Traditional and Herbal Medicines for COVID-19

Traditional and Herbal Medicines for COVID-19 explores promising ways to manage COVID-19, post-COVID, and long-COVID conditions. The management plans are based on anti-virus activity, anti-inflammatory activity, and diverse health benefits of traditional and herbal medicines through a comprehensive summarization of scientific literature by experts in the field. It presents views of the origin of SARS-CoV-2 and emerging variants and pathogenesis, and it proposes renewed strategies of diagnostics, vaccines, and therapies.

Features

- Provides an in-depth analysis to illustrate the impact of traditional and herbal medicines on crucial protein targets responsible for the progress of SARS-CoV-2 infection and symptoms.
- Presents knowledge of SARS-CoV-2 and variants.
- Explores strategies to manage COVID-19, post-COVID, and long-COVID by applying traditional herbal medicines.
- Illustrates molecular aspects of anti-coronavirus activity from traditional herbal medicines.
- Features information on molecular mechanisms of target proteins involving COVID-19 infection and symptoms.

Traditional and Herbal Medicines for COVID-19 serves as an ideal reference for researchers and experts in the fields of virology, epidemiology, drug discovery, and traditional herbal medicine. This book aligns with supporting the Sustainable Development Goals (SDGs) 2030 by the United Nations to establish "Good Health and Well-Being."

Traditional and Herbal Medicines for COVID-19

Edited by
Jen-Tsung Chen

CRC Press
Taylor & Francis Group
Boca Raton London New York

CRC Press is an imprint of the
Taylor & Francis Group, an **informa** business

Designed cover image: Shutterstock

First edition published 2025
by CRC Press
2385 NW Executive Center Drive, Suite 320, Boca Raton FL 33431

and by CRC Press
4 Park Square, Milton Park, Abingdon, Oxon, OX14 4RN

CRC Press is an imprint of Taylor & Francis Group, LLC

Library of Congress Cataloging-in-Publication Data
Names: Chen, Jen-Tsung, editor.
Title: Traditional and herbal medicines for COVID-19 / edited by Jen-Tsung Chen.
Description: First edition. | Boca Raton FL : CRC Press, 2025. | Includes bibliographical references and index. | Summary: "Traditional and Herbal Medicines for COVID-19 explores promising ways to manage COVID-19, post-COVID, and long-COVID conditions based on anti-virus activity, anti-inflammatory, and diverse health benefits of traditional and herbal medicines through a comprehensive summarization of scientific literature by scientists and experts. It presents the current view on the origin of SARS-CoV-2, the emerging variants, and its pathogenesis and proposes renewed strategies of diagnostics, vaccines, and therapies. This book provides an in-depth analysis to illustrate the impact of traditional and herbal medicines on crucial protein targets that are responsible for the progress of SARS-CoV-2 infection and symptoms covering angiotensinogen-converting enzyme 2 (ACE-2), Toll-like receptors (TLRs), transmembrane serine protease 2 (TMPRSS2), papain-like protease (PLpro), main protease (Mpro), 3-chymotrypsin-like protease (3CLpro), RNA-dependent RNA polymerase (RdRp), spike glycoprotein, and envelope protein"-- Provided by publisher.
Identifiers: LCCN 2024017720 (print) | LCCN 2024017721 (ebook) | ISBN 9781032589442 (hardback) | ISBN 9781032590301 (paperback) | ISBN 9781003452621 (ebook)
Subjects: MESH: COVID-19 Drug Treatment--methods | Phytotherapy--methods | Antiviral Agents--pharmacology | Anti-Inflammatory Agents--pharmacology | Plants, Medicinal
Classification: LCC RM666.H33 (print) | LCC RM666.H33 (ebook) | NLM WC 506.2 | DDC 615.3/21--dc23/eng/20240613
LC record available at https://lccn.loc.gov/2024017720
LC ebook record available at https://lccn.loc.gov/2024017721

ISBN: 9781032589442 (hbk)
ISBN: 9781032590301 (pbk)
ISBN: 9781003452621 (ebk)

DOI: 10.1201/9781003452621

Typeset in Times
by KnowledgeWorks Global Ltd.

Contents

About the Editor

Jen-Tsung Chen is a professor of cell biology at the National University of Kaohsiung in Taiwan. He also teaches genomics, proteomics, plant physiology, and plant biotechnology. Dr. Chen's research interests include: bioactive compounds, chromatography techniques, plant molecular biology, plant biotechnology, bioinformatics, and systems pharmacology. He is an active editor of academic books and journals that advance the exploration of multidisciplinary knowledge involving plant physiology, plant biotechnology, nanotechnology, ethnopharmacology, and systems biology. He serves as an editorial board member of reputed journals, including *Plant Methods*, *GM Crops & Food*, *Plant Nano Biology*, *Biomolecules*, and *International Journal of Molecular Sciences*, and serves as a guest editor of *Frontiers in Plant Science*, *Frontiers in Pharmacology*, *Journal of Fungi*, and *Current Pharmaceutical Design*. Dr. Chen has published books in collaboration with Springer Nature, CRC Press/Taylor & Francis Group, and In-TechOpen, and he handles book projects for other international publishers on diverse topics such as drug discovery, herbal medicine, medicinal biotechnology, nanotechnology, bioengineering, plant functional genomics, plant speed breeding, and CRISPR-based plant genome editing. In 2023, Dr. Chen was included in the World's Top 2% Scientists by Stanford University.

Contributors

Sheikh Naem Islam Abhi
Department of Pharmacy
Faculty of Biological Science Islamic
    University
Kushtia, Bangladesh

Arunachalam Abitha
Department of Human Genetics
    and Molecular Genetics
Bharathiar University
Coimbatore, Tamil Nadu, India

Raja Ahmed
Department of Bioengineering
    and Technology
Gauhati University
Guwahati, Assam, India

Suraiya Akhtar
Department of Bioengineering and
    Technology
Gauhati University
Guwahati, Assam, India

Athanasius Alexiou
Department of Science and Engineering
Novel Global Community and Educational
    Foundation
Australia

Arumugam Vijaya Anand
Department of Human Genetics
    and Molecular Genetics
Bharathiar University
Coimbatore, Tamil Nadu, India

Sadia Aziz
Department of Biological Sciences
International Islamic University
Islamabad, Pakistan

Sofia Banu
Department of Bioengineering and
    Technology
Gauhati University
Guwahati, Assam, India

Khaleda Begum
Department of Bioengineering and Technology
Gauhati University
Guwahati, Assam, India

Ch. V. S. S. S. Bharathi
Department of Genetics and Plant Breeding
Palli Siksha Bhavana (Institute of Agriculture)
Visva-Bharati
Sriniketan, West Bengal, India

Divya Raj Bharti
Department of Genetics and Plant Breeding
Palli Siksha Bhavana (Institute of Agriculture)
Visva-Bharati
Sriniketan, West Bengal, India

Shabana Bibi
Department of Biosciences
Shifa Tameer-e-Millat University
Islamabad, Pakistan

Prakash S. Bisen
School of Studies in Biotechnology
Jiwaji University
Gwalior, India

Partha Biswas
Laboratory of Pharmaceutical Biotechnology
    and Bioinformatics
Department of Genetic Engineering and
    Biotechnology
Jashore University of Science and Technology
Jashore, Bangladesh

Mònica Bulló
Department of Biochemistry & Biotechnology
University Rovira i Virgili (URV)
Spain
Nutrition and Metabolic Health Research Group
Institute of Health Pere Virgili (IISPV)
Spain
CIBER Physiology of Obesity and Nutrition
    (CIBEROBN)
Carlos III Health Institute
Spain

Cagla Celik
Pharmacy Services Program
Vocational School of Health Services
Hitit University
Corum, Turkey

Amartya Chakraborty
Department of Bioengineering and Technology
Gauhati University
Guwahati, Assam, India

Ankita Chakraborty
Integrative Biochemistry and Immunology
    Laboratory (IBIL)
Department of Animal Science
Kazi Nazrul University
Asansol, West Bengal, India

Jen-Tsung Chen
Department of Life Sciences
National University of Kaohsiung
Kaohsiung, Taiwan

Ankur Das
Department of Bioengineering and Technology
Gauhati University
Guwahati, Assam, India

Kanchana Das
Toxicology Research Unit
Department of Zoology
The University of Burdwan
Purba Bardhaman, West Bengal, India

Sandip Debnath
Department of Genetics and Plant Breeding
Palli Siksha Bhavana (Institute of Agriculture)
Visva-Bharati
Sriniketan, West Bengal, India

Chenniappan Anchana Devi
Postgraduate & Research Department
    of Biotechnology
Women's Christian College
Chennai, Tamil Nadu, India

Dipta Dey
Department of Biochemistry and Molecular
    Biology, Life Science Faculty
Bangabandhu Sheikh Mujibur Rahman Science
    and Technology University
Gopalgonj, Bangladesh

Jayakumar Dharshini
Department of Human Genetics
    and Molecular Genetics
Bharathiar University
Coimbatore, Tamil Nadu, India

Burtram Clinton Fielding
University of the Western Cape
Cape Town, South Africa

Abhratanu Ganguly
Department of Animal Science
Kazi Nazrul University
Asansol, West Bengal, India

Abhijit Ghosh
Independent Researcher &
    Registered Medical Practitioner
Japandanga, West Bengal, India

Arun Bahadur Gurung
Department of Biological Sciences
Indian Institute of Science Education and
    Research Kolkata
Mohanpur, West Bengal, India

Md Hasibul Hasan
Department of Food Engineering
Bangabandhu Sheikh Mujibur Rahman
    Science and Technology University
Gopalgonj, Bangladesh

Md Nazmul Hasan
Laboratory of Pharmaceutical
    Biotechnology and Bioinformatics
Department of Genetic Engineering and
    Biotechnology
Jashore University of Science
    and Technology
Jashore, Bangladesh

Rashid Hussain
Department of Biosciences
Shifa Tameer-e-Millat University
Islamabad, Pakistan

Kalibulla Syed Ibrahim
Department of Biotechnology
PSG College of Arts and Science
Coimbatore, Tamil Nadu, India

Nilay Ildiz
Bandirma Onyedi Eylul University
Vocational School of Health Services
Medical Imaging Department
Bandirma, Turkey

Md Imtiaz
Laboratory of Pharmaceutical Biotechnology
 and Bioinformatics
Department of Genetic Engineering
 and Biotechnology
Jashore University of Science
 and Technology
Jashore, Bangladesh

Faryal Jahan
Shifa College of Pharmaceutical Sciences
Shifa Tameer-e-Millat University
Islamabad, Pakistan

Chetna Jhagta
Department of Medicinal Chemistry
Government College of Pharmacy
Rohru, India

Pankaj Jinta
Department of Pharmacognosy
Government College of Pharmacy
Rohru, India

Marudhachalam Kamalesh
Department of Human Genetics
 and Molecular Genetics
Bharathiar University
Coimbatore, Tamil Nadu, India

Aditya Khatri
All India Institute of Medical
 Sciences Bhopal
Madhya Pradesh, India

Ayse Baldemir Kılıc
Department of Pharmaceutical Botany
Faculty of Gulhane Pharmacy
University of Health Sciences
Keçiören, Turkey

Ashwin Kotnis
Department of Biochemistry
All India Institute of Medical Sciences Bhopal
Madhya Pradesh, India

Kruthi Ashok Kumar
Centre for Biomaterials, Cellular and
 Molecular Theragnostic
Vellore Institute of Technology, VIT
Vellore, Tamil Nadu, India

Rajesh Kumari
Department of Zoology
University of Allahabad
Prayagraj, Uttar Pradesh, India

Swastika Maitra
Centre for Global Health Research
Saveetha Medical College and Hospital
Chennai, India
Shifa College of Pharmaceutical
 Sciences
Shifa Tameer-e-Millat University
Islamabad, Pakistan

Sucheta Mandal
Department of Botany
Banwarilal Bhalotia College
Asansol, West Bengal, India

Moutushi Mandi
Toxicology Research Unit
Department of Zoology
The University of Burdwan
Purba Bardhaman, West Bengal,
 India

Mani Manoj
Department of Human Genetics and
 Molecular Genetics
Bharathiar University
Coimbatore, Tamil Nadu, India

Asim Mehmood
Department of Biosciences,
Sahiwal Campus
COMSATS University Islamabad
Sahiwal, Pakistan

Vineet Mehta
Department of Pharmacology
Government College of Pharmacy
Rohru, India

Swetha M. Menon
Department of Human Genetics
    and Molecular Genetics
Bharathiar University
Coimbatore, Tamil Nadu, India

Md Moinuddin
Department of Pharmacy
Faculty of Biological Science
    Islamic University
Kushtia, Bangladesh

Suprabhat Mukherjee
Integrative Biochemistry and Immunology
    Laboratory (IBIL)
Department of Animal Science
Kazi Nazrul University
Asansol, West Bengal, India

Priyanka Nagu
Department of Pharmaceutics
Government College of Pharmacy
Rohru, India

Sayantani Nanda
Department of Animal Science
Kazi Nazrul University
Asansol, West Bengal, India

Abhiruj Navabhatra
Department of Pharmacology
College of Pharmacy
Rangsit University
Pathum Thani, Thailand

Tanisha Neog
Department of Bioengineering
    and Technology
Gauhati University
Guwahati, Assam, India

Ramesh Nivedha
Department of Human Genetics
    and Molecular Genetics
Bharathiar University
Coimbatore, Tamil Nadu, India

Abdullah Al Noman
School of Pharmacy
Brac University
Dhaka, Bangladesh

Ismail Ocsoy
Department of Analytical
    Chemistry
Faculty of Pharmacy
Erciyes University
Kayseri, Turkey

Anand Kumar Pandey
Department of Biotechnology
    Engineering
Institute of Engineering and
    Technology
Bundelkhand University
Jhansi, Uttar Pradesh, India

Anwar Parvez
Department of Pharmacy
Faculty of Allied Health Sciences
Daffodil International University
Dhaka, Bangladesh

Adityaraj Patidar
All India Institute of Medical
    Sciences Bhopal
Madhya Pradesh, India

Eeshan Pawar
All India Institute of Medical Sciences
    Bhopal
Madhya Pradesh, India

Biswajit Pramanik
Department of Genetics and
    Plant Breeding
Palli Siksha Bhavana
    (Institute of Agriculture)
Visva-Bharati
Sriniketan, West Bengal, India

Natarajan Pushpa
Postgraduate & Research Department
    of Microbiology
Cauvery College for Women
Tiruchirappalli, Tamil Nadu, India

Md Habibur Rahaman
Department of Pharmacy
Faculty of Biological Science Islamic
    University
Kushtia, Bangladesh

Prem Rajak
Department of Animal Science
Kazi Nazrul University
Asansol, West Bengal, India

Shivani Rana
Department of Pharmacognosy
Govt. College of Pharmacy
Rohru, India

Balasubramaniam Rashmika
Department of Human Genetics
    and Molecular Genetics
Bharathiar University
Coimbatore, Tamil Nadu, India

Soumya Rathore
PG Department of Food Technology
Raja Balwant Singh Engineering
    Technical Campus
Bichpuri, Agra, India

Anindya Sundar Ray
Department of Botany
Visva-Bharati University
Santiniketan, West Bengal, India

Panneerselvam Saamisha
Department of Biotechnology
Bharathiar University
Coimbatore, Tamil Nadu, India

Saurabh Sarkar
Science Research Laboratory
Department of Zoology
Gushkara Mahavidyalaya
Gushkara, West Bengal, India

Abhijit Sarma
Department of Bioengineering
    and Technology
Gauhati University
Guwahati, Assam, India

Dewald Schoeman
University of the Western Cape
Cape Town, South Africa

Arumugam Sembiyaa
Department of Human Genetics
    and Molecular Genetics
Bharathiar University
Coimbatore, Tamil Nadu, India

Maria Shah
Department of Biosciences
COMSATS University Islamabad
Islamabad, Pakistan

Muhammad Furqan Shah
Department of Biosciences
COMSATS University
Islamabad, Pakistan

Amina Shamas
Department of Bioinformatics and Biosciences
Capital University of Science & Technology
Islamabad, Pakistan

Minaxi Sharma
Department of Applied Biology
University of Science and Technology
    Meghalaya
Baridua, India

Labib Shahriar Siam
Laboratory of Pharmaceutical Biotechnology
    and Bioinformatics
Department of Genetic Engineering
    and Biotechnology
Jashore University of Science and Technology
Jashore, Bangladesh

Rupanjali Singh
Department of Biotechnology
    Engineering
Institute of Engineering and Technology
Bundelkhand University
Jhansi, Uttar Pradesh, India

Ramaraj Sivamano
Department of Human Genetics
    and Molecular Genetics
Bharathiar University, Coimbatore
Tamil Nadu, India

Kandi Sridhar
Department of Food Technology
Karpagam Academy of Higher Education
    (Deemed to be University)
Coimbatore, India

Gunna Sureshbabu Suruthi
Department of Human Genetics
    and Molecular Genetics
Bharathiar University
Coimbatore, Tamil Nadu, India

Karuppannan Swathi
Department of Biotechnology
Bharathiar University
Coimbatore, Tamil Nadu, India

Sathyalingam Sathya Trisha
Department of Human Genetics
    and Molecular Genetics
Bharathiar University
Coimbatore, Tamil Nadu, India

Shanmugam Velayuthaprabhu
Department of Biotechnology
Bharathiar University
Coimbatore, Tamil Nadu, India

Ravindra Verma
School of Studies in Biotechnology
Jiwaji University
Gwalior, India

Shalja Verma
Department of Biotechnology Engineering
Institute of Engineering and Technology
Bundelkhand University
Jhansi, Uttar Pradesh, India
Department of Biosciences and Bioengineering
Indian Institute of Technology
Roorkee, Uttarakhand, India

Viroj Wiwanitkit
Chandigarh University
Punjab, India

Sora Yasri
KM Center
Bangkok Thailand

Bancha Yingngam
Department of Pharmaceutical Chemistry
    and Technology
Faculty of Pharmaceutical Sciences
Ubon Ratchathani University
Ubon Ratchathani, Thailand

Sadi Yusufbeyoglu
Department of Pharmaceutical Botany
Faculty of Gulhane Pharmacy
University of Health Sciences
Keçiören, Turkey

Preface

In recent years, one of the most severe pandemics in human history, coronavirus disease 2019 (COVID-19), greatly damaged global public health as well as economic activity. This disease was caused by a fast-mutated RNA virus, severe acute respiratory syndrome coronavirus 2 (SARS-CoV-2), belonging to the family Coronaviridae, which contains members that led to severe acute respiratory syndrome (SARS) and Middle East respiratory syndrome (MERS) outbreaks. The world is continued to be threatened by its new variants with changing symptoms and immune escape capacity and, therefore, there is a high demand for exploring treatments, medications, and vaccines for efficient and precise management of COVID-19 in the post-pandemic era, highlighting ways to combat new variants as well as post-COVID and long-COVID conditions.

This book provides an updated overview of COVID-19, by presenting advances in the origin of SARS-CoV-2 and its pathogenesis, and additionally, summarizing renewed strategies of diagnostics, vaccines, and therapies. An in-depth analysis was given to illustrate the molecular structure of crucial protein targets either from the host cell or the virus responsible for the progress of SARS-CoV-2 infection and symptoms. The current achievements in uncovering the molecular mechanisms and the hub proteins involved in COVID-19 are comprehensively presented, including parts of angiotensinogen-converting enzyme 2 (ACE-2), toll-like receptors (TLRs), transmembrane serine protease 2 (TMPRSS2), papain-like protease (PLpro), main protease (Mpro), 3-chymotrypsin-like protease (3CLpro), RNA-dependent RNA polymerase (RdRp), and additionally, the current findings of spike (S) glycoprotein and the envelope (E) protein of coronaviruses are discussed in separate chapters.

The text explores the potential roles of traditional and herbal medicines, which have been well-recognized for curing and benefiting human health for thousands of years around the world, in the management of COVID-19, chiefly based on scientific research using modern technologies performed under *in silico*, *in vitro*, *in vivo*, and clinical conditions. By systematically refining, this book introduces the structure-activity relationship of promising bioactive compounds derived from traditional herbal medicines such as traditional Chinese medicine that have been applied to clinical treatments for COVID-19 and further discusses their potential anti-SARS-CoV-2 activity and the underlying mechanisms.

Crucial groups of bioactive compounds in herbal medicines, including flavonoids, alkaloids, and terpenoids, are discussed in separate chapters for their impacts on antibacterial, anticoagulant, anti-inflammatory, antioxidant, and antivirus activities, which have been investigated using reliable approaches such as multiple and integrated omics with high-throughput technologies, systems/network pharmacology, and clinical experiments. These achievements might inspire future studies designed for the development of herbal remedies combating SARS-CoV-2 and the management of COVID-19.

A diverse group of scientists and experts contributed to this book, which is highly recommended to be a reference for students, lecturers, and professors in cross-filed life sciences and epidemiology. The knowledge is valuable for supporting the Sustainable Development Goals (SDGs) 2030 by the United Nations to establish "Good Health and Well-Being". In ending, I'd like to express my sincere appreciation to all the authors for their insightful and valuable chapters. I am very grateful for the instruction and help from the publisher.

Jen-Tsung Chen
Kaohsiung, Taiwan

1 An Updated Overview of Coronavirus Disease 2019

Dewald Schoeman and Burtram Clinton Fielding

1.1 INTRODUCTION: CHARACTERISTICS OF SARS-CoV-2

1.1.1 HUMAN CORONAVIRUSES (HCoVs) AND SARS-CoV-2

Coronaviruses (CoVs) are a group of zoonotic viruses found worldwide that traditionally infect a wide range of animals and can cause a variety of illnesses ranging from respiratory tract infections (RTIs) and gastrointestinal tract infections (GITs) to neurological diseases [1, 2]. Seven CoVs have been able to cross the species barrier and have demonstrated the ability to infect humans. Four of these human CoVs (hCoVs), HCoV-229E, HCoV-NL63, HCoV-OC43, and HCoV-HKU1, commonly circulate within the human population and are less virulent, typically causing less severe, seasonal RTIs [3]. The other three, severe acute respiratory syndrome coronavirus 1 (SARS-CoV-1, formerly SARS-CoV), Middle East respiratory syndrome coronavirus (MERS-CoV), and severe acute respiratory syndrome coronavirus 2 (SARS-CoV-2), are more virulent and are generally associated with more severe RTIs, having caused deadly outbreaks in the past 20 years [4].

All CoVs belong to the family *Coronaviridae* in the order Nidovirales and are taxonomically classified into four genera: alpha (α), beta (β), gamma (γ), and delta (δ). Along with SARS-CoV-1, MERS-CoV, HCoV-OC43, and HCoV-HKU1, SARS-CoV-2 is classified under the genus β-CoVs, whereas HCoV-229E and HCoV-NL63 belong to the genus α-CoVs [5]. SARS-CoV-2 was named after SARS-CoV-1 due to its high genetic sequence similarity and is the second hCoV, after SARS-CoV-1, to have caused a global pandemic [6, 7]. SARS-CoV-2 is the causative agent of the coronavirus disease 2019 (COVID-19) [8].

1.1.2 GENOME AND STRUCTURE OF SARS-CoV-2

Each SARS-CoV-2 virion has an average diameter of approximately 108 nm and is spherical or ellipsoidal in morphology [9, 10]. Inside each virion is the approximate 30 kilobase (kb) positive-sense single-stranded ribonucleic acid (RNA) genome, which encodes the four structural proteins spike (S) protein, membrane (M) protein, nucleocapsid (N) protein, and envelope (E) protein, along with the 14 open-reading frames (ORFs) that encode 27 proteins [11, 12]. The 5'-terminus of the genome contains the ORF1a and ORF1ab genes that encode the polypeptides (pp) 1a and pp1ab, respectively, whereas the genes that code for the four structural proteins and the eight accessory proteins (ORF3a, ORF3b, p6, ORF7a, ORF7b, ORF8b, ORF9b, and ORF14) are located at the 3'-terminus (Figure 1.1). This genome organization is conserved and common to all CoVs. The genome is capped at the 5' end and polyadenylated at the 3' end [13]. Although SARS-CoV-1 and SARS-CoV-2 are quite similar at the amino acid level, there are a few differences in some proteins. Notable examples in SARS-CoV-2 include the absence of ORF8a, a somewhat longer ORF8b compared to SARS-CoV-1, and a considerably shorter ORF3b [12].

The S proteins cover the external surface of the virion where most of the S proteins adopt a prefusion state and a smaller fraction adopt the post-fusion state [9, 14]. Hinges in the stalk region of the S protein allow the head to tilt, and this flexibility, coupled with the extensive glycosylation, enables the virus to scan the surface of host cells before binding to the host-cell receptor angiotensin-converting enzyme 2 (ACE2) [9, 15]. The outer membrane of the virus consists of a combination of

DOI: 10.1201/9781003452621-1

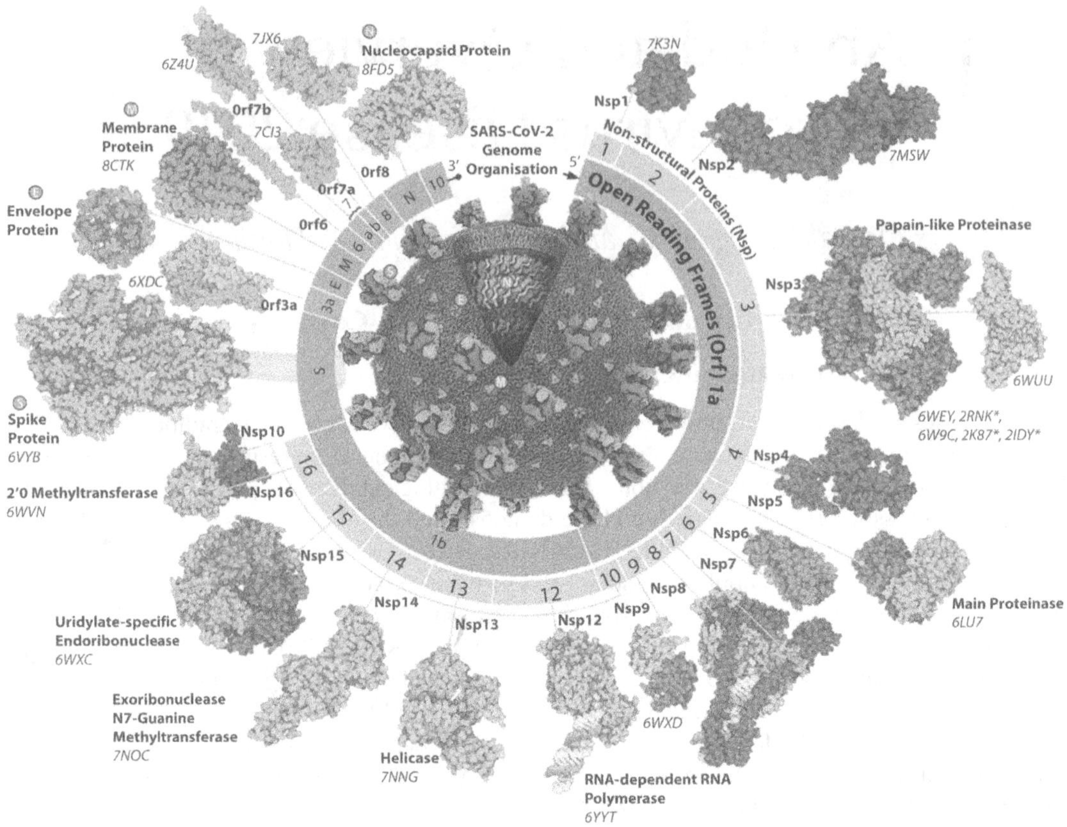

FIGURE 1.1 A graphical representation of the SARS-CoV-2 genome and protein architecture. Polyproteins pp1a and pp1ab (blue) comprised of the different non-structural proteins (nsps), structural proteins (teal/cyan), open-reading frames (ORFs; green), double-stranded RNA complex (red: template strand, yellow: product strand), and transmembrane portions of the spike (S) protein shown in cartoon form (light teal). Protein Data Bank IDs (PDB IDs) for the experimentally resolved structure is also shown [16].

the M and E proteins [9]. Enclosed by this outer membrane, in the lumen of the virus, is the viral genome complexed with the N protein in ribonucleoprotein (RNP) complexes that is responsible for the packaging of the viral RNA genome. Each virion is estimated to contain between 30 and 35 RNPs [9].

Computational tools have made a valuable contribution to SARS-CoV-2 research as evident by the most recent use of integrative modelling to elucidate the molecular architecture and dynamics of the SARS-CoV-2 envelope to near atomistic scale [17, 18]. The authors combined existing experimental data at different resolutions to obtain a consistent model that provides both structural and functional insights that would not otherwise be possible to obtain from a single experimental method. The viral envelope of the model incorporates a lipid bilayer of up to six types of phospholipids that reflects the composition of the endoplasmic reticulum (ER). The outer surface of the envelope contains the transmembrane domains (TMDs) of M protein dimers, whereas the endodomains (EDs) of the dimers are tightly packed on the inner surface. The stability of the viral envelope appears to remain unaffected by varying stoichiometries of the structural proteins E (pentamer), S (trimer), and M (dimer), thus supporting *in vivo* experimental evidence that demonstrated the possibility of having a range of different stoichiometries. The lack of a full-length experimentally resolved structure of M underpins the value of this model as the authors showed that the TMDs of M dimers preferentially formed filament-like assemblies and there were no contacts between adjacent filaments. Conversely, the EDs of M dimers are tightly packed and bind neighbours in two directions, reflecting a tendency to form a well-ordered lattice which is consistent with previously reported EM data for SARS-CoV-1 and other CoVs [19]. Such data on the complex assemblies of the M dimers and their interactions with other structural proteins play an important role in contributing to our understanding of the molecular mechanisms behind viral assembly.

1.1.3 GENES AND PROTEINS

1.1.3.1 Spike (S) Protein

The S glycoprotein is arguably the most important of the SARS-CoV-2 structural proteins and has received a lot of attention. It is a 600 kilodalton (kDa) type I membrane protein that assembles into a homotrimer with a large ectodomain exposed on the virion surface, while its transmembrane segment anchors it to the virion surface. It is estimated that each SARS-CoV-2 virion contains approximately 15–40 S proteins distributed randomly across the surface of the virion, and most of them adopt a prefusion conformation [14]. The full-length S protein contains two subunits, S1 and S2, which mediate binding to the host-cell receptor ACE2 and facilitates fusion between the viral particle and the host cell membrane, respectively [20–22]. The receptor-binding domain (RBD) of S1 recognizes and binds to the ACE2 receptor, whereas its amino (N)-terminal domain (NTD) appears to play a role in the overall conformation of the S protein [23, 24]. Mutations in the NTD have also been linked to SARS-CoV-2 evading the immune system [24]. Similar to SARS-CoV-1, the SARS-CoV-2 S protein undergoes proteolytic cleavage prior to membrane fusion, and proteases such as furin, transmembrane protease serine 2 (TMPRSS2), and cathepsin L are involved in this process [22, 25–27]. The SARS-CoV-1 and -2 S proteins are both highly conserved but in SARS-CoV-2, the S protein contains a unique furin cleavage site ([681]PRRAR[685]) that is suggested to contribute to its increased ability to infect host cells [26]. The importance of the furin cleavage site was demonstrated in mice where viral replication was diminished, or even abrogated, in the absence of the furin cleavage site [28].

Although it is not considered the main determinant of pathogenicity, the SARS-CoV-2 S protein characteristically plays a role in forming syncytia (giant, multinucleated cells) by fusing infected cells with neighbouring uninfected cells [29, 30]. It has also been reported to contribute to vascular damage and to induce coagulation, raising the concern of using spike-based vaccines and the potential risk of thrombotic complications [31–34]. The S protein is extensively glycosylated, and each monomer is reported to contain 22 N-linked glycosylation sites with some O-linked glycosylation on the S1 subunit as well [9, 35–37]. This has been suggested to play an important role in evading the immune system by serving as a glycan shield to hide epitopes from recognition by neutralizing antibodies [37, 38]. In support of this, the majority of neutralizing antibodies appear to be generated against the RBD, which is glycosylated to a lesser extent [39].

Several variants arose over the course of the COVID-19 pandemic, and although not all persisted, five strains of SARS-CoV-2 were designated as variants of concern (VOCs) in the order that they emerged: Alpha, Beta, Gamma, Delta, and Omicron [40, 41]. Each variant became dominant by outcompeting the previous and were designated as VOCs as they demonstrated an increase in transmissibility and could evade natural immunity or infect vaccinated individuals [42, 43]. Mutations accrued by the S protein have been shown to confer to SARS-CoV-2 the ability to evade the immune system [44–46]. Figure 1.2 depicts the five notable VOCs and the position of their defining mutations along the S protein [40].

As early as March 2020, the three-dimensional (3D) structure of the isolated, recombinant trimeric S protein was resolved by cryo-electron microscopy (cryo-EM) and has been deposited in the Protein Data Bank (PDB) as the full-length S protein in its prefusion state (PDB ID: 6VSB) and ectodomain closed or open state (PDB ID: 6VXX and 6VYB) [47, 48]. These structures described two conformations of the S protein: one (6VXX) in which the RBD of each monomer is closed

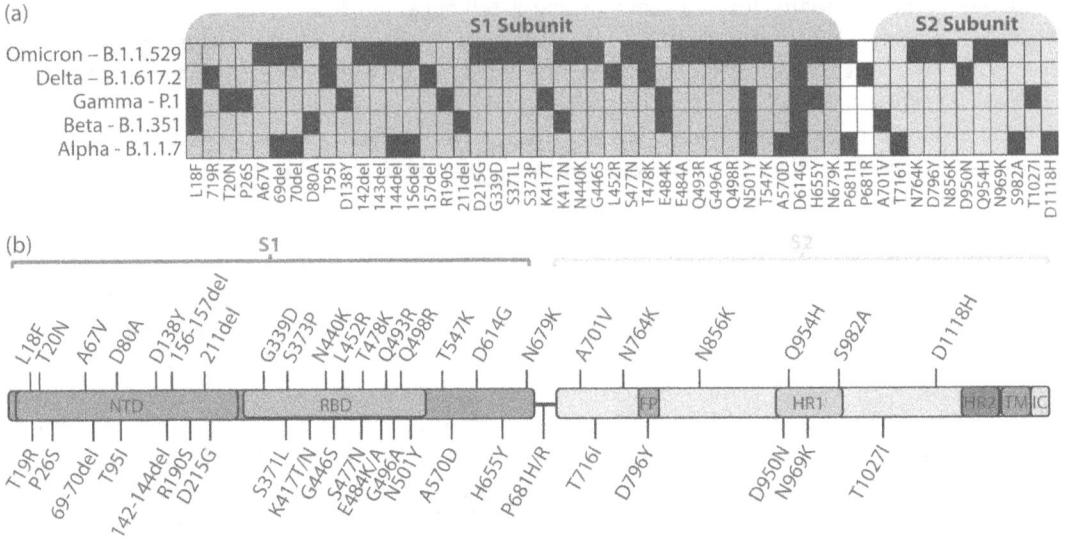

FIGURE 1.2 The mutations accrued in the Spike (S) protein of the different SARS-CoV-2 variants of concern (VOCs). (a) Mutations that define the different VOCs and their relative position in the S1/S2 Subunits of the S protein. (b) A visual representation of the SARS-CoV-2 S protein indicating the relative positions of all the reported mutations. **NTD**: N-terminal domain; **RBD**: receptor-binding domain; **FP**: fusion peptide; **HR1**: heptad repeat 1; **HR2**: heptad repeat 2; **TM**: transmembrane region; **IC**: intracellular domain [40].

(or 'down') and the other (6VSB, 6VYB) containing a partially open (or 'up') monomer with one exposed RBD. Over the course of the pandemic, more experimental structures of the S protein were resolved, including those of the mutated S proteins for each of the VOCs and these can also be found in the PDB.

1.1.3.2 Membrane (M) Protein

The M protein, also the most abundant of the structural proteins, is a transmembrane protein comprising of 222 residues with three TMDs [19]. It is well-known for its central role in orchestrating the assembly of new viral particles, where it plays a role in the membrane curvature of the viral envelope and the subsequent morphology of viral particles, but also appears to be involved in viral maturation, secretion, and budding [49–51]. In a completely assembled viral particle, its N- and C-termini are positioned toward the interior and exterior of the virus, respectively [52, 53]. During viral assembly, the C-terminus facilitates binding with both the N and E proteins, while the TMD mediates binding to the S protein [54]. Commensurate with its function in defining the shape of the viral envelope, the M protein is also capable of homotypic interactions with itself [19, 55, 56]. There is a high level of sequence similarity between the M proteins of SARS-CoV-1 and -2, and SARS-CoV-2 M shares six of its eight N-linked glycosylation sites with that of SARS-CoV-1 [57].

The M protein is considered to function as a homodimer but cryo-EM studies of different CoVs suggest that it might adopt two distinct conformations, depending on its role and protein-protein interactions (PPIs) with other structural proteins [49, 58]. Aside from its structural role, M is also reported to antagonize type I and III interferon (IFN; IFN-I, IFN-III) signalling, important aspects of the antiviral response [49, 59]. The absence of an experimentally resolved structure or a suitable structural homolog early on in the pandemic prompted the use of bioinformatics tools, such as I-TASSER, Robetta, trRosetta, and SWISS-MODEL to predict the 3D structure of the SARS-CoV-2 M protein [60, 61]. Such *in silico* predicted structures were important early on in the pandemic as they could be used to ascertain potential structural/functional differences between the

SARS-CoV-1 and -2 M proteins, for mutational analysis, and other tools could be used for screening of antiviral drugs [60, 62]. AlphaFold was also used to predict the full-length structure of the M protein, and shortly thereafter, cryo-EM resolved structures were reported: M complexed with Fab fragments (PDB IDs: 7VGR and 7VGS) and the structure of M in lipid nanodiscs (PDB ID: 8CTK) [62, 63].

1.1.3.3   Nucleocapsid (N) Protein

The N protein is a major protein component inside SARS-CoV-2 that is important for the transcription and replication of viral mRNA, binds the viral RNA genome, and packages it into RNP structures [64]. It can be described as a multidomain RNA-binding protein that consists of 419 residues with an NTD and carboxy (C)-terminal domain (CTD) that is conserved among CoVs [65, 66]. The protein is positively charged, hydrophobic, unstable, and is comprised largely of coils (~55%) [67]. The NTD and CTD are both flanked and linked by intrinsically disordered regions (IDRs): the N-arm and C-tail flank the NTD and CTD, respectively, and the central linker region (CLR) links the NTD and CTD [66] (Figure 1.3). The NTD, which can bind both single-stranded (ss) RNA and double-stranded (ds) RNA, initiates binding to the viral RNA [68, 69]. The CTD can also bind RNA and it stabilizes the protein through dimerization but can also self-oligomerize into trimers, tetramers, and octamers [70–72]. The binding of viral RNA induces a more ordered conformation that permits dimerization, dimer-dimer interactions, and ultimately the assembly of multimeric structures to accomplish genome packaging [73]. Interaction between N and M anchors the RNP complexes to the viral membrane to facilitate packaging of the viral genome in virions [74]. The IDRs of N do not adopt fixed structures like the NTD and CTD, but are inherently flexible and variable and contribute to the functions of N [75, 76]. The CLR contains a cluster of charged residues

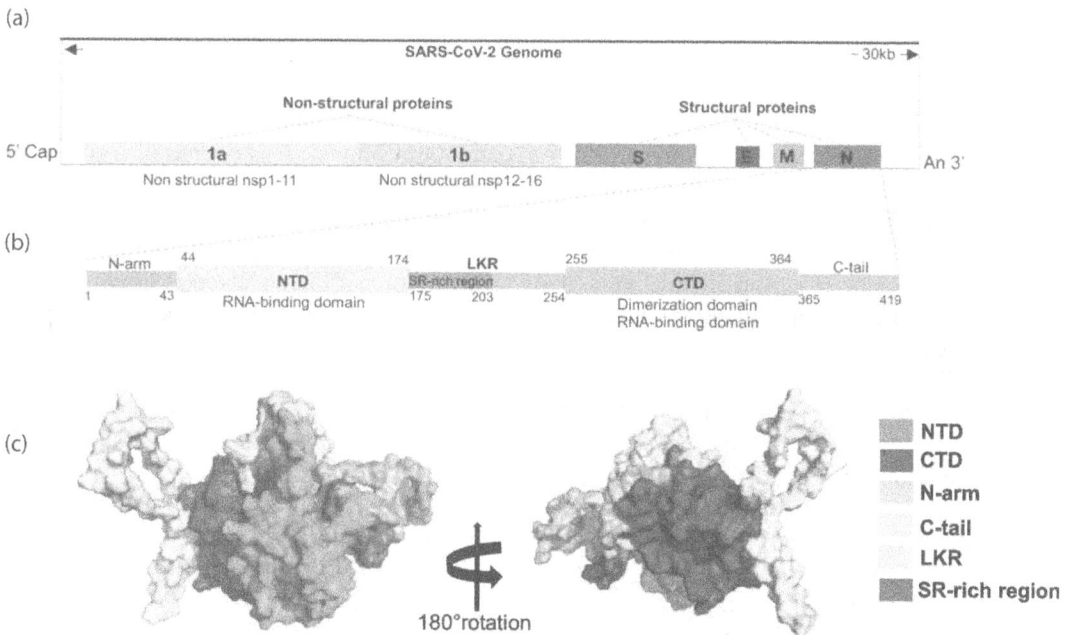

FIGURE 1.3   The genome organization of the SARS-CoV-2 virus and a structural overview of the SARS-CoV-2 Nucleocapsid (N) protein. (a) The genome organization of SARS-CoV-2. (b) Schematic representation of the domains of the SARS-CoV-2 N protein. The N-terminal domain (NTD), C-terminal domain (CTD), and three intrinsically disordered regions (IDRs), i.e. N-arm, linker region (LKR), and C-tail. The LKR is shown to contain the charged Ser/Arg (SR)-rich region. (c) Deletion analyses of the SARS-CoV-2 N protein as visualised in PyMol v2.4. [93].

and its phosphorylation state regulates the binding between N and other proteins such as viral non-structural protein 3 (nsp3), cyclin-dependent kinase 1 (Cdk1), glycogen synthase 3, and the different isoforms of the 14-3-3 proteins [77–80]. The N-arm binds to the Ras-GTPase-activating protein binding protein 1 (G3BP1) and disrupts the formation of stress granules, an obstacle for viral replication [81]. Based on the corresponding region in the N protein of other CoVs, the C-tail is thought to be involved in its interaction with M [82].

In addition to its properties as a structural protein, N is also capable of host–pathogen interactions, regulating progression of the host cell cycle, and apoptosis [83, 84]. It appears to exhibit a complex interplay with the host's immune response, evident by its ability to inhibit the host cell's RNA interference defence and type I IFN signalling, but it can also induce type III IFNs, trigger hyperactivation of nuclear factor kappa-light-chain-enhancer of activated B cells (NF-κB), and activate the NLR family pyrin domain containing 3 (NLRP3) inflammasome [85–88]. This dual role appears to depend on the expression level of N: at low levels, N suppresses the immune system, whereas higher levels evoke a response [89]. Interestingly, the stimulatory effect of NF-κB has been attributed to N's ability to undergo liquid–liquid phase separation (LLPS) in which N binds to viral RNA and forms liquid/solid-like condensates, depending on the N:RNA ratio [90, 91]. Binding of viral RNA to N promotes LLPS, which recruits signalling complexes that ultimately enhance the activation of NF-κB, and the CTD of N plays a critical role in this process [88].

Cryo-EM resolved monomeric (PDB ID: 8FD5) and dimeric (PDB ID: 8FG2) structures of the full-length SARS-CoV-2 N protein has been deposited in the PDB along with several structures of its NTD and CTD, including a dodecameric CTD (PDB ID: 7F2E) [92–94].

1.1.3.4   Envelope (E) Protein

The E protein is the smallest of the CoV structural proteins, and probably the most enigmatic since it is not as well characterized as the other structural proteins. The E protein of SARS-CoV-2 has 75 residues, one less than SARS-CoV-1, sharing a 94% sequence similarity due to a single residue deletion and three residue mutations [95–97]. Although there is very little sequence similarity between the seven hCoV E proteins, each typically consists of a short N-terminus, followed by a hydrophobic TMD, and end in a long C-terminus [96]. As a structural protein, E is primarily involved in the assembly of new virions and facilitate their release from infected cells [98–100]. Despite its small size, the protein contains several domains and motifs that confer additional properties to it: the TMD is important for its viroporin activity, the C-terminus contains a PDZ-binding motif (PBM) necessary for PPIs, and a BH3-like domain upstream of its PBM involved in binding to the anti-apoptotic B-cell lymphoma-extra large (Bcl-xL) protein [101, 102]. These domains and motifs demonstrate the major role that E plays in virulence and viral pathogenesis, whereby their absence significantly attenuates the disease severity [103–107].

The mechanism by which E facilitates viral release is not yet clear, but its presence enhances the release of new viral particles [98]. The E protein participates in lysosomal deacidification, the proposed mechanism by which CoVs egress from the cell, which also serves to prevent premature cleavage of the S protein, thereby promoting the release of infectious virions [99, 108, 109]. Mutants of the E protein have revealed that distinct motifs are associated with some of its functions: its amyloidization motif and PBM were critical for the release of virus-like particles (VLPs), and the viroporin activity, C40/C43, or K63 dibasic motif were required for lysosomal deacidification [99]. Numerous host proteins have also been reported to interact with the SARS-CoV-2 E protein through its PBM, highlighting the importance of it both in the viral life cycle and as a therapeutic target [110–117]. No full-length structure of the hCoV E protein has been experimentally resolved yet. Only the dimeric (PDB ID: 8U1T) and pentameric (PDB ID: 7K3G) structures of the SARS-CoV-2 E TMD have been resolved by solid-state nuclear magnetic resonance (NMR) spectroscopy and deposited in the PDB, along with a few structures of host proteins complexed to peptides of the

E protein [118, 119]. Several studies have, however, generated full-length structures *in silico* using different computational tools [120–126].

1.1.3.5 Accessory Proteins

The accessory proteins of CoVs are highly variable and virus-specific, with limited conservation. Due to this lack of homology to other CoV accessory proteins, even within individual species, their molecular functions remain largely unknown. Nevertheless, they are thought to contribute primarily to modulating the host's responses to the infection and are considered determinants of viral pathogenicity [127–130].

The largest of the accessory proteins is ORF3a. It comprises 275 residues and is a transmembrane protein that functions as a viroporin [131, 132]. The cryo-EM structure of 3a reveals that it has three transmembrane helices followed by two anti-parallel β-sheets in its cytoplasmic domain, which forms a β-sandwich [133]. As a dimer, the transmembrane helices form an ion channel which is lined interiorly with polar/charged residues and allows for cations to be transported, showing a preference for Ca^{2+} and K^+ over Na^+ ions. It contains a cysteine-rich domain, a tyrosine-based sorting motif (YXXΦ), and a diacidic EXD motif, all of which contribute to its function in various aspects of viral pathogenesis, such as activating NF-κB signalling and NLRP3 inflammasomes to promote interleukin (IL)-1β secretion, and to induce lysosomal damage and cell death by necrosis [134]. The much shorter, 41-amino acid ORF3c is also predicted to be a transmembrane protein [135]. It does not significantly affect the replication of SARS-CoV-2 and has been shown to suppress IFN-β signalling by inhibiting retinoic acid-inducible gene I (RIG-I) and melanoma differentiation-associated protein 5 (MDA5) [136, 137]. The ORF3d protein is a 51-residue accessory protein that, as of yet, has no determined function [138].

ORF6, a 61-residue protein, is reported to be localized to the ER and membrane vesicles, such as autophagosomes and lysosomes, within the cell [139]. Currently, it is known to function as an IFN antagonist [140]. ORF7a is a type-I transmembrane protein that has a length of 121 residues and is 85% identical to the ORF7a of SARS-CoV-1 [141]. It comprises a 15-amino acid N-terminal signal peptide, followed by an ectodomain, a transmembrane region, and a cytoplasmic di-lysine motif (KRKTE) for ER localization [142]. The protein is polyubiquitinated at K119, possibly when the virus exploits the host ubiquitin system, which may subsequently enable ORF7a to inhibit the type I IFN response [143]. The high efficiency with which it binds to $CD14^+$ monocytes has suggested a possible role for recruiting monocytes to the lungs during COVID-19. This binding also decreases antigen presentation but induces inflammatory cytokine expression, suggestive of an immunomodulatory role in viral pathogenesis [144].

The crystal structure of ORF8 shows that it exists as an asymmetric dimer with a monomeric core of 60 amino acids. Its N-terminal contains a signal for ER localization and the remaining β-strand core resembles that of immunoglobulin (Ig)-like folds but lacks the C-terminal transmembrane domain found in other Ig superfamily proteins. The eight anti-parallel β-sheets of each monomer are stabilized by three intramolecular disulphide bonds [145]. Specific interfaces of ORF8 suggest that it is capable of oligomerization, but its biological function and the significance of the interfaces remain to be determined. Exogenous overexpression of the protein has shown that it can disrupt IFN signalling, and it downregulates major-histocompatibility complex (MHC) I expression [146, 147].

ORF9b is 97 amino acids long and structurally resembles ORF9b of SARS-CoV-1 by superimposition (RMSD: 1.14 Å) [142, 148]. Structural analysis reveals that it exists as a dimer primarily composed of β-sheets. The β-strands from each monomer interlock to form a twisted anti-parallel β-sheet at the interface. The hydrophobic cavity at the centre of the dimer appears to accommodate lipids and based on its orthologue structure in SARS-CoV-1, it suggests that the protein might function in a role similar to viral assembly [149]. Its localization to the mitochondrial membrane alludes to its function in supressing the IFN response by interacting with the adaptor protein translocase of outer mitochondrial membrane 70 (TOM70) [150].

1.2   VIRAL REPLICATION

Before attachment, the SARS-CoV-2 S protein primarily adopts a closed conformation in which all its RBDs are buried and inaccessible to ACE2. Conversely, in its open state, the RBD faces upward, and the receptor-binding motif (RBM) is exposed to facilitate interaction with ACE2 [47, 48]. The S protein then recognizes ACE2 on the host-cell surface and the binding affinity between ACE2 and the RBD of S is critical in determining SARS-CoV-2 infectivity [23] (Figure 1.4). Hinges on the S protein allow it to flexibly orient the RBD and facilitates effective engagement with ACE2 [9, 15, 151]. The RBM, within the RBD, interacts directly with ACE2 and this interaction is 10-fold higher in SARS-CoV-2 than in SARS-CoV-1 due to a more compact conformation of the RBD [23, 47]. Specific mutations in the binding region allow for ionic and π–π interactions between S and ACE2 that results in SARS-CoV-2 S having a higher affinity for ACE2 [48]. After binding to ACE2, TMPRSS2 cleaves the S1 and S2 subunits at the furin cleavage site located between the two

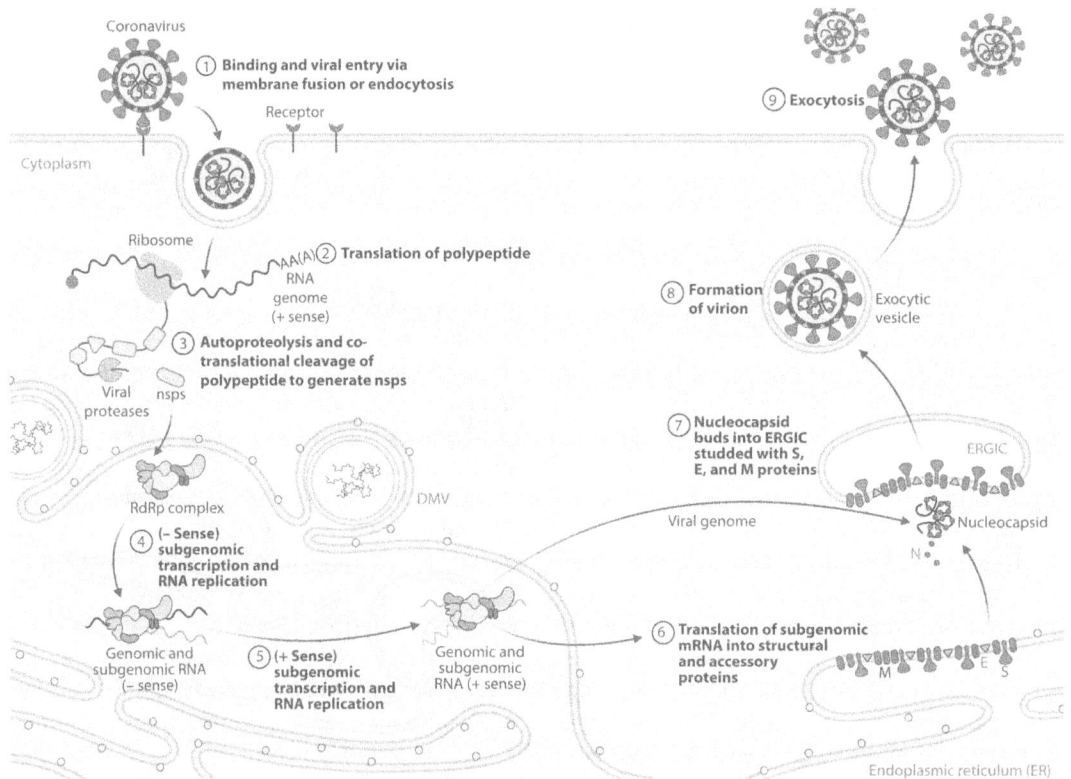

FIGURE 1.4   The typical coronavirus (CoV) replication cycle inside an infected cell. The CoV spike (S) protein binds to the angiotensin-converting enzyme 2 (ACE2) receptor on host cells and the CoV RNA genome is deposited into the host cytoplasm after membrane fusion or receptor-mediated endocytosis (1). Host translation machinery translates the positive-sense viral RNA (2) to produce polyproteins that are sequentially cleaved by viral proteases to generate the components of the RNA-dependent RNA polymerase (RdRp) complex (3). The RdRp complex generates both negative-sense subgenomic RNAs (sgRNAs) and genome-length RNAs using the original viral genome as a template (4). These, in turn, serve as templates for synthesizing more positive-sense full-length genomes, to be incorporated into viral progeny and subgenomic mRNAs (sg-mRNAs) (5). The sg-mRNAs are translated into structural and accessory proteins (6). The Nucleocapsid (N) protein binds to positive-sense genomic RNA and buds into the endoplasmic reticulum-Golgi intermediate compartment (ERGIC), which contains the structural proteins S, E, and M already translated from positive-sense sgRNAs (6 and 7). The enveloped viral progeny can be exported from the cell by exocytosis (8 and 9). (This figure was created and adapted using BioRender.com [183].)

subunits on the S protein [22]. Although other proteases such as cathepsin B and L are also involved in proteolytic cleavage, viral entry relies mainly on TMPRSS2 cleavage as inhibition thereof was sufficient to prevent viral entry [22]. Following cleavage, the S protein changes conformation from a prefusion state to a post-fusion state and exposes the fusion peptide (FP) with the FP domain. The S2 subunit can now mediate fusion between the viral envelope and host-cell membrane through its FP domain [152]. The hydrophobic amino acids of the FP now become exposed and can insert into the membrane of the host cell [153]. The two heptapeptide repeat (HR) domains contain the sequence motif 'HPPHCPC', where 'H' corresponds to a hydrophobic residue, 'P' to a polar residue, and 'C' to a charged residue, which contribute to fusion by mediating the interactions between the adjacent S trimers [154]. When the HR1 and HR2 domains of each S monomer interact within the S trimer, they form a six-helix bundle structure, which increases the proximity between the viral envelope and host-cell membrane to facilitate membrane fusion by a yet to be characterized process [155, 156].

Although the S protein primarily interacts with ACE2, the virus can infect other cells and potentially utilize alternative entry mechanisms using different surface receptors. These include neuropilin-1 (NRP-1), cluster of differentiation 147 (CD147), Toll-like receptor 4 (TLR4), and Arginine-Glycine-Aspartic acid (RGD)-recognizing integrins [157–161]. Alternatively, the virus can enter the host cell by exploiting its innate endocytosis pathway rather than using TMPRSS2. Briefly, if the virus remains attached to the host-cell surface for a prolonged period, it triggers receptor-mediated endocytosis and the particle becomes enveloped in an endosome. The endosome acidifies to activate cathepsins B and L which cleave the S protein, and results in membrane fusion between the viral envelope and endosomal membrane [52]. Regardless of the entry mechanism, the viral genome is released into the host cell after membrane fusion, and viral replication begins [22, 162].

The ~30 kb long genome is translated in the cytoplasm by host ribosomes [163]. Ribosomal frameshifting is exploited to translate the first two-thirds of the genome to produce the polypeptides pp1a and pp1ab [164]. Together, these two polypeptides comprise 16 non-structural proteins (nsps) that are produced when the precursor polyproteins pp1a and pp1ab are self-cleaved by the viral proteases papain-like protease (PL^{pro}) and the main protease (M^{pro}) [165, 166]. Many of the resulting nsps form part of the viral replication-transcription complex (RTC), which is essential for viral replication, and each one has its own function(s) [11, 167]. The protease PL^{pro} cleaves the first four nsps (nsps1–4) whereas M^{pro} cleaves the remaining nsps5–16 after asymmetric dimerization [168, 169]. While the core enzymatic functions of RNA synthesis, proofreading and modification is attributed to nsp12–16, the nsps2–11 appear to support the viral RTC in various ways, including serving as cofactors necessary for replication and remodelling intracellular membranes to form replication organelles, or double-membrane vesicles (DMVs) that accommodate viral RNA synthesis [170, 171]. Nsp1 is rapidly released to shut down the translation of host mRNAs [172]. Nsp12, also known as the RNA-dependent RNA polymerase (RdRp), is responsible for synthesizing new viral RNA in conjunction with its cofactors nsp7, nsp8, and nsp14; nsp8 possesses primase activity whereas nsp14 exhibits a 3'–5' exonuclease activity and provides a unique RNA proofreading function [127, 173, 174]. The capping machinery is not completely characterized yet but is currently understood to consist of nsps 10, 13, 14, and 16 each of which performs a different function [170, 175]. The successive steps in viral replication are both temporally and spatially coordinated and occur simultaneously to ensure the optimal and efficient production of viral progeny.

The viral RTC is described in much greater detail in other papers but will be summarized here [175, 176]. At this point, the RTC has proteolytically matured, and the translation of genomic RNA (gRNA) is underway. Viral RNA synthesis occurs inside DMVs and starts by using the gRNA as the template to produce a negative-sense single-stranded complement of the full-length genome as well as negative-sense subgenomic RNAs (sgRNAs). In turn, the negative-sense gRNA and sgRNAs serve as templates to synthesize new positive-sense gRNA and subgenomic mRNAs (sg-mRNAs) [177]. As the process continues, the full-length, positive-sense gRNA can be used to produce additional replicase polyproteins, serve as a template for additional negative-sense gRNA, or be packaged into the progeny virions. The sg-mRNAs are used to produce the structural proteins that function in the assembly of new virions. Accessory proteins are also produced from some of the sg-mRNAs, and they function predominantly to modulate the immune response to the virus [178]. All sg-mRNAs share the same 3'-terminal sequence, and they carry a common leader sequence at the 5' end that is identical to the gRNA 5'-terminal sequence, a feature that facilitates recognition and binding by the viral polymerase for efficient replication and transcription. The 5'-leader sequence is formed through a discontinuous step, when the RTC terminates at the 3'-proximal quarter of the gRNA template and relocates near the 5' end where synthesis of negative-sense strands resume [179]. The sg-mRNAs produced from the discontinuous transcription step form a characteristic nested set of positive-sense sg-mRNAs that can either serve as templates for transcription or be exported through a molecular pore in the DMV and translocated to the ER/Golgi where they can be translated into the desired proteins [180, 181].

The 'core' of the RTC exists as a heterotrimer comprised of the RdRp and nsps 7 and 8, with RdRp functioning as the main component [182]. The replicating RNA binds to the conserved polymerase motifs of RdRp (A–G), where the active site is located at motifs A–E, RNA synthesis occurs at motif C, and motifs F and G position the RNA template. A single copy of nsp7 and two copies of nsp8 bind to RdRp, both of which function as processivity factors. Many of the SARS-CoV-2 RdRp structural features are consistent with what is known about SARS-CoV-1, but a previously unresolved β-hairpin was found at the N-terminal and predicted to stabilize the overall structure [182]. The structure of the RTC allows for the positively charged RNA template and incoming nucleoside triphosphate (NTP) substrates to be positioned in a central hydrophilic cavity so that RNA synthesis can occur in a template-directed manner. As nucleotides are incorporated into the nascent RNA strand, a dsRNA intermediate is formed which then leaves the complex through the front side of the RdRp [182]. RNA viruses are known for having high mutation rates, but the RTC confers to CoVs a mutation rate that is one order of magnitude lower than most other RNA viruses [183]. This high-fidelity replication and proofreading are attributed to the RTC's ability to remove erroneously incorporated nucleotides from the nascent RNA strand, by nsps 13 and 14. This appears to be accomplished when nsp13 pushes the mismatched RNA duplex backward into the RdRp, exposing the mismatched nucleotide to nsp14, which can then be removed by nsp14 [184, 185]. The viral mRNAs are then finally processed when they are polyadenylated at the 3' end and capped at the 5' end by nsps 13, 14, and 16 [186].

Viral progeny are produced at the endoplasmic reticulum-Golgi intermediate compartment (ERGIC) where the membrane-associated structural proteins S, M, and E, have assembled and inserted in the membrane of the ER [180]. The newly synthesized genome binds to dimerized N proteins and is densely packaged into RNPs that resemble a beads-on-a-string conformation [180]. This is proposed to allow the unusually large viral genome to be packaged into viral progeny while still maintaining the high steric flexibility required to be incorporated into the virions. How genome packaging between N and RNA is driven is not entirely clear, but it appears that the T20 region in the genome and the N protein's IDRs might play a role in this [187, 188]. The condensed RNPs then move to the cytoplasmic side of the ERGIC where they interact with C-terminus of the M protein [74, 180]. It is not entirely clear yet whether M or E induces membrane curvature and how this is accomplished, but both M and E serve as an assembly hub, and once all the structural proteins have been assembled at the ERGIC, a positively curved membrane is formed with the S proteins in the prefusion form located on the luminal side and the RNPs on the cytosolic side [180]. Vesicles containing new viral progeny, sometimes with multiple viruses per vesicle, start to form during the budding process [181]. While it is generally understood that CoVs bud and egress via exocytosis, recent evidence showed that new virions are released from infected cells through the lysosomal trafficking pathway [109]. This coincides with recent evidence that the CoV E protein deacidifies lysosomes during egress to protect the S protein and, by extension, newly formed virions from the acidic lysosomal environment [99, 108, 189].

1.3   EPIDEMIOLOGY AND RESERVOIR

The origin of SARS-CoV-2, as it stands, remains unclear. Despite having made some significant advancements on COVID-19 over the past 4 years, not much progress has been made regarding the origin of the virus [190, 191]. It is of utmost importance to identify the origin and potential hosts of viruses, such as SARS-CoV-2, that are capable of causing pandemics. This will allow us to implement proper epidemiological and public health measures so that we can prevent or, at best, mitigate future pandemics [190, 192]. Coronaviruses are zoonotic viruses with animal reservoirs, which suggests that SARS-CoV-2 most likely has an animal reservoir [193]. Most of the first COVID-19 cases were epidemiologically linked to the Huanan Seafood Wholesale Market, which sells seafood and live animals [194, 195]. Some of the initial reported cases had no link to the seafood market and suggested that the market might not have been the initial source of exposure to the virus [196]. A later study, however, found that even such cases were geographically centred around the market. The study also found that animals sold at the market were susceptible to live SARS-CoV-2 and environmental samples, spatially associated with the vendors selling the live animals, were positive for SARS-CoV-2 [197].

An early study reported finding a 1378 bp gene fragment in the SARS-CoV-2 S gene that is also found in several naturally occurring bat CoVs, and bats, such as the horseshoe bat *Rhinolophus affinis*, are reservoirs for many CoVs [198–200]. The SARS-CoV-2 full-length genome is 96.3% identical to the CoV 'RaTG13' found in *R. affinis*, and the ORFs of SARS-CoV-2, including the hypervariable S protein and ORF8, are also >90% identical to RaTG13, corroborating that SARS-CoV-2 likely originated in bats [201]. However, the RBD of RaTG13 and SARS-CoV-2 S share less than 90% sequence similarity, and other studies have found sequences more closely related to other circulating bat CoVs [202–204]. More recently, the bat CoV BANAL-52 and SARS-CoV-2 was found to share a higher genomic identity (96.8%) and that the nucleotide conservation of BANAL-52 in the NTD and RBD of its S protein is higher than that of RaTG13 [205]. The RBD of BANAL-52 differs by one residue and could bind to human ACE2 receptors, and lentiviral vectors with this RBD could infect and replicate in host cells. Recombination analysis revealed that SARS-CoV-2 has a mosaic genome to which RaTG13 and BANAL-52 contributed, but no sequence from a pangolin CoV was immediately associated with a recombination event at the origin of SARS-CoV-2 [206, 207]. Interestingly, the authors pointed out that these S proteins lack a furin cleavage site, which relates to another recent study.

Others have also noted the genomic similarity between SARS-CoV-2 and that of bat CoVs, but remarked that there were stretches of the genome that showed notable similarity with viruses from other animals [208]. One study reported conserved molecular signatures in the S protein, suggestive of recombination between sarbecovirus lineages, possibly even between SARS-CoV-2 and a pangolin CoV (PgCoV) [209]. A recent review of the SARS-CoV-2 S protein sequence and the origins of amino acid insertions concluded that it is plausible for homologous recombination to have occurred between a bat CoV and an unknown intermediate animal host [210]. A recent study demonstrated that the wild-type Pg CoV GD (PgCoV GD) could replicate in primary human cells and mice expressing human (h) ACE2 (hACE2) and demonstrated a capacity for airborne spread [211]. Despite the lack of a furin cleavage site in its S protein, PgCoV GD could infect and grow as efficient as SARS-CoV-2 in cell lines overexpressing furin. In fact, the SARS-CoV-1 S protein lacked a furin cleavage site but was able to replicate efficiently in primary human airway epithelia and efficiently transmit between civets, humans, and other animals [212]. It was previously shown that PgCoV GD shares 91.2% sequence similarity with SARS-CoV-2 and its RBD residues are 96.8% similar, and it can infect human cells [213]. However, the relationship between PgCoV GD and SARS-CoV-2 is unclear since their whole genome similarity is low, and this still makes *Rhinolophus* bats currently the most probable origin of SARS-CoV-2 [214].

Pangolins have been reported to contain a diversity of sarbecoviruses that are genetically related to SARS-CoV-2, including those that circulate in horseshoe bats [201, 204, 205, 213]. Given the significant homology between the RBDs of PgCoV GD, BANAL-52, and SARS-CoV-2, in conjunction with the recombination potential of bat CoVs, it might not be inconceivable for a bat CoV to have recombined with a PgCoV. This might point to the possibility of SARS-CoV-2 originating from an ancestral bat CoV (reservoir) that recombined with a PgCoV (intermediate host), which could have happened in the months leading up to the reported outbreak in December 2019, as some suggested that the virus may have already circulated prior to the actual outbreak [215, 216].

1.4 TRANSMISSION

As a respiratory virus, one of the major, direct routes of transmission for SARS-CoV-2 is via respiratory droplets from coughs and sneezes, and human-to-human transmission has already been established [217, 218]. Airborne transmission through aerosols is another transmission route, even containing replication-competent viral particles [219, 220]. Interestingly, one study found that in a contained environment with minimal air flow, mere speech can also serve as a route of transmission [221]. While an infectious person can produce both respiratory droplets and aerosols, the droplets will quickly fall to the ground while the air remains enriched with aerosols, which not only suggests that aerosol transmission is the predominant route of transmission for SARS-CoV-2, but also highlights the importance of maintaining adequate distance to minimize transmission of the virus between persons [222].

The virus was also reported to be transmitted indirectly by means of contaminated surfaces, with many objects and surfaces such as glass, stainless steel, plastic, clothing, cardboard, and banknotes being tested [223, 224]. However, many studies were either conducted under ideal experimental conditions, which do not reflect real-world conditions, or measured viral RNA rather than viable virus [223, 225, 226]. In cases where the replication-competent virus was investigated, levels of viable SARS-CoV-2 on contaminated surfaces would decrease rapidly and occur at extremely low levels or become undetectable within 2 days [225, 227–229]. Similarly, both viral RNA and replicative virus were recovered from faecal samples of COVID-19 patients, but the presence of RNA alone in faeces is not indicative of active viral shedding and the levels of viable replicative virus are an important consideration in establishing a new infection [230–235]. It is worth noting that a single study reported finding viral RNA in a single semen sample, but several others have found no viral RNA in semen samples, demonstrating that there is, as of yet, no indication of SARS-CoV-2 being sexually transmissible [236–240].

Vertical maternal-foetal transmission of SARS-CoV-2 was a major concern since the start of the pandemic and is still a matter of debate. Several studies have reported the presence of SARS-CoV-2 RNA or virions in the placenta, amniotic fluid, or umbilical blood of COVID-19 positive pregnant mothers, but no viral particles were detected in foetal or neonatal tissues [241–248]. A few studies also found no viral RNA or particles in the amniotic fluid or cord blood of COVID-19-positive pregnant mothers [249–252]. However, the presence of anti-SARS-CoV-2 IgM antibodies in neonates suggests that the neonate must have been exposed to the virus *in utero* and mounted an immune response against it, likely transmitted from the mother [244, 246, 252]. This, coupled with IgM antibodies not being able to cross the placenta and neonates testing positive for only a brief time after birth, suggests that vertical transmission of SARS-CoV-2 during pregnancy is possible [246, 252, 253]. Studies should, however, be cautious about interpreting a positive polymerase chain reaction (PCR) result as the virus being present in neonates since mere fragments of viral RNA can remain in tissues and be amplified to give a false impression that viral transmission occurred between mother and foetus [253–255]. Negative PCR results from nasopharyngeal swabs of neonates, coupled with the expression of ACE2 and TMPRSS2 in neonatal tissues, suggests that organs and tissues other than the lungs might be a target for SARS-CoV-2 replication in neonates, and that samples for PCR testing also be collected from other tissues [245, 256, 257].

The novelty of SARS-CoV-2 has made its transmissibility an area of interest in the early stages of the pandemic and is given by the basic reproductive number (R_0). This is defined as the mean number of secondary infections that are the result of an infected person in a susceptible population and is generally useful in understanding whether a pathogen poses an epidemic threat [226, 258]. The approximations of R_0 value for SARS-CoV-2 have increased from the wild-type strain, responsible for the initial outbreak, ($R_0 \approx 3$) to the alpha variant ($R_0 \approx 4.5$) to the delta variant ($R_0 \approx 8$) [259, 260]. Another study reported different R_0 values for each variant, but the values nonetheless reflected a similar trend: that the transmissibility of variants increased and Omicron exhibited the largest R_0 value [261]. The increased transmissibility of each variant has been attributed to the increasing number of accrued mutations in the S protein [262, 263].

1.5   CLINICAL FEATURES

Persons of all ages are susceptible to COVID-19 for which a variety of symptoms have been reported, and the severity can range from asymptomatic to mild, moderate, and severe [264, 265]. The most common symptoms included fever, dry cough, sore throat, runny nose or nasal congestion, fatigue, sputum production, difficulty breathing, and chills [266–268]. Other symptoms such as headache, myalgia, loss of smell or taste, shortness of breath, nausea, vomiting, abdominal pain, and diarrhoea have also been reported [268–271]. Persons with comorbidities were generally more prone to severe disease and more likely to be hospitalized or succumb to the infection [272–275]. The clinical presentation of COVID-19 does appear to differ between age groups. Most children and younger persons tend to be asymptomatic or only develop a mild form of the disease, whereas men older than 60 years of age, with comorbidities, are generally more likely to develop a severe form that requires hospitalization, or they may even die from the infection. In persons older than 80 years, the mortality rate increases significantly [276–278].

Immunocompromised persons are generally at higher risk for developing severe COVID-19 due to their compromised immune system. However, based on the association between a hyperactivated immune response and severe COVID-19, persons with a weakened immune system may be less likely to develop severe COVID-19 due to a cytokine storm [279, 280]. Their compromised immune system does not make them less susceptible to the SARS-CoV-2 infection, rather, their weakened immune system is less likely to trigger a cytokine storm that would otherwise culminate in a severe form of the disease [281, 282]. After all, immunosuppressive treatments such as dexamethasone and anti-cytokine drugs improve patient outcome and prognosis [283–285]. Immunocompromised persons are, nevertheless, at risk of contracting COVID-19 with longer recovery times or hospitalizations and higher mortality rates compared to immunocompetent individuals [286–289].

Asymptomatic persons have been an ongoing area of research during the pandemic [290]. An asymptomatic person is generally characterized as someone who has tested positive for the SARS-CoV-2 virus but exhibits no clinical symptoms consistent with COVID-19, making it fundamentally challenging to distinguish asymptomatics from healthy individuals [291, 292]. The concern is that such persons may still be able to transmit the virus and infect others, placing the elderly and those with comorbidities at risk [293, 294]. Indeed, a joint review of several articles concluded that the absolute and relative risk of transmission by asymptomatic individuals is probable [295]. There exists a small degree of variability in symptoms, but for the most part, their symptoms are subclinical with favourable outcomes [296]. Several studies compared the differences in demographics, age, laboratory findings, epidemiology, and clinical features in an attempt to understand the differences between symptomatic and asymptomatic persons [296–298]. Studies have found lower antibody responses that are short-lived in asymptomatic persons, whereas symptomatic persons generated stronger, more sustained antibody responses [299–302]. Asymptomatics also exhibited a reduced inflammatory response as evident by lower levels of circulating cytokines and chemokines, a likely reason behind the decreased antibody responses [300, 303]. The T-cell dynamics of asymptomatic persons exhibit reduced proportions of SARS-CoV-2-specific T cells, and as one case study showed, regulatory T cells may play a role in regulating the immune response to SARS-CoV-2 by suppressing other T cells early in the infection, possibly mitigating the risk of an excessive immune response

[304, 305]. This apparent tighter regulation might explain the poor, or lack of, clinical presentation of asymptomatic persons [306]. Thus, it would appear as though asymptomatic persons mount some measure of a humoral response but that their antiviral cellular response is much more functional and effective [307].

Early studies have noted that men appear to be at a higher risk for developing severe COVID-19 than women [276, 308, 309]. Differences in response to stress and the influence of sex-specific hormones have been suggested to play a role in this [310]. One study reported positive PCR results from some semen samples and, coupled with a higher expression of both *ACE2* and *TMPRSS2* in testes compared to ovaries, the authors suggested that men might exhibit a delayed clearance of the virus [237, 311]. Biopsies from deceased men have found SARS-CoV-2 viral particles in testicular tissue, indicating that the testis may be a target tissue for the virus. This supports the possibility of a delayed viral clearance in males as the testes are considered an immune-privileged site [312, 313]. However, studies have yet to demonstrate whether SARS-CoV-2 can be sexually transmitted. Other reports have echoed the impact of differences in *ACE2* and *TMPRSS2* expression levels; some also proposed that behavioural and/or social differences may be a factor, while others have offered the different effects that testosterone and oestrogen have on the immune response as an explanation [314–317]. Serum levels of pro-inflammatory cytokines were higher in men than in women, which has been associated with more adverse outcomes and a higher mortality for COVID-19 [318]. It would, therefore, appear as though there is, in fact, a male bias in COVID-19 mortality, but more research is needed to determine whether this is due to a single factor or a combination of factors. It is also worth mentioning that studies have proposed the possibility of a negative impact on reproductive health by COVID-19, whether directly through infection of testicular cells or indirectly from an overactivated immune response [319, 320].

Over the course of the pandemic, it became increasingly apparent that persons recovering from COVID-19 also experience post-infectious complications. The most notable complications are multi-system inflammatory syndrome in children (MIS-C) and post-acute sequelae of SARS-CoV-2 infection (PASC), more commonly known as 'long COVID'. Children infected with SARS-CoV-2 tend to experience a milder form of COVID-19 [321, 322]. However, within a few weeks of recovering from COVID-19, some children developed extreme systemic inflammation resembling Kawasaki disease or toxic shock syndrome [323, 324]. While the clinical features can often vary, persistent fever was a defining feature and some additional symptoms such as diarrhoea, abdominal pain, vomiting, myalgia, fatigue, rashes, enlarged lymph nodes, pink eye, and strawberry tongue were occasionally present [323, 325, 326]. It is considered a serious condition as children are often hospitalized even though death is reported in fewer than 2% of cases [323]. A recent study also reported that very few children experience long-term consequences associated with COVID-19 [327].

MIS-C is considered a consequence of the SARS-CoV-2 infection due to the temporal association between the condition and the infection, although this is neither established nor well understood [328–331]. Several mechanisms have been proposed, and research is ongoing, but a post-infectious mechanism seems plausible [329, 332]. These mechanisms include antibody-dependent enhancement (ADE), blocking of type I and III IFN responses by SARS-CoV-2 leading to a delayed cytokine storm, and a superantigen-like motif in the S protein leading to the clonal expansion of a subset of T cells [329–331, 333, 334]. Regardless of the exact cause or mechanism, the inflammatory nature of the condition typically calls for treatment centred around an anti-inflammatory effect. This includes corticosteroids, intravenous immunoglobulin (IVIG), or cytokine blockers such as tocilizumab, anakinra, or infliximab, depending on the severity [335–338].

Long COVID-19 refers to a group of health problems that develop or persist after recovering from COVID-19. Symptoms and their duration are highly varied, lasting weeks, months, or years, and are often debilitating, with some resolving and others worsening over time [339–341]. While it is generally understood as a consequence of severe COVID-19, it may also occur in mild COVID-19 cases that did not require hospitalization [342, 343]. Symptoms impact many different organ systems, and the severity can range from mild to incapacitating (Figure 1.5). These include,

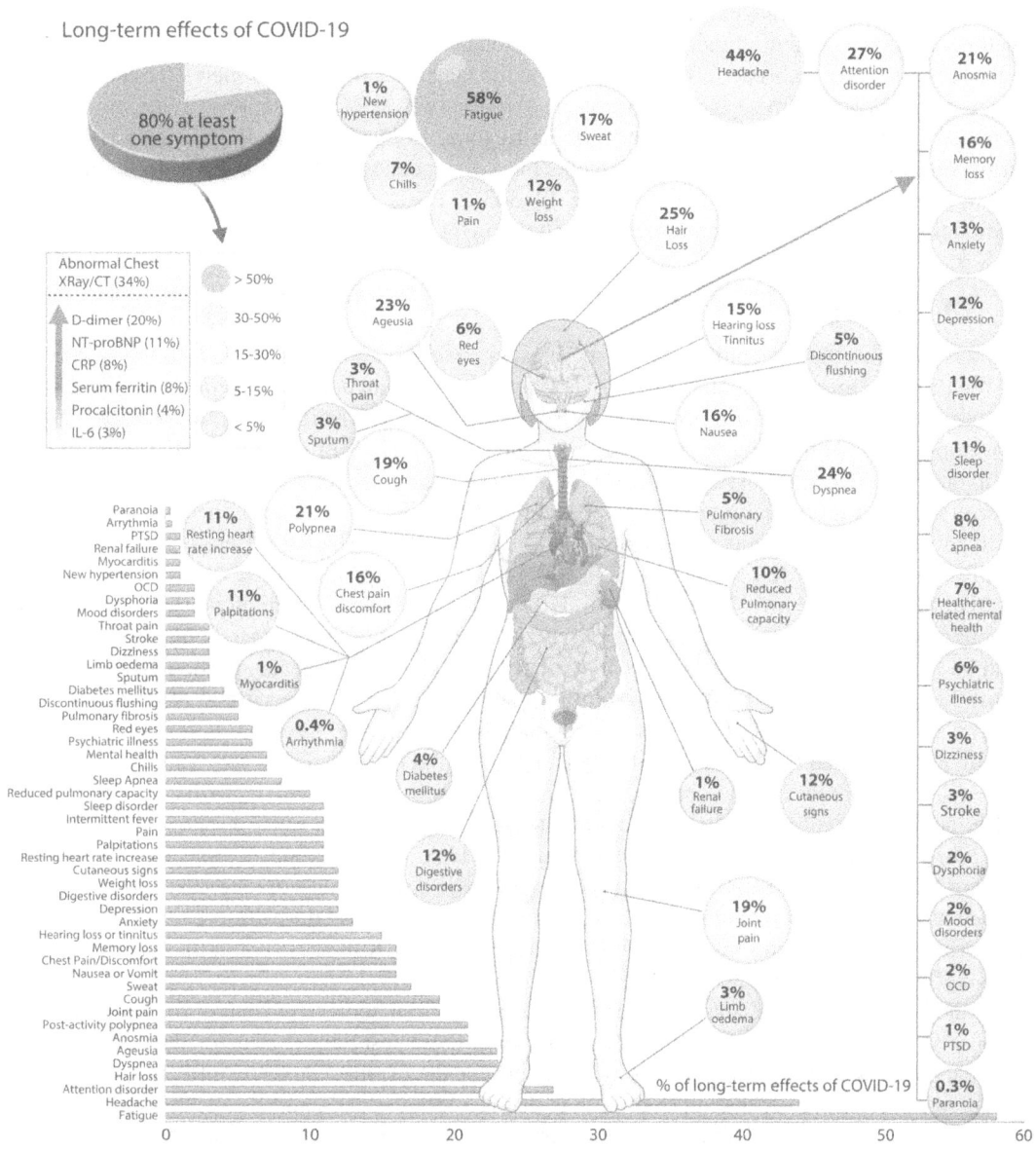

FIGURE 1.5 The long-term effects associated with COVID-19 or long COVID [345].

but are not limited to, fatigue, hair loss, muscle pain, shortness of breath, chest pain, red eyes, throat pain, post-exertional malaise, weight loss, general pain, gastrointestinal disorders, arrhythmia, and palpitations [344]. Neurological and cognitive symptoms are hallmarks of long COVID, and reported symptoms include headache, dizziness, nausea, depression, anxiety, cognitive dysfunction ('brain fog'), and dysautonomia, all of which can impact daily living to different extents [339, 345, 346]. There is no standard diagnostic tool for long COVID, and diagnosis generally relies on a patient's history with suspected or confirmed COVID-19 symptoms [347]. Interestingly, women appear to be at higher risk than men, with older age, obesity, type II diabetes, chronic obstructive pulmonary disease (COPD) like asthma, and specific autoantibodies identified as risk factors [348, 349]. Mechanistic studies are still in the early stages, but some have proposed persisting viral proteins/RNA, organ damage, reactivation of dormant viruses, and aberrant T-cell responses and T-cell exhaustion as potential mechanisms behind long COVID [350–356].

There is currently no broad, effective treatment for long COVID, but treatments for certain components have shown promise or have been effective for subsets of populations. Symptom-specific pharmacological options include β-blockers, naltrexone, IVIG, the drug BC007, and H_1 and H_2 antihistamines [357–360]. Certain supplements have also been reported to assist with the management of long COVID-associated symptoms [361–363]. This is certainly an area of active, ongoing research that warrants more attention as few people suffering from long COVID fully recover [341].

1.6 DIAGNOSTICS

Symptomatic presentation could initially serve as a reasonable diagnosis for COVID-19. However, sole reliance on clinical presentation might not always lead to an accurate diagnosis as the disease presented similar to other viral respiratory tracts infections, such as influenza, or became increasingly impractical with growing infection rates [364, 365]. This, in addition to the potential for asymptomatic carriers, necessitated the use of *in vitro* diagnostic tools [365].

Reverse transcription polymerase chain reaction (RT-PCR) is considered the gold standard for the diagnosis of COVID-19 as it is the most sensitive [365–367]. Several different genes, including E, RdRp, ORF1ab, and N, typically serve as the targets, and a positive PCR result is indicative of an active SARS-CoV-2 infection [368–370]. However, some challenges and concerns arose as the pandemic progressed. Emerging variants raised concerns about false-negative PCR results, but this could be mitigated by using primer sets that amplify multiple target genes, particularly genes that tend to have low mutation rates [371]. Another concern was over the accuracy of detecting viral RNA from different respiratory tract samples, such as throat swabs, nasopharyngeal swabs, saliva, sputum, and bronchial fluid, since viral load is typically higher in the lower respiratory tract (LRT) [201, 365, 372]. Moreover, stool, blood, and urine samples have also been reported to contain viral nucleic acid but might be less sensitive since these are not primary sites of active viral replication [373, 374]. Yet another concern was that an individual could test positive again up to several weeks or months later by PCR. However, there is a very low probability of recovering a replication-competent viral load sufficient to cause a productive infection several days after recovery, regardless of severity of the infection [365, 375, 376]. In such cases, a PCR-positive result most likely reflects genomic fragments rather than an actively replicating virus [377].

Antibody, or serological, testing relies on detecting IgM and/or IgG antibodies produced in response to a SARS-CoV-2 infection [378]. As such, these tests are useful for determining past infections and vaccine efficacy but are not suitable for diagnosing an active infection [379–384]. The advantage of antibody testing over PCR is that it is simpler and can be done in a much shorter time, making it particularly suitable for rapid testing and testing in low-income countries [385, 386]. Several serological assays have been developed to detect antibodies produced against the S and/or N proteins, each with certain benefits and caveats [383]. Antibodies against the N protein are the first to be produced because N is highly immunogenic and abundantly expressed during infection [387, 388]. However, since the S protein is considerably more exposed than N is in a natural infection, and the majority of the initial vaccines were based on the S protein, anti-N antibodies would not be a suitable marker for determining the effectiveness of S-based vaccines. This makes anti-S antibodies an apt marker for neutralizing antibodies that can be used to determine the effectiveness of such vaccines. However, serological assays that rely solely on anti-S antibodies would not be able to distinguish vaccinated persons from those who have recovered from a natural infection, favouring the use of diagnostics that detect anti-N antibodies in such cases [379].

Antigen testing, or immunoassays, detect the S or N proteins, or the S protein RBD of the virus, and can reflect an early, active infection as these proteins are actively produced during viral replication [386, 389]. Rapid antigen test kits that can detect the N protein of the Alpha, Beta, Gamma, Delta, and Omicron variants are also commercially available [390–392]. Like antibody tests, immunoassays are simpler, and results can be obtained quickly [364]. However, it should be noted that while rapid antigen tests offer more convenient and rapid screening for COVID-19, they are generally less sensitive than PCR, increasing the probability of false-negative results [393, 394].

Additionally, in-hospital tests such as X-rays and computed tomography (CT) scans can also be used to assist with a COVID-19 diagnosis. However, such tests only reveal characteristic lung abnormalities associated with COVID-19 and do not specifically detect the presence of SARS-CoV-2. Therefore, they are not advised to be used as a primary diagnostic tool but rather in conjunction with other tests [364, 395]. Nevertheless, chest CTs and X-rays of COVID-19 patients often show bilateral, multilobar ground-glass opacification with a peripheral or posterior distribution, lung consolidation, and peripheral and multifocal location of pulmonary infiltrates [396–398]. Other imaging techniques such as lung ultrasound and positron emission tomography CT (PET/CT) have occasionally also been employed [395, 399, 400].

1.7 PATHOGENESIS

The lungs are the most affected by COVID-19 since it is the target tissue of SARS-CoV-2 and exhibits a high expression of ACE2 [401]. Histopathology reports of post-mortem autopsies have demonstrated diffuse alveolar damage (DAD) as the predominant pattern of lung injury, evident by characteristic features such as interstitial and intra-alveolar oedema, dying pneumocytes, hyaline membranes, microvascular thrombosis, congested capillaries, and hyperplasia of type 2 alveolar pneumocytes [402]. Alveolar cell death or damage disrupts the alveolar epithelium, causing an imbalance in the activation of coagulation and the inhibition of fibrinolysis, another key feature of DAD [403]. The result is a fibrin-rich exudate, which forms hyaline membranes that function to prevent fluid accumulation by sealing the alveoli, but limits oxygen exchange at the same time [404]. The tissue distribution of ACE2 has demonstrated that SARS-CoV-2 can also infect other organ systems and cause organ-specific pathology [405–407]. The virus predominantly exhibits a tropism for the respiratory tract, but one of the hallmarks of severe COVID-19 is the cytokine storm–a systemic hyperinflammatory response characterized by elevated levels of inflammatory cytokines and chemokines, including IL-1β, IL-2, IL-6, IL-10, and tumour necrosis factor alpha (TNFα). In patients with severe COVID-19, the cytokine storm is often associated with a worse prognosis and can lead to acute respiratory distress syndrome (ARDS) which may rapidly progress to respiratory failure. The cytokine storm can cause widespread vasodilation that, coupled with high levels of chemokines in the lung, can also lead to lymphocytic and monocytic infiltrates of the lungs which, in turn, has been associated with alveolar injury [408]. The mechanism that drives the cytokine storm, and its associated immunopathology, is not fully understood but there has been some progress on this front.

Both the IFN-I, consisting of IFN-α and IFN-β, and IFN-III response, consisting of IFN-λ, play an important part in the antiviral response. However, SARS-CoV-2 is very effective at inhibiting the IFN response, as evident by the many viral proteins that can inhibit this response at different levels, likely due to its sensitivity to IFN [409–413]. Indeed, in a study of critically ill COVID-19 patients, peripheral blood responses displayed a reduced IFN response coupled with a pro-inflammatory response [414]. Comparing the *in vitro* transcriptional responses of SARS-CoV-2 and other viruses demonstrated the inability of SARS-CoV-2 to induce a robust IFN-I and -III response, but high chemokine levels were observed for recruiting immune cells [415]. Interestingly, a study of the IFN response in the upper (U) respiratory tract (RT) and LRT of COVID-19 found no difference between the IFN levels in the URT of COVID-19 patients and healthy controls. The bronchoalveolar lavage fluid (BALF) of severe COVID-19 patients, however, contained elevated levels of IFN-I and -III and pro-inflammatory cytokines [416]. This suggests that SARS-CoV-2 replicates efficiently in the URT by suppressing the IFN response, but by the time the virus reaches the LRT, the surge of upregulated IFNs causes an excessive immune response which does more harm than good [416–418]. Indeed, severe COVID-19 patients display enhanced IFN-α and IFN-λ levels 2 weeks after mild-moderate patients, who exhibit an early transient IFN response [419]. This illustrates that

SARS-CoV-2 likely combines a delayed, or dysregulated, IFN response with an excessive inflammatory response that can develop into a cytokine storm and progress to severe COVID-19 and even death [420–424].

Another possible contributing factor to the cytokine storm, and subsequently SARS-CoV-2 pathogenesis, may be the abortive infection of immune cells. Most immune cells in the blood express little to no ACE2, but tissue-resident macrophages, such as alveolar macrophages, highly express ACE2 [425, 426]. This expression can mediate viral entry into immune cells, but other receptors and coreceptors have also been reported to mediate viral entry [157, 427–429]. Several lines of evidence have detected viral replication and products in immune cells indicative of an infection, and this may disrupt the normal functioning of macrophages [430–434]. These dysfunctional, viral-infected macrophages may infiltrate the lungs in a hyperactivated state, a phenomenon known as 'macrophage activation syndrome' (MAS), which can contribute to the production of inflammatory cytokines and lead to ARDS [435]. Infection of macrophages with SARS-CoV-2 leads to pyroptosis, a type of inflammatory cell death in which cellular contents such as damage-associated molecular patterns (DAMPs) and inflammatory cytokines such as IL-1β, are released and evokes an inflammatory response [426, 436]. This leads to an abortive form of infection in which immune-related genes are expressed and attract lymphocytes to the site of infection as a physiological response to infection control [437]. However, it also triggers excessive inflammatory cytokine production that has been reported to correlate with the development of severe COVID-19 [432, 437]. Indeed, monocytes infected with SARS-CoV-2 from healthy donors and COVID-19 patients have demonstrated that the virus induces formation and activation of the inflammasome, which results in the production of IL-1β [438]. Furthermore, deceased COVID-19 patients were found to have macrophages with activated inflammasomes in their lungs [432]. Therefore, it appears that MAS may also contribute to the progression of severe COVID-19.

While the major concern and focus was on the mutations of the S protein, and the resulting variants, there were somewhat neglected mutations in other viral proteins that might also have an impact on viral pathogenesis. The mutation G215C in the CLR of the Delta variant's N protein was reported to become dominant in the variant, outcompeting the other G215 variants. This mutation was suggested to stabilize helices in the CLR region and comparisons to the wild-type N protein showed a significantly more compact structure that led to a more enhanced coassembly between N and nucleic acids. The authors proposed that this mutation might imbue Delta with a greater efficiency to assemble viral progeny and, thereby, enhance its infectivity [439]. While this will have to be validated by further research, it does highlight the importance of tracking mutations in other structural proteins as well to obtain a more complete understanding of how SARS-CoV-2 evolves and finding other aspects to target.

Some studies also pointed out that the pathogenicity of variants seemed to decline as they emerged. The RBD of the Omicron variant reportedly showed a lower pathogenicity than other variants but its antigenicity was increased [440]. The authors reported a decrease in the order of pathogenicity: Wuhan > Gamma > Delta > Omicron, which was also associated with a decreased induction of IL-6 by Omicron. Conversely, the stability of the structural proteins was found to decrease in the reverse order: Omicron > Gamma > Delta > Wuhan. The E protein of Omicron was reportedly the most unstable and suggested to be responsible for a decreased viral load by impeding viral maturation, which in turn caused a less severe form of the disease and led to faster patient recovery. It would be interesting to corroborate this *in vitro* and *in vivo* since the authors used computational tools for this analysis, but it nevertheless points to the possibility of the virus adapting to the human population and becoming more attenuated. Another study noted a potential decrease in the pathogenicity of the Omicron variant that they attributed to a mutation in the E protein [441]. Mutational analysis, based on the National Genomics Data Center (NGDC), revealed that the mutations T9I and T11A in the Omicron E protein were associated with less severe disease in patients compared to the wild-type virus. Mice infected with variants containing these mutations had lower levels of inflammatory cytokines and no severe damage was observed based on pulmonary histopathology. Interestingly, the mutation P71L was reported to confer higher disease severity to the virus by being more lethal to cells, producing higher viral loads, and inducing highers level of inflammatory cytokine production. This study too highlights the need to include the effects of mutations on multiple proteins to obtain a holistic understanding of the pathogenicity of emerging variants, rather than just focussing on the mutations of a single protein.

1.8 PREVENTION AND TREATMENT

The fight against SARS-CoV-2 proved challenging and required a combination of measures. At the onset of the pandemic, the focus was to limit the spread of the virus, and public health interventions such as restricting large gatherings, social distancing, wearing masks, and frequently sanitizing surfaces were implemented to achieve this [222, 442]. These were somewhat effective in delaying the spread of the virus but could not completely break the chain of transmission. Several countries also imposed flight restrictions and hard lockdowns as SARS-CoV-2 continued to spread globally. The rollout of the COVID-19 vaccine(s) was a major milestone in the fight against the virus and was developed to limit the spread of the virus and prevent symptomatic, often severe, illness. Several vaccines have been developed, some of which have been made publicly available while others are still in development [443].

Many of the vaccines were based solely on the S protein of the wild-type that was responsible for the initial outbreak in late 2019, and since variants emerged through key mutations in the S protein, an increasing number of breakthrough infections were observed in which vaccinated persons became infected with the virus [444, 445]. Booster doses then became necessary to increase antibody titres by stimulating existing vaccine-induced memory [445, 446]. Persons could opt for boosting pre-existing immunity with the same vaccine that they initially received (homologous) or a vaccine different from the one they initially received (heterologous) [447]. Persons with naturally acquired immunity from the SARS-CoV-2 infection were also encouraged to be vaccinated, and this hybrid immunity was reported to provide a more effective protection than natural immunity or vaccine-induced immunity alone [445, 448–450]. Despite breakthrough infections, in which vaccinated persons still contract COVID-19, vaccinated persons generally experienced less severe symptoms after being infected [451].

Several studies attempted to provide a duration of immunity, but the lack of a baseline antibody titre for which an individual is certain to be protected from infection made it difficult to accurately determine how long a person would be immune to the disease. Nevertheless, most studies reported sustained elevated antibody titres as a measure of immunity for approximately 6 months [445, 452]. It should be noted, however, that waning immunity is not unique to SARS-CoV-2. Immunity to infections naturally decline over time, whether naturally acquired or vaccine-induced [453–455]. Vaccines have since been updated or released as modified monovalent versions based on recent variants or released as bivalent vaccines which are designed to provide protection against the SARS-CoV-2 wild type and/or the latest circulating variant [445, 456–458].

1.8.1 VACCINES

1.8.1.1 mRNA Vaccines: mRNA-1273 (Moderna) and BNT162b2 (Pfizer-BioNTech)

Moderna and Pfizer were the first companies to roll out their COVID-19 vaccines, also marking the release of the first mRNA vaccine. The vaccines were composed of modified mRNA, encapsulated in a lipid nanoparticle, that encoded the full-length S protein of SARS-CoV-2 [459, 460]. Both have been shown to induce a robust immune response comprised of antibodies and $CD4^+$ and $CD8^+$ T cells [461–463]. However, the emergence of variants demonstrated that the initial two-dose regimen for both vaccines would not confer durable protection, as neutralizing antibodies against the wild-type and variants Alpha, Beta, Gamma, and Delta declined after 6 months, and only weakly neutralizing antibodies were elicited against Omicron [464]. The varying availability of vaccine doses, coupled with updated vaccine recommendations, resulted in the combined use ('mixing and matching') of some vaccines. An adenoviral vector prime, boosted with an mRNA vaccine, demonstrated superior induction of humoral and cellular immune responses compared to homologous prime-boost strategies [447, 465, 466]. Results were similar for mRNA-1273 and BNT162b2 as well as for ChAdOx1 and Ad26.COV2.S used as primary and secondary vaccine modalities, respectively.

Both Moderna and the Pfizer vaccines have since been updated to monovalent and bivalent vaccines to confer protection against the Omicron variant [467]. However, the results were not as significant as anticipated showing that, in a small group of people, the bivalent vaccine-induced immunity against Omicron started waning after 3 months [468, 469]. The reduced effectiveness of the updated vaccines was attributed to imprinting–a phenomenon in which the immune systems of the recipients of the bivalent vaccine were primed against the wild-type virus by repeated previous doses of the vaccine based on the wild-type virus. Thus, their immune systems were likely more responsive, or primed, to the epitopes shared between the wild-type and Omicron variant rather than to the new Omicron-specific epitopes [470, 471].

1.8.1.2   Viral Vector Vaccine: Ad26.COV2.S (Johnson & Johnson) and AZD1222/ChAdOx1 (AstraZeneca)

The COVID-19 vaccines manufactured by Johnson & Johnson and Oxford-AstraZeneca are both based on a human adenoviral vector that has been modified to produce the SARS-CoV-2 S protein inside the cells of recipients [472]. The Ad26.COV2.S vaccine is designed to be a one-shot vaccine, whereas the AZD1222 vaccine is given in two doses [473]. Neutralizing antibodies against most variants induced by one dose of Ad26.COV2.S were weaker than those induced by the mRNA vaccines, but were moderately neutralizing against the Delta and Omicron variants and lasted for 6 months [464, 474]. Recipients of the two-dose AZD1222 vaccine elicited antibody responses against the SARS-CoV-2 S protein as well as polyfunctional CD4$^+$ and CD8$^+$ T-cell responses that were primarily skewed towards a T_H1 profile [475, 476]. The two-dose AZD1222 vaccine regimen was able to induce effective neutralizing antibody activity against the Delta variant, but failed to induce a similar response against the Omicron variant [477].

An overall lower immunogenicity was observed for a homologous ChAdOx1-ChAdOx1 prime-boost scheme, which might be due to immunity generated against the vector. Conversely, the Ad26.COV2.S vaccine encodes a prefusion-stabilized form of the S protein that has an additional stabilization at the furin cleavage site, which likely results in a different type of quality and quantity response in the heterologous Ad26.COV2.S-mRNA primer-boost scheme [445]. Post-vaccination clotting complications were reported in recipients of both the ChAdOx1 vaccine and the Ad26.COV2.S vaccine, first for the former and then the latter, causing the use of both vaccines to be halted until the risks were reassessed [478–480]. After reassessment, the vaccines were cleared for use again noting that there was a plausible relationship between the use of the vaccine and thrombosis occurring in vaccinated persons.

1.8.1.3   Inactivated Virus Vaccine: CoronaVac (Sinovac Biotech)

Sinovac Biotech manufactures CoronaVac, a whole inactivated SARS-CoV-2 virus COVID-19 vaccine [481]. The CN2 strain of the virus was obtained from the bronchoalveolar lavage fluid of a patient, after which it was purified, adapted to Vero cells, and inactivated by β-propiolactone. A few studies have found that CoronaVac did not induce a durable, effective neutralizing antibody response, and the antibody responses were consistently lower compared to BNT162b2 [482–485]. However, one study compared both the humoral and cellular responses of CoronaVac to that of BNT162b2 and found that CoronaVac elicited significantly higher CD4$^+$ and CD8$^+$ T-cell responses against the structural proteins after one month than BNT162b2 did [486]. Another study showed that a CoronaVac homologous prime-boost vaccination scheme was considerably less effective than a heterologous prime-boost scheme with either BNT162b2 or ChAdOx1 [485]. These two studies demonstrate the risk of relying solely on antibody production as a measure of vaccine effectiveness as well as the importance of considering heterologous prime-boost as an effective vaccination strategy.

1.8.1.4 Mucosal Vaccines

There has been quite some interest in mucosal vaccines as a way of inducing immunity against SARS-CoV-2 given that the oral and nasal cavities, lined with mucosal surfaces, serve as the primary portals of entry for the virus [487–489]. Advantages to mucosal immunity include inducing a strong mucosal and systemic immune response and being non-invasive as vaccines can be administered via oral, sublingual, or intranasal routes, depending on the desired site for inducing the immune response [490, 491]. Studies on mucosal vaccines designed against SARS-CoV-2 are limited, but provided insight into whether mucosal vaccination could be a safe and effective means of inducing durable systemic and mucosal immunity. Studies on mucosal vaccines for SARS-CoV-1 and MERS-CoV have shown promise that this type of vaccine might offer effective protection [488]. India and China have both recently approved the use of their own intranasal mucosal vaccines named iNCOVACC and Ad5-nCOV, respectively. Recent trial results of iNCOVACC showed that it induces a neutralizing humoral response against the wild-type virus that also cross-reacts with the Omicron variant, as well as T-cell memory and high serum IgA titres [492, 493]. A small phase I trial of aerosolized Ad5-nCOV showed that a two-dose regimen of aerosolized Ad5-nCOV elicited a strong neutralizing antibody response and a booster dose with the aerosolized vaccine 28 days after a single intramuscular dose induced strong IgG and neutralizing antibody responses. The authors also remarked that different compositions of antibody response were produced depending on the route of administration [494]. This is an area that warrants more research and could prove beneficial given the portals of entry and site of replication of the SARS-CoV-2 virus.

Different vaccines seemed to produce different immune responses. Interestingly, the more 'virus-like' designed vaccines, such as CoronaVac, Ad26.COV2.S, and ChAdOx1, seemed to favour the cell-mediated T-cell response. In fact, individuals who received a heterologous Ad26.COV2.S primer-boost scheme with Ad26.COV2.S as their first vaccine and either mRNA-1273 or BNT162b2 as their booster were reported to generate the most potent T-cell response [495]. This may not be all that surprising as their design likely intentionally resembles a viral infection and, therefore, produces the desired response. It should also be noted that using the mRNA vaccines as boosters seemed to generate a more productive immune response, inducing both antibody and T-cell responses after being primed with a viral-vector vaccine [465, 466, 495, 496]. From this, while homologous vaccination regimens did confer protective immunity, it would seem that the heterologous regimens produced a more robust immune response in general, regardless of the combination of vaccines used. It may be that this combination of different vaccine designs induces a more rounded immune response. Nevertheless, the mRNA vaccines also induced effective immune responses on their own [497, 498].

1.8.1.5 Importance of T-Cell Immunity in COVID-19

Neutralizing antibody levels correlate with protection against symptomatic infection as they function to stop infection of susceptible host cells [499, 500]. However, as demonstrated over the course of the pandemic, variants emerged and accrued mutations that allowed them to evade neutralizing antibodies [501, 502]. And yet many studies often neglect to report on the T-cell response and cellular immunity and have consistently reported on vaccine-induced antibody responses as the seemingly most important measure of vaccine effectiveness [443]. T cells might not on their own prevent viruses from infecting cells as antibodies do, but CD8+ T cells function to limit viral replication and CD4+ T cells promote both effective CD8+ T-cell functions and antibody production by B cells–aspects which appear to be grossly overlooked in the context of COVID-19. Early studies have demonstrated the power of T cells: that they are resilient and retained their effectiveness against the initial variants [503–505]. Cell-mediated immunity, encompassing B cells and T cells, has been shown to persist for at least 6–8 months, demonstrating durability against the Delta variant [506, 507]. Another study found that vaccines induced durable CD8+ and CD4+ T-cell responses and exhibited extensive cross-reactivity against both the Delta and Omicron variants, demonstrating that cell-mediated immunity provides robust protection even against the Omicron variant despite neutralizing antibody responses waning substantially over time [508]. Indeed, while the vaccine-induced humoral response to SARS-CoV-2 undoubtedly plays an important role in providing protection, the role of T-cell immunity in COVID-19 cannot be overstated. The results of several studies clearly demonstrate that T-cell immunity against SARS-CoV-2 is very resilient and highly durable, and the importance thereof underpins the need for it be included in correlates of protection [509].

Historically, hCoV infections are treated by managing symptoms and providing supportive care in the form of ventilation support and fluids [510, 511]. However, COVID-19 has evoked a renewed interest in the treatment of hCoV infections, leading to the investigation of several different lines of treatment for COVID-19 and potential antiviral drugs against SARS-CoV-2.

1.8.2 IMMUNOGLOBULIN THERAPY: MONOCLONAL ANTIBODIES AND COVID-19 CONVALESCENT PLASMA

Binding of the SARS-CoV-2 S protein to the ACE2 receptor facilitates viral entry into host cells, therefore disrupting this interaction, is one potential avenue of treatment for COVID-19. Bamlanivimab/etesevimab is a combination of neutralizing monoclonal antibodies (mAbs) that bind to the S protein and disrupt the interaction between S and ACE2. On its own bamlanivimab was shown to be ineffective, but in combination with etesevimab, the two reportedly decrease viral load, and in patients with mild-moderate COVID-19, no deaths were observed [512, 513]. Casirivimab/imdevimab is another combination of recombinant mAbs that bind to the RBD of the S protein and also block the interaction between SARS-CoV-2 S and ACE2. The combination of mAbs appeared to be a more effective treatment against variants as combination treatment efficiently inhibited viral entry for all variants [514, 515]. The treatment of mild-moderate and severe cases of COVID-19 with mAbs appears to be relatively effective, but concerns have been raised about the efficacy of anti-S mAbs in treating VOCs as the antibodies appear to become increasingly less effective [514, 516–519].

The excessive inflammatory response often seen in severe COVID-19 cases also makes for a potential target to lessen the severity of such cases. Tocilizumab is a recombinant humanized mAb that blocks IL-6 signalling by binding IL-6 receptors (IL-6R) and prevents IL-6 from binding to the receptor. Studies have shown that its use reduces mortality and the need for mechanical ventilation [520, 521]. Similar concerns have been raised about the efficacy of tocilizumab for different variants. However, since tocilizumab targets the immune response rather than the variants, there is a smaller probability of its efficacy decreasing, and studies has shown that it remains effective regardless of variants [522, 523]. Sarilumab is another human mAb against the IL-6R, but previous trials reported no significant improvement of clinical outcomes for patients treated with sarilumab [524, 525]. Tocilizumab demonstrated superior efficacy over sarilumab, but the evaluation of sarilumab was limited to small trials and might be more beneficial in targeted patients and/or in combination with tocilizumab similar to the use of anti-S mAbs.

COVID-19 convalescent plasma (CCP) therapy provides passive immunity to patients with COVID-19 by transfusion with plasma from patients who have recovered from COVID-19, and has been demonstrated to be effective, particularly in immunocompromised patients [526]. It has even been reported to significantly reduce COVID-19-induced ARDS in mechanically ventilated patients [527]. Although there was a concern that variants may become resistant to CCP, studies showed that plasma from donors with hybrid immunity could provide effective passive immunity, minimizing the risk of resistance to CCP from variants [528, 529]. This demonstrates the importance of variant-matched CCP in that it is the only antibody therapy that could keep up with variants, since every variant elicits an immune response that neutralizes it during convalescence [530, 531].

1.8.3 Antivirals Agents

While the SARS-CoV-1 outbreak could provide some guidance, many studies relied on virtual screenings to identify potentially novel antiviral compounds or to repurpose existing drugs that could demonstrate anti-SARS-CoV-2 activity [532–535]. Remdesivir, a broad-spectrum antiviral, is an adenosine analogue that is metabolized to a structural analogue of adenosine triphosphate (ATP), which competes with ATP for incorporation by the viral RdRp and inhibits viral replication by delaying genome replication [536]. *In vitro* studies have demonstrated that remdesivir retains its antiviral activity, irrespective of the variant [537, 538]. While only a few clinical studies exist, they have observed the same [539, 540]. *In vitro* studies have demonstrated the potential of remdesivir-resistant mutations, which raised some concerns, but such resistance has not yet been widely observed in the clinical setting [541–544]. More recently, the potential combinatory effect of remdesivir and baricitinib to treat COVID-19 more effectively was investigated and was indeed found to produce favourable clinical outcomes [545]. It should, however, be noted that baricitinib causes immunosuppression and can therefore delay viral clearance, but that this combination could potentially be effective at treating the infection [546].

Molnupiravir is a nucleoside derivative that possesses antiviral properties by misdirecting viral RdRp to introduce errors in the genome during viral replication [547]. A few studies found positive outcomes, demonstrating promise for the use of molnupiravir in COVID-19. In mild COVID-19 cases, it significantly reduced the need for hospitalization or death, but more studies are needed to confirm whether it was also beneficial in moderate to severe cases [547, 548]. In high-risk, vaccinated adults, molnupiravir reduced the viral load and patients recovered faster, but the chances of hospitalization or death were not reduced in this population [549, 550]. A very recent study investigated several SARS-CoV-2 sequences and, concerningly, found that the clinical use of molnupiravir induced mutations in the virus, raising concerns that using the drug might allow the virus to acquire additional, unwanted mutations [551].

A combination of the protease inhibitors nirmatrelvir and ritonavir, commonly known as Paxlovid, works synergistically in that nirmatrelvir exerts antiviral activity while ritonavir inhibits the metabolism of nirmatrelvir, thereby enhancing its bioavailability [552]. Studies have found that for non-hospitalized patients, Paxlovid reduced the risk of hospitalization, the risk of in-hospital disease progression, as well as the risk of death [553, 554]. However, concerning reports of virological rebound have recently emerged in which a small proportion of patients, both those who have used Paxlovid and those who have not, experienced a recurrence of COVID-19 symptoms and some have tested positive again for COVID-19 [555–557]. The Centers for Disease Control and Prevention (CDC) confirmed that viral rebound could occur with or without the use of Paxlovid and reported that there was no association with COVID-19-related hospitalization or death [558].

1.8.4 Immunomodulatory Agents

Baricitinib/tofacitinib are existing drugs that are used as inhibitors of Janus activated kinases (JAKs), central players in inflammatory diseases, which potentially makes them good candidates for repurposing to reduce COVID-19-induced inflammation [559–561]. Their use has been beneficial in hospitalized COVID-19 patients as well as elderly persons as they have been shown to modulate the immune response [560, 562]. The use of baricitinib in COVID-19 patients with moderate pneumonia showed promise as it reduced the need for intensive care unit (ICU) admission, significantly decreased the 2-week case fatality rate, and significantly increased the discharge rate [563]. A clinical trial in which hospitalized COVID-19 patients were treated with a combination of baricitinib and remdesivir reported a reduction to recovery time within 29 days [560].

Dexamethasone and methylprednisolone are both corticosteroid drugs for which the Oxford University RECOVERY Trial demonstrated that dexamethasone provided no benefit to those without respiratory support, but in patients on oxygen alone or receiving invasive mechanical ventilation, it reduced the 28-day mortality [564]. Based on this, treatment guidelines were updated to include dexamethasone, but not in COVID-19 patients that do not require oxygen [565–567]. The bioinformatics company AdvaitaBio analysed transcriptomics data from COVID-19 patients and identified methylprednisolone as a potential drug to mitigate the cytokine storm experienced by patients suffering from severe COVID-19 [415, 568]. Interestingly, a study comparing the efficacy of methylprednisolone and dexamethasone in COVID-19 ICU patients reported both are effective in decreasing mortality, but that methylprednisolone was superior to dexamethasone [569]. Both methylprednisolone and dexamethasone nevertheless currently form part of standard care for severe COVID-19 patients.

The IL-1 receptor antagonist (IL-1Ra) Anakinra is another immunomodulatory agent that has been investigated for its potential benefits in COVID-19. Based on available evidence, treatment with Anakinra reduces the need for mechanical ventilation as well as the risk of mortality for non-intubated patients hospitalized with COVID-19 [570]. Another study found that treating hospitalized COVID-19 patients, which also have an increased risk of respiratory failure, with Anakinra significantly improved the outcome of these patients [571]. Anakinra has even been reported to improve the clinical outcome of a COVID-19 patient who was refractory to tocilizumab and lopinavir/ritonavir [572].

1.8.5 OTHER DRUGS

Ivermectin is an antiparasitic drugs that has drawn a lot of controversy after showing potential anti-SARS-CoV-2 activity *in vitro* [573, 574]. However, authors of the initial study cautioned that ivermectin warranted further investigation, and doses higher than those used *in vitro* and those previously approved would be needed for use in humans to treat COVID-19 [575]. A small randomized clinical trial of 476 patients found no significant difference between patients who received a 5-day course of ivermectin and those who did not [576]. A larger double-blind, randomized trial also reported no significant advantage to using ivermectin, noting that ivermectin treatment did not lower the incidence of hospital admission [577]. Based on the current evidence, it is still unclear whether ivermectin treatment has any therapeutic benefit for COVID-19 [578]. A few recent clinical trials attempted to determine whether higher doses of ivermectin had any benefit in treating COVID-19, evaluating different outcomes in different populations, but none of them reported any benefits [579–582]. Some trials are still ongoing, but the current evidence base, although slightly improved, is still limited.

Chloroquine and hydroxychloroquine are antimalarial compounds that, in the context of COVID-19, were expected to have therapeutic benefits by inhibiting entry of the virus. Hydroxychloroquine inhibits endosomal acidification and prevents the protease cathepsin L from being activated. However, since proteolytic cleavage of SARS-CoV-2 S is more dependent on TMPRSS2, inhibition of cathepsin L by hydroxychloroquine has a limited therapeutic effect on the virus. The use of both a TMPRSS2 inhibitor and hydroxychloroquine, however, partially restored the inhibitory effect of hydroxychloroquine on SARS-CoV-2 entry [583]. In a clinical setting, many studies have not succeeded in demonstrating a therapeutic benefit for using hydroxychloroquine in COVID-19, after evaluating different outcomes [584–588]. In fact, some authors advised against the use of hydroxychloroquine, noting an increased risk for adverse events [585, 586].

Camostat mesylate has been shown to inhibit serine proteases such as TMPRSS2, which is necessary for virus–host membrane fusion and subsequent entry of SARS-CoV-2 into host cells, making it a probable candidate for COVID-19 treatment [22, 589]. An initial study reported that it may be clinically beneficial after a favourable outcome in patients with organ dysfunction [590]. A subsequent study also found clinical benefits in critically ill COVID-19 patients with pneumonia [591]. Conversely, a double-blind, randomized clinical trial with mild-moderate COVID-19 patients reported no clinical benefit [592]. The authors suggested evaluating its efficacy in high-risk patients and based on the previous two studies of severe COVID-19 patients, their recommendation seems plausible. In support of this, a trial conducted in an ambulatory setting enrolled symptomatic patients that were not at high risk of hospitalization. They too reported no clinical benefit to camostat mesylate, suggesting that camostat may probably not be clinically beneficial for all COVID-19 patients [593].

1.9   CONCLUSION

In just 4 years, COVID-19 research has led to some significant advances in both science and society: the first mRNA vaccine used in humans, a concerted global effort of public health interventions, implementing virtual platforms for various applications such online education, the use of computational tools in resolving protein structures, and drug discovery. There are, however, still some unanswered questions: what is the exact reservoir and/or intermediate host for SARS-CoV-2? Will SARS-CoV-2 become extinct like SARS-CoV-1 and MERS-CoV, or will it become significantly attenuated and circulate seasonally? Can we design a pan-CoV vaccine to protect against potential future variants and/or emerging CoVs, and how to avoid vaccine imprinting? How long does long COVID last? In the event of another large-scale CoV outbreak, will we have CoV-specific antiviral treatment? How do we regain the public's trust? Four years later and there is still a lot we do not know or understand. The COVID-19 pandemic has made its mark in history and it will take some time to recover from the consequences of the pandemic. Should there be another similar outbreak, and from experience with SARS, MERS, and COVID-19 there may very well be, we should be better prepared.

REFERENCES

1. MacLachlan, N.J.; Dubovi, E.J., Eds. Chapter 24 - Coronaviridae. In *Fenner's Veterinary Virology* (Fifth Edition, pp. 435–461), Academic Press: Boston, 2017; https://doi.org/10.1016/B978-0-12-800946-8.00024-6.
2. Bonilauri, P.; Rugna, G. Animal coronaviruses and SARS-COV-2 in animals, what do we actually know? *Life (Basel)* 2021, 11, doi:10.3390/life11020123.
3. Su, S.; Wong, G.; Shi, W.; Liu, J.; Lai, A.C.K.; Zhou, J.; Liu, W.; Bi, Y.; Gao, G.F. Epidemiology, genetic recombination, and pathogenesis of coronaviruses. *Trends Microbiol* 2016, 24, 490–502, doi:10.1016/j.tim.2016.03.003.
4. Schoeman, D.; Gordon, B.; Fielding, B.C. Coronaviruses. *Encyclopedia of Infection and Immunity* 2022, 241–258, doi:10.1016/b978-0-12-818731-9.00052-5.
5. Malik, Y.A. Properties of coronavirus and SARS-CoV-2. *Malays J Pathol* 2020, 42, 3–11.
6. Liu, Y.V.; Massare, M.J.; Barnard, D.L.; Kort, T.; Nathan, M.; Wang, L.; Smith, G. Chimeric severe acute respiratory syndrome coronavirus (SARS-CoV) S glycoprotein and influenza matrix 1 efficiently form virus-like particles (VLPs) that protect mice against challenge with SARS-CoV. *Vaccine* 2011, 29, 6606–6613, doi:10.1016/j.vaccine.2011.06.111.
7. Coronaviridae Study Group of the International Committee on Taxonomy of Viruses. The species severe acute respiratory syndrome-related coronavirus: Classifying 2019-nCoV and naming it SARS-CoV-2. *Nat Microbiol* 2020, 5, 536–544, doi:10.1038/s41564-020-0695-z.
8. Zhu, N.; Zhang, D.; Wang, W.; Li, X.; Yang, B.; Song, J.; Zhao, X.; Huang, B.; Shi, W.; Lu, R., et al. A novel coronavirus from patients with pneumonia in China, 2019. *N Engl J Med* 2020, 382, 727–733, doi:10.1056/NEJMoa2001017.
9. Yao, H.P.; Song, Y.T.; Chen, Y.; Wu, N.P.; Xu, J.L.; Sun, C.J.; Zhang, J.X.; Weng, T.H.; Zhang, Z.Y.; Wu, Z.G., et al. Molecular architecture of the SARS-CoV-2 virus. *Cell* 2020, 183, 730–738.e13, doi:10.1016/j.cell.2020.09.018.
10. Liu, C.; Mendonca, L.; Yang, Y.; Gao, Y.; Shen, C.; Liu, J.; Ni, T.; Ju, B.; Liu, C.; Tang, X., et al. The architecture of inactivated SARS-CoV-2 with postfusion spikes revealed by cryo-EM and cryo-ET. *Structure* 2020, 28, 1218–1224.e4, doi:10.1016/j.str.2020.10.001.
11. V'Kovski, P.; Kratzel, A.; Steiner, S.; Stalder, H.; Thiel, V. Coronavirus biology and replication: Implications for SARS-CoV-2. *Nat Rev Microbiol* 2021, 19, 155–170, doi:10.1038/s41579-020-00468-6.
12. Wu, A.; Peng, Y.; Huang, B.; Ding, X.; Wang, X.; Niu, P.; Meng, J.; Zhu, Z.; Zhang, Z.; Wang, J., et al. Genome composition and divergence of the novel coronavirus (2019-nCoV) originating in China. *Cell Host Microbe* 2020, 27, 325–328, doi:10.1016/j.chom.2020.02.001.

13. Miao, Z.; Tidu, A.; Eriani, G.; Martin, F. Secondary structure of the SARS-CoV-2 5'-UTR. *RNA Biol* 2021, 18, 447–456, doi:10.1080/15476286.2020.1814556.

14. Ke, Z.; Oton, J.; Qu, K.; Cortese, M.; Zila, V.; McKeane, L.; Nakane, T.; Zivanov, J.; Neufeldt, C.J.; Cerikan, B., et al. Structures and distributions of SARS-CoV-2 spike proteins on intact virions. *Nature* 2020, 588, 498–502, doi:10.1038/s41586-020-2665-2.

15. Turonova, B.; Sikora, M.; Schurmann, C.; Hagen, W.J.H.; Welsch, S.; Blanc, F.E.C.; von Bulow, S.; Gecht, M.; Bagola, K.; Horner, C., et al. In situ structural analysis of SARS-CoV-2 spike reveals flexibility mediated by three hinges. *Science* 2020, 370, 203–208, doi:10.1126/science.abd5223.

16. Berman, H.M.; Westbrook, J.; Feng, Z.; Gilliland, G.; Bhat, T.N.; Weissig, H.; Shindyalov, I.N.; Bourne, P.E. The protein data bank. *Nucleic Acids Res* 2000, 28, 235–242, doi:10.1093/nar/28.1.235.

17. Pezeshkian, W.; Grunewald, F.; Narykov, O.; Lu, S.; Arkhipova, V.; Solodovnikov, A.; Wassenaar, T.A.; Marrink, S.J.; Korkin, D. Molecular architecture and dynamics of SARS-CoV-2 envelope by integrative modeling. *Structure* 2023, 31, 492–503.e7, doi:10.1016/j.str.2023.02.006.

18. Casalino, L.; Dommer, A.C.; Gaieb, Z.; Barros, E.P.; Sztain, T.; Ahn, S.-H.; Trifan, A.; Brace, A.; Bogetti, A.T.; Clyde, A., et al. AI-driven multiscale simulations illuminate mechanisms of SARS-CoV-2 spike dynamics. *Int J High Perform Comput Appl* 2021, 35, 432–451, doi:10.1177/10943420211006452.

19. Neuman, B.W.; Kiss, G.; Kunding, A.H.; Bhella, D.; Baksh, M.F.; Connelly, S.; Droese, B.; Klaus, J.P.; Makino, S.; Sawicki, S.G., et al. A structural analysis of M protein in coronavirus assembly and morphology. *J Struct Biol* 2011, 174, 11–22, doi:10.1016/j.jsb.2010.11.021.

20. Huang, Y.; Yang, C.; Xu, X.F.; Xu, W.; Liu, S.W. Structural and functional properties of SARS-CoV-2 spike protein: Potential antivirus drug development for COVID-19. *Acta Pharmacologica Sinica* 2020, 41, 1141–1149, doi:10.1038/s41401-020-0485-4.

21. Xia, X.H. Domains and functions of spike protein in SARS-Cov-2 in the context of vaccine design. *Viruses* 2021, 13, doi:10.3390/v13010109.

22. Hoffmann, M.; Kleine-Weber, H.; Schroeder, S.; Krüger, N.; Herrler, T.; Erichsen, S.; Schiergens, T.S.; Herrler, G.; Wu, N.-H.; Nitsche, A., et al. SARS-CoV-2 cell entry depends on ACE2 and TMPRSS2 and is blocked by a clinically proven protease inhibitor. *Cell* 2020, 181, 271–280.e8, https://doi.org/10.1016/j.cell.2020.02.052.

23. Wan, Y.; Shang, J.; Graham, R.; Baric, R.S.; Li, F. Receptor recognition by the novel coronavirus from Wuhan: An analysis based on decade-long structural studies of SARS coronavirus. *J Virol* 2020, 94, doi:10.1128/JVI.00127-20.

24. Klinakis, A.; Cournia, Z.; Rampias, T. N-terminal domain mutations of the spike protein are structurally implicated in epitope recognition in emerging SARS-CoV-2 strains. *Comput Struct Biotechnol J* 2021, 19, 5556–5567, doi:10.1016/j.csbj.2021.10.004.

25. Ou, T.L.; Mou, H.H.; Zhang, L.Z.; Ojha, A.; Choe, H.; Farzan, M. Hydroxychloroquine-mediated inhibition of SARS-CoV-2 entry is attenuated by TMPRSS2. *PLoS Pathog* 2021, 17, doi:10.1371/journal.ppat.1009212.

26. Coutard, B.; Valle, C.; de Lamballerie, X.; Canard, B.; Seidah, N.G.; Decroly, E. The spike glycoprotein of the new coronavirus 2019-nCoV contains a furin-like cleavage site absent in CoV of the same clade. *Antiviral Res* 2020, 176, 104742, doi:10.1016/j.antiviral.2020.104742.

27. Zhu, C.; He, G.; Yin, Q.; Zeng, L.; Ye, X.; Shi, Y.; Xu, W. Molecular biology of the SARs-CoV-2 spike protein: A review of current knowledge. *J Med Virol* 2021, 93, 5729–5741, doi:10.1002/jmv.27132.

28. Johnson, B.A.; Xie, X.; Bailey, A.L.; Kalveram, B.; Lokugamage, K.G.; Muruato, A.; Zou, J.; Zhang, X.; Juelich, T.; Smith, J.K., et al. Loss of furin cleavage site attenuates SARS-CoV-2 pathogenesis. *Nature* 2021, 591, 293–299, doi:10.1038/s41586-021-03237-4.

29. Buchrieser, J.; Dufloo, J.; Hubert, M.; Monel, B.; Planas, D.; Rajah, M.M.; Planchais, C.; Porrot, F.; Guivel-Benhassine, F.; Van der Werf, S., et al. Syncytia formation by SARS-CoV-2-infected cells. *EMBO J* 2020, 39, e106267, doi:10.15252/embj.2020106267.

30. Boson, B.; Legros, V.; Zhou, B.J.; Siret, E.; Mathieu, C.; Cosset, F.L.; Lavillette, D.; Denolly, S. The SARS-CoV-2 envelope and membrane proteins modulate maturation and retention of the spike protein, allowing assembly of virus-like particles. *J Biol Chem* 2021, 296, 100111, doi:10.1074/jbc.RA120.016175.

31. Lei, Y.; Zhang, J.; Schiavon, C.R.; He, M.; Chen, L.; Shen, H.; Zhang, Y.; Yin, Q.; Cho, Y.; Andrade, L., et al. SARS-CoV-2 spike protein impairs endothelial function via downregulation of ACE 2. *Circ Res* 2021, 128, 1323–1326, doi:10.1161/CIRCRESAHA.121.318902.

32. Zhang, S.; Liu, Y.Y.; Wang, X.F.; Yang, L.; Li, H.S.; Wang, Y.Y.; Liu, M.D.; Zhao, X.Y.; Xie, Y.H.; Yang, Y., et al. SARS-CoV-2 binds platelet ACE2 to enhance thrombosis in COVID-19. *J Hematol Oncol* 2020, 13, 120, doi:10.1186/s13045-020-00954-7.

33. McFadyen, J.D.; Stevens, H.; Peter, K. The emerging threat of (micro)thrombosis in COVID-19 and its therapeutic implications. *Circ Res* 2020, 127, 571–587, doi:10.1161/Circresaha.120.317447.

34. Zheng, Y.; Zhao, J.X.; Li, J.Q.; Guo, Z.M.; Sheng, J.J.; Ye, X.L.; Jin, G.W.; Wang, C.R.; Chai, W.G.; Yan, J.Y., et al. SARS-CoV-2 spike protein causes blood coagulation and thrombosis by competitive binding to heparan sulfate. *Int J Biol Macromol* 2021, 193, 1124–1129, doi:10.1016/j.ijbiomac.2021.10.112.

35. Zhao, P.; Praissman, J.L.; Grant, O.C.; Cai, Y.F.; Xiao, T.S.; Rosenbalm, K.E.; Aoki, K.; Kellman, B.P.; Barouch, D.H.; Lewis, N.E., et al. Virus-receptor interactions of glycosylated SARS-CoV-2 spike and human ACE2 receptor. *Glycobiology* 2020, 30, 1133–1134.

36. Shajahan, A.; Supekar, N.T.; Gleinich, A.S.; Azadi, P. Deducing the N- and O-glycosylation profile of the spike protein of novel coronavirus SARS-CoV-2. *Glycobiology* 2020, 30, 981–988, doi:10.1093/glycob/cwaa042.

37. Watanabe, Y.; Allen, J.D.; Wrapp, D.; McLellan, J.S.; Crispin, M. Site-specific glycan analysis of the SARS-CoV-2 spike. *Science* 2020, 369, 330–333, doi:10.1126/science.abb9983.

38. Casalino, L.; Gaieb, Z.; Goldsmith, J.A.; Hjorth, C.K.; Dommer, A.C.; Harbison, A.M.; Fogarty, C.A.; Barros, E.P.; Taylor, B.C.; McLellan, J.S., et al. Beyond shielding: The roles of glycans in the SARS-CoV-2 spike protein. *ACS Cent Sci* 2020, 6, 1722–1734, doi:10.1021/acscentsci.0c01056.

39. Greaney, A.J.; Loes, A.N.; Crawford, K.H.D.; Starr, T.N.; Malone, K.D.; Chu, H.Y.; Bloom, J.D. Comprehensive mapping of mutations in the SARS-CoV-2 receptor-binding domain that affect recognition by polyclonal human plasma antibodies. *Cell Host Microbe* 2021, 29, 463–476.e6, doi:10.1016/j.chom.2021.02.003.

40. Magazine, N.; Zhang, T.; Wu, Y.; McGee, M.C.; Veggiani, G.; Huang, W. Mutations and evolution of the SARS-CoV-2 spike protein. *Viruses* 2022, 14, doi:10.3390/v14030640.

41. European Centre for Disease Prevention and Control (ECDC). SARS-CoV-2 variants of concern as of 01 December 2023. Available online: https://www.ecdc.europa.eu/en/covid-19/variants-concern (accessed on 04 December 2023).

42. Carabelli, A.M.; Peacock, T.P.; Thorne, L.G.; Harvey, W.T.; Hughes, J.; Consortium, C.-G.U.; Peacock, S.J.; Barclay, W.S.; de Silva, T.I.; Towers, G.J., et al. SARS-CoV-2 variant biology: Immune escape, transmission and fitness. *Nat Rev Microbiol* 2023, 21, 162–177, doi:10.1038/s41579-022-00841-7.

43. Centers for Disease Control and Prevention (CDC). SARS-CoV-2 Variant Classifications and Definitions. Available online: https://www.cdc.gov/coronavirus/2019-ncov/variants/variant-classifications.html (accessed on 04 December 2023).

44. Mohammadi, M.; Shayestehpour, M.; Mirzaei, H. The impact of spike mutated variants of SARS-CoV2 [Alpha, Beta, Gamma, Delta, and Lambda] on the efficacy of subunit recombinant vaccines. *Braz J Infect Dis* 2021, 25, 101606, doi:10.1016/j.bjid.2021.101606.

45. Sahni, C.; Basu Roy Chowdhury, P.; Devadas, D.; Ashish, A.; Singh, N.K.; Yadav, A.; Kaur, M.; Mishra, S.; Vishwakarma, S.; Singh, R. SARS-CoV-2 mutations responsible for immune evasion leading to breakthrough infection. *Cureus* 2022, 14, e29544, doi:10.7759/cureus.29544.

46. Souza, P.F.N.; Mesquita, F.P.; Amaral, J.L.; Landim, P.G.C.; Lima, K.R.P.; Costa, M.B.; Farias, I.R.; Belem, M.O.; Pinto, Y.O.; Moreira, H.H.T., et al. The spike glycoprotein of SARS-CoV-2: A review of how mutations of spike glycoproteins have driven the emergence of variants with high transmissibility and immune escape. *Int J Biol Macromol* 2022, 208, 105–125, doi:10.1016/j.ijbiomac.2022.03.058.

47. Wrapp, D.; Wang, N.; Corbett, K.S.; Goldsmith, J.A.; Hsieh, C.L.; Abiona, O.; Graham, B.S.; McLellan, J.S. Cryo-EM structure of the 2019-nCoV spike in the prefusion conformation. *Science* 2020, 367, 1260–1263, doi:10.1126/science.abb2507.

48. Walls, A.C.; Park, Y.J.; Tortorici, M.A.; Wall, A.; McGuire, A.T.; Veesler, D. Structure, function, and antigenicity of the SARS-CoV-2 spike glycoprotein. *Cell* 2020, 181, 281–292.e6, doi:10.1016/j.cell.2020.02.058.

49. Wong, N.A.; Saier, M.H. The SARS-coronavirus infection cycle: A survey of viral membrane proteins, their functional interactions and pathogenesis. *Int J Mol Sci* 2021, 22, doi:10.3390/ijms22031308.

50. Collins, L.T.; Elkholy, T.; Mubin, S.; Hill, D.; Williams, R.; Ezike, K.; Singhal, A. Elucidation of SARS-Cov-2 budding mechanisms through molecular dynamics simulations of M and E protein complexes. *J Phys Chem Lett* 2021, 12, 12249–12255, doi:10.1021/acs.jpclett.1c02955.

51. Zhang, Z.K.; Nomura, N.; Muramoto, Y.; Ekimoto, T.; Uemura, T.; Liu, K.H.; Yui, M.; Kono, N.; Aoki, J.; Ikeguchi, M., et al. Structure of SARS-CoV-2 membrane protein essential for virus assembly. *Nat Commun* 2022, 13, doi:10.1038/s41467-022-32019-3.

52. Tang, T.; Bidon, M.; Jaimes, J.A.; Whittaker, G.R.; Daniel, S. Coronavirus membrane fusion mechanism offers a potential target for antiviral development. *Antiviral Res* 2020, 178, 104792, doi:10.1016/j.antiviral.2020.104792.

53. Fu, Y.Z.; Wang, S.Y.; Zheng, Z.Q.; Yi, H.; Li, W.W.; Xu, Z.S.; Wang, Y.Y. SARS-CoV-2 membrane glycoprotein M antagonizes the MAVS-mediated innate antiviral response. *Cell Mol Immunol* 2021, 18, 613–620, doi:10.1038/s41423-020-00571-x.

54. Eskandarzade, N.; Ghorbani, A.; Samarfard, S.; Diaz, J.; Guzzi, P.H.; Fariborzi, N.; Tahmasebi, A.; Izadpanah, K. Network for network concept offers new insights into host- SARS-CoV-2 protein interactions and potential novel targets for developing antiviral drugs. *Comput Biol Med* 2022, 146, doi:10.1016/j.compbiomed.2022.105575.

55. Tseng, Y.T.; Chang, C.H.; Wang, S.M.; Huang, K.J.; Wang, C.T. Identifying SARS-CoV membrane protein amino acid residues linked to virus-like particle assembly. *PLoS One* 2013, 8, e64013, doi:10.1371/journal.pone.0064013.

56. Tseng, Y.T.; Wang, S.M.; Huang, K.J.; Lee, A.I.; Chiang, C.C.; Wang, C.T. Self-assembly of severe acute respiratory syndrome coronavirus membrane protein. *J Biol Chem* 2010, 285, 12862–12872, doi:10.1074/jbc.M109.030270.

57. Dawood, A.A. Glycosylation, ligand binding sites and antigenic variations between membrane glycoprotein of COVID-19 and related coronaviruses. *Vacunas* 2021, 22, 1–9, https://doi.org/10.1016/j.vacun.2020.09.005.

58. Neuman, B.W.; Kiss, G.; Kunding, A.H.; Bhella, D.; Baksh, M.F.; Connelly, S.; Droese, B.; Klaus, J.P.; Makino, S.; Sawicki, S.G., et al. A structural analysis of M protein in coronavirus assembly and morphology. *J Struct Biol* 2011, 174, 11–22, doi:10.1016/j.jsb.2010.11.021.

59. Zheng, Y.; Zhuang, M.W.; Han, L.L.; Zhang, J.; Nan, M.L.; Zhan, P.; Kang, D.W.; Liu, X.Y.; Gao, C.J.; Wang, P.H. Severe acute respiratory syndrome coronavirus 2 (SARS-CoV-2) membrane (M) protein inhibits type I and III interferon production by targeting RIG-I/MDA-5 signaling. *Signal Transduct Target Ther* 2020, 5, doi:10.1038/s41392-020-00438-7.

60. Mahtarin, R.; Islam, S.; Islam, M.J.; Ullah, M.O.; Ali, M.A.; Halim, M.A. Structure and dynamics of membrane protein in SARS-CoV-2. *J Biomol Struct Dyn* 2022, 40, 4725–4738, doi:10.1080/07391102.2020.1861983.

61. Thomas, S. The structure of the membrane protein of SARS-CoV-2 resembles the sugar transporter SemiSWEET. *Pathog Immun* 2020, 5, 342–363, doi:10.20411/pai.v5i1.377.

62. Marques-Pereira, C.; Pires, M.N.; Gouveia, R.P.; Pereira, N.N.; Caniceiro, A.B.; Rosario-Ferreira, N.; Moreira, I.S. SARS-CoV-2 membrane protein: From genomic data to structural new insights. *Int J Mol Sci* 2022, 23, doi:10.3390/ijms23062986.

63. Zhang, Z.; Nomura, N.; Muramoto, Y.; Ekimoto, T.; Uemura, T.; Liu, K.; Yui, M.; Kono, N.; Aoki, J.; Ikeguchi, M., et al. Structure of SARS-CoV-2 membrane protein essential for virus assembly. *Nat Commun* 2022, 13, 4399, doi:10.1038/s41467-022-32019-3.

64. Ye, Q.Z.; Lu, S.; Corbett, K.D. Structural basis for SARS-CoV-2 nucleocapsid protein recognition by single-domain antibodies. *Front Immunol* 2021, 12, doi:10.3389/fimmu.2021.719037.

65. Schiavina, M.; Pontoriero, L.; Uversky, V.N.; Felli, I.C.; Pierattelli, R. The highly flexible disordered regions of the SARS-CoV-2 nucleocapsid N protein within the 1-248 residue construct: Sequence-specific resonance assignments through NMR. *Biomol NMR Assign* 2021, 15, 219–227, doi:10.1007/s12104-021-10009-8.

66. Peng, Y.; Du, N.; Lei, Y.Q.; Dorje, S.; Qi, J.X.; Luo, T.R.; Gao, G.F.; Song, H. Structures of the SARS-CoV-2 nucleocapsid and their perspectives for drug design. *EMBO J* 2020, 39, doi:10.15252/embj.2020105938.

67. Gao, T.Y.; Gao, Y.D.; Liu, X.X.; Nie, Z.L.; Sun, H.L.; Lin, K.; Peng, H.X.; Wang, S.K. Identification and functional analysis of the SARS-COV-2 nucleocapsid protein. *BMC Microbiol* 2021, 21, doi:10.1186/s12866-021-02107-3.

68. Dinesh, D.C.; Chalupska, D.; Silhan, J.; Koutna, E.; Nencka, R.; Veverka, V.; Boura, E. Structural basis of RNA recognition by the SARS-CoV-2 nucleocapsid phosphoprotein. *PLoS Pathog* 2020, 16, doi:10.1371/journal.ppat.1009100.

69. Morse, M.; Sefcikova, J.; Rouzina, I.; Beuning, P.J.; Williams, M.C. Structural domains of SARS-CoV-2 nucleocapsid protein coordinate to compact long nucleic acid substrates. *Nucleic Acids Res* 2023, 51, 290–303, doi:10.1093/nar/gkac1179.

70. Zinzula, L.; Basquin, J.; Bohn, S.; Beck, F.; Klumpe, S.; Pfeifer, G.; Nagy, I.; Bracher, A.; Hartl, F.U.; Baumeister, W. High-resolution structure and biophysical characterization of the nucleocapsid phosphoprotein dimerization domain from the Covid-19 severe acute respiratory syndrome coronavirus 2. *Biochem Biophys Res Commun* 2021, 538, 54–62, doi:10.1016/j.bbrc.2020.09.131.

71. Zeng, W.H.; Liu, G.F.; Ma, H.; Zhao, D.; Yang, Y.R.; Liu, M.Y.; Mohammed, A.; Zhao, C.C.; Yang, Y.; Xie, J.J., et al. Biochemical characterization of SARS-CoV-2 nucleocapsid protein. *Biochem Biophys Res Commun* 2020, 527, 618–623, doi:10.1016/j.bbrc.2020.04.136.

72. Chen, C.Y.; Chang, C.K.; Chang, Y.W.; Sue, S.C.; Bai, H.I.; Riang, L.; Hsiao, C.D.; Huang, T.H. Structure of the SARS coronavirus nucleocapsid protein RNA-binding dimerization domain suggests a mechanism for helical packaging of viral RNA. *J Mol Biol* 2007, 368, 1075–1086, doi:10.1016/j.jmb.2007.02.069.

73. Zhao, H.Y.; Di, W.; Nguyen, A.; Li, Y.; Adao, R.C.; Valkov, E.; Patterson, G.H.; Piszczek, G.; Schuck, P. Energetic and structural features of SARS-CoV-2 N-protein co-assemblies with nucleic acids. *iScience* 2021, 24, doi:10.1016/j.isci.2021.102523.

74. Lu, S.; Ye, Q.Z.; Singh, D.; Cao, Y.; Diedrich, J.K.; Yates, J.R.; Villa, E.; Cleveland, D.W.; Corbett, K.D. The SARS-CoV-2 nucleocapsid phosphoprotein forms mutually exclusive condensates with RNA and the membrane-associated M protein. *Nat Commun* 2021, 12, doi:10.1038/s41467-020-20768-y.

75. Dunker, A.K.; Silman, I.; Uversky, V.N.; Sussman, J.L. Function and structure of inherently disordered proteins. *Curr Opin Struct Biol* 2008, 18, 756–764, doi:10.1016/j.sbi.2008.10.002.

76. Barik, S. Genus-specific pattern of intrinsically disordered central regions in the nucleocapsid protein of coronaviruses. *Comput Struct Biotechnol J* 2020, 18, 1884–1890, doi:10.1016/j.csbj.2020.07.005.

77. Ni, X.C.; Han, Y.Z.; Zhou, R.J.; Zhou, Y.M.; Lei, J. Structural insights into ribonucleoprotein dissociation by nucleocapsid protein interacting with non-structural protein 3 in SARS-CoV-2. *Commun Biol* 2023, 6, doi:10.1038/s42003-023-04570-2.

78. Koetzner, C.A.; Hurst-Hess, K.R.; Kuo, L.L.; Masters, P.S. Analysis of a crucial interaction between the coronavirus nucleocapsid protein and the major membrane-bound subunit of the viral replicase-transcriptase complex. *Virology* 2022, 567, doi:10.1016/j.virol.2021.12.004.

79. Carlson, C.R.; Asfaha, J.B.; Ghent, C.M.; Howard, C.J.; Hartooni, N.; Safari, M.; Frankel, A.D.; Morgan, D.O. Phosphoregulation of phase separation by the SARS-CoV-2 N protein suggests a biophysical basis for its dual functions. *Mol Cell* 2020, 80, 1092–1103.e4, doi:10.1016/j.molcel.2020.11.025.

80. Tugaeva, K.V.; Hawkins, D.E.D.P.; Smith, J.L.R.; Bayfield, O.W.; Ker, D.S.; Sysoev, A.A.; Klychnikov, O.I.; Antson, A.A.; Sluchanko, N.N. The mechanism of SARS-CoV-2 nucleocapsid protein recognition by the human 14-3-3 proteins. *J Mol Biol* 2021, 433, doi:10.1016/j.jmb.2021.166875.

81. Biswal, M.; Lu, J.W.; Song, J.K. SARS-CoV-2 nucleocapsid protein targets a conserved surface groove of the NTF2-like domain of G3BP1. *J Mol Biol* 2022, 434, doi:10.1016/j.jmb.2022.167516.

82. Masters, P.S. Coronavirus genomic RNA packaging. *Virology* 2019, 537, 198–207, doi:10.1016/j.virol.2019.08.031.

83. McBride, R.; van Zyl, M.; Fielding, B.C. The coronavirus nucleocapsid is a multifunctional protein. *Viruses* 2014, 6, 2991–3018, doi:10.3390/v6082991.

84. Kwarteng, A.; Asiedu, E.; Sakyi, S.A.; Asiedu, S.O. Targeting the SARS-CoV2 nucleocapsid protein for potential therapeutics using immuno-informatics and structure-based drug discovery techniques. *Biomed Pharmacother* 2020, 132, doi:10.1016/j.biopha.2020.110914.

85. Mu, J.F.; Xu, J.Y.; Zhang, L.K.; Shu, T.; Wu, D.; Huang, M.H.; Ren, Y.J.; Li, X.F.; Geng, Q.; Xu, Y., et al. SARS-CoV-2-encoded nucleocapsid protein acts as a viral suppressor of RNA interference in cells. *Sci China Life Sci* 2020, 63, 1413–1416, doi:10.1007/s11427-020-1692-1.

86. Mu, J.F.; Fang, Y.H.; Yang, Q.; Shu, T.; Wang, A.; Huang, M.H.; Jin, L.; Deng, F.; Qiu, Y.; Zhou, X. SARS-CoV-2 N protein antagonizes type I interferon signaling by suppressing phosphorylation and nuclear translocation of STAT1 and STAT2. *Cell Discov* 2020, 6, doi:10.1038/s41421-020-00208-3.

87. Wu, J.L.; Kuan, I.I.; Guo, J.Y.; Hsu, W.C.; Tang, W.C.; Chan, H.J.; Chen, Y.J.; Chen, B.C.; Wu, H.C.; Liao, J.C. SARS-CoV-2 N protein mediates intercellular nucleic acid dispersion, a feature reduced in Omicron. *iScience* 2023, 26, doi:10.1016/j.isci.2023.105995.

88. Wu, Y.X.; Ma, L.; Cai, S.H.; Zhuang, Z.; Zhao, Z.Y.; Jin, S.H.; Xie, W.H.; Zhou, L.L.; Zhang, L.; Zhao, J.C., et al. RNA-induced liquid phase separation of SARS-CoV-2 nucleocapsid protein facilitates NF-κB hyper-activation and inflammation. *Signal Transduct Target Ther* 2021, 6, doi:10.1038/s41392-021-00575-7.

89. Zhao, Y.H.; Sui, L.Y.; Wu, P.; Wang, W.F.; Wang, Z.D.; Yu, Y.; Hou, Z.J.; Tan, G.Y.; Liu, Q.; Wang, G.Q. A dual-role of SARS-CoV-2 nucleocapsid protein in regulating innate immune response. *Signal Transduct Target Ther* 2021, 6, doi:10.1038/s41392-021-00742-w.

90. Chen, H.; Cui, Y.; Han, X.L.; Hu, W.; Sun, M.; Zhang, Y.; Wang, P.H.; Song, G.T.; Chen, W.; Lou, J.Z. Liquid-liquid phase separation by SARS-CoV-2 nucleocapsid protein and RNA. *Cell Res* 2020, 30, 1143–1145, doi:10.1038/s41422-020-00408-2.

91. Jack, A.; Ferro, L.S.; Trnka, M.J.; Wehri, E.; Nadgir, A.; Nguyenla, X.; Fox, D.; Costa, K.; Stanley, S.; Schaletzky, J., et al. SARS-CoV-2 nucleocapsid protein forms condensates with viral genomic RNA. *PloS Biol* 2021, 19, doi:10.1371/journal.pbio.3001425.

92. Casasanta, M.A.; Jonaid, G.M.; Kaylor, L.; Luqiu, W.Y.; DiCecco, L.A.; Solares, M.J.; Berry, S.; Dearnaley, W.J.; Kelly, D.F. Structural insights of the SARS-CoV-2 nucleocapsid protein: Implications for the inner-workings of rapid antigen tests. *Microsc Microanal* 2023, 29, 649–657, doi:10.1093/micmic/ozac036.

93. Wu, W.B.; Cheng, Y.; Zhou, H.; Sun, C.Z.; Zhang, S.J. The SARS-CoV-2 nucleocapsid protein: Its role in the viral life cycle, structure and functions, and use as a potential target in the development of vaccines and diagnostics. *Virol J* 2023, 20, doi:10.1186/s12985-023-01968-6.

94. Jia, Z.H.; Liu, C.; Chen, Y.W.; Jiang, H.; Wang, Z.J.; Yao, J.L.; Yang, J.; Zhu, J.X.; Zhang, B.Q.; Yuchi, Z.G. Crystal structures of the SARS-CoV-2 nucleocapsid protein C-terminal domain and development of nucleocapsid-targeting nanobodies. *FEBS J* 2022, 289, 3813–3825, doi:10.1111/febs.16239.

95. Schoeman, D.; Fielding, B.C. Is there a link between the pathogenic human coronavirus envelope protein and immunopathology? A review of the literature. *Front Microbiol* 2020, 11, doi:10.3389/fmicb.2020.02086.

96. Duart, G.; García-Murria, M.J.; Grau, B.; Acosta-Cáceres, J.M.; Martínez-Gil, L.; Mingarro, I. SARS-CoV-2 envelope protein topology in eukaryotic membranes. *Open Biol* 2020, 10, doi:10.1098/rsob.200209.

97. Yang, R.; Wu, S.J.; Wang, S.; Rubino, G.; Nickels, J.D.; Cheng, X.L. Refinement of SARS-CoV-2 envelope protein structure in a native-like environment by molecular dynamics simulations. *Front Mol Biosci* 2022, 9, doi:10.3389/fmolb.2022.1027223.

98. Xu, R.D.; Shi, M.F.; Li, J.; Song, P.; Li, N. Construction of SARS-CoV-2 virus-like particles by mammalian expression system. *Front Bioeng Biotechnol* 2020, 8, doi:10.3389/fbioe.2020.00862.

99. Miura, K.; Suzuki, Y.; Ishida, K.; Arakawa, M.; Wu, H.; Fujioka, Y.; Emi, A.; Maeda, K.; Hamajima, R.; Nakano, T., et al. Distinct motifs in the E protein are required for SARS-CoV-2 virus particle formation and lysosomal deacidification in host cells. *J Virol* 2023, doi:10.1128/jvi.00426-23.

100. Bracquemond, D.; Muriaux, D. Betacoronavirus assembly: Clues and perspectives for elucidating SARS-CoV-2 particle formation and egress. *mBio* 2021, 12, doi:10.1128/mBio.02371-21.

101. Schoeman, D.; Fielding, B.C. Coronavirus envelope protein: Current knowledge. *Virol J* 2019, 16, doi:10.1186/s12985-019-1182-0.

102. Yang, Y.; Xiong, Z.Y.; Zhang, S.; Yan, Y.; Nguyen, J.; Ng, B.; Lu, H.F.; Brendese, J.; Yang, F.; Wang, H., et al. Bcl-xL inhibits T-cell apoptosis induced by expression of SARS coronavirus E protein in the absence of growth factors. *Biochem J* 2005, 392, 135–143, doi:10.1042/Bj20050698.

103. Honrubia, J.M.; Gutierrez-Alvarez, J.; Sanz-Bravo, A.; González-Miranda, E.; Muñoz-Santos, D.; Castaño-Rodriguez, C.; Wang, L.; Villarejo-Torres, M.; Ripoll-Gómez, J.; Esteban, A., et al. SARS-CoV-2-mediated lung edema and replication are diminished by cystic fibrosis transmembrane conductance regulator modulators. *mBio* 2023, 14, doi:10.1128/mbio.03136-22.

104. Xia, B.Q.; Shen, X.R.; He, Y.; Pan, X.Y.; Liu, F.L.; Wang, Y.; Yang, F.P.; Fang, S.; Wu, Y.; Duan, Z.L., et al. SARS-CoV-2 envelope protein causes acute respiratory distress syndrome (ARDS)-like pathological damages and constitutes an antiviral target. *Cell Res* 2021, 31, 847–860, doi:10.1038/s41422-021-00519-4.

105. DeDiego, M.L.; Alvarez, E.; Almazán, F.; Rejas, M.T.; Lamirande, E.; Roberts, A.; Shieh, W.J.; Zaki, S.R.; Subbarao, K.; Enjuanes, L. A severe acute respiratory syndrome coronavirus that lacks the E gene is attenuated in vitro and in vivo. *J Virol* 2007, 81, 1701–1713, doi:10.1128/Jvi.01467-06.

106. Nieto-Torres, J.L.; DeDiego, M.L.; Verdiá-Báguena, C.; Jimenez-Guardeño, J.M.; Regla-Nava, J.A.; Fernandez-Delgado, R.; Castaño-Rodriguez, C.; Alcaraz, A.; Torres, J.; Aguilella, V.M., et al. Severe acute respiratory syndrome coronavirus envelope protein ion channel activity promotes virus fitness and pathogenesis. *PloS Pathog* 2014, 10, doi:10.1371/journal.ppat.1004077.

107. Jimenez-Guardeño, J.M.; Nieto-Torres, J.L.; DeDiego, M.L.; Regla-Nava, J.A.; Fernandez-Delgado, R.; Castaño-Rodriguez, C.; Enjuanes, L. The PDZ-binding motif of severe acute respiratory syndrome coronavirus envelope protein is a determinant of viral pathogenesis. *PloS Pathog* 2014, 10, doi:10.1371/journal.ppat.1004320.

108. Westerbeck, J.W.; Machamer, C.E. The infectious bronchitis coronavirus envelope protein alters golgi pH to protect the spike protein and promote the release of infectious virus. *J Virol* 2019, 93, doi:10.1128/JVI.00015-19.

109. Ghosh, S.; Dellibovi-Ragheb, T.A.; Kerviel, A.; Pak, E.; Qiu, Q.; Fisher, M.; Takvorian, P.M.; Bleck, C.; Hsu, V.W.; Fehr, A.R., et al. β-coronaviruses use lysosomes for egress instead of the biosynthetic secretory pathway. *Cell* 2020, 183, 1520–1535.e14, doi:10.1016/j.cell.2020.10.039.

110. Gordon, D.E.; Jang, G.M.; Bouhaddou, M.; Xu, J.W.; Obernier, K.; White, K.M.; O'Meara, M.J.; Rezelj, V.V.; Guo, J.F.Z.; Swaney, D.L., et al. A SARS-CoV-2 protein interaction map reveals targets for drug repurposing. *Nature* 2020, 583, 459–468, doi:10.1038/s41586-020-2286-9.

111. Avila-Flores, A.; Sánchez-Cabezón, J.J.; Ochoa-Echeverría, A.; Checa, A.I.; Rosas-García, J.; Téllez-Araiza, M.; Casado, S.; Liébana, R.; Santos-Mendoza, T.; Mérida, I. Identification of host PDZ-based interactions with the SARS-CoV-2 E protein in human monocytes. *Int J Mol Sci* 2023, 24, doi:10.3390/ijms241612793.

112. Zhu, Y.L.; Alvarez, F.; Wolff, N.; Mechaly, A.; Brule, S.; Neitthoffer, B.; Etienne-Manneville, S.; Haouz, A.; Boeda, B.; Caillet-Saguy, C. Interactions of severe acute respiratory syndrome coronavirus 2 protein E with cell junctions and polarity PSD-95/Dlg/ZO-1-containing proteins. *Front Microbiol* 2022, 13, doi:10.3389/fmicb.2022.829094.

113. Planès, R.; Bert, J.B.; Tairi, S.; BenMohamed, L.; Bahraoui, E. SARS-CoV-2 envelope (E) protein binds and activates TLR2 pathway: A novel molecular target for COVID-19 interventions. *Viruses* 2022, 14, doi:10.3390/v14050999.

114. Caillet-Saguy, C.; Durbesson, F.; Rezelj, V.V.; Gogl, G.; Tran, Q.D.; Twizere, J.C.; Vignuzzi, M.; Vincentelli, R.; Wolff, N. Host PDZ-containing proteins targeted by SARS-CoV-2. *FEBS J* 2021, 288, 5148–5162, doi:10.1111/febs.15881.

115. Vann, K.R.; Acharya, A.; Jang, S.M.; Lachance, C.; Zandian, M.; Holt, T.A.; Smith, A.L.; Pandey, K.; Durden, D.L.; El-Gamal, D., et al. Binding of the SARS-CoV-2 envelope E protein to human BRD4 is essential for infection. *Structure* 2022, 30, 1224–1232.e5, doi:10.1016/j.str.2022.05.020.

116. Chen, I.P.; Longbotham, J.E.; McMahon, S.; Suryawanshi, R.K.; Khalid, M.M.; Taha, T.Y.; Tabata, T.; Hayashi, J.M.; Soveg, F.W.; Carlson-Stevermer, J., et al. Viral E protein neutralizes BET protein-mediated post-entry antagonism of SARS-CoV-2. *Cell Rep* 2022, 40, doi:10.1016/j.celrep.2022.111088.

117. Tang, Z.H.; Xu, Y.Y.; Tan, Y.; Shi, H.; Jin, P.P.; Li, Y.Q.; Teng, J.L.; Liu, H.L.; Pan, H.Y.; Hu, Q.Y., et al. CD36 mediates SARS-CoV-2-envelope-protein-induced platelet activation and thrombosis. *Nat Commun* 2023, 14, doi:10.1038/s41467-023-40824-7.

118. Zhang, R.; Qin, H.; Prasad, R.; Fu, R.; Zhou, H.X.; Cross, T.A. Dimeric transmembrane structure of the SARS-CoV-2 E protein. *Commun Biol* 2023, 6, 1109, doi:10.1038/s42003-023-05490-x.

119. Mandala, V.S.; McKay, M.J.; Shcherbakov, A.A.; Dregni, A.J.; Kolocouris, A.; Hong, M. Structure and drug binding of the SARS-CoV-2 envelope protein transmembrane domain in lipid bilayers. *Nat Struct Mol Biol* 2020, 27, 1202–1208, doi:10.1038/s41594-020-00536-8.

120. Schoeman, D.; Cloete, R.; Fielding, B.C. The flexible, extended coil of the PDZ-binding motif of the three deadly human coronavirus E proteins plays a role in pathogenicity. *Viruses* 2022, 14, doi:10.3390/v14081707.

121. Sarkar, M.; Saha, S. Structural insight into the role of novel SARS-CoV-2 E protein: A potential target for vaccine development and other therapeutic strategies. *PLoS One* 2020, 15, doi:10.1371/journal.pone.0237300.

122. Heo, L.; Feig, M. Modeling of severe acute respiratory syndrome coronavirus 2 (SARS-CoV-2) proteins by machine learning and physics-based refinement. *bioRxiv* 2020, doi:10.1101/2020.03.25.008904.

123. Kuzmin, A.; Orekhov, P.; Astashkin, R.; Gordeliy, V.; Gushchin, I. Structure and dynamics of the SARS-CoV-2 envelope protein monomer. *Proteins* 2022, 90, 1102–1114, doi:10.1002/prot.26317.

124. Dey, D.; Borkotoky, S.; Banerjee, M. Identification of tretinoin as a SARS-CoV-2 envelope (E) protein ion channel inhibitor. *Comput Biol Med* 2020, 127, doi:10.1016/j.compbiomed.2020.104063.

125. Park, S.H.; Siddiqi, H.; Castro, D.V.; De Angelis, A.A.; Oom, A.L.; Stoneham, C.A.; Lewinski, M.K.; Clark, A.E.; Croker, B.A.; Carlin, A.F., et al. Interactions of SARS-CoV-2 envelope protein with amilorides correlate with antiviral activity. *PloS Pathog* 2021, 17, doi:10.1371/journal.ppat.1009519.

126. De Maio, F.; Lo Cascio, E.; Babini, G.; Sali, M.; Della Longa, S.; Tilocca, B.; Roncada, P.; Arcovito, A.; Sanguinetti, M.; Scambia, G., et al. Improved binding of SARS-CoV-2 envelope protein to tight junction-associated PALS1 could play a key role in COVID-19 pathogenesis. *Microbes Infect* 2020, 22, 592–597, doi:10.1016/j.micinf.2020.08.006.

127. Perlman, S.; Netland, J. Coronaviruses post-SARS: Update on replication and pathogenesis. *Nat Rev Microbiol* 2009, 7, 439–450, doi:10.1038/nrmicro2147.

128. Masters, P.S. The molecular biology of coronaviruses. *Adv Virus Res* 2006, 66, 193–292, doi:10.1016/S0065-3527(06)66005-3.

129. Arshad, N.; Laurent-Rolle, M.; Ahmed, W.S.; Hsu, J.C.; Mitchell, S.M.; Pawlak, J.; Sengupta, D.; Biswas, K.H.; Cresswell, P. SARS-CoV-2 accessory proteins ORF7a and ORF3a use distinct mechanisms to down-regulate MHC-I surface expression. *Proc Natl Acad Sci U S A* 2023, 120, e2208525120, doi:10.1073/pnas.2208525120.

130. Hassan, S.S.; Choudhury, P.P.; Dayhoff, G.W. 2nd; Aljabali, A.A.A.; Uhal, B.D.; Lundstrom, K.; Rezaei, N.; Pizzol, D.; Adadi, P.; Lal, A., et al. The importance of accessory protein variants in the pathogenicity of SARS-CoV-2. *Arch Biochem Biophys* 2022, 717, 109124, doi:10.1016/j.abb.2022.109124.

131. Issa, E.; Merhi, G.; Panossian, B.; Salloum, T.; Tokajian, S. SARS-CoV-2 and ORF3a: Nonsynonymous mutations, functional domains, and viral pathogenesis. *mSystems* 2020, 5, doi:10.1128/mSystems.00266-20.

132. Azad, G.K.; Khan, P.K. Variations in Orf3a protein of SARS-CoV-2 alter its structure and function. *Biochem Biophys Rep* 2021, 26, 100933, doi:10.1016/j.bbrep.2021.100933.

133. Kern, D.M.; Sorum, B.; Mali, S.S.; Hoel, C.M.; Sridharan, S.; Remis, J.P.; Toso, D.B.; Kotecha, A.; Bautista, D.M.; Brohawn, S.G. Cryo-EM structure of SARS-CoV-2 ORF3a in lipid nanodiscs. *Nat Struct Mol Biol* 2021, 28, 573–582, doi:10.1038/s41594-021-00619-0.

134. Redondo, N.; Zaldivar-Lopez, S.; Garrido, J.J.; Montoya, M. SARS-CoV-2 accessory proteins in viral pathogenesis: Knowns and unknowns. *Front Immunol* 2021, 12, 708264, doi:10.3389/fimmu.2021.708264.

135. Firth, A.E. A putative new SARS-CoV protein, 3c, encoded in an ORF overlapping ORF3a. *J Gen Virol* 2020, 101, 1085–1089, doi:10.1099/jgv.0.001469.

136. Müller, M.; Herrmann, A.; Fujita, S.; Uriu, K.; Kruth, C.; Strange, A.; Kolberg, J.E.; Schneider, M.; Ito, J.; Müller, M.A., et al. ORF3c is expressed in SARS-CoV-2-infected cells and inhibits innate sensing by targeting MAVS. *EMBO Rep* 2023, 24, e57137, doi:10.15252/embr.202357137.

137. Stewart, H.; Lu, Y.; O'Keefe, S.; Valpadashi, A.; Cruz-Zaragoza, L.D.; Michel, H.A.; Nguyen, S.K.; Carnell, G.W.; Lukhovitskaya, N.; Milligan, R., et al. The SARS-CoV-2 protein ORF3c is a mitochondrial modulator of innate immunity. *iScience* 2023, 26, 108080, https://doi.org/10.1016/j.isci.2023.108080.

138. Nelson, C.W.; Ardern, Z.; Goldberg, T.L.; Meng, C.; Kuo, C.-H.; Ludwig, C.; Kolokotronis, S.-O.; Wei, X. Dynamically evolving novel overlapping gene as a factor in the SARS-CoV-2 pandemic. *Elife* 2020, 9, e59633, doi:10.7554/eLife.59633.

139. Lee, J.G.; Huang, W.; Lee, H.; van de Leemput, J.; Kane, M.A.; Han, Z. Characterization of SARS-CoV-2 proteins reveals Orf6 pathogenicity, subcellular localization, host interactions and attenuation by Selinexor. *Cell Biosci* 2021, 11, 58, doi:10.1186/s13578-021-00568-7.

140. Miorin, L.; Kehrer, T.; Sanchez-Aparicio, M.T.; Zhang, K.; Cohen, P.; Patel, R.S.; Cupic, A.; Makio, T.; Mei, M.; Moreno, E., et al. SARS-CoV-2 Orf6 hijacks Nup98 to block STAT nuclear import and antagonize interferon signaling. *Proc Natl Acad Sci U S A* 2020, 117, 28344–28354, doi:10.1073/pnas.2016650117.

141. Yoshimoto, F.K. The proteins of severe acute respiratory syndrome coronavirus-2 (SARS CoV-2 or n-COV19), the cause of COVID-19. *Protein J* 2020, 39, 198–216, doi:10.1007/s10930-020-09901-4.

142. Arya, R.; Kumari, S.; Pandey, B.; Mistry, H.; Bihani, S.C.; Das, A.; Prashar, V.; Gupta, G.D.; Panicker, L.; Kumar, M. Structural insights into SARS-CoV-2 proteins. *J Mol Biol* 2021, 433, 166725, https://doi.org/10.1016/j.jmb.2020.11.024.

143. Cao, Z.; Xia, H.; Rajsbaum, R.; Xia, X.; Wang, H.; Shi, P.Y. Ubiquitination of SARS-CoV-2 ORF7a promotes antagonism of interferon response. *Cell Mol Immunol* 2021, 18, 746–748, doi:10.1038/s41423-020-00603-6.

144. Zhou, Z.; Huang, C.; Zhou, Z.; Huang, Z.; Su, L.; Kang, S.; Chen, X.; Chen, Q.; He, S.; Rong, X., et al. Structural insight reveals SARS-CoV-2 ORF7a as an immunomodulating factor for human CD14+ monocytes. *iScience* 2021, 24, 102187, https://doi.org/10.1016/j.isci.2021.102187.

145. Flower, T.G.; Buffalo, C.Z.; Hooy, R.M.; Allaire, M.; Ren, X.; Hurley, J.H. Structure of SARS-CoV-2 ORF8, a rapidly evolving immune evasion protein. *Proc Natl Acad Sci U S A* 2021, 118, doi:10.1073/pnas.2021785118.

146. Li, J.Y.; Liao, C.H.; Wang, Q.; Tan, Y.J.; Luo, R.; Qiu, Y.; Ge, X.Y. The ORF6, ORF8 and nucleocapsid proteins of SARS-CoV-2 inhibit type I interferon signaling pathway. *Virus Res* 2020, 286, 198074, doi:10.1016/j.virusres.2020.198074.

147. Zhang, Y.; Chen, Y.; Li, Y.; Huang, F.; Luo, B.; Yuan, Y.; Xia, B.; Ma, X.; Yang, T.; Yu, F. The ORF8 protein of SARS-CoV-2 mediates immune evasion through down-regulating MHC-I. *Proc Natl Acad Sci U S A* 2021, 118, e2024202118.

148. Jiang, H.W.; Zhang, H.N.; Meng, Q.F.; Xie, J.; Li, Y.; Chen, H.; Zheng, Y.X.; Wang, X.N.; Qi, H.; Zhang, J., et al. SARS-CoV-2 Orf9b suppresses type I interferon responses by targeting TOM70. *Cell Mol Immunol* 2020, 17, 998–1000, doi:10.1038/s41423-020-0514-8.

149. Meier, C.; Aricescu, A.R.; Assenberg, R.; Aplin, R.T.; Gilbert, R.J.; Grimes, J.M.; Stuart, D.I. The crystal structure of ORF-9b, a lipid binding protein from the SARS coronavirus. *Structure* 2006, 14, 1157–1165, doi:10.1016/j.str.2006.05.012.

150. Gao, X.; Zhu, K.; Qin, B.; Olieric, V.; Wang, M.; Cui, S. Crystal structure of SARS-CoV-2 Orf9b in complex with human TOM70 suggests unusual virus-host interactions. *Nat Commun* 2021, 12, 2843, doi:10.1038/s41467-021-23118-8.

151. Kuhn, C.C.; Basnet, N.; Bodakuntla, S.; Alvarez-Brecht, P.; Nichols, S.; Martinez-Sanchez, A.; Agostini, L.; Soh, Y.M.; Takagi, J.; Biertumpfel, C., et al. Direct Cryo-ET observation of platelet deformation induced by SARS-CoV-2 spike protein. *Nat Commun* 2023, 14, 620, doi:10.1038/s41467-023-36279-5.

152. Tang, X.; Wu, C.; Li, X.; Song, Y.; Yao, X.; Wu, X.; Duan, Y.; Zhang, H.; Wang, Y.; Qian, Z., et al. On the origin and continuing evolution of SARS-CoV-2. *Natl Sci Rev* 2020, 7, 1012–1023, doi:10.1093/nsr/nwaa036.

153. Millet, J.K.; Whittaker, G.R. Physiological and molecular triggers for SARS-CoV membrane fusion and entry into host cells. *Virology* 2018, 517, 3–8, doi:10.1016/j.virol.2017.12.015.

154. Chambers, P.; Pringle, C.R.; Easton, A.J. Heptad repeat sequences are located adjacent to hydrophobic regions in several types of virus fusion glycoproteins. *J Gen Virol* 1990, 71 *(Pt 12)*, 3075–3080, doi:10.1099/0022-1317-71-12-3075.

155. Xia, S.; Liu, M.; Wang, C.; Xu, W.; Lan, Q.; Feng, S.; Qi, F.; Bao, L.; Du, L.; Liu, S., et al. Inhibition of SARS-CoV-2 (previously 2019-nCoV) infection by a highly potent pan-coronavirus fusion inhibitor targeting its spike protein that harbors a high capacity to mediate membrane fusion. *Cell Research* 2020, 30, 343–355, doi:10.1038/s41422-020-0305-x.

156. Du, L.; He, Y.; Zhou, Y.; Liu, S.; Zheng, B.J.; Jiang, S. The spike protein of SARS-CoV–a target for vaccine and therapeutic development. *Nat Rev Microbiol* 2009, 7, 226–236, doi:10.1038/nrmicro2090.

157. Cantuti-Castelvetri, L.; Ojha, R.; Pedro, L.D.; Djannatian, M.; Franz, J.; Kuivanen, S.; van der Meer, F.; Kallio, K.; Kaya, T.; Anastasina, M., et al. Neuropilin-1 facilitates SARS-CoV-2 cell entry and infectivity. *Science* 2020, 370, 856–860, doi:10.1126/science.abd2985.

158. Wang, K.; Chen, W.; Zhang, Z.; Deng, Y.; Lian, J.Q.; Du, P.; Wei, D.; Zhang, Y.; Sun, X.X.; Gong, L., et al. CD147-spike protein is a novel route for SARS-CoV-2 infection to host cells. *Signal Transduct Target Ther* 2020, 5, 283, doi:10.1038/s41392-020-00426-x.

159. Shirato, K.; Kizaki, T. SARS-CoV-2 spike protein S1 subunit induces pro-inflammatory responses via toll-like receptor 4 signaling in murine and human macrophages. *Heliyon* 2021, 7, e06187, doi:10.1016/j.heliyon.2021.e06187.

160. Sigrist, C.J.; Bridge, A.; Le Mercier, P. A potential role for integrins in host cell entry by SARS-CoV-2. *Antiviral Res* 2020, 177, 104759, doi:10.1016/j.antiviral.2020.104759.

161. Park, E.J.; Myint, P.K.; Appiah, M.G.; Darkwah, S.; Caidengbate, S.; Ito, A.; Matsuo, E.; Kawamoto, E.; Gaowa, A.; Shimaoka, M. The spike glycoprotein of SARS-CoV-2 binds to beta1 integrins expressed on the surface of lung epithelial cells. *Viruses* 2021, 13, doi:10.3390/v13040645.

162. Xu, H.; Zhong, L.; Deng, J.; Peng, J.; Dan, H.; Zeng, X.; Li, T.; Chen, Q. High expression of ACE2 receptor of 2019-nCoV on the epithelial cells of oral mucosa. *Int J Oral Sci* 2020, 12, 8, doi:10.1038/s41368-020-0074-x.

163. Kim, D.; Kim, S.; Park, J.; Chang, H.R.; Chang, J.; Ahn, J.; Park, H.; Park, J.; Son, N.; Kang, G., et al. A high-resolution temporal atlas of the SARS-CoV-2 translatome and transcriptome. *Nat Commun* 2021, 12, 5120, doi:10.1038/s41467-021-25361-5.

164. Finkel, Y.; Mizrahi, O.; Nachshon, A.; Weingarten-Gabbay, S.; Morgenstern, D.; Yahalom-Ronen, Y.; Tamir, H.; Achdout, H.; Stein, D.; Israeli, O., et al. The coding capacity of SARS-CoV-2. *Nature* 2021, 589, 125–130, doi:10.1038/s41586-020-2739-1.

165. Osipiuk, J.; Azizi, S.A.; Dvorkin, S.; Endres, M.; Jedrzejczak, R.; Jones, K.A.; Kang, S.; Kathayat, R.S.; Kim, Y.; Lisnyak, V.G., et al. Structure of papain-like protease from SARS-CoV-2 and its complexes with non-covalent inhibitors. *Nat Commun* 2021, 12, 743, doi:10.1038/s41467-021-21060-3.

166. Chen, J.; Malone, B.; Llewellyn, E.; Grasso, M.; Shelton, P.M.M.; Olinares, P.D.B.; Maruthi, K.; Eng, E.T.; Vatandaslar, H.; Chait, B.T., et al. Structural basis for helicase-polymerase coupling in the SARS-CoV-2 replication-transcription complex. *Cell* 2020, 182, 1560–1573.e13, doi:10.1016/j.cell.2020.07.033.

167. Yadav, R.; Chaudhary, J.K.; Jain, N.; Chaudhary, P.K.; Khanra, S.; Dhamija, P.; Sharma, A.; Kumar, A.; Handu, S. Role of structural and non-structural proteins and therapeutic targets of SARS-CoV-2 for COVID-19. *Cells* 2021, 10, doi:10.3390/cells10040821.

168. Klemm, T.; Ebert, G.; Calleja, D.J.; Allison, C.C.; Richardson, L.W.; Bernardini, J.P.; Lu, B.G.; Kuchel, N.W.; Grohmann, C.; Shibata, Y., et al. Mechanism and inhibition of the papain-like protease, PLpro, of SARS-CoV-2. *EMBO J* 2020, 39, e106275, doi:10.15252/embj.2020106275.

169. Jin, Z.; Du, X.; Xu, Y.; Deng, Y.; Liu, M.; Zhao, Y.; Zhang, B.; Li, X.; Zhang, L.; Peng, C., et al. Structure of M(pro) from SARS-CoV-2 and discovery of its inhibitors. *Nature* 2020, 582, 289–293, doi:10.1038/s41586-020-2223-y.

170. Snijder, E.J.; Decroly, E.; Ziebuhr, J. The nonstructural proteins directing coronavirus RNA synthesis and processing. *Adv Virus Res* 2016, 96, 59–126, doi:10.1016/bs.aivir.2016.08.008.

171. Roingeard, P.; Eymieux, S.; Burlaud-Gaillard, J.; Hourioux, C.; Patient, R.; Blanchard, E. The double-membrane vesicle (DMV): A virus-induced organelle dedicated to the replication of SARS-CoV-2 and other positive-sense single-stranded RNA viruses. *Cell Mol Life Sci* 2022, 79, 425, doi:10.1007/s00018-022-04469-x.

172. Thoms, M.; Buschauer, R.; Ameismeier, M.; Koepke, L.; Denk, T.; Hirschenberger, M.; Kratzat, H.; Hayn, M.; Mackens-Kiani, T.; Cheng, J., et al. Structural basis for translational shutdown and immune evasion by the Nsp1 protein of SARS-CoV-2. *Science* 2020, 369, 1249–1255, doi:10.1126/science.abc8665.

173. Ogando, N.S.; Zevenhoven-Dobbe, J.C.; van der Meer, Y.; Bredenbeek, P.J.; Posthuma, C.C.; Snijder, E.J. The enzymatic activity of the nsp14 exoribonuclease is critical for replication of MERS-CoV and SARS-CoV-2. *J Virol* 2020, 94, e01246-20, doi:10.1128/jvi.01246-20.

174. Gao, Y.; Yan, L.; Huang, Y.; Liu, F.; Zhao, Y.; Cao, L.; Wang, T.; Sun, Q.; Ming, Z.; Zhang, L., et al. Structure of the RNA-dependent RNA polymerase from COVID-19 virus. *Science* 2020, 368, 779–782, doi:10.1126/science.abb7498.

175. Lou, Z.; Rao, Z. The life of SARS-CoV-2 inside cells: Replication-transcription complex assembly and function. *Annu Rev Biochem* 2022, 91, 381–401, doi:10.1146/annurev-biochem-052521-115653.

176. Malone, B.; Urakova, N.; Snijder, E.J.; Campbell, E.A. Structures and functions of coronavirus replication-transcription complexes and their relevance for SARS-CoV-2 drug design. *Nat Rev Mol Cell Biol* 2022, 23, 21–39, doi:10.1038/s41580-021-00432-z.

177. Wang, D.; Jiang, A.; Feng, J.; Li, G.; Guo, D.; Sajid, M.; Wu, K.; Zhang, Q.; Ponty, Y.; Will, S., et al. The SARS-CoV-2 subgenome landscape and its novel regulatory features. *Mol Cell* 2021, 81, 2135–2147.e5, doi:10.1016/j.molcel.2021.02.036.

178. Kim, D.; Lee, J.Y.; Yang, J.S.; Kim, J.W.; Kim, V.N.; Chang, H. The architecture of SARS-CoV-2 transcriptome. *Cell* 2020, 181, 914–921.e10, doi:10.1016/j.cell.2020.04.011.

179. Brant, A.C.; Tian, W.; Majerciak, V.; Yang, W.; Zheng, Z.M. SARS-CoV-2: From its discovery to genome structure, transcription, and replication. *Cell Biosci* 2021, 11, 136, doi:10.1186/s13578-021-00643-z.

180. Klein, S.; Cortese, M.; Winter, S.L.; Wachsmuth-Melm, M.; Neufeldt, C.J.; Cerikan, B.; Stanifer, M.L.; Boulant, S.; Bartenschlager, R.; Chlanda, P. SARS-CoV-2 structure and replication characterized by in situ cryo-electron tomography. *Nat Commun* 2020, 11, 5885, doi:10.1038/s41467-020-19619-7.

181. Mendonca, L.; Howe, A.; Gilchrist, J.B.; Sheng, Y.; Sun, D.; Knight, M.L.; Zanetti-Domingues, L.C.; Bateman, B.; Krebs, A.S.; Chen, L., et al. Correlative multi-scale cryo-imaging unveils SARS-CoV-2 assembly and egress. *Nat Commun* 2021, 12, 4629, doi:10.1038/s41467-021-24887-y.

182. Hillen, H.S.; Kokic, G.; Farnung, L.; Dienemann, C.; Tegunov, D.; Cramer, P. Structure of replicating SARS-CoV-2 polymerase. *Nature* 2020, 584, 154–156, doi:10.1038/s41586-020-2368-8.

183. Hartenian, E.; Nandakumar, D.; Lari, A.; Ly, M.; Tucker, J.M.; Glaunsinger, B.A. The molecular virology of coronaviruses. *J Biol Chem* 2020, 295, 12910–12934, doi:10.1074/jbc.REV120.013930.

184. Lin, S.; Chen, H.; Chen, Z.; Yang, F.; Ye, F.; Zheng, Y.; Yang, J.; Lin, X.; Sun, H.; Wang, L., et al. Crystal structure of SARS-CoV-2 nsp10 bound to nsp14-ExoN domain reveals an exoribonuclease with both structural and functional integrity. *Nucleic Acids Res* 2021, 49, 5382–5392, doi:10.1093/nar/gkab320.

185. Malone, B.; Chen, J.; Wang, Q.; Llewellyn, E.; Choi, Y.J.; Olinares, P.D.B.; Cao, X.; Hernandez, C.; Eng, E.T.; Chait, B.T., et al. Structural basis for backtracking by the SARS-CoV-2 replication-transcription complex. *Proc Natl Acad Sci U S A* 2021, 118, doi:10.1073/pnas.2102516118.

186. Romano, M.; Ruggiero, A.; Squeglia, F.; Maga, G.; Berisio, R. A structural view of SARS-CoV-2 RNA replication machinery: RNA synthesis, proofreading and final capping. *Cells* 2020, 9, doi:10.3390/cells9051267.

187. Syed, A.M.; Taha, T.Y.; Tabata, T.; Chen, I.P.; Ciling, A.; Khalid, M.M.; Sreekumar, B.; Chen, P.Y.; Hayashi, J.M.; Soczek, K.M., et al. Rapid assessment of SARS-CoV-2-evolved variants using virus-like particles. *Science* 2021, 374, 1626–1632, doi:10.1126/science.abl6184.

188. Cubuk, J.; Alston, J.J.; Incicco, J.J.; Singh, S.; Stuchell-Brereton, M.D.; Ward, M.D.; Zimmerman, M.I.; Vithani, N.; Griffith, D.; Wagoner, J.A., et al. The SARS-CoV-2 nucleocapsid protein is dynamic, disordered, and phase separates with RNA. *Nat Commun* 2021, 12, 1936, doi:10.1038/s41467-021-21953-3.

189. Wang, W.A.; Carreras-Sureda, A.; Demaurex, N. SARS-CoV-2 infection alkalinizes the ERGIC and lysosomes through the viroporin activity of the viral envelope protein. *J Cell Sci* 2023, 136, doi:10.1242/jcs.260685.

190. Chams, N.; Chams, S.; Badran, R.; Shams, A.; Araji, A.; Raad, M.; Mukhopadhyay, S.; Stroberg, E.; Duval, E.J.; Barton, L.M., et al. COVID-19: A multidisciplinary review. *Front Public Health* 2020, 8, 383, doi:10.3389/fpubh.2020.00383.

191. Domingo, J.L. What we know and what we need to know about the origin of SARS-CoV-2. *Environ Res* 2021, 200, 111785, https://doi.org/10.1016/j.envres.2021.111785.

192. Frutos, R.; Pliez, O.; Gavotte, L.; Devaux, C.A. There is no "origin" to SARS-CoV-2. *Environ Res* 2022, 207, 112173, https://doi.org/10.1016/j.envres.2021.112173.

193. Kane, Y.; Wong, G.; Gao, G.F. Animal models, zoonotic reservoirs, and cross-species transmission of emerging human-infecting coronaviruses. *Annu Rev Anim Biosci* 2023, 11, 1–31, doi:10.1146/annurev-animal-020420-025011.

194. Deng, S.Q.; Peng, H.J. Characteristics of and public health responses to the coronavirus disease 2019 outbreak in China. *J Clin Med* 2020, 9, doi:10.3390/jcm9020575.

195. Jiang, S.; Du, L.; Shi, Z. An emerging coronavirus causing pneumonia outbreak in Wuhan, China: Calling for developing therapeutic and prophylactic strategies. *Emerg Microbes Infect* 2020, 9, 275–277, doi:10.1080/22221751.2020.1723441.

196. Li, Q.; Guan, X.; Wu, P.; Wang, X.; Zhou, L.; Tong, Y.; Ren, R.; Leung, K.S.M.; Lau, E.H.Y.; Wong, J.Y., et al. Early transmission dynamics in Wuhan, China, of novel coronavirus-infected pneumonia. *N Engl J Med* 2020, 382, 1199–1207, doi:10.1056/NEJMoa2001316.

197. Worobey, M.; Levy, J.I.; Malpica Serrano, L.; Crits-Christoph, A.; Pekar, J.E.; Goldstein, S.A.; Rasmussen, A.L.; Kraemer, M.U.G.; Newman, C.; Koopmans, M.P.G., et al. The huanan seafood wholesale market in Wuhan was the early epicenter of the COVID-19 pandemic. *Science* 2022, 377, 951–959, doi:10.1126/science.abp8715.

198. Banerjee, A.; Kulcsar, K.; Misra, V.; Frieman, M.; Mossman, K. Bats and coronaviruses. *Viruses* 2019, 11, doi:10.3390/v11010041.

199. Li, W.; Shi, Z.; Yu, M.; Ren, W.; Smith, C.; Epstein, J.H.; Wang, H.; Crameri, G.; Hu, Z.; Zhang, H., et al. Bats are natural reservoirs of SARS-like coronaviruses. *Science* 2005, 310, 676–679, doi:10.1126/science.1118391.

200. Hao, P.; Zhong, W.; Song, S.; Fan, S.; Li, X. Is SARS-CoV-2 originated from laboratory? A rebuttal to the claim of formation via laboratory recombination. *Emerg Microb Infect* 2020, 9, 545–547, doi:10.1080/22221751.2020.1738279.

201. Zhou, P.; Yang, X.L.; Wang, X.G.; Hu, B.; Zhang, L.; Zhang, W.; Si, H.R.; Zhu, Y.; Li, B.; Huang, C.L., et al. A pneumonia outbreak associated with a new coronavirus of probable bat origin. *Nature* 2020, 579, 270–273, doi:10.1038/s41586-020-2012-7.

202. Nie, J.; Li, Q.; Zhang, L.; Cao, Y.; Zhang, Y.; Li, T.; Wu, J.; Liu, S.; Zhang, M.; Zhao, C., et al. Functional comparison of SARS-CoV-2 with closely related pangolin and bat coronaviruses. *Cell Discov* 2021, 7, 21, doi:10.1038/s41421-021-00256-3.

203. Zhou, H.; Ji, J.; Chen, X.; Bi, Y.; Li, J.; Wang, Q.; Hu, T.; Song, H.; Zhao, R.; Chen, Y., et al. Identification of novel bat coronaviruses sheds light on the evolutionary origins of SARS-CoV-2 and related viruses. *Cell* 2021, 184, 4380–4391.e14, doi:10.1016/j.cell.2021.06.008.

204. Zhou, H.; Chen, X.; Hu, T.; Li, J.; Song, H.; Liu, Y.; Wang, P.; Liu, D.; Yang, J.; Holmes, E.C., et al. A novel bat coronavirus closely related to SARS-CoV-2 contains natural insertions at the S1/S2 cleavage site of the spike protein. *Curr Biol* 2020, 30, 2196–2203.e3, doi:10.1016/j.cub.2020.05.023.

205. Temmam, S.; Vongphayloth, K.; Baquero, E.; Munier, S.; Bonomi, M.; Regnault, B.; Douangboubpha, B.; Karami, Y.; Chretien, D.; Sanamxay, D., et al. Bat coronaviruses related to SARS-CoV-2 and infectious for human cells. *Nature* 2022, 604, 330–336, doi:10.1038/s41586-022-04532-4.

206. Singh, D.; Yi, S.V. On the origin and evolution of SARS-CoV-2. *Exp Mol Med* 2021, 53, 537–547, doi:10.1038/s12276-021-00604-z.

207. Turakhia, Y.; Thornlow, B.; Hinrichs, A.; McBroome, J.; Ayala, N.; Ye, C.; Smith, K.; De Maio, N.; Haussler, D.; Lanfear, R., et al. Pandemic-scale phylogenomics reveals the SARS-CoV-2 recombination landscape. *Nature* 2022, 609, 994–997, doi:10.1038/s41586-022-05189-9.

208. Mitra, A. Investigations into the origin of SARS-CoV-2: An update. *Curr Sci* 2021, 121, 77–84.

209. Khadka, B.; Gupta, R.S. Conserved molecular signatures in the spike protein provide evidence indicating the origin of SARS-CoV-2 and a Pangolin-CoV (MP789) by recombination (s) between specific lineages of Sarbecoviruses. *PeerJ* 2021, 9, e12434.

210. Cattaneo, A.M. Reviewing findings on the polypeptide sequence of the SARS-CoV-2 S-protein to discuss the origins of the virus. *Future Virol* 2022, 0, doi:10.2217/fvl-2021-0233.

211. Hou, Y.J.; Chiba, S.; Leist, S.R.; Meganck, R.M.; Martinez, D.R.; Schafer, A.; Catanzaro, N.J.; Sontake, V.; West, A.; Edwards, C.E., et al. Host range, transmissibility and antigenicity of a pangolin coronavirus. *Nat Microbiol* 2023, 8, 1820–1833, doi:10.1038/s41564-023-01476-x.

212. Wang, M.; Yan, M.; Xu, H.; Liang, W.; Kan, B.; Zheng, B.; Chen, H.; Zheng, H.; Xu, Y.; Zhang, E., et al. SARS-CoV infection in a restaurant from palm civet. *Emerg Infect Dis* 2005, 11, 1860–1865, doi:10.3201/eid1112.041293.

213. Xiao, K.; Zhai, J.; Feng, Y.; Zhou, N.; Zhang, X.; Zou, J.J.; Li, N.; Guo, Y.; Li, X.; Shen, X., et al. Isolation of SARS-CoV-2-related coronavirus from Malayan pangolins. *Nature* 2020, 583, 286–289, doi:10.1038/s41586-020-2313-x.

214. Lytras, S.; Hughes, J.; Martin, D.; Swanepoel, P.; de Klerk, A.; Lourens, R.; Kosakovsky Pond, S.L.; Xia, W.; Jiang, X.; Robertson, D.L. Exploring the natural origins of SARS-CoV-2 in the light of recombination. *Genome Biol Evol* 2022, 14, doi:10.1093/gbe/evac018.

215. Reis, J.; Faou, A.L.; Buguet, A.; Sandner, G.; Spencer, P. Covid-19: Early cases and disease spread. *Ann Glob Health* 2022, 88, 83, doi:10.5334/aogh.3776.

216. Nishiura, H.; Linton, N.M.; Akhmetzhanov, A.R. Initial cluster of novel coronavirus (2019-nCoV) infections in Wuhan, China is consistent with substantial human-to-human transmission. *J Clin Med* 2020, 9, 488.

217. Chan, J.F.; Yuan, S.; Kok, K.H.; To, K.K.; Chu, H.; Yang, J.; Xing, F.; Liu, J.; Yip, C.C.; Poon, R.W., et al. A familial cluster of pneumonia associated with the 2019 novel coronavirus indicating person-to-person transmission: A study of a family cluster. *Lancet* 2020, 395, 514–523, doi:10.1016/S0140-6736(20)30154-9.

218. Jayaweera, M.; Perera, H.; Gunawardana, B.; Manatunge, J. Transmission of COVID-19 virus by droplets and aerosols: A critical review on the unresolved dichotomy. *Environ Res* 2020, 188, 109819, doi:10.1016/j.envres.2020.109819.

219. Lednicky, J.A.; Lauzardo, M.; Alam, M.M.; Elbadry, M.A.; Stephenson, C.J.; Gibson, J.C.; Morris, J.G., Jr. Isolation of SARS-CoV-2 from the air in a car driven by a COVID patient with mild illness. *Int J Infect Dis* 2021, 108, 212–216, doi:10.1016/j.ijid.2021.04.063.

220. Santarpia, J.L.; Herrera, V.L.; Rivera, D.N.; Ratnesar-Shumate, S.; Reid, S.P.; Ackerman, D.N.; Denton, P.W.; Martens, J.W.S.; Fang, Y.; Conoan, N., et al. The size and culturability of patient-generated SARS-CoV-2 aerosol. *J Expo Sci Environ Epidemiol* 2022, 32, 706–711, doi:10.1038/s41370-021-00376-8.

221. Stadnytskyi, V.; Bax, C.E.; Bax, A.; Anfinrud, P. The airborne lifetime of small speech droplets and their potential importance in SARS-CoV-2 transmission. *Proc Natl Acad Sci U S A* 2020, 117, 11875–11877, doi:10.1073/pnas.2006874117.

222. Wang, C.C.; Prather, K.A.; Sznitman, J.; Jimenez, J.L.; Lakdawala, S.S.; Tufekci, Z.; Marr, L.C. Airborne transmission of respiratory viruses. *Science* 2021, 373, doi:10.1126/science.abd9149.

223. Baker, C.A.; Gibson, K.E. Persistence of SARS-CoV-2 on surfaces and relevance to the food industry. *Curr Opin Food Sci* 2022, 47, 100875, doi:10.1016/j.cofs.2022.100875.

224. Xu, J.; Xu, C.; Chen, R.; Yin, Y.; Wang, Z.; Li, K.; Shi, J.; Chen, X.; Huang, J.; Hong, J., et al. Stability of SARS-CoV-2 on inanimate surfaces: A review. *Microbiol Res* 2023, 272, 127388, doi:10.1016/j.micres.2023.127388.

225. Paton, S.; Spencer, A.; Garratt, I.; Thompson, K.A.; Dinesh, I.; Aranega-Bou, P.; Stevenson, D.; Clark, S.; Dunning, J.; Bennett, A., et al. Persistence of severe acute respiratory syndrome coronavirus 2 (SARS-CoV-2) virus and viral RNA in relation to surface type and contamination concentration. *Appl Environ Microbiol* 2021, 87, e0052621, doi:10.1128/AEM.00526-21.

226. Meyerowitz, E.A.; Richterman, A. SARS-CoV-2 transmission and prevention in the era of the delta variant. *Infect Dis Clin North Am* 2022, 36, 267–293, doi:10.1016/j.idc.2022.01.007.

227. van Doremalen, N.; Bushmaker, T.; Morris, D.H.; Holbrook, M.G.; Gamble, A.; Williamson, B.N.; Tamin, A.; Harcourt, J.L.; Thornburg, N.J.; Gerber, S.I., et al. Aerosol and surface stability of SARS-CoV-2 as compared with SARS-CoV-1. *N Engl J Med* 2020, 382, 1564–1567, doi:10.1056/NEJMc2004973.

228. Rocha, A.L.S.; Pinheiro, J.R.; Nakamura, T.C.; da Silva, J.D.S.; Rocha, B.G.S.; Klein, R.C.; Birbrair, A.; Amorim, J.H. Fomites and the environment did not have an important role in COVID-19 transmission in a Brazilian mid-sized city. *Sci Rep* 2021, 11, 15960, doi:10.1038/s41598-021-95479-5.

229. Harvey, A.P.; Fuhrmeister, E.R.; Cantrell, M.E.; Pitol, A.K.; Swarthout, J.M.; Powers, J.E.; Nadimpalli, M.L.; Julian, T.R.; Pickering, A.J. Longitudinal monitoring of SARS-CoV-2 RNA on high-touch surfaces in a community setting. *Environ Sci Technol Lett* 2021, 8, 168–175, doi:10.1021/acs.estlett.0c00875.

230. Lavania, M.; Joshi, M.S.; Ranshing, S.S.; Potdar, V.A.; Shinde, M.; Chavan, N.; Jadhav, S.M.; Sarkale, P.; Mohandas, S.; Sawant, P.M., et al. Prolonged shedding of SARS-CoV-2 in feces of COVID-19 positive patients: Trends in genomic variation in first and second wave. *Front Med* 2022, 9, 835168, doi:10.3389/fmed.2022.835168

231. Xiao, F.; Sun, J.; Xu, Y.; Li, F.; Huang, X.; Li, H.; Zhao, J.; Huang, J.; Zhao, J. Infectious SARS-CoV-2 in feces of patient with severe COVID-19. *Emerg Infect Dis* 2020, 26, 1920–1922, doi:10.3201/eid2608.200681.

232. Natarajan, A.; Zlitni, S.; Brooks, E.F.; Vance, S.E.; Dahlen, A.; Hedlin, H.; Park, R.M.; Han, A.;
     Schmidtke, D.T.; Verma, R., et al. Gastrointestinal symptoms and fecal shedding of SARS-CoV-2 RNA
     suggest prolonged gastrointestinal infection. *Med* 2022, 3, 371–387e379, doi:10.1016/j.medj.2022.04.001.
233. Joshi, M.; Mohandas, S.; Prasad, S.; Shinde, M.; Chavan, N.; Yadav, P.D.; Lavania, M. Lack of evidence
     of viability and infectivity of SARS-CoV-2 in the fecal specimens of COVID-19 patients. *Front Public
     Health* 2022, 10, 1030249, doi:10.3389/fpubh.2022.1030249.
234. Foladori, P.; Cutrupi, F.; Segata, N.; Manara, S.; Pinto, F.; Malpei, F.; Bruni, L.; La Rosa, G. SARS-
     CoV-2 from faeces to wastewater treatment: What do we know? A review. *Sci Total Environ* 2020, 743,
     140444, https://doi.org/10.1016/j.scitotenv.2020.140444.
235. Amirian, E.S. Potential fecal transmission of SARS-CoV-2: Current evidence and implications for pub-
     lic health. *Int J Infect Dis* 2020, 95, 363–370, doi:10.1016/j.ijid.2020.04.057.
236. Gacci, M.; Coppi, M.; Baldi, E.; Sebastianelli, A.; Zaccaro, C.; Morselli, S.; Pecoraro, A.; Manera, A.;
     Nicoletti, R.; Liaci, A., et al. Semen impairment and occurrence of SARS-CoV-2 virus in semen after
     recovery from COVID-19. *Hum Reprod* 2021, 36, 1520–1529, doi:10.1093/humrep/deab026.
237. Li, D.; Jin, M.; Bao, P.; Zhao, W.; Zhang, S. Clinical characteristics and results of semen tests
     among men with coronavirus disease 2019. *JAMA Netw Open* 2020, 3, e208292, doi:10.1001/jama
     networkopen.2020.8292.
238. Pan, X.; Chen, D.; Xia, Y.; Wu, X.; Li, T.; Ou, X.; Zhou, L.; Liu, J. Asymptomatic cases in a family cluster
     with SARS-CoV-2 infection. *Lancet Infect Dis* 2020, 20, 410–411, doi:10.1016/S1473-3099(20)30114-6.
239. Song, C.; Wang, Y.; Li, W.; Hu, B.; Chen, G.; Xia, P.; Wang, W.; Li, C.; Diao, F.; Hu, Z., et al. Absence of
     2019 novel coronavirus in semen and testes of COVID-19 patientsdagger. *Biol Reprod* 2020, 103, 4–6,
     doi:10.1093/biolre/ioaa050.
240. Guo, L.; Zhao, S.; Li, W.; Wang, Y.; Li, L.; Jiang, S.; Ren, W.; Yuan, Q.; Zhang, F.; Kong, F., et al.
     Absence of SARS-CoV-2 in semen of a COVID-19 patient cohort. *Andrology* 2021, 9, 42–47, doi:10.1111/
     andr.12848.
241. Algarroba, G.N.; Rekawek, P.; Vahanian, S.A.; Khullar, P.; Palaia, T.; Peltier, M.R.; Chavez, M.R.;
     Vintzileos, A.M. Visualization of severe acute respiratory syndrome coronavirus 2 invading the
     human placenta using electron microscopy. *Am J Obstet Gynecol* 2020, 223, 275–278, doi:10.1016/j.
     ajog.2020.05.023.
242. Penfield, C.A.; Brubaker, S.G.; Limaye, M.A.; Lighter, J.; Ratner, A.J.; Thomas, K.M.; Meyer, J.A.;
     Roman, A.S. Detection of severe acute respiratory syndrome coronavirus 2 in placental and fetal mem-
     brane samples. *Am J Obstet Gynecol MFM* 2020, 2, 100133, doi:10.1016/j.ajogmf.2020.100133.
243. Valk, J.E.; Chong, A.M.; Uhlemann, A.C.; Debelenko, L. Detection of SARS-CoV-2 in placental but not
     fetal tissues in the second trimester. *J Perinatol* 2021, 41, 1184–1186, doi:10.1038/s41372-020-00877-8.
244. Allotey, J.; Chatterjee, S.; Kew, T.; Gaetano, A.; Stallings, E.; Fernandez-Garcia, S.; Yap, M.; Sheikh,
     J.; Lawson, H.; Coomar, D., et al. SARS-CoV-2 positivity in offspring and timing of mother-to-child
     transmission: Living systematic review and meta-analysis. *BMJ* 2022, 376, e067696, doi:10.1136/
     bmj-2021-067696.
245. Sessa, R.; Masciullo, L.; Filardo, S.; Di Pietro, M.; Brandolino, G.; Brunelli, R.; Galoppi, P.; Terrin, G.;
     Viscardi, M.F.; Anastasi, E., et al. SARS-CoV-2 vertical transmission in a twin-pregnant woman: A case
     report. *Int J Inf Dis* 2022, 125, 192–194, doi:10.1016/j.ijid.2022.10.019.
246. Fenizia, C.; Biasin, M.; Cetin, I.; Vergani, P.; Mileto, D.; Spinillo, A.; Gismondo, M.R.; Perotti, F.;
     Callegari, C.; Mancon, A., et al. Analysis of SARS-CoV-2 vertical transmission during pregnancy. *Nat
     Commun* 2020, 11, 5128, doi:10.1038/s41467-020-18933-4.
247. Richtmann, R.; Torloni, M.R.; Oyamada Otani, A.R.; Levi, J.E.; Crema Tobara, M.; de Almeida Silva,
     C.; Dias, L.; Miglioli-Galvao, L.; Martins Silva, P.; Macoto Kondo, M. Fetal deaths in pregnancies
     with SARS-CoV-2 infection in Brazil: A case series. *Case Rep Womens Health* 2020, 27, e00243,
     doi:10.1016/j.crwh.2020.e00243.
248. Maeda, M.F.Y.; Brizot, M.L.; Gibelli, M.; Ibidi, S.M.; Carvalho, W.B.; Hoshida, M.S.; Machado, C.M.;
     Sabino, E.C.; Oliveira da Silva, L.C.; Jaenisch, T., et al. Vertical transmission of SARS-CoV2 during
     pregnancy: A high-risk cohort. *Prenat Diagn* 2021, 41, 998–1008, doi:10.1002/pd.5980.
249. Chen, H.; Guo, J.; Wang, C.; Luo, F.; Yu, X.; Zhang, W.; Li, J.; Zhao, D.; Xu, D.; Gong, Q., et al.
     Clinical characteristics and intrauterine vertical transmission potential of COVID-19 infection in nine
     pregnant women: A retrospective review of medical records. *Lancet* 2020, 395, 809–815, doi:10.1016/
     S0140-6736(20)30360-3.
250. Schwartz, D.A. An analysis of 38 pregnant women with COVID-19, their newborn infants, and mater-
     nal-fetal transmission of SARS-CoV-2: maternal coronavirus infections and pregnancy outcomes. *Arch
     Pathol Lab Med* 2020, 144, 799–805, doi:10.5858/arpa.2020-0901-SA.

251. Yan, J.; Guo, J.; Fan, C.; Juan, J.; Yu, X.; Li, J.; Feng, L.; Li, C.; Chen, H.; Qiao, Y., et al. Coronavirus disease 2019 in pregnant women: A report based on 116 cases. *Am J Obstet Gynecol* 2020, 223, 111. e1–111.e14, doi:10.1016/j.ajog.2020.04.014.

252. Dong, L.; Pei, S.; Ren, Q.; Fu, S.; Yu, L.; Chen, H.; Chen, X.; Yin, M. Evaluation of vertical transmission of SARS-CoV-2 in utero: Nine pregnant women and their newborns. *Placenta* 2021, 111, 91–96, doi:10.1016/j.placenta.2021.06.007.

253. Zeng, L.; Xia, S.; Yuan, W.; Yan, K.; Xiao, F.; Shao, J.; Zhou, W. Neonatal early-onset infection with SARS-CoV-2 in 33 neonates born to mothers with COVID-19 in Wuhan, China. *JAMA Pediatr* 2020, 174, 722–725, doi:10.1001/jamapediatrics.2020.0878.

254. Vivanti, A.J.; Vauloup-Fellous, C.; Prevot, S.; Zupan, V.; Suffee, C.; Do Cao, J.; Benachi, A.; De Luca, D. Transplacental transmission of SARS-CoV-2 infection. *Nat Commun* 2020, 11, 3572, doi:10.1038/s41467-020-17436-6.

255. Vivanti, A.J.; Mattern, J.; Vauloup-Fellous, C.; Jani, J.; Rigonnot, L.; El Hachem, L.; Le Gouez, A.; Desconclois, C.; Ben M'Barek, I.; Sibiude, J., et al. Retrospective description of pregnant women infected with severe acute respiratory syndrome coronavirus 2, France. *Emerg Infect Dis* 2020, 26, 2069–2076, doi:10.3201/eid2609.202144.

256. Sessa, R.; Filardo, S.; Masciullo, L.; Di Pietro, M.; Angeloni, A.; Brandolino, G.; Brunelli, R.; D'Alisa, R.; Viscardi, M.F.; Anastasi, E., et al. SARS-CoV-2 infection in pregnancy: Clues and proof of adverse outcomes. *Int J Environ Res Public Health* 2023, 20, doi:10.3390/ijerph20032616.

257. Beesley, M.A.; Davidson, J.R.; Panariello, F.; Shibuya, S.; Scaglioni, D.; Jones, B.C.; Maksym, K.; Ogunbiyi, O.; Sebire, N.J.; Cacchiarelli, D., et al. COVID-19 and vertical transmission: Assessing the expression of ACE2/TMPRSS2 in the human fetus and placenta to assess the risk of SARS-CoV-2 infection. *BJOG* 2022, 129, 256–266, doi:10.1111/1471-0528.16974.

258. Delamater, P.L.; Street, E.J.; Leslie, T.F.; Yang, Y.T.; Jacobsen, K.H. Complexity of the basic reproduction number (R_0). *Emerg Infect Dis* 2019, 25, 1–4, doi:10.3201/eid2501.171901.

259. Billah, M.A.; Miah, M.M.; Khan, M.N. Reproductive number of coronavirus: A systematic review and meta-analysis based on global level evidence. *PLoS One* 2020, 15, e0242128, doi:10.1371/journal.pone.0242128.

260. Xia, F.; Yang, X.; Cheke, R.A.; Xiao, Y. Quantifying competitive advantages of mutant strains in a population involving importation and mass vaccination rollout. *Infect Dis Model* 2021, 6, 988–996, doi:10.1016/j.idm.2021.08.001.

261. Manathunga, S.S.; Abeyagunawardena, I.A.; Dharmaratne, S.D. A comparison of transmissibility of SARS-CoV-2 variants of concern. *Virol J* 2023, 20, 59, doi:10.1186/s12985-023-02018-x.

262. Dhawan, M.; Sharma, A.; Priyanka; Thakur, N.; Rajkhowa, T.K.; Choudhary, O.P. Delta variant (B.1.617.2) of SARS-CoV-2: Mutations, impact, challenges and possible solutions. *Hum Vaccin Immunother* 2022, 18, 2068883, doi:10.1080/21645515.2022.2068883.

263. Karimizadeh, Z.; Dowran, R.; Mokhtari-Azad, T.; Shafiei-Jandaghi, N.Z. The reproduction rate of severe acute respiratory syndrome coronavirus 2 different variants recently circulated in human: A narrative review. *Eur J Med Res* 2023, 28, 94, doi:10.1186/s40001-023-01047-0.

264. Wu, Z.Y.; McGoogan, J.M. Characteristics of and important lessons from the coronavirus disease 2019 (COVID-19) outbreak in China summary of a report of 72 314 cases from the Chinese Center for Disease Control and Prevention. *JAMA* 2020, 323, 1239–1242, doi:10.1001/jama.2020.2648.

265. Guan, W.; Ni, Z.; Hu, Y.; Liang, W.; Ou, C.; He, J.; Liu, L.; Shan, H.; Lei, C.; Hui, D.S.C., et al. Clinical characteristics of coronavirus disease 2019 in China. *N Engl J Med* 2020, 382, 1708–1720, doi:10.1056/NEJMoa2002032.

266. WHO. Report of the WHO-China Joint Mission on Coronavirus Disease 2019 (COVID-19). Available online: https://www.who.int/docs/default-source/coronaviruse/who-china-joint-mission-on-covid-19-final-report.pdf (accessed on 15 November 2023).

267. Grant, M.C.; Geoghegan, L.; Arbyn, M.; Mohammed, Z.; McGuinness, L.; Clarke, E.L.; Wade, R.G. The prevalence of symptoms in 24,410 adults infected by the novel coronavirus (SARS-CoV-2; COVID-19): A systematic review and meta-analysis of 148 studies from 9 countries. *PLoS One* 2020, 15, e0234765, doi:10.1371/journal.pone.0234765.

268. Docherty, A.B.; Harrison, E.M.; Green, C.A.; Hardwick, H.E.; Pius, R.; Norman, L.; Holden, K.A.; Read, J.M.; Dondelinger, F.; Carson, G., et al. Features of 20 133 UK patients in hospital with covid-19 using the ISARIC WHO clinical characterisation protocol: Prospective observational cohort study. *BMJ* 2020, 369, m1985, doi:10.1136/bmj.m1985.

269. Huang, C.L.; Wang, Y.M.; Li, X.W.; Ren, L.L.; Zhao, J.P.; Hu, Y.; Zhang, L.; Fan, G.H.; Xu, J.Y.; Gu, X.Y., et al. Clinical features of patients infected with 2019 novel coronavirus in Wuhan, China. *Lancet* 2020, 395, 497–506, doi:10.1016/S0140-6736(20)30183-5.

270. Wang, D.W.; Hu, B.; Hu, C.; Zhu, F.F.; Liu, X.; Zhang, J.; Wang, B.B.; Xiang, H.; Cheng, Z.S.; Xiong, Y., et al. Clinical characteristics of 138 hospitalized patients with 2019 novel coronavirus-infected pneumonia in Wuhan, China. *JAMA* 2020, 323, 1061–1069, doi:10.1001/jama.2020.1585.

271. Krishnan, A.; Hamilton, J.P.; Alqahtani, S.A.; Woreta, T.A. COVID-19: An overview and a clinical update. *World J Clin Cases* 2021, 9, 8–23, doi:10.12998/wjcc.v9.i1.8.

272. Chen, N.S.; Zhou, M.; Dong, X.; Qu, J.M.; Gong, F.Y.; Han, Y.; Qiu, Y.; Wang, J.L.; Liu, Y.; Wei, Y., et al. Epidemiological and clinical characteristics of 99 cases of 2019 novel coronavirus pneumonia in Wuhan, China: A descriptive study. *Lancet* 2020, 395, 507–513, doi:10.1016/S0140-6736(20)30211-7.

273. Ng, W.H.; Tipih, T.; Makoah, N.A.; Vermeulen, J.G.; Goedhals, D.; Sempa, J.B.; Burt, F.J.; Taylor, A.; Mahalingam, S. Comorbidities in SARS-CoV-2 patients: A systematic review and meta-analysis. *mBio* 2021, 12, doi:10.1128/mBio.03647-20.

274. Karagiannidis, C.; Mostert, C.; Hentschker, C.; Voshaar, T.; Malzahn, J.; Schillinger, G.; Klauber, J.; Janssens, U.; Marx, G.; Weber-Carstens, S., et al. Case characteristics, resource use, and outcomes of 10 021 patients with COVID-19 admitted to 920 German hospitals: An observational study. *Lancet Respir Med* 2020, 8, 853–862, doi:10.1016/S2213-2600(20)30316-7.

275. Mungroo, M.R.; Khan, N.A.; Siddiqui, R. Novel coronavirus: Current understanding of clinical features, diagnosis, pathogenesis, and treatment options. *Pathogens* 2020, 9, doi:10.3390/pathogens9040297.

276. Jin, J.M.; Bai, P.; He, W.; Wu, F.; Liu, X.F.; Han, D.M.; Liu, S.; Yang, J.K. Gender differences in patients with COVID-19: Focus on severity and mortality. *Front Public Health* 2020, 8, 152, doi:10.3389/fpubh.2020.00152.

277. Lu, X.; Zhang, L.; Du, H.; Zhang, J.; Li, Y.Y.; Qu, J.; Zhang, W.; Wang, Y.; Bao, S.; Li, Y., et al. SARS-CoV-2 infection in children. *N Engl J Med* 2020, 382, 1663–1665, doi:10.1056/NEJMc2005073.

278. Wu, Z.; McGoogan, J.M. Characteristics of and important lessons from the coronavirus disease 2019 (COVID-19) outbreak in China: Summary of a report of 72 314 cases from the Chinese Center for Disease Control and Prevention. *JAMA* 2020, 323, 1239–1242, doi:10.1001/jama.2020.2648.

279. Minotti, C.; Tirelli, F.; Barbieri, E.; Giaquinto, C.; Donà, D. How is immunosuppressive status affecting children and adults in SARS-CoV-2 infection? A systematic review. *J Infect* 2020, 81, E61–E66, doi:10.1016/j.jinf.2020.04.026.

280. Schoeman, D.; Fielding, B.C. Leptin deficiency, caused by malnutrition, makes you susceptible to SARS-CoV-2 infection but could offer protection from severe COVID-19. *mSphere* 2021, 6, e00031-21, doi:10.1128/mSphere.00031-21.

281. Mascolo, S.; Romanelli, A.; Carleo, M.A.; Esposito, V. Could HIV infection alter the clinical course of SARS-CoV-2 infection? When less is better. *J Med Virol* 2020, 92, 1777–1778, doi:10.1002/jmv.25881.

282. Romanelli, A.; Mascolo, S. Immunosuppression drug-related and clinical manifestation of coronavirus disease 2019: A therapeutic hypothesis. *Am J Transplant* 2020, 20, 1947–1948, doi:10.1111/ajt.15905.

283. Ahmed, W.S.; Anand, A.; Stanbrook, M.B. Dexamethasone in hospitalized patients with Covid-19 - Preliminary report. *Can J Respir Crit Care Sleep Med* 2021, 5, 270–272, doi:10.1080/24745332.2021.1920865.

284. Horby, P.; Lim, W.S.; Emberson, J.R.; Mafham, M.; Bell, J.L.; Linsell, L.; Staplin, N.; Brightling, C.; Ustianowski, A.; Elmahi, E., et al. Dexamethasone in hospitalized patients with Covid-19. *N Engl J Med* 2021, 384, 693–704, doi:10.1056/NEJMoa2021436.

285. Hojyo, S.; Uchida, M.; Tanaka, K.; Hasebe, R.; Tanaka, Y.; Murakami, M.; Hirano, T. How COVID-19 induces cytokine storm with high mortality. *Inflamm Regen* 2020, 40, doi:10.1186/s41232-020-00146-3.

286. Fung, M.; Babik, J.M. COVID-19 in immunocompromised hosts: What we know so far. *Clin Infect Dis* 2021, 72, 340–350, doi:10.1093/cid/ciaa863.

287. Belsky, J.A.; Tullius, B.P.; Lamb, M.G.; Sayegh, R.; Stanek, J.R.; Auletta, J.J. COVID-19 in immuno-compromised patients: A systematic review of cancer, hematopoietic cell and solid organ transplant patients. *J Infect* 2021, 82, 329–338, doi:10.1016/j.jinf.2021.01.022.

288. Nevejan, L.; Ombelet, S.; Laenen, L.; Keyaerts, E.; Demuyser, T.; Seyler, L.; Soetens, O.; Van Nedervelde, E.; Naesens, R.; Geysels, D., et al. Severity of COVID-19 among hospitalized patients: Omicron remains a severe threat for immunocompromised hosts. *Viruses* 2022, 14, 2736, doi:10.3390/v14122736.

289. Abasian, L.; Jafari, F.; SeyedAlinaghi, S.; Yazdi, N.A.; Daraei, M.; Ahmadinejad, N.; Ghiasvand, F.; Khalili, H.; Seifi, A.; Maydani, M., et al. The comparison of clinical, laboratory, and radiological findings in immunocompromised and immunocompetent patients with COVID-19: A case-control study. *Immun Inflamm Dis* 2023, 11, e806, doi:10.1002/iid3.806.

290. Muller, C.P. Do asymptomatic carriers of SARS-COV-2 transmit the virus? *Lancet Reg Health Eur* 2021, 4, 100082, doi:10.1016/j.lanepe.2021.100082.

291. NIH. Clinical Spectrum of SARS-CoV-2 Infection. Available online: https://www.covid19treatment-guidelines.nih.gov/overview/clinical-spectrum/ (accessed on 20 November 2023).

292. Parasher, A. COVID-19: Current understanding of its pathophysiology, clinical presentation and treatment. *Postgrad Med J* 2021, 97, 312–320, doi:10.1136/postgradmedj-2020-138577.

293. Wang, B.; Li, R.; Lu, Z.; Huang, Y. Does comorbidity increase the risk of patients with COVID-19: Evidence from meta-analysis. *Aging (Albany NY)* 2020, 12, 6049–6057, doi:10.18632/aging.103000.

294. Tian, S.; Hu, N.; Lou, J.; Chen, K.; Kang, X.; Xiang, Z.; Chen, H.; Wang, D.; Liu, N.; Liu, D., et al. Characteristics of COVID-19 infection in Beijing. *J Infect* 2020, 80, 401–406, doi:10.1016/j.jinf.2020.02.018.

295. Mugglestone, M.A.; Ratnaraja, N.V.; Bak, A.; Islam, J.; Wilson, J.A.; Bostock, J.; Moses, S.E.; Price, J.R.; Weinbren, M.; Loveday, H.P., et al. Presymptomatic, asymptomatic and post-symptomatic transmission of SARS-CoV-2: Joint British Infection Association (BIA), Healthcare Infection Society (HIS), Infection Prevention Society (IPS) and Royal College of Pathologists (RCPath) guidance. *BMC Infect Dis* 2022, 22, 453, doi:10.1186/s12879-022-07440-0.

296. Xu, T.; Huang, R.; Zhu, L.; Wang, J.; Cheng, J.; Zhang, B.; Zhao, H.; Chen, K.; Shao, H.; Zhu, C., et al. Epidemiological and clinical features of asymptomatic patients with SARS-CoV-2 infection. *J Med Virol* 2020, 92, 1884–1889, doi:10.1002/jmv.25944.

297. AlJishi, J.M.; Alhajjaj, A.H.; Alkhabbaz, F.L.; AlAbduljabar, T.H.; Alsaif, A.; Alsaif, H.; Alomran, K.S.; Aljanobi, G.A.; Alghawi, Z.; Alsaif, M., et al. Clinical characteristics of asymptomatic and symptomatic COVID-19 patients in the Eastern Province of Saudi Arabia. *J Infect Public Health* 2021, 14, 6–11, doi:10.1016/j.jiph.2020.11.002.

298. El-Ghitany, E.M.; Hashish, M.H.; Farghaly, A.G.; Omran, E.A.; Osman, N.A.; Fekry, M.M. Asymptomatic versus symptomatic SARS-CoV-2 infection: A cross-sectional seroprevalence study. *Trop Med Health* 2022, 50, 98, doi:10.1186/s41182-022-00490-9.

299. Lei, Q.; Li, Y.; Hou, H.Y.; Wang, F.; Ouyang, Z.Q.; Zhang, Y.; Lai, D.Y.; Banga Ndzouboukou, J.L.; Xu, Z.W.; Zhang, B., et al. Antibody dynamics to SARS-CoV-2 in asymptomatic COVID-19 infections. *Allergy* 2021, 76, 551–561, doi:10.1111/all.14622.

300. Long, Q.X.; Tang, X.J.; Shi, Q.L.; Li, Q.; Deng, H.J.; Yuan, J.; Hu, J.L.; Xu, W.; Zhang, Y.; Lv, F.J., et al. Clinical and immunological assessment of asymptomatic SARS-CoV-2 infections. *Nat Med* 2020, 26, 1200–1204, doi:10.1038/s41591-020-0965-6.

301. Sui, Z.; Dai, X.; Lu, Q.; Zhang, Y.; Huang, M.; Li, S.; Peng, T.; Xie, J.; Zhang, Y.; Wu, C., et al. Viral dynamics and antibody responses in people with asymptomatic SARS-CoV-2 infection. *Signal Transduct Target Ther* 2021, 6, 181, doi:10.1038/s41392-021-00596-2.

302. Milani, G.P.; Dioni, L.; Favero, C.; Cantone, L.; Macchi, C.; Delbue, S.; Bonzini, M.; Montomoli, E.; Bollati, V.; Consortium, U. Serological follow-up of SARS-CoV-2 asymptomatic subjects. *Sci Rep* 2020, 10, 20048, doi:10.1038/s41598-020-77125-8.

303. Yin, S.W.; Zhou, Z.; Wang, J.L.; Deng, Y.F.; Jing, H.; Qiu, Y. Viral loads, lymphocyte subsets and cytokines in asymptomatic, mildly and critical symptomatic patients with SARS-CoV-2 infection: A retrospective study. *Virology Journal* 2021, 18, 126, doi:10.1186/s12985-021-01597-x.

304. Yang, J.; Zhang, E.; Zhong, M.; Yang, Q.; Hong, K.; Shu, T.; Zhou, D.; Xiang, J.; Xia, J.; Zhou, X., et al. Longitudinal characteristics of T cell responses in asymptomatic SARS-CoV-2 infection. *Virol Sin* 2020, 35, 838–841, doi:10.1007/s12250-020-00277-4.

305. Mazzoni, A.; Maggi, L.; Capone, M.; Spinicci, M.; Salvati, L.; Colao, M.G.; Vanni, A.; Kiros, S.T.; Mencarini, J.; Zammarchi, L., et al. Cell-mediated and humoral adaptive immune responses to SARS-CoV-2 are lower in asymptomatic than symptomatic COVID-19 patients. *Eur J Immunol* 2020, 50, 2013–2024, https://doi.org/10.1002/eji.202048915.

306. Yu, S.; Di, C.; Chen, S.; Guo, M.; Yan, J.; Zhu, Z.; Liu, L.; Feng, R.; Xie, Y.; Zhang, R., et al. Distinct immune signatures discriminate between asymptomatic and presymptomatic SARS-CoV-2pos subjects. *Cell Res* 2021, 31, 1148–1162, doi:10.1038/s41422-021-00562-1.

307. Le Bert, N.; Clapham, H.E.; Tan, A.T.; Chia, W.N.; Tham, C.Y.L.; Lim, J.M.; Kunasegaran, K.; Tan, L.W.L.; Dutertre, C.A.; Shankar, N., et al. Highly functional virus-specific cellular immune response in asymptomatic SARS-CoV-2 infection. *J Exp Med* 2021, 218, doi:10.1084/jem.20202617.

308. Fabião, J.; Sassi, B.; Pedrollo, E.F.; Gerchman, F.; Kramer, C.K.; Leitão, C.B.; Pinto, L.C. Why do men have worse COVID-19-related outcomes? A systematic review and meta-analysis with sex adjusted for age. *Braz J Med Biol Res* 2022, 55, e11711, doi:10.1590/1414-431X2021e11711.

309. Peckham, H.; de Gruijter, N.M.; Raine, C.; Radziszewska, A.; Ciurtin, C.; Wedderburn, L.R.; Rosser, E.C.; Webb, K.; Deakin, C.T. Male sex identified by global COVID-19 meta-analysis as a risk factor for death and ITU admission. *Nat Commun* 2020, 11, 6317, doi:10.1038/s41467-020-19741-6.

310. Pradhan, A.; Olsson, P.E. Sex differences in severity and mortality from COVID-19: Are males more vulnerable? *Biol Sex Differ* 2020, 11, 53, doi:10.1186/s13293-020-00330-7.

311. Fagerberg, L.; Hallstrom, B.M.; Oksvold, P.; Kampf, C.; Djureinovic, D.; Odeberg, J.; Habuka, M.; Tahmasebpoor, S.; Danielsson, A.; Edlund, K., et al. Analysis of the human tissue-specific expression by genome-wide integration of transcriptomics and antibody-based proteomics. *Mol Cell Proteomics* 2014, 13, 397–406, doi:10.1074/mcp.M113.035600.

312. Achua, J.K.; Chu, K.Y.; Ibrahim, E.; Khodamoradi, K.; Delma, K.S.; Iakymenko, O.A.; Kryvenko, O.N.; Arora, H.; Ramasamy, R. Histopathology and ultrastructural findings of fatal COVID-19 infections on testis. *World J Mens Health* 2021, 39, 65–74, doi:10.5534/wjmh.200170.

313. Ma, X.; Guan, C.; Chen, R.; Wang, Y.; Feng, S.; Wang, R.; Qu, G.; Zhao, S.; Wang, F.; Wang, X., et al. Pathological and molecular examinations of postmortem testis biopsies reveal SARS-CoV-2 infection in the testis and spermatogenesis damage in COVID-19 patients. *Cell Mol Immunol* 2021, 18, 487–489, doi:10.1038/s41423-020-00604-5.

314. Mukherjee, S.; Pahan, K. Is COVID-19 gender-sensitive? *J Neuroimmune Pharmacol* 2021, 16, 38–47, doi:10.1007/s11481-020-09974-z.

315. Bienvenu, L.A.; Noonan, J.; Wang, X.; Peter, K. Higher mortality of COVID-19 in males: Sex differences in immune response and cardiovascular comorbidities. *Cardiovasc Res* 2020, 116, 2197–2206, doi:10.1093/cvr/cvaa284.

316. Okwan-Duodu, D.; Lim, E.C.; You, S.; Engman, D.M. TMPRSS2 activity may mediate sex differences in COVID-19 severity. *Signal Transduct Target Ther* 2021, 6, 100, doi:10.1038/s41392-021-00513-7.

317. Aleksanyan, Y.; Weinman, J.P. Women, men and COVID-19. *Soc Sci Med* 2022, 294, 114698, doi:10.1016/j.socscimed.2022.114698.

318. Hu, H.; Pan, H.; Li, R.; He, K.; Zhang, H.; Liu, L. Increased circulating cytokines have a role in COVID-19 severity and death with a more pronounced effect in males: A systematic review and meta-analysis. *Front Pharmacol* 2022, 13, doi:10.3389/fphar.2022.802228.

319. Ly, J.; Campos, R.K.; Hager-Soto, E.E.; Camargos, V.N.; Rossi, S.L. Testicular pathological alterations associated with SARS-CoV-2 infection. *Front Reprod Health* 2023, 5, 1229622, doi:10.3389/frph.2023.1229622.

320. Dubin, J.M.; Bennett, N.E.; Halpern, J.A. The adverse impact of COVID-19 on men's health. *Curr Opin Urol* 2022, 32, 146–151, doi:10.1097/MOU.0000000000000966.

321. Parri, N.; Lenge, M.; Buonsenso, D.; Coronavirus Infection in Pediatric Emergency Departments Research Group. Children with Covid-19 in Pediatric Emergency Departments in Italy. *N Engl J Med* 2020, 383, 187–190, doi:10.1056/NEJMc2007617.

322. She, J.; Liu, L.; Liu, W. COVID-19 epidemic: Disease characteristics in children. *J Med Virol* 2020, 92, 747–754, doi:10.1002/jmv.25807.

323. Ahmed, M.; Advani, S.; Moreira, A.; Zoretic, S.; Martinez, J.; Chorath, K.; Acosta, S.; Naqvi, R.; Burmeister-Morton, F.; Burmeister, F., et al. Multisystem inflammatory syndrome in children: A systematic review. *EClinicalMedicine* 2020, 26, 100527, doi:10.1016/j.eclinm.2020.100527.

324. Belot, A.; Levy-Bruhl, D.; French Covid-19 Pediatric Inflammation Consortium. Multisystem inflammatory syndrome in children in the United States. *N Engl J Med* 2020, 383, 1793–1794, doi:10.1056/NEJMc2026136.

325. Henderson, L.A.; Canna, S.W.; Friedman, K.G.; Gorelik, M.; Lapidus, S.K.; Bassiri, H.; Behrens, E.M.; Ferris, A.; Kernan, K.F.; Schulert, G.S., et al. American College of Rheumatology Clinical Guidance for multisystem inflammatory syndrome in children associated with SARS-CoV-2 and hyperinflammation in pediatric COVID-19: Version 1. *Arthritis Rheumatol* 2020, 72, 1791–1805, doi:10.1002/art.41454.

326. Sokunbi, O.; Akinbolagbe, Y.; Akintan, P.; Oyeleke, G.; Kusimo, O.; Owowo, U.; Olonade, E.; Ojo, O.; Ikhazobor, E.; Amund, O., et al. Clinical presentation and short-term outcomes of multisystemic inflammatory syndrome in children in Lagos, Nigeria during the COVID-19 pandemic: A case series. *EClinicalMedicine* 2022, 49, 101475, doi:10.1016/j.eclinm.2022.101475.

327. Dun-Dery, F.; Xie, J.; Winston, K.; Burstein, B.; Gravel, J.; Emsley, J.; Sabhaney, V.; Zemek, R.; Berthelot, S.; Beer, D., et al. Post-COVID-19 condition in children 6 and 12 months after infection. *JAMA Netw Open* 2023, 6, e2349613, doi:10.1001/jamanetworkopen.2023.49613.

328. ECDC. Rapid risk assessment: Paediatric inflammatory multisystem syndrome and SARS-CoV-2 infection in children. Available online: https://www.ecdc.europa.eu/en/publications-data/paediatric-inflammatory-multisystem-syndrome-and-sars-cov-2-rapid-risk-assessment (accessed on 18 November 2023).

329. Panaro, S.; Cattalini, M. The spectrum of manifestations of severe acute respiratory syndrome-coronavirus 2 (SARS-CoV2) infection in children: What we can learn from multisystem inflammatory syndrome in children (MIS-C). *Front Med* 2021, 8, doi:10.3389/fmed.2021.747190.

330. Noval Rivas, M.; Porritt, R.A.; Cheng, M.H.; Bahar, I.; Arditi, M. Multisystem inflammatory syndrome in children and long COVID: The SARS-CoV-2 viral superantigen hypothesis. *Front Immunol* 2022, 13, 941009, doi:10.3389/fimmu.2022.941009.

331. Benezech, S.; Khoryati, L.; Cognard, J.; Netea, S.A.; Khan, T.; Moreews, M.; Saker, K.; De Guillebon, J.M.; Khaldi-Plassart, S.; Pescarmona, R., et al. Pre-Covid-19, SARS-CoV-2-negative multisystem inflammatory syndrome in children. *N Engl J Med* 2023, 389, 2105–2107, doi:10.1056/NEJMc2307574.

332. Rowley, A.H. Understanding SARS-CoV-2-related multisystem inflammatory syndrome in children. *Nat Rev Immunol* 2020, 20, 453–454, doi:10.1038/s41577-020-0367-5.

333. Fialkowski, A.; Gernez, Y.; Arya, P.; Weinacht, K.G.; Kinane, T.B.; Yonker, L.M. Insight into the pediatric and adult dichotomy of COVID-19: Age-related differences in the immune response to SARS-CoV-2 infection. *Pediatr Pulmonol* 2020, 55, 2556–2564, https://doi.org/10.1002/ppul.24981.

334. Alunno, A.; Carubbi, F.; Rodríguez-Carrio, J. Storm, typhoon, cyclone or hurricane in patients with COVID-19? Beware of the same storm that has a different origin. *RMD Open* 2020, 6, doi:10.1136/rmdopen-2020-001295.

335. Son, M.B.F.; Murray, N.; Friedman, K.; Young, C.C.; Newhams, M.M.; Feldstein, L.R.; Loftis, L.L.; Tarquinio, K.M.; Singh, A.R.; Heidemann, S.M., et al. Multisystem inflammatory syndrome in children - Initial therapy and outcomes. *N Engl J Med* 2021, 385, 23–34, doi:10.1056/NEJMoa2102605.

336. McArdle, A.J.; Vito, O.; Patel, H.; Seaby, E.G.; Shah, P.; Wilson, C.; Broderick, C.; Nijman, R.; Tremoulet, A.H.; Munblit, D., et al. Treatment of multisystem inflammatory syndrome in children. *N Engl J Med* 2021, 385, 11–22, doi:10.1056/NEJMoa2102968.

337. Sperotto, F.; Friedman, K.G.; Son, M.B.F.; VanderPluym, C.J.; Newburger, J.W.; Dionne, A. Cardiac manifestations in SARS-CoV-2-associated multisystem inflammatory syndrome in children: A comprehensive review and proposed clinical approach. *Eur J Pediatr* 2021, 180, 307–322, doi:10.1007/s00431-020-03766-6.

338. Zhu, Y.P.; Shamie, I.; Lee, J.C.; Nowell, C.J.; Peng, W.; Angulo, S.; Le, L.N.; Liu, Y.; Miao, H.; Xiong, H., et al. Immune response to intravenous immunoglobulin in patients with Kawasaki disease and MIS-C. *J Clin Invest* 2021, 131, doi:10.1172/JCI147076.

339. Davis, H.E.; Assaf, G.S.; McCorkell, L.; Wei, H.; Low, R.J.; Re'em, Y.; Redfield, S.; Austin, J.P.; Akrami, A. Characterizing long COVID in an international cohort: 7 months of symptoms and their impact. *EClinicalMedicine* 2021, 38, 101019, doi:10.1016/j.eclinm.2021.101019.

340. Jason, L.A.; Islam, M.; Conroy, K.; Cotler, J.; Torres, C.; Johnson, M.; Mabie, B. COVID-19 symptoms over time: Comparing long-haulers to ME/CFS. *Fatigue* 2021, 9, 59–68, doi:10.1080/21641846.2021.1922140.

341. Tran, V.T.; Porcher, R.; Pane, I.; Ravaud, P. Course of post COVID-19 disease symptoms over time in the ComPaRe long COVID prospective e-cohort. *Nat Commun* 2022, 13, 1812, doi:10.1038/s41467-022-29513-z.

342. Townsend, L.; Dowds, J.; O'Brien, K.; Sheill, G.; Dyer, A.H.; O'Kelly, B.; Hynes, J.P.; Mooney, A.; Dunne, J.; Ni Cheallaigh, C., et al. Persistent poor health after COVID-19 is not associated with respiratory complications or initial disease severity. *Ann Am Thorac Soc* 2021, 18, 997–1003, doi:10.1513/AnnalsATS.202009-1175OC.

343. Kim, Y.; Bae, S.; Chang, H.H.; Kim, S.W. Long COVID prevalence and impact on quality of life 2 years after acute COVID-19. *Sci Rep* 2023, 13, 11207, doi:10.1038/s41598-023-36995-4.

344. Lopez-Leon, S.; Wegman-Ostrosky, T.; Perelman, C.; Sepulveda, R.; Rebolledo, P.A.; Cuapio, A.; Villapol, S. More than 50 long-term effects of COVID-19: A systematic review and meta-analysis. *Sci Rep* 2021, 11, 16144, doi:10.1038/s41598-021-95565-8.

345. Spudich, S.; Nath, A. Nervous system consequences of COVID-19. *Science* 2022, 375, 267–269, doi:10.1126/science.abm2052.

346. Stefanou, M.I.; Palaiodimou, L.; Bakola, E.; Smyrnis, N.; Papadopoulou, M.; Paraskevas, G.P.; Rizos, E.; Boutati, E.; Grigoriadis, N.; Krogias, C., et al. Neurological manifestations of long-COVID syndrome: A narrative review. *Ther Adv Chronic Dis* 2022, 13, 20406223221076890, doi:10.1177/20406223221076890.

347. Srikanth, S.; Boulos, J.R.; Dover, T.; Boccuto, L.; Dean, D. Identification and diagnosis of long COVID-19: A scoping review. *Prog Biophys Mol Biol* 2023, 182, 1–7, doi:10.1016/j.pbiomolbio.2023.04.008.

348. Chen, C.; Haupert, S.R.; Zimmermann, L.; Shi, X.; Fritsche, L.G.; Mukherjee, B. Global prevalence of post-coronavirus disease 2019 (COVID-19) condition or long COVID: A meta-analysis and systematic review. *J Infect Dis* 2022, 226, 1593–1607, doi:10.1093/infdis/jiac136.

349. Su, Y.; Yuan, D.; Chen, D.G.; Ng, R.H.; Wang, K.; Choi, J.; Li, S.; Hong, S.; Zhang, R.; Xie, J., et al. Multiple early factors anticipate post-acute COVID-19 sequelae. *Cell* 2022, 185, 881–895.e20, doi:10.1016/j.cell.2022.01.014.

350. Goh, D.; Lim, J.C.T.; Fernaindez, S.B.; Joseph, C.R.; Edwards, S.G.; Neo, Z.W.; Lee, J.N.; Caballero, S.G.; Lau, M.C.; Yeong, J.P.S. Case report: Persistence of residual antigen and RNA of the SARS-CoV-2 virus in tissues of two patients with long COVID. *Front Immunol* 2022, 13, 939989, doi:10.3389/fimmu.2022.939989.

351. Cheung, C.C.L.; Goh, D.; Lim, X.; Tien, T.Z.; Lim, J.C.T.; Lee, J.N.; Tan, B.; Tay, Z.E.A.; Wan, W.Y.; Chen, E.X., et al. Residual SARS-CoV-2 viral antigens detected in GI and hepatic tissues from five recovered patients with COVID-19. *Gut* 2022, 71, 226–229, doi:10.1136/gutjnl-2021-324280.

352. Castanares-Zapatero, D.; Chalon, P.; Kohn, L.; Dauvrin, M.; Detollenaere, J.; Maertens de Noordhout, C.; Primus-de Jong, C.; Cleemput, I.; Van den Heede, K. Pathophysiology and mechanism of long COVID: A comprehensive review. *Ann Med* 2022, 54, 1473–1487, doi:10.1080/07853890.2022.2076901.

353. Chen, B.; Julg, B.; Mohandas, S.; Bradfute, S.B.; Force, R.M.P.T. Viral persistence, reactivation, and mechanisms of long COVID. *Elife* 2023, 12, doi:10.7554/eLife.86015.

354. Cruz, T.; Mendoza, N.; Lledo, G.M.; Perea, L.; Albacar, N.; Agusti, A.; Sellares, J.; Sibila, O.; Faner, R. Persistence of a SARS-CoV-2 T-cell response in patients with long COVID and lung sequelae after COVID-19. *ERJ Open Res* 2023, 9, doi:10.1183/23120541.00020-2023.

355. Shakiba, M.H.; Gemund, I.; Beyer, M.; Bonaguro, L. Lung T cell response in COVID-19. *Front Immunol* 2023, 14, 1108716, doi:10.3389/fimmu.2023.1108716.

356. Altmann, D.M.; Whettlock, E.M.; Liu, S.; Arachchillage, D.J.; Boyton, R.J. The immunology of long COVID. *Nat Rev Immunol* 2023, 23, 618–634, doi:10.1038/s41577-023-00904-7.

357. Desai, A.D.; Boursiquot, B.C.; Melki, L.; Wan, E.Y. Management of arrhythmias associated with COVID-19. *Curr Cardiol Rep* 2020, 23, 2, doi:10.1007/s11886-020-01434-7.

358. Bolton, M.J.; Chapman, B.P.; Van Marwijk, H. Low-dose naltrexone as a treatment for chronic fatigue syndrome. *BMJ Case Rep* 2020, 13, doi:10.1136/bcr-2019-232502.

359. Hohberger, B.; Harrer, T.; Mardin, C.; Kruse, F.; Hoffmanns, J.; Rogge, L.; Heltmann, F.; Moritz, M.; Szewczykowski, C.; Schottenhamml, J., et al. Case report: Neutralization of autoantibodies targeting G-protein-coupled receptors improves capillary impairment and fatigue symptoms after COVID-19 infection. *Front Med (Lausanne)* 2021, 8, 754667, doi:10.3389/fmed.2021.754667.

360. Glynne, P.; Tahmasebi, N.; Gant, V.; Gupta, R. Long COVID following mild SARS-CoV-2 infection: Characteristic T cell alterations and response to antihistamines. *J Investig Med* 2022, 70, 61–67, doi:10.1136/jim-2021-002051.

361. Wood, E.; Hall, K.H.; Tate, W. Role of mitochondria, oxidative stress and the response to antioxidants in myalgic encephalomyelitis/chronic fatigue syndrome: A possible approach to SARS-CoV-2 'long-haulers'? *Chronic Dis Transl Med* 2021, 7, 14–26, doi:10.1016/j.cdtm.2020.11.002.

362. Tosato, M.; Ciciarello, F.; Zazzara, M.B.; Pais, C.; Savera, G.; Picca, A.; Galluzzo, V.; Coelho-Junior, H.J.; Calvani, R.; Marzetti, E., et al. Nutraceuticals and dietary supplements for older adults with long COVID-19. *Clin Geriatr Med* 2022, 38, 565–591, doi:10.1016/j.cger.2022.04.004.

363. Rathi, A.; Jadhav, S.B.; Shah, N. A randomized controlled trial of the efficacy of systemic enzymes and probiotics in the Resolution of post-COVID fatigue. *Medicines (Basel)* 2021, 8, doi:10.3390/medicines8090047.

364. Kevadiya, B.D.; Machhi, J.; Herskovitz, J.; Oleynikov, M.D.; Blomberg, W.R.; Bajwa, N.; Soni, D.; Das, S.; Hasan, M.; Patel, M., et al. Diagnostics for SARS-CoV-2 infections. *Nat Mater* 2021, 20, 593–605, doi:10.1038/s41563-020-00906-z.

365. Peeling, R.W.; Heymann, D.L.; Teo, Y.Y.; Garcia, P.J. Diagnostics for COVID-19: Moving from pandemic response to control. *Lancet* 2022, 399, 757–768, doi:10.1016/S0140-6736(21)02346-1.

366. Liu, R.; Han, H.; Liu, F.; Lv, Z.H.; Wu, K.L.; Liu, Y.L.; Feng, Y.; Zhu, C.L. Positive rate of RT-PCR detection of SARS-CoV-2 infection in 4880 cases from one hospital in Wuhan, China, from Jan to Feb 2020. *Clinica Chimica Acta* 2020, 505, 172–175, doi:10.1016/j.cca.2020.03.009.

367. Corman, V.M.; Landt, O.; Kaiser, M.; Molenkamp, R.; Meijer, A.; Chu, D.K.W.; Bleicker, T.; Brünink, S.; Schneider, J.; Schmidt, M.L., et al. Detection of 2019 novel coronavirus (2019-nCoV) by real-time RT-PCR. *Euro Surveill* 2020, 25, 23–30, doi:10.2807/1560-7917.Es.2020.25.3.2000045.

368. Gdoura, M.; Abouda, I.; Mrad, M.; Ben Dhifallah, I.; Belaiba, Z.; Fares, W.; Chouikha, A.; Khedhiri, M.; Layouni, K.; Touzi, H., et al. SARS-CoV2 RT-PCR assays: In vitro comparison of 4 WHO approved protocols on clinical specimens and its implications for real laboratory practice through variant emergence. *Virol J* 2022, 19, doi:10.1186/s12985-022-01784-4.

369. Linkowska, K.; Bogiel, T.; Lamperska, K.; Marszalek, A.; Starzynski, J.; Szylberg, L.; Szwed-Kowalska, A.; Pawlowska, M.; Grzybowski, T. Commercially available SARS-CoV-2 RT-qPCR diagnostic tests need obligatory internal validation. *Sci Rep* 2023, 13, doi:10.1038/s41598-023-34220-w.

370. Tavakoli, F.; Yavarian, J.; Jandaghi, N.Z.S.; Sadeghi, K.; Ghavami, N.; Salimi, V.; Mokhtari-Azad, T. Evaluation of SARS-CoV-2 genome detection by real-time PCR assay using pooled specimens. *Future Virol* 2022, 17, 455–461, doi:10.2217/fvl-2021-0014.

371. Mentes, A.; Papp, K.; Visontai, D.; Stéger, J.; Csabai, I.; Medgyes-Horváth, A.; Pipek, O.A.; Grp, V.T.W. Identification of mutations in SARS-CoV-2 PCR primer regions. *Sci Rep* 2022, 12, doi:10.1038/s41598-022-21953-3.

372. Zou, L.R.; Ruan, F.; Huang, M.X.; Liang, L.J.; Huang, H.T.; Hong, Z.S.; Yu, J.X.; Kang, M.; Song, Y.C.; Xia, J.Y., et al. SARS-CoV-2 viral load in upper respiratory specimens of infected patients. *N Engl J Med* 2020, 382, 1177–1179, doi:10.1056/NEJMc2001737.

373. Wang, W.L.; Xu, Y.L.; Gao, R.Q.; Lu, R.J.; Han, K.; Wu, G.Z.; Tan, W.J. Detection of SARS-CoV-2 in different types of clinical specimens. *JAMA* 2020, 323, 1843–1844, doi:10.1001/jama.2020.3786.

374. Zhang, W.; Du, R.H.; Li, B.; Zheng, X.S.; Yang, X.L.; Hu, B.; Wang, Y.Y.; Xiao, G.F.; Yan, B.; Shi, Z.L., et al. Molecular and serological investigation of 2019-nCoV infected patients: Implication of multiple shedding routes. *Emerg Microbes Infect* 2020, 9, doi:10.1080/22221751.2020.1729071.

375. Xiang, F.; Wang, X.R.; He, X.L.; Peng, Z.H.; Yang, B.H.; Zhang, J.C.; Zhou, Q.; Ye, H.; Ma, Y.L.; Li, H., et al. Antibody detection and dynamic characteristics in patients with coronavirus disease 2019. *Clin Infect Dis* 2020, 71, 1930–1934, doi:10.1093/cid/ciaa461.

376. Zhao, J.J.; Yuan, Q.; Wang, H.Y.; Liu, W.; Liao, X.J.; Su, Y.Y.; Wang, X.; Yuan, J.; Li, T.D.; Li, J.X., et al. Antibody responses to SARS-CoV-2 in patients with novel coronavirus disease 2019. *Clin Infect Dis* 2020, 71, 2027–2034, doi:10.1093/cid/ciaa344.

377. Alexandersen, S.; Chamings, A.; Bhatta, T.R. SARS-CoV-2 genomic and subgenomic RNAs in diagnostic samples are not an indicator of active replication. *Nat Commun* 2020, 11, 6059, doi:10.1038/s41467-020-19883-7.

378. Carter, L.J.; Garner, L.V.; Smoot, J.W.; Li, Y.Z.; Zhou, Q.Q.; Saveson, C.J.; Sasso, J.M.; Gregg, A.C.; Soares, D.J.; Beskid, T.R., et al. Assay techniques and test development for COVID-19 diagnosis. *ACS Cent Sci* 2020, 6, 591–605, doi:10.1021/acscentsci.0c00501.

379. Poehlein, E.; Rane, M.S.; Frogel, D.; Kulkarni, S.; Gainus, C.; Profeta, A.; Robertson, M.; Nash, D. Presence of SARS-CoV-2 antibodies following COVID-19 diagnosis: A longitudinal study of patients at a major urgent care provider in New York. *Diagn Microbiol Infect Dis* 2022, 103, doi:10.1016/j.diagmicrobio.2022.115720.

380. Hou, H.Y.; Wang, T.; Zhang, B.; Luo, Y.; Mao, L.; Wang, F.; Wu, S.J.; Sun, Z.Y. Detection of IgM and IgG antibodies in patients with coronavirus disease 2019. *Clin Transl Immunol* 2020, 9, doi:10.1002/cti2.1136.

381. Sethuraman, N.; Jeremiah, S.S.; Ryo, A. Interpreting diagnostic tests for SARS-CoV-2. *JAMA* 2020, 323, 2249–2251, doi:10.1001/jama.2020.8259.

382. Yang, Z.R.; Jiang, Y.W.; Li, F.X.; Liu, D.; Lin, T.F.; Zhao, Z.Y.; Wei, C.; Jin, Q.Y.; Li, X.M.; Jia, Y.X., et al. Efficacy of SARS-CoV-2 vaccines and the dose-response relationship with three major antibodies: A systematic review and meta-analysis of randomised controlled trials. *Lancet Microbe* 2023, 4, E236–E246, doi:10.1016/S2666-5247(22)00390-1.

383. Hajissa, K.; Mussa, A.; Karobari, M.I.; Abbas, M.A.; Ibrahim, I.K.; Assiry, A.A.; Iqbal, A.; Alhumaid, S.; Mutair, A.A.; Rabaan, A.A., et al. The SARS-CoV-2 antibodies, their diagnostic utility, and their potential for vaccine development. *Vaccines* 2022, 10, doi:10.3390/vaccines10081346.

384. Bordi, L.; Sberna, G.; Piscioneri, C.N.; Cocchiara, R.A.; Miani, A.; Grammatico, P.; Mariani, B.; Parisi, G. Longitudinal dynamics of SARS-CoV-2 anti-receptor binding domain IgG antibodies in a wide population of health care workers after BNT162b2 vaccination. *Int J Infect Dis* 2022, 122, 174–177, doi:10.1016/j.ijid.2022.05.061.

385. Scohy, A.; Anantharajah, A.; Bodéus, M.; Kabamba-Mukadi, B.; Verroken, A.; Rodriguez-Villalobos, H. Low performance of rapid antigen detection test as frontline testing for COVID-19 diagnosis. *J Clin Virol* 2020, 129, doi:10.1016/j.jcv.2020.104455.

386. Drain, P.K.; Ampajwala, M.; Chappel, C.; Gvozden, A.B.; Hoppers, M.; Wang, M.; Rosen, R.; Young, S.; Zissman, E.; Montano, M. A rapid, high-sensitivity SARS-CoV-2 nucleocapsid immunoassay to aid diagnosis of acute COVID-19 at the point of care: A clinical performance study. *Infect Dis Ther* 2021, 10, 753–761, doi:10.1007/s40121-021-00413-x.

387. West, R.; Kobokovich, A.; Connell, N.; Gronvall, G.K. COVID-19 antibody tests: A valuable public health tool with limited relevance to individuals. *Trends Microbiol* 2021, 29, 214–223, doi:10.1016/j.tim.2020.11.002.

388. Smits, V.A.J.; Hernandez-Carralero, E.; Paz-Cabrera, M.C.; Cabrera, E.; Hernandez-Reyes, Y.; Hernandez-Fernaud, J.R.; Gillespie, D.A.; Salido, E.; Hernandez-Porto, M.; Freire, R. The nucleocapsid protein triggers the main humoral immune response in COVID-19 patients. *Biochem Biophys Res Commun* 2021, 543, 45–49, doi:10.1016/j.bbrc.2021.01.073.

389. Drain, P.K. Rapid diagnostic testing for SARS-CoV-2. *N Engl J Med* 2022, 386, 264–272, doi:10.1056/NEJMcp2117115.

390. Jungnick, S.; Hobmaier, B.; Mautner, L.; Hoyos, M.; Haase, M.; Baiker, A.; Lahne, H.; Eberle, U.; Wimmer, C.; Hepner, S., et al. In vitro rapid antigen test performance with the SARS-CoV-2 variants of concern B.1.1.7 (Alpha), B.1.351 (Beta), P.1 (Gamma), and B.1.617.2 (Delta). *Microorganisms* 2021, 9, doi:10.3390/microorganisms9091967.

391. Bekliz, M.; Adea, K.; Essaidi-Laziosi, M.; Sacks, J.A.; Escadafal, C.; Kaiser, L.; Eckerle, I. SARS-CoV-2 antigen-detecting rapid tests for the delta variant. *Lancet Microbe* 2022, 3, E90–E90, doi:10.1016/S2666-5247(21)00302-5.

392. Deerain, J.; Druce, J.; Tran, T.; Batty, M.; Yoga, Y.; Fennell, M.; Dwyer, D.E.; Kok, J.; Williamson, D.A. Assessment of the analytical sensitivity of 10 lateral flow devices against the SARS-CoV-2 Omicron variant. *J Clinl Microbiol* 2022, 60, doi:10.1128/jcm.02479-21.

393. Eshghifar, N.; Busheri, A.; Shrestha, R.; Beqaj, S. Evaluation of analytical performance of seven rapid antigen detection kits for detection of SARS-CoV-2 virus. *International J Gen Med* 2021, 14, 435–440, doi:10.2147/Ijgm.S297762.

394. Krüttgen, A.; Cornelissen, C.G.; Dreher, M.; Hornef, M.W.; Imöhl, M.; Kleines, M. Comparison of the Sars-CoV-2 rapid antigen test to the real star Sars-CoV-2 RT PCR kit. *J Virol Methods* 2021, 288, doi:10.1016/j.jviromet.2020.114024.

395. Quarato, C.M.I.; Mirijello, A.; Maggi, M.M.; Borelli, C.; Russo, R.; Lacedonia, D.; Foschino Barbaro, M.P.; Scioscia, G.; Tondo, P.; Rea, G., et al. Lung ultrasound in the diagnosis of COVID-19 pneumonia: Not always and not only what is COVID-19 "Glitters." *Front Med (Lausanne)* 2021, 8, 707602, doi:10.3389/fmed.2021.707602.

396. Kanne, J.P. Chest CT findings in 2019 novel coronavirus (2019-nCoV) infections from Wuhan, China: Key points for the radiologist. *Radiology* 2020, 295, 16–17, doi:10.1148/radiol.2020200241.

397. Churruca, M.; Martínez-Besteiro, E.; Counago, F.; Landete, P. COVID-19 pneumonia: A review of typical radiological characteristics. *World J Radiol* 2021, 13, 327–343, doi:10.4329/wjr.v13.i10.327.

398. Rousan, L.A.; Elobeid, E.; Karrar, M.; Khader, Y. Chest x-ray findings and temporal lung changes in patients with COVID-19 pneumonia. *BMC Pulm Med* 2020, 20, doi:10.1186/s12890-020-01286-5.

399. Poggiali, E.; Dacrema, A.; Bastoni, D.; Tinelli, V.; Demichele, E.; Ramos, P.M.; Marcianò, T.; Silva, M.; Vercelli, A.; Magnacavallo, A. Can lung US help critical care clinicians in the early diagnosis of novel coronavirus (COVID-19) pneumonia? *Radiology* 2020, 295, E6–E6, doi:10.1148/radiol.2020200847.

400. Yeh, R.; Elsakka, A.; Wray, R.; Johnston, R.P.; Gangai, N.C.; Yarmohammadi, H.; Schoder, H.; Pandit-Taskar, N. FDG PET/CT imaging features and clinical utility in COVID-19. *Clin Imaging* 2021, 80, 262–267, doi:10.1016/j.clinimag.2021.08.002.

401. Verdecchia, P.; Cavallini, C.; Spanevello, A.; Angeli, F. The pivotal link between ACE2 deficiency and SARS-CoV-2 infection. *Eur J Intern Med* 2020, 76, 14–20, doi:10.1016/j.ejim.2020.04.037.

402. Carsana, L.; Sonzogni, A.; Nasr, A.; Rossi, R.S.; Pellegrinelli, A.; Zerbi, P.; Rech, R.; Colombo, R.; Antinori, S.; Corbellino, M., et al. Pulmonary post-mortem findings in a series of COVID-19 cases from northern Italy: A two-centre descriptive study. *Lancet Infect Dis* 2020, 20, 1135–1140, doi:10.1016/S1473-3099(20)30434-5.

403. Sebag, S.C.; Bastarache, J.A.; Ware, L.B. Therapeutic modulation of coagulation and fibrinolysis in acute lung injury and the acute respiratory distress syndrome. *Curr Pharm Biotechnol* 2011, 12, 1481–1496, doi:10.2174/138920111798281171.

404. Iba, T.; Levy, J.H.; Levi, M.; Thachil, J. Coagulopathy in COVID-19. *J Thromb Haemost* 2020, 18, 2103–2109, doi:10.1111/jth.14975.

405. Liu, J.; Li, Y.; Liu, Q.; Yao, Q.; Wang, X.; Zhang, H.; Chen, R.; Ren, L.; Min, J.; Deng, F., et al. SARS-CoV-2 cell tropism and multiorgan infection. *Cell Discovery* 2021, 7, 17, doi:10.1038/s41421-021-00249-2.

406. Louis, T.J.; Qasem, A.; Abdelli, L.S.; Naser, S.A. Extra-pulmonary complications in SARS-CoV-2 infection: A comprehensive multi organ-system review. *Microorganisms* 2022, 10, doi:10.3390/microorganisms10010153.

407. Rabaan, A.A.; Smajlovic, S.; Tombuloglu, H.; Cordic, S.; Hajdarevic, A.; Kudic, N.; Al Mutai, A.; Turkistani, S.A.; Al-Ahmed, S.H.; Al-Zaki, N.A., et al. SARS-CoV-2 infection and multi-organ system damage: A review. *Biomol Biomed* 2023, 23, 37–52, doi:10.17305/bjbms.2022.7762.

408. Xu, Z.; Shi, L.; Wang, Y.; Zhang, J.; Huang, L.; Zhang, C.; Liu, S.; Zhao, P.; Liu, H.; Zhu, L., et al. Pathological findings of COVID-19 associated with acute respiratory distress syndrome. *Lancet Respir Med* 2020, 8, 420–422, doi:10.1016/S2213-2600(20)30076-X.

409. Zandi, M.; Shafaati, M.; Kalantar-Neyestanaki, D.; Pourghadamyari, H.; Fani, M.; Soltani, S.; Kaleji, H.; Abbasi, S. The role of SARS-CoV-2 accessory proteins in immune evasion. *Biomed Pharmacother* 2022, 156, 113889, doi:10.1016/j.biopha.2022.113889.
410. Mantlo, E.; Bukreyeva, N.; Maruyama, J.; Paessler, S.; Huang, C. Antiviral activities of type I interferons to SARS-CoV-2 infection. *Antiviral Res* 2020, 179, 104811, doi:10.1016/j.antiviral.2020.104811.
411. Hoagland, D.A.; Møller, R.; Uhl, S.A.; Oishi, K.; Frere, J.; Golynker, I.; Horiuchi, S.; Panis, M.; Blanco-Melo, D.; Sachs, D., et al. Leveraging the antiviral type I interferon system as a first line of defense against SARS-CoV-2 pathogenicity. *Immunity* 2021, 54, 557–570.e5, doi:10.1016/j.immuni.2021.01.017.
412. Hung, I.F.; Lung, K.C.; Tso, E.Y.; Liu, R.; Chung, T.W.; Chu, M.Y.; Ng, Y.Y.; Lo, J.; Chan, J.; Tam, A.R., et al. Triple combination of interferon beta-1b, lopinavir-ritonavir, and ribavirin in the treatment of patients admitted to hospital with COVID-19: An open-label, randomised, phase 2 trial. *Lancet* 2020, 395, 1695–1704, doi:10.1016/S0140-6736(20)31042-4.
413. Zhang, Q.; Bastard, P.; Liu, Z.; Le Pen, J.; Moncada-Velez, M.; Chen, J.; Ogishi, M.; Sabli, I.K.D.; Hodeib, S.; Korol, C., et al. Inborn errors of type I IFN immunity in patients with life-threatening COVID-19. *Science* 2020, 370, doi:10.1126/science.abd4570.
414. Hadjadj, J.; Yatim, N.; Barnabei, L.; Corneau, A.; Boussier, J.; Smith, N.; Pere, H.; Charbit, B.; Bondet, V.; Chenevier-Gobeaux, C., et al. Impaired type I interferon activity and inflammatory responses in severe COVID-19 patients. *Science* 2020, 369, 718–724, doi:10.1126/science.abc6027.
415. Blanco-Melo, D.; Nilsson-Payant, B.E.; Liu, W.C.; Uhl, S.; Hoagland, D.; Moller, R.; Jordan, T.X.; Oishi, K.; Panis, M.; Sachs, D., et al. Imbalanced host response to SARS-CoV-2 drives development of COVID-19. *Cell* 2020, 181, 1036–1045.e39, doi:10.1016/j.cell.2020.04.026.
416. Broggi, A.; Ghosh, S.; Sposito, B.; Spreafico, R.; Balzarini, F.; Lo Cascio, A.; Clementi, N.; De Santis, M.; Mancini, N.; Granucci, F., et al. Type III interferons disrupt the lung epithelial barrier upon viral recognition. *Science* 2020, 369, 706–712, doi:10.1126/science.abc3545.
417. King, C.; Sprent, J. Dual nature of type I interferons in SARS-CoV-2-induced inflammation. *Trends Immunol* 2021, 42, 312–322, doi:10.1016/j.it.2021.02.003.
418. Hatton, C.F.; Botting, R.A.; Duenas, M.E.; Haq, I.J.; Verdon, B.; Thompson, B.J.; Spegarova, J.S.; Gothe, F.; Stephenson, E.; Gardner, A.I., et al. Delayed induction of type I and III interferons mediates nasal epithelial cell permissiveness to SARS-CoV-2. *Nat Commun* 2021, 12, 7092, doi:10.1038/s41467-021-27318-0.
419. Lucas, C.; Wong, P.; Klein, J.; Castro, T.B.R.; Silva, J.; Sundaram, M.; Ellingson, M.K.; Mao, T.; Oh, J.E.; Israelow, B., et al. Longitudinal analyses reveal immunological misfiring in severe COVID-19. *Nature* 2020, 584, 463–469, doi:10.1038/s41586-020-2588-y.
420. Eskandarian Boroujeni, M.; Sekrecka, A.; Antonczyk, A.; Hassani, S.; Sekrecki, M.; Nowicka, H.; Lopacinska, N.; Olya, A.; Kluzek, K.; Wesoly, J., et al. Dysregulated interferon response and immune hyperactivation in severe COVID-19: Targeting STATs as a novel therapeutic strategy. *Front Immunol* 2022, 13, 888897, doi:10.3389/fimmu.2022.888897.
421. Lopes-Pacheco, M.; Silva, P.L.; Cruz, F.F.; Battaglini, D.; Robba, C.; Pelosi, P.; Morales, M.M.; Caruso Neves, C.; Rocco, P.R.M. Pathogenesis of multiple organ injury in COVID-19 and potential therapeutic strategies. *Front Physiol* 2021, 12, 593223, doi:10.3389/fphys.2021.593223.
422. Kuri-Cervantes, L.; Pampena, M.B.; Meng, W.; Rosenfeld, A.M.; Ittner, C.A.G.; Weisman, A.R.; Agyekum, R.S.; Mathew, D.; Baxter, A.E.; Vella, L.A., et al. Comprehensive mapping of immune perturbations associated with severe COVID-19. *Sci Immunol* 2020, 5, doi:10.1126/sciimmunol.abd7114.
423. Harrison, A.G.; Lin, T.; Wang, P. Mechanisms of SARS-CoV-2 transmission and pathogenesis. *Trends Immunol* 2020, 41, 1100–1115, doi:10.1016/j.it.2020.10.004.
424. Paludan, S.R.; Mogensen, T.H. Innate immunological pathways in COVID-19 pathogenesis. *Sci Immunol* 2022, 7, eabm5505, doi:10.1126/sciimmunol.abm5505.
425. Song, X.; Hu, W.; Yu, H.; Zhao, L.; Zhao, Y.; Zhao, X.; Xue, H.-H.; Zhao, Y. Little to no expression of angiotensin-converting enzyme-2 on most human peripheral blood immune cells but highly expressed on tissue macrophages. *Cytometry Part A* 2023, 103, 136–145, https://doi.org/10.1002/cyto.a.24285.
426. Sefik, E.; Qu, R.; Junqueira, C.; Kaffe, E.; Mirza, H.; Zhao, J.; Brewer, J.R.; Han, A.; Steach, H.R.; Israelow, B., et al. Inflammasome activation in infected macrophages drives COVID-19 pathology. *Nature* 2022, 606, 585–593, doi:10.1038/s41586-022-04802-1.
427. Thepaut, M.; Luczkowiak, J.; Vives, C.; Labiod, N.; Bally, I.; Lasala, F.; Grimoire, Y.; Fenel, D.; Sattin, S.; Thielens, N., et al. DC/L-SIGN recognition of spike glycoprotein promotes SARS-CoV-2 trans-infection and can be inhibited by a glycomimetic antagonist. *PloS Pathog* 2021, 17, e1009576, doi:10.1371/journal.ppat.1009576.

428. Gu, Y.; Cao, J.; Zhang, X.; Gao, H.; Wang, Y.; Wang, J.; He, J.; Jiang, X.; Zhang, J.; Shen, G., et al. Receptome profiling identifies KREMEN1 and ASGR1 as alternative functional receptors of SARS-CoV-2. *Cell Res* 2022, 32, 24–37, doi:10.1038/s41422-021-00595-6.

429. Wang, S.; Qiu, Z.; Hou, Y.; Deng, X.; Xu, W.; Zheng, T.; Wu, P.; Xie, S.; Bian, W.; Zhang, C., et al. AXL is a candidate receptor for SARS-CoV-2 that promotes infection of pulmonary and bronchial epithelial cells. *Cell Res* 2021, 31, 126–140, doi:10.1038/s41422-020-00460-y.

430. Ramos da Silva, S.; Ju, E.; Meng, W.; Paniz Mondolfi, A.E.; Dacic, S.; Green, A.; Bryce, C.; Grimes, Z.; Fowkes, M.; Sordillo, E.M., et al. Broad severe acute respiratory syndrome coronavirus 2 cell tropism and immunopathology in lung tissues from fatal coronavirus disease 2019. *J Infect Dis* 2021, 223, 1842–1854, doi:10.1093/infdis/jiab195.

431. Pontelli, M.C.; Castro, I.A.; Martins, R.B.; La Serra, L.; Veras, F.P.; Nascimento, D.C.; Silva, C.M.; Cardoso, R.S.; Rosales, R.; Gomes, R., et al. SARS-CoV-2 productively infects primary human immune system cells in vitro and in COVID-19 patients. *J Mol Cell Biol* 2022, 14, doi:10.1093/jmcb/mjac021.

432. Junqueira, C.; Crespo, A.; Ranjbar, S.; de Lacerda, L.B.; Lewandrowski, M.; Ingber, J.; Parry, B.; Ravid, S.; Clark, S.; Schrimpf, M.R., et al. FcgammaR-mediated SARS-CoV-2 infection of monocytes activates inflammation. *Nature* 2022, 606, 576–584, doi:10.1038/s41586-022-04702-4.

433. Lv, J.; Wang, Z.; Qu, Y.; Zhu, H.; Zhu, Q.; Tong, W.; Bao, L.; Lv, Q.; Cong, J.; Li, D., et al. Distinct uptake, amplification, and release of SARS-CoV-2 by M1 and M2 alveolar macrophages. *Cell Discov* 2021, 7, 24, doi:10.1038/s41421-021-00258-1.

434. Rodrigues, T.S.; de Sa, K.S.G.; Ishimoto, A.Y.; Becerra, A.; Oliveira, S.; Almeida, L.; Goncalves, A.V.; Perucello, D.B.; Andrade, W.A.; Castro, R., et al. Inflammasomes are activated in response to SARS-CoV-2 infection and are associated with COVID-19 severity in patients. *J Exp Med* 2021, 218, doi:10.1084/jem.20201707.

435. Roy, R.K.; Sharma, U.; Wasson, M.K.; Jain, A.; Hassan, M.I.; Prakash, H. Macrophage activation syndrome and COVID 19: Impact of MAPK driven immune-epigenetic programming by SARS-Cov-2. *Front Immunol* 2021, 12, 763313, doi:10.3389/fimmu.2021.763313.

436. Jorgensen, I.; Miao, E.A. Pyroptotic cell death defends against intracellular pathogens. *Immunol Rev* 2015, 265, 130–142, doi:10.1111/imr.12287.

437. Shrivastava, G.; Leon-Juarez, M.; Garcia-Cordero, J.; Meza-Sanchez, D.E.; Cedillo-Barron, L. Inflammasomes and its importance in viral infections. *Immunol Res* 2016, 64, 1101–1117, doi:10.1007/s12026-016-8873-z.

438. Ferreira, A.C.; Soares, V.C.; de Azevedo-Quintanilha, I.G.; Dias, S.; Fintelman-Rodrigues, N.; Sacramento, C.Q.; Mattos, M.; de Freitas, C.S.; Temerozo, J.R.; Teixeira, L., et al. SARS-CoV-2 engages inflammasome and pyroptosis in human primary monocytes. *Cell Death Discov* 2021, 7, 43, doi:10.1038/s41420-021-00428-w.

439. Zhao, H.; Nguyen, A.; Wu, D.; Li, Y.; Hassan, S.A.; Chen, J.; Shroff, H.; Piszczek, G.; Schuck, P. Plasticity in structure and assembly of SARS-CoV-2 nucleocapsid protein. *PNAS Nexus* 2022, 1, pgac049, doi:10.1093/pnasnexus/pgac049.

440. Barh, D.; Tiwari, S.; Rodrigues Gomes, L.G.; Ramalho Pinto, C.H.; Andrade, B.S.; Ahmad, S.; Aljabali, A.A.A.; Alzahrani, K.J.; Banjer, H.J.; Hassan, S.S., et al. SARS-CoV-2 variants show a gradual declining pathogenicity and pro-inflammatory cytokine stimulation, an increasing antigenic and anti-inflammatory cytokine induction, and rising structural protein instability: A minimal number genome-based approach. *Inflammation* 2023, 46, 297–312, doi:10.1007/s10753-022-01734-w.

441. Wang, Y.; Pan, X.; Ji, H.; Zuo, X.; Xiao, G.F.; Li, J.; Zhang, L.K.; Xia, B.; Gao, Z. Impact of SARS-CoV-2 envelope protein mutations on the pathogenicity of Omicron XBB. *Cell Discov* 2023, 9, 80, doi:10.1038/s41421-023-00575-7.

442. Lai, C.C.; Shih, T.P.; Ko, W.C.; Tang, H.J.; Hsueh, P.R. Severe acute respiratory syndrome coronavirus 2 (SARS-CoV-2) and coronavirus disease-2019 (COVID-19): The epidemic and the challenges. *Int J Antimicrob Agents* 2020, 55, 105924, doi:10.1016/j.ijantimicag.2020.105924.

443. Sadeghalvad, M.; Mansourabadi, A.H.; Noori, M.; Nejadghaderi, S.A.; Masoomikarimi, M.; Alimohammadi, M.; Rezaei, N. Recent developments in SARS-CoV-2 vaccines: A systematic review of the current studies. *Rev Med Virol* 2023, 33, e2359, doi:10.1002/rmv.2359.

444. Martinez-Flores, D.; Zepeda-Cervantes, J.; Cruz-Resendiz, A.; Aguirre-Sampieri, S.; Sampieri, A.; Vaca, L. SARS-CoV-2 vaccines based on the spike glycoprotein and implications of new viral variants. *Front Immunol* 2021, 12, 701501, doi:10.3389/fimmu.2021.701501.

445. Lapuente, D.; Winkler, T.H.; Tenbusch, M. B-cell and antibody responses to SARS-CoV-2: Infection, vaccination, and hybrid immunity. *Cell Mol Immunol* 2024, 21, 144–158, doi:10.1038/s41423-023-01095-w.

446. Diaz-Dinamarca, D.A.; Diaz, P.; Barra, G.; Puentes, R.; Arata, L.; Grossolli, J.; Riveros-Rodriguez, B.; Ardiles, L.; Santelises, J.; Vasquez-Saez, V., et al. Humoral immunity against SARS-CoV-2 evoked by heterologous vaccination groups using the CoronaVac (Sinovac) and BNT162b2 (Pfizer/BioNTech) vaccines in Chile. *Front Public Health* 2023, 11, 1229045, doi:10.3389/fpubh.2023.1229045.

447. Klemis, V.; Schmidt, T.; Schub, D.; Mihm, J.; Marx, S.; Abu-Omar, A.; Ziegler, L.; Hielscher, F.; Guckelmus, C.; Urschel, R., et al. Comparative immunogenicity and reactogenicity of heterologous ChAdOx1-nCoV-19-priming and BNT162b2 or mRNA-1273-boosting with homologous COVID-19 vaccine regimens. *Nat Commun* 2022, 13, 4710, doi:10.1038/s41467-022-32321-0.

448. Levi, R.; Azzolini, E.; Pozzi, C.; Ubaldi, L.; Lagioia, M.; Mantovani, A.; Rescigno, M. One dose of SARS-CoV-2 vaccine exponentially increases antibodies in individuals who have recovered from symptomatic COVID-19. *J Clin Invest* 2021, 131, doi:10.1172/JCI149154.

449. Wise, J. Covid-19: People who have had infection might only need one dose of mRNA vaccine. *BMJ* 2021, 372, n308, doi:10.1136/bmj.n308.

450. Pooley, N.; Abdool Karim, S.S.; Combadiere, B.; Ooi, E.E.; Harris, R.C.; El Guerche Seblain, C.; Kisomi, M.; Shaikh, N. Durability of vaccine-induced and natural immunity against COVID-19: A narrative review. *Infect Dis Ther* 2023, 12, 367–387, doi:10.1007/s40121-022-00753-2.

451. Painter, M.M.; Johnston, T.S.; Lundgreen, K.A.; Santos, J.J.S.; Qin, J.S.; Goel, R.R.; Apostolidis, S.A.; Mathew, D.; Fulmer, B.; Williams, J.C., et al. Prior vaccination promotes early activation of memory T cells and enhances immune responses during SARS-CoV-2 breakthrough infection. *Nat Immunol* 2023, 24, 1711–1724, doi:10.1038/s41590-023-01613-y.

452. Israel, A.; Shenhar, Y.; Green, I.; Merzon, E.; Golan-Cohen, A.; Schaffer, A.A.; Ruppin, E.; Vinker, S.; Magen, E. Large-scale study of antibody titer decay following BNT162b2 mRNA vaccine or SARS-CoV-2 infection. *Vaccines* 2021, 10, doi:10.3390/vaccines10010064.

453. Wendelboe, A.M.; Van Rie, A.; Salmaso, S.; Englund, J.A. Duration of immunity against pertussis after natural infection or vaccination. *Pediatr Infect Dis J* 2005, 24, S58–61, doi:10.1097/01.inf.0000160914.59160.41.

454. Gu, X.X.; Plotkin, S.A.; Edwards, K.M.; Sette, A.; Mills, K.H.G.; Levy, O.; Sant, A.J.; Mo, A.; Alexander, W.; Lu, K.T., et al. Waning Immunity and Microbial Vaccines-Workshop of the National Institute of Allergy and Infectious Diseases. *Clin Vaccine Immunol* 2017, 24, doi:10.1128/CVI.00034-17.

455. Hernandez-Suarez, C.; Murillo-Zamora, E. Waning immunity to SARS-CoV-2 following vaccination or infection. *Front Med* 2022, 9, 972083, doi:10.3389/fmed.2022.972083.

456. Lin, D.Y.; Xu, Y.; Gu, Y.; Zeng, D.; Wheeler, B.; Young, H.; Sunny, S.K.; Moore, Z. Effectiveness of bivalent boosters against severe Omicron infection. *N Engl J Med* 2023, 388, 764–766, doi:10.1056/NEJMc2215471.

457. Li, W.; Zhao, T.; Tao, B.; Zhao, L.; Xiao, H.; Ding, X.; Li, C.; Chen, L.; Cheng, H.; Lou, Y., et al. Monovalent Omicron COVID-19 vaccine triggers superior neutralizing antibody responses against Omicron subvariants than Delta and Omicron bivalent vaccine. *Hum Vaccin Immunother* 2023, 19, 2264589, doi:10.1080/21645515.2023.2264589.

458. Patel, N.; Trost, J.F.; Guebre-Xabier, M.; Zhou, H.; Norton, J.; Jiang, D.; Cai, Z.; Zhu, M.; Marchese, A.M.; Greene, A.M., et al. XBB.1.5 spike protein COVID-19 vaccine induces broadly neutralizing and cellular immune responses against EG.5.1 and emerging XBB variants. *Sci Rep* 2023, 13, 19176, doi:10.1038/s41598-023-46025-y.

459. Kowalski, P.S.; Rudra, A.; Miao, L.; Anderson, D.G. Delivering the messenger: Advances in technologies for therapeutic mRNA delivery. *Mol Ther* 2019, 27, 710–728, doi:10.1016/j.ymthe.2019.02.012.

460. Kaplonek, P.; Cizmeci, D.; Fischinger, S.; Collier, A.R.; Suscovich, T.; Linde, C.; Broge, T.; Mann, C.; Amanat, F.; Dayal, D., et al. mRNA-1273 and BNT162b2 COVID-19 vaccines elicit antibodies with differences in Fc-mediated effector functions. *Sci Transl Med* 2022, 14, eabm2311, doi:10.1126/scitranslmed.abm2311.

461. Sahin, U.; Muik, A.; Derhovanessian, E.; Vogler, I.; Kranz, L.M.; Vormehr, M.; Baum, A.; Pascal, K.; Quandt, J.; Maurus, D., et al. COVID-19 vaccine BNT162b1 elicits human antibody and T(H)1 T cell responses. *Nature* 2020, 586, 594–599, doi:10.1038/s41586-020-2814-7.

462. Mateus, J.; Dan, J.M.; Zhang, Z.; Rydyznski Moderbacher, C.; Lammers, M.; Goodwin, B.; Sette, A.; Crotty, S.; Weiskopf, D. Low-dose mRNA-1273 COVID-19 vaccine generates durable memory enhanced by cross-reactive T cells. *Science* 2021, 374, eabj9853, doi:10.1126/science.abj9853.

463. Fodor, E.; Olmos Calvo, I.; Kuten-Pella, O.; Hamar, E.; Bukva, M.; Madár, Á.; Hornyák, I.; Hinsenkamp, A.; Hetényi, R.; Földes, F., et al. Comparison of immune activation of the COVID vaccines: ChAdOx1, BNT162b2, mRNA-1273, BBIBP-CorV, and Gam-COVID-Vac from serological human samples in Hungary showed higher protection after mRNA-based immunization. *Eur Rev Med Pharmacol Sci* 2022, 26, 5297–5306, doi:10.26355/eurrev_202207_29321.

464. Zhang, G.F.; Meng, W.; Chen, L.; Ding, L.; Feng, J.; Perez, J.; Ali, A.; Sun, S.; Liu, Z.; Huang, Y., et al. Neutralizing antibodies to SARS-CoV-2 variants of concern including Delta and Omicron in subjects receiving mRNA-1273, BNT162b2, and Ad26.COV2.S vaccines. *J Med Virol* 2022, 94, 5678–5690, doi:10.1002/jmv.28032.

465. Vogel, E.; Kocher, K.; Priller, A.; Cheng, C.C.; Steininger, P.; Liao, B.H.; Korber, N.; Willmann, A.; Irrgang, P.; Held, J., et al. Dynamics of humoral and cellular immune responses after homologous and heterologous SARS-CoV-2 vaccination with ChAdOx1 nCoV-19 and BNT162b2. *EBioMedicine* 2022, 85, 104294, doi:10.1016/j.ebiom.2022.104294.

466. Khoo, N.K.H.; Lim, J.M.E.; Gill, U.S.; de Alwis, R.; Tan, N.; Toh, J.Z.N.; Abbott, J.E.; Usai, C.; Ooi, E.E.; Low, J.G.H., et al. Differential immunogenicity of homologous versus heterologous boost in Ad26. COV2.S vaccine recipients. *Med* 2022, 3, 104–118.e4, doi:10.1016/j.medj.2021.12.004.

467. Shin, D.H.; Smith, D.M.; Choi, J.Y. SARS-CoV-2 Omicron variant of concern: Everything you wanted to know about Omicron but were afraid to ask. *Yonsei Med J* 2022, 63, 977–983, doi:10.3349/ymj.2022.0383.

468. Barda, N.; Lustig, Y.; Indenbaum, V.; Zibly, D.; Joseph, G.; Asraf, K.; Weiss-Ottolenghi, Y.; Amit, S.; Kliker, L.; Abd Elkader, B., et al. Immunogenicity of Omicron BA.1-adapted BNT162b2 vaccines: Randomized trial, 3-month follow-up. *Clin Microbiol Infect* 2023, 29, 918–923, doi:10.1016/j.cmi.2023.03.007.

469. Collier, A.Y.; Miller, J.; Hachmann, N.P.; McMahan, K.; Liu, J.; Bondzie, E.A.; Gallup, L.; Rowe, M.; Schonberg, E.; Thai, S., et al. Immunogenicity of BA.5 bivalent mRNA vaccine boosters. *N Engl J Med* 2023, 388, 565–567, doi:10.1056/NEJMc2213948.

470. Offit, P.A. Bivalent Covid-19 vaccines — A cautionary tale. *N Engl J Med* 2023, 388, 481–483, doi:10.1056/NEJMp2215780.

471. Reynolds, C.J.; Pade, C.; Gibbons, J.M.; Otter, A.D.; Lin, K.M.; Munoz Sandoval, D.; Pieper, F.P.; Butler, D.K.; Liu, S.; Joy, G., et al. Immune boosting by B.1.1.529 (Omicron) depends on previous SARS-CoV-2 exposure. *Science* 2022, 377, eabq1841, doi:10.1126/science.abq1841.

472. Deng, S.; Liang, H.; Chen, P.; Li, Y.; Li, Z.; Fan, S.; Wu, K.; Li, X.; Chen, W.; Qin, Y., et al. Viral vector vaccine development and application during the COVID-19 pandemic. *Microorganisms* 2022, 10, doi:10.3390/microorganisms10071450.

473. Vanaparthy, R.; Mohan, G.; Vasireddy, D.; Atluri, P. Review of COVID-19 viral vector-based vaccines and COVID-19 variants. *Infez Med* 2021, 29, 328–338, doi:10.53854/liim-2903-3.

474. Gray, G.; Collie, S.; Goga, A.; Garrett, N.; Champion, J.; Seocharan, I.; Bamford, L.; Moultrie, H.; Bekker, L.G. Effectiveness of Ad26.COV2.S and BNT162b2 vaccines against Omicron variant in South Africa. *N Engl J Med* 2022, 386, 2243–2245, doi:10.1056/NEJMc2202061.

475. Swanson, P.A. 2nd; Padilla, M.; Hoyland, W.; McGlinchey, K.; Fields, P.A.; Bibi, S.; Faust, S.N.; McDermott, A.B.; Lambe, T.; Pollard, A.J., et al. AZD1222/ChAdOx1 nCoV-19 vaccination induces a polyfunctional spike protein-specific T(H)1 response with a diverse TCR repertoire. *Sci Transl Med* 2021, 13, eabj7211, doi:10.1126/scitranslmed.abj7211.

476. Asano, M.; Okada, H.; Itoh, Y.; Hirata, H.; Ishikawa, K.; Yoshida, E.; Matsui, A.; Kelly, E.J.; Shoemaker, K.; Olsson, U., et al. Immunogenicity and safety of AZD1222 (ChAdOx1 nCoV-19) against SARS-CoV-2 in Japan: A double-blind, randomized controlled phase 1/2 trial. *Int J Infect Dis* 2022, 114, 165–174, doi:10.1016/j.ijid.2021.10.030.

477. Chua, J.X.; Durrant, L.G.; Chok, Y.L.; Lai, O.M. Susceptibility to SARS-CoV-2 omicron following ChAdOx1 nCoV-19 and BNT162b2 versus CoronaVac vaccination. *iScience* 2022, 25, 105379, doi:10.1016/j.isci.2022.105379.

478. Greinacher, A.; Thiele, T.; Warkentin, T.E.; Weisser, K.; Kyrle, P.A.; Eichinger, S. Thrombotic thrombocytopenia after ChAdOx1 nCov-19 vaccination. *N Engl J Med* 2021, 384, 2092–2101, doi:10.1056/NEJMoa2104840.

479. See, I.; Su, J.R.; Lale, A.; Woo, E.J.; Guh, A.Y.; Shimabukuro, T.T.; Streiff, M.B.; Rao, A.K.; Wheeler, A.P.; Beavers, S.F., et al. US case reports of cerebral venous sinus thrombosis with thrombocytopenia after Ad26.COV2.S vaccination, March 2 to April 21, 2021. *JAMA* 2021, 325, 2448–2456, doi:10.1001/jama.2021.7517.

480. Liu, Y.; Shao, Z.; Wang, H. SARS-CoV-2 vaccine-induced immune thrombotic thrombocytopenia. *Thromb Res* 2022, 209, 75–79, doi:10.1016/j.thromres.2021.12.002.

481. Gao, Q.; Bao, L.; Mao, H.; Wang, L.; Xu, K.; Yang, M.; Li, Y.; Zhu, L.; Wang, N.; Lv, Z., et al. Development of an inactivated vaccine candidate for SARS-CoV-2. *Science* 2020, 369, 77–81, doi:10.1126/science.abc1932.

482. Muena, N.A.; Garcia-Salum, T.; Pardo-Roa, C.; Avendano, M.J.; Serrano, E.F.; Levican, J.; Almonacid, L.I.; Valenzuela, G.; Poblete, E.; Strohmeier, S., et al. Induction of SARS-CoV-2 neutralizing antibodies by CoronaVac and BNT162b2 vaccines in naive and previously infected individuals. *EBioMedicine* 2022, 78, 103972, doi:10.1016/j.ebiom.2022.103972.

483. Yavuz, E.; Gunal, O.; Basbulut, E.; Sen, A. SARS-CoV-2 specific antibody responses in healthcare workers after a third booster dose of CoronaVac or BNT162b2 vaccine. *J Med Virol* 2022, 94, 3768–3775, doi:10.1002/jmv.27794.

484. Çağlayan, D.; Süner, A.F.; Şiyve, N.; Güzel, I.; Irmak, Ç.; Işik, E.; Appak, Ö.; Çelik, M.; Öztürk, G.; Alp Çavuş, S., et al. An analysis of antibody response following the second dose of CoronaVac and humoral response after booster dose with BNT162b2 or CoronaVac among healthcare workers in Turkey. *J Med Virol* 2022, 94, 2212–2221, https://doi.org/10.1002/jmv.27620.

485. Vargas, L.; Valdivieso, N.; Tempio, F.; Simon, V.; Sauma, D.; Valenzuela, L.; Beltran, C.; Castillo-Delgado, L.; Contreras-Benavides, X.; Acevedo, M.L., et al. Serological study of CoronaVac vaccine and booster doses in Chile: Immunogenicity and persistence of anti-SARS-CoV-2 spike antibodies. *BMC Med* 2022, 20, 216, doi:10.1186/s12916-022-02406-0.

486. Mok, C.K.P.; Cohen, C.A.; Cheng, S.M.S.; Chen, C.; Kwok, K.O.; Yiu, K.; Chan, T.O.; Bull, M.; Ling, K.C.; Dai, Z., et al. Comparison of the immunogenicity of BNT162b2 and CoronaVac COVID-19 vaccines in Hong Kong. *Respirology* 2022, 27, 301–310, doi:10.1111/resp.14191.

487. Pilapitiya, D.; Wheatley, A.K.; Tan, H.X. Mucosal vaccines for SARS-CoV-2: Triumph of Hope over experience. *EBioMedicine* 2023, 92, 104585, doi:10.1016/j.ebiom.2023.104585.

488. Moreno-Fierros, L.; García-Silva, I.; Rosales-Mendoza, S. Development of SARS-CoV-2 vaccines: Should we focus on mucosal immunity? *Expert Opin Biol Ther* 2020, 20, 831–836, doi:10.1080/14712598.2020.1767062.

489. Gong, W.; Parkkila, S.; Wu, X.; Aspatwar, A. SARS-CoV-2 variants and COVID-19 vaccines: Current challenges and future strategies. *Int Rev Immunol* 2023, 42, 393–414, doi:10.1080/08830185.2022.2079642.

490. Lycke, N. Recent progress in mucosal vaccine development: Potential and limitations. *Nat Rev Immunol* 2012, 12, 592–605, doi:10.1038/nri3251.

491. Li, M.; Wang, Y.; Sun, Y.; Cui, H.; Zhu, S.J.; Qiu, H.J. Mucosal vaccines: Strategies and challenges. *Immunol Lett* 2020, 217, 116–125, doi:10.1016/j.imlet.2019.10.013.

492. Singh, C.; Verma, S.; Reddy, P.; Diamond, M.; Curiel, D.; Patel, C.; Jain, M.K.; Redkar, S.V.; Bhate, A.S.; Gundappa, V., et al. Immunogenicity and tolerability of BBV154 (iNCOVACC®), an intranasal SARS-CoV-2 vaccine, compared with intramuscular Covaxin® in healthy adults: A randomised, open-label, phase 3 clinical trial. *The Lancet (Preprint)* 2023, doi:10.2139/ssrn.4342771.

493. Singh, C.; Verma, S.; Reddy, P.; Diamond, M.S.; Curiel, D.T.; Patel, C.; Jain, M.K.; Redkar, S.V.; Bhate, A.S.; Gundappa, V., et al. Phase III Pivotal comparative clinical trial of intranasal (iNCOVACC) and intramuscular COVID 19 vaccine (Covaxin®). *NPJ Vaccines* 2023, 8, 125, doi:10.1038/s41541-023-00717-8.

494. Wu, S.; Huang, J.; Zhang, Z.; Wu, J.; Zhang, J.; Hu, H.; Zhu, T.; Zhang, J.; Luo, L.; Fan, P., et al. Safety, tolerability, and immunogenicity of an aerosolised adenovirus type-5 vector-based COVID-19 vaccine (Ad5-nCoV) in adults: Preliminary report of an open-label and randomised phase 1 clinical trial. *Lancet Infect Dis* 2021, 21, 1654–1664, doi:10.1016/S1473-3099(21)00396-0.

495. Atmar, R.L.; Lyke, K.E.; Deming, M.E.; Jackson, L.A.; Branche, A.R.; El Sahly, H.M.; Rostad, C.A.; Martin, J.M.; Johnston, C.; Rupp, R.E., et al. Homologous and heterologous Covid-19 booster vaccinations. *N Engl J Med* 2022, 386, 1046–1057, doi:10.1056/NEJMoa2116414.

496. Tenbusch, M.; Schumacher, S.; Vogel, E.; Priller, A.; Held, J.; Steininger, P.; Beileke, S.; Irrgang, P.; Brockhoff, R.; Salmanton-Garcia, J., et al. Heterologous prime-boost vaccination with ChAdOx1 nCoV-19 and BNT162b2. *Lancet Infect Dis* 2021, 21, 1212–1213, doi:10.1016/S1473-3099(21)00420-5.

497. Accorsi, E.K.; Britton, A.; Shang, N.; Fleming-Dutra, K.E.; Link-Gelles, R.; Smith, Z.R.; Derado, G.; Miller, J.; Schrag, S.J.; Verani, J.R. Effectiveness of homologous and heterologous Covid-19 boosters against Omicron. *N Engl J Med* 2022, 386, 2433–2435, doi:10.1056/NEJMc2203165.

498. Sette, A.; Sidney, J.; Crotty, S. T cell responses to SARS-CoV-2. *Annu Rev Immunol* 2023, 41, 343–373, doi:10.1146/annurev-immunol-101721-061120.

499. Khoury, D.S.; Cromer, D.; Reynaldi, A.; Schlub, T.E.; Wheatley, A.K.; Juno, J.A.; Subbarao, K.; Kent, S.J.; Triccas, J.A.; Davenport, M.P. Neutralizing antibody levels are highly predictive of immune protection from symptomatic SARS-CoV-2 infection. *Nat Med* 2021, 27, 1205–1211, doi:10.1038/s41591-021-01377-8.

500. Gilbert, P.B.; Montefiori, D.C.; McDermott, A.B.; Fong, Y.; Benkeser, D.; Deng, W.; Zhou, H.; Houchens, C.R.; Martins, K.; Jayashankar, L., et al. Immune correlates analysis of the mRNA-1273 COVID-19 vaccine efficacy clinical trial. *Science* 2022, 375, 43–50, doi:10.1126/science.abm3425.

501. Hoffmann, M.; Krüger, N.; Schulz, S.; Cossmann, A.; Rocha, C.; Kempf, A.; Nehlmeier, I.; Graichen, L.; Moldenhauer, A.S.; Winkler, M.S., et al. The Omicron variant is highly resistant against antibody-mediated neutralization: Implications for control of the COVID-19 pandemic. *Cell* 2022, 185, 447–456.e11, doi:10.1016/j.cell.2021.12.032.

502. Jalkanen, P.; Kolehmainen, P.; Hakkinen, H.K.; Huttunen, M.; Tahtinen, P.A.; Lundberg, R.; Maljanen, S.; Reinholm, A.; Tauriainen, S.; Pakkanen, S.H., et al. COVID-19 mRNA vaccine induced antibody responses against three SARS-CoV-2 variants. *Nat Commun* 2021, 12, 3991, doi:10.1038/s41467-021-24285-4.

503. Geers, D.; Shamier, M.C.; Bogers, S.; den Hartog, G.; Gommers, L.; Nieuwkoop, N.N.; Schmitz, K.S.; Rijsbergen, L.C.; van Osch, J.A.T.; Dijkhuizen, E., et al. SARS-CoV-2 variants of concern partially escape humoral but not T cell responses in COVID-19 convalescent donors and vaccine recipients. *Sci Immunol* 2021, 6, eabj1750, doi:10.1126/sciimmunol.abj1750.

504. Tarke, A.; Sidney, J.; Methot, N.; Yu, E.D.; Zhang, Y.; Dan, J.M.; Goodwin, B.; Rubiro, P.; Sutherland, A.; Wang, E., et al. Impact of SARS-CoV-2 variants on the total CD4(+) and CD8(+) T cell reactivity in infected or vaccinated individuals. *Cell Rep Med* 2021, 2, 100355, doi:10.1016/j.xcrm.2021.100355.

505. Redd, A.D.; Nardin, A.; Kared, H.; Bloch, E.M.; Pekosz, A.; Laeyendecker, O.; Abel, B.; Fehlings, M.; Quinn, T.C.; Tobian, A.A.R. CD8+ T-cell responses in COVID-19 convalescent individuals target conserved epitopes from multiple prominent SARS-CoV-2 circulating variants. *Open Forum Infect Dis* 2021, 8, ofab143, doi:10.1093/ofid/ofab143.

506. Goel, R.R.; Painter, M.M.; Apostolidis, S.A.; Mathew, D.; Meng, W.; Rosenfeld, A.M.; Lundgreen, K.A.; Reynaldi, A.; Khoury, D.S.; Pattekar, A., et al. mRNA vaccines induce durable immune memory to SARS-CoV-2 and variants of concern. *Science* 2021, 374, abm0829, doi:10.1126/science.abm0829.

507. Dan, J.M.; Mateus, J.; Kato, Y.; Hastie, K.M.; Yu, E.D.; Faliti, C.E.; Grifoni, A.; Ramirez, S.I.; Haupt, S.; Frazier, A., et al. Immunological memory to SARS-CoV-2 assessed for up to 8 months after infection. *Science* 2021, 371, doi:10.1126/science.abf4063.

508. Liu, J.; Chandrashekar, A.; Sellers, D.; Barrett, J.; Jacob-Dolan, C.; Lifton, M.; McMahan, K.; Sciacca, M.; VanWyk, H.; Wu, C., et al. Vaccines elicit highly conserved cellular immunity to SARS-CoV-2 Omicron. *Nature* 2022, 603, 493–496, doi:10.1038/s41586-022-04465-y.

509. Wherry, E.J.; Barouch, D.H. T cell immunity to COVID-19 vaccines. *Science* 2022, 377, 821–822, doi:10.1126/science.add2897.

510. Fisher, D.; Heymann, D. Q&A: The novel coronavirus outbreak causing COVID-19. *BMC Med* 2020, 18, 57, doi:10.1186/s12916-020-01533-w.

511. Wang, T.; Du, Z.; Zhu, F.; Cao, Z.; An, Y.; Gao, Y.; Jiang, B. Comorbidities and multi-organ injuries in the treatment of COVID-19. *Lancet* 2020, 395, e52, doi:10.1016/S0140-6736(20)30558-4.

512. Dougan, M.; Nirula, A.; Azizad, M.; Mocherla, B.; Gottlieb, R.L.; Chen, P.; Hebert, C.; Perry, R.; Boscia, J.; Heller, B., et al. Bamlanivimab plus etesevimab in mild or moderate Covid-19. *N Engl J Med* 2021, 385, 1382–1392, doi:10.1056/NEJMoa2102685.

513. Gottlieb, R.L.; Nirula, A.; Chen, P.; Boscia, J.; Heller, B.; Morris, J.; Huhn, G.; Cardona, J.; Mocherla, B.; Stosor, V., et al. Effect of bamlanivimab as monotherapy or in combination with etesevimab on viral load in patients with mild to moderate COVID-19: A randomized clinical trial. *JAMA* 2021, 325, 632–644, doi:10.1001/jama.2021.0202.

514. Falcone, M.; Tiseo, G.; Valoriani, B.; Barbieri, C.; Occhineri, S.; Mazzetti, P.; Vatteroni, M.L.; Suardi, L.R.; Riccardi, N.; Pistello, M., et al. Efficacy of Bamlanivimab/Etesevimab and Casirivimab/Imdevimab in preventing progression to severe COVID-19 and role of variants of concern. *Infect Dis Ther* 2021, 10, 2479–2488, doi:10.1007/s40121-021-00525-4.

515. Hoffmann, M.; Arora, P.; Gross, R.; Seidel, A.; Hornich, B.F.; Hahn, A.S.; Kruger, N.; Graichen, L.; Hofmann-Winkler, H.; Kempf, A., et al. SARS-CoV-2 variants B.1.351 and P.1 escape from neutralizing antibodies. *Cell* 2021, 184, 2384–2393.e12, doi:10.1016/j.cell.2021.03.036.

516. Yu, S.Y.; Choi, M.; Cheong, C.; Ryoo, S.; Huh, K.; Yoon, Y.K.; Choi, J.; Kim, S.B. Clinical efficacy and safety of SARS-CoV-2-neutralizing monoclonal antibody in patients with COVID-19: A living systematic review and meta-analysis. *J Microbiol Immunol Infect* 2023, 56, 909–920, doi:10.1016/j.jmii.2023.07.009.

517. Tai, Y.L.; Lee, M.D.; Chi, H.; Chiu, N.C.; Lei, W.T.; Weng, S.L.; Liu, L.Y.; Chen, C.C.; Huang, S.Y.; Huang, Y.N., et al. Effects of bamlanivimab alone or in combination with etesevimab on subsequent hospitalization and mortality in outpatients with COVID-19: A systematic review and meta-analysis. *PeerJ* 2023, 11, e15344, doi:10.7717/peerj.15344.

518. Mahase, E. Covid-19: Has the spread of omicron BA.2 made antibody treatments redundant? *BMJ* 2022, 377, o1009, doi:10.1136/bmj.o1009.

519. VanBlargan, L.A.; Errico, J.M.; Halfmann, P.J.; Zost, S.J.; Crowe, J.E. Jr.; Purcell, L.A.; Kawaoka, Y.; Corti, D.; Fremont, D.H.; Diamond, M.S. An infectious SARS-CoV-2 B.1.1.529 Omicron virus escapes neutralization by therapeutic monoclonal antibodies. *Nat Med* 2022, 28, 490–495, doi:10.1038/s41591-021-01678-y.

520. Snow, T.A.C.; Saleem, N.; Ambler, G.; Nastouli, E.; Singer, M.; Arulkumaran, N. Tocilizumab in COVID-19: A meta-analysis, trial sequential analysis, and meta-regression of randomized-controlled trials. *Intensive Care Med* 2021, 47, 641–652, doi:10.1007/s00134-021-06416-z.

521. Malgie, J.; Schoones, J.W.; Pijls, B.G. Decreased mortality in coronavirus disease 2019 patients treated with tocilizumab: A rapid systematic review and meta-analysis of observational studies. *Clin Infect Dis* 2021, 72, e742–e749, doi:10.1093/cid/ciaa1445.

522. Laffont-Lozes, P.; Laureillard, D.; Loubet, P.; Stephan, R.; Chiaruzzi, M.; Clemmer, E.; Martin, A.; Roger, C.; Muller, L.; Claret, P.G., et al. Effect of tocilizumab on mortality in patients with SARS-CoV-2 pneumonia caused by Delta or Omicron variants: A propensity-matched analysis in Nimes University Hospital, France. *Antibiotics (Basel)* 2023, 12, doi:10.3390/antibiotics12010088.

523. Oliynyk, O.; Barg, W.; Oliynyk, Y.; Dubrov, S.; Gurianov, V.; Rorat, M. Lack of difference in tocilizumab efficacy in the treatment of severe COVID-19 caused by different SARS-CoV-2 variants. *J Pers Med* 2022, 12, doi:10.3390/jpm12071103.

524. Yu, S.Y.; Koh, D.H.; Choi, M.; Ryoo, S.; Huh, K.; Yeom, J.S.; Yoon, Y.K. Clinical efficacy and safety of interleukin-6 receptor antagonists (tocilizumab and sarilumab) in patients with COVID-19: A systematic review and meta-analysis. *Emerg Microbes Infect* 2022, 11, 1154–1165, doi:10.1080/22221751.2022.2059405.

525. Sivapalasingam, S.; Lederer, D.J.; Bhore, R.; Hajizadeh, N.; Criner, G.; Hosain, R.; Mahmood, A.; Giannelou, A.; Somersan-Karakaya, S.; O'Brien, M.P., et al. Efficacy and safety of sarilumab in hospitalized patients with coronavirus disease 2019: A randomized clinical trial. *Clin Infect Dis* 2022, 75, e380–e388, doi:10.1093/cid/ciac153.

526. Senefeld, J.W.; Franchini, M.; Mengoli, C.; Cruciani, M.; Zani, M.; Gorman, E.K.; Focosi, D.; Casadevall, A.; Joyner, M.J. COVID-19 convalescent plasma for the treatment of immunocompromised patients: A systematic review and meta-analysis. *JAMA Netw Open* 2023, 6, e2250647, doi:10.1001/jamanetworkopen.2022.50647.

527. Misset, B.; Piagnerelli, M.; Hoste, E.; Dardenne, N.; Grimaldi, D.; Michaux, I.; De Waele, E.; Dumoulin, A.; Jorens, P.G.; van der Hauwaert, E., et al. Convalescent plasma for Covid-19-induced ARDS in mechanically ventilated patients. *N Engl J Med* 2023, 389, 1590–1600, doi:10.1056/NEJMoa2209502.

528. Lang, K. Is convalescent plasma still useful as a covid treatment? *BMJ* 2023, 383, p2185, doi:10.1136/bmj.p2185.

529. Wang, M.; Zhou, B.; Fan, Q.; Zhou, X.; Liao, X.; Lin, J.; Ma, Z.; Dong, J.; Wang, H.; Ge, X., et al. Omicron variants escape the persistent SARS-CoV-2-specific antibody response in 2-year COVID-19 convalescents regardless of vaccination. *Emerg Microbes Infect* 2023, 12, 2151381, doi:10.1080/22221751.2022.2151381.

530. Franchini, M.; Focosi, D.; Percivalle, E.; Beccaria, M.; Garuti, M.; Arar, O.; Pecoriello, A.; Spreafico, F.; Greco, G.; Bertacco, S., et al. Variant of concern-matched COVID-19 convalescent plasma usage in seronegative hospitalized patients. *Viruses* 2022, 14, doi:10.3390/v14071443.

531. Vickers, M.A.; Sariol, A.; Leon, J.; Ehlers, A.; Locher, A.V.; Dubay, K.A.; Collins, L.; Voss, D.; Odle, A.E.; Holida, M., et al. Exponential increase in neutralizing and spike specific antibodies following vaccination of COVID-19 convalescent plasma donors. *Transfusion* 2021, 61, 2099–2106, doi:10.1111/trf.16401.

532. Sokouti, B. A review on in silico virtual screening methods in COVID-19 using anticancer drugs and other natural/chemical inhibitors. *Explor Target Antitumor Ther* 2023, 4, 994–1026, doi:10.37349/etat.2023.00177.

533. Jang, W.D.; Jeon, S.; Kim, S.; Lee, S.Y. Drugs repurposed for COVID-19 by virtual screening of 6,218 drugs and cell-based assay. *Proc Natl Acad Sci U S A* 2021, 118, doi:10.1073/pnas.2024302118.

534. Luttens, A.; Gullberg, H.; Abdurakhmanov, E.; Vo, D.D.; Akaberi, D.; Talibov, V.O.; Nekhotiaeva, N.; Vangeel, L.; De Jonghe, S.; Jochmans, D., et al. Ultralarge virtual screening identifies SARS-CoV-2 main protease inhibitors with broad-spectrum activity against coronaviruses. *J Am Chem Soc* 2022, 144, 2905–2920, doi:10.1021/jacs.1c08402.

535. Chan, W.K.B.; Olson, K.M.; Wotring, J.W.; Sexton, J.Z.; Carlson, H.A.; Traynor, J.R. In silico analysis of SARS-CoV-2 proteins as targets for clinically available drugs. *Sci Rep* 2022, 12, 5320, doi:10.1038/s41598-022-08320-y.

536. Kokic, G.; Hillen, H.S.; Tegunov, D.; Dienemann, C.; Seitz, F.; Schmitzova, J.; Farnung, L.; Siewert, A.; Hobartner, C.; Cramer, P. Mechanism of SARS-CoV-2 polymerase stalling by remdesivir. *Nat Commun* 2021, 12, 279, doi:10.1038/s41467-020-20542-0.

537. Pitts, J.; Li, J.; Perry, J.K.; Du Pont, V.; Riola, N.; Rodriguez, L.; Lu, X.; Kurhade, C.; Xie, X.; Camus, G., et al. Remdesivir and GS-441524 retain antiviral activity against Delta, Omicron, and other emergent SARS-CoV-2 variants. *Antimicrob Agents Chemother* 2022, 66, e0022222, doi:10.1128/aac.00222-22.

538. Vangeel, L.; Chiu, W.; De Jonghe, S.; Maes, P.; Slechten, B.; Raymenants, J.; Andre, E.; Leyssen, P.; Neyts, J.; Jochmans, D. Remdesivir, molnupiravir and nirmatrelvir remain active against SARS-CoV-2 Omicron and other variants of concern. *Antiviral Res* 2022, 198, 105252, doi:10.1016/j.antiviral.2022.105252.

539. Dobrowolska, K.; Zarebska-Michaluk, D.; Brzdek, M.; Rzymski, P.; Rogalska, M.; Moniuszko-Malinowska, A.; Kozielewicz, D.; Hawro, M.; Rorat, M.; Sikorska, K., et al. Retrospective analysis of the effectiveness of remdesivir in COVID-19 treatment during periods dominated by Delta and Omicron SARS-CoV-2 variants in clinical settings. *J Clin Med* 2023, 12, doi:10.3390/jcm12062371.

540. Giannitsioti, E.; Mavroudis, P.; Speggos, I.; Katsoulidou, A.; Pantazis, N.; Loupis, T.; Daniil, I.; Rekleiti, N.; Damianidou, S.; Louka, C., et al. Real life treatment experience and outcome of consecutively hospitalised patients with SARS-CoV-2 pneumonia by Omicron-1 vs Delta variants. *Infect Dis (Lond)* 2023, 55, 706–715, doi:10.1080/23744235.2023.2232445.

541. Gandhi, S.; Klein, J.; Robertson, A.J.; Pena-Hernandez, M.A.; Lin, M.J.; Roychoudhury, P.; Lu, P.; Fournier, J.; Ferguson, D.; Mohamed Bakhash, S.A.K., et al. De novo emergence of a remdesivir resistance mutation during treatment of persistent SARS-CoV-2 infection in an immunocompromised patient: A case report. *Nat Commun* 2022, 13, 1547, doi:10.1038/s41467-022-29104-y.

542. Stevens, L.J.; Pruijssers, A.J.; Lee, H.W.; Gordon, C.J.; Tchesnokov, E.P.; Gribble, J.; George, A.S.; Hughes, T.M.; Lu, X.; Li, J., et al. Mutations in the SARS-CoV-2 RNA-dependent RNA polymerase confer resistance to remdesivir by distinct mechanisms. *Sci Transl Med* 2022, 14, eabo0718, doi:10.1126/scitranslmed.abo0718.

543. Takashita, E.; Yamayoshi, S.; Simon, V.; van Bakel, H.; Sordillo, E.M.; Pekosz, A.; Fukushi, S.; Suzuki, T.; Maeda, K.; Halfmann, P., et al. Efficacy of antibodies and antiviral drugs against Omicron BA.2.12.1, BA.4, and BA.5 subvariants. *N Engl J Med* 2022, 387, 468–470, doi:10.1056/NEJMc2207519.

544. Takashita, E.; Kinoshita, N.; Yamayoshi, S.; Sakai-Tagawa, Y.; Fujisaki, S.; Ito, M.; Iwatsuki-Horimoto, K.; Halfmann, P.; Watanabe, S.; Maeda, K., et al. Efficacy of antiviral agents against the SARS-CoV-2 Omicron subvariant BA.2. *N Engl J Med* 2022, 386, 1475–1477, doi:10.1056/NEJMc2201933.

545. Kalil, A.C.; Patterson, T.F.; Mehta, A.K.; Tomashek, K.M.; Wolfe, C.R.; Ghazaryan, V.; Marconi, V.C.; Ruiz-Palacios, G.M.; Hsieh, L.; Kline, S., et al. Baricitinib plus remdesivir for hospitalized adults with Covid-19. *N Engl J Med* 2021, 384, 795–807, doi:10.1056/NEJMoa2031994.

546. Akbarzadeh-Khiavi, M.; Torabi, M.; Rahbarnia, L.; Safary, A. Baricitinib combination therapy: A narrative review of repurposed Janus kinase inhibitor against severe SARS-CoV-2 infection. *Infection* 2022, 50, 295–308, doi:10.1007/s15010-021-01730-6.

547. Jayk Bernal, A.; Gomes da Silva, M.M.; Musungaie, D.B.; Kovalchuk, E.; Gonzalez, A.; Delos Reyes, V.; Martin-Quiros, A.; Caraco, Y.; Williams-Diaz, A.; Brown, M.L., et al. Molnupiravir for Oral treatment of Covid-19 in nonhospitalized patients. *N Engl J Med* 2022, 386, 509–520, doi:10.1056/NEJMoa2116044.

548. Singh, A.K.; Singh, A.; Singh, R.; Misra, A. Molnupiravir in COVID-19: A systematic review of literature. *Diabetes Metab Syndr* 2021, 15, 102329, doi:10.1016/j.dsx.2021.102329.

549. Butler, C.C.; Hobbs, F.D.R.; Gbinigie, O.A.; Rahman, N.M.; Hayward, G.; Richards, D.B.; Dorward, J.; Lowe, D.M.; Standing, J.F.; Breuer, J., et al. Molnupiravir plus usual care versus usual care alone as early treatment for adults with COVID-19 at increased risk of adverse outcomes (PANORAMIC): An open-label, platform-adaptive randomised controlled trial. *Lancet* 2023, 401, 281–293, doi:10.1016/s0140-6736(22)02597-1.

550. Fischer, W.A.; Eron, J.J.; Holman, W.; Cohen, M.S.; Fang, L.; Szewczyk, L.J.; Sheahan, T.P.; Baric, R.; Mollan, K.R.; Wolfe, C.R., et al. A phase 2a clinical trial of molnupiravir in patients with COVID-19 shows accelerated SARS-CoV-2 RNA clearance and elimination of infectious virus. *Sci Transl Med* 2022, 14, eabl7430, doi:10.1126/scitranslmed.abl7430.

551. Sanderson, T.; Hisner, R.; Donovan-Banfield, I.; Hartman, H.; Lochen, A.; Peacock, T.P.; Ruis, C. A molnupiravir-associated mutational signature in global SARS-CoV-2 genomes. *Nature* 2023, 623, 594–600, doi:10.1038/s41586-023-06649-6.

552. Akinosoglou, K.; Schinas, G.; Gogos, C. Oral Antiviral treatment for COVID-19: A comprehensive review on Nirmatrelvir/Ritonavir. *Viruses* 2022, 14, doi:10.3390/v14112540.

553. Wong, C.K.H.; Au, I.C.H.; Lau, K.T.K.; Lau, E.H.Y.; Cowling, B.J.; Leung, G.M. Real-world effectiveness of molnupiravir and nirmatrelvir plus ritonavir against mortality, hospitalisation, and in-hospital outcomes among community-dwelling, ambulatory patients with confirmed SARS-CoV-2 infection during the omicron wave in Hong Kong: An observational study. *Lancet* 2022, 400, 1213–1222, doi:10.1016/S0140-6736(22)01586-0.

554. Hammond, J.; Leister-Tebbe, H.; Gardner, A.; Abreu, P.; Bao, W.; Wisemandle, W.; Baniecki, M.; Hendrick, V.M.; Damle, B.; Simon-Campos, A., et al. Oral Nirmatrelvir for high-risk, nonhospitalized adults with Covid-19. *N Engl J Med* 2022, 386, 1397–1408, doi:10.1056/NEJMoa2118542.

555. Anderson, A.S.; Caubel, P.; Rusnak, J.M.; Investigators, E.-H.T. Nirmatrelvir-ritonavir and viral load rebound in Covid-19. *N Engl J Med* 2022, 387, 1047–1049, doi:10.1056/NEJMc2205944.

556. Edelstein, G.E.; Boucau, J.; Uddin, R.; Marino, C.; Liew, M.Y.; Barry, M.; Choudhary, M.C.; Gilbert, R.F.; Reynolds, Z.; Li, Y., et al. SARS-CoV-2 virologic rebound with nirmatrelvir-ritonavir therapy: An observational study. *Ann Intern Med* 2023, 176, 1577–1585, doi:10.7326/M23-1756.

557. Charness, M.E.; Gupta, K.; Stack, G.; Strymish, J.; Adams, E.; Lindy, D.C.; Mohri, H.; Ho, D.D. Rebound of SARS-CoV-2 infection after nirmatrelvir-ritonavir treatment. *N Engl J Med* 2022, 387, 1045–1047, doi:10.1056/NEJMc2206449.

558. Harrington, P.R.; Cong, J.; Troy, S.B.; Rawson, J.M.O.; O'Rear, J.J.; Valappil, T.I.; McGarry Connelly, S.; Farley, J.; Birnkrant, D. Evaluation of SARS-CoV-2 RNA rebound after nirmatrelvir/ritonavir treatment in randomized, double-blind, placebo-controlled trials - United States and international sites, 2021–2022. *MMWR Morb Mortal Wkly Rep* 2023, 72, 1365–1370, doi:10.15585/mmwr.mm 7251a2.

559. Seif, F.; Aazami, H.; Khoshmirsafa, M.; Kamali, M.; Mohsenzadegan, M.; Pornour, M.; Mansouri, D. JAK inhibition as a new treatment strategy for patients with COVID-19. *Int Arch Allergy Immunol* 2020, 181, 467–475, https://doi.org/10.1159/000508247.

560. Stebbing, J.; Sanchez Nievas, G.; Falcone, M.; Youhanna, S.; Richardson, P.; Ottaviani, S.; Shen, J.X.; Sommerauer, C.; Tiseo, G.; Ghiadoni, L., et al. JAK inhibition reduces SARS-CoV-2 liver infectivity and modulates inflammatory responses to reduce morbidity and mortality. *Sci Adv* 2021, 7, doi:10.1126/ sciadv.abe4724.

561. Bronte, V.; Ugel, S.; Tinazzi, E.; Vella, A.; De Sanctis, F.; Cane, S.; Batani, V.; Trovato, R.; Fiore, A.; Petrova, V., et al. Baricitinib restrains the immune dysregulation in patients with severe COVID-19. *J Clin Invest* 2020, 130, 6409–6416, doi:10.1172/JCI141772.

562. Marconi, V.C.; Ramanan, A.V.; de Bono, S.; Kartman, C.E.; Krishnan, V.; Liao, R.; Piruzeli, M.L.B.; Goldman, J.D.; Alatorre-Alexander, J.; de Cassia Pellegrini, R., et al. Efficacy and safety of baricitinib for the treatment of hospitalised adults with COVID-19 (COV-BARRIER): A randomised, double-blind, parallel-group, placebo-controlled phase 3 trial. *Lancet Respir Med* 2021, 9, 1407–1418, doi:10.1016/ S2213-2600(21)00331-3.

563. Cantini, F.; Niccoli, L.; Nannini, C.; Matarrese, D.; Natale, M.E.D.; Lotti, P.; Aquilini, D.; Landini, G.; Cimolato, B.; Pietro, M.A.D., et al. Beneficial impact of baricitinib in COVID-19 moderate pneumonia; multicentre study. *J Infect* 2020, 81, 647–679, doi:10.1016/j.jinf.2020.06.052.

564. Horby, P.; Lim, W.S.; Emberson, J.R.; Mafham, M.; Bell, J.L.; Linsell, L.; Staplin, N.; Brightling, C.; Ustianowski, A.; Elmahi, E., et al. Dexamethasone in hospitalized patients with Covid-19. *N Engl J Med* 2021, 384, 693–704, doi:10.1056/NEJMoa2021436.

565. Bhimraj A, M.R.; Shumaker, A.H.; Baden, L.; Cheng, V.C.; Edwards, K.M.; Gallagher, J.C.; Gandhi, R.T.; Muller, W.J.; Nakamura, M.M.; O'Horo, J.C.; Shafer, R.W.; Shoham, S.; Murad, M.H.; Mustafa, R.A.; Sultan, S.; Falck-Ytter, Y. Infectious Diseases Society of America Guidelines on the Treatment and Management of Patients with COVID-19. Available online: https://www.idsociety.org/practice-guideline/covid-19-guideline-treatment-and-management/ (accessed on 05 December 2023).

566. NIH. Systemic Corticosteroids. Available online: https://www.covid19treatmentguidelines.nih.gov/ therapies/immunomodulators/systemic-corticosteroids/ (accessed on 05 December 2023).

567. WHO. Coronavirus disease (COVID-19): Corticosteroids, including dexamethasone. Available online: https://www.who.int/news-room/questions-and-answers/item/coronavirus-disease-covid-19-dexameth-asone (accessed on 05 December 2023).

568. Draghici, S.; Nguyen, T.M.; Sonna, L.A.; Ziraldo, C.; Vanciu, R.; Fadel, R.; Morrison, A.; Kenney, R.M.; Alangaden, G.; Ramesh, M., et al. COVID-19: Disease pathways and gene expression changes predict methylprednisolone can improve outcome in severe cases. *Bioinformatics* 2021, 37, 2691–2698, doi:10.1093/bioinformatics/btab163.

569. Ko, J.J.; Wu, C.; Mehta, N.; Wald-Dickler, N.; Yang, W.; Qiao, R. A comparison of methylprednisolone and dexamethasone in intensive care patients with COVID-19. *J Intensive Care Med* 2021, 36, 673–680, doi:10.1177/0885066621994057.

570. Barkas, F.; Filippas-Ntekouan, S.; Kosmidou, M.; Liberopoulos, E.; Liontos, A.; Milionis, H. Anakinra in hospitalized non-intubated patients with coronavirus disease 2019: A systematic review and meta-analysis. *Rheumatology (Oxford)* 2021, 60, 5527–5537, doi:10.1093/rheumatology/keab447.

571. Kyriazopoulou, E.; Poulakou, G.; Milionis, H.; Metallidis, S.; Adamis, G.; Tsiakos, K.; Fragkou, A.; Rapti, A.; Damoulari, C.; Fantoni, M., et al. Early treatment of COVID-19 with anakinra guided by soluble urokinase plasminogen receptor plasma levels: A double-blind, randomized controlled phase 3 trial. *Nat Med* 2021, 27, 1752–1760, doi:10.1038/s41591-021-01499-z.

572. Figuero-Pérez, L.; Olivares-Hernández, A.; Escala-Cornejo, R.A.; Terán-Brage, E.; López-Gutiérrez, Á.; Cruz-Hernández, J.J. Anakinra as a potential alternative in the treatment of severe acute respiratory infection associated with SARS-CoV-2 refractory to tocilizumab. *Reumatología Clínica (English Edition)* 2021, 17, 559–561, https://doi.org/10.1016/j.reumae.2020.06.008.

573. Caly, L.; Druce, J.D.; Catton, M.G.; Jans, D.A.; Wagstaff, K.M. The FDA-approved drug ivermectin inhibits the replication of SARS-CoV-2 in vitro. *Antiviral Res* 2020, 178, 104787, doi:10.1016/j.antiviral.2020.104787.

574. Kinobe, R.T.; Owens, L. A systematic review of experimental evidence for antiviral effects of ivermectin and an in silico analysis of ivermectin's possible mode of action against SARS-CoV-2. *Fundam Clin Pharmacol* 2021, 35, 260–276, doi:10.1111/fcp.12644.

575. Bray, M.; Rayner, C.; Noel, F.; Jans, D.; Wagstaff, K. Ivermectin and COVID-19: A report in antiviral research, widespread interest, an FDA warning, two letters to the editor and the authors' responses. *Antiviral Res* 2020, 178, 104805, doi:10.1016/j.antiviral.2020.104805.

576. López-Medina, E.; López, P.; Hurtado, I.C.; Dávalos, D.M.; Ramirez, O.; Martínez, E.; Díazgranados, J.A.; Oñate, J.M.; Chavarriaga, H.; Herrera, S., et al. Effect of ivermectin on time to resolution of symptoms among adults with mild COVID-19: A randomized clinical trial. *JAMA* 2021, 325, 1426–1435, doi:10.1001/jama.2021.3071.

577. Reis, G.; Silva, E.A.S.M.; Silva, D.C.M.; Thabane, L.; Milagres, A.C.; Ferreira, T.S.; dos Santos, C.V.Q.; Campos, V.H.S.; Nogueira, A.M.R.; de Almeida, A.P.F.G., et al. Effect of early treatment with ivermectin among patients with Covid-19. *N Engl J Med* 2022, 386, 1721–1731, doi:10.1056/NEJMoa2115869.

578. Popp, M.; Reis, S.; Schiesser, S.; Hausinger, R.I.; Stegemann, M.; Metzendorf, M.I.; Kranke, P.; Meybohm, P.; Skoetz, N.; Weibel, S. Ivermectin for preventing and treating COVID-19. *Cochrane Database Syst Rev* 2022, 6, CD015017, doi:10.1002/14651858.CD015017.pub3.

579. Buonfrate, D.; Chesini, F.; Martini, D.; Roncaglioni, M.C.; Ojeda Fernandez, M.L.; Alvisi, M.F.; De Simone, I.; Rulli, E.; Nobili, A.; Casalini, G., et al. High-dose ivermectin for early treatment of COVID-19 (COVER study): A randomised, double-blind, multicentre, phase II, dose-finding, proof-of-concept clinical trial. *Int J Antimicrob Agents* 2022, 59, 106516, doi:10.1016/j.ijantimicag.2021.106516.

580. Naggie, S.; Boulware, D.R.; Lindsell, C.J.; Stewart, T.G.; Slandzicki, A.J.; Lim, S.C.; Cohen, J.; Kavtaradze, D.; Amon, A.P.; Gabriel, A., et al. Effect of higher-dose ivermectin for 6 days vs placebo on time to sustained recovery in outpatients with COVID-19: A randomized clinical trial. *JAMA* 2023, 329, 888–897, doi:10.1001/jama.2023.1650.

581. Schilling, W.H.K.; Jittamala, P.; Watson, J.A.; Ekkapongpisit, M.; Siripoon, T.; Ngamprasertchai, T.; Luvira, V.; Pongwilai, S.; Cruz, C.; Callery, J.J., et al. Pharmacometrics of high-dose ivermectin in early COVID-19 from an open label, randomized, controlled adaptive platform trial (PLATCOV). *Elife* 2023, 12, doi:10.7554/eLife.83201.

582. Manomaipiboon, A.; Pholtawornkulchai, K.; Poopipatpab, S.; Suraamornkul, S.; Maneerit, J.; Ruksakul, W.; Phumisantiphong, U.; Trakarnvanich, T. Efficacy and safety of ivermectin in the treatment of mild to moderate COVID-19 infection: A randomized, double-blind, placebo-controlled trial. *Trials* 2022, 23, 714, doi:10.1186/s13063-022-06649-3.

583. Ou, T.; Mou, H.; Zhang, L.; Ojha, A.; Choe, H.; Farzan, M. Hydroxychloroquine-mediated inhibition of SARS-CoV-2 entry is attenuated by TMPRSS2. *PloS Pathog* 2021, 17, e1009212, doi:10.1371/journal.ppat.1009212.

584. Abella, B.S.; Jolkovsky, E.L.; Biney, B.T.; Uspal, J.E.; Hyman, M.C.; Frank, I.; Hensley, S.E.; Gill, S.; Vogl, D.T.; Maillard, I., et al. Efficacy and safety of hydroxychloroquine vs placebo for pre-exposure SARS-CoV-2 prophylaxis among health care workers: A randomized clinical trial. *JAMA Intern Med* 2021, 181, 195–202, doi:10.1001/jamainternmed.2020.6319.

585. Kumar, J.; Jain, S.; Meena, J.; Yadav, A. Efficacy and safety of hydroxychloroquine/chloroquine against SARS-CoV-2 infection: A systematic review and meta-analysis. *J Infect Chemother* 2021, 27, 882–889, doi:10.1016/j.jiac.2021.02.021.

586. Tanni, S.E.; Bacha, H.A.; Naime, A.; Bernardo, W.M. Use of hydroxychloroquine to prevent SARS-CoV-2 infection and treat mild COVID-19: A systematic review and meta-analysis. *J Bras Pneumol* 2021, 47, e20210236, doi:10.36416/1806-3756/e20210236.

587. Jung, S.Y.; Kim, M.S.; Kim, M.C.; Choi, S.H.; Chung, J.W.; Choi, S.T. Effect of hydroxychloroquine pre-exposure on infection with SARS-CoV-2 in rheumatic disease patients: A population-based cohort study. *Clin Microbiol Infect* 2021, 27, 611–617, doi:10.1016/j.cmi.2020.12.003.

588. WHO Solidarity Trial Consortium. Remdesivir and three other drugs for hospitalised patients with COVID-19: Final results of the WHO solidarity randomised trial and updated meta-analyses. *Lancet* 2022, 399, 1941–1953, doi:10.1016/S0140-6736(22)00519-0.

589. Hoffmann, M.; Hofmann-Winkler, H.; Smith, J.C.; Kruger, N.; Arora, P.; Sorensen, L.K.; Sogaard, O.S.; Hasselstrom, J.B.; Winkler, M.; Hempel, T., et al. Camostat mesylate inhibits SARS-CoV-2 activation by TMPRSS2-related proteases and its metabolite GBPA exerts antiviral activity. *EBioMedicine* 2021, 65, 103255, doi:10.1016/j.ebiom.2021.103255.

590. Hofmann-Winkler, H.; Moerer, O.; Alt-Epping, S.; Brauer, A.; Buttner, B.; Muller, M.; Fricke, T.; Grundmann, J.; Harnisch, L.O.; Heise, D., et al. Camostat mesylate May reduce severity of coronavirus disease 2019 sepsis: A first observation. *Crit Care Explor* 2020, 2, e0284, doi:10.1097/CCE.0000000000000284.

591. Sakr, Y.; Bensasi, H.; Taha, A.; Bauer, M.; Ismail, K.; the UAE-Jena Research Group. Camostat mesylate therapy in critically ill patients with COVID-19 pneumonia. *Intensive Care Med* 2021, 47, 707–709, doi:10.1007/s00134-021-06395-1.

592. Kim, Y.S.; Jeon, S.H.; Kim, J.; Koh, J.H.; Ra, S.W.; Kim, J.W.; Kim, Y.; Kim, C.K.; Shin, Y.C.; Kang, B.D., et al. A double-blind, randomized, placebo-controlled, phase II clinical study to evaluate the efficacy and safety of camostat mesylate (DWJ1248) in adult patients with mild to moderate COVID-19. *Antimicrob Agents Chemother* 2023, 67, e0045222, doi:10.1128/aac.00452-22.

593. Tobback, E.; Degroote, S.; Buysse, S.; Delesie, L.; Van Dooren, L.; Vanherrewege, S.; Barbezange, C.; Hutse, V.; Romano, M.; Thomas, I., et al. Efficacy and safety of camostat mesylate in early COVID-19 disease in an ambulatory setting: A randomized placebo-controlled phase II trial. *Int J Infect Dis* 2022, 122, 628–635, doi:10.1016/j.ijid.2022.06.054.

2 Exploring Natural Antivirals

The Potential Role of Medicinal Plants in COVID-19 Prevention and Treatment

Biswajit Pramanik, Divya Raj Bharti,
Ch. V. S. S. S. Bharathi, and Sandip Debnath

2.1 INTRODUCTION

2.1.1 THE GLOBAL IMPACT OF COVID-19

Severe acute respiratory syndrome coronavirus 2 (SARS-CoV-2) or nCoV-19, a coronavirus upsurge, was first detected in the city named Wuhan, the capital of the People's Republic of China in December, 2019. This was a fatal endemic virus that caused Middle East respiratory syndrome (MERS). This virus was discovered in individuals with pneumonia in January 2020, and in February 2020, the disease was renamed as Coronavirus Disease 2019 (COVID-19) (Pramanik et al. 2023). As of September 2020, the World Health Organization (WHO) had recorded more than 969,585 deaths and about 31,499,146 confirmed cases of COVID-19 worldwide (Adetunji et al. 2021). SARS-CoV-2 has the largest single-stranded RNA and the genetic material belonging to the order Nidovirales, which is enlisted under the family *Coronaviridae*, according to the binomial classification (Khan et al. 2020). Angiotensin converting enzyme 2 (ACE-2) is the principal receptor protein of SARS-CoV-2, enabling its binding to host cells through the spike (S) proteins. This binding eventually results in the rearranging shapes of the S2 subunit in the C-terminal of the S proteins, forming the intracellular gateway for the virus (Jiang et al. 2020).

For this virus, no therapy was authorized to be an effective one. To address this condition, researchers and physicians used alternative therapy modalities. Herbs have been utilized to treat a variety of disorders, including infections, since ancient times. Currently, efforts are underway to create vaccinations. However, some medications are still under development, while others are now undergoing various stages of clinical research. These medications' therapeutic effectiveness is debatable.

Because herbal therapy has been successful in treating other viral illnesses, research into it is now underway. SARS-CoV-2 treatment may benefit from phyto-sourced compounds that specifically block the targeted ACE-2 receptor, altering the necessary restriction in the activity of the enzyme, which significantly reduces viral entrance into host cells. There are several medicinal plants which possess such potent antiviral activities, among which *Onopordum acanthium*, *Crataegus laevigata, Quercus infectoria*, and *Berberis integerrima* Bunge are the most well known. These herbs have been shown to suppress ACE. This results from the substantial amount of phenolic chemicals, which have antioxidant effects. Oxidative stress and viral infection co-occur, which promotes viral replication. Additionally, it was shown that some plant species from several families like *Oleaceae, Magnoliaceae, Lauraceae, Labiatae, Nelumbonaceae, Polygonaceae*, etc. impede the event of binding of the S protein and ACE, which was eventually attributed to higher concentration of emodin.

DOI: 10.1201/9781003452621-2

The transmembrane serine-proteinase serine type 2 (TMPRSS2) is another target of certain phytochemicals, according to newly available knowledge. According to findings of Matsuyama et al. (2020), nCoV-19 also employs TMPRSS2 for clearing the pathway of entrance into a cell. As a result, it is thought that natural remedies that can inhibit TMPRSS2 expression are useful in treating SARS-CoV-2. Kaempferol, for example, dramatically depleted the expression of TMPRSS2. Additionally, a combination of luteolin, quercetin, and kaempferol has been shown in another investigation to show the ability to also inhibit TMPRSS 2. The SARS-CoV-2 genome codes for a protein called papain-like proteinase (PLPro). It is essential for the reproduction of viruses. Additionally, it reportedly interferes with the host's natural defenses.

2.1.2 SHORTCOMINGS OF SYNTHETIC PHARMACEUTICALS

Antiviral medications, chloroquine/hydroxychloroquine, dexamethasone, and convalescent plasma transfusion (CPT) are a few promising therapeutic agents for COVID-19 treatment; however, the majority of them still provide erratic outcomes. As a COVID-19 therapy, many antiviral medications are being studied. One of the significant synthetic antiviral medications utilized for COVID-19 is remdesivir, which is an adenosine analogue that combines with viral RNA chains to cause an early termination (Wang et al. 2020a). The huge death toll from SARS-CoV-2 throughout the globe has contributed to the virus's appearance as a worldwide pandemic. Since the virus possesses the ability to adapt to varied environments and is spreading quickly, immediate action is required to reverse the situation. A large population who contract the infection will require the already depleted healthcare system. Therefore, the quick discovery of natural therapeutic agents may offer an alternative to the synthetic medications now used to treat COVID-19 infections.

Dexamethasone and remdesivir, two common treatments for the new COVID-19 pandemic, were found by Vetter et al. (2020) to be exceedingly weak and ineffective. Consequently, the medications against the infection of COVID-19 still becomes a major issue to be solved. According to them, different herbal remedies derived from numerous decoctures obtained from several plants like *Artesimia* spp. are abundant in bioactive ingredients and have significant biological activity, which eventually make them suitable for use in medication formulations. Further, these methods regarding the conventional procedures of remediation were generally practised in low-income countries. Also, the determination of toxicity indices, safety concerns, and efficacy of those plant decoctions still require rigorous investigation.

Remdesivir may have a better benefit–risk profile than a placebo in cases of severe infection of this disease. Although it possessed a promising impact in numerous *in vitro* experiments, significant practical exposure is further required. Remdesivir was actually the subject of a multicenter, randomized, double-blind, placebo-controlled clinical study which was terminated early due to several serious side effects in both the treatment and control groups. Additionally, the research highlighted zero difference in the time of occurrence of outcome exposed to clinical improvements. Further research revealed that remdesivir did not significantly lower the patients' mortality risk (8% vs. 11.6% in the placebo group), and the National Institute of Allergy and Infectious Diseases (NIAID) terminated the experiment due to the trial's poor benefit outcomes and ethical problems (Wang et al. 2020b).

Another systematic study of the deployment of lopinavir/ritonavir for the treatment of COVID-19 was also carried out, where a fixed-dose of the antiviral combination was administered against HIV infection. HIV protease was inhibited by this antiviral drug, which created an inhibitor enzyme complex, though Gag-pol polyproteins may not be cleaved by this complex. Notably, the substances demonstrated antiviral efficacy against both SARS-CoV and MERS-CoV infections in *in vitro* research. In clinical research, the administration of lopinavir/ritonavir reduced the need of steroids, mechanical ventilations, the rate of intubations, and patient mortality rate in SARS cases. In a randomized controlled experiment carried out by several other researchers, it was demonstrated that none of the artificial medications could significantly lower viral clearance in both mild and severe

COVID-19 patients. These studies also revealed that taking both drugs together might result in certain unfavorable side effects, typically in the digestive system (Cao et al. 2020).

Chloroquine and hydroxychloroquine are two more drug-administered therapies which were researched, to date. The former can raise the endosomal pH necessary for the virus to bind to the cell membrane of the host, which in *in vitro* condition was able to inhibit viral infection. The combination of hydroxychloroquine and azithromycin is effective in eradicating nasopharyngeal viral carriage of SARS-CoV-2 in the infected patients in 3–6 days, according to Gautret et al. (2020) in an open-label non-randomized clinical experiment. However, another comprehensive review suggested that there haven't been any high-quality, randomized controlled trials that can demonstrate the efficacy as well as safety of chloroquine or hydroxychloroquine. Additionally, the combination of macrolide, chloroquine, and hydroxychloroquine had an adverse side effect during therapy and might not minimize the risk of COVID-19 infection. The use of corticosteroids may lessen the pro-inflammatory cytokines implicated since severe COVID-19 instances are frequently accompanied by cytokine release syndrome. In recovery trial research on the use of corticosteroids for COVID-19, dexamethasone significantly decreased the risk of mortality in severe patients on a ventilator or moderate cases receiving supplementary oxygen treatment. Also, ingestion of dexamethasone had no impact in mild to moderate patients not requiring oxygen treatment, and only benefited severe instances. Nevertheless, a multifaceted investigation is required to establish dependability. In their systematic review of CPT for the treatment of COVID-19, Rajendran et al. (2020) made the case that CPT, in addition to antiviral/antimicrobial medications, could be a useful therapeutic option with promising evidence regarding safety, improvement of clinical symptoms, and decreased mortality. Additionally, CPT considerably raises antibody levels over time and lowers viral load. All study participants tolerated CPT without any issues. However, only a few case reports and series were used to collect this data. No high-quality randomized controlled study has been conducted to determine the effectiveness and safety of the implementation of these synthetic antiviral drugs.

2.1.3 The Quest for Natural Alternatives

It is anticipated that natural items will outperform synthetic medications made to suppress PLPro. These naturally occurring substances include Terrestriamide, Cinnamic amides (N-trans-Feruloyloctopamine), N-trans-Feruloyltyramine, N-trans-Caffeoyltyramine, and Terrestrimine, which are all isolated from the fruits of *Tribulus terrestris* L. Additionally, the flavonoid-rich plant *Cullen corylifolium* (L) Medik was reported to be involved in the inhibition process of SARS-CoV-2, and *Paulownia tomentosa* (Thunb) Steud-sourced flavonoids also demonstrated suppressive effects on the virus (Adetunji et al. 2021).

It has been shown that several medicinal herbs can inhibit the 3-chymotrypsin-like protease (3CL PRO). Since 3CL PRO is involved in the replication process of SARS-CoV-2, anti-COVID-19 therapies are likely to target it. *Angelica keiskei* (Miq) Koidz-derived coumarins and alkylated chalcones are two prime examples of these natural substances. Additionally, phlorotannins isolated from *Ecklonia cava* significantly inhibited 3CL PRO in the same virus. Considering these, flavonoids obtained from various sources come out to be very promising. In this chapter, the pathophysiology of SARS-CoV-2 coupled with different herbs-derived antiviral effects will be discussed in detail.

2.2 THE NEED FOR NATURAL PHARMACEUTICALS

2.2.1 Medicinal Plants: An Untapped Resource

A few nations with prolonged experience in the field of traditional medicine invention as well as usage, such as China and India, began investigating traditional therapies as a complementary addition to the conventional treatment of COVID-19 as the entire global medical profession was looking for a potent vaccine or drug (Rastogi et al. 2022). The Malaysian population took great interest while experimenting with the implementation of complementary herbal medicines based on their regional traditional knowledge, because of its status as a tropical multiracial nation rich in flora and fauna. The biomedical research institute in Malaysia's herbal medicine research division fielded many inquiries regarding the efficient application of complementary remedies, including single medicinal plants, conventional treatments, processed and supplementary herbal products, dietary products, and numerous medical gadgets against COVID-19, during the Movement Control Order that the Government of Malaysia implemented in March 2020 in an effort to slow the widely spreading disease.

China-originated medicinal plants were found to be much more effective against an array of infections in the respiratory system because of the presence of numerous bioactive molecules fulfilled with potent biological activities. Therefore, these Chinese medicinal herbs might be of great use in the treatment of nCoV-19. To assess the effectiveness of these herbs against immunological responses and inflammation, similar to those caused by coronaviruses, pharmacological analysis and molecular docking were used.

2.2.2 HISTORICAL AND CONTEMPORARY USE OF ANTIVIRAL HERBS IN COVID-19 MANAGEMENT

Coronavirus 2 was reportedly causing pneumonia and producing severe acute respiratory syndrome (SARS). Currently, no vaccination or antivirus medication is available to treat the symptoms other than in supportive care. Although the biological characteristics and mode of action of the active ingredients were not clearly defined, several authors observed that traditional China-originated medicine is being used to cure many SARS CoV-2 patients in China (Yang et al. 2020). The absence of effective antiviral vaccinations or medications to combat COVID-19 has resulted in major devastation throughout the globe. Hence, the search for a strong medicinal plant with antiviral and immune-stimulating components would help patients with viral infections through therapeutic assistance. Dexamethasone and hydroxychloroquine, two common synthetic medications for COVID-19 infections, have demonstrated a high prevalence of adverse effects, especially in individuals with compromised immune systems. Since many of these medications have demonstrated egregious inefficiency and a variety of side effects, much focus is now being placed on discovering the substitutions possessing pure biomolecules with antiviral capabilities. The antiviral capabilities are mainly caused by several polyphenols (amentoflavone, hesperetin, papyriflavonol A, luteolin), anthraquinones (aloe-emodin, emodin), terpenes (curcumin, betulinic acid), alkaloids (lycorine, jubanine H, 7-methoxycryptopleurine), etc. These biochemical compounds are generally deployed in the developmental process of these candidate drugs. Again, using additional allopathic immune booster medications may be expensive; instead, the body will receive rich elements from medicinal plants and Ayurvedic products altering the adverse side effects.

As per the research of Boukhatem and Setzer (2020), the treatment options available against COVID-19 were time-consuming when the duration of development was considered. These included various practices involving small-molecule medications, interferon therapies, and vaccines. Additionally, there is a risk of viral resistance developing, as well as adverse side effects, including dormancy as well as re-emergence of the viral components in response to currently practiced diverse synthetic drugs. Consequently, therapeutic plants comprising various phytochemical extracts like essential oils, might be considered as rich sources of bioactive molecules to serve as potent antiviral substances. The researchers mainly concentrated on the bio-physiological processes adapted by the plants against the entire process of viral establishment from entry to discharge through assembly, and replication upon exposure to the antiviral agents. They also focussed on the mode of action of the substances as well as modulations in the receptor specified to the viral components.

There were no recognized antiviral medications that are effective against COVID-19 infection, until the discovery of a vaccine. However, several traditional medicines have positive effects against COVID-19. Though the exact processes are yet unknown, propagatory progress of this virus can be checked by using these medicinal herbs of Indian, Chinese, African, and Iranian origin. Therefore, more research will be required to determine the toxicity, dose, biological activity, and concentrations of these therapeutic plants. A few medicinal plants native to Nigeria were carefully chosen by the authors for their potential pharmacological efficacy against SARS-CoV-2. In severe cases, this disease can enhance the permeability of entire pulmonary system through alveolar damage, inhibition of the activities of ACE-2, and trigger to the c-Jun N-terminal kinase pathway, which eventually leads to the failure of multiple organs. On the other hand, Chinese herbal remedies might be taken into consideration for potential treatment and therapeutic studies against COVID-19.

According to the reports of Nadeem (2020), COVID-19-related pathophysiological diseases have a significant death rate. *Alangium salviifolium*-sourced bioactive substances were potent enough to prevent the entry of virus into the host cells and, thus, can be utilized as antiviral medications, as reported by the aforesaid scientists. Molecular simulation and docking were carried out with the aid of bioinformatic tools. Based on their research, it was shown that the plant's bioactive substance had a high affinity for the trimeric S protein of the virus, which eventually modulated the prohibitory effects against the S proteins.

Plants that contain protease inhibitors can be used as significant antiviral agents in the procurement of COVID-19. Numerous bioinformatical technologies were employed to study the impeding actions of the secondary metabolites produced from several medicinal plants against the virus. Following molecular docking, curcumin was found to be a potent protease inhibitor.

A majority of the China-originated medicinal herb preparations were studied for their potent antiviral activity against nCOV-19. Strong antiviral agents have been shown to be present in *Lycoris radiata* and its active component, lycorine. Traditional herbal remedies have been suggested as a potential therapeutic answer to the current pandemic after research by Benarba and Pandiella (2020) showed that the artificial medications currently used for supplementary treatment for COVID-19 might not provide the required reaction upon viral infection. According to the scientists, these substances may have an inhibitory impact on viral entrance or growth. Among all the naturally abundant herbal remedies, some have been detected to inhibit the serine proteases used by the virus to enter cells, including TMPRRS2, ACE-2 receptor, papain- and chymotrypsin-like proteases.

According to Ikpa et al. (2020), the recent coronavirus epidemic necessitates the rapid development of therapeutic drugs to treat SARS. The researchers hypothesized that regional medicinal herbs *i.e. Allium cepa, Allium sativum, Aloe vera, Zingiber officinale, Garcinia kola, Olea europaea, Euphorbia hirta*, etc. would provide helpful strategies for limiting the COVID-19 epidemic. Bioactive components from these plants can suppress viral replication, indicating prospective therapeutic candidates against the virus.

Sagar and Kumar (2020) suggested that *Tinospora cordifolia* inhibited viral attachment and replication, making it an effective antiviral agent against SARS-CoV-2. With the implementation of *in silico* methods, the scientists examined the plants' resistance to the surface glycoproteins, receptors of the virus, RNA-dependent RNA polymerase, and viral protease. Furthermore, they found that the plant's active biomolecules, including berberine, magnoflorine, tinocordiside, and isocolumbin, had excellent binding capacity, establishing the plant as a strong contender in the therapeutic process of COVID-19 infection.

Using the AutoDock 4.2 tool, a molecular docking system targeting the protein *in silico*, Laksmiani et al. (2020) conducted research with the goal of identifying several naturally occurring active compounds with inhibitory action against COVID-19. Several plants like *Caesalpinia sappan* L., *Citrus* sp., etc., were assessed for targeting proteins integrated within a number of active substances, which included hesperetin, andrographolide, curcumin, luteolin, and quercetin.

The findings of Shawky et al. (2020) suggested that essential proteins as well as enzymes could be targeted well by several significant constituents obtained from medicinal plants viz. *Chrysanthemum coronarium, Epilobium hirsutum, Psidium guajava, Glycyrrhiza glabra, Euphorbia* sp. and many more. These comprised cyanidin 3-(600-malonylglucoside), caftaric acid, rutin, luteolin 7-rutinoside, caffeic acid, etc. The scientists studied the modulatory features of several pathogenic viral components and enzymes such as RNA-dependent RNA polymerase (RdRp), 3CL PRO, and PLPro using molecular docking. Reports revealed that these plants with an array of powerful biomolecules, including isorhamnetin and quercetin, could regulate the mitigation pathways against the viral infection through inducing various immunological and inflammatory response.

According to Mani et al. (2020), the evolving nature of COVID-19 has drawn attention of using conventional Indian medicinal herbs to stop the transmission of the virus. Many active macromolecules found in many types of medicinal plants have been shown to have the potential to stop pathogenic illnesses like COVID-19. The proposal of applying the medicinal constituents in clinical items like soaps, masks, and sanitizers to help fight viral infections was one of the most promising aspects suggested by the authors. These could, therefore, advance the evolution of sustainable natural medicine and vaccine development, coupled with overall economic development. Furthermore, plenty of research is still ongoing for determining the physiological properties of numerous Indian medicinal herbs to provide potential COVID-19 control candidates.

According to Chen et al. (2020), COVID-19's physiological manifestations were not only confined within severe viral pneumonia type of symptoms, but also spread towards a few numbers of asymptomatic infections, which eventually led to mortality, in severe cases. They prioritized the vital role of Chinese medicine in the treatment of many different ailments, hence the effectiveness of this treatment was assessed against COVID-19. According to their findings, Chinese medicinal herbs include a significant number of bioactive components that have strong antiviral activity against COVID-19.

According to Shahrajabian et al. (2020), China-originated herbs are mediated immunity boosters that have provided significant advantages for the procurement of coronaviruses by both preventing the entry of the virus as well as regulating crucial enzymes. The majority of these herbs, including dandelion, *Artemisia annua, Lycoris radiate, Isatis indigotica, Torreya nucifera*, etc., are used to prepare herbal teas. The scientists hypothesized that combining synthetic drugs with conventional Chinese drugs would lead to better outcomes, but active ingredients need to be discovered in order to understand the biological aspects in the mode of action coupled with particular drug interactions. Coronavirus infections are on the rise all over the world, and many scientists are looking for an alternate remedy, instead of the previously found effective therapies or vaccines, from the rich variety of regular African medications. The effectiveness of phytomedicines against COVID-19 requires more research to determine and establish the mechanisms of action. Some of the proposed methods of action include blocking cell membrane-derived vesicles for transportation, reducing endosome acidity and preventing the generation of pro-inflammatory cytokines. As concluding remarks, the authors advised several precautions against the non-systematic use and ingestion of the extracts of medicinal herbs under the pretext of treating COVID-19.

Numerous techniques have been proposed for vaccine and medication formulations or designs in response to Pandey et al.'s (2020) demonstration that public health disasters brought on by infectious illnesses like the coronavirus might inflict significant worldwide socio-economic destruction. Due to their high degree of accessibility, cost, and acceptance, conventional herbal medicines could be administered for immuno-suppressing and antiviral action against several contagious illnesses *viz.* tuberculosis, influenza, malaria, and COVID-19. Many modern nations have been successful in integrating the use of traditional medicines with their healthcare systems to treat a variety of infectious illnesses. COVID-19 principally causes respiratory illnesses. The lacuna in the proper effective medications against the virus created a definite urge among researchers to look into several phytochemicals to find new biomolecules possessing antibacterial, antifungal, and antiviral action. Using molecular docking techniques, they were able to assess 10 components, and they came to the conclusion that these molecules can be used to create therapeutic formulations to treat COVID-19.

By using nanofiber-based respiratory masks and bioactive substances from traditional medicinal plants, Balachandar et al. (2020) demonstrated that it was possible to construct a viral inhibitory system. It was recommended to create three-layered fiber filtration masks using biomolecules from strong medicinal herbs to deactivate viruses. Biomolecules with strong antiviral properties were found in various traditional medicinal herbs. They went through the many-faceted processes behind their immunological and antiviral modulatory activities.

According to Azam et al. (2020), Bangladesh's high population density and limited opportunities for social isolation contributed to the COVID-19 surge that caused severe economic and health damage. This was made worse by the costly expense of medication; therefore, individuals turned to folk medicine as an alternate form of treatment for COVID-19. Due to the lack of an effective vaccine or medication to stop the transmission of the virus, Mirzaie et al. (2020) underlined the significance of traditional medications in managing the pandemic. The substance's active components can interact with essential viral proteins and enzymes to prevent viral replication. In their conclusion, the scientists recommended that more research be done to determine the biological relationship between the virus and plant-derived biochemicals. It has been proven that it is urgently necessary to find powerful antiviral compounds to combat the unique coronavirus illness. Numerous healing plants, including *Z. officinale, G. kola, O. europaea, Aloe vera, A. sativum, A. cepa, E. hirta, C. longa*, etc., have strong antiviral substances to stop the viral multiplication. Drug repurposing has been carried out on a variety of biomolecules to manage as well as treat the COVID symptoms. Recently, novel Nigerian medicinal herbs were researched for their potent therapeutic benefits against COVID-19.

One of the most effective methods for battling COVID-19, according to Srivastava et al. (2020), is viral inactivation using organic chemicals. Although the mechanism of action has not yet been completely defined, several parts of Africa, China, and India have approved traditional medicines. To clarify the inhibitory pathways from blocking the entry of the virus throughout its entire life cycle, comprised of assembly and replication, molecular biology techniques implementing herbal remedies extracted from *Aloe vera*, garlic, ginger, onions, and *G. kola* can be used.

2.3 MECHANISMS OF HERBAL ANTIVIRALS AGAINST CORONAVIRUSES

2.3.1 COVID LIFE CYCLE

SARS-CoV-2, the COVID-19, virus is classified as a β-coronavirus and has a genome length of approximately 30 kilobase pairs (kb). It also has a 3′poly(A) tail and a 5′-cap structure. Its homology percentages with SARS-CoV and MERS-CoV are 80% and 50%, respectively (Kim et al. 2020). It is an enveloped virus with a spherical shape and single-stranded positive-sense (+) genomic RNA. Its surface is composed of ultra-structural S proteins that have an appearance similar to a crown. This virus's genome encodes non-structural, accessory, and structural proteins. The main structural proteins are the nucleocapsid (N), spike protein (S), membrane protein (M), and envelope protein (E) (Haake et al. 2020).

The viral infection initiates with the binding of the S protein to the host's ACE-2 receptor. Two subunits, S1 and S2, formed from the breakdown of S protein by the action of the cellular serine proteases TMPRSS2 take charge of the invading cells (Hoffmann et al. 2020). The S protein is generally utilized by the virus for binding with ACE-2 while attacking the respiratory system. Afterwards, viral guide RNA (gRNA) translates polypeptides pp1a and pp1ab, which eventually produce 16 non-structural proteins (Nsps). Notable to mention, the Nsp formation at the C-terminal is processed by 3CL PRO/the main protease (Mpro) while that of the other end, *i.e.* N-terminus, is executed by PLPro. Nsps play a crucial role in the spread of viruses as 12 of the 16 Nsps contain RdRp, an enzyme that is necessary for the replication of viruses. However, the genomic RNA of the virus may be utilized as a template for replication as well as transcription, resulting in progeny genomic RNA that can encode additional sub-gRNAs and the structural proteins S, E, M, and N, as well as accessory proteins 1ab, 3a, 6, 7a, 7b, 8, and 10. The newly generated gRNA enters the next replication cycle and either transmits to other asymptomatic individuals or infects more cells after being further integrated inside the virus particles and ousted from the cell (Poduri et al. 2020).

When developing a pharmacological therapy to combat this virus, a number of functional molecules involved in its replication could be important targets (V'kovski et al. 2021). Bioactive metabolites derived from medicinal plants have the ability to inhibit and/or interfere with the cellular and molecular targets of the coronavirus. We can approach it two ways.

2.3.1.1 Disturb SARS-CoV-2 Replication and Proliferation

During the virus's life cycle, the S protein is necessary for it to enter the host. It also has a major impact on tissue phagocytosis and host tropism of the coronavirus. De Wilde et al.'s (2018) research has shown that mutagenic changed in the MERS-CoV S protein may contribute to the viral zoonoses, while the SARS-CoV-2 S protein mutations in N479L and T487S can greatly boost the virus's affinity to human ACE-2 receptor. This suggests that for the coronavirus to infect humans, the S protein and the host are essential. The S protein of SARS-CoV-2 was shown to be inhibited by a number of metabolites derived from medicinal plants, including (a) hesperidin, seselin, 6-gingerol, and humulene epoxide (Amparo et al. 2021; Nivetha et al. 2021); (b) ACE-2 receptor inhibition was observed by, apigenin, luteolin, quercetin, hesperidin, kaempferol and curcumin (Goyal et al. 2020; Manoharan et al. 2020); and (c) hesperidin, seselin, epigallocatechin-3-gallate (EGCG), curcumin, 6-gingerol, and humulene epoxide disrupted the S protein-receptor interaction (Henss et al. 2021). The synthesis of potential SARS-CoV-2 entry inhibitors may take these metabolites into account.

Following endocytosis, the gRNA is translated into non-functional polypeptides, which the Mpro/3CL PRO and PLPro enzymes' proteolytic activity cleaves into functional proteins. The main protease, Mpro/3CL PRO enzyme, has been shown to be inhibited by, 6-gingerol, humulene epoxide, isotine, hesperidin, seselin, feralolide catechin, epicatechin gallate (ECG), EGCG, curcumin, quercetin, and kaempferol (Amparo et al. 2021; Khaerunnisa et al. 2020; Nivetha et al. 2021); however, 6-gingerol and humulene epoxide inhibited the PLPro enzyme (Amparo et al. 2021), leading to proteolysis inhibition and non-infective non-functional proteins. Furthermore, 6LU7 and 7JTL, which are necessary for the proteolysis pathway, were blocked by guaiol and gingerenone A (Pandey et al. 2021). The subsequent stage entails the replication of genomic RNA from 16 different nonstructural protein types (nsp 1–16). RdRp is in charge of this process, which is blocked by humulene epoxide, hesperidin, luteolin, quercetin, and ECG (Amparo et al. 2021; Goyal et al. 2020). Recent studies on the development of drugs against the SARS-CoV-2 virus indicate that RdRp enzyme inhibitors and ACE-2 receptor blockers are the most promising options.

Moreover, curcumin and 6-gingerol hinder the replication process (Khaerunnisa et al. 2020). The assembly of genomic RNA and proteins, as well as the initiation of the exocytosis pathway, causes the host cell to undergo apoptosis after translation and post-translational maturation. A bioactive molecule derived from plants called kaempferol-3-O-(6″-O-E-p-coumaroyl)-β-d-glucopyranoside suppressed ORF 3a, a viral protein involved in both the release mechanism and apoptosis of SARS-CoV-2 (Ren et al. 2020).

2.3.1.2 Affect the Host Immune/Inflammatory Response

COVID-19 is a lung infection with thrombotic inflammatory symptoms, spreading to all organs as the disease progresses. The severity of illness is attributed to the body's inflammatory response to the SARS-CoV-2 infection. High concentrations of cytokines and lymphocytes in the kidneys, heart, spleen, lymph nodes, and lungs, as well as mononuclear cell infiltration, may contribute to the disease's severity (Liu et al. 2022).

Qingfei Paidu Tang (QFPD), a China-sourced medicinal recipe for COVID-19 treatment, may prevent SARS-CoV-2 replication by directly attacking the virus's 3C protein and human ACE-2. By focusing on ribosomal proteins necessary for virus reproduction, it preserves immunological equilibrium, lowers inflammation, and prevents the translation of viral mRNA (Zhao 2020).

2.4 THE ANTIVIRAL PROPERTIES OF MEDICINAL PLANTS

2.4.1 EXAMPLES OF MEDICINAL PLANTS IN COVID-19 TREATMENT

The SARS CoV-2 global pandemic forced us to probe natural bioresources. Today, the world looks for substitute medications with antiviral properties, notably those made from natural substances derived from medicinal plants. The Chinese Health Commission has recognized the use of traditional and conventional medicine in tandem to treat coronavirus patients (Yang et al., 2020). The study outlines various medicinal plants evaluated for their medicinal effects, antiviral effectiveness against coronavirus, and immune-modulating activity, which are shown in Table 2.1.

TABLE 2.1

Uses of Several Medicinal Plants and Their Modes of Action

Plant Name	Modes of Action		
	Antiviral	Anti-inflammatory	Immunomodulation
Androgaphis paniculata (Burm. f.) Nees	Possibility of binding to ACE-2 active residues, which mediate the host viral interface.	In LPS/IL-4-activated murine macrophages, andrographolide decreases the amounts of inflammatory cytokines TNFα, IL-12, IL-1β, IL-6, and IL-18.	Andrographolide greatly enhances the immune response, controls the generation of NK cells and cytokines, and promotes the development of cytotoxic T-lymphocytes.
Allium sativum L.	Inhibiting the synthesis of amino acids, aqueous bulb extracts, and essential oils bound the viral protease enzyme.	Allicin inhibits TNF-induced IP10, CXCL8, and IL-1β production.	Crushed garlic extract boosts CD4+ T-cell proliferation and IFNγ production.
Rhus chinensis Mill.	Tetra-O-galloyl-β-d-glucose was said to limit SARS CoV replication, as ethanol extract of gall had the ability to impede the SARS-CoV virus into host cells.	*In vivo*, aqueous extract suppresses the release of the inflammatory cytokine IL-6.	Aqueous extract acts as an immunomodulator and inhibits the expression of IL-10.
Cassia tora L	Ethanol seed extract inhibits 3CL protease and SARS CoV replication, as demonstrated by cell-based tests on Vero E6 cells.	Phosphorylated cAMP response is induced by ethanol seed extract.	–
Cibotium barometz (L.) J. Sm.	SARS-CoV 3CL protease activity, spike protein levels, and viral replication are all inhibited by methanol and ethanol extract of the dried rhizome.	The methanol extract of rhizome reduced the expression of COX-2 and iNOS and inhibited NO and IL-6.	–
Chrysanthemum indicum L.	This herb was found inhibitory to SARS CoV-2 virus.	Ethanol extract from flower and bud suppressed TNF-α, IL-6, and IL-1β production.	Ethanol extract markedly elevated the levels of IgG and IgM, splenic cell-produced antibodies, and delayed-type hypersensitivity reaction.

(Continued)

TABLE 2.1 *(Continued)*
Uses of Several Medicinal Plants and Their Modes of Action

Plant Name	Modes of Action		
	Antiviral	**Anti-inflammatory**	**Immunomodulation**
Phyllanthus emblica L.	Phyllaemblicin B and phyllaemblinol bind to helicase (Nsp13), spike protein, ACE-2 protein, and 3CL PRO, inhibiting RNA-dependent RNA polymerase.	Aqueous fruit extract suppresses COX-2, iNOS, IL-16, IL-6, and reduces TNF-α and IL-1 β.	Antibody-dependent cellular cytotoxicity and NK cell activity are both increased by the aqueous fruit extract.
Glycyrrhiza glabra L.	Glycyrrhizin inhibits SARS-CoV replication.	Glycyrrhizic acid, liquiritin, and liquiritigenin inhibited iNOS, COX-2, TNF-α, IL-1β and IL-6 while root extract also suppressed the expression levels of TNF-α, IL-1β and IL-6.	Glycyrrhizin suppressed IL-4 production and stimulated allogenic T-cell proliferation, IFN-γ and IL-10 production.
Ocimum sanctum	Tulsinol (A, B, C, D, E, F, G) and dihydroeugenol-B inhibited the COVID-19 main protease and papain like protease.	When detected *in silico*, luteolin-7-O-glucuronide and chlorogenic acid covalently bound to the active residue Cys145 of the primary protease of SARS-CoV-2, irreversibly inhibiting the viral enzyme.	It possesses ACE-2 blocking properties with immune-modulatory feature
Salvia officinalis L.	Fruit essential oils were found active against SARS-CoV and HSV-1 replication.	In the mouse carrageenan model, flavonoids extracted from fresh leaves and flowers decrease inflammation. Its primary component, caffeic acid, lowers the level of IL-6.	The polysaccharide fractions exhibited immune-suppressive properties.
Cinnamomum zeylanicum L.	Bark butanol extract suppressed the SARS-CoV virus.	Methanol bark extract exhibited anti-inflammatory activity in *in vivo* animal models	Polyphenolic fractions of bark extract stimulate lymphocytes proliferation, immunoglobulin production, and IL-1β production. The oil and bark extract have immunosuppressive potential.
Laurus nobilis L.	*In vitro*, essential oils prevented the replication of SARS-CoV and HSV-1.	Mice treated with hydroalcoholic extracts of leaves and seeds had anti-inflammatory properties.	By reducing hematocrits such haemoglobin (HGB) and HCT and elevating white blood cell counts, essential oils from leaves have an immune-stimulating effect.

(Continued)

TABLE 2.1 *(Continued)*
Uses of Several Medicinal Plants and Their Modes of Action

Plant Name	Modes of Action		
	Antiviral	Anti-inflammatory	Immunomodulation
Azadirachta indica A. Juss	Nimbolide, or terpenoid lactone, is a crucial pathogenic characteristic of COVID-19 and is beneficial at controlling ARDS.	Through the suppression of TNF-α, quercetin from leaf methanol extract showed anti-inflammatory properties.	Nimbin, a constituent of neem oil, stimulates immune responses by potentiating phagocytic activity, antigen-presenting ability of macrophages, and enhancing IL-1, IFN-Γ, and TNF-α production. Flowers aqueous stimulated both specific and non-specific immune responses, humoral and cell-mediated response, and Nimbin exhibited immunomodulatory activity by potentiate phagocytic activity, antigen-presenting ability of macrophages and enhances IL-1, IFN-Γ, and TNF-α production
Tinospora cordifolia (Willd.) Miers	Berberine, isocolumbin, magnoflorine, and tinocordiside interfere with the viral attachment and replication due to binding efficacy against surface glycoprotein and receptor binding domain and main protease.	Stem extract from chloroform inhibited IL-6, IL-1β, and PGE2.	IFN-γ, TNF-α, and IL-1β are produced in response to aqueous and methanolic stem extract. Syringin and cordifolioside A also have immunomodulatory properties.
Syzygium aromaticum (L.) Merr.	Enhanced binding affinity for human proinflammatory mediators, human ACE-2, furin proteins, spike, SARSCoV-2 main proteases, and other host and viral macromolecular targets.	Eugenol inhibits the production of IL-4 and IL-5 as well as the downregulation of TNFα and IL-6, two pro-inflammatory cytokines.	Eugenol found immunomodulatory activity.
Piper nigrum L.	Phenolic compounds Kadsurenin L and methysticin found in *Piper nigrum* was found to inhibit the COVID-19 main protease.	Piperine prevents NF-κB activation, which in turn prevents the generation of pro-inflammatory cytokines such as IL-1β, IL-6, IL-10, and TNFα.	It acts as immunomodulator.
Nigella sativa L.	In an *in silico* investigation, nigellidine and α-hederin were found to inhibit 3CL PRO/Mpro COVID-19 and 3CL PRO/Mpro SARS-coronavirus.	Thymoquinone and fixed oil have been shown to down-regulate COX, 5-LO, and 5-HETE and to reduce the production of IL-6, TNFα, and NO.	By promoting CD4+ cells, the aqueous seed extract enhances humoral as well as cellular immunity.

(Continued)

TABLE 2.1 *(Continued)*
Uses of Several Medicinal Plants and Their Modes of Action

	Modes of Action		
Plant Name	Antiviral	Anti-inflammatory	Immunomodulation
Withania somnifera (L.) Dunal	Withanone prevents or lessens COVID-19 entrance and subsequent contagiousness. It is possible that quercetin and steroidal lactones suppressed SARS COV PLPro and 3CL PRO.	TNF-α, IL-1β, IL-8, and IL-12 were all suppressed by aqueous root extract.	Ashwagandha included in a herbal mineral formulation that markedly raised the CD4+ and CD8+.
Camellia sinensis (L.) Kuntze	Leaf ethanol extract contains hydrolysable tannins that exhibit antiviral properties against influenza A by inhibiting viral replication and RNA-dependent RNA polymerase.	The ethanol leaf extract reduced NO, COX-2, IL-6, IL-1b, and TNF-α levels in RAW 264.7 cells.	Aqueous leaf extract raised the levels of IL-17A, IL-8, and HBD-2 in immuno-compromised rats.
Curcuma longa L.	Vitamin C, curcumin, and glycyrrhizic acid showed inhibitions of COVID-19 Mpro. Cucumin impeded the replication and budding of the human respiratory syncytial virus.	Curcuminoids suppress leukotrienes, prostaglandins, TNF, IL-12, IL-6, and interferon-inducible protein.	Aqueous extract elevation NO, IL-2, IL-6, IL-10, IL-12, IFN-γ, and TNF-α.
Zingiber Officinale Roscoe	6-gingerol binds to the active sites of R7Y COVID-19, the main protease required for SARS-Cov-2 replication and reproduction.	6-Gingerol prevents the secretion of pro-inflammatory cytokines such as TNFα, IL-1β, and IL-12.	6-Gingerol has immune-modulating activities. The humoral immune response was recovered by ginger essential oil.

Source: Adapted from Yashvardhini et al. 2021; Mukherjee et al. 2022.

2.4.2 Phytochemicals as COVID-19 Inhibitors

Phytochemicals are plant-based bioactive compounds produced by plants for their protection. The development of prophylactic and therapeutic drugs for COVID-19 has made use of phytochemicals as well as a few medicinal plants with recognized antiviral, antimicrobial, and antifungal properties (Gyebi et al. 2021). Naringin, quercetin, capsaicin, psychotrine and gallic acid are just a few of the compounds that were isolated from medicinal plants using Gas Chromatography/Mass Spectroscopy (GC/MS) and High-Performance Liquid Chromatography (HPLC) in a recent study by Alrasheid et al. (2021). These compounds are significant sources for novel antiviral drugs that target COVID-19. Resveratrol, myricetin, apigenin, kaempferol, and quercetin have all demonstrated strong anticoronavirus properties in a different study (Wahedi et al. 2021). Lycorine, an indolizidine alkaloid that was extracted from the Lycoris radiata (L'Hér) herb, is one example of an alkaloid that has demonstrated antiviral properties against coronaviruses (Gyebi et al. 2021).

Proteins like Nsp15 and 3CL PRO in SARS-CoV-2 have potential applications in developing medications against COVID-19. Nsp15 is crucial for viral replication, while 3CL PRO processes translated polyproteins. SARS-CoV-2 exposure triggers various biochemical processes in the human body, leading to the development of various small molecule medications to combat the virus. According to reports, the essential amino acids in 3CL PRO's active binding site include CYS145 and HIS41, while THR341 is important in Nsp15 (Al-Janabi et al. 2021; Kumar et al. 2020; Sinha et al. 2021). Researchers, therefore, generally agree that to prevent protein activity in viral replication, prospective novel antiviral activity medications must interact with these essential amino acids.

In an *in silico* perspective by Jamhour et al. (2022), of the 36 phytoconstituents, six compounds–quercetin, rhamnetin, campesterol, stigmasterol, and rutin–have shown exceptional molecular docking results. The chemicals in question exhibit the ability to bind to the P-glycoprotein substrate of 3CL PRO and Nsp15, as demonstrated by their lowest binding energies (LBE) and inhibition constant (Ki).

Their capacity to form strong hydrogen bonds (H-bonds) with essential amino acids (HIS41 and/or CYS145) of the primary protease 3CL PRO is noteworthy. Stigmasterol is the most dominant hydroxyl group while forming the H-bonds, followed by campesterol, quercetin, and rutin. The minimal function in this regard can be found in rhamnetin and catechin gallate. On the other hand, quercetin occupies the last position while detecting the potentiality of the inhibitors on H-bond formation with the principal residues in Nsp15. Rutin and campesterol were predominantly impeded here, while rutin can be listed among the most effective natural substances for both enzymes, accordingly succeeded by stigmasterol and campesterol when measured against the reference control.

Remarkably, rutin has been discovered to be a highly effective putative inhibitor for COVID-19 proteins due to its superior LBE score, Ki value, and capacity to establish strong H-bonds with COVID-19 proteins compared to other substances. This observation suggests that these compounds might serve as effective agents for impeding the viral replication of COVID-19 by inhibiting the efficacies of the two crucial proteins. To validate the computational outcomes of this research, further investigations through *in vitro* and *in vivo* studies are imperative.

2.5 UNDERSTANDING PHYTOCHEMICALS

2.5.1 SPECIFIC PHYTOCHEMICALS WITH ANTIVIRAL ACTIVITY AGAINST COVID-19

Nature holds answers to many questions, including those about the recent pandemic. Different plants from different genera produce a range of phytochemicals found to be effective in the antiviral activity against COVID-19. The array of phytochemicals which are antiviral include flavonoids, phenolic compounds, alkaloids, terpenoids, saponins, lignans, and other compounds. Pal et al. (2022) added lectins and polysaccharides to the list of phytochemicals. Phytochemicals are effective fighters of SARS-CoV-2 as they disrupt different functions of the viral particle which include entry, replication, assembly, and various host targets (Mlozi 2022).

2.5.1.1 Flavonoids

Flavonoids are a set of secondary metabolites synthesized by plants which are a skeleton of carbon atoms having two aromatic rings (Españo et al. 2021). Many recent studies reported that different flavonoids such as kaempferol, quercetin, naringenin, and hesperetin effectively act contrary to COVID-19 (Ghidoli et al. 2021). Quercetin which is found in capers, buckwheat, and onions is a good antiviral compound especially for SARS-COV-2. Hesperetin which is found more in citrus fruits is an effective antiviral compound and is also an important antioxidant compound. Kaempferol is most widely found in *Brassicaceae* plants acts as an antioxidant, anti-inflammatory, anticancer, and antiviral compound. Naringenin is found more in grapefruit, tangerines, tomatoes, and oranges and was also reported to be a strong antiviral agent (Ghidoli et al. 2021). According to Españo et al. 2021, flavonoids such as brazilin, EGCG, isorhamnetin, theaflavin 3,3'-di-O-gallate, herbacetin, and pectolinarin are found to be antiviral. Another compound belonging to flavonoid class, luteolin extracted from *Veronica* sp. is a strong inhibitor for the entry of corona virus. Honey is also rich in flavonoids such as chrysin, galangin, caffeic acid, hesperidin, isoquercetin, rutin, quercetin, and pinocembrin, which are found to be antiviral *in silico* (Al-Hatamleh et al. 2020).

2.5.1.2 Alkaloids

Alkaloids constitute another important class of secondary metabolites synthesized in the amino-acid pathway, which consists of a heterocyclic ring containing nitrogen. This chemical inhibits SARS-COV-2 infection effectively. Berbamine, cepharanthine, hernandezine, isoliensinine, liensi-nine, neferine, quinine, and tetrandrine are the alkaloid chemicals reported as effective inhibitors of SARS-COV-2 (Españo et al. 2021). Berberine also finds its place in this list of chemicals found in medicinal plants (*Berberis vulgaris*) also possessing antiviral activity that could be effective against this virus (Behl et al. 2021).

2.5.1.3 Phenolic Compounds

Curcumin, an important phenolic compound that is a regular Indian traditional medicine, is recorded as a potent inhibitor of COVID-19. The action of this chemical can be enhanced by increasing the bioavailability by combining the chemical with piperine (Ghidoli et al. 2021). Polyphenols consti-tute a class of phenols obtained largely from plants. Tannins, polyphenols such as chebulagic acid, tannic acid, and punicalagin also reported anti-COVID-19 activity (Españo et al. 2021). As given by Davella et al. (2022), phenolic compounds obtained from Piper species, such as Kadsurenin L and methysticin, inhibit the main protease of COVID-19. Polyphenolic acids derived from the basil plant, primarily luteolin-7-O-glucuronide and chlorogenic acid, inhibit the viral activity by inhibit-ing the enzyme's main protease and Cys145 (Mohapatra et al. 2020). Various natural polyphenols including indigo, betulinic acid, aloe emodin, luteolin, quercetin or gallates, and quinomethyl triter-penoids are potential anti SARS-CoV-2 agents by inhibiting viral proteins (Chojnacka et al. 2020). Green tea rich in polyphenols such as EGCG. ECG, and gallocatechin-3-gallate can also help to curb the activity of the coronavirus (Ghosh et al. 2021).

2.5.1.4 Other Phytochemicals

Among plant products, lectins, and other secondary metabolites, baicalin and glycyrrhizin have shown promising activity against the coronavirus. An alkaloid derivative homoharringtonine derived from *Cephalotaxus fortunei* is effective in inhibiting viral replication of SARS-CoV-2. Strong reduction in viral entry has been achieved by two compounds luteolin (flavonoid) and res-veratrol (terpenoid) obtained from *Veronica* sp. (Pal et al. 2022). Glycyrrhizin inhibits the virus by effecting the main protease of the virus (Mpro). Compounds derived from Neem, nimbolin A, nimocin, and cycloartenols act as antiviral compounds specific to COVID-19 by binding to the enve-lope and M proteins of SARS-CoV-2 (Borkotoky and Banerjee 2020). Azeem et al. (2022) reported several chemicals such as pedalitin (a flavonoid), baicalin (betaxanthins), ichangin (a terpenoid), kaempferol (flavonoid), silvestrol, betaxanthin, gallic acid, hortensin, fomecin A, quercetin, limo-noic acid, and epigallocatechin displayed more potential as they have reported good binding inter-action and S-scores. Emirik (2022) conducted molecular docking on compounds present in turmeric and found that they have an efficient binding and form stable ligand-receptor complex. Españo et al. (2021) found that platycodin D (saponin), tannic acid (tannins), digoxin, ouabin, sennoside B, acetoside (glycosides), resveratrol and pterostilbene (stilbenoid compound), and hypericin (anthra-quinone) are said to be resistant to SARS-CoV-2. Sulphur compounds such as sulforaphane can be used against COVID-19, but is not confirmed (Bousquet et al. 2020). Various other phytochemicals inhibiting the main protease activity of the virus include withanoside V and somniferine from ash-wagandha; tinocordiside and berberine from Giloy; ursolic acid, vicenin, Isorientin 4′-O-glucoside 2″-O-p hydroxybenzoate (Saxena 2022).

2.5.2 The Role of Bioinformatics and Computational Tools

2.5.2.1 Introduction to Bioinformatics and Computational Approaches Used in COVID-19

Bioinformatics is an important branch of science, especially in recent times, which combines different branches of science such as biological science with statistics, mathematics, computer science, and information engineering. It assists present studies by helping to understand, integrate, and analyze biological data with the help of various software tools and databases. With emphasis on COVID-19, bioinformatics involves two important aspects: genomics, which involves mapping and editing the genome, and proteomics of the pathogen, involving viral protein analysis. Various databases include Genbank, DNA Data Bank of Japan (DDBJ), European Nucleotide Archive (ENA), European Bioinformatics Institute (EMBL-EBI), Protein Data Bank (PDB), InterPro, Kyoto Encyclopedia of Genes and Genomes (KEGG), UniProt Knowledgebase (UniProtKB), Ensembl, TrEMBL, Prosite and many more. Different software packages are essential tools for analysis such as sequence search tools (BLAST [Basic Local Alignment Search Tool] or FASTA [FAST-All]), structure view tools like Rasmol, sequence alignment tools (ClustalW), and structure prediction tools like PROSPECT. Assembling virus genomes has also been facilitated with the use of a Web-based software tool named Genome Detective which helps to track new mutations in the viral genome and facilitate the development of new drug or vaccine (Poojary 2020).

Computational biology refers to the application of mathematics to the theoretical and experimental study of biology. Paired with bioinformatics, especially in the case of COVID-19, computational biology helps to validate mathematical models and in understanding the relationships in biological systems regarding efficiency of the drug target and molecular interactions with the virus. It also involves analysis of molecular docking to enhance understanding of the binding affinities of molecules, thus aiding in drug discovery (AutoDock software, for example). ADMET profiling evaluates pharmacokinetic properties of the compounds, and the compounds that satisfy these parameters will be considered as a potential drug. Molecular dynamics (MD) simulation is done to determine the stability of the complex that is docked (Mahmud et al. 2021). These techniques help to accelerate drug discovery, whereas different computational approaches, reverse vaccinology, immunoinformatics, and structural vaccinology help to identify suitable vaccines. Reverse vaccinology (RV) helps to discover novel vaccine compounds with the help of genome or proteome of pathogens. Immunoinformatics applies computational methods to design epitope-based peptide vaccines, whereas structural vaccinology involves binding epitopes and immunogenic domains to develop potential vaccines (L. Ma et al. 2021).

Other computer technologies include Computer Aided Drug Design (CADD), also known as *in silico* screening, which involves homology modelling, molecular docking, and high throughput experiments. This technology helps to understand various biological processes to carryout drug discovery and, also with the help of algorithms, it develops digital repositories to study various chemical interactions. It is used for structure-based and network-based drug design to discover potential drug chemicals against viral spike proteins. Various other tools include artificial intelligence (AI), Internet of Things (IoT), cloud computing (CC), Big Data analytics (BDA), and supercomputers (Poojary 2020).

2.5.2.2 Various Studies Involving Bioinformatics in Selection of Phytochemicals against SARS-CoV-2

Based on studies conducted by Azeem et al. (2022) on the stability of drug entries and active-site targeting against receptors of SARS-CoV-2, it is reported that molecular docking, when paired with an MD approach, are great providers of information. Zhang et al. (2020) considered Chinese herbal formulations for the study and conducted *in silico* processing, ADME screening and protein-molecular docking, which were reported to be antiviral. Nag et al. (2021) conducted various studies (using seven different phytochemicals, allicin, capsaicin, gingerol, piperine, and zingiberene) such as ADME profiling and molecular docking, and the results of these are validated by MD simulation. PCA and molecular alignment tools were also used and reported that curcumin and piperin are structurally similar. Here are a few studies involving use of CADD against SARS-CoV-2: Prasanth et al. (2020) identified potential inhibitors in Cinnamon; Fantini et al. (2020) studied the effects of hydroxychloroquine and chloroquine against COVID-19; Jo et al. (2019) screened various flavonoids against COVID-19; Panda et al. (2020) conducted drug designs based on structure and immunoinformatic approaches against COVID-19. Using MD simulation, bioflavonoids and tannins are considered to have the best binding affinity for Mpro and, thus, inhibit replication of the viral particles (Ristovski et al. 2022). In studies carried out by Basu et al. (2020), three natural chemicals, hesperidin, emodin, and chrysin are selected out of five phytochemicals as effective antivirals using studies on molecular docking. With the help of network analysis on phylogenetic and sequence similarity, phytochemicals such as δ-viniferin, hesperidin, myricitrin, nympholide A, afzelin, Taiwanhomoflavone A, lactucopicrin 15-oxalate, biorobin, and phyllaemblicin B were characterized to be strong antivirals providing resistance to COVID-19 (Joshi et al. 2020).

2.6 ENHANCING THE BIOLOGICAL ACTIVITY OF MEDICINAL PLANTS

2.6.1 Unravelling Mechanisms of Action against the Virus

As discussed earlier in the chapter, there are a plethora of phytochemicals used against the COVID-19 virus, but each one has a different mode of action (Table 2.2). To know their mechanisms of action we have to get an idea of the structure of SARS-CoV-2 and different proteins and enzymes responsible for its replication and infection and utilized as important drug targets. The first step of infection is an interaction between the spike glycoprotein (S) of the virus and ACE-2 of the host cell. After the entry of the viral particle, its genome is transferred to the genome of host cell with the help of transmembrane proteins. This genome (RNA) replicates in the host cell with the help of the enzyme RdRp. Proteins are then synthesized, resulting in the formation of new viral particles. Various other proteases such as 3CL PRO, Mpro, and PLPro also serve as key targets. Phytochemicals which interact with ACE-2, S glycoprotein, and transmembrane proteins help to obstruct the entry point for viruses; if they interact with RdRp and the proteases mentioned previously, replication and assembling of the viral genome is affected (Khanal et al. 2020).

Various studies have been carried out on wide range of phytochemicals using molecular docking approach. Based on docking scores and binding energy potential antiviral chemicals along with their mechanism of action specific to various viral proteins and enzymes has been given in the Table 2.3.

TABLE 2.2
Mechanism of Various Antiviral Phytochemicals

Phytochemicals	Mechanisms of Action
Flavonoids	Inhibits the activity of reverse transcriptase.
Phenolic compounds	Affects the surface protein of the viral particles.
Alkaloids	Affects binding and growth of the viral pathogens.
Terpenes	Inhibits replication of virus.
Lectin	Prevents integration and reverse transcription, glucohydrolases activity of virus.
Carbohydrate	Blocks attachment and inhibits viral replication.

Source: Modified from Pal et al. (2022).

TABLE 2.3

Phytochemicals and Their Interaction with Viral Proteins and Enzymes

Phytochemicals	Inhibitors of Component	References
Luteolin-7-glucoside, demethoxycurcumin, apigenin-7-glucoside, oleuropein, curcumin, catechin, and epicatechin-gallate	Mpro inhibitors	Khaerunnisa et al. 2020
Polyphenols (gallic acid, quercetin, caffeine, resveratrol, naringenin, benzoic acid, oleuropein and ellagic acid)	RdRp inhibitor	Abdo et al. 2021
Hesperidin, nabiximols, pectolinarin, EGCG, and rhoifolin	Mpro and S protein	Tallei et al. 2020
Amentoflavone, guggulsterone, puerarin, and piperine	Mpro	Tripathi et al. 2020
Flavonoids (herbacetin, rhoifolin, and pectolinarin)	3CL PRO	Jo et al. 2019
Alkaloids and terpenoids	3CL PRO	Gyebi et al. 2021

2.6.2 FUTURE RESEARCH AND DRUG DEVELOPMENT

Being a pandemic, COVID-19 created great havoc in the sphere of public health. Various drugs were reported as effective antiviral compounds that later became ineffective due to the evolution of various viral mutations. Hence, there is a need to focus on the successful strategies which can help protect us in the future. Development of potential antiviral chemicals is the need of the hour to prevent higher mortality. There are several challenges which serve as hurdles in curbing the disease and its aftermath. Insufficient data available on the viral genome and drug chemicals lag the antiviral development; sharing available data is slow and there is no proper streamlining of the information which results in improper and duplicate research, which results in a waste of time, costs, and materials. The short time available to conduct research and the long process of drug research, discovery, and development are various challenges in drug discovery. They can be achieved by using new era tools including AI.

A rapid coordinated response to situations regarding the pandemic, a more focused response, and proper collaboration between research institutes, hospitals, and pharmaceuticals is needed. Gene therapies along with cell-based therapy help to transform future health research. Due to the shortage of time available to discover antiviral drugs, repurposing drugs can be a great tool to resolve the issues at clinical level. Technological tools such as computational calculation and AI help to resolve the issues effectively in a short span of time with an accurate response. AI can be applied to various areas of drug discovery and development. Deep learning models are supported with greater accuracy because to its autonomous feature extraction capability. Its ability helps synthesize potential drug development of broad-spectrum antivirals to combat future dangerous coronaviruses like COVID-19 molecules and reduce the chances of failure in trails. Transfer learning, which is a branch of AI, utilizes the data available and transfers this knowledge to the task at hand, thus alleviating the problem of scarce therapies available for a new virus (Keshavarzi Arshadi et al. 2020). Future research should also focus on the discovery and development of broad-spectrum antivirals against COVID-19 and other future pathogenic coronaviruses (Ghosh et al. 2020).

2.6.3 IMPLICATIONS FOR DRUG DEVELOPMENT

Medicinal plants are most recently often used to obtain drugs for diseases like cancer and infections (Katiyar et al. 2012). Many adverse effects were reported from synthetic medicines and were costly, hence, phytochemicals are regarded as the safest tools (Mohanty et al. 2022). India, being an ancient civilization, has a deep and profound knowledge of herbal compounds from plants which can be used for various diseases. As discussed previously in the chapter, different phytochemicals are reported to be effective against SARS-CoV-2 that can be further isolated and tested under clinical trials to utilize them as drugs.

Following are the different steps in the process of drug discovery from medicinal plants:

i. Primarily desirable chemicals should be studied from the literature available from traditional scripts and other available sources and then the desirable lead compounds from the selected plants have to be detected, isolated, and purified.

ii. The selected lead compounds have to be primarily screened against the targets using various techniques, especially high throughput screening (HTS). Later, various biochemical and pharmacological tests are carried out to detect the compounds effective for specific targets.

iii. Sometimes, the selected compounds may not be selective enough. To improve the selectivity, the structure of the lead compound has to be modified.

iv. After improving selectivity of the compound, a series of testing is carried out both *in vivo* and *in vitro* and, if the result is positive, they are forwarded for safety tests. If it passes these safety tests, then the lead compound can be regarded as a potential drug.

Various approaches are utilized in each step of drug discovery. Ethnopharmacology is an approach used to select plants using the history of traditional medicines. It involves observation, description, and experimental investigation of indigenous drugs. ADMET profiling has also to be carried out to evaluate the potential usage (Katiyar et al. 2012). Plant extracts contain a mixture of chemicals and to identify and isolate desired bioactive compounds, a technique named bioactivity-guided fractionation is followed. It can be defined as a method for screening and validating plant extracts for bioactive substances that could serve as sources of novel biobased pharmaceuticals (Mani et al. 2020). Reverse pharmacology is another technique which involves screening of compounds with functional activity and can be used further for drug discovery. HTS is the latest method used to evaluate plant extracts and biological activity of samples which uses analytical tools and fractionation techniques for screening these natural compounds. An alternative for HTS is fragment-based drug discovery (FBDD), which involves drug designing based on structure using X-ray crystallography or nuclear magnetic resonance (NMR) spectroscopy for identification of lead drug molecules. Polypharmacology and network pharmacology are other approaches used to improve the efficiency of the selected drug molecules. Polypharmacology is a multitarget approach which develops the drugs for multiple targets and, hence, has more efficiency. The polypharmacological approach uses the profile of five natural compounds, EGCG, curcumin, quercetin, berberine, and resveratrol. Network pharmacology is based on concepts of system biology, network analysis, redundancy, pleiotropy, and connectivity. It helps to improve the efficiency of the drug by evaluating its toxicity level and other side effects. The quality of the extracted drugs has also to be evaluated by carrying out chromatographic fingerprinting (Mohanty et al. 2022).

Other technologies involved in drug discovery include the "Omics" technologies which involves genomics, proteomics, and metabolomics. Coming to genomics, DNA barcoding offers rapid and accurate identification of the plant species; genomic analysis, genome sequencing, helps to evaluate the systems at various levels, and the CRISPR-Cas9 technique helps to facilitate the identification of multiple drug targets. Proteomics helps to identify target proteins before their use as medicines. The main advantage of using proteomics is to understand the mechanism of action of natural products in drug discovery. Metabolomics also plays a vital role in drug development by elucidating the biological mechanism of action and effects of natural products on the individual. It also helps to identify and evaluate metabolites derived from natural products (Dzobo 2022). There is a dire need to combine these tools to succeed in releasing natural plant products as drugs.

2.6.4 Complementary Therapeutic Strategies

Various therapeutic strategies involved to curb the pathogenicity of the virus, SARS-CoV-2, other than phytochemicals that can be used as drugs, include drug repurposing in which other pharmacotherapeutic agents are tested for their potentiality against COVID-19, interferons, traditional medicines used across countries, and vaccines. Other alternate interventions include yoga, herbal formulations, nutritional supplements, exercise, and aromatherapy. Detailed notes of these complementary and alternate therapeutic approaches are discussed next.

2.6.4.1 Drug Repurposing

Drug repurposing is a technique which identifies a novel clinical use for an already approved medication that is approved for a different indication to expedite the drug-discovery process. Due to the extreme time constraints in the case of a viral pandemic, it is probably difficult to carryout vivid research and discover a novel drug. Hence, this technique comes into play to save time and other resources. In this method, various drugs such as hydroxychloroquine and chloroquine, remdesivir, favipiravir, lopinavir/ritonavir, arbidol, ribavirin, disulfiram, baricitinib, and ivermectin were considered to study their effectiveness against SARS-CoV-2.

The above mentioned drugs have been reported against the virus through different mechanisms. Hydroxychloroquine and chloroquine increase the pH of the cell and modify the ACE-2 receptor glycosylation and has inhibitory effect on ACE-2 receptor, respectively. Remdesivir is an antiviral drug and it inhibits the RdRp enzyme, thus affecting the viral replication and exhibits significant effect on the viral inhibition. It has been recommended only for emergency usage in hospitalized COVID-19 patients. Favipiravir also acts against the replication of the viral genome by binding to the RdRp. Lopinavir acts as antiretroviral protease inhibitor and ritonavir, also being a protease inhibitor when administered along with lopinavir, has enhanced bioavailability and antiviral effect. Arbidol prevents the infection of the viral particles by interfering with their entry and also induces the production of interferons exhibiting its broad range of antiviral effects. Ribavirin is a synthetic drug which terminates the replication of DNA and RNA viruses; in case of SARS-CoV-2, it inhibits PLPro activity, resulting in a decline in replication activity of the virus. The same mechanism can be observed in case of disulfiram and ivermectin. Baricitinib is another drug used primarily for rheumatoid arthritis which reduces the viral entry.

2.6.4.2 Immunomodulators

Immunomodulators are drugs that modify the immune system's response by either stimulating or suppressing the formation of serum antibodies (Bascones-Martinez et al. 2014). Convalescent plasma, interferons, cyclosporine A, and corticosteroids are various immunomodulators studied for the treatment of COVID-19.

Passive immunity can be achieved by administering convalescent plasma from the recovered patients to the infected patients. It contains a good number of antibodies which provide immunity against the virus by neutralizing the virus or by antibody-dependent cellular cytotoxicity involving complement activation or phagocytosis. This method has been reported to be positive for various viral diseases such as measles, mumps, polio, and influenza. Few clinical studies have been conducted; however, and further research is needed to know the efficacy of this therapy.

Interferons are the protein molecules produced and released by the host immune system to combat the virus. Alfacon-1 is an interferon reported to have a suppressive effect on SARS-CoV-2. It was administered to COVID-19 patients along with corticosteroids. Cyclosporine A is another type of immunosuppressant used for autoimmune disorders and transplantations which affects the replication of the virus. Corticosteroids, methylprednisolone, and dexamethasone, all anti-inflammatory agents, were studied for treatment of COVID-19. Some reviews on the use of corticosteroids revealed that their usage is harmful through drug complications and reduction in viral clearance. Most of the studies remained inconclusive. Whereas dexamethasone has no role in curing the disease, studies reported it improved the survival rate of severely ill patients.

2.6.4.3 Vaccines

Vaccines can be the best tool in cases of viral disease for their long-lasting effect, cost-effective approach, and control of the spread of the disease. Various types of vaccines have been developed against the virus of which mRNA vaccines, vectored vaccines, and protein-based vaccines are reported to be more effective (Soraci et al. 2022). There are two types of mRNA vaccines: (1) a nonreplicating RNA vaccine, which encodes the antigen based on the disease of interest and (2) a self-replicating RNA vaccine, which bears the gene's encoded viral antigen and a replication complex that allows replication on its own in the host cell. mRNA vaccines are considered to be the best as they trigger strong immune responses and their activity can be enhanced by combining safe agents which form complexes, such as lipid- and polymer-based nanoparticles. Protein vaccines are made of desired purified antigens which can trigger the immune response in the host cell. These are considered to be a safe option as they do not bear any infectious components. Vector vaccines do not contain antigens by themselves, but use the host-cell machinery to produce antigens. In the case of SARS-CoV-2, vector vaccines are modified viral vectors which encode the surface S proteins. There are two types of vector vaccines: replicating vaccines in which the vaccine enters the host cell to produce whole viral particles, and nonreplicating vaccines that can produce the viral antigens directly within the host cells. Due to the evolving mutations in the virus, even vaccines are becoming ineffective; hence, studies have to be focused on developing measures for long-lasting and broad-range effects.

2.6.4.4 Other Therapeutic Agents

Several monoclonal antibodies in clinical use for therapy and other diagnostic works are studied for their action against COVID-19. Antibodies such as tocilizumab and sarilumab are under clinical trials.

2.6.4.5 Traditional Medicinal Approaches and Nutritional Supplements

Traditional Chinese medicine (TCM) and Indian Ayurveda are the major complementary and alternate approaches in most parts of the world. In various parts of the world with socio-economic constraints and restricted access to various research and clinical facilities, there is a high usage of these interventions to reduce mortality and improve the survival rate. Traditional medicine in China involves the use of various herbal formulas to prevent viral infection. China's National Health Commission (NHC) recommended 15 TCMs, seven oral formulations, and eight injectable formulations, which consist of Qingwen, Shufeng Jiedu, Shenfu, Shengmai, Shenmai, and others. Most of them are anti-inflammatory in nature. Trails must be conducted to know their efficacy (Paudyal et al. 2022).

The majority of the studies focused on traditional Chinese medicine rather than Indian Ayurveda which have similar effective herbal solutions. The Indian Ministry of AYUSH recommended use of Kadha (an Ayurvedic remedy containing curcumin, ginger, cumin seeds, fennel seeds, honey, and cloves), turmeric, garlic, carom seeds, Indian gooseberry, Tulsi leaves, ashwagandha, and Giloy, which were reported as protease inhibitors of SARS-CoV-2. The study conducted by Agrawal et al. (2023) in India suggested that the most popular complementary and alternative medicine (CAM) interventions include the use of Kadha, followed by Ayurvedic medicines and herbal tea. It also reported that most people have focused on eating food rich in vitamins and minerals to improve their immunity.

Nutritional supplements are another intervention which improves immunity because good immunity comes from a good nutritional status of the body. This includes taking various supplements such as vitamin D3, vitamin C, zinc, iron, and selenium. Vitamin D3 is reported to reduce the mortality rate, whereas the antioxidant vitamin C boosts and activates natural immunity. Zinc was reported to clear viral particles from the nasopharynx and zinc cations along with the zinc ionophore pyrithione inhibited RNA polymerase of the coronavirus. Iron and selenium help improve the immune response of the body (Saxena 2022).

2.7 CONCLUSION

2.7.1 The Way Forward in the Fight against COVID-19

This chapter has discussed various principles to be followed for managing the devastating effects of COVID-19 by employing the extracts of numerous medicinal herbs. The use of herbs as possible therapeutic agents and alternate therapies for treating coronavirus have also been demonstrated by several evidence-based studies. The majority of these medicinal herbs demonstrate their antiviral activity against this viral infection, even though many studies are still being conducted to determine the main mechanism. Numerous medicinal plants have antiviral properties, but more research is still needed to understand their structures, understand their pharmacology, and standardize the bioactive components of those plants. It is necessary to conduct evidence-based experiments using accepted methodologies as well as to thoroughly analyze and research formulation development as well as the smallest effective dosage. The use of metabolomics may also be essential in locating pharmacologically active substances that can be used to synthesize naturally derived medications for the efficient treatment of COVID-19 infection.

REFERENCES

Abdo, N., Moheyeldin, O., Shehata, M. G., & El Sohaimy, S. (2021). Inhibition of COVID-19 RNA-Dependent RNA Polymerase by Natural Bioactive Compounds: Molecular Docking Analysis. Egyptian Journal of Chemistry, 64(4), 1989–2001. https://doi.org/10.21608/ejchem.2021.45739.2947

Adetunji, C. O., Akram, M., Olaniyan, O. T., Ajayi, O. O., Inobeme, A., Olaniyan, S., … & Adetunji, J. B. (2021). Targeting SARS-CoV-2 novel corona (COVID-19) virus infection using medicinal plants. In: Dua, K., Nammi, S., Chang, D., Chellappan, D.K., Gupta, G., Collet, T. (eds) Medicinal Plants for Lung Diseases: A Pharmacological and Immunological Perspective, 461–495.

Agrawal, A., Sharma, A., Mathur, M., Sharma, A., Modi, G., & Patel, T. (2023). Perspective toward complementary & alternative medicines in the prevention of COVID-19 infection. Indian Journal of Community Medicine, 48(3): 401–406. https://doi.org/10.4103/ijcm.ijcm_282_22

Al-Hatamleh, M. A. I., Hatmal, M. M., Sattar, K., Ahmad, S., Mustafa, M. Z., Bittencourt, M. D. C., & Mohamud, R. (2020). Antiviral and Immunomodulatory Effects of Phytochemicals from Honey against COVID-19: Potential Mechanisms of Action and Future Directions. Molecules, 25(21), 5017. https://doi.org/10.3390/molecules25215017

Al-Janabi, A. S., Elzupir, A. O., & Yousef, T. A. (2021). Synthesis, anti-bacterial evaluation, DFT study and molecular docking as a potential 3-chymotrypsin-like protease (3CLpro) of SARS-CoV-2 inhibitors of a novel Schiff bases. Journal of Molecular Structure, 1228, 129454.

Alrasheid, A. A., Babiker, M. Y., & Awad, T. A. (2021). Evaluation of certain medicinal plants compounds as new potential inhibitors of novel corona virus (COVID-19) using molecular docking analysis. In Silico Pharmacology, 9, 1–7.

Amparo, T. R., Seibert, J. B., Silveira, B. M., Costa, F. S. F., Almeida, T. C., & Braga, S. F. P., et al. (2021). Brazilian Essential oils as source for the discovery of new anti-COVID-19 drug: A review guided by in silico study. Phytochemistry Reviews, 20, 1013–1032. https://doi.org/10.1007/s11101-021-09754-4

Azam, M. N. K., Al Mahamud, R., Hasan, A., Jahan, R., & Rahmatullah, M. (2020). Some home remedies used for treatment of COVID-19 in Bangladesh. Journal of Medicinal Plants Studies, 8(4), 27–32.

Azeem, M., Mustafa, G., & Mahrosh, H. S. (2022). Virtual screening of phytochemicals by targeting multiple proteins of severe acute respiratory syndrome coronavirus 2: Molecular docking and molecular dynamics simulation studies. International Journal of Immunopathology and Pharmacology. https://doi.org/10.1177/03946320221142793

Balachandar, V., Mahalaxmi, I., Kaavya, J., Vivekanandhan, G., Ajithkumar, S., Arul, N., … & Devi, S. M. (2020). COVID-19: Emerging protective measures. European Review for Medical and Pharmacological Sciences, 24(6), 3422–3425.

Bascones-Martinez, A., Mattila, R., Gomez-Font, R., & Meurman, J. H. (2014). Immunomodulatory drugs: Oral and systemic adverse effects. Medicina oral, patologia oral y cirugia bucal, 19(1), e24. http://dx.doi.org/doi:10.4317/medoral.19087

Basu, A., Sarkar, A., & Maulik, U. (2020). Molecular docking study of potential phytochemicals and their effects on the complex of SARS-CoV2 spike protein and human ACE2. Scientific Reports, 10(1), 17699. https://doi.org/10.1038/s41598-020-74715-4

Behl, T., Rocchetti, G., Chadha, S., Zengin, G., Bungau, S., Kumar, A., Mehta, V., Uddin, M. S., Khullar, G., Setia, D., Arora, S., Sinan, K. I., Ak, G., Putnik, P., Gallo, M., & Montesano, D. (2021). Phytochemicals from plant foods as potential source of antiviral agents: An overview. Pharmaceuticals, 14(4), 381. https://doi.org/10.3390/ph14040381

Benarba, B., & Pandiella, A. (2020). Medicinal plants as sources of active molecules against COVID-19. Frontiers in Pharmacology, 11, 1189. https://doi.org/10.1080/07391102.2020.1758788

Borkotoky, S., & Banerjee, M. (2020). A computational prediction of SARS-Cov-2 structural protein inhibitors from *Azadirachta indica* (neem). Journal of Biomolecular Structure & Dynamics, 39(21): 4111–4121. https://doi.org/10.1080/07391102.2020.1774419

Boukhatem, M. N., & Setzer, W. N. (2020). Aromatic herbs, medicinal plant-derived essential oils, and phytochemical extracts as potential therapies for coronaviruses: Future perspectives. Plants, 9(6), 800.

Bousquet, J., Anto, J., Czarlewski, W., Haahtela, T., Fonseca, S., & Iaccarino, G. (2020). Sulforaphane: From death rate heterogeneity in countries to candidate for prevention of severe COVID-19. Authorea 14:100498. https://doi.org/10.22541/au.159493397.79345039

Cao, B., Wang, Y., Wen, D., Liu, W., Wang, J., Fan, G., ... & Wang, C. (2020). A trial of lopinavir–ritonavir in adults hospitalized with severe Covid-19. New England Journal of Medicine, 382(19), 1787–1799.

Chen, H., Xie, Z., Zhu, Y., Chen, Q., & Xie, C. (2020). Chinese medicine for COVID-19: A protocol for systematic review and meta-analysis. Medicine, 99(25), e20660.

Ikpa, Chiyere, B. C., Maduka Tochukwu, O. D., Enyoh, C. E., & JM, I. U. (2020). Potential plants for treatment and management of COVID-19 in Nigeria. Academic Journal of Chemistry, 5(6), 69–80.

Chojnacka, K., Witek-Krowiak, A., Skrzypczak, D., Mikula, K., & Młynarz, P. (2020). Phytochemicals containing biologically active polyphenols as an effective agent against Covid-19-inducing coronavirus. Journal of Functional Foods, 73, 104146. https://doi.org/10.1016/j.jff.2020.104146

Davella, R., Gurrapu, S., & Mamidala, E. (2022). Phenolic compounds as promising drug candidates against COVID-19 - An integrated molecular docking and dynamics simulation study. Materials Today: Proceedings, 51(1), 522–527. https://doi.org/10.1016/j.matpr.2021.05.595

De Wilde, A. H., Snijder, E. J., Kikkert, M., & van Hemert, M. J. (2018). Host factors in coronavirus replication. Current Topics in Microbiology and Immunology, 49, 1–42.

Dzobo, K. (2022). The Role of Natural Products as Sources of Therapeutic Agents for Innovative Drug Discovery. Comprehensive Pharmacology (pp. 408–422). Elsevier. https://doi.org/10.1016/b978-0-12-820472-6.00041-4

Emirik, M. (2022). Potential therapeutic effect of turmeric contents against SARS-CoV-2 compared with experimental COVID-19 therapies: in silico study. Journal of Biomolecular Structure and Dynamics, 40(5), 2024–2037. https://doi.org/10.1080/07391102.2020.1835719

Espaňo, E., Kim, J., Lee, K., & Kim, J.-K. (2021). Phytochemicals for the treatment of COVID-19. *Journal of Microbiology*, 59(11), 959–977. https://doi.org/10.1007/s12275-021-1467-z

Fantini, J., Di Scala, C., Chahinian, H., & Yahi, N. (2020). Structural and molecular modelling studies reveal a new mechanism of action of chloroquine and hydroxychloroquine against SARS-CoV-2 infection. International journal of antimicrobial agents, 55(5), 105960. https://doi.org/10.1016/j.ijantimicag.2020.105960

Gautret, P., Lagier, J. C., Parola, P., Meddeb, L., Mailhe, M., Doudier, B., ... & Raoult, D. (2020). Hydroxychloroquine and azithromycin as a treatment of COVID-19: Results of an open-label non-randomized clinical trial. International Journal of Antimicrobial Agents, 56(1), 105949.

Ghidoli, M., Colombo, F., Sangiorgio, S., Landoni, M., Giupponi, L., Nielsen, E., & Pilu, R. (2021). Food containing bioactive flavonoids and other phenolic or sulfur phytochemicals with antiviral effect: Can we design a promising diet against COVID-19? Frontiers in Nutrition, 8. https://doi.org/10.3389/fnut.2021.661331

Ghosh, A. K., Brindisi, M., Shahabi, D., Chapman, M. E., & Mesecar, A. D. (2020). Drug development and medicinal chemistry efforts toward SARS-Coronavirus and Covid-19 therapeutics. ChemMedChem, 15(11), 907–932. https://doi.org/10.1002/cmdc.202000223

Ghosh, R., Chakraborty, A., Biswas, A., & Chowdhuri, S. (2021). Identification of alkaloids from *Justicia adhatoda* as potent SARS CoV-2 main protease inhibitors: An *in silico* perspective. Journal of Molecular Structure, 1229, 129489. https://doi.org/10.1016/j.molstruc.2020.129489

Ghosh, R., Chakraborty, A., Biswas, A., & Chowdhuri, S. (2021) Evaluation of green tea polyphenols as novel corona virus (SARS Cov-2) main protease (Mpro) inhibitors – An in silico docking and molecular dynamics simulation study. Journal of Biomolecular Structure and Dynamics, 39(12), 4362–4374. https://doi.org/10.1080/07391102.2020.1779818

Goyal, R. K., Majeed, J., Tonk, R., Dhobi, M., Patel, B., & Sharma, K., et al. (2020). Current targets and drug candidates for prevention and treatment of SARSCoV-2 (COVID-19) infection. Reviews in Cardiovascular Medicine, 21, 365–384. https://doi.org/10.31083/j.rcm.2020.03.118

Gyebi, G. A., Ogunro, O. B., Adegunloye, A. P., Ogunyemi, O. M., & Afolabi, S. O. (2021). Potential inhibitors of coronavirus 3-chymotrypsin-like protease (3CLpro): An in silico screening of alkaloids and terpenoids from African medicinal plants. Journal of Biomolecular Structure and Dynamics, 39(9), 3396–3408. https://doi.org/10.1080/07391102.2020.1764868

Haake, C., Cook, S., Pusterla, N., & Murphy, B. (2020). Coronavirus infections in companion animals: Virology, epidemiology, clinical and pathologic features. Viruses, 12, 1023. https://doi.org/10.3390/v12091023

Henss, L., Auste, A., Schürmann, C., Schmidt, C., Von Rhein, C., & Mühlebach, M. D., et al. (2021). The green tea catechin epigallocatechin gallate inhibits SARSCoV-2 infection. Journal of General Virology, 102, 001574. https://doi.org/10.1099/jgv.0.001574

Hoffmann, M., Kleine-Weber, H., Schroeder, S., Krüger, N., Herrler, T., Erichsen, S., … & Pöhlmann, S. (2020). SARS-CoV-2 cell entry depends on ACE2 and TMPRSS2 and is blocked by a clinically proven protease inhibitor. Cell, 181(2), 271–280.

Jamhour, R. M., Al-Nadaf, A. H., Wedian, F., Al-Mazaideh, G. M., Mustafa, M., Huneif, M. A., … & Alakhras, F. (2022). Phytochemicals as a potential inhibitor of COVID-19: An in-silico perspective. Russian Journal of Physical Chemistry A, 96(7), 1589–1597.

Jiang, S., Hillyer, C., & Du, L. (2020). Neutralizing antibodies against SARS-CoV-2 and other human coronaviruses. Trends in Immunology, 41(5), 355–359.

Jo, S., Kim, S., Shin, D. H., & Kim, M.-S. (2019). Inhibition of SARS-CoV 3CL protease by flavonoids. Journal of Enzyme Inhibition and Medicinal Chemistry, 35(1), 145–151. https://doi.org/10.1080/14756366.2019.1690480

Joshi, M. G., Kshersagar, J., Desai, S. R., & Sharma, S. (2020). Antiviral properties of placental growth factors: A novel therapeutic approach for COVID-19 treatment. Placenta, 99, 117–130. https://doi.org/10.1016/j.placenta.2020.07.033

Katiyar, C., Kanjilal, S., Gupta, A., & Katiyar, S. (2012). Drug discovery from plant sources: An integrated approach. AYU (An International Quarterly Journal of Research in Ayurveda), 33(1), 10. https://doi.org/10.4103/0974-8520.100295

Keshavarzi Arshadi, A., Webb, J., Salem, M., Cruz, E., Calad-Thomson, S., Ghadirian, N., Collins, J., Diez-Cecilia, E., Kelly, B., Goodarzi, H., & Yuan, J. S. (2020). Artificial intelligence for COVID-19 drug discovery and vaccine development. Frontiers in Artificial Intelligence, 3, 65. https://doi.org/10.3389/frai.2020.00065

Khaerunnisa, S., Kurniawan, H., Awaluddin, R., Suhartati, S., & Soetjipto, S. (2020). Potential inhibitor of COVID-19 main protease (Mpro) from several medicinal plant compounds by molecular docking study. Preprints. https://doi.org/10.20944/preprints202003.0226.v1

Khanal, L. N., Pokharel, Y. R., Sharma, K., & Kalauni, S. K. (2020). Plant-derived secondary metabolites as potential mediators against COVID-19: A review. Prithvi Academic Journal, 3(1), 1–18. https://doi.org/10.3126/paj.v3i1.31282

Khan, S., Liu, J., & Xue, M. (2020). Transmission of SARS-CoV-2, required developments in research and associated public health concerns. Frontiers in Medicine, 7, 310.

Kim, D., Lee, J. Y., Yang, J. S., Kim, J. W., Kim, V. N., & Chang, H. (2020). The architecture of SARS-CoV-2 transcriptome. Cell, 181(4), 914–921.

Kumar, Y., Singh, H., & Patel, C. N. (2020). In silico prediction of potential inhibitors for the main protease of SARS-CoV-2 using molecular docking and dynamics simulation based drug-repurposing. Journal of Infection and Public Health, 13(9), 1210–1223.

Laksmiani, N. P. L., Larasanty, L. P. F., Santika, A. A. G. J., Prayoga, P. A. A., Dewi, A. A. I. K., & Dewi, N. P. A. K. (2020). Active compounds activity from the medicinal plants against SARS-CoV-2 using in silico assay. Biomedical and Pharmacology Journal, 13(2), 873–881.

Liu, Y. X., Zhou, Y. H., Jiang, C. H., Liu, J., & Chen, D. Q. (2022). Prevention, treatment and potential mechanism of herbal medicine for Corona viruses: A review. Bioengineered, 13(3), 5480–5508.

Mahmud, S., Hasan, M. R., Biswas, S., Paul, G. K., Afrose, S., Mita, M. A., Sultana Shimu, M. S., Promi, M. M., Hani, U., Rahamathulla, M., Khan, M. A., Zaman, S., Uddin, M. S., Rahmatullah, M., Jahan, R., Alqahtani, A. M., Saleh, M. A., & Emran, T. B. (2021). Screening of potent phytochemical inhibitors against SARS-CoV-2 main protease: An integrative computational approach. Frontiers in Bioinformatics, 1, 717141. https://doi.org/10.3389/fbinf.2021.717141

Ma, L., Li, H., Lan, J., Hao, X., Liu, H., Wang, X., & Huang, Y. (2021). Comprehensive analyses of bioinformatics applications in the fight against COVID-19 pandemic. Computational Biology and Chemistry, 95, 107599. https://doi.org/10.1016/j.compbiolchem.2021.107599

Mani, J. S., Johnson, J. B., Steel, J. C., Broszczak, D. A., Neilsen, P. M., Walsh, K. B., & Naiker, M. (2020). Natural product-derived phytochemicals as potential agents against coronaviruses: A review. Virus Research, 284, 197989.

Manoharan, Y., Haridas, V., Vasanthakumar, K. C., Muthu, S., Thavoorullah, F. F., & Shetty, P. (2020). Curcumin: A wonder drug as a preventive measure for COVID19 management. Indian Journal of Clinical Biochemistry, 35, 373–375. https://doi.org/10.1007/s12291-020-00902-9

Mathur, S., & Hoskins, C. (2017). Drug development: Lessons from nature. Biomedical Reports, 6(6), 612–614. https://doi.org/10.3892/br.2017.909

Matsuyama, S., Nao, N., Shirato, K., Kawase, M., Saito, S., Takayama, I., ... & Takeda, M. (2020). Enhanced isolation of SARS-CoV-2 by TMPRSS2-expressing cells. Proceedings of the National Academy of Sciences, 117(13), 7001–7003.

Mirzaie, A., Halaji, M., Dehkordi, F. S., Ranjbar, R., & Noorbazargan, H. (2020). A narrative literature review on traditional medicine options for treatment of corona virus disease 2019 (COVID-19). Complementary Therapies in Clinical Practice, 40, 101214.

Mohapatra, P. K., Chopdar, K. S., Dash, G. C., & Raval, M. K. (2020). In Silico Screening of Phytochemicals of Ocimum Sanctum Against Main Protease of SARS-CoV-2. ChemRxiv. https://doi.org/10.26434/chemrxiv.12599915.v1

Mohanty, M., Mishra, B., Singh, A. K., Mohapatra, P. R., Gupta, K., Patro, B. K., ... & Bhuniya, S. (2022). Comparison of clinical presentation and vaccine effectiveness among omicron and non-omicron SARS coronavirus-2 patients. Cureus, 14(12). https://doi.org/10.7759/cureus.32354

Mukherjee, P. K., Efferth, T., Das, B., Kar, A., Ghosh, S., Singha, S., ... & Haldar, P. K. (2022). Role of medicinal plants in inhibiting SARS-CoV-2 and in the management of post-COVID-19 complications. Phytomedicine, 98, 153930.

Nadeem, M. K. (2020). In-silico study to elucidate corona virus by plant phytoderivatives that hits as a fusion inhibitor targeting HR1 domain in spike protein which conformational changes efficiently inhibit entry COVID-19. *Translational Biomedicine*, 11(3), 1.

Nag, A., Paul, S., Banerjee, R., & Kundu, R. (2021). In silico study of some selective phytochemicals against a hypothetical SARS-Cov-2 spike RBD using molecular docking tools. Computers in Biology and Medicine, 137, 104818. https://doi.org/10.1016/j.compbiomed.2021.104818

Nivetha, R., Bhuvaragavan, S., & Janarthanan, S. (2021). Inhibition of multiple SARS-CoV-2 roteins by an antiviral biomolecule, Seselin from Aeglemarmelos deciphered using molecular docking analysis. Research Square. https://doi.org/10.21203/rs.3.rs-31134/v1

Pal, S., Chowdhury, T., Paria, K., Manna, S., Parveen, S., Singh, M., Sharma, P., Islam, S. S., Abu Imam Saadi, S. M., & Mandal, S. M. (2022). Brief survey on phytochemicals to prevent COVID-19. Journal of the Indian Chemical Society, 99(1), 100244. https://doi.org/10.1016/j.jics.2021.100244

Panda, P. K., Arul, M. N., Patel, P., Verma, S. K., Luo, W., Rubahn, H. G., ... & Ahuja, R. (2020). Structure-based drug designing and immunoinformatics approach for SARS-CoV-2. Science advances, 6(28), eabb8097. https://doi.org/10.1126/sciadv.abb8097

Pandey, P., Basnet, A., & Mali, A. (2020). Quest for COVID-19 cure: Integrating traditional herbal medicines in the modern drug paradigm. Applied Science and Technology Annals, 1(1), 63–71.

Pandey, P., Singhal, D., Khan, F., & Arif, M. (2021). An in silico screening on Piper Nigrum, Syzygium Aromaticum and Zingiber Officinale roscoe derived compounds against Sars-Cov-2: A drug repurposing approach. Biointerface Research in Applied Chemistry. 11. https://doi.org/10.33263/BRIAC114.1112211134

Paudyal, V., Sun, S., Hussain, R., Abutaleb, M. H., & Hedima, E. W. (2022). Complementary and alternative medicines use in COVID-19: A global perspective on practice, policy and research. Research in Social and Administrative Pharmacy 18(3), 2524–2528. https://doi.org/10.1016/j.sapharm.2021.05.004

Poduri, R., Joshi, G., & Jagadeesh, G. (2020). Drugs targeting various stages of the SARS-CoV-2 life cycle: Exploring promising drugs for the treatment of Covid-19. Cellular Signalling, 74, 109721.

Poojary, S. (2020). Role of bioinformatics, computational biology and computer technologies in combating COVID-19 virus – A review. International Journal of Biotech Trends and Technology, 10(2), 26–30. https://doi.org/10.14445/22490183/ijbtt-v10i2p605

Pramanik, S., Seth, D., & Debnath, S. (2023). Insights into In Silico Methods to Explore Plant Bioactive Substances in Combating SARS-CoV-2. In Ethnopharmacology and Drug Discovery for COVID-19: Anti-SARS-CoV-2 Agents from Herbal Medicines and Natural Products (pp. 243–264). Singapore: Springer Nature Singapore.

Prasanth, D. S. N. B. K., Murahari, M., Chandramohan, V., Panda, S. P., Atmakuri, L. R., & Guntupalli, C. (2020). In silico identification of potential inhibitors from Cinnamon against main protease and spike glycoprotein of SARS CoV-2. Journal of Biomolecular Structure and Dynamics, 39(13), 4618–4632. https://doi.org/10.1080/07391102.2020.1779129

Rajendran, K., Krishnasamy, N., Rangarajan, J., Rathinam, J., Natarajan, M., & Ramachandran, A. (2020). Convalescent plasma transfusion for the treatment of COVID-19: Systematic review. Journal of Medical Virology, 92(9), 1475–1483.

Rastogi, S., Pandey, D. N., & Singh, R. H. (2022). COVID-19 pandemic: A pragmatic plan for ayurveda intervention. Journal of Ayurveda and Integrative Medicine, 13(1), 100312.

Rathinavel, T., Palanisamy, M., Srinivasan, P., Subramanian, A., & Thangaswamy, S. (2020). Phytochemical 6-Gingerol – A promising drug of choice for COVID-19. International Journal of Advanced Science and Engineering, 06, 1482–1489. https://doi.org/10.29294/ijase.6.4.2020.1482-1489

Ray, M., Sable, M. N., Sarkar, S., & Hallur, V. (2021). Essential interpretations of bioinformatics in COVID-19 pandemic. Meta Gene, 27, 100844. https://doi.org/10.1016/j.mgene.2020.100844

Ren, Y., Shu, T., Wu, D., Mu, J., Wang, C., & Huang, M., et al. (2020). The ORF3a protein of SARS-CoV-2 induces apoptosis in cells. Cellular & Molecular Immunology 17, 881–883. https://doi.org/10.1038/s41423-020-0485-9

Ristovski, J. T., Matin, M. M., Kong, R., Kusturica, M. P., & Zhang, H. (2022). In vitro testing and computational analysis of specific phytochemicals with antiviral activities considering their possible applications against COVID-19. South African Journal of Botany, 151, 248–258. https://doi.org/10.1016/j.sajb.2022.02.009

Sagar, V., & Kumar, A. H. (2020). Efficacy of natural compounds from Tinospora cordifolia against SARS-CoV-2 protease, surface glycoprotein and RNA polymerase. Biology, Engineering, Medicine and Science Reports, 6(1), 06–08.

Saxena, B. (2022). The Role of Complementary and Alternative Medicines in the Treatment and Management of COVID-19. In Complementary Therapies. IntechOpen. https://doi.org/10.5772/intechopen.100422

Shahrajabian, M. H., Sun, W., Shen, H., & Cheng, Q. (2020). Chinese Herbal medicine for SARS and SARS-CoV-2 treatment and prevention, encouraging using herbal medicine for COVID-19 outbreak. Acta Agriculturae Scandinavica, Section B—Soil & Plant Science, 70(5), 437–443.

Shawky, E., Nada, A. A., & Ibrahim, R. S. (2020). Potential role of medicinal plants and their constituents in the mitigation of SARS-CoV-2: Identifying related therapeutic targets using network pharmacology and molecular docking analyses. RSC Advances, 10(47), 27961–27983.

Sinha, S. K., Shakya, A., Prasad, S. K., Singh, S., Gurav, N. S., Prasad, R. S., & Gurav, S. S. (2021). An in-silico evaluation of different Saikosaponins for their potency against SARS-CoV-2 using NSP15 and fusion spike glycoprotein as targets. Journal of Biomolecular Structure and Dynamics, 39(9), 3244–3255.

Soraci, L., Lattanzio, F., Soraci, G., Gambuzza, M. E., Pulvirenti, C., Cozza, A., Corsonello, A., Luciani, F., & Rezza, G. (2022). COVID-19 vaccines: Current and future perspectives. Vaccines, 10(4), 608. https://doi.org/10.3390/vaccines10040608

Srivastava, A. K., Kumar, A., & Misra, N. (2020). On the inhibition of COVID-19 protease by Indian herbal plants: An in silico investigation. arXiv preprint arXiv:2004.03411.

Tallei, T. E., Tumilaar, S. G., Niode, N. J., Fatimawali, Kepel, B. J., Idroes, R., Effendi, Y., Sakib, S. A., & Emran, T. B. (2020). Potential of Plant Bioactive Compounds as SARS-CoV-2 Main Protease (Mpro) and Spike (S) Glycoprotein Inhibitors: A Molecular Docking Study. In C. Riganti (Ed.), Scientifica (Vol. 2020, pp. 1–18). Hindawi Limited. https://doi.org/10.1155/2020/6307457

Tripathi, A., Tyagi, V. K., Vivekanand, V., Bose, P., & Suthar, S. (2020). Challenges, opportunities and progress in solid waste management during COVID-19 pandemic. Case Studies in Chemical and Environmental Engineering, 2, 100060. https://doi.org/10.1016/j.cscee.2020.100060

Vetter, P., Kaiser, L., Calmy, A., Agoritsas, T., & Huttner, A. (2020). Dexamethasone and remdesivir: finding method in the COVID-19 madness. The Lancet Microbe, 1(8), e309–e310. https://doi.org/10.1016/S2666-5247(20)30173-7

V'kovski, P., Kratzel, A., Steiner, S., Stalder, H., & Thiel, V. (2021). Coronavirus biology and replication: Implications for SARS-CoV-2. Nature Reviews Microbiology, 19, 155–170. https://doi.org/10.1038/s41579-020-00468-6

Wahedi, H. M., Ahmad, S., & Abbasi, S. W. (2021). Stilbene-based natural compounds as promising drug candidates against COVID-19. Journal of Biomolecular Structure and Dynamics, 39(9), 3225–3234.

Wang, L. Y., Cui, J. J., Ouyang, Q. Y., Zhan, Y., Guo, C. X., & Yin, J. Y. (2020b). Remdesivir and COVID-19. The Lancet, 396(10256), 953–954. https://doi.org/10.1016/S0140-6736(20)32019-5

Wang, M., Cao, R., Zhang, L., Yang, X., Liu, J., Xu, M., ... & Xiao, G. (2020a). Remdesivir and chloroquine effectively inhibit the recently emerged novel coronavirus (2019-nCoV) in vitro. Cell Research, 30(3), 269–271.

Yang, Y., Islam, M. S., Wang, J., Li, Y., & Chen, X. (2020). Traditional Chinese medicine in the treatment of patients infected with 2019-new coronavirus (SARS-CoV-2): A review and perspective. International Journal of Biological Sciences, 16(10), 1708.

Yashvardhini, N., Samiksha, S., & Jha, D. (2021). Pharmacological intervention of various Indian medicinal plants in combating COVID-19 infection. Biomedical Research and Therapy, 8(7), 4461–4475. http://bmrat.org/index.php/BMRAT/article/view/685

Zhang, D., Wu, K., Zhang, X., Deng, S., & Peng, B. (2020). In silico screening of Chinese herbal medicines with the potential to directly inhibit 2019 novel coronavirus. Journal of Integrative Medicine, 18(2), 152–158. https://doi.org/10.1016/j.joim.2020.02.005

Zhao, J. (2020). Investigating mechanism of Qing-Fei-Pai-Du-Tang for treatment of COVID-19 by network pharmacology. Chinese Traditional and Herbal Drugs, 829–835.

3 Anti-Virus and Anti-SARS-CoV-2 Activities of Traditional Herbal Medicines

Cagla Celik, Nilay Ildiz, Ayse Baldemir Kılıc,
Sadi Yusufbeyoglu, and Ismail Ocsoy

3.1 INTRODUCTION

Viruses are pathogens associated with many diseases, including difficult-to-treat diseases such as cancer, Alzheimer's disease and type 1 diabetes [1, 2]. Increased travel and interaction in developing countries significantly increase contamination and spread. In particular, viruses for which there are no vaccines or effective drug treatments pose a serious regional public health threat. Similar situations have occurred with outbreaks of the dengue virus, measles virus, influenza virus and severe acute respiratory syndrome (SARS) virus [3, 4]. Although there are many approved antiviral drugs, the development of antiviral drugs is still very important. Drug resistance is developing against enzyme inhibitors and, as a result, drugs become ineffective [5–7]. The lack of effective drugs and protective vaccines makes it difficult to combat viral pathogens. The world has seen the best example of this situation in the pandemic infection called coronavirus disease that emerged in 2019.

Coronavirus disease 2019 (COVID-19), also known as severe acute respiratory syndrome, is a disease caused by sever acute respiratory syndrome coronavirus 2 (SARS-CoV-2). It first appeared in Wuhan, China, in December 2019 and spread around the world as a pandemic. SARS-CoV-2 affects the angiotensin-converting enzyme 2 (ACE2) receptor in the lungs. Although SARS-CoV-2 affects the respiratory system, it can have devastating effects on the entire body, including the heart, intestines, kidneys and bile [8, 9]. The presence of this pathogen stimulates the body's immune system and triggers the production of cytokines. This can lead to a cytokine storm and symptoms can be very serious [9]. Symptoms may include fever, fatigue, dry cough and, depending on the severity, excessive inflammation, lung damage, lung tissue edema and coagulation problems. Cytokine storms can cause serious damage and may lead to death [10, 11].

The search continues for the development of effective treatments and vaccines. There is a need to develop anti-SARS-CoV-2 drugs. Initially, broad-spectrum antibiotics, various antiviral drugs and immunosuppressive corticosteroids have been used for treatment [12]. In particular, the therapeutic efficacy of other antiviral drugs has been tested for the treatment of COVID-19. Drugs such as ivermectin, chloroquine, hydroxychloroquine, lopinavir, camostat mesylate, ritonavir, favipiravir and remdesivir have been used to eradicate the virus. They have been tested in clinical trials but have not shown the expected effect [13]. Currently, the lack of SARS-CoV-2-specific and effective drugs perpetuates the severity of the disease [14]. The high rate of spread of COVID-19 is a cause for concern. For this reason, some people have begun to prefer herbal medicines for the treatment and protection of COVID-19.

Herbal medicines have been used for many years in different cultures in complementary and alternative medicine. Today, phytotherapy is also very important in modern medicine. Plants have been preferred as a source of medicines, especially in the treatment of cancer. High-efficacy therapeutic drugs such as vinblastine, paclitaxel and vincristine have been discovered [15]. In addition, a variety of herbal biomolecules such as piperidines, picrosides, fillantins and curcuminoids have

DOI: 10.1201/9781003452621-3

been used in traditional Ayurvedic medicine [16]. Artemisinin is an antimalarial drug isolated from the plant *Artemisia annua*. This discovery depends on Traditional Chinese Medicine (TCM) [17]. People believe that herbal medicines are more environmentally friendly, accessible, reliable and affordable than laboratory-synthesized drugs. This perception has increased the tendency to use herbal medicines in recent years. The efficacy of traditional herbal medicines is due to the presence of biomolecules with therapeutic activity in different parts of the plant. There are compounds with pharmacological effects in many parts of plants, such as roots, leaves and flowers [18]. The COVID-19 outbreak occurred at a time when patients preferred herbal medicines for the treatment of a variety of diseases. Similarly, herbal medicines have become increasingly popular in different cultures to prevent and treat the disease. The lack of a specific drug for SARS-CoV-2 tested in modern medicine has increased the tendency to use traditional herbal medicines. Herbal medicines have the potential to directly affect the virus, signaling pathways, virus-host interaction, host receptors and the immune system [11]. The preference for herbal medicines is most pronounced in Asian cultures. These products are widely used in many countries such as Korea, Japan, Thailand, Vietnam, India, Indonesia, Bangladesh and especially China. In the COVID-19 treatment strategy of TCM, it is observed that physicians prescribe herbal drugs alone or in combination with modern synthetic drugs [11]. The National Health Commission of the People's Republic of China has published guidelines for the diagnosis and treatment of COVID-19. TCM is recommended in this guideline [19]. The sixth edition of this guideline recommends the use of Jinhua Qinggan (JHQG) granules, Huoxiang Zhengqi (HXZQ) capsules, Shufeng Jiedu (SFJD) capsules, or Lianhua Qingwen (LHQW) capsules depending on the stage of the disease [20]. On 14 April 2020, a Chinese official held a press conference and shared critical information on this issue. He announced that three patented herbal medicines of TCM, including LHQW capsules and JHQG granules, had been approved as indicated for the treatment of COVID-19 [21]. It is indicated for patients with mild disease. The guideline also includes information on traditional treatment, covering more than 74,187 (91.5%) of confirmed cases in Chinese people. Data from the treatment of these cases, most of which had mild symptoms, showed that patients using TCM-recommended herbal medicines shortened the duration of the disease and prevented its progression by relieving symptoms [22, 23]. In ancient Chinese practices, various herbal medicines are used for antiviral activity. In particular, herbal medicines were used effectively to prevent SARS triggered by SARS coronavirus (SARS-CoV) and to reduce the mortality ratio in 2003. Because of their effectiveness in preventing the spread of the SARS virus in the past, Chinese medicine again used herbal medicines as a weapon during the COVID-19 outbreak. In the absence of effective drugs and vaccines, the trend toward herbal medicine has increased in this culture. The guidelines published by China for COVID-19 argue that TCM is effective in the prevention and treatment of diseases in various age groups, including children and the elderly [24–27]. In addition, their hypothesis on this issue is that there will be no adverse effects [28, 29]. Herbal medicines have a high potential for use in the treatment of various diseases, especially infectious diseases, due to their low toxicity and easy availability [30]. In addition, biomolecules in herbal medicines can be screened and targeted by computational drug design technology [31, 32]. This chapter aims to summarize the antiviral herbal medicines and their mechanisms of action that are most commonly used in Asian countries. It will provide general information on the use of antiviral herbal medicines against various viruses, especially SARS-CoV-2.

3.2   COMMON HERBAL MEDICINES IN COVID-19

3.2.1   TYPES OF CHINESE HERBAL MEDICINE

TCM is based on a philosophy that combines many therapeutic experiences in the control and treatment of disease. TCM has been used to prevent and treat infectious diseases for more than 2000 years. The general approach of Chinese medicine is 'multi-component, multi-target, multi-path'. Treatment is individualized according to the physical condition and living conditions of the individual [33–35]. Chinese medicine characterizes COVID-19 disease as an epidemic and attaches importance to the stage of the disease. It is recommended to take different herbal medicines according to the stage of the disease [36].

In China, the traditional treatment recommendations of Chinese medicine for COVID-19 patients have been officially published. Thus, the use of herbal medicines such as *A. annua* (antimalarial herbal medicine), Qingfei Paidu (QFPD) Tang, Gancao, Huoxiang Zhengqi and Cangshu has been officially announced [20]. Guidelines for effective treatment of COVID-19 with TCM have been published in China include 17 herbal medicine formulas. In these guidelines, the Angong Niuhuang Pill or Su He Xiang Pill and the Shen Fu Tang herbal formula are most commonly recommended when the disease is severe. In the recovery phase, the Li Zhong Pill with Xiang Sha Liu Junzi Tang formula can be used together. Only in severe cases can the Angong Niuhuang Pill, Zhi Bao Dan, Zi Xue San and Su He Xiang Pill be prescribed alone. In addition, four different QFPD Tang's herbal formulas, prepared with different combinations of 21 different herbs, can be used according to the stage of the disease. TCM herbal medicines recommended by the COVID-19 guidelines are shown in Table 3.1 [20].

TCM recommendations for the treatment of COVID-19 in children play an important role in the guidelines. Xiang Su San and Yin Qiao San are recommended for mild cases. For moderate cases, the guidelines recommend San Ren Tang and Ma Xing Shi Gan Tang, and for severe cases, Buhuan Jin Zhengqi San, Ganlu Xiaodu Dan and Xuanbai Chengqi Tang. Children in recovery can use Yu Ping Feng San and Liu Junzi Tang in combination. When all the guidelines are analyzed, *Poria Sclerotium, Ophiopogonis Radix, Citri Reticulatae Pericarpium* are among the most recommended drugs.

From another point of view, it is a much better strategy to prevent the disease rather than to get the disease. One of the principles of TCM is prophylaxis [38]. In Chinese medicine, oral intake of various herbal medicines has been recommended for prophylaxis [39]. Prescriptions for the prevention of SARS-CoV-2 infection based on the 'natural factor' theory have been proposed in Chinese medicine. Scientific pharmacological studies on the efficacy of the recommended herbal medicine ingredients have shown that these substances are effective against the virus. *Radix astragali* (Huangqi), *Radix saposhnikoviae* (Fangfeng), *Radix glycyrrhizae* (Gancao), *Lonicerae japonicae flos* (Jinyinhua), *Rhizoma atractylodis macrocephalae* (Baizhu) and *Forsythiae Fructus* (Lianqiao)

TABLE 3.1

TCM that Recommended by the COVID-19 Guidelines [37]

Stage of Disease	Symptom	Recommended Chinese Patent Medicine
Medical observation period	Fatigue with gastrointestinal discomfort	Huo Xiang Zheng Qi Shui
	Fatigue with fever	Ian Hua Qing Wen capsule, Shu Feng Jie Du capsule, Jin Hua Qing Gan capsule
Clinical treatment period	Mild cases	Qing Fei Pai Du Tang
	General cases	Qing Fei Pai Du Tang
	Severe cases	Xi Yan Ping injection, Xue Bi Jing injection, Re Du Nine injection, Tan Re Qing injection, Xing Nao Jing injection, Qing Fei Pai Du Tang
	Critical cases	Xue Bi Jing injection, Re Du Nine injection, Tan Re Qing injection, Shen Fu injection, Sheng Mai injection, Shen Mai injection

are the most recommended herbs in TCM [40]. Among these plants, Huangqi, Baizhu and Fangfeng are included in the herbal medicine known as Yupingfeng Powder [38]. Lau et al., have suggested that Yupingfeng powder has anti-inflammatory, antiviral and immune supportive properties against the threat of influenza. Thus, its protective and therapeutic efficacy in the fight against SARS has been demonstrated [38]. In addition, glycyrrhizic acid in Gancao (Chinese name of *Radix glycyrrhizae*) has also been suggested to be effective in preventing infection [41]. Yinqiao powder, which is used for prophylaxis, also contains Lianqiao and Jinyinhua [42]. All these studies support the efficacy of herbal medicines recommended by TCM for the prevention of COVID-19. The lack of a specific WHO-recommended drug for the treatment and prevention of COVID-19 has increased the tendency of most medical professionals in China toward TCM treatment strategies [43].

3.2.2   TYPES OF ASIAN HERBAL MEDICINE

Apart from the fact that the most common traditional medicine recommendations are in China, a traditional medicine guideline for the prophylaxis and effective treatment of COVID-19 has been published by official institutions in South Korea [44]. This guideline includes 15 herbal formulas for the treatment of mild disease. There are three herbal formulas for patients with more severe symptoms and two herbal formulas for the recovery phase [45]. At the same time, herbal medicines with local names such as Qing-Fei-Pai-Du-Tang, Huo-Xiang-ZhengqiSan and Yin-Qiao-San were recommended by experts for the treatment of COVID-19. Of these prescriptions, 30% were for the use of Qing-Fei-Pai-Du-Tang [46]. South Korea has previously used Youngyopaedoc San plus Saengmaek San and Youngyopaedoc San plus Bojungikgi Tang to treat people infected with SARS-CoV2. Youngyopaedoc San plus Bojungikgi Tang and Youngyopaedoc San plus Bulhwangeumjeonggi San were also prescribed for mild-stage patients [45]. In Japan, Maoto and the traditional Japanese (Kampo) medicine Kakkontokasenkyushin have been used for the treatment of viral infection [47, 48]. In another study, ninjin'yoeito (NYT), a Japanese Kampo medicine, was found to be an effective drug in the treatment of severe COVID-19 patients [49].

In Thailand, *Andrographis paniculata* extract has been officially approved for the treatment of COVID-19. In addition, taking *A. paniculata* together with Maoto and Maotoka-senshinren is effective in viral inhibition [50].

3.2.3   TYPES OF SAUDI ARABIA HERBAL MEDICINE

Herbal medicines are widely used as part of traditional medicine in the Middle East, as well as in the Far East, to boost immunity and protect against disease. Among these, especially in Saudi Arabia, various dietary supplements and herbal medicines with pharmacological effects are used [51, 52]. A survey showed that Saudis, especially during the COVID-19 pandemic, used various herbs, especially cinnamon and garlic [51, 52]. In addition, the clinical effect of *Nigella sativa* oil (NSO) has been demonstrated [53]. One study showed that COVID-19 patients with mild symptoms recovered faster with *N. sativa* than with usual care alone. *N. sativa* is a popular herb used in traditional medicines in various cultures, not only in Saudi Arabia [54]. *N. sativa* extract contains a variety of antiviral molecules, such as thymoquinone [55]. Computational molecular drug studies have shown that the plant's major compounds, nigelid and alpha-hederin, can inhibit SAR-CoV-2 by docking to the main protease (Mpro) [56]. Although the mechanism of action has not been adequately clarified, recent studies have demonstrated their pharmacological efficacy. These results are seen as promising that the use of herbal medicine will increase in the future.

3.2.4   TYPES OF BANGLADESHI HERBAL MEDICINE

Physicians in Bangladesh recommend the use of masks, home isolation, and personal hygiene to prevent the spread of the disease. It is also recommended to consume supplementary food and some medicinal herbs to support the immune system. Although the consumption of mixed ginger, black cumin, honey and cloves in hot water is widely recommended as an herbal treatment, its effectiveness against SARS-CoV-2 needs to be proven [57].

3.2.5 TYPES OF INDIAN HERBAL MEDICINE

Ayurvedic and Siddha medicine are very common in the Indian community. The basic principle of Ayurvedic medicine is to evaluate the human body in its entirety, thus, treating the body as a whole. This approach can be effective in the treatment of COVID-19. For instance, plant-derived biomolecules such as curcumin, silymarin, palmatine and syringin are used in Ayurvedic medicine. These biomolecules have antiviral activity and they can be used as a potential agent against SARS-CoV-2 [58–60]. A study showed that Divya-Swasari-Vati, a calcium-supported herbal medicine used in India for respiratory diseases, exhibits antiviral activity [61].

The other common system is Siddha medicine in India. Siddha medicine focuses on the radical treatment of disease rather than symptomatic treatment. Accordingly, the use of mineral, herbal and animal medicines in treatment is recommended. In the case of high fever, the local name of Kabasura Kudineer's Aiya suram drug is recommended, and Aiya noigal is used for respiratory diseases. It has also been shown in clinical trials to reduce viral loading [62].

3.2.6 TYPES OF AFRICAN HERBAL MEDICINE

Due to the varying levels of economic development in different countries, modern medicines and vaccines are not equally available in all countries. Therefore, an adequate supply of vaccines and medicines may not be possible in these regions. Herbal medicines are used by more than 80% of the African population due to their availability and affordability. For this reason, herbal medicines are playing a critical role in the struggle against SARS-CoV-2 in African countries [63].

More than 68,000 different plants grow in the African region. Studies by Attah et al. [64], Nair et al. [65] and Asprilio and Wilar [66] suggest that there are bioactive components with antiviral activity in plants from the region. Some of the antiviral compounds found in the plants have been shown to alleviate COVID-19. Experimentally, they have also been shown to inhibit SARS-CoV-2 under *in vitro* conditions. However, there are no *in vivo* or clinical trials.

For this purpose, ginger, lemon, garlic, and the plant known locally as 'damakase' are used. Artemis afra tea is also preferred. However, the efficacy of these plants needs to be investigated [67]. In Ethiopia, these plants are also preferred for the treatment of COVID-19 [68].

COVID organics (CVO), COROCUR, IHP Tea Detox, STAR Yellow, and COVIDEX are examples of the most commonly used herbal medicines in Africa. CVO is a liquid drug that essentially contains the plant, *Artemesia annua*. It is licensed in Madagascar for the treatment of COVID-19. COROCUR is a dry powder containing the herb *Thymus vulgaris*. It is recommended as an adjuvant in the treatment of COVID-19. IHP Tea Detox is an herbal medicine containing the plants *A. paniculata*; *Garcinia kola* and *Psidium guajava*, which is recommended to be consumed in tea form. It is effective in the treatment of COVID-19. STAR Yellow is in liquid form and consists of 10 herbs including *Cephalotaxus mannii* and *Allium sativum*. It is claimed to prevent the transmission of COVID-19 in feces. COVIDEX is made from three plants including *Zanthoxylum gilleti* and *Warburgia ugandensis*. Like COROCUR, it is recommended as an adjuvant treatment. None of these drugs has been validated by randomized controlled clinical trials. Therefore, their efficacy and safety are unclear [69].

3.2.7 TYPES OF MEXICAN AMERICAN HERBAL MEDICINE

Herbal medicine has an important place in Mexico. It is one of the five most biodiverse countries in the world. It is the second-richest country in medicinal plants after China. In a study, 100 compounds isolated from medicinal plants grown in Mexico were examined. The binding rates of these

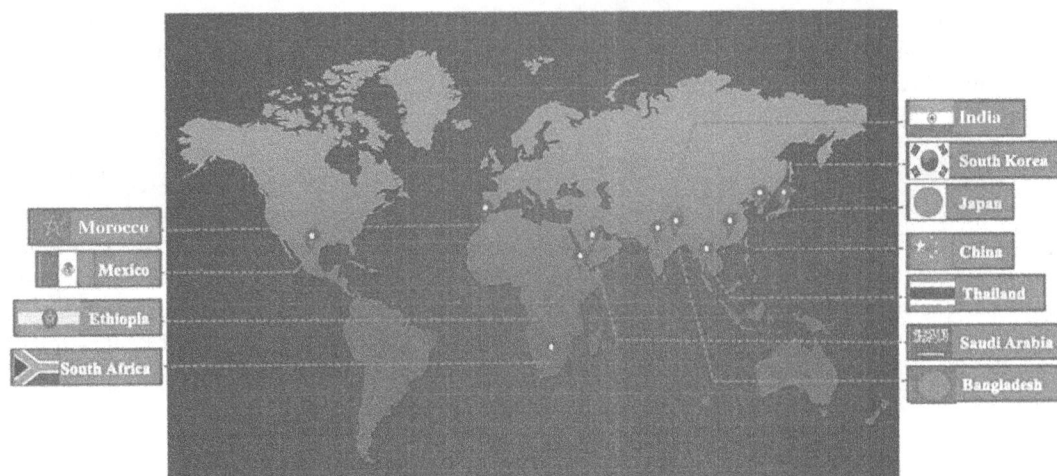

FIGURE 3.1 Countries that use traditional herbal medicines.

bioactive compounds to the SARS-CoV-2 were investigated and kaempferol, quercetin, emodin anthrone, esculin, luteolin, and cichorium were found to be effective. These compounds may have antiviral activity, but to become drug candidates, safety and efficacy need to be supported by further studies [70]. Figure 3.1 shows that countries where traditional herbal medicine is commonly used [37].

3.3 SYMPTOMATIC AND RADICAL EFFECTS OF HERBAL MEDICINES IN COVID-19

In silico and *in vitro* studies have shown that various plant-derived bioactive components have inhibitory effects on proteins in SARS or Middle East respiratory syndrome (MERS) coronavirus [71–74]. The gene sequence of SARS-CoV-2 is highly similar to the virus identified in other members of this family, SARS-CoV or MERS-CoV [75]. Therefore, TCM products used for the treatment of viral pneumonia caused by these viruses may also be effective against SARS-CoV-2. Herbal formulas recommended by TCM for the eradication of the SARS virus include Yin Qiao San [76, 77], Shuanghuanglian [78], Yu Ping Feng San [79], LHQW capsule, Sang Ju Yin [79], Ma Xin Gan Shi Tang [44]. These herbal formulas act as anti-inflammatory, antiviral, and immunomodulatory agents.

These herbal medicines recommended for the eradication of SARS-CoV are currently being used in the inhibition of SARS-CoV-2. One of these medicines, the LHQW capsule, has been clinically shown to reduce symptoms of fever, cough, and fatigue in patients. It has also been shown to be effective in preventing viral replication [80–84]. In addition, Shuanghuanglian, a TCM herbal product containing forsythia and honeysuckle extracts, has been shown to inhibit the virus [85]. Another traditional herbal medicine, LHQW, prepared from 13 plants, has antiviral effects by inhibiting the proliferation of the virus. Another study showed that this drug has a therapeutic effect through immunomodulation and reduction of cytokine storms [86, 87]. The QFPD drug in the TCM guideline has been shown to block pathogen entry and replication by targeting the 3-chymotrypsin-like protease (3CLpro) protein and the ACE2 receptor in the host cell [86, 87]. SFJD herbal medicine is prepared from eight plants. It is used in the treatment of COVID-19 with the recommendation of TCM. It has anti-inflammatory, antiviral, antitumor, and antibacterial properties [88]. In the clinical trial, SFJD was combined with standard treatment. Compared with standard treatment alone, the combined treatment reduced cough and fatigue symptoms. It was even found to be much more beneficial to use SFJD in the first few days of COVID-19 symptoms [89].

In Beijing Ditan Hospital in China, Yindan Jiedu Granules (YDJDG) was used and its efficacy in treatment was compared with lopinavir–ritonavir antiviral drug treatment. The use of YDJDG was shown to be effective for exudative lung lesions and high fever [90]. Shenhuang Granule (SHG), another Chinese herbal medicine, is composed of six different herbs and was found to be effective in severe patients [91]. Another study by Hetrick et al., investigated the efficacy of the Respiratory Detox Shot taken orally. They showed that it could inhibit SARS-CoV-2, including viruses in lung disease [92].

In summary, *in vitro* and clinical studies have suggested that these Chinese herbal medicines are effective in reducing fever and mortality [93, 94]. Unfortunately, clinical trials should be large-scale, randomized, double-blind, placebo-controlled studies. After these trials, safety should be investigated. Open-label, randomized, controlled trials are available for some herbal preparations. One of these, *N. sativa*, has been shown to have immunomodulatory and antiviral effects. It has also been shown to decrease symptoms and reduce the healing process in patients with mild symptoms in an open-label, randomized, controlled clinical trial conducted in Saudi Arabia [53]. In a controlled clinical trial of Kabasura Kudineer used in India with the same method, it was shown to reduce the viral load in COVID-19 patients [62].

3.4 COMBINATION OF MODERN TREATMENTS WITH HERBAL MEDICINES IN COVID-19

The combination of herbal medicines with modern medicines may be effective in the management of COVID-19. In a study, Huashi Baidu, an herbal medicine, was used in combination with current medical treatment strategies. An effect superior to that of modern drugs alone was observed in the eradication of SARS-CoV-2 [55]. However, the incidence of adverse effects with herbal medicines is not higher than with modern medicines [95].

In a study by Liu et al., LHQW and arbidol were administered in combination. Compared with the use of LHQW alone, it was found to reduce the period of disease and lung inflammation [96]. In another clinical trial, the use of LHQW in combination with TCM herbal medicines was found to reduce symptoms such as fatigue, fever, cough and recovery time [86].

Injectable preparations of Chinese medicine are also used for immunosuppressive effects. In a study by Wang et al., preparations such as the Shen Fu injection, Re Du Ning injection, and Shenmai injection were reported to affect inflammatory factors. In particular, the preparations were found to have the potential to inhibit factors involved in cytokine storms such as IL-6, IL-1β and TNFα [97–100]. In addition, Xiyanping injection may also alleviate symptoms in mild to moderate SARS-CoV-2 infected patients [101]. These results indicate that injectable preparations of TCM are effective against inflammation. Further *in vitro* experiments and *in vivo* trials could support these results. TCM herbal formulas used in the treatment of COVID-19 are listed in Table 3.2.

TABLE 3.2
TCM Formulas Used in the Treatment of SARS-CoV Infection

TCM Formula	Therapeutics Effect	Reference
Yin Qiao San	Treatment of upper-respiratory tract infection. Improvement of the function of upper-respiratory mucosal immune system.	[102, 103]
Yu Ping Feng San	Antiviral, anti-inflammatory, and immunoregulatory effects.	[79, 104,105]
Sang Ju Yin and Yu Ping Feng San	Antiviral and immunoregulatory effects.	[106]
Lian Hua Qing Wen capsule	Antiviral, anti-inflammatory, and immunoregulatory effects.	[77, 107]
Shuanghuanglian	Anti-SARS-CoV-2 activity, immunosuppressive effects.	[108–110]
Ma Xing Shi Gan Tang	Anti-SARS-CoV activity.	[111, 112]

3.5   COMMON HERBAL MEDICINES IN DNA AND RNA VIRUSES

3.5.1   HERBAL MEDICINES IN DNA VIRAL INFECTIONS

3.5.1.1   Hepatitis B Virus (HBV)

The hepatitis B virus (HBV) is an enveloped, double-stranded DNA virus. Also, it belongs to the *Hepadnaviridae* family. It is known to be destructive to the liver and to trigger inflammatory pathways in liver cells. Chronic infection can lead to very serious diseases such as cirrhosis and hepatocellular carcinoma [113]. The antivirals entecavir and lamivudine are the drugs recommended by modern medicine for the treatment of HBV infection. In addition, interferon-alpha (IFN-a) is the preferred immunomodulator [114]. Although there are different treatment strategies for the infection, some drugs have serious disadvantages. These disadvantages are the high cost, serious side effects and drug resistance [115]. For this reason, herbal sources have been used in recent years to develop new antiviral products. *Terminalia bellirica*, *Enicostemma axillare*, *Phyllanthus amarus*, *Hybanthus enneaspermus*, *Bombyx mori* L. and *Boehmeria nivea* are plants used for this purpose that have shown an inhibitory effect on virus entry and replication [116–119]. *Pulsatilla chinensis*, *Curcuma longa* L. and *Liriope platyphylla* plants are effective in targeting the host cell factor [120–122]. There is also the prescription drug Xiao-Chai-Hu-Tang (XCHT) from TCM. The medicine, known in Japan as Sho-saiko-to, has been used for many years. It is a prescription herbal medicine recommended by TCM for chronic liver disease. It contains seven different plants: Chaihu (*Bupleurum chinense* DC.), Gancao (*Glycyrrhiza uralensis* Fisch.), Huangqin (*Scutellaria baicalensis* Georgi), Banxia (*Pinellia ternata* [Thunb.]), Renshen (Panax ginseng [C.A. Mey], Makino), Shengjiang (*Zingiber officinale* Roscoe) and Dazao (*Ziziphus jujuba* Mill.). In a study conducted in a rat model of CCl4-induced liver fibrosis, the herbal drug XCHT was pharmacologically shown to be effective in the treatment of liver fibrosis [123, 124]. Researchers have also shown that this herbal formula is effective against liver inflammation [125, 126]. In a study by Takahashi et al., XCHT was shown to reduce liver necroinflammation and fibrosis in a mouse model [127]. It has also been shown to be efficacious in children. In pediatric patients with chronic HBV infection, it has been shown to induce HBeAg clearance and improve liver function [128]. Fuzheng Huayu (FZHY) is another Chinese formula used for this purpose and has 319 recipes. It is prepared from six different Chinese plants: Jiaogulan (*Gynostemma aggregatum* C.Y.Wu & S.K.Chen), Taoren (*Prunus persica* [L.] Batsch), Dongchongxiacao (*Cordyceps sinensis* [Berk.] Sacc), Songhuafen (pollen pini), Danshen (*Salvia miltiorrhiza* Bunge) and Wuweizi (*Schisandra chinensis* [Turcz.] Baill.) [97]. This drug has been approved by the Chinese Food and Drug Administration (CFDA) in tablet form. It is also the first TCM drug to complete a US Food and Drug Administration (FDA) phase II clinical trial for the treatment of liver disease [98]. A study by Tian et al., showed that the FZHY formula can be used in the treatment of liver fibrosis with its antiviral effect in chronic hepatitis B patients [99]. In a randomized, double-blind, placebo-controlled trial, this herbal formula demonstrated therapeutic efficacy in HBV-induced cirrhosis and improved Child-Pugh scores [100]. In addition, Dahuang zhechong and anluohuaxianwan have also been shown to be effective in HBV-induced cirrhosis in randomized trials [129, 130].

3.5.1.2   Herpes Simplex Virus

There are two types of herpes simplex viruses (HSV)–type 1 and type 2 (HSV-1 and HSV-2)–and they are members of the *Herpesviridae* family. They are double-stranded, enveloped DNA viruses with a worldwide distribution. The virus causes infection in the mouth and genital area. HSV-1 infection is very common and is estimated to occur in 45–98% of the world's population [131]. The virus, which can infect for life, lives in sensory neurons. It can be activated and infected when stimulated by a weak immune system, febrile illness or menstrual cycle [132]. Because of the high prevalence of HSV-1 viral infections worldwide, many antiviral medicines have been developed. In addition, research has shown that plant-derived molecules may also be effective. *C. longa* L.,

Houttuynia cordata, *Pistacia vera* L., *Prunus dulcis* and *Aloe vera* plants have been shown to inhibit viral binding, viral DNA synthesis and expression of viral proteins [133–140].

Long-Dan-Xie-Gan-Tan (LDXGT) and Yin-Chen-Hao-Tang are herbal medicines used for HSV infections. LDXGT, called Ryutan-shakan-to in traditional Kampo medicine, is one of the most popular herbal formulas in TCM. According to the *Pharmacopoeia of the People's Republic of China*, it is prepared from 10 different herbs. A water extract of LDXGT was prepared and tested *in vitro* against HSV-1 and HSV-2 and was found to reduce infectivity [141].

Yin Chen Hao Tang (YCHT) is one of the most commonly used recommended formulations in TCM. It is prepared from three different Chinese herbs, *Artemisia capillaris* Thunb. (*Compositae* family), *Gardenia jasminoides* Ellis (*Rubiaceae* family) and *Rheum officinale* Baillon (*Polygonaceae* family). The efficacy of the water extract of this recipe, which is actually used in the treatment of jaundice, against HSV viruses has also been investigated and has been shown to inhibit HSV-1 and HSV-2 *in vitro*. Although it was observed that the inhibition was higher for HSV-2, it is thought that this effect is based on direct inactivation of viral infectivity [141].

3.5.1.3 Human Papilloma Virus (HPV)

Human papillomaviruses (HPVs) are generally sexually transmitted viruses. They are non-enveloped, circular DNA genomic viruses and can cause very high-risk infections. For example, the types HPV16 and HPV18 are associated with cervical cancer [142].

Plant-derived polyphenols have the potential to be effective agents in the fight against cervical cancer [143]. *A. vera*, *C. longa* L., *Ficus carica*, *Pinellia pedatisecta*, *Azadirachta indica*, and *Emblica officinalis* are effective plants in the downregulation of oncogenes and the prevention of viral entry into the cell [144, 145]. TCM also includes formulas used in the treatment of HPV. Paiteling (PTL) is a patented liquid TCM medicine developed by the Chinese Academy of Sciences. It is widely prescribed in China for the treatment of warts. HPV infection has been prevalent in China for many years [146]. PTL is composed of 13 different herbs, including *Smilax glabra* Roxb. (15%), *Scleromitrion diffusum* (Willd.) R.J.Wang (10%), *Phellodendron amurense* Rupr. (10%) and *Taraxacum mongolicum* Hand.-Mazz. (10%). Many clinical studies have shown that PTL is effective in the treatment of cancer with high efficacy when used alone or in combination with other drugs. In a study by Liu et al., the Liquid Chromatography with tandem mass spectrometry (LC-MS-MS) content was analyzed and 36 biomolecules including phenyls, flavonoids, terpenoids and alkaloids were identified. Pharmacological studies conducted to investigate efficacy showed that PTL components inhibited HPV virus replication [147, 148].

Youdujing (YDJ) is one of the formulas of Chinese medicine and contains nine different plants: Arnebiae Radix, Polygoni cuspidati Rhizoma et Radix, Phellodendri amurensis Cortex, Sophorae flavescentis Radix, Isatidis Folium, Isatidis Radix, Curcumae Rhizoma, Bolbostemmatis Rhizoma and Glycyrrhizae Radix et Rhizoma. YDJ is a hospital-prepared and administered drug that has been shown to prevent HPV transmission and DNA replication in HPV-related cervical cancer [149, 150].

3.5.2 Herbal Medicines in RNA Viral Infections

3.5.2.1 Human Immunodeficiency Virus (HIV)

Human immunodeficiency viruses are enveloped viruses of the *Retroviridae* family. The virus is two types known as HIV-1 and HIV-2. HIV is transmitted through sexual contact, the sharing of contaminated needles/sharp instruments, and the exchange of blood and body fluids [151]. HIV causes acquired immunodeficiency syndrome (AIDS). Research on the infectious agent has been ongoing for more than 30 years; however, effective vaccines and drugs are still not available [152]. Problems with modern drugs include toxicity, patient compliance and drug resistance. Herbal medicines are also preferred [153]. *Griffithsia* sp., *Nostoc ellipsosporum*, *Siliquariaspongia mirabilis*, *A. annua*, *Syzygium claviflorum* and *Stelletta clavosa* are plants with proven anti-HIV properties in this area [154–157]. Bioactive molecules contained in these plants inhibit the maturation and entry of the virus.

Traditional herbal medicines are also used in the treatment of HIV in various cultures. Indian Ayurvedic medicine uses the herbal medicine called Tulsi or holy basil for its anti-HIV activity. In addition, *Rheum palmatum* L., *Polygonum cuspidatum* and *Euphorbia kansui*, which are registered in TCM, are widely used in China. *P. cuspidatum* Sieb. and Zucc are popular Chinese medicines also known as Huzhang. *E. kansui* has been used in TCM for centuries to treat cancer and fluid retention and has been shown to have anti-HIV properties. These drugs act by inhibiting HIV replication [158].

3.5.2.2 Influenza Viruses

Influenza viruses are RNA viruses in the *Orthomyxoviridae* family. There are three types: type A, B and C. Type A has subtypes according to hemagglutinin and neuraminidase [159]. Infected people may have symptoms such as sore throat, fever, sneezing, headache and muscle aches. Infection can progress to very serious illnesses, such as pneumonia [160, 161]. Influenza can be a seasonal infection or pandemic. It is a major health problem with a high mortality rate [81].

Litchi chinensis, green tea, Aglaia, *Mycale hentscheli* and *C. longa* L. are preferred in traditional treatment strategies. They act as RNA polymerase inhibitors and hemagglutinin inhibitors for H1N1, H3N2 and H6N1. Yi-Zhi-Hao pellets is an herbal formula registered in TCM. It is used in the treatment of influenza on the recommendation of TCM. It contains *Isatis Tinctoria*, *I. Radix* (local name is Ban-Lan-Gen), *I. Folium* (local name is Da-Qing-Ye) and *Artemisia rupestris* L., *Artemisia rupestris* (local name is Yi-Zhi-Hao) [162]. Another TCM formula is Ma Xing Shi Gan (MXSGD) decoction. It is prepared from four herbal sources including Ephedra Herba (Ma Huang), Glycyrrhizae Radix Et Rhizoma (Gan Cao), *Armeniacae Semen Amarum* (Ku Xing Ren) and gypsum fibrosum (Shi Gao). It is used to treat cough and asthma symptoms caused by lung inflammation [163]. Yu Ping Feng San is a traditional herbal formula of Chinese medicine. It is prepared using Astragali Radix (Huangqi), Saposhnikoviae Radix (Fangfeng) and Atractylodis Macrocephalae Rhizoma (Baizhu). Although its mechanism of action is not clear, it is suggested that it has antiviral activity by acting on inflammatory responses [164]. Yinqiaosan, Xinjiaxiangruyin and Gui Zhi Ma Huang are other TCM formulations with anti-influenza activity. Comparison of these drugs with oseltamivir also showed that they reduced lung inflammation [165].

3.5.2.3 Hepatitis C Virus

Hepatitis C virus (HCV) is an enveloped RNA virus of the *Flaviviridae* family. HCV is transmitted through contact with blood. It can be transmitted by intravenous injection, blood transfusion and sharing toothbrushes [166]. The standard treatment is IFN-a and the antiviral ribavirin. The new protease inhibitors boceprevir and telaprevir have also been approved. However, there are disadvantages such as existing drug resistance, serious side effects and patient compliance [167]. For this reason, herbal medicines containing various phytochemicals are used depending on the life cycle of HCV. *Trichilia dregeana*, *Detarium microcarpum*, *Phragmanthera capitata*, *Bupleurum kaoi* and *Alloeocomatella polycladia* are natural sources used for this purpose [168–171]. Their activity is based on the prevention of viral entry into the host cell and the inhibition of helicase enzyme activity. Kuan-Sin-Yin is a TCM formula prepared from seven different Chinese herbs in the form of a decoction. This formula includes *G. uralensis* Fisch., *Ligustrum lucidum*, *Agastache rugosa*, *Astragalus membranaceus*, *Poria cocos*, *Atractylodes macrocephala* Koidz. and *Codonopsis pilosula* (Franch.) Nannf. It has been shown to reduce viral load after 6 weeks of use [172]. In Taiwan, several traditional Chinese herbal formulas are widely used, including Jia-Wei-Xiao-Yao-San (10 different plants), XCHT (seven different plants), Yin-Chen-Wu-Ling-San (six different plants) and Ping-Wei-San (six different plants). In addition to these multi-herb formulas, there are also single herbal medicines. Dan-Shen is prepared from Radix Salviae Miltiorrhizae; Hai-Piao-Xiao from Endoconcha Sepiae; Hu-Zhang from Rhizoma Polygoni Cuspidati. In a study comparing patients who used these drugs with those who did not, important results were obtained. Mortality and cirrhosis rates decreased in patients using TCM. In this sense, it is suggested that the use of TCM has effects that may help standard treatment [173].

3.6 CONCLUSION

Traditional herbal medicines are widely used to treat viral infections. Especially during the COVID-19 pandemic, people have shown a positive attitude toward the use of herbal medicines due to the lack of an effective therapeutic agents developed by modern medicine. This chapter summarizes the antiviral efficacy of TCM-approved herbal medicines in various Asian countries, particularly China. In addition to the recommendations of Chinese traditional medicine against SARS-CoV-2, herbal medicines used in Saudi Arabia, Bangladesh, Morocco, India, Africa and Mexico are also mentioned. The contents of single- and multi-component herbal medicines are included. *In vitro*, *in vivo* and clinical studies show that herbal medicines are promising in the fight against viral infections. It is predicted that as their efficacy is proven through advanced clinical trials, they will become even more preferred in the future [174].

REFERENCES

1. Ball MJ, Lukiw WJ, Kammerman EM, Hill JM. Intracerebral propagation of Alzheimer's disease: Strengthening evidence of a herpes simplex virus etiology. Alzheimers Dement. 2013;9:169–75. [PMC free article] [PubMed] [Google Scholar]
2. Hober D, Sane F, Jaidane H, Riedweg K, Goffard A, Desailloud R. Immunology in the clinic review series; Focus on type 1 diabetes and viruses: Role of antibodies enhancing the infection with coxsackievirus-B in the pathogenesis of type 1 diabetes. Clin Exp Immunol. 2012;168:47–51. [PMC free article] [PubMed] [Google Scholar]
3. Christou L. The global burden of bacterial and viral zoonotic infections. Clin Microbiol Infect. 2011;17:326–30. [PMC free article] [PubMed] [Google Scholar]
4. Cascio A, Bosilkovski M, Rodriguez-Morales AJ, Pappas G. The socio-ecology of zoonotic infections. Clin Microbiol Infect. 2011;17:336–42. [PubMed] [Google Scholar]
5. Sheu TG, Deyde VM, Okomo-Adhiambo M, Garten RJ, Xu X, Bright RA, et al. Surveillance for neuraminidase inhibitor resistance among human influenza A and B viruses circulating worldwide from 2004 to 2008. Antimicrob Agents Chemother. 2008;52:3284–92. [PMC free article] [PubMed] [Google Scholar]
6. Geretti AM, Armenia D, Ceccherini-Silberstein F. Emerging patterns and implications of HIV-1 integrase inhibitor resistance. Curr Opin Infect Dis. 2012;25:677–86. [PubMed] [Google Scholar]
7. Locarnini SA, Yuen L. Molecular genesis of drug-resistant and vaccine-escape HBV mutants. Antivir Ther. 2010;15:451–61. [PubMed] [Google Scholar]
8. Wang Z, Yang L. Chinese herbal medicine: Fighting SARSCoV-2 infection on all fronts. J Ethnopharmacol. 2021;270:113869.
9. Illian DN, Siregar ES, Sumaiyah S, Utomo AR, Nuryawan A, Basyuni M. Potential compounds from several Indonesian plants to prevent SARS-CoV-2 infection: A mini-review of SARS-CoV-2 therapeutic targets. Heliyon. 2021;7(1):e06001. https://doi.org/10.1016/j.heliyon.2021.e06001
10. Teuwen L-A, Geldhof V, Pasut A, Carmeliet P. COVID19: The vasculature unleashed. Nat Rev Immunol. 2020;20(7):389–91. https://doi.org/10.1038/s41577-020-0343-0
11. Leung EL-H, Pan H-D, Huang Y-F, Fan X-X, Wang W-Y, He F, Cai J, Zhou H, Liu L. The scientific foundation of Chinese herbal medicine against COVID-19. Engineering. 2020;6(10):1099–107. https://doi.org/10.1016/j.eng.2020.08.009
12. Susskind D, Vines D. The economics of the COVID-19 pandemic: An assessment. Oxford Rev Econ Pol. 2020;36(Supplement_1):S1–13. https://doi.org/10.1093/oxrep/graa036
13. Akindele AJ, Agunbiade FO, Sofidiya MO, Awodele O, Sowemimo A, Ade-Ademilua O, Akinleye MO, Ishola IO, Orabueze I. COVID-19 pandemic: A case for phytomedicines. Nat Prod Commun. 2020;15(8):1–9. https://doi.org/10.1177/1934578X20945086
14. Zhao J, Zhao S, Ou J, Zhang J, Lan W, Guan W, Wu X, Yan Y, Zhao W, Wu J. COVID-19: Vaccine development updates. Front Immunol. 2020;11:3435. https://doi.org/10.3389/fimmu.2020.602256

15. Cragg GM, Newman DJ. Natural product drug discovery in the next millennium. Pharm Biol. 2001;39(supl):8–17. https://doi.org/10.1076/phbi.39.s1.8.0009

16. Patwardhan B, Mashelkar RA. Traditional medicine inspired approaches to drug discovery: Can ayurveda show the way forward? Drug Discov Today. 2009;14(15):804–11. https://doi.org/10.1016/j.drudis.2009.05.009

17. Sucher NJ. The application of Chinese medicine to novel drug discovery. Expert Opin Drug Disc. 2013;8(1):21–34. https://doi.org/10.1517/17460441.2013.739602

18. Shahrajabian MH, Sun W, Cheng Q. Traditional herbal medicine for the prevention and treatment of cold and fu in the autumn of 2020, overlapped with COVID-19. Nat Prod Commun. 2020;15(8):1–10. https://doi.org/10.1177/1934578X20951431

19. Han YY, Zhao MR, Shi Y, et al. Application of integrative medicine protocols on treatment of coronavirus disease 2019. Chin Tradit Herb Drugs. 2020;2020:2–18.

20. National Health Commission of the People's Republic of China. Notice on the issuance of guidelines of diagnosis and treatment for 2019-nCoV infected pneumonia (version 6). 6 ed; http://www.nhc.gov.cn/yzygj/s7653p/202002/8334a8326dd94d329df351d7da8aefc2.shtml?from=timeline.%202020. 2020

21. Yang Y. Use of herbal drugs to treat COVID-19 should be with caution. Lancet. 2020;395(10238):1689–90.

22. Xiao M, Tian J, Zhou Y, et al. Efficacy of Huoxiang Zhengqi dropping pills and Lianhua Qingwen granules in treatment of COVID-19: A randomized controlled trial. Pharmacol Res. 2020;161:105126.

23. Province S. Clinical features and treatment of COVID19 patients in northeast Chongqing, https://pubmed.ncbi.nlm.nih.gov/32198776/

24. Hung I, To K, Lee CK, et al. Hyperimmune IV immunoglobulin treatment: A multicenter double-blind randomized controlled trial for patients with severe 2009 influenza A(H1N1) infection. Chest. 2013;144(2):464–73. https://doi.org/10.1378/chest.12-2907

25. Zhu YL, Yang BB, Wu F. Understanding of COVID-19 in children from different perspectives of traditional Chinese medicine and western medicine. Chin Tradit Herb Drugs. 2020;51(4): 883–87.

26. Ni L, Zhou L, Zhou M, et al. Combination of western medicine and Chinese traditional patent medicine in treating a family case of COVID-19. Front Med. 2020;14(2):210–14. https://doi.org/10.1007/s11684-020-0757-x

27. Chan KW, Wong VT, Tang SCW. COVID-19: An update on the epidemiological, clinical, preventive and therapeutic evidence and guidelines of integrative Chinese–Western medicine for the management of 2019 novel coronavirus disease. Am J Chin Med. 2020;48(3):737–62.

28. Shahrajabian MH, Biotechnology Research Institute CAOA. A review of ginseng species in different regions as a multipurpose herb in traditional Chinese medicine, modern herbology and pharmacological science. J Med Plants Res. 2019;10(13):213–26.

29. Sun W, Shahrajabian MH, Cheng Q. Anise (Pimpinella anisum L), a dominant spice and traditional medicinal herb for both food and medicinal purposes. Cogent Biology. 2019;5(1):1673688. https://doi.org/10.1080/23312025.2019.1673688

30. Xia S, Zhong Z, Gao B, et al. The important herbal pair for the treatment of COVID-19 and its possible mechanisms. Chin Med. 2021;16(1):1–16.

31. Chia WY, Kok H, Chew KW, et al. Can algae contribute to the war with covid-19? Bioengineered. 2021;12(1):1226–37. https://doi.org/10.1080/21655979.2021.1910432

32. Peter AP, Wayne CK, Show PL, et al. Potential pathway that could treat coronaviruses (COVID-19). Curr Biochem Eng. 2020;6(1):3–4. https://doi.org/10.2174/221271190699200228100507

33. Yuan H, Ma Q, Ye L, et al. The traditional medicine and modern medicine from natural products. Molecules. 2016;21(5):559. https://doi.org/10.3390/molecules21050559

34. Tong T, Wu YQ, Ni WJ, et al. The potential insights of traditional Chinese medicine on treatment of COVID-19. Chin Med. 2020;15(1):51. https://doi.org/10.1186/s13020-020-00326-w

35. Jiang W. Therapeutic wisdom in traditional Chinese medicine: A perspective from modern science. Trends Pharmacol Sci. 2005;26(11):558–63.

36. Miao Q, Cong XD, Wang B, et al. Understanding and thinking of novel coronavirus pneumonia in traditional Chinese medicine. J Trad Chin Med. 2020;61:286–88.

37. Yang Y, Islam MS, Wang J, Li Y, Chen X. Traditional Chinese medicine in the treatment of patients infected with 2019-new coronavirus (SARS-CoV-2): A review and perspective. Int J Biol Sci. 2020;16(10):1708.

38. Lau TF, Leung PC, Wong EL, et al. Using herbal medicine as a means of prevention experience during the SARS crisis. Am J Chin Med. 2005;33(3):345–56. https://doi.org/10.1142/S0192415X05002965

39. Needham J, Gwei-djen L. Hygiene and preventive medicine in ancient China. J Hist Med Allied Sci. 1962;17(4):429–78.

40. Luo H, Tang QL, Shang YX, et al. Can Chinese medicine be used for prevention of corona virus disease 2019 (COVID-19)? A review of historical classics, research evidence and current prevention programs. Chin J Integr Med. 2020;26(4):243–50. https://doi.org/10.1007/s11655-020-3192-6
41. Hoever G, Baltina L, Michaelis M, et al. Antiviral activity of glycyrrhizic acid derivatives against SARS-coronavirus. J Med Chem. 2005;48(4):1256–59. https://doi.org/10.1021/jm0493008
42. Wang C, Cao B, Liu QQ, et al. Oseltamivir compared with the Chinese traditional therapy maxingshiganyinqiaosan in the treatment of H1N1 influenza: A randomized trial. Ann Intern Med. 2011;155(4):217–25. https://doi.org/10.7326/0003-4819-155-4-201108160-00005
43. Pu J, Mei H, Lei L, et al. Knowledge of medical professionals, their practices, and their attitudes toward traditional Chinese medicine for the prevention and treatment of coronavirus disease 2019: A survey in Sichuan, China. PLOS ONE. 2021;16(3):e234855. https://doi.org/10.1371/journal.pone.0234855
44. Ang L, Lee HW, Choi JY, et al. Herbal medicine and pattern identification for treating COVID-19: A rapid review of guidelines. Integr Med Res. 2020;9(2):100407.
45. Lee B, Lee JA, Kim K, et al. A consensus guideline of herbal medicine for coronavirus disease 2019. Integr Med Res. 2020;9(3):100470. https://doi.org/10.1016/j.imr.2020.100470
46. Kim D, Chu H, Min BK, et al. Telemedicine center of Korean medicine for treating patients with COVID-19: A retrospective analysis. Integr Med Res. 2020;9(3):100492. https://doi.org/10.1016/j.imr.2020.100492
47. Takayama S, Arita R, Ono R, et al. Treatment of COVID-19-related olfactory disorder promoted by kakkontokasenkyushin'i: A case series. Tohoku J Exp Med. 2021;254(2):71–80. https://doi.org/10.1620/tjem.254.71
48. Takayama S, Namiki T, Ito T, et al. A multi-center, randomized controlled trial by the integrative management in Japan for epidemic disease (IMJEDI study-RCT) on the use of Kampo medicine, kakkonto with shosaikotokakikyosekko, in mild-to-moderate COVID-19 patients for symptomatic relief and prevention of severe stage: A structured summary of a study protocol for a randomized controlled trial. Trials. 2020;21(1):1–3. https://doi.org/10.1186/s13063-019-3906-2
49. Aomatsu N, Shigemitsu K, Nakagawa H, et al. Efficacy of Ninjin'yoeito in treating severe coronavirus disease 2019 in patients in an intensive care unit. Neuropeptides. 2021;90:102201.
50. Kuchta K, Cameron S, Lee M, et al. Which East Asian herbal medicines can decrease viral infections? Phytochem Rev. 2021:1–19. https://doi.org/10.1007/s11101-021-09756-2
51. AlNajrany SM, Asiri Y, Sales I, et al. The commonly utilized natural products during the COVID-19 pandemic in Saudi Arabia: A cross-sectional online survey. Int J Environ Res Public Health. 2021;18(9):4688. https://doi.org/10.3390/ijerph18094688
52. ALkharashi NA. The consumption of nutritional supplements and herbal products for the prevention and treatment of COVID-19 infection among the Saudi population in Riyadh. Clin Nutr Open Sci. 2021;39:11–20. https://doi.org/10.1016/j.nutos.2021.09.001
53. Koshak AE, Koshak EA, Mobeireek AF, et al. Nigella sativa for the treatment of COVID-19: An open-label randomized controlled clinical trial. Complement Ther Med. 2021;61:102769.
54. Ahmad A, Husain A, Mujeeb M, et al. A review on therapeutic potential of Nigella sativa: A miracle herb. Asian Pac J Trop Biomed. 2013;3(5):337–52. https://doi.org/10.1016/S2221-1691(13)60075-1
55. Rahman MT. Potential benefits of combination of Nigella sativa and Zn supplements to treat COVID-19. J Herbs Med. 2020;23:100382.
56. Bouchentouf S, Missoum N. Identification of Compounds from Nigella Sativa as new potential inhibitors of 2019 novel Coronasvirus (Covid-19): Molecular docking study, https://doi.org/10.26434/chemrxiv.12055716.v1. 2020.
57. Ahmed I, Hasan M, Akter R, et al. Behavioral preventive measures and the use of medicines and herbal products among the public in response to covid-19 in Bangladesh: A cross-sectional study. PLOS ONE. 2020;15(12):e243706. https://doi.org/10.1371/journal.pone.0243706
58. Saraswat J, Singh P, Patel R. A computational approach for the screening of potential antiviral compounds against SARS-CoV-2 protease: Ionic liquid vs herbal and natural compounds. J Mol Liq. 2021;326:115298.
59. Charan J, Bhardwaj P, Dutta S, et al. Use of complementary and alternative medicine (CAM) and home remedies by COVID-19 patients: A telephonic survey. Indian J Clin Biochem. 2021;36(1):108–11. https://doi.org/10.1007/s12291-020-00931-4
60. Banerjee S, Kar A, Mukherjee PK, et al. Immunoprotective potential of ayurvedic herb kalmegh (*Androgrpahis paniculata*) against respiratory viral infections–LC–MS/MS and network pharmacology analysis. Phytochem Analysis. 2021;32(4):629–39. https://doi.org/10.1002/pca.3011
61. Balkrishna A, Verma S, Solleti SK, et al. Calcio-herbal medicine Divya-Swasari-Vati ameliorates SARS-CoV-2 spike protein-induced pathological features and inflammation in humanized zebrafish model by moderating IL-6 and TNF-α cytokines. J Inflamm Res. 2020;13:1219.

62. Natarajan S, Anbarasi C, Sathiyarajeswaran P, et al. Kabasura Kudineer (KSK), a poly-herbal Siddha medicine, reduced SARS-CoV-2 viral load in asymptomatic COVID-19 individuals as compared to vitamin C and zinc supplementation: Findings from a prospective, exploratory, open-labeled, comparative, randomized controlled trial, Tamil Nadu, India. Trials. 2021;22(1):1–11. https://doi.org/10.1186/s13063-020-04976-x

63. Dandara C, Dzobo K, Chirikure S. COVID-19 pandemic and Africa: From the situation in Zimbabwe to a case for precision herbal medicine. OMICS. 2021;25(4):209–12.

64. Attah AF, Fabgbemi AA, Olubiyi O, Dada-Adegbola H, Oluwadorun A, Elujoba A, Babalola CP. Therapeutic potentials of antiviral plants used in traditional medicine with covid in focus: A Nigerian perspective. Front Pharmacol. 2021;12. https://doi.org/10.3389/fphar.2021.596855

65. Nair MS, Huang Y, Fidock DA, Polyak SJ, Wagoner J, Towler MJ, Weathers PJ. Artemisia annua L. Extracts inhibit the in vitro replication of SARS-CoV-2 and two of its variants. J Ethnopharmacol. 2021;274:114016. https://doi.org/10.1016/j.jep.2021.114016

66. Asprilio K, Wilar G. Emergence of ethnomedical COVID-19 treatment: A literature review. Infect Drug Resist. 2021;14:4277–89.

67. Nie C, Trimpert J, Moon S, Haag, R., Gilmore, K., Kaufer, B. B., & Seeberger, P. H. In vitro efficacy of artemisia extracts against SARS-CoV-2. Virology Journal. 2021; *18*, 1–7.

68. Chali BU, Melaku T, Berhanu N, et al. Traditional medicine practice in the context of COVID-19 pandemic: Community claim in Jimma zone, Oromia, Ethiopia. Infect Drug Resist. 2021;14:3773.

69. Titanji, VP. The case for Phytomedicines in Africa with particular focus on Cameroon. J Cameroon Acad Sci. 2021;17(2):163–75.

70. Rivero-Segura NA, Gomez-Verjan JC. In silico screening of natural products isolated from Mexican herbal medicines against COVID-19. Biomolecules. 2021;11(2):216.

71. Wen CC, Kuo YH, Jan JT, et al. Specific plant terpenoids and lignoids possess potent antiviral activities against severe acute respiratory syndrome coronavirus. J Med Chem. 2007;50(17):4087–95. https://doi.org/10.1021/jm070295s

72. Ryu YB, Park SJ, Kim YM, et al. SARS-CoV 3CLpro inhibitory effects of quinone-methide triterpenes from Tripterygium regelii. Bioorg Med Chem Lett. 2010;20(6):1873–76. https://doi.org/10.1016/j.bmcl.2010.01.152

73. Park JY, Kim JH, Kim YM, et al. Tanshinones as selective and slow-binding inhibitors for SARS-CoV cysteine proteases. Bioorg Med Chem. 2012;20(19):5928–35. https://doi.org/10.1016/j.bmc.2012.07.038.

74. Park JY, Kim JH, Kwon JM, et al. Dieckol, a SARS-CoV 3CL(pro) inhibitor, isolated from the edible brown algae Ecklonia cava. Bioorg Med Chem. 2013;21(13):3730–37. https://doi.org/10.1016/j.bmc.2013.04.026.

75. Zhou P, Yang XL, Wang XG, et al. A pneumonia outbreak associated with a new coronavirus of probable bat origin. Nature. 2020;579(7798):270–73. https://doi.org/10.1038/s41586-020-2012-7.

76. Liu L-S, Lei, N., Lin, Q., Wang, W. L., Yan, H. W., & Duan, X. H. The effects and mechanism of Yinqiao powder on upper respiratory tract infection. Int J Biotech Well Indus. 2015; *4*(2):57.

77. Ding Y, Zeng L, Li R, et al. The Chinese prescription Lianhuaqingwen capsule exerts anti-influenza activity through the inhibition of viral propagation and impacts immune function. BMC Complement Altern Med. 2017;17(1):130. https://doi.org/10.1186/s12906-017-1585-7.

78. Su H, Yao S, Zhao W, et al. Anti-SARS-CoV-2 activities in vitro of Shuanghuanglian preparations and bioactive ingredients. Acta Pharmacol Sin. 2020;41(9):1167–77. https://doi.org/10.1038/s41401-020-0483-6

79. Gao J, Li J, Shao X, et al. Antiinflammatory and immunoregulatory effects of total glucosides of Yupingfeng powder. Chin Med J (Engl). 2009;122(14):1636–41.

80. Zeng M, Li L, Wu Z. Traditional Chinese medicine Lianhua Qingwen treating corona virus disease 2019 (COVID-19): Meta-analysis of randomized controlled trials. PLOS ONE. 2020;15(9):e238828.

81. Cheng Dezhong LY. Clinical effectiveness and case analysis in 54 NCP patients treated with Lianhuaqingwen Granules. World Chin Med. 2020;15(15):150–4.

82. Lv R, Wang, W, & Li, X Clinical observation on 63 cases of suspected cases of new coronavirus pneumonia treated by Chinese medicine Lianhua Qingwen. J Tradit Chin Med. 2020; *61*(08): 655–59,

83. Yao K, Liu, M Y, Li, X, Huang, J H, & Cai, H B. Retrospective clinical analysis on treatment of novel coronavirus-infected pneumonia with traditional Chinese medicine Lianhua Qingwen. Chin J Exp Tradit Med Formul. 2020; 1–7.

84. Li H, Yang L, Liu FF, et al. Overview of therapeutic drug research for COVID-19 in China. Acta Pharmacol Sin. 2020;41(9):1133–40. https://doi.org/10.1038/s41401-020-0438-y

85. Li T, Lu H, Zhang W. Clinical observation and management of COVID-19 patients. Emerg Microbes Infect. 2020;9(1):687–90.
86. Hu K, Guan WJ, Bi Y, et al. Efficacy and safety of Lianhuaqingwen capsules, A repurposed Chinese herb, in patients with coronavirus disease 2019: A multicenter, prospective, randomized controlled trial. Phytomedicine. 2020;85:153242.
87. Runfeng L, Yunlong H, Jicheng H, et al. Lianhuaqingwen exerts anti-viral and anti-inflammatory activity against novel coronavirus (SARS-CoV-2). Pharmacol Res. 2020;156:104761.
88. Pan, X, Dong, L, Yang, L, Chen, D, & Peng, C. Potential drugs for the treatment of the novel coronavirus pneumonia (COVID-19) in China. Virus Research, 2020; 286, 198057.
89. Lu X, Yujing S, Jie SU, et al. Shufeng Jiedu, a promising herbal therapy for moderate COVID-19: Antiviral and anti-inflammatory properties, pathways of bioactive compounds, and a clinical real-world pragmatic study. Phytomedicine. 2021;85:153390.
90. Liu J, Jiang Y, Liu Y, et al. Yindan Jiedu granules, a traditional Chinese medicinal formulation, as a potential treatment for coronavirus disease 2019. Front Pharmacol. 2021;11:2449.
91. Feng J, Fang B, Zhou D, et al. Clinical effect of traditional Chinese medicine Shenhuang granule in critically ill patients with COVID-19: A single-centered, retrospective, observational study. J Microbiol Biotechnol. 2021 31(3):380–86.
92. Hetrick B, Yu D, Olanrewaju AA, et al. A traditional medicine, respiratory detox shot (RDS), inhibits the infection of SARS-CoV, SARS-CoV-2, and the influenza A virus in vitro. Cell Biosci. 2021;11(1):1–12.
93. Wang Y, Liu Y, Lv Q, et al. Effect and safety of Chinese herbal medicine granules in patients with severe coronavirus disease 2019 in Wuhan, China: A retrospective, single-center study with propensity score matching. Phytomedicine. 2020;85:153404.
94. Shi N, Guo L, Liu B, et al. Efficacy and safety of Chinese herbal medicine versus lopinavir-ritonavir in adult patients with coronavirus disease 2019: A non-randomized controlled trial. Phytomedicine. 2021;81:153367.
95. Liang S, Zhang Y, Shen C, et al. Chinese herbal medicine used with or without conventional Western therapy for COVID-19: An evidence review of clinical studies. Front Pharmacol. 2021;11:2321.
96. Liu L, Shi F, Tu P, et al. Arbidol combined with the Chinese medicine Lianhuaqingwen capsule versus arbidol alone in the treatment of COVID-19. Medicine (Baltimore). 2021;100(4):e24475.
97. Liu P, Liu C, Hu YY. Effect of Fuzheng Huayu recipe in treating posthepatitic cirrhosis. Chin J Integr Med Chin J Integr Med. 1996;16(8):459–62.
98. Hong M, Li S, Tan HY, Wang N, Tsao SW, Feng Y. Current status of herbal medicines in chronic liver disease therapy: The biological effects, molecular targets and future prospects. Int J Mol Sci. 2015;16(12):28705–745.
99. Tian YL, Zhu XY, Yin WW, Zang ZD, Wang L, Fu XL. Supplemental capsule therapy for improving liver fibrosis markers in patients with chronic hepatitis B following unsatisfactory outcome of nucleos (t) ide analogue monotherapy. Chin J Hepatol. 2013;21(7):514–8.
100. Chen Q, Wu F, Wang M, Dong S, Liu Y, Lu Y, Su S. Transcriptional profiling and miRNA-target network analysis identify potential biomarkers for efficacy evaluation of Fuzheng-Huayu formula-treated hepatitis B caused liver cirrhosis. Int J Mol Sci. 2016;17(6):883.
101. Zhang XY, Lv L, Zhou YL, et al. Efficacy and safety of Xiyanping injection in the treatment of COVID-19: A multicenter, prospective, open-label and randomized controlled trial. Phytother Res. 2021;35(8):4401–10. https://doi.org/10.1002/ptr.7141
102. Liu LS, Lei N, Lin Q, Wang WL, Yan HW, Duan XH. The effects and mechanism of Yinqiao powder on upper respiratory tract infection. Int J Biotechnol Wellness Ind. 2015;4:57–60.
103. Fu YJ, Yan YQ, Qin HQ, Wu S, Shi SS, Zheng X, et al. Effects of different principles of traditional Chinese medicine treatment on TLR7/NF-κB signaling pathway in influenza virus infected mice. Chin Med. 2018;13:42.
104. Lau JT, Leung PC, Wong EL, Fong C, Cheng KF, Zhang SC, et al. The use of an herbal formula by hospital care workers during the severe acute respiratory syndrome epidemic in Hong Kong to prevent severe acute respiratory syndrome transmission, relieve influenza-related symptoms, and improve quality of life: A prospective cohort study. J Altern Complement Med. 2005;11:49–55.
105. Du CY, Zheng KY, Bi CW, Dong TT, Lin H, Tsim KW. Yu Ping Feng San, an ancient Chinese herbal decoction, induces Gene expression of anti-viral proteins and inhibits neuraminidase activity. Phytother Res. 2015;29:656–61.
106. Poon PM, Wong CK, Fung KP, Fong CY, Wong EL, Lau JT, et al. Immunomodulatory effects of A traditional Chinese medicine with potential antiviral activity: A self-control study. Am J Chin Med. 2006;34:13–21.

107. Dong L, Xia JW, Gong Y, Chen Z, Yang H-H, Zhang J, et al. Effect of Lianhuaqingwen capsules on airway inflammation in patients with acute exacerbation of chronic obstructive pulmonary disease. Evid Based Complement Alternat Med. 2014;2014:1–11.

108. Science CAo. Researchers in Shanghai Institute of Drugs and Wuhan Virus Institute discovered that the Chinese patent medicine Shuanghuanglian oral liquid can inhibit the 2019-new coronavirus.; 2020.

109. Gao Y, Fang L, Cai R, Zong C, Chen X, Lu J, et al. Shuang-Huang-Lian exerts anti-inflammatory and anti-oxidative activities in lipopolysaccharide-stimulated murine alveolar macrophages. Phytomedicine. 2014;21:461–9.

110. Zhang H, Chen Q, Zhou W, Gao S, Lin H, Ye S, et al. Chinese medicine injection Shuanghuanglian for treatment of acute upper respiratory tract infection: A systematic review of randomized controlled trials. Evid Based Complement Alternat Med. 2013;2013:987326.

111. Xiao GL, Song K, Yuan CJ et al. A literature report on the treatment of SARS by stages with traditional Chinese medicine. J Emerg Chin Med Hunan. 2005;4:53–5.

112. Bao L, & Ma, J Research progress of Da Yuan Yin on the treatment of infectious diseases. J Emerg Tradit Chin Med. 2010;2:263–87.

113. Selzer L, Zlotnick A. Assembly and release of hepatitis B virus. Cold Spring Harb Perspect Med. 2015;5:a021394. [Google Scholar] [CrossRef][Green Version]

114. Serigado JM, Izzy M, Kalia H. Novel therapies and potential therapeutic targets in the management of chronic hepatitis B. Eur J Gastroenterol Hepatol. 2017;29:987–93. [Google Scholar] [CrossRef]

115. Farrell GC, Teoh NC. Management of chronic hepatitis b virus infection: A new era of disease control. Intern Med J. 2006;36:100–13. [Google Scholar] [CrossRef]

116. Anbalagan S, Sankareswaran M, Rajendran P, Karthikeyan M, Priyadharshini K, Hamza H. In vitro screening of anti-HBV properties of selected Indian medicinal plants from Kolli hills, Namakkal district of Tamil Nadu, India. World J Pharm Sci. 2015;4:909–15. [Google Scholar]

117. Lee C-D, Ott M, Thyagarajan SP, Shafritz DA, Burk RD, Gupta S. Phyllanthus amarus down-regulates hepatitis B virus mRNA transcription and replication. Eur J Clin Investig. 1996;26:1069–76. [Google Scholar] [CrossRef]

118. Jacob JR, Mansfield K, You JE, Tennant BC, Kim YH. Natural iminosugar derivatives of 1-deoxynojirimycin inhibit glycosylation of hepatitis viral envelope proteins. J Microbiol. 2007;45:431–40. [Google Scholar]

119. Wei J, Lin L, Su X, Qin S, Xu Q, Tang Z, Deng Y, Zhou Y, He S. Anti-hepatitis B virus activity of Boehmeria nivea leaf extracts in human HepG2.2.15 cells. Biomed Rep. 2014;2:147–51. [Google Scholar] [CrossRef] [PubMed]

120. Yao D, Li H, Gou Y, Zhang H, Vlessidis AG, Zhou H, Evmiridis NP, Liu Z. Betulinic acid-mediated inhibitory effect on hepatitis B virus by suppression of manganese superoxide dismutase expression. FEBS J. 2009;276:2599–614. [Google Scholar] [CrossRef] [PubMed]

121. Kim HJ, Yoo HS, Kim JC, Park CS, Choi MS, Kim M, Choi H, Min JS, Kim YS, Yoon SW, et al. Antiviral effect of Curcuma longa Linn extract against hepatitis B virus replication. J Ethnopharmacol. 2009;124:189–96. [Google Scholar] [CrossRef] [PubMed]

122. Huang T-J, Tsai Y-C, Chiang S-Y, Wang G-J, Kuo Y-C, Chang Y-C, Wu Y-Y, Wu Y-C. Anti-viral effect of a compound isolated from Liriope platyphylla against hepatitis B virus in vitro. Virus Res. 2014;192:16–24. [Google Scholar] [CrossRef]

123. Hu R, Jia WY, Xu SF, Zhu ZW, Xiao Z, Yu SY, Li J. Xiaochaihutang inhibits the activation of hepatic stellate cell line T6 through the Nrf2 pathway. Front Pharmacol. 2019;9:1516.

124. Li J, Hu R, Xu S, Li Y, Qin Y, Wu Q, Xiao Z. Xiaochaihutang attenuates liver fibrosis by activation of Nrf2 pathway in rats. Biomed Pharmacother. 2017;96:847–53.

125. Kusunose M, Qiu B, Cui T, Hamada A, Yoshioka S, Ono M, Nishioka Y. Effect of Sho-Saiko-to extract on hepatic inflammation and fibrosis in dimethylnitrosamine induced liver injury rats. Biol Pharm Bull. 2002;25(11):1417–21.

126. Shimizu I, Ma YR, Mizobuchi Y, Liu F, Miura T, Nakai Y, Ito S. Effects of Sho-Saiko-to, a Japanese herbal medicine, on hepatic fibrosis in rats. Hepatology. 1999;29(1):149–60.

127. Takahashi Y, Soejima Y, Kumagai A, Watanabe M, Uozaki H, Fukusato T. Inhibitory effects of Japanese herbal medicines Sho-Saiko-to and Juzen-Taiho-to on nonalcoholic steatohepatitis in mice. PLOS ONE. 2014;9(1):e87279.

128. Lee JK, Kim JH, Shin HK. Therapeutic effects of the oriental herbal medicine Sho-Saiko-to on liver cirrhosis and carcinoma. Hepatol Res. 2011;41(9):825–37.

129. Wei F, Lang Y, Gong D, Fan Y. Effect of Dahuang zhechong formula on liver fibrosis in patients with chronic hepatitis B: A meta-analysis. Complement Ther Med. 2015;23(1):129–38.

130. Miao L, Yang WN, Dong XQ, Zhang ZQ, Xie SB, Zhang DZ, Wang GQ. Combined Anluohuaxianwan and entecavir treatment significantly improve the improvement rate of liver fibrosis in patients with chronic hepatitis B virus infection. Chin J Hepatol. 2019;27(7):521–6.

131. Fatahzadeh M, Schwartz RA. Human herpes simplex virus infections: Epidemiology, pathogenesis, symptomatology, diagnosis, and management. J Am Acad Dermatol. 2007;57:737–63. [Google Scholar] [CrossRef] [PubMed]

132. Fatahzadeh M, Schwartz RA. Human herpes simplex labialis. Clin Exp Dermatol. 2007;32:625–30.

133. Zandi K, Ramedani E, Mohammadi K, Tajbakhsh S, Deilami I, Rastian Z, Fouladvand M, Yousefi F, Farshadpour F. Evaluation of antiviral activities of curcumin derivatives against HSV-1 in Vero cell line. Nat Prod Commun. 2010;5:1935–38. [Google Scholar] [CrossRef][Green Version]

134. Kutluay SB, Doroghazi J, Roemer ME, Triezenberg SJ. Curcumin inhibits herpes simplex virus immediate-early gene expression by a mechanism independent of p300/CBP histone acetyltransferase activity. Virology. 2008;373:239–47. [Google Scholar] [CrossRef][Green Version]

135. Musarra-Pizzo M, Pennisi R, Ben-Amor I, Smeriglio A, Mandalari G, Sciortino MT. In vitro anti-HSV-1 activity of polyphenol-rich extracts and pure polyphenol compounds derived from pistachios kernels (Pistacia vera L.). Plants. 2020;9:267. [Google Scholar] [CrossRef][Green Version]

136. Rezazadeh F, Moshaverinia M, Motamedifar M, Alyaseri M. Assessment of anti HSV-1 activity of Aloe vera gel extract: An in vitro study. J Dent. 2016;17:49–54. [Google Scholar]

137. Croft KD. The chemistry and biological effects of flavonoids and phenolic acidsa. Ann NY Acad Sci. 1998;854:435–42. [Google Scholar] [CrossRef]

138. Hung P-Y, Ho B-C, Lee S-Y, Chang S-Y, Kao C-L, Lee S-S, Lee C-N. Houttuynia cordata targets the beginning stage of herpes simplex virus infection. PLOS ONE. 2015;10:e0115475. [Google Scholar] [CrossRef][Green Version]

139. Bisignano C, Mandalari G, Smeriglio A, Trombetta D, Pizzo MM, Pennisi R, Sciortino MT. Almond skin extracts abrogate HSV-1 replication by blocking virus binding to the cell. Viruses. 2017;9:178. [Google Scholar] [CrossRef]

140. Musarra-Pizzo M, Ginestra G, Smeriglio A, Pennisi R, Sciortino MT, Mandalari G. The antimicrobial and antiviral activity of polyphenols from almond (Prunus dulcis L.) skin. Nutrients. 2019;11:2355. [Google Scholar] [CrossRef] [PubMed][Green Version]

141. Cheng HY, Lin LT, Huang HH, Yang CM, Lin CC. Yin Chen Hao Tang, a Chinese prescription, inhibits both herpes simplex virus type-1 and type-2 infections in vitro. Antivir Res. 2008;77(1):14–19.

142. Brentjens MH, Yeung-Yue KA, Lee PC, Tyring SK. Human papillomavirus: A review. Dermatol Clin. 2002;20:315–31. [Google Scholar] [CrossRef]

143. Di Domenico F, Foppoli C, Coccia R, Perluigi M. Antioxidants in cervical cancer: Chemopreventive and chemotherapeutic effects of polyphenols. Biochim Biophys Acta. 2012;1822:737–47. [Google Scholar] [CrossRef][Green Version]

144. Divya CS, Pillai MR. Antitumor action of curcumin in human papillomavirus associated cells involves downregulation of viral oncogenes, prevention of NFkB and AP-1 translocation, and modulation of apoptosis. Mol Carcinog. 2006;45:320–32. [Google Scholar] [CrossRef]

145. Talwar G, Dar SA, Rai MK, Reddy K, Mitra D, Kulkarni SV, Doncel GF, Buck CB, Schiller JT, Muralidhar S, et al. A novel polyherbal microbicide with inhibitory effect on bacterial, fungal and viral genital pathogens. Int J Antimicrob Agents. 2008;32:180–5. [Google Scholar] [CrossRef]

146. Hu Y, Lu Y, Qi X, Chen X, Liu K, Zhou X, Hu Y. Clinical efficacy of paiteling in the treatment of condyloma acuminatum infected with different subtypes of HPV. Dermatol Ther. 2019;32(5):e13065.

147. Liu Y, Zheng P, Jiao T, Zhang M, Wu Y, Zhang X, Zhao Z. Paiteling induces apoptosis of cervical cancer cells by down-regulation of the E6/E7-Pi3k/Akt pathway: A network pharmacology. J Ethnopharmacol. 2023;305:116062.

148. Jin G, Liu Y, Wang S, Zhang X, Wang S, Zhou K, Zhao Z. Efficacy and safety of Chinese patent medicine paiteling in condyloma acuminatum: A systematic review and meta-analysis. J Ethnopharmacol. 2023;318:116894.

149. Xiao J, Wu J, Yu B. Therapeutic efficacy of Youdujing preparation in treating cervical high-risk human papilloma virus infection patients. Chin J Integr Med. 2012;32(9):1212–15.

150. Chen X, Hu X, Liu L, Liang X, Xiao J. Extracts derived from a traditional Chinese herbal formula triggers necroptosis in ectocervical Ect1/E6E7 cells through activation of RIP1 kinase. J Ethnopharmacol. 2019;239:111922.

151. Shaw GM, Hunter E. HIV transmission. Cold Spring Harb Perspect Med. 2012;2:11.

152. Burton DR, Desrosiers RC, Doms RW, Koff WC, Kwong PD, Moore JP, et al. HIV vaccine design and the neutralizing antibody problem. Nat Immunol. 2004;5:233–6.

153. Evans A, Lee R, Mammen-Tobin A, Piyadigamage A, Shann S, Waugh M. HIV revisited: The global impact of the HIV/AIDS epidemic. Skinmed. 2004;3:149–56.

154. Lusvarghi S, Bewley CA. Griffithsin: An antiviral lectin with outstanding therapeutic potential. Viruses. 2016;8:296. [Google Scholar] [CrossRef]

155. Gandhi MJ, Boyd MR, Yi L, Yang GG, Vyas GN. Properties of Cyanovirin-N (CV-N): Inactivation of HIV-1 by sessile Cyanovirin-N (SCV-N). Dev Biol. 2000;102:141–8. [Google Scholar]

156. Plaza A, Gustchina E, Baker HL, Kelly M, Bewley CA, Mirabamides AD, Depsipeptides from the spongesiliquariaspongia mirabilis that inhibit HIV-1 fusion. J Nat Prod. 2007;70:1753–60. [Google Scholar] [CrossRef]

157. Kashiwada Y, Hashimoto F, Cosentino LM, Chen C-H, Garrett APE, Lee K-H. Betulinic acid and dihydrobetulinic acid derivatives as potent anti-HIV Agents1. J Med Chem. 1996;39:1016–17. [Google Scholar] [CrossRef]

158. Cary DC, Peterlin BM. Natural products and HIV/AIDS. AIDS Res Hum. 2018;34(1):31–38.

159. Medina RA, García-Sastre A. Influenza A viruses: New research developments. Nat Rev Genet. 2011;9:590–603. [Google Scholar] [CrossRef]

160. Eccles R. Understanding the symptoms of the common cold and influenza. Lancet Infect Dis. 2005;5:718–25.

161. Rello J, Pop-Vicas A. Clinical review: Primary influenza viral pneumonia. Crit Care. 2009;13:235.

162. Yin J, Ma L, Wang H, Yan H, Hu J, Jiang W, Li Y. Chinese herbal medicine compound Yi-Zhi-Hao pellet inhibits replication of influenza virus infection through activation of heme oxygenase-1. Acta Pharmaceutica Sinica B. 2017;7(6):630–7.

163. Xi L, Le-Ping L, Xin-Yi X, Ling LI, Guo-Min Z. A network pharmacology study on the effects of Ma Xing Shi Gan decoction on influenza. Digit Chin Med. 2020;3(3):163–79.

164. Du CY, Zheng KY, Bi CW, Dong TT, Lin H, Tsim KW. Yu Ping Feng San, an ancient Chinese herbal decoction, induces gene expression of anti-viral proteins and inhibits neuraminidase activity. Phytother Res. 2015;29(5):656–61.

165. Qin HQ, Shi SS, Fu YJ, Yan YQ, Wu S, Tang XL, Jiang ZY. Effects of Gui Zhi Ma Huang Ge Ban Tang on the TLR7 pathway in influenza virus infected mouse lungs in a cold environment. Evid Based Complementary Altern Med. 2023;2018(1), 5939720.

166. El-Serag HB. Epidemiology of viral hepatitis and hepatocellular carcinoma. Gastroenterology. 2012;142:1264–73.

167. Welsch C, Jesudian A, Zeuzem S, Jacobson I. New direct-acting antiviral agents for the treatment of hepatitis C virus infection and perspectives. Gut. 2012;61(Suppl 1):i36–46.

168. Galani BRT, Sahuc M-E, Njayou FN, Deloison G, Mkounga P, Feudjou WF, Brodin P, Rouillé Y, Nkengfack AE, Moundipa PF, et al. Plant extracts from Cameroonian medicinal plants strongly inhibit hepatitis c virus infection in vitro. Front Microbiol. 2015;6:488. [Google Scholar] [CrossRef] [PubMed] [Green Version]

169. Lin L-T, Chung C-Y, Hsu W-C, Chang S-P, Hung T-C, Shields J, Russell RS, Lin C-C, Liang-Tzung L, Yen M-H, et al. Saikosaponin b2 is a naturally occurring terpenoid that efficiently inhibits hepatitis C virus entry. J Hepatol. 2015;62:541–48. [Google Scholar] [CrossRef]

170. Calland N, Sahuc M-E, Belouzard S, Pène V, Bonnafous P, Mesalam AA, Deloison G, Descamps V, Sahpaz S, Wychowski C, et al. Polyphenols inhibit hepatitis c virus entry by a new mechanism of action. J Virol. 2015;89:10053–63. [Google Scholar] [CrossRef][Green Version]

171. Yamashita A, Salam KA, Furuta A, Matsuda Y, Fujita O, Tani H, Fujita Y, Fujimoto Y, Ikeda M, Kato N, et al. Inhibition of hepatitis C virus replication and viral helicase by ethyl acetate extract of the marine feather star Alloeocomatella polycladia. Mar Drugs. 2012;10:744–61. [Google Scholar] [CrossRef][Green Version]

172. Liu CY, Ko PH, Yen HR, Cheng CH, Li YH, Liao ZH, Hsu CH. The Chinese medicine Kuan-Sin-Yin improves liver function in patients with chronic hepatitis C: A randomised and placebo-controlled trial. Complement Ther Med. 2016;27:114–22.

173. Tsai FJ, Cheng CF, Chen CJ, Lin CY, Wu YF, Li TM, Lin YJ. Effects of Chinese herbal medicine therapy on survival and hepatic outcomes in patients with hepatitis C virus infection in Taiwan. Phytomedicine. 2019;57:30–38.

174. Diceinson GT. Epidemic and endemic influenza. Can Med Assoc J. 1962;86:588–9. [Google Scholar]

4 Molecular Insights into Antiviral Activity from Traditional and Herbal Medicine for the Management of SARS-CoV-2 Variants

*Marudhachalam Kamalesh, Ramaraj Sivamano,
Sathyalingam Sathya Trisha, Arunachalam Abitha,
Jen-Tsung Chen, Kalibulla Syed Ibrahim,
Shanmugam Velayuthaprabhu, and
Arumugam Vijaya Anand*

4.1 INTRODUCTION

The virus that drives the infectious condition comprehended as severe acute respiratory syndrome coronavirus 2 (SARS-CoV-2) first surfaced in China in late 2019 and went on to cause the coronavirus disease 2019 (COVID-19; Hu *et al.,* 2021). SARS-CoV-2 belongs to the *Coronaviridae* family and contains a single-stranded, positive-sense RNA genome (Malik, 2020). The viral genome codes for four main structural proteins: membrane (M), spike (S), envelope (E), and nucleocapsid (N) proteins. The trimeric S protein on the viral membranal surface facilitates access into host cells by attaching to angiotensin-converting enzyme 2 (ACE2) receptors (Saxena *et al.,* 2020). SARS-CoV-2 has a rapid mutation rate, leading to the occurrence of various variants, which show increased immune evasion abilities and transmissibility (Arumugam *et al.,* 2020; Bhotla *et al.,* 2020; Shanmugam *et al.,* 2020; Bhotla *et al.,* 2021; Meyyazhagan *et al.,* 2022; Pushparaj *et al.,* 2022; Carabelli *et al.,* 2023). Traditional medicines use natural ingredients and have been used all over the world for hundreds or even thousands of years. They have developed into orderly-regulated systems of medicine (Yuan *et al.,* 2016).

Conventional herbal therapies have been investigated as possible alternative therapies against SARS-CoV-2 due to their antimicrobial, antiphlogistic, and immunomodulatory effects. Compounds like glycyrrhizin from *Glycyrrhiza glabra*, artemisinin from *Artemisia annua*, curcumin from *Curcuma longa*, and eugenol from *Syzygium aromaticum* have exhibited antiviral effects through mechanisms like inhibiting viral replication, blocking ACE2 receptors, reducing cytokine storms, and targeting viral proteases (Armanini *et al.,* 2020; Fuzimoto, 2021; Kharisma *et al.,* 2021; Li *et al.,* 2022). *In vitro* investigations on extracts from specific herbal extracts showed potent inhibition of SARS-CoV-2 replication at a certain concentration. Animal models like hamsters and mice have been used to assess the antiviral efficacies of herbal compounds through assays like viral titer reduction, cytopathic effect inhibition, and viral load quantification (Poochi *et al.,* 2020; Anand *et al.,* 2021; Bhotla *et al.,* 2021; Jan *et al.,* 2021; Al-Kuraishy *et al.,* 2022; Chandra Manivannan *et al.,* 2022; Abidharini *et al.,* 2023; Bharathi *et al.,* 2023; Boro *et al.,* 2023; Kaviya *et al.,* 2023; Latha *et al.,* 2023; Ramya *et al.,* 2023).

DOI: 10.1201/9781003452621-4

The synergistic effects of multiple components present in medicinal herbal plants have exhibited greater efficacy as opposed to SARS-CoV-2. Computational methodologies like molecular docking and homology modeling have accelerated antiviral drug discovery for SARS-CoV-2 by enabling high-throughput exposure of libraries of natural and synthetic compounds (Khodadadi *et al.*, 2020). Bioinformatics approaches have identified phytochemicals derived from herbal plants as potential inhibitors targeting major viral proteins, including the main protease 3-chymotrypsin-like protease (3CLpro), papain-like protease (PLpro), and the spike glycoprotein (Nallusamy *et al.*, 2021; Yan and Gao, 2021). Network pharmacology approaches integrating drug-target-disease interactions have provided insights into repurposing US Food and Drug Administration (FDA)-endorsed drugs with potential antiviral effects on SARS-CoV-2 (Liu *et al.*, 2021). This chapter focuses on exploring the antiviral properties of traditional and herbal medicine, with a specific focus on their molecular mechanisms and potential applications in managing SARS-CoV-2 variants. Although the study described here demonstrates both *in vivo* and *in vitro* activity, their findings may aid in the effective screening of plant bioactive compounds with the possibility for growth as drugs that can combat SARS-CoV-2 variants. As a result, the goal of this study is to establish the efficacy of employing natural sources as a SARS-CoV-2 control method.

4.2   OVERVIEW AND IMPLICATIONS OF SARS-CoV-2 VARIANTS

The variants of SARS-CoV-2 result from mutations during replication in its genome. Coronavirus has a single-stranded RNA as its genome (ssRNA). They are less likely to undergo mutations (i.e., antigenic shift and antigenic drift) that result in changes in their infectivity, transmission, and disease severity (Ramesh *et al.*, 2021). The original virus variant contains one or a few mutations (Janik *et al.*, 2021). Different nomenclatures were used to divide SARS-CoV-2 variants; they include the Nextstrain, Pango, and GISAID varieties. These nomenclatures refer to genetic lineages. In naming non-scientific groups, the help of the Greek alphabet has also been used (Parra-Lucares *et al.*, 2022). The genome of SARS-CoV-2 involves six functional open reading frames (ORFs) and four structural proteins that include S, M, E, and N. The S proteins considered the major structural proteins, are arranged into homotrimers noted on the surface of a virus, which is vital for the virus to enter the cell (Yi *et al.*, 2021).

B.1.1.7

The first sample was found in a survey study in the UK on September 20, 2020 (Yi *et al.*, 2021). The variant is categorized into variants of concern (VOC) and comprehended as **Alpha**, (Galloway *et al.*, 2021). They have also been registered in about 30 countries, including Brazil and South Africa. This variant contains 14 missense mutations, 6 silent mutations, and 3 deleterious mutations in S protein (Dubey *et al.*, 2022). The important mutations include N501Y, P681H, and H69-V70del (Golubchik *et al.*, 2021). The variant is valued as highly infectious and contains a high reproduction number of ~90%, which shows that individuals infected are more infectious than those infected with other variants (Dubey *et al.*, 2022). It shows a risen nasopharyngeal viral load than the wild-type strain, which also has a 35% increased risk of death (Aleem *et al.*, 2024). An investigation discovered that the B.1.1.7 variant shows an elevated death rate compared to the pre-existing virus in unvaccinated people. From this study, older-age people with comorbidities have a complete risk of death within 28 days, and males have a more increased risk of death than females (Davies *et al.*, 2021). The mutation in D614G (S protein) is considered dominant because it is seen in all variants of SARS-CoV-2 (Flores-Vega *et al.*, 2022).

B.1.351

It is also called **Beta**. It was originally recognized in October 2020 and has become dominant in South Africa's Eastern Cape province (Mwenda *et al.*, 2021). It had a higher occurrence rate during the second wave (Mistry *et al.,* 2022). The transmission rate of this variant is increasing; however, there is no sign of increased virulence or severity of the disease (Gómez *et al.,* 2021). The mutations are K417N, E484K, and N501Y, which alter the receptor binding domain (RBD). It helps the virus hide from the immune system, and its affinity towards the ACE2 receptor is increased by 19 times (Flores-Vega *et al.,* 2022). Among the 16 countries in the European Union, about 350 patients have been noticed with this variant. Some were travel-related, and patients with no epidemiological link have also been documented (Janik *et al.,* 2021). The other signature mutations include D80A, A701V, ΔL242, D614G, ΔA243, E484K, D215G, K417N, and R246I. The variants and their new mutations reduce the protective efficacy of the known vaccines. The Pfizer-BioNTech vaccine is shown to be 75% potent after two doses (Flores-Vega *et al.,* 2022). Studies showed that neutralization by sera of a vaccinated individual is reduced or completely ineffective (Greaney *et al.,* 2021).

B.1.617.2

During February 2021, this variant rendered the ferocious outbreak of the second wave. The **Delta** variant has evolved into the dominant variant in India and the UK. The lineage also contains many variants, like B.1.617.2.1 and AY.1. Some mutations, such as P681R, D614G, E484Q, L452R, D111D, and G142D in S protein, led to the development of this variant. All variants of B.1.617 emerged between October 2020 and February 2021 (Dubey *et al.,* 2022). This variant can evade antibody neutralization that is produced during infection and is more pertinent following vaccination. It shows a decrease in antibody titer in comparison with the alpha variant (Ramesh *et al.,* 2021). The patients affected by this variant experienced several complications that led to hospitalizations, intensive care unit (ICU) requirements, and even death. The BNT162b2 vaccine provided by Pfizer-BioNTech for the Delta variant showed 88% efficacy. Other vaccines, such as the ChAdOx1 nCoV-19 vaccine provided by Oxford-AstraZeneca, revealed 67% efficacy (Parra-Lucares *et al.,* 2022). The variants showed a reduction in intramolecular and intermolecular interactions with ACE2 receptors. There was also an increase in the replication capability of this variant (Cherian *et al.,* 2021).

B.1.1.529

The **Omicron** variant was detected by a traveler from Botswana and reported in several South African patients. This variant had the power to escalate the rate of daily cases from 280 to 800 within weeks. The variant is placed under the category of variant under monitoring (VUM) and then reclassified as VOC by the World Health Organization (WHO). The transmissibility rate was higher, which also increased the rate of hospitalization (Parra-Lucares *et al.,* 2022). The sublineage of this variant is BA.1 to 3. There are 30 mutations in the S protein, which has several clinical complications such as risk of reinfection, immune resistance, and lower pathogenicity when compared with other variants, and immune resistance. This is because of the evolution of the virus in patients who have immunosuppression (Kumar *et al.,* 2022). Omicron contains mutations not only in the S protein but also in the M protein, E protein, N protein, and ORF. The third dosage of BNT162b2 injection improved the neutralizing capacity of antibody titer by 25-fold and shows greater efficacy (Aleem *et al.,* 2024).

4.3 TRADITIONAL MEDICINE AND HERBAL REMEDIES

Throughout history, plants have been considered an important source of medicine. Even though there is tremendous development in the current pharmaceutical investigation, the use of therapeutic herbs in our daily lives has not been reduced (Shahrajabian *et al.,* 2020). Traditional Chinese Medicine (TCM) uses acupuncture and herbal medicines that help stimulate the immune system.

To prevent adverse reactions and side effects, Chinese herbs must be used correctly. During the earlier SARS infection in 2003, the use of TCM reduced the hospitalization period for the patients, reduced the side effects caused by steroids, and improved disease symptoms (Umashankar *et al.,* 2021). Indian tribal and rural populations use the traditional preparation of herbs for treating diseases as they contain phytochemical components that have improved healing properties (Shahrajabian *et al.,* 2020).

4.4   HERBS THAT HAVE ANTIVIRAL PROPERTIES

Tylophora indica

This plant is found in India and is generally known by the name 'Antamool.' It has been used for almost 1000 years. It represents a significant part of Indian traditional medicine and is found in the eastern and southern parts of India. The important therapeutic ingredients are present in the leaves and roots and are used for treating whooping cough, asthma, cancer, anaphylaxis, and other respiratory infections. The alkaloids present in the plants, like tylophorinidine, tylophorinine, and tylophorine, prevent viral RNA replication (Shahzad *et al.,* 2015). It acts against the coronavirus by blocking the 3a channel protein (Umashankar *et al.,* 2021).

Glycyrrhiza glabra (licorice)

This plant is a type of herb native to the Mediterranean area in Europe and southwestern Asia. It is extensively utilized in China, India, Greece, and Africa for treating various diseases that are related to ulcers and arthritis. It was used as a drug in the prehistoric period. It is commonly known by the name 'Yashtimadhu' ('sweet root' in Sanskrit) (Wang *et al.,* 2015). It shows antimicrobial and anti-fungal activities against certain bacteria and fungi. The compounds contributing to antiviral efficacy include glycyrrhizin (glycyrrhizic acid), $18\alpha/\beta$-glycyrrhetinic acid, flavonoids, and triterpenes. The root of this plant contains phytochemicals such as umbelliferone, liquiritin, glucuronic acid, liquiritigenin, resins, and herniarin (coumarin compounds). The other key components of this plant are its anti-inflammatory properties and autophagy-enhancing capability, which can help patients fight against coronavirus (Abraham and Florentine, 2021).

Camellia sinensis

The herb is used in the preparation of tea and beverages that act as medicines. This procedure has been followed for almost 5000 years and has been widely used in India. It belongs to the family *Thaeceae*. The special property is that it comprises various phytochemicals and polyphenols (epigallocatechin gallate, epigallocatechin, epicatechin, and most importantly, catechin) (Rubab *et al.,* 2020). Polyphenols have been recognized for their effectiveness in treating viral diseases. It is used for its chemoprotective, wound healing, antimicrobial, antioxidant, antidiarrheal, and anti-viral properties. The phytochemicals can also inhibit the activity of some enzymes belonging to the protease family, e.g., matrix metalloproteinases (MMPs). Owing to the properties mentioned previously, the plant is also used *in silico* for the management of COVID-19 (Kanbarkar and Mishra, 2021).

Syzygium aromaticum

This plant grows in tropical climates and belongs to the family *Myrtaceae*. It has been utilized as a medication in TCM and Ayurveda for more than 2000 years (Kumar *et al.,* 2012). It is popularly used as a medicinal spice. Cloves, which are dried flower buds of this plant, contain all the essential oils. It is also used for culinary purposes. It is rich in eugenol (a virucidal agent). The phytoconstituents present in the clove include β-caryophyllene and eugenyl acetate. In addition, they contain several sesquiterpenes such as α-copaene, γ- and δ-cadinene, and α-cubebene (Gopalakrishnan *et al.*, 1984). The aromatherapy procedure followed in Asia uses cloves for treating respiratory illnesses such as bronchitis, cough, cold, sinusitis, and asthma. The flavonoids (rhamnetin, β-caryophyllene, and kaempferol) present in cloves have anti-inflammatory properties. Liquid extract from the cloves is used to treat pyelonephritis (kidney inflammation) seen in COVID-19 cases (Vicidomini *et al.*, 2021).

Piper longum

This is known as the Indian long pepper and grows mainly in tropical regions (rainforests). The plant grows in countries including Nepal, Malaysia, Indonesia, India, and Sri Lanka. The use of the plant began around the 16th century. It contains a special protein known as ribosomal inactivating proteins that can prevent viral infections (Doshi *et al.*, 2013). It can also revive tissues and cells in people who are infected, thereby enhancing a person's immunity. The Rasayana treatment, which involves pepper, reduces the symptoms of COVID-19. It can induce both specific and non-specific immunity. It consists of various metabolites like essential oils, flavonoids, alkaloids, steroids, anthraglycosides, coumarins, anthraquinones, arbutin, sterols, etc. (Subramaniam *et al.*, 2021). The important alkaloids include pipernonaline, piperlongumine, longamide, methyl piperine, pipercide, etc. Flavonoids such as epicatechin, naringenin, apigenin, etc., are also isolated from this plant. These compounds can help in the development of pharmaceutical products that can fight against COVID-19 (Manoj *et al.*, 2004).

Ocimum tenuiflorum (Ocimum sanctum)

It is generally known as Tulsi, and it has been used for 1000 years. It is also widely known as the 'Queen of Herbs' and has an astringent taste with a strong aroma. Extracts from this plant are used in Ayurveda remedies and are treated as the 'elixir of life' (Pattanayak *et al.*, 2010). It shows virucidal and prophylactic activities. Components such as linalool, ursolic acid, and apigenin purified from *Ocimum basilicum* showed antiviral activities against DNA viruses. Terpenoid, crude, and phenol-rich extracts from *O. sanctum* also show virucidal activities by decreasing viral genome copy number (Cohen, 2014). Dihydrodieuginol B and tulsinol A-G are inhibitors of SARS-CoV-2 proteases. It also has ACE2 receptor-blocking activity (Shree *et al.*, 2022) (Figure 4.1).

4.5 MOLECULAR TARGETS AND MECHANISMS OF ACTION

COVID-19 is transferred via droplets, aerosols, and direct interaction with contaminated surfaces. Transmission through the fecal-oral route has also been reported in some cases. Earlier symptoms were coughing and lung opacities, and it progressed to severe pneumonia. The virus reproduces in both respiratory tracts (upper and lower). The initiation of viral infection starts when the S protein and receptors on the target host cell surface interact with each other. The S protein is divided as S1 and S2, by TMPRSS2. The carboxyl-terminal domain present in S protein subunits binds to the ACE2 or CDil147 receptor on the cell surface of the host cell. The S2 subunit then forms an anti-parallel six-helical structure, facilitating the access of the viral nucleic acid into the target host cell (Harrison *et al.*, 2020). Vigorous reproduction of the virus facilitates its access into the lung cells, resulting in symptoms including myalgia, headache, fever, and other respiratory signs (Cevik *et al.*, 2020).

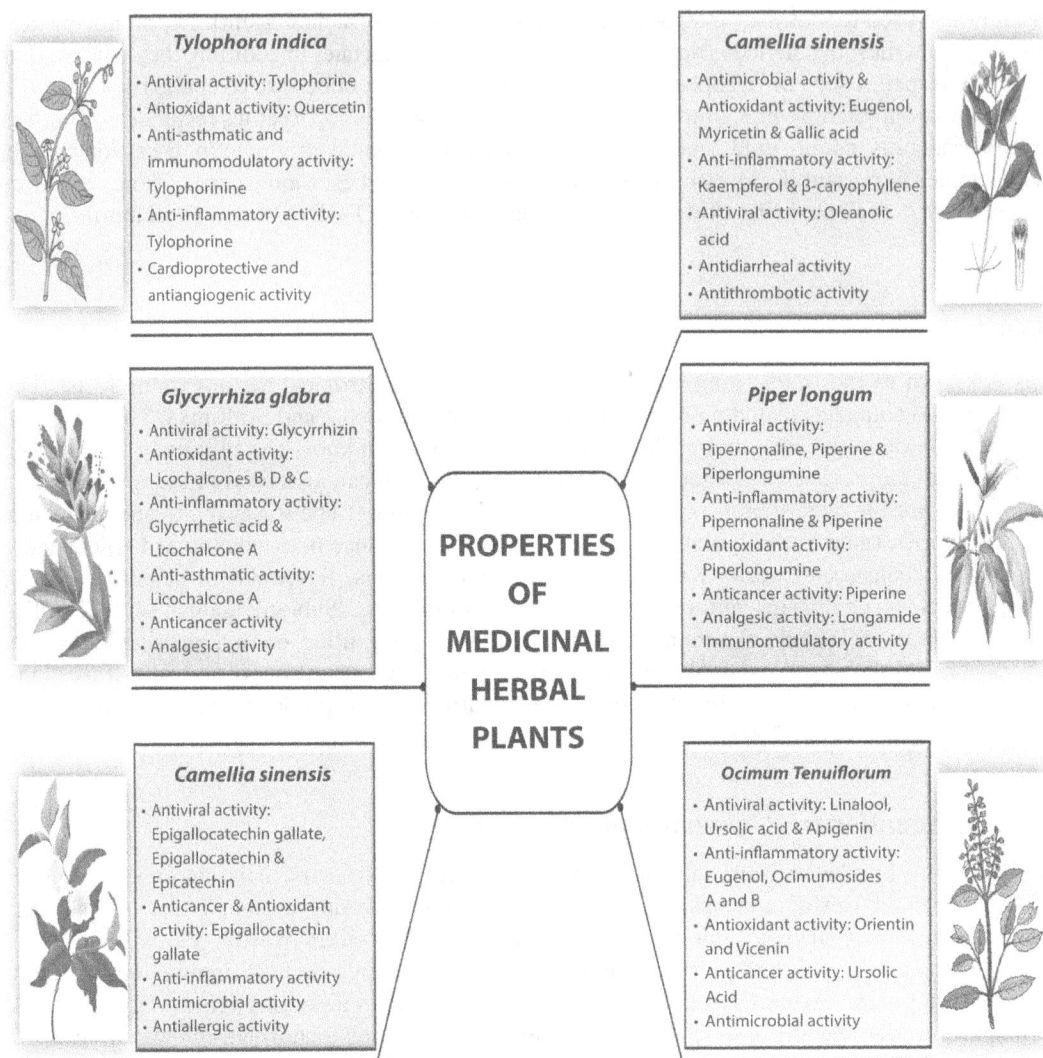

FIGURE 4.1 Properties of medicinal herbal plants.

The discovery of the target protein is the foundation of the pathogenesis study. The most considered target proteins are ACE2 and TMPRSS2 in endothelial cells, 3CLpro, S protein, PLpro, and RNA-dependent RNA polymerase (RdRp) of coronavirus (Harrison *et al.,* 2020). Glycyrrhizin, which is found in the roots of licorice plants, has potential antiviral activity. The molecular targets of this compound are that it affects the signal transduction pathways of casein kinase 2, nuclear factor κB, protein kinase C, and activator protein 1. The metabolites of this compound carry out the upregulation activity of the nitrous oxide synthase enzyme in macrophages. Lycorine, which is another compound from the same plant, acts as a safe antiviral agent (Khare *et al.,* 2020). Chalcone (xanthoangelol E), present in the leaf of *Angelica keiskei,* has inhibitory activity against proteins such as PLpro and 3CLpro. The inhibition is non-competitive against PLpro and competitive inhibition against 3CLpro (IC_{50} values: 11.4 and 1.2 µM). In another study, flavonoid compounds that are separated from the seeds of *Psoralea corylifolia* exhibited inhibition against the target protein PLpro in a dosage-dependent manner (Verma *et al.,* 2020). Compounds present in the Haritaki and tea extracts hinder the activity of 3CLpro (the target protein) of SARS-CoV-2. Molecules such as hesperidin, thearubigin, and quercetin-3-O-rutinoside interfere with the active spot of the target protein 3CLpro and prevent its action (Shahzad *et al.,* 2015). Another compound, Spiroketalenol, separated from the roots of *Tanacetum vulgare* is found to reduce the activity of viral glycoproteins (Upadhyay *et al.,* 2020). Theaflavin extracted from the plant *C. sinensis* interacts with the proteases present in the catalytic site of SARS-COV-2 (Cys145 and His41). Several other interactions form hydrogen bonds with side-chain amino acids (Met165, Glu166, His41, and Leu141). It shows no side effects. Hence, it can reduce the prophylaxis of the virus (Kanbarkar and Mishra, 2021). Eugenol from clove has shown inhibitory potency against SARS-CoV-2 at high nanomolar concentrations. It can hamper the activity of the protease enzyme 3CLpro in the virus. Its propanoic tail attaches to the His41 amino acid present in the catalytic residue, thus affecting the virus's catalytic activity (Rizzuti *et al.,* 2021). About 32 phytochemicals isolated from *P. longum* have shown drug-like properties. The most important are fargesin, I-asarinin, aristololactam, piperundecalidine, lignans, pluviatilol, and machilin-F. These compounds interact with the target proteins by carbon-hydrogen bonds and pi-donor bonds between the active spot amino acids like tyrosine, methionine, glutamine, lysine, asparagine, threonine, leucine, histidine, and aspartic acid (Lakhera *et al.,* 2021).

4.6 EXPERIMENTAL EVIDENCE AND STUDIES

Many experimental studies were performed to demonstrate the antiviral effects of traditional medicines and herbal compounds that can combat SARS-CoV-2 variants. These medicines mainly target the S proteins through which the virus enters the cell (Ohishi *et al.,* 2022). The main approach to countering the virus's effects is by reducing the cytokine storm (Venugopal *et al.,* 2022). These traditional medicines hinder viral entry by inhibiting their replication process, acting on ACE2 receptors, SARS-CoV helicase, and many more (Ugwah-Oguejiofor and Adebisi, 2021). The herbal extracts from *Areca, Quercus* bark, and *Polygala* root were performed using an antiviral action assay to estimate their antiviral impact on SARS-CoV-2. The extracts from herbal drugs with various concentrations were diluted in a 2% fetal calf serum minimum essential medium. Twenty-four hours before infection, Vero E6 cells were put in 96-well plates. Before an hour of infection with the virus, these were treated with diluted herbal drug extract, and the infection was allowed for two hours. The inoculant was then removed. Phosphate-buffered saline was used to wash these inoculates; the extracts from herbal drugs were added until the 48th hour of infection. Then the cells were quantified utilizing reverse transcription polymerase chain reaction (RT-PCR). This showed a subsequent reduction in viral titer in a concentration-dependent way (Ngwe Tun *et al.,* 2022). It was found that these herbal medicine extracts without cytotoxicity inhibited 75% to 100% of viral RNA levels and contagious virus titers of SARS-CoV-2 when given at a concentration of 50 µg/ml (Shang *et al.,* 2020). These extracts show an earlier intervention effect against the virus than before the virus enters the cell *in vitro* (Ngwe Tun *et al.,* 2022). Along with *Areca*, it showed inhibitory effects on human immunodeficiency virus (HIV) proteases (Patel *et al.,* 2021). Artemisinin, derived from *A. annua*, has a wide range of antiviral actions. It was found that this compound suppressed SARS-CoV-2 (Cao *et al.,* 2020; Nair *et al.,* 2021). Also, both artemisinin and the plant *A. annua* have lower levels of inflammatory markers, including interleukin (IL)-6 and tumor necrosis factor-alpha (TNF-α; Desrosiers *et al.,* 2020). *O. sanctum* (Tulasi) has antiviral and antiphlogistic activity (Malabadi *et al.,* 2021). The efficiency of ursolic acid, eugenol, β-caryophyllene, and oleanolic acid, an important constituent of Tulasi, was used to check the binding of SARS-CoV-2 spike S1 with ACE2. This was done using the ACE2: SARS-CoV-2 spike inhibitor screening assay kit. This is a 96-nickel-treated well coated with ACE2 solution (Paidi *et al.,* 2021a). Later, the well was treated with an immune buffer and blocking buffer to various concentrations, and SARS-CoV-2 Spike S1 was added, followed by treatment with an anti-mouse Fc-HRP substrate. A Perkin Elmer multi-mode microplate reader, the Victor X5, was used to monitor the chemiluminescence.

An *in silico*-structural investigation was then performed to study the interaction between SARS-CoV-2 spike S1 and human ACE2, followed by a thermal shift assay. This *in vitro* analysis revealed that the binding of the S1 protein to the ACE2 receptor can be prevented by different levels of eugenol, thereby preventing infection (Paidi *et al.*, 2021b). Additionally, a 4-week *in vivo* study conducted among 24 healthy individuals demonstrated that the administration of 300 mg of holy basil extract elevated the levels of IL-4 and interferon (IFN)-γ, in progress with an upsurge in the rates of natural killer cells and T-helper cells (Mondal *et al.*, 2011).

4.7   ANIMAL MODEL STUDIES

The world, especially the clinical and scientific communities, was on a chase to find effective preventive and treatment measures for COVID-19 (Johansen *et al.*, 2020). The animal model served as one of the essential factors in learning the prophylaxis, pathogenesis, and treatment of the diseases (Roberts *et al.*, 2008). A protocol for how animal models must be chosen is stated in the FDA Animal Efficacy Rule; it states that the model should have the same receptors as humans, and the outcome and severity should match human infections (Pandey *et al.*, 2021). The animal models that mimicked the human infection for a particular disease provided a deeper insight into understanding the disease, which is invaluable (Roberts *et al.*, 2008; de Vries *et al.*, 2021). The *in vivo* and *in vitro* experimentations contributed to finding the viral pathway (Takayama, 2020). During the pandemic, the need for vaccines was critical; therefore, they were produced in an unprecedented manner, and animal models played an important role in their success (de Vries *et al.*, 2021). These tests on an animal model were done for both synthetic and herbal drugs.

G. glabra (Yashtimadhu) is a medicinal plant from which glycyrrhizin is extracted, which is utilized as an active compound for treating COVID-19 (Table 4.1 Luo *et al.*, 2020). Out of many compounds extracted from plants, it was found that the most active compound that inhibits SARS and its associated virus was glycyrrhizin (Cinatl *et al.*, 2003). The mouse animal model was used to find the efficacy of glycyrrhizin. Various assays, like cytotoxicity assay, ELISA, cell viability assay, virucidal and virustatic tests, and cytopathic reduction assay, were employed *in vivo* (Khuntia *et al.*, 2022). It was found that glycyrrhizin inhibited viral penetration, absorption, and replication against SARS-CoV (Table 4.1 Conn et al., 1995; Table 4.1 Khuntia et al., 2022).

Curcumin, which is obtained from turmeric, shows anti-SARS-CoV properties at concentrations of 20 μM and 40 μM. Its mechanism of action is to hinder viral replication by exhibiting protease activity. Studies have revealed that this compound interacts with ACE2 protein and S protein, thus blocking viral entry and preventing infection (Table 4.1 Thimmulappa *et al.*, 2021). It also decreases IL-6 and TNFα release, therefore decreasing the cytokine storm (Table 4.1 Roshdy *et al.*, 2020). Water-soluble curcumin was orally given to the mice infected with pneumonia, resulting in a decrease in the cytokine storm (Table 4.1 Zhang *et al.*, 2019). Thus, it shows its potential to act against SARS-CoV (Thimmulappa *et al.*, 2021).

Ganoderma lucidum is a therapeutic mushroom traditionally utilized in Japanese and Chinese medicine that is widely used to treat many diseases like hepatitis, gastric cancer, hypertension, etc. (Table 4.1 Pan *et al.*, 2013). Recent *in vivo* investigations on hamsters have demonstrated that the extracts of *G. lucidum* are a potential compound that have effects against SARS-CoV-2. Various assays, like cell-based techniques and *in vitro* enzymatic assays combined with computer modeling, were performed. It was found that it restricted the pathway of RdRp and virus 3CLpro (Table 4.1 Jan *et al.*, 2021).

Perilla (*Perilla frutescens*) is a therapeutic plant extensively utilized by Asian populations to treat abnormal stomach functions, heat discharge, etc. (Tang *et al.*, 2021). Studies have revealed its potential to treat SARS-CoV-2 *in vivo* in the Syrian hamster. It was found that adults aged 18 or older with a severe or moderate COVID infection showed high levels of inflammatory cytokines (IL-6) in the blood plasma (Table 4.1 Evans *et al.*, 2022). Cell-based assays performed by examining the virus-induced immunopathology in the lungs and viral loads revealed that perilla

TABLE 4.1

List of Derived Compounds on Animal Model Studies

Traditional/ Herbal Medicine	Active Compounds	Animal Model	Assays/Experiments	Key Mechanisms	Reference
Glycyrrhiza glabra (Yashtimadhu)	Glycyrrhizin	Mice	Cytotoxicity assay, ELISA, cell viability assay, virucidal and virustatic test, cytopathic reduction assay	Inhibited viral penetration, absorption, and replication against influenza and SARS-CoV	Conn *et al.* (1995); Luo *et al.* (2020); Khuntia *et al.* (2022)
Curcuma longa (Turmeric)	Curcumin	Mice	Cytotoxicity assay, plaque-reduction assay, hemagglutination inhibition assay, ELISA, protease inhibition assay, ACE2 and S protein interaction, cytokine reduction	Inhibited viral replication and inhibition of 3CLpro, blocked viral entry, decreased cytokine storm	Zhang *et al.* (2019); Roshdy *et al.* (2020); Thimmulappa *et al.* (2021)
Ganoderma lucidum (Lingzhi or Reishi)	Terpenoids	Hamster	Cell-based assay, *in vitro* enzymatic assay, computer modeling	Restricted RdRP and virus 3CLpro pathway	Pan *et al.* (2013); Jan *et al.* (2021)
Perilla frutescens (Perilla)	Perilla leaf extract	Syrian Hamster	Cell-based assays, examination of virus-induced immunopathology, viral loads	Controlled cytokine levels, exhibited *in vivo* therapeutic activity and immunomodulatory effects	Pan *et al.* (2013); Evans *et al.* (2022); Chin *et al.* (2022)
Nerium oleander (Oleandrin)	Oleandrin (*Nerium oleander*)	Golden Syrian Hamster	Concentration testing (130 µg/ml) for 5 consecutive days before infection sublingual route	Lessening of viral titer in nasal turbinates of golden Syrian hamsters	Hong *et al.* (2014); Plante *et al.* (2021)

extract controlled the cytokine levels and exerted *in vivo* therapeutic activity and immunomodulatory effects (Table 4.1 Chin *et al.*, 2022), thus reducing the symptoms and exhibiting antiviral activity.

Various tests on many medicinal plants were conducted on non-human primate (NHP) models as well. Oleandrin, derived from *Nerium oleander* is a cardiac glycoside that is also lipid-soluble and is used as an active principal agent for clinical trials in phase I and II of cancer (Table 4.1 Hong *et al.*, 2014). Oleandrin was tested on infected SARS-CoV-2-affected African green monkey kidney Vero cells; doses between 0.005 µg/ml and 100 µg/ml resulted in reduced viral production (Table 4.1 Plante *et al.*, 2021).

Animal models used for studies should be lenient to infection and should be able to produce infection similar to humans for that particular disease of study (Pandey *et al.*, 2021). The potential for using animal models is wide. Hamsters express ACE2 receptors, which produce a similar infection as that of humans (Table 4.1 Chin *et al.*, 2022). The older monkey can develop a rigorous form of the COVID-19 infection. They also produce antibodies as neutralizing activity and T-cell actions against the virus as an immunological effect. These models were mainly used as therapeutic agents and for vaccine testing (Bestion *et al.*, 2022). Animal models can be small, like mice, hamsters, etc., or large, like NHPs. The limitations of using small animals are that the intrinsic biological differences between humans and small animals are large. Due to the short life span of animals like hamsters, it is difficult to study the long prognosis of the diseases (Agostini *et al.*, 2018). A study revealed that NHPs contributed only 5% of the animal model study. The main limitations of large animal models like NHPs are their financial, ethical, and technical issues (Bestion *et al.*, 2022) (Table 4.1).

4.8 SYNERGISTIC EFFECTS OF COMBINING COMPOUNDS

For much of history, the approach to developing medicines and treating any disease was 'one drug, one target, one disease.' This pharmaceutical approach was good until the previous decade when this mono-substance approach shifted to adopting combination therapies. This approach involved the treatment of disease using many combinations of active compounds. Recent studies have revealed that combination therapy used to treat complex diseases like cancer, AIDS, atherosclerosis, etc., showed greater therapeutic effectiveness. As they show complex pathophysiological and etiological mechanisms, it is hard to treat them with drugs that show a single-target approach (Devita *et al.*, 1975; Zhou *et al.*, 2016). Traditional medicines, which involve plant-based preparations, are successful in boosting immunity and are tolerant to infections caused by viruses (Prasad *et al.*, 2020). The multi-component character exhibited by therapeutic herbs causes them to become specialized in treating various complicated diseases by exhibiting synergistic effects (Yang *et al.*, 2014). Its availability and low cost make it more popular among indigenous people. This can ensure that many compounds present in plants, like flavonoids, alkaloids, tannins, lignins, and many other phytochemicals produced by plants, help treat viral diseases like SARS-CoV (Matsabisa *et al.*, 2022).

Quinine, an alkaloid extracted from *Cinchona officinalis bark*, is widely employed to treat malaria. A study revealed that hydroxychloroquine and chloroquine, which are the structural analogs of quinine, when combined with azithromycin, were found to be more effective against SARS-CoV-2. The outcomes demonstrated that this combination was more effective and reduced the viral load of SARS-CoV-2 (Gentile *et al.*, 2020). PHELA is an herbal mixture of four exotic African therapeutic plants, like *Rotheca myricoides, Clerodendrum glabrum, Senna occidentalis,* and *Gladiolus dalenii. In vitro* studies found that when given at a dose of 0.005–0.03 mg/mL, 90% of the infection was inhibited for SARS-CoV-2 and SARS-CoV. This also inhibited 100% of the infection induced by MERS-CoV when given at concentrations of 0.1–0.6 mg/mL. PHELA showed ~0.01 mg/ml *in vitro* IC_{50} average on SARS-CoV, SARS-CoV-2 (Wuhan strain), and MERS-CoV. *In silico* docking tests revealed that PHELA has powerful binding energy interchanges with the proteins present on SARS-CoV-2. PHELA exhibited inhibitory activity even when given at a low microgram concentration. The test also showed that when given at the lowest concentration, it showed its inhibitory action by 30% against the wild-type Wuhan-SARS-CoV-2. This suggests that PHELA has more possibility to act as a COVID-19 therapeutic (Matsabisa *et al.*, 2022).

Myricetin obtained from *Myrica rubra* is a flavonoid compound, and scutellarein obtained from *Asplenium belangeri* and *Scutellaria baicalensis* is a flavone compound. The combination of these flavonoid and flavone compounds was effective against SARS-CoV. It was observed that they inhibited the ATPase action of SARS-CoV helicase nsP13 (Yang *et al.*, 2020). Many plants have been tested to see if they pose anti-inflammatory and antiviral properties by exhibiting dual inhibition. Cryptochlorogenic acid and curcumin extracted from *C. longa and Moringa oleifera* inhibit SARS-CoV-2 infection by inhibiting NF-κB1 activity (Kharisma *et al.*, 2021). In a previous era, significant research was not carried out on plants due to the absence of advanced molecular testing and high-tech analytical methods. In this generation, due to the availability of advanced biological techniques, it is possible to study the multi-component nature exhibited by plants, and significant research is ongoing with many medicinal plants used in TCM (Yang *et al.*, 2014) (Figure 4.2).

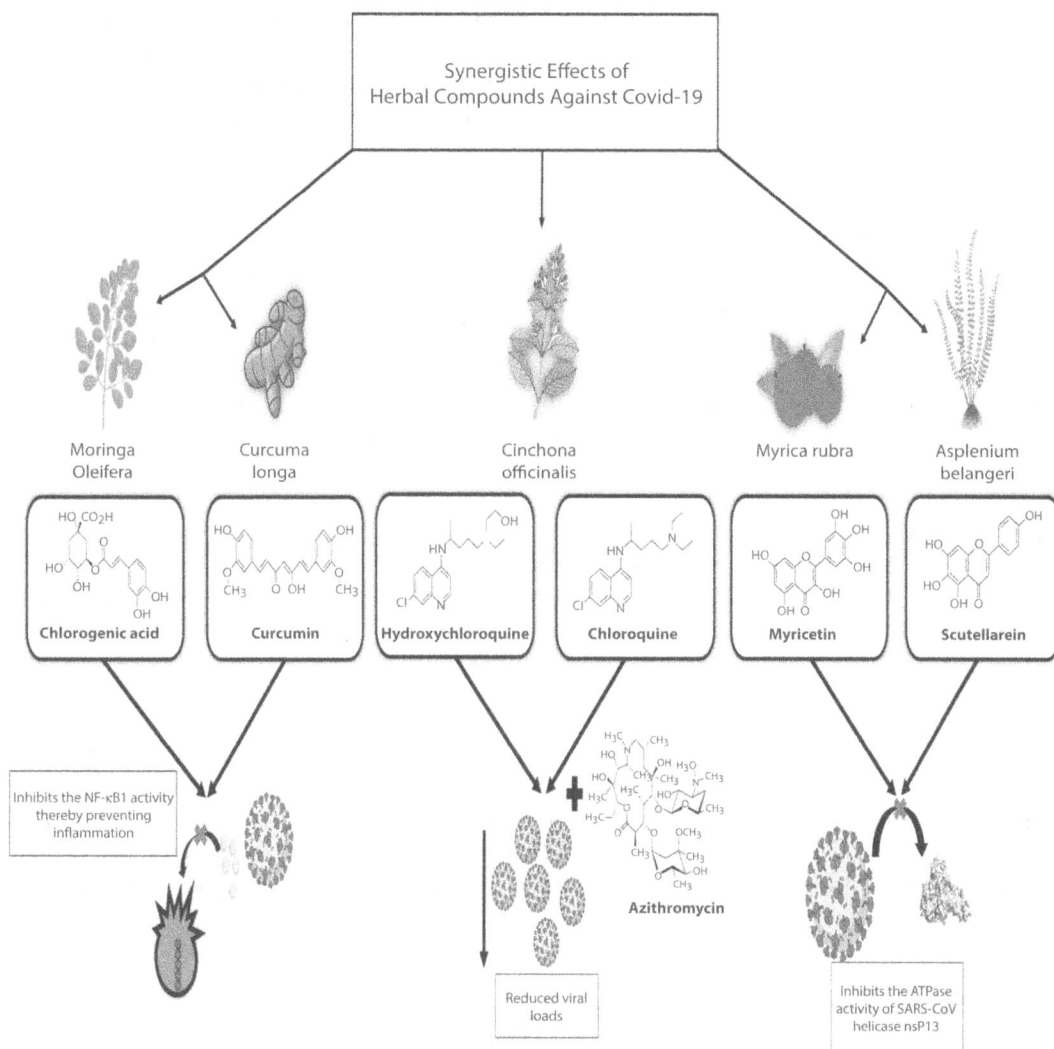

FIGURE 4.2   Synergistic effects of herbal compounds against COVID-19.

4.9   BIOINFORMATICS AND TARGETED-DRUG APPROACH

Finding drug-target interactions (DTIs) is a significant difficulty in drug research. DTIs become more potent by accelerating the costly and time-consuming experimental labor by using prediction (Wen *et al.*, 2017). Experimental methods for DTI identification are very costly and time-consuming. Therefore, various patterns have been designed to predict the possibility of medicine targets for large-scale production (Chen *et al.*, 2016). The current drug targets are cellular proteins whose main aim is to treat or interpret the disease by interacting with chemical compounds. Research has shown that traditional therapeutic target drugs contain ~130 protein families, like transporters, G-protein-coupled receptors (GPCRs), nuclear hormone receptors, ion channels, and enzymes. Determining compound-protein interactions or essential DTIs can be highly challenging. Predicted interactions provide crucial additional evidence for experimental investigations, providing support and understanding of these complicated relationships (Ymanishi *et al.*, 2008). The DTI model is used to identify the availability of commercial antiviral medicines that could discompose SARS-CoV-2 viral elements, like RdRp, helicase, and proteinase (Beck *et al.*, 2020). In this context, a bioinformatics analysis was conducted on the proteins encoded by the new and unexplored coronavirus genes. The proteins of SARS-CoV-2 were systematically corresponded with those of other viruses (Wu *et al.*, 2020).

POTENTIAL THERAPEUTIC DRUG TARGETS

The first approach involves testing for broad-spectrum antivirals. The metabolic characteristics, potential efficacy, dosage uses, and side effects of these antivirals are well-established, as they have been authorized for combating viral infections. The benefits of these treatments are evident due to their prior approval for viral infections (Chen *et al.*, 2020). The second approach entails utilizing current molecular databases to identify molecules with therapeutic effects on coronaviruses (Wu *et al.*, 2020). The development of novel therapies has shown to benefit greatly from the application of structure-based drug discovery. Numerous ongoing investigations have been conducted to treat COVID-19, with a primary focus on the viral proteases and S protein (Tan *et al.*, 2005). There is little effective targeted therapy currently. The third strategy involves using the clinical traits and genomic information of several coronaviruses to develop new therapeutic targets from scratch (Omrani *et al.*, 2014). This revolutionary approach seeks to pinpoint new targets by delving deeper into the biology and pathophysiology of the virus.

METHODS

In bioinformatics analysis, performing *in silico* tests on existing medications allows for the rapid identification of possible therapeutic properties. Researchers have used differential expression investigation to recognize genes that are differently represented in COVID-19 and linked with ACE2. Furthermore, these investigations have revealed clear connections between drug–gene interactions and these differentially expressed genes. A network technique has been employed to combine these drug–gene interactions, providing an enhanced understanding (Cava *et al.*, 2020).

MOLECULAR DOCKING

Sequences available for homology modeling enabled the investigation of several medicines using *in silico* docking to target two major COVID-19 proteins: 3CLpro and spike glycoprotein. This kind of technique is being used to study modern medication molecules. Simultaneously, relevant studies endeavor to identify plausible plant-based medicines using screening to impede pivotal proteins linked to COVID-19. Additionally, specialized research has used molecular docking methods to examine 83 chemical structures from substances used in TCM. Among these substances, chemicals derived from black tea called theaflavin were found to be strong inhibitors that might reduce the action of the SARS-CoV 3CLpro. These results point to the possibility of using compounds of theaflavin as COVID-19 inhibitors (Sakkiah *et al.*, 2021).

HOMOLOGY MODELING

The SARS-CoV-2 PLpro has been examined as a potential medicinal target for antiviral medicines because of its vital function in viral replication and its interaction with the innate immune response. To find target medications for the treatment of SARS-CoV-2, computational techniques including homology modeling, molecular dynamics, and molecular docking studies have been used (Arwansyah *et al.*, 2021). The S protein's main sequence was taken from the Uniport database and used as a template sequence for homology modeling. Furthermore, the complex of the RBD of the SARS-CoV-2 S protein with ACE2 and the mutant trimeric S protein of SARS-CoV-2 were chosen as template structures. Aligning the templates with the target sequence through homology modeling allowed for insights into the structural characteristics (Shahrajabian *et al.*, 2020).

4.10 CHALLENGES AND LIMITATIONS

Traditional medicine is a broad term that involves various forms of native medicine. The explosion of COVID-19, which was driven by SARS-CoV-2, has caused more than 1 million deaths worldwide and more than 50 million confirmed cases. There are no effective antiviral drugs authorized by the FDA (Yang *et al.*, 2020). The richness of bioactive phytoconstituents in plants makes them important for drug discovery, the origin of possible alternative therapies, and also for utilizing herbal treatment for COVID-19. Natural products are valid for managing SARS-CoV-2 and their mode of action for treatment (Jayamohan *et al.*, 2021).

PLANTS WITH ANTIVIRAL ACTIONS

Several plant-derived compounds have demonstrated antiviral properties. These include flavonoids such as rutin, acaciin, hyperoside, and quercetin; phenolic compounds like p-hydroxyacetophenone and chlorogenic acid; terpenes such as (3R,4R,6S)-3,6-dihydroxy-1-menthene; coumarins including scopoletin and scoparone; and other compounds like apigenin, luteolin, acacetin, and aristolactam. Apigenin-7,4/-dimethyl ether is a known compound used to treat and prevent SARS (Chen and Zhang, 2021). Researchers found that amentoflavone, extracted from Torreya nucifera, shows potential in combating SARS-CoV infection. Additionally, other compounds such as myricetin and scutellarein have been identified as having effects against SARS-CoV, with scutellarein specifically targeting the virus's helicase (Jayamohan *et al.*, 2021). In research studies, various compounds of phytochemicals and antiviral compounds found in Indian herbs like haldi, Tulsi, black pepper, Giloy, cloves, ginger, cardamom, ashwagandha, and kemon were examined to see if they could cure COVID-19 infections (Pandey *et al.*, 2020). Chinese plants and herbs are being used as traditional medicine and are also considered to boost the immune system in humans. There are seven coronaviruses understood to infect humans; the three viruses most treacherous among them are MERS, SARS, and SARS-CoV-2 (COVID-19) (Arwansyah *et al.*, 2021).

EXISTING LIMITATIONS IN RESEARCH ON SARS-CoV

Sequence identification is a critical first step in developing probes and primers for nucleic acid (NA)-based RT-PCR testing. Regular sequencing reveals critical information about the virus's genetic makeup, guiding its evolution and assisting in the development of possible vaccinations and therapeutics. Sequencing is currently accepted as a clinical diagnostic tool, and next-generation sequencing (NGS) is critical for detecting and monitoring viral outbreaks. Long-read and short-read methods can be used to broadly classify sequencing techniques. Sequence data in NA fragments shorter than 1000 base pairs is produced by short-read technology (Popoola *et al.*, 2022). Due to the differences in sequencing technology, researchers can select the best method for their specific needs, including genome assembly, variant identification, or other analytical purposes. Pseudo-viruses based on adenoviruses, and ChAdOx type 42, have emerged as effective vaccinations in the fight against SARS-CoV-2 (Nitulescu *et al.*, 2020). According to research, infection with SARS-CoV-2 triggers specific antibody responses, including IgA, IgG, and IgM antibodies. These antibodies target epitopes in the C-terminal and N-terminal domains of the immunogenic protein N (Budweiser *et al.*, 2021). Also, a positive link has been shown between the severity of the disease and the levels of these antibodies. Prolonged symptom duration is linked to elevated levels of IgM and IgA antibodies (Dobaño *et al.*, 2021). Diastolic blood pressure, age, SpO_2 (oxygen saturation), body temperature, major symptoms, creatinine levels, comorbidities, hemoglobin, and C-reactive protein (CRP) exhibited significant differences between patients without treatment limitations (Rahman *et al.*, 2022). Logistic regression, combined with receiver operating characteristic (ROC) analyses, was employed to derive a predictive measure for treatment limitations. The measure demonstrated an accuracy of 82.5% for predicting ICU admission, 82.0% for predicting the need for intubation, and 83.4% for predicting the need for cardiopulmonary resuscitation when compared to existing criteria (Dobaño *et al.*, 2021).

4.11 FUTURE DIRECTIONS AND IMPLICATIONS

SARS-CoV-2 has induced a greater number of casualties around the world, making it a severe general health risk. The absence of vaccinations or treatments continues to create issues, thereby causing the output of new drugs. Computer-aided drug design has helped to speed up research in the drug and outcome approach by reducing time and cost (Budweiser *et al.*, 2021). In West Africa, scientific reports show that the usage of plants that combat SARS-CoV-2 or SARS-CoV; many reports suggest traditional medicines and natural products may play an essential part in fighting the current pandemic (Rahman *et al.*, 2022). Plants are used for discovery in drug efforts for present and future purposes that aim to find the result of anti-COVID-19 chemical therapeutics and also the potential for processing the plants to develop forms of phytomedicines for the nations from which development in plants originates (Popoola *et al.*, 2022).

THE SIGNIFICANCE OF ONGOING RESEARCH AND THE OUTCOME OF NEW THERAPEUTICS

SARS-CoV-2 has garnered significant attention from researchers, prompting efforts to identify treatments that can effectively inhibit the transmission of the disease (Nitulescu *et al.*, 2020). Initial clinical trials have highlighted the therapeutic potential of various drugs, including favipiravir, an antiviral drug that impedes viral reproduction, and hydroxychloroquine, an antimalarial medicine known to impact the virus entering the endosomal pathway (Tu *et al.*, 2020). Concerning COVID-19, ACE2 is supposed to act as an access receptor in human cells. ACE2 expression is downregulated during SARS-COV-2, which recreates a vital part of pathogenic COVID-19. Importantly, blocking the ACE2 receptor and targeting the S protein interaction in the SARS-CoV-2 infection are important therapeutics, and they are widely used for curing and preventing COVID-19 infection (Datta *et al.*, 2020).

The advances obtained in understanding SARS-CoV-2 pave the way for the discovery of new possible treatment targets and medicines. However, more comprehensive preclinical and clinical research is required to completely investigate effective and safe COVID-19 therapies (Li *et al.*, 2020). Specific antiviral drugs were used in the beginning to treat various human coronaviruses. In certain cases, their efficacy was diminished because human coronaviruses differ structurally from SARS-CoV-2. The pathogenicity of SARS-CoV-2 is more elevated than that of other human coronaviruses (Shamsi *et al.*, 2021).

Based on protein–protein docking experiments, it has been determined that the SARS-CoV-2 S protein has a substantial binding relationship with human ACE2; however, this binding is less than that reported with the SARS-CoV S protein (Neves *et al.*, 2012). The results of the most current study, which describes the clinical features of SARS-CoV-2 infection, support the idea that the virus can transmit quickly from person to person. As anticipated in earlier studies, this consistency is confirmed by the reported protein–protein docking results of the human ACE2 and SARS-CoV-2 spike RBD (Huang *et al.*, 2020).

4.12 CONCLUSION

COVID-19 took millions of lives, leaving the world in distress. The pandemic has underscored the critical requirement for efficacious antiviral therapeutics, especially with the emergence of variants like Omicron. As the virus continues to evolve into new variants, research efforts are keen to develop treatments that can inhibit viral transmission and disease severity. Traditional herbal medicines used for centuries represent a promising direction for antiviral drug discovery and development. Many medicinal plants contain beneficial phytochemicals that have been demonstrated to exert antiviral effects on SARS-CoV-2 through mechanisms like blocking viral entry, inhibiting replication enzymes, and suppressing protease activity. This chapter summarizes the investigations that have revealed the antiviral possibility of compounds obtained from herbs like licorice, turmeric, tea, cloves, black pepper, and Tulsi in both *in vitro* experiments and certain animal models. Combination therapies utilizing traditional medicines also warrant further exploration, as synergistic effects may boost efficacy compared to single compounds. This comprehensive chapter gives substantial evidence from studies, instilling confidence in considering these plants as versatile, multi-objective, and multi-purpose entities with potential therapeutic benefits. Continued research is critical to establishing the safety, optimal dosing, and clinical effectiveness of herbal medicines against SARS-CoV-2 variants. Global collaborations leveraging bioinformatics, genomics, and emerging biotechnologies can accelerate screening and validation. Efforts are also needed to make traditional medicines accessible and affordable, particularly in lower-resource settings bearing the greatest disease burden. Investigation into herbal antivirals should be an integral constituent of the continuous public health reaction to COVID-19, alongside vaccine development, pharmaceutical research, and non-pharmaceutical interventions.

ACKNOWLEDGEMENT

All the authors are thankful to their respective Universities and Institutions for their valuable support. The corresponding author is thankful to Skill Development Centre, BEICH, RUSA2.0 (SDC Project order No. BU/RUSA/SDC/BEICH 2020/15; 28.12.2020), Bharathiar University for providing financial support.

DATA AVAILABILITY

The authors confirm that the data supporting the findings of this study are available within the book/chapter.

REFERENCES

Abidharini, J. D., Souparnika, B. R., Elizabeth, J., Vishalini, G., Nihala, S., Shanmugam, V., … Anand, A. V. Herbal formulations in fighting against the SARS-CoV-2 infection. In: *Ethnopharmacology and Drug Discovery for COVID-19: Anti-SARS-CoV-2 Agents from Herbal Medicines and Natural Products.* Singapore: Springer Nature Singapore. 2023, pp. 85–113.

Abraham, J., & Florentine, S. Licorice (Glycyrrhiza glabra) extracts-suitable pharmacological interventions for COVID-19? A review. *Plants.* 2021, *10*(12), 2600.

Agostini, M. L., Andres, E. L., Sims, A. C., Graham, R. L., Sheahan, T. P., Lu, X., … Denison, M. R. Coronavirus susceptibility to the antiviral remdesivir (GS-5734) is mediated by the viral polymerase and the proofreading exoribonuclease. *MBio.* 2018, *9*(2), 10–1128.

Aleem, A., Akbar Samad, A. B., & Vaqar, S. Emerging variants of SARS-CoV-2 and novel therapeutics against coronavirus (COVID-19). 2023 May 8. In: StatPearls [Internet]. Treasure Island (FL): StatPearls Publishing. 2024.

Al-Kuraishy, H. M., Al-Fakhrany, O. M., Elekhnawy, E., Al-Gareeb, A. I., Alorabi, M., De Waard, M., & Batiha, G. E. S. Traditional herbs against COVID-19: Back to old weapons to combat the new pandemic. *European Journal of Medical Research.* 2022, *27*(1), 186.

Anand, A. V., Balamuralikrishnan, B., Kaviya, M., Bharathi, K., Parithathvi, A., Arun, M., … Dhama, K. Medicinal plants, phytochemicals, and herbs to combat viral pathogens including SARS-CoV-2. *Molecules.* 2021, *26*(6), 1775.

Armanini, D., Fiore, C., Bielenberg, J., Sabbadin, C., & Bordin, L. Coronavirus-19: Possible therapeutic implications of spironolactone and dry extract of Glycyrrhiza glabra L.(licorice). *Frontiers in Pharmacology.* 2020, *11*, 558418.

Arumugam, V. A., Thangavelu, S., Fathah, Z., Ravindran, P., Sanjeev, A. M. A., Babu, S., … Tiwari, R. COVID-19 and the world with co-morbidities of heart disease, hypertension and diabetes. *Journal of Pure and Applied Microbiology*. 2020, *14*(3), 1623–1638.

Arwansyah, A., Arif, A. R., Ramli, I., Kurniawan, I., Sukarti, S., Nur Alam, M., … Manguntungi, B. Molecular modelling on SARS-CoV-2 papain-like protease: An integrated study with homology modelling, molecular docking, and molecular dynamics simulations. *SAR and QSAR in Environmental Research*. 2021, *32*(9), 699–718.

Beck, B. R., Shin, B., Choi, Y., Park, S., & Kang, K. Predicting commercially available antiviral drugs that may act on the novel coronavirus (SARS-CoV-2) through a drug-target interaction deep learning model. *Computational and Structural Biotechnology Journal*. 2020, *18*, 784–790.

Bestion, E., Halfon, P., Mezouar, S., & Mège, J. L. Cell and animal models for SARS-CoV-2 research. *Viruses*. 2022, *14*(7), 1507.

Bharathi, K., Sivasangar Latha, A., Jananisri, A., Bavyataa, V., Rajan, B., Balamuralikrishnan, B., … Anand, A. V. Antiviral properties of South Indian plants against SARS-CoV-2. In: *Ethnopharmacology and Drug Discovery for COVID-19: Anti-SARS-CoV-2 Agents from Herbal Medicines and Natural Products*. Springer. 2023, pp. 447–478.

Bhotla, H. K., Balasubramanian, B., Arumugam, V. A., Pushparaj, K., Easwaran, M., Baskaran, R., … Meyyazhagan, A. Insinuating cocktailed components in biocompatible-nanoparticles could act as an impressive neo-adjuvant strategy to combat COVID-19. *Natural Resources for Human Health*. 2021, *1*(1), 3–7.

Bhotla, H. K., Balasubramanian, B., Meyyazhagan, A., Pushparaj, K., Easwaran, M., Pappusamy, M., … Alfalih, A. M. Opportunistic mycoses in COVID-19 patients/survivors: Epidemic inside a pandemic. *Journal of Infection and Public Health*. 2021, *14*(11), 1720–1726.

Bhotla, H. K., Kaul, T., Balasubramanian, B., Easwaran, M., Arumugam, V. A., Pappusamy, M., … Meyyazhagan, A. Platelets to surrogate lung inflammation in COVID-19 patients. *Medical Hypotheses*. 2020, *143*, 110098.

Boro, A., Souparnika, B. R., Atchaya, S., Ramya, S., Senthilkumar, N., Velayuthaprabhu, S., Rengarajan, R. L., & Vijaya Anand A. Antiviral activities of the flavonoids compound against COVID-19 and other viruses causing ailments. In: Chen, JT. (eds). *Bioactive Compound Against SARS-CoV-2*. CRC Press. 2023, pp. 68–81.

Budweiser, S., Baş, Ş., Jörres, R. A., Engelhardt, S., von Delius, S., Lenherr, K., … Bauer, A. Patients' treatment limitations as predictive factor for mortality in COVID-19: Results from hospitalized patients of a hotspot region for SARS-CoV-2 infections. *Respiratory Research*. 2021, *22*(1), 1–11.

Cao, R., Hu, H., Li, Y., Wang, X., Xu, M., Liu, J., … Zhong, W. Anti-SARS-CoV-2 potential of artemisinins in vitro. *ACS Infectious Diseases*. 2020, *6*(9), 2524–2531.

Carabelli, A. M., Peacock, T. P., Thorne, L. G., Harvey, W. T., Hughes, J., de Silva, T. I., … Robertson, D. L., COVID-19 Genomics UK Consortium. SARS-CoV-2 variant biology: Immune escape, transmission and fitness. *Nature Reviews Microbiology*. 2023, *21*(3), 162–177.

Cava, C., Bertoli, G., & Castiglioni, I. A protein interaction map identifies existing drugs targeting SARS-CoV-2. *BMC Pharmacology and Toxicology*. 2020, *21*, 1–11.

Cevik, M., Kuppalli, K., Kindrachuk, J., & Peiris, M. Virology, transmission, and pathogenesis of SARS-CoV-2. *BMJ*. 2020, *371*, m3862.

Chandra Manivannan, A., Malaisamy, A., Eswaran, M., Meyyazhagan, A., Arumugam, V. A., Rengasamy, K. R. R., … Liu, W.-C. Evaluation of clove phytochemicals as potential antiviral drug candidates targeting SARS-CoV-2 main protease: Computational Docking, molecular dynamics simulation, and pharmacokinetic profiling. *Frontiers in Molecular Biosciences*. 2022, *9*, 918101.

Chen, Y., Liu, Q., & Guo, D. Emerging coronaviruses: Genome structure, replication, and pathogenesis. *Journal of Medical Virology*. 2020, *92*(4), 418–423.

Chen, X., Yan, C. C., Zhang, X., Zhang, X., Dai, F., Yin, J., & Zhang, Y. Drug–target interaction prediction: Databases, web servers and computational models. *Briefings in Bioinformatics*. 2016, *17*(4), 696–712.

Chen, M., & Zhang, X. E. Construction and applications of SARS-CoV-2 pseudoviruses: A mini review. *International Journal of Biological Sciences*. 2021, *17*(6), 1574.

Cherian, S., Potdar, V., Jadhav, S., Yadav, P., Gupta, N., Das, M., … Team, N. I. C. SARS-CoV-2 spike mutations, L452R, T478K, E484Q and P681R, in the second wave of COVID-19 in Maharashtra, India. *Microorganisms*. 2021, *9*(7), 1542.

Chin, Y. F., Tang, W. F., Chang, Y. H., Chang, T. Y., Lin, W. C., Lin, C. Y., … Horng, J. T. Orally delivered perilla (Perilla frutescens) leaf extract effectively inhibits SARS-CoV-2 infection in a Syrian hamster model. *Journal of Food and Drug Analysis*. 2022, *30*(2), 252.

Cinatl, J., Morgenstern, B., Bauer, G., Chandra, P., Rabenau, H., & Doerr, H. Glycyrrhizin, an active component of liquorice roots, and replication of SARS-associated coronavirus. *The Lancet*. 2003, *361*(9374), 2045–2046.

Cohen, M. M. Tulsi-Ocimum sanctum: A herb for all reasons. *Journal of Ayurveda and Integrative Medicine*. 2014, *5*(4), 251.

Conn, C. A., McClellan, J. L., Maassab, H. F., Smitka, C. W., Majde, J. A., & Kluger, M. J. Cytokines and the acute phase response to influenza virus in mice. *American Journal of Physiology-Regulatory, Integrative and Comparative Physiology*. 1995, *268*(1), R78–R84.

Datta, P. K., Liu, F., Fischer, T., Rappaport, J., & Qin, X. SARS-CoV-2 pandemic and research gaps: Understanding SARS-CoV-2 interaction with the ACE2 receptor and implications for therapy. *Theranostics*. 2020, *10*(16), 7448.

Davies, N. G., Abbott, S., Barnard, R. C., Jarvis, C. I., Kucharski, A. J., Munday, J. D., … Edmunds, W. J. Estimated transmissibility and impact of SARS-CoV-2 lineage B. 1.1. 7 in England. *Science*. 2021, *372*(6538), eabg3055.

de Vries, R. D., Rockx, B., Haagmans, B. L., Herfst, S., Koopmans, M. P., & de Swart, R. L. Animal models of SARS-CoV-2 transmission. *Current Opinion in Virology*. 2021, *50*, 8–16.

Desrosiers, M. R., Mittleman, A., & Weathers, P. J. Dried leaf Artemisia annua improves bioavailability of artemisinin via cytochrome P450 inhibition and enhances artemisinin efficacy downstream. *Biomolecules*. 2020, *10*(2), 254.

Devita, V. T. Jr, Young, R. C., & Canellos, G. P. Combination versus single agent chemotherapy: A review of the basis for selection of drug treatment of cancer. *Cancer*. 1975, *35*(1), 98–110.

Dobaño, C., Santano, R., Jiménez, A., Vidal, M., Chi, J., Melero, N. R., … & Izquierdo, L. Immunogenicity and crossreactivity of antibodies to the nucleocapsid protein of SARS-CoV-2: Utility and limitations in seroprevalence and immunity studies. *Translational Research*. 2021, *232*, 60–74.

Doshi, G. M., Une, H. D., & Shanbhag, P. P. Rasayans and non-rasayans herbs: Future immunodrug–targets. *Pharmacognosy Reviews*. 2013, *7*(14), 92.

Dubey, A., Choudhary, S., Kumar, P., & Tomar, S. Emerging SARS-CoV-2 variants: Genetic variability and clinical implications. *Current Microbiology*. 2022, *79*, 1–18.

Evans, R. A., Leavy, O. C., Richardson, M., Elneima, O., McAuley, H. J., Shikotra, A., … Berridge, A. Clinical characteristics with inflammation profiling of long COVID and association with 1-year recovery following hospitalisation in the UK: A prospective observational study. *The Lancet Respiratory Medicine*. 2022, *10*(8), 761–775.

Flores-Vega, V. R., Monroy-Molina, J. V., Jiménez-Hernández, L. E., Torres, A. G., Santos-Preciado, J. I., & Rosales-Reyes, R. SARS-CoV-2: Evolution and emergence of new viral variants. *Viruses*. 2022, *14*(4), 653.

Fuzimoto, A. D. An overview of the anti-SARS-CoV-2 properties of Artemisia annua, its antiviral action, protein-associated mechanisms, and repurposing for COVID-19 treatment. *Journal of Integrative Medicine*. 2021, *19*(5), 375–388.

Galloway, S. E., Paul, P., MacCannell, D. R., Johansson, M. A., Brooks, J. T., MacNeil, A., … Dugan, V. G. Emergence of SARS-CoV-2 b. 1.1. 7 lineage-united states, December 29, 2020–January 12, 2021. *Morbidity and Mortality Weekly Report*. 2021, *70*(3), 95.

Gentile, D., Fuochi, V., Rescifina, A., & Furneri, P. M. New anti SARS-Cov-2 targets for quinoline derivatives chloroquine and hydroxychloroquine. *International Journal of Molecular Sciences*. 2020, *21*(16), 5856.

Golubchik, T., Lythgoe, K. A., Hall, M., Ferretti, L., Fryer, H. R., MacIntyre-Cockett, G., … Bonsall, D. Early analysis of a potential link between viral load and the N501Y mutation in the SARS-COV-2 spike protein. *MedRxiv*. 2021, *01*, 2021–01.

Gómez, C. E., Perdiguero, B., & Esteban, M. Emerging SARS-CoV-2 variants and impact in global vaccination programs against SARS-CoV-2/COVID-19. *Vaccines*. 2021, *9*(3), 243.

Gopalakrishnan, N., Narayanan, C. S., & Mathew, A. G. Sesquiterpene hydrocarbons from clove oil. *Lebensmittel-Wissenschaft+ Technologie*. 1984, *17*(1), 42–43.

Greaney, A. J., Loes, A. N., Crawford, K. H., Starr, T. N., Malone, K. D., Chu, H. Y., & Bloom, J. D. Comprehensive mapping of mutations in the SARS-CoV-2 receptor-binding domain that affect recognition by polyclonal human plasma antibodies. *Cell Host & Microbe*. 2021, *29*(3), 463–476.

Harrison, A. G., Lin, T., & Wang, P. Mechanisms of SARS-CoV-2 transmission and pathogenesis. *Trends in Immunology*. 2020, *41*(12), 1100–1115.

Hong, D. S., Henary, H., Falchook, G. S., Naing, A., Fu, S., Moulder, S., … Kurzrock, R. First-in-human study of pbi-05204, an oleander-derived inhibitor of akt, fgf-2, nf-κB and p70s6k, in patients with advanced solid tumors. *Investigational New Drugs*. 2014, *32*, 1204–1212.

Hu, B., Guo, H., Zhou, P., & Shi, Z. L. Characteristics of SARS-CoV-2 and COVID-19. *Nature Reviews Microbiology.* 2021, *19*(3), 141–154.

Huang, C., Wang, Y., Li, X., Ren, L., Zhao, J., Hu, Y., … Cao, B. Clinical features of patients infected with 2019 novel coronavirus in Wuhan, China. *The Lancet.* 2020, *395*(10223), 497–506.

Jan, J. T., Cheng, T. J. R., Juang, Y. P., Ma, H. H., Wu, Y. T., Yang, W. B., … Wong, C. H. Identification of existing pharmaceuticals and herbal medicines as inhibitors of SARS-CoV-2 infection. *Proceedings of the National Academy of Sciences.* 2021, *118*(5), e2021579118.

Janik, E., Niemcewicz, M., Podogrocki, M., Majsterek, I., & Bijak, M. The emerging concern and interest of SARS-CoV-2 variants. *Pathogens.* 2021, *10*(6), 633.

Jayamohan, H., Lambert, C. J., Sant, H. J., Jafek, A., Patel, D., Feng, H., … Gale, B. K. SARS-CoV-2 pandemic: A review of molecular diagnostic tools including sample collection and commercial response with associated advantages and limitations. *Analytical and Bioanalytical Chemistry.* 2021, *413*, 49–71.

Johansen, M. D., Irving, A., Montagutelli, X., Tate, M. D., Rudloff, I., Nold, M. F., … Hansbro, P. M. Animal and translational models of SARS-CoV-2 infection and COVID-19. *Mucosal Immunology.* 2020, *13*(6), 877–891.

Kanbarkar, N., & Mishra, S. Matrix metalloproteinase inhibitors identified from Camellia sinensis for COVID-19 prophylaxis: An in-silico approach. *Advances in Traditional Medicine.* 2021, *21*(1), 173–188.

Kaviya, M., Geofferina, I. P., Poornima, P., Rajan, A. P., Balamuralikrishnan, B., Arun, M., … Anand, A. V. Dietary plants, spices, and fruits in curbing SARS-CoV-2 virulence. In: *Ethnopharmacology and Drug Discovery for COVID-19: Anti-SARS-CoV-2 Agents from Herbal Medicines and Natural Products.* Singapore: Springer Nature Singapore. 2023, pp. 265–316.

Khare, P., Sahu, U., Pandey, S. C., & Samant, M. Current approaches for target-specific drug discovery using natural compounds against SARS-CoV-2 infection. *Virus Research.* 2020, *290*, 198169.

Kharisma, V. D., Agatha, A., Ansori, A. N. M., Widyananda, M. H., Rizky, W. C., Dings, T. G. A., … Zainul, R. Herbal combination from Moringa oleifera Lam. and Curcuma longa L. as SARS-CoV-2 antiviral via dual inhibitor pathway: A viroinformatics approach. *Journal of Pharmacy & Pharmacognosy Research.* 2021, *10*(1), 138–146.

Khodadadi, E., Maroufi, P., Khodadadi, E., Esposito, I., Ganbarov, K., Espsoito, S., … Kafil, H. S. Study of combining virtual screening and antiviral treatments of the Sars-CoV-2 (Covid-19). *Microbial Pathogenesis.* 2020, *146*, 104241.

Khuntia, B. K., Sharma, V., Wadhawan, M., Chhabra, V., Kidambi, B., Rathore, S., … Sharma, G. Antiviral potential of Indian medicinal plants against influenza and SARS-CoV: A systematic review. *Natural Product Communications.* 2022, *17*(3), 1934578X221086988.

Kumar, S., Thambiraja, T. S., Karuppanan, K., & Subramaniam, G. Omicron and Delta variant of SARS-CoV-2: A comparative computational study of spike protein. *Journal of Medical Virology.* 2022, *94*(4), 1641–1649.

Kumar, K. S., Yadav, A., Srivastava, S., Paswan, S., & Sankar Dutta, A. Recent trends in Indian traditional herbs Syzygium aromaticum and its health benefits. *Journal of Pharmacognosy and Phytochemistry.* 2012, *1*(1), 13–22.

Lakhera, S., Devlal, K., Ghosh, A., & Rana, M. In silico investigation of phytoconstituents of medicinal herb 'Piper Longum'against SARS-CoV-2 by molecular docking and molecular dynamics analysis. *Results in Chemistry.* 2021, *3*, 100199.

Latha, A. S., Anita, P. P., Sidhic, N., James, E., Kaviya, M., Bharathi, K., … Anand, A. V. Phytonutrients and secondary metabolites to cease SARS-CoV-2 loop. In: *Bioactive Compounds Against SARS-CoV-2.* CRC Press. 2023, pp. 111–124.

Li, H., Zhou, Y., Zhang, M., Wang, H., Zhao, Q., & Liu, J. Updated approaches against SARS-CoV-2. *Antimicrobial Agents and Chemotherapy.* 2020, *64*(6), 10–1128.

Li, Y., Liu, Z., Zeng, M., El Kadiri, A., Huang, J., Kim, A., … Yu, L. L. Chemical compositions of clove (Syzygium aromaticum (L.) merr. & L.) extracts and their potentials in suppressing SARS-CoV-2 spike protein–ACE2 binding, inhibiting ACE2, and scavenging free radicals. *Journal of Agricultural and Food Chemistry.* 2022, *70*(45), 14403–14413.

Liu, D. Y., Liu, J. C., Liang, S., Meng, X. H., Greenbaum, J., Xiao, H. M., … Deng, H. W. Drug repurposing for COVID-19 treatment by integrating network pharmacology and transcriptomics. *Pharmaceutics.* 2021, *13*(4), 545.

Luo, P., Liu, D., & Li, J. Pharmacological perspective: Glycyrrhizin may be an efficacious therapeutic agent for COVID-19. *International Journal of Antimicrobial Agents.* 2020, *55*(6), 105995.

Malabadi, R. B., Meti, N. T., & Chalannavar, R. K. Role of herbal medicine for controlling coronavirus (SARS-CoV-2) disease (COVID-19). *International Journal of Research and Scientific Innovations.* 2021, *8*(2), 135–165.

Malik, Y. A. Properties of coronavirus and SARS-CoV-2. *The Malaysian Journal of Pathology*. 2020, *42*(1), 3–11.

Manoj, P., Soniya, E. V., Banerjee, N. S., & Ravichandran, P. Recent studies on well-known spice, Piper longum Linn. *Natural Product Research*. 2004, *3*, 222–227.

Matsabisa, M. G., Alexandre, K., Ibeji, C. U., Tripathy, S., Erukainure, O. L., Malatji, K., … Chabalala, H. P. In vitro study on efficacy of PHELA, an African traditional drug against SARS-CoV-2. *Scientific Reports*. 2022, *12*(1), 10305.

Meyyazhagan, A., Pushparaj, K., Balasubramanian, B., Kuchi Bhotla, H., Pappusamy, M., Arumugam, V. A., … Di Renzo, G. C. COVID-19 in pregnant women and children: Insights on clinical manifestations, complexities, and pathogenesis. *International Journal of Gynecology & Obstetrics*. 2022, *156*(2), 216–224.

Mistry, P., Barmania, F., Mellet, J., Peta, K., Strydom, A., Viljoen, I. M., … Pepper, M. S. SARS-CoV-2 variants, vaccines, and host immunity. *Frontiers in Immunology*. 2022, *12*, 809244.

Mondal, S., Varma, S., Bamola, V. D., Naik, S. N., Mirdha, B. R., Padhi, M. M., … Mahapatra, S. C. Double-blinded randomized controlled trial for immunomodulatory effects of Tulsi (Ocimum sanctum Linn.) leaf extract on healthy volunteers. *Journal of Ethnopharmacology*. 2011, *136*(3), 452–456.

Mwenda, M., Saasa, N., Sinyange, N., Busby, G., Chipimo, P. J., Hendry, J., … Bridges, D. J. Detection of, B. 1.351 SARS-CoV-2 variant strain—Zambia, December 2020. *Morbidity and Mortality Weekly Report*. 2021, *70*(8), 280.

Nair, M. S., Huang, Y., Fidock, D. A., Polyak, S. J., Wagoner, J., Towler, M. J., & Weathers, P. J. Artemisia annua L. extracts inhibit the in vitro replication of SARS-CoV-2 and two of its variants. *Journal of Ethnopharmacology*. 2021, *274*, 114016.

Nallusamy, S., Mannu, J., Ravikumar, C., Angamuthu, K., Nathan, B., Nachimuthu, K., … Neelakandan, K. Exploring phytochemicals of traditional medicinal plants exhibiting inhibitory activity against main protease, spike glycoprotein, RNA-dependent RNA polymerase and non-structural proteins of SARS-CoV-2 through virtual screening. *Frontiers in Pharmacology*. 2021, *12*, 667704.

Neves, M. A., Totrov, M., & Abagyan, R. Docking and scoring with ICM: The benchmarking results and strategies for improvement. *Journal of Computer-Aided Molecular Design*. 2012, *26*, 675–686.

Ngwe Tun, M. M., Toume, K., Luvai, E., Nwe, K. M., Mizukami, S., Hirayama, K., … Morita, K. The discovery of herbal drugs and natural compounds as inhibitors of SARS-CoV-2 infection in vitro. *Journal of Natural Medicines*. 2022, *76*(2), 402–409.

Nitulescu, G. M., Paunescu, H., Moschos, S. A., Petrakis, D., Nitulescu, G., Ion, G. N. D., … Tsatsakis, A. Comprehensive analysis of drugs to treat SARS- CoV- 2 infection: Mechanistic insights into current COVID- 19 therapies. *International Journal of Molecular Medicine*. 2020, *46*(2), 467–488.

Ohishi, T., Hishiki, T., Baig, M. S., Rajpoot, S., Saqib, U., Takasaki, T., & Hara, Y. Epigallocatechin gallate (EGCG) attenuates severe acute respiratory coronavirus disease 2 (SARS-CoV-2) infection by blocking the interaction of SARS-CoV-2 spike protein receptor-binding domain to human angiotensin-converting enzyme 2. *PLOS ONE*. 2022, *17*(7), e0271112.

Omrani, A. S., Saad, M. M., Baig, K., Bahloul, A., Abdul-Matin, M., Alaidaroos, A. Y., … Albarrak, A. M. Ribavirin and interferon alfa-2a for severe Middle East respiratory syndrome coronavirus infection: A retrospective cohort study. *The Lancet Infectious Diseases*. 2014, *14*(11), 1090–1095.

Paidi, R. K., Jana, M., Mishra, R. K., Dutta, D., Raha, S., & Pahan, K. ACE-2-interacting domain of SARS-CoV-2 (AIDS) peptide suppresses inflammation to reduce fever and protect lungs and heart in mice: Implications for COVID-19 therapy. *Journal of Neuroimmune Pharmacology*. 2021a, *16*, 59–70.

Paidi, R. K., Jana, M., Raha, S., McKay, M., Sheinin, M., Mishra, R. K., & Pahan, K. Eugenol, a component of holy basil (Tulsi) and common spice clove, inhibits the interaction between SARS-CoV-2 spike S1 and ACE2 to induce therapeutic responses. *Journal of Neuroimmune Pharmacology*. 2021b, *16*, 743–755.

Pan, K., Jiang, Q., Liu, G., Miao, X., & Zhong, D. Optimization extraction of Ganoderma lucidum polysaccharides and its immunity and antioxidant activities. *International Journal of Biological Macromolecules*. 2013, *55*, 301–306.

Pandey, K., Acharya, A., Mohan, M., Ng, C. L., Reid, S. P., & Byrareddy, S. N. Animal models for SARS-CoV-2 research: A comprehensive literature review. *Transboundary and Emerging Diseases*. 2021, *68*(4), 1868–1885.

Pandey, P., Basnet, A., & Mali, A. Quest for COVID-19 cure: Integrating traditional herbal medicines in the modern drug paradigm. *Applied Science and Technology Annals*. 2020, *1*(1), 63–71.

Parra-Lucares, A., Segura, P., Rojas, V., Pumarino, C., Saint-Pierre, G., & Toro, L. Emergence of SARS-CoV-2 variants in the world: How could this happen? *Life*. 2022, *12*(2), 194.

Patel, B., Sharma, S., Nair, N., Majeed, J., Goyal, R. K., & Dhobi, M. Therapeutic opportunities of edible antiviral plants for COVID-19. *Molecular and Cellular Biochemistry*. 2021, *476*, 2345–2364.

Pattanayak, P., Behera, P., Das, D., & Panda, S. K. Ocimum sanctum Linn. A reservoir plant for therapeutic applications: An overview. *Pharmacognosy Reviews*. 2010, *4*(7), 95.

Plante, K. S., Dwivedi, V., Plante, J. A., Fernandez, D., Mirchandani, D., Bopp, N., ... Newman, R. A. Antiviral activity of oleandrin and a defined extract of Nerium oleander against SARS-CoV-2. *Biomedicine & Pharmacotherapy*. 2021, *138*, 111457.

Poochi, S. P., Easwaran, M., Balasubramanian, B., Anbuselvam, M., Meyyazhagan, A., Park, S., ... Kaul, T. Employing bioactive compounds derived from Ipomoea obscura (L.) to evaluate potential inhibitor for SARS-CoV-2 main protease and ACE2 protein. *Food Frontiers*. 2020, *1*(2), 168–179.

Popoola, T. D., Segun, P. A., Ekuadzi, E., Dickson, R. A., Awotona, O. R., Nahar, L., ... Fatokun, A. A. West African medicinal plants and their constituent compounds as treatments for viral infections, including SARS-CoV-2/COVID-19. *DARU Journal of Pharmaceutical Sciences*. 2022, *30*(1), 191–210.

Prasad, A., Muthamilarasan, M., & Prasad, M. Synergistic antiviral effects against SARS-CoV-2 by plant-based molecules. *Plant Cell Reports*. 2020, *39*, 1109–1114.

Pushparaj, K., Bhotla, H. K., Arumugam, V. A., Pappusamy, M., Easwaran, M., Liu, W. C., ... Balasubramanian, B. Mucormycosis (black fungus) ensuing COVID-19 and comorbidity meets-magnifying global pandemic grieve and catastrophe begins. *Science of the Total Environment*. 2022, *805*, 150355.

Rahman, M. M., Islam, M. R., Akash, S., Mim, S. A., Rahaman, M. S., Emran, T. B., ... Wilairatana, P. In silico investigation and potential therapeutic approaches of natural products for COVID-19: Computer-aided drug design perspective. *Frontiers in Cellular and Infection Microbiology*. 2022, *12*, 929430.

Ramesh, S., Govindarajulu, M., Parise, R. S., Neel, L., Shankar, T., Patel, S., ... Moore, T. Emerging SARS-CoV-2 variants: A review of its mutations, its implications and vaccine efficacy. *Vaccines*. 2021, *9*(10), 1195.

Ramya, S., Boro, A., Geofferina, I. P., Jananisri, A., Vishalini, G., Balamuralikrishna, B., ... Anand, A. V. The potential contribution of vitamin K as a nutraceutical to scale down the mortality rate of COVID-19. In: *Bioactive Compounds Against SARS-CoV-2*. CRC Press. 2023, pp. 140–159.

Rizzuti, B., Ceballos-Laita, L., Ortega-Alarcon, D., Jimenez-Alesanco, A., Vega, S., Grande, F., ... Velazquez-Campoy, A. Sub-micromolar inhibition of SARS-CoV-2 3CLpro by natural compounds. *Pharmaceuticals*. 2021, *14*(9), 892.

Roberts, A., Lamirande, E. W., Vogel, L., Jackson, J. P., Paddock, C. D., Guarner, J., ... Subbarao, K. Animal models and vaccines for SARS-CoV infection. *Virus Research*. 2008, *133*(1), 20–32.

Roshdy, W. H., Rashed, H. A., Kandeil, A., Mostafa, A., Moatasim, Y., Kutkat, O., ... Ali, M. A. EGYVIR: An immunomodulatory herbal extract with potent antiviral activity against SARS-CoV-2. *PLOS ONE*. 2020, *15*(11), e0241739.

Rubab, S., Rizwani, G. H., Bahadur, S., Shah, M., Alsamadany, H., Alzahrani, Y., ... Zaman, W. Neuropharmacological potential of various morphological parts of Camellia sinensis L *Saudi Journal of Biological Sciences*. 2020, *27*(1), 567–573.

Sakkiah, S., Guo, W., Pan, B., Ji, Z., Yavas, G., Azevedo, M., ... Hong, H. Elucidating interactions between SARS-CoV-2 trimeric spike protein and ACE2 using homology modeling and molecular dynamics simulations. *Frontiers in Chemistry*. 2021, *8*, 622632.

Saxena, S. K., Kumar, S., Baxi, P., Srivastava, N., Puri, B., & Ratho, R. K. Chasing COVID-19 through SARS-CoV-2 spike glycoprotein. *Virusdisease*. 2020, *31*, 399–407.

Shahrajabian, M. H., Sun, W., Shen, H., & Cheng, Q. Chinese herbal medicine for SARS and SARS-CoV-2 treatment and prevention, encouraging using herbal medicine for COVID-19 outbreak. *Acta Agriculture Scandinavica, Section B-Soil & Plant Science*. 2020, *70*(5), 437–443.

Shahzad, A., Sharma, S. & Siddiqui, S. A. (eds). *Biotechnological Strategies for the Conservation of Medicinal and Ornamental Climbers*. Springer. 2015.

Shamsi, A., Mohammad, T., Anwar, S., Amani, S., Khan, M. S., Husain, F. M., ... Hassan, M. I. Potential drug targets of SARS-CoV-2: From genomics to therapeutics. *International Journal of Biological Macromolecules*. 2021, *177*, 1–9.

Shang, J., Wan, Y., Luo, C., Ye, G., Geng, Q., Auerbach, A., & Li, F. Cell entry mechanisms of SARS-CoV-2. *Proceedings of the National Academy of Sciences*. 2020, *117*(21), 11727–11734.

Shanmugam, R., Thangavelu, S., Fathah, Z., Yatoo, M. I., Tiwari, R., Pandey, M. K., ... Arumugam, V. A. SARS-CoV-2/COVID-19 pandemic-an update. *Journal of Experimental Biology and Agricultural Sciences*. 2020, *8*, S219–S245.

Shree, P., Mishra, P., Selvaraj, C., Singh, S. K., Chaube, R., Garg, N., & Tripathi, Y. B. Targeting COVID-19 (SARS-CoV-2) main protease through active phytochemicals of ayurvedic medicinal plants—Withania somnifera (Ashwagandha), Tinospora cordifolia (Giloy) and Ocimum sanctum (Tulsi)–a molecular docking study. *Journal of Biomolecular Structure and Dynamics*. 2022, *40*(1), 190–203.

Subramaniam, K., Subramanian, S. K., Bhargav, S., Parameswari, R., Praveena, R., Ravikumar, R., … Kumar, V. M. Review on potential antiviral and immunomodulatory properties of Piper Longum. IOP Conference Series. *Materials Science and Engineering*. 2021, *1145*(1), 012099.

Takayama, K. In vitro and animal models for SARS-CoV-2 research. *Trends in Pharmacological Sciences*. 2020, *41*(8), 513–517.

Tan, Y. J., Lim, S. G., & Hong, W. Characterization of viral proteins encoded by the SARS-coronavirus genome. *Antiviral Research*. 2005, *65*(2), 69–78.

Tang, W. F., Tsai, H. P., Chang, Y. H., Chang, T. Y., Hsieh, C. F., Lin, C. Y., … Horng, J. T. Perilla (Perilla frutescens) leaf extract inhibits SARS-CoV-2 via direct virus inactivation. *Biomedical Journal*. 2021, *44*(3), 293–303.

Thimmulappa, R. K., Mudnakudu-Nagaraju, K. K., Shivamallu, C., Subramaniam, K. T., Radhakrishnan, A., Bhojraj, S., & Kuppusamy, G. Antiviral and immunomodulatory activity of curcumin: A case for prophylactic therapy for COVID-19. *Heliyon*. 2021, *7*(2), e06350.

Tu, Y. F., Chien, C. S., Yarmishyn, A. A., Lin, Y. Y., Luo, Y. H., Lin, Y. T., … Chiou, S. H. A review of SARS-CoV-2 and the ongoing clinical trials. *International Journal of Molecular Sciences*. 2020, *21*(7), 2657.

Ugwah-Oguejiofor, C. J., & Adebisi, I. M. Potential medicinal plant remedies and their possible mechanisms against COVID-19: A review. *Ife Journal of Science*. 2021, *23*(1), 161–194.

Umashankar, V., Deshpande, S. H., Hegde, H. V., Singh, I., & Chattopadhyay, D. Phytochemical moieties from Indian traditional medicine for targeting dual hotspots on SARS-CoV-2 spike protein: An integrative in-silico approach. *Frontiers in Medicine*. 2021, *8*, 672629.

Upadhyay, S., Tripathi, P. K., Singh, M., Raghavendhar, S., Bhardwaj, M., & Patel, A. K. Evaluation of medicinal herbs as a potential therapeutic option against SARS-CoV-2 targeting its main protease. *Phytotherapy Research*. 2020, *34*(12), 3411–3419.

Venugopal, V., deenadayalan, B., Poonguzhali, S., & Maheshkumar, K. Potential role of functional foods in the management of SARS-CoV-2 omicron variant. *Open Health*. 2022, *3*(1), 141–144.

Verma, S., Twilley, D., Esmear, T., Oosthuizen, C. B., Reid, A. M., Nel, M., & Lall, N. Anti-SARS-CoV natural products with the potential to inhibit SARS-CoV-2 (COVID-19). *Frontiers in Pharmacology*. 2020, *11*, 561334.

Vicidomini, C., Roviello, V., & Roviello, G. N. Molecular basis of the therapeutical potential of clove (Syzygium aromaticum L.) and clues to its anti-COVID-19 utility. *Molecules*. 2021, *26*(7), 1880.

Wang, L., Yang, R., Yuan, B., Liu, Y., & Liu, C. The antiviral and antimicrobial activities of licorice, a widely-used Chinese herb. *Acta Pharmaceutica Sinica B*. 2015, *5*(4), 310–315.

Wen, M., Zhang, Z., Niu, S., Sha, H., Yang, R., Yun, Y., & Lu, H. Deep-learning-based drug–target interaction prediction. *Journal of Proteome Research*. 2017, *16*(4), 1401–1409.

Wu, C., Liu, Y., Yang, Y., Zhang, P., Zhong, W., Wang, Y., … Li, H. Analysis of therapeutic targets for SARS-CoV-2 and discovery of potential drugs by computational methods. *Acta Pharmaceutica Sinica B*. 2020, *10*(5), 766–788.

Yan, F., & Gao, F. An overview of potential inhibitors targeting non-structural proteins 3 (PLpro and Mac1) and 5 (3CLpro/Mpro) of SARS-CoV-2. *Computational and Structural Biotechnology Journal*. 2021, *19*, 4868–4883.

Yang, Y., Islam, M. S., Wang, J., Li, Y., & Chen, X. Traditional Chinese medicine in the treatment of patients infected with 2019-new coronavirus (SARS-CoV-2): A review and perspective. *International Journal of Biological Sciences*. 2020, *16*(10), 1708.

Yang, Y., Zhang, Z., Li, S., Ye, X., Li, X., & He, K. Synergy effects of herb extracts: Pharmacokinetics and pharmacodynamic basis. *Fitoterapia*. 2014, *92*, 133–147.

Yi, H., Wang, J., Wang, J., Lu, Y., Zhang, Y., Peng, R., … Chen, Z. The emergence and spread of novel SARS-CoV-2 variants. *Frontiers in Public Health*. 2021, *9*, 696664.

Ymanishi, Y., Araki, M., Gutteridge, A., Honda, W., & Kanehisa, M. Prediction of drug–target interaction networks from the integration of chemical and genomic spaces. *Bioinformatics*. 2008, *24*(13), i232–i240.

Yuan, H., Ma, Q., Ye, L., & Piao, G. The traditional medicine and modern medicine from natural products. *Molecules*. 2016, *21*(5), 559.

Zhang, B., Swamy, S., Balijepalli, S., Panicker, S., Mooliyil, J., Sherman, M. A., … Suresh, M. V. Direct pulmonary delivery of solubilized curcumin reduces severity of lethal pneumonia. *The FASEB Journal*. 2019, *33*(12), 13294.

Zhou, X., Seto, S. W., Chang, D., Kiat, H., Razmovski-Naumovski, V., Chan, K., & Bensoussan, A. Synergistic effects of Chinese herbal medicine: A comprehensive review of methodology and current research. *Frontiers in Pharmacology*. 2016, *7*, 201.

5 Structure-Activity Relationship of Antiviral and Anti-SARS-CoV-2 Compounds from Traditional Herbal Medicines

Anand Kumar Pandey, Shalja Verma,
Soumya Rathore, and Rupanjali Singh

5.1 INTRODUCTION

The structure-activity relationship (SAR) represents the association between the chemical entities present in a compound and its biological activity in *in vivo* systems. It plays a lead in the discovery of drugs from initial screening to the final optimization of lead compounds (Guha, 2013). Practically, the identification of SAR in a group of compounds and their bioactive properties for revealing structural details that are characteristic of a particular bioactivity can provide scope for structural modifications for optimization or enhancement of specific bioactivity. This encourages the exploration of chemical space based on identified SAR for a specific collection of compounds. Moreover, the development of a chemical series will require the optimization of different biological and physiochemical properties simultaneously with a target to enhance potency, decrease toxic effects, and achieve significant bioavailability along with the required specific properties (Atanasov et al., 2021). Experimental high-throughput screening of different chemical series, along with *in silico* studies to develop SARs and eventually utilize it to screen the best counterparts from millions of chemical libraries, have been a robust approach in the field of drug discovery (Doytchinova, 2022). Numerous studies have developed SARs for specific drug targets of different diseases and have utilized these established SARs to further screen effective drug candidates (Kim et al., 2015; Chia et al., 2018; Ivanov et al., 2020; Li et al., 2023; Wei et al., 2024; Yap & Gan, 2024).

Traditional herbal medicines and their components have been used as effective antiviral formulations. Herbal plants including *Allium sativum, Citrus sinensis, Mentha piperita, Curcuma longa* and *Nigella sativa* have been used to prepare herbal drinks to manage severe acute respiratory syndrome coronavirus-2 (SARS-CoV-2) infections. Further studies have investigated the antiviral potentials of *Camellia sinensis, Aesculus hippocastanum L., Echinacea purpurea, Eucalyptus globulus* Labill., *Glycyrrhiza glabra* L., *Filipendula ulmaria* (L.) maxim., *Gingko biloba* L., etc. (Ahmad et al., 2021). Identification of their mechanism of action illuminates the interaction of drug-like components with virus-specific proteins that take part in either entry of the virus, its replication, formation of assembly, virus release, or virus-host interactions (Demeke et al., 2021).

SAR-based studies of components of traditional herbal medicines can provide high efficacy, low toxicity, natural drugs or their combinations, which can target a broad spectrum of viruses by acting on conserved regions of viral drug targets that mediate essential viral mechanisms (Jan et al., 2021). Several investigations have been conducted to elaborate on the effective components of traditional herbal medicines and to study their SAR to target different viral infections (Lin et al., 2014; Zhang et al., 2020; Jan et al., 2021). Licorice root is utilized for treating a variety of diseases including hepatitis B. Glycyrrhizin or glycyrrhetinic acid, the lead components of licorice root, displays

DOI: 10.1201/9781003452621-5

antiviral effects. Wang et al., 2012 investigated 57 derivatives of glycyrrhetinic acid for their antiviral properties against the hepatitis B virus and established the SAR providing essential scaffold features required in the drug to act against the virus (Wang et al., 2012). Betulinic acid, obtained from the bark of various plants, possesses effective antiviral activity against Human Immunodeficiency Virus (HIV). Bevirimat, a known anti-HIV agent, is obtained from betulinic acid. Zhao et al., 2021, in their study on the SAR of betulinic acid derivatives, identified 27 derivatives having high effectiveness compared to bevirimat against HIV and provided a three-dimensional (3D) quantitative SAR model that can be utilized for investigating further potential drug candidates (Zhao et al., 2021). A wide range of viruses have been targeted by traditional herbal medicines and their compounds, and SAR models of effective compounds have been developed to open up further scope for drug discovery (Najmi et al., 2022).

The present chapter illustrates the importance of SAR in drug discovery and different approaches used for SAR establishment with a lead focus on antiviral traditional herbal medicines. Pharmacophore-based identification of novel antiviral compounds and disclosure of effective chemical moieties that contribute to the bioactivity of the compound using *in silico* analysis as well as experimental techniques have been elaborated followed by a discussion of herbal antiviral against SARS-CoV-2 drug targets and their associated SAR studies.

5.2   SAR IN DRUG DISCOVERY

A chemical compound can be described based on its structure, physiochemical properties, and bioactivity. The structure of the compound is displayed by the chemical subgroups or descriptors that constitute the compound, whereas physiochemical properties are analyzed by experiments or calculated by computational methods. The biological activity or responses of a chemical can be evaluated by different assays (Bertoni et al., 2021). According to chemistry principles, physiochemical properties can be understood by the structure of a compound and understanding of these properties can help in understanding bioactivity and the effect of that compound on the organism. Though the structure and physiochemical properties of chemical compounds are linked and the relation between the two is clear, the bioactivity of a compound, which is a response induced by the compound in a biological system, is affected by different factors depending on the complexity of the considered biological system (Guha, 2013; Wawer et al., 2014; Lewandowski et al., 2020).

Conceptually, a similarity in structural features of a cluster of compounds indicates similarity in their physiochemical properties and bioactivities given that the biological system being studied remains the same. The generation of a SAR thus requires chemical, biological, computational, and statistical analysis for the characterization of compounds and then extraction of patterns using chemometric approaches to establish SAR (Figure 5.1) (Guha, 2013; Lo et al., 2018).

Therefore, the establishment of the SAR model in a collection of compounds emphasizes finding out the association between specific chemical features of compounds and their physiochemical nature and biological activities concerned with a particular disease. A diverse range of compounds are studied for their chemical substructures that interact with a disease-target protein and convey specific biological effects. Information about these substructures can mediate the modification or derivatization of potential compounds to increase their efficiency. Further, these substructures can be utilized to screen or filter new compound libraries to find the most potent counterpart (McKinney, 2000; Temml & Kutil, 2021).

A wide range of methodologies has been used to develop SAR models for specific diseases, while a broad classification divides them into two classes including the statistical method and the pharmacophore-based method (Sliwoski et al., 2013).

The selection of a method for SAR and its quantification is critical for establishing an accurate link between the chemical structures and activities of compounds. Though the association between the chemical descriptors and bioactivities of compounds may be nonlinear or linear, usually a linear model like partial least square or multivariate linear regression is used for a quantitative

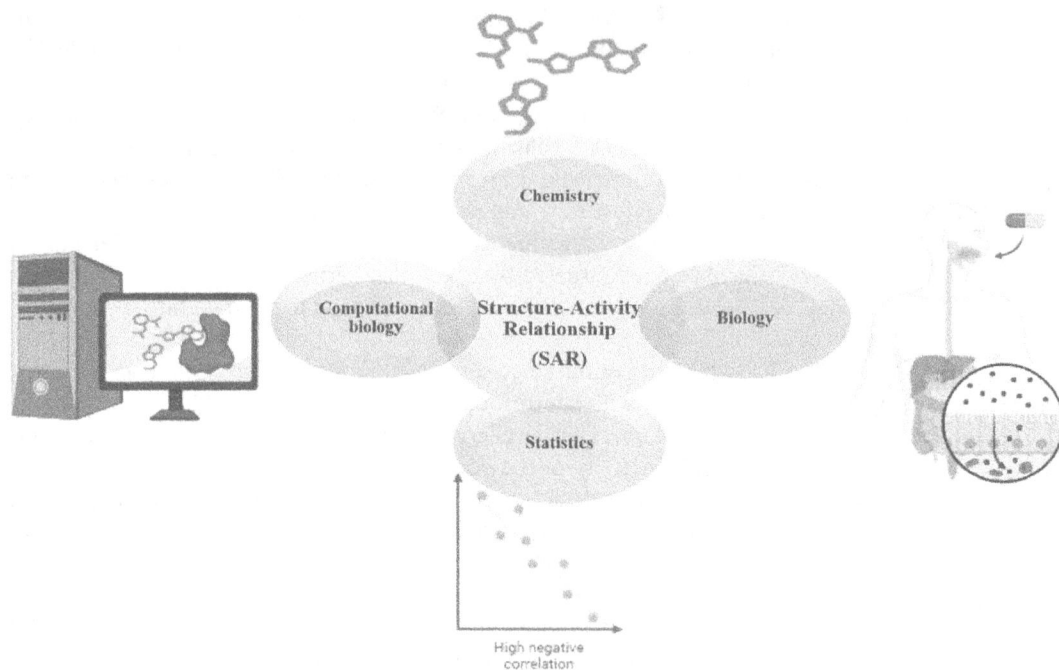

FIGURE 5.1 Structure-activity relationship–a multidisciplinary approach to drug discovery.

SAR (QSAR) model construction, which makes the accuracy of the model questionable (Niazi & Mariam, 2023). The polynomial neural network used for nonlinear model generation combines the essential features of the partial least square or multivariate linear regression and the artificial neural network, thus providing characteristic nonlinearity of the artificial neural network along with required regression (Tetko et al., 2000).

New approaches not only consider building models to predict the selectivity and potency of compounds but also focus on the prediction of absorption, distribution, metabolism, elimination, and toxicity (ADMET) properties, thus predicting pharmacodynamic as well as pharmacokinetic properties. ADMET properties play an essential role in eliminating compounds from the race of drug development, thus QSAR can be used to evaluate effective properties like solubility, permeability, and metabolism by cytochrome P450 (Cumming et al., 2013).

An automated QSAR method of variable selection using the principle of k-nearest neighbor (kNN) has been introduced by Zheng and Tropsha. The chemical activity in this kNN-QSAR method was estimated as the activity mean value of its kNN on the basis of Euclidean distance in a space of multidimensional descriptor. An optimum descriptor subset was identified by the method of stimulated annealing where the equation of QSAR-containing descriptors selected randomly is equated with a similar equation that is disturbed by replacing and removing a few descriptors at random. If the modified equation seems better on the basis of the Metropolis criterion than the previous, it is accepted, otherwise it is rejected (Zheng & Tropsha, 1999).

In today's scenario, artificial intelligence (AI) and deep learning have been frequently employed in drug discovery where deep neural networks are used for the prediction of properties as statistical engines. It indicates nonlinear complex statistical models containing neural nets with several hidden layers that perform prediction. The lead limitation associated with deep neural network models is their explainability as they are built as a black box and the algorithm design does not permit understanding of prediction. This limits the model improvement in a focused way. Recently, some methods like SHAP and LIME have been introduced that provide an explanation of the neural network model having fixed input size and are based on general backpropagation and gradient methods (Ali et al., 2023). An explainable AI approach integrated with a gradients-based method to interpret graph-based neural networks has been devised. This explainable AI approach is used for ADMET data interpretation (Jiménez-Luna et al., 2020). Harren et al., have developed a method that can explain the deep neural network model for SAR trained on bioactivity parameters like inhibition constant and IC50 values for target proteins (Harren et al., 2022).

The statistical method includes two-dimensional descriptors that do not consider stereochemistry and can avoid SAR depending on the chirality of compound-target interaction. Three-dimensional approaches are thus more informative as they can directly interpret the nature of the interaction between a target protein and ligand that results in SAR (Guha, 2013).

Three-dimensional QSAR (3DQSAR) correlates features localized spatially in a chemical compound collection with bioactivity. As the descriptors used for the representation of chemical structure generally encode structural characteristics in a location-dependent manner, which are responsible for the activity, the structures of a series of compounds are required to be aligned with each other (Cherkasov et al., 2014). The types of descriptors of 3DQSAR are divided into two types including surface-based and lattice-based descriptors. The most frequently studied lattice-based method is Comparative Molecular Field Analysis (CoMFA) in which the alignment of chemical structure is highly important but slightly subjective. The alignment of chemical structures is accurate and easy, provided information on the action mechanism at the molecular level is available, which defines the rules for molecular alignments. The availability of a ligand-bound complex crystal structure of a target protein can assist in the identification of identical or similar features in a series of compounds like the location of steric interaction or hydrogen bond (Morales-Bayuelo et al., 2015). Thus, alignment is a process that is guided by available knowledge. Other comparable 3DQSAR methods have also been devised that are alignment-free, e.g. WHIM, EVA and Comparative Molecular Moment Analysis (CoMMA) but are not frequently used.

5.3 PHARMACOPHORE AND BIOACTIVITY OF COMPOUNDS

Conventionally, the pharmacophore term was vaguely used for common functional or structural elements in a compound set that are essential for their bioactivity against a target protein. The formal International Union of Pure and Applied Chemistry (IUPAC) definition states that a pharmacophore is a group of electronic and steric features that are required to confirm optimum supra-molecular interactions with a particular target protein structure to initiate or block the function or biological response of the target. This definition represents pharmacophores as a precise picture of stereoelectronic molecular properties (Seidel et al., 2020).

Molecules having different structures but containing similar patterns of pharmacophores can be identified by the specific binding site of the same biological target and, hence, can display similar biological activities. Therefore, the model of pharmacophore is the essence of the information available about the SAR procured from the investigation of large amounts of inactive and active compounds against a particular drug target (Giordano et al., 2022).

Three-dimensional pharmacophores display the location and nature of chemical features of a compound taking part in the interaction to its target as geometrical entities. For matching pharmacophore, the model features have to match in spatial arrangement and the type, but not in fundamental chemical structure, thus, displaying the characteristic ability of scaffold hopping in pharmacophore, which is revealed by the high diversity of compound structures obtained as hits in virtual screening (Meyenburg et al., 2022). The set of features selected for the pharmacophore model should be optimal, as a very general set of features can provide fewer selective hits and a highly restrictive large number of features can be problematic when the matching compounds are structurally unrelated but are active against the same target protein. Different feature types usually include hydrogen bond acceptors (carboxylates, amines, ketones, fluorine, alcohols, etc.), hydrogen bond donors (amides, amines, alcohols), aromatic groups that mediate pi stacking and cation-pi interactions, positive ions

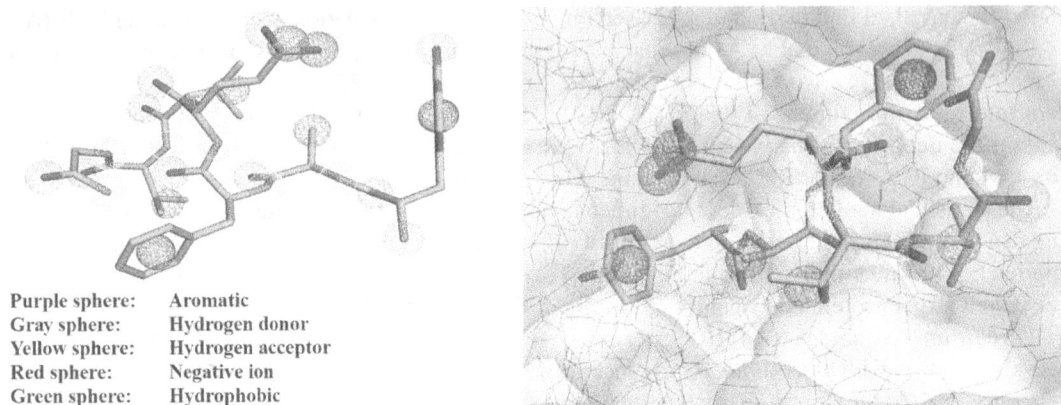

Purple sphere: Aromatic
Gray sphere: Hydrogen donor
Yellow sphere: Hydrogen acceptor
Red sphere: Negative ion
Green sphere: Hydrophobic

FIGURE 5.2   Different pharmacophoric features of ligand (sticks) responsible for interaction with the target protein (surface + wire).

like ammonium or metal ions mediating ionic or cation-pi interaction, negative ions like carboxylate ions, hydrophobic groups like halogens, alkyl groups, alicycles, and nonpolar or weak aromatic rings forming hydrophobic contacts (Figure 5.2) (Seidel et al., 2020; Zhu et al., 2023).

However, a compound that can fit the model of pharmacophore may not bind to the target receptor due to steric clashes, thus, models of pharmacophore should also take into account the spatial constraints due to the architecture of the binding site. This is possible by the incorporation of exclusion volumes where the ligand is not allowed to fill the space after alignment with the model of pharmacophore (Klebe, 2013). Two approaches are used for the development of the pharmacophore model, including a structure-based and ligand-based approach. The structure-based approach for modeling pharmacophore requires an experimental complex structure of a ligand–protein complex that can be used as a guide to generate a pharmacophore model. The structural information about the protein structure encourages the identification of important interactions between the protein and ligand. But if the protein structure in the ligand unbound state is available, development of an efficient pharmacophore model becomes challenging. Computational models and refinement of the model can be beneficial in the absence of complex structures for the development of the pharmacophore model (Opo et al., 2021; Luo et al., 2021).

A ligand-based approach for pharmacophore model development utilizes sufficient known active compounds that bind at the specific binding site of a receptor in a specific orientation. As the active conformation of ligands of the training set is usually unknown, it is crucial to generate different ligand conformers in a manner that at least one conformation is the precise approximation of the active conformation. Then, chemical features common to all known active ligands have to be identified. Several pharmacophoric models can be generated and can be ranked based on a fitness function (Flores & Gerstein, 2011; Pal et al., 2019).

Eventually, the developed pharmacophore model from either of the approaches is utilized to screen huge libraries to identify the most potent novel compounds with effective bioactive properties.

5.4   TRADITIONAL HERBAL MEDICINES AND ANTIVIRAL THERAPIES

Traditional medicine enhances the skills, knowledge, and practices based on experiences, theories, and beliefs indigenous to diverse cultures utilized to maintain health and improve, prevent, or treat mental and physical illness. Traditional medicine systems of different areas have different practices and philosophies that are affected by the environment, prevailing conditions, and respective geographic area in which they were evolved. These systems follow a holistic approach to body-mind equilibrium, environment, and life where health is emphasized rather than the disease. Traditional medicine does not focus only on the disease or ailment, rather it focuses on the complete condition of the patient. The utilization of herbs to maintain health is a fundamental part of all systems of traditional medicine (Musa et al., 2022). Herbal plants with antiviral effects are universally present in all traditional medicine systems and have been proven effective against a variety of viruses (Ben-Shabat et al., 2019).

Coronaviruses are single-stranded positive sense-enveloped RNA viruses of the *Coronaviridae* family. This family contains a variety of species that infect the gastrointestinal tract and upper respiratory tract in birds and mammals. It usually causes the common cold, but some highly virulent strains lead to pneumonia and severe acute respiratory syndrome. To date, the known coronaviruses infecting humans are HCoV-OC43, HCoV-229E, HKU1, HCoV-NL63, SARS-CoV, MERS-CoV, and SARS-CoV2. Though the recent pandemic caused by SARS-CoV2 has accelerated research on treatments for coronaviruses, no specific treatments have been identified (Payne, 2017). Several herbal medicines are effective against coronaviruses. Saikosaponins or triterpene glycosides found in medicinal plants like *Bupleurum spp.*, *Scrophularia scorodonia*, and *Heteromorpha spp.* are effective against HCoV-22E9 and prevent attachment and penetration of virus (Cheng et al., 2006). Further, *Lindera aggregate*, *Lycoris radiata*, *Pyrrosia lingua*, and *Artemisia annua* have been reported for their antiviral effect against SARS-CoV (Li et al., 2005). Scutellarein, myricetin, and several phenolics of *Torreya nucifera* and *Isatis indigotica* have the potential to inhibit the 3CL protease and nsP13 helicase (Baker et al., 2023). The extract of *Houttuynia cordata* in turn shows antiviral effects against SARS-CoV by inhibiting RNA-dependent RNA-polymerase and 3CL protease (Lau et al., 2008). Several herbs of Indian tradition have shown antiviral effects against SARS-CoV-2 including *Tinospora cordifolia*, *Ocimum tenuiflorum*, *Cinnamomum cassia*, *Achyranthes bidentata*, *Embelin ribes*, *Cydonia oblonga*, *Momordica charantia*, *Zingiber officinale*, *Withania somnifera*, Kabusura kudineer, and camphor (Singh et al., 2021).

The hepatitis B virus (Family *Hepadnaviridae*) has circular relaxed DNA as a partially double-stranded genome. This enveloped virus causes hepatitis B infection and is transmitted via contaminated body fluids or blood. The chronic infection of hepatitis B develops hepatocellular carcinoma and cirrhosis. Although vaccines to prevent infection of this virus are available, the areas deprived of vaccination programs are affected by the disease and require efficient treatments (Datta et al., 2012). Therapeutics including nucleotides and nucleoside analogs are known but total eradication from the host is difficult. Exhaustive research on finding natural herbal treatments against hepatitis B identified several herbs and their constituents. Amide alkaloids of *Piper longum*, isochlorogenic acid A of *Laggera alata*, and dehydrocheilanthifoline component of *Corydalis saxicola* are known to show anti-hepatitis B effect (Ali et al., 2021). Curcumin has also been reported to inhibit the replication and expression of the hepatitis B viral gene, as it downregulates the peroxisome proliferator-activated receptor gamma coactivator 1 alpha (PGC-1α) hepatitis B transcription co-activator (Rechtman et al., 2010).

Dengue virus (DENV), a single-stranded positive sense-enveloped RNA virus belonging to *Flaviviridae* family, is prominent in parts of Southeast Asia and has *Aedes aegypti* as its vector. It has four serotypes ranging from DENV1–4, which can cause high fever (accompanied by myalgias, headache, nausea, joint pain, skin rashes, and vomiting), mild febrile presentation, and severe hemorrhagic disease like dengue shock syndrome or dengue hemorrhagic fever. Though dengue is an old disease, effective treatments are still limited and prevention measures include vector mosquito control and isolating infected individuals (Bhatt et al., 2020). Natural compounds have been promising candidates for the treatment of DENV. Baicalein flavone is effective in hindering the adsorption of DENV and viral replication after entering the host (Zandi et al., 2012). Narasin, quercetin, and extract from seaweeds possess exceptional anti-DENV properties (Chavda et al., 2022). Tannins like punicalagin and chebulagic acid from *Terminalia chebula* have been reported to show antiviral effects against different viruses, including DENV, by hindering the attachment and then fusing the virus during its entry into the host cell (Thomas et al., 2022). Moreover, Ayurveda, the Indian traditional medicine system, has put forward extracts of *Acorus calamus* L., *Carica papaya* L., *Andrographis paniculata* Nees, *Boesenbergia rotunda* (L.) Mansf., *Alternanthera philoxeroides* (Mart.) Griseb., *Azadirachta indica* A. Juss., *Cissampelos pareira* L., etc., as effective herbal treatments against DENV as they control hyperthermia and enhance the ability of the immune system (Singh & Rawat, 2017).

Hepatitis C virus, a positive sense single-stranded RNA flavivirus transmits via blood-to-blood transmission. It has a high mutation rate and, thus, development of treatment against it is highly challenging (Li & Lo, 2015). Several natural products with antiviral effects against this virus have been investigated. Flavonolignans of *Silybum marianum* are known for their anti-hepatitis C effect as they reduce the virus load (Polyak et al., 2013). Curcumin, a known antiviral, also inhibits the replication of the virus by down-regulating the sterol regulatory element binding protein 1 (SREBP-1)-Akt pathway (Kim et al., 2009). Further, griffithsin, epigallocatechin-3-gallate, ladanein, and tellimagrandin have been investigated for their antiviral potential for preventing entry of the hepatitis C virus (Tamura et al., 2010; Meuleman et al., 2011; Calland et al., 2012). Tannins like punicalagin and chebulagic acid inhibit the entry of hepatitis C as they inhibit the attachment and penetration of the virus to the host cell, inactivate virus particles present in free form, and disrupt cell-to-cell viral transmission after infection (Lin et al., 2011).

Natural treatments for herpes simplex virus 1 and 2 types have also been investigated, considering the severe mucocutaneous lesions in genital and perioral or oral areas. This enveloped double-stranded DNA virus of the *Herpesviridae* family establishes infection in sensory neurons and can become reactivated by different stimuli (Zhu & Viejo-Borbolla, 2021). Traditional herbal plants such as *Cassia javanica* containing ent-epiafzelechin-(4α→8)-epiafzelechin shows an inhibitory effect against viral replication (Cheng et al., 2006). Hippomanin-A, 1,3,4,6-tetra-O-galloyl-beta-d-glucose, geraniin, and excoecarianin extracted from *Phyllanthus urinaria* are reported to obstruct the infection of the herpes simplex virus (Yang et al., 2007b; Yang et al., 2007a). Moreover, the meliacine component of *Melia azedarach* stimulates the production of IFN-g and TNF-α and decreases the shedding of HSV-2, thus, improving pathogenesis induced by the virus in the mouse model having vaginal herpetic infection (Petrera & Coto, 2009). The glucoevatromonoside component of *Digitalis lanata* alters the electro-chemical gradient of the cell and inhibits propagation of the herpes simplex virus 1 and 2 (Bertól et al., 2011). Extract of blackberry, *Rhododendron ferrugineum* L., and proanthocyanidin containing *Myrothamnus flabellifolius* Welw. extract also displays inhibition against the herpes simplex virus (Gescher et al., 2011). Further, flavonoids like houttuynoids A–E extracted from *H. cordata* display anti-herpes virus activity (T. Li et al., 2017).

HIV, from the *Retroviridae* family, is an enveloped virus that attacks the immune cells, by performing a reverse transcription of ssRNA and integrating it into chromosomal DNA. HIV transmission occurs by the exchange of infected body fluids or blood and causes acquired immunodeficiency syndrome (AIDS), which leads to progressive immune system failure because of the depletion of CD4+ T lymphocyte, which results in deadly opportunistic infections and also malignancies. Though several therapies have been developed to fight HIV, however, drug resistance, toxicity, and restricted accessibility in areas with poor resources make the treatment challenging (Fanning, 2010). Several herbal treatments have been evaluated for anti-HIV properties including extracts of *Artemisia afra* and *A. annua* (Lubbe et al., 2012). The *Calophyllum* species containing coumarins have been identified to convey anti-HIV effects by decreasing the levels of nuclear factor-kappa B, leading to inhibition of HIV replication (Spino et al., 1998).

Influenza viruses, the single-stranded negative sense RNA viruses categorized in the *Orthomyxoviridae* family, infect the respiratory system, resulting in fever, sore throat, headache, sneezing, and pain in joints and muscles. These viruses can be severe enough to cause pneumonia (Taubenberger & Morens, 2008). Herbal treatments against these viruses have been devised and include extracts of elderberry, root extracts of *Pelargonium sidoides*, and dandelion (He et al., 2011). *P. sidoides* root extract inhibits the activity of neuraminidase and hemagglutinin, while dandelion extract reduces the activity of polymerase and RNA levels of nucleoprotein

(Ganjhu et al., 2015). *Illicium oligandrum* root extract containing spirooliganone B also has anti-influenza properties (Ma et al., 2013). Chalcones extracted from *Glycyrrhiza inflata*, homoiso-flavonoids of *Caesalpinia sappan*, and xanthones of *Polygala karensium* have been identified as neuraminidase inhibitors and show effective anti-influenza effects (Đào et al., 2011; Jeong et al., 2012; Đào et al., 2012).

Many other viruses have been targeted by natural herbal medicines and several investigations are underway. Hence, traditional herbal medicines are being explored exhaustively against a variety of viruses and can be utilized to develop effective natural drugs.

5.5 SARS-CoV-2 AND ITS LEAD DRUG TARGETS

SARS-CoV-2 is a single-stranded positive sense-enveloped RNA virus of the family *Coronaviridae*. It has 29.9 kilobase (kb) genome, which encodes four structural, 16 nonstructural, and nine accessory proteins. The four structural proteins include envelope, spike, nucleocapsid, and membrane proteins and nonstructural proteins account for NSP1 to NSP16. The inhibitors against SARS-CoV-2 have been investigated to target virus entry by inhibiting spike protein, proteolytic cleavage by blocking papain-like or main protease, RNA synthesis as NSP12, 13, 14, 15 and 16 inhibitors, and viral assembly inhibiting the nucleocapsid (Shengman et al., 2022).

The spike protein of SARS-CoV-2, a fusing glycoprotein of class I type, is homotrimeric and is present on the surface of the virion to mediate the entry of the virus into the host cell. This protein is cleaved by the protease of the host into the S1 subunit (containing receptor binding domain) and S2 subunit, having a shield of O-linked and N-linked glycans. Rearrangements of conformations lead to the binding of the receptor binding domain with angiotensin-converting enzyme-2 (ACE-2) of the host, and subsequent transitions in structure and cleavage of protein result in post-fusing conformation that fuses the membrane of the virus with the host membrane (Huang et al., 2020).

Papain-like protease or NSP3, a cysteine protease, cleaves polyprotein pp1a and pp1ab and releases NSP3, NSP2, and NSP1. It also removes the ubiquitin-like interferon-stimulated gene-15 and ubiquitin of the host from signaling proteins to reduce the innate immune system response. The active site contains a conventional catalytic triad of C111, H272, and D286 and the cleavage occurs at the LXGG|XX motif lying adjacent to NSP1-2, NSP2-3, NSP3-4, and the tail of ubiquitin and ubiquitin-like protein at C terminal. A β-hair pin loop called blocking loop 2 regulates the substrate's access to the active site (Osipiuk et al., 2021).

Another cysteine protease of SARS-CoV-2, i.e., main protease or NSP5 or 3C-like protease, cleaves the pp1ab and pp1a to produce NSP4, NSP5, NSP6, NSP7, NSP8, NSP9, NSP10, NSP11, NSP12, NSP13, NSP14, NSP15, and NSP16. The lead viral enzymes NSP13 and NSP12, which are responsible for viral replication, become matured by the action of the main protease. The homodimer of the main protease possesses a preference for glutamine at the P1 position for hydrolysis in the motif Q|(S/A/N). A catalytic dyad of C145 and H41 in both homodimers mediates the cleavage by this protease. This dyad catalyzes the formation of carbon-sulfur covalent bonds. This bond is formed between the thiolate of C145 and the carbonyl of the P1 glutamine of the main chain of the substrate (Jin et al., 2020).

NSP12 or RNA-dependent RNA-polymerase mediates the replication and transcription of viral RNA and is highly conserved. After proteolytic cleavage by the main protease, this polymerase comes together with NSP7, NSP8, NSP9, NSP10, NSP13, NSP14, NSP15, and NSP16 to form replication and transcription complex that catalyzes the unwinding of template, synthesizes RNA, and proofreads the RNA and its capping. The individual components of this complex play a vital role in the replication and transcription mechanism where NSP13 conveys helicase activity or template unwinding, NSP14 is an exoribonuclease that mediates RNA proofreading, uridine cleavage is mediated by endoribonuclease NSP15, and RNA capping involves NSP9, NSP12 NiRAN, NSP13 ATPase, NSP14 guanine N7 methyltransferase, and NSP16 2'-O-methyltransferase (Jiang et al., 2021).

Nucleocapsid is a multivalent flexible protein that has multiple functions in the packaging of the virus genome and antiviral innate immunity suppression. It is the most abundant and dynamic protein in SARS-CoV-2, which makes it challenging to identify a highly potent inhibitor (Cubuk et al., 2021).

All the above-mentioned proteins of SARS-CoV-2 play essential roles in viral machinery and can be effective drug targets to treat the deadly infection of COVID-19. Numerous studies have been conducted to target these proteins and to inhibit the viral propagation in the host. Several have opted for the SAR-based approach to investigate potential inhibitors against these mentioned drug targets and some of the most effective studies are discussed in the next section.

5.6 SAR STUDIES TO IDENTIFY ANTIVIRALS FROM HERBAL MEDICINES AGAINST SARS-CoV-2

Herbal medicines can provide safe and efficient treatment even for a disease like COVID-19, however, the characteristic composition of herbal medicines has been less explored. The phytochemicals present in these medicines have specific functional groups that form their structure and mediate their chemistry. The chemical interactions of the functional groups of compounds present in herbal medicines with that of specific drug targets mediate the bioactivities of these compounds, thus establishing the SAR (Figure 5.3). Several studies have evaluated traditional herbal medicines effective against SARS-CoV-2 for their active components to provide structures that can inhibit the target proteins and can be used as a basis for developing novel natural drugs with the fewest toxicities. Some of the lead studies are elaborated in this section.

Nguyen et al. (2021), in their study on polyphenols and black garlic extract, revealed IC50 values of 9–197 μM and 137 μg/mL, respectively, for the main protease of SARS-CoV-2. The structural

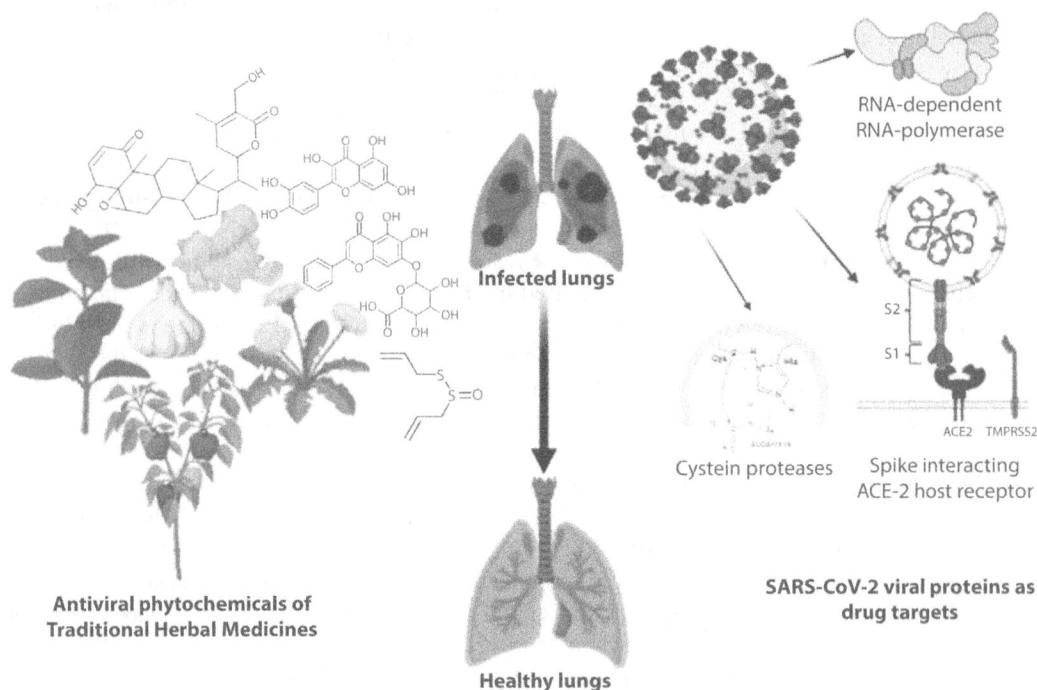

FIGURE 5.3 Phytochemicals present in traditional herbal medicines act as antivirals against drug targets of SARS-CoV-2.

analysis disclosed the SAR between the polyphenols and their *in vitro* effectiveness. The hydroxyl at C5′, C4′, C3′ in ring B, at C3 of ring C, C7 of ring A, C2=C3 of ring C, and C8 glycosylation in ring A displayed the main protease inhibition and can be considered as a pharmacophore for further studies (Nguyen et al., 2021).

Ghosh et al., 2021, in their SAR study, evaluated 300 derivatives of hesperidin, withaferin-A, and baicalin phytochemicals developed by a machine learning and neural networking-based approach to study the unexplored chemical structure space regions. The withaferin-A derivatives displayed significant binding energy and interactions with the catalytic residues of the main protease of SARS-CoV-2 and effectively predicted IC50 values. Further, molecular docking and molecular dynamic (MD) simulation-based analysis provided an effective withaferin-A derivative (mol 61) for main protease inhibition. This study provided a SAR of different derivatives against COVID-19 and provided a potential structural scaffold of Withaferin-A derivatives with effective inhibition activity for the main protease. Therefore, withaferin-A, a component of *W. somnifera* or Ashwagandha (an herbal plant), and its identified derivative mol 61 can be further utilized as a structural scaffold for the investigation of main protease inhibitor against SARS-CoV-2 (Ghosh et al., 2021).

Vincent et al., 2020, in their bioinformatics approach, studied 145 phytochemicals of Kabasura kudineer, an herbal AYUSH formulation used to treat COVID-19, to inhibit the main protease of SARS-CoV-2. Molecular docking results provided effective structural interactions and binding energies of the components of the herbal formulation, thus developing a SAR between the structures of components and their activities (Vincent et al., 2020).

Jan et al. (2021), in their *in vitro* and *in vivo* analysis of 190 herbal medicines and 2855 small molecules, identified 15 small molecules that were active against SARS-CoV-2. Out of the herbal extracts, the extracts of *Perilla frutescens*, *Ganoderma lucidum,* and *Mentha haplocalyx* were effective in both *in vitro* as well as in *in vivo* experiments. These extracts were active in cell-based assays that were developed to evaluate RNA-dependent-RNA polymerase and 3CL protease inhibition thus, provided the SARS-CoV-2 inhibition mechanism of these extracts against the virus. This study provides scope for the development of a SAR between the chemical scaffolds present in the extracts that can work against the respective viral enzymes and can result in the development of therapeutics (Jan et al., 2021b).

Amin et al. (2021), in their QSAR-based main protease inhibitor identification study, classified different descriptors of inhibitors and their inhibitory activities. The structural fingerprints enhancing the inhibitory activity were mapped for optimizing the lead molecule. Baicalein, a flavone found in Chinese herbal medicine made from root extract of *Scutellaria baicalensis*. This study revealed that structural inhibitors derived from baicalein were effective for lead optimization. The pockets S1′, S1, S2, and S4 of the main protease possess intrinsic flexibility and indicate that hydrophobic substituents can mediate effective main protease inhibition (Amin et al., 2021).

Kumar et al., 2020 identified a potential scaffold from the known antiviral against the main protease. Indinavir was found effective due to the presence of hydroxyethylamine pharmacophore. The library of compounds containing this pharmacophore was investigated and a potent molecule 16 was selected as lead for main protease inhibition at I and II domains and at the linker of domains II and III (Kumar et al., 2020).

Ayipo et al. (2022), in their analysis of the identification of inhibitors against the NSP12 or RNA-dependent RNA polymerase and NSP3 macrodomain, identified 2,3,4,5,6-pentahydroxyhexyl o-hydroxybenzoate as the most potential natural inhibitor that can target both NSP3 and NSP12. The SAR analysis of this compound showed that the entire compound is a good fit in the active sites of both NSP3 macrodomain and NSP12. The 2-hydroxy benzoate group has been known to be present in several drugs mediating antioxidant, anticancer, antifungal, anti-inflammatory, etc., effects. The aliphatic or phenolic hydroxyl group is flexible for mediating antiviral effects due to its lipophilic nature and effective binding with the target, forming inter-molecular hydrogen bonding. The hydroxyl group at the ortho position is more favorable for hydrogen bonding compared to the meta and para positions. Further, the polyols promote pharmacodynamic effects. In the ligand, the aliphatic hydroxyl displays a tendency for a high penalty of desolvation that affects binding. Though phenolic and primary hydroxyl groups have a tendency for rapid production of metabolites that are toxic, the flexible chemistry helps in the modulation of this limitation (Ayipo et al., 2022).

Ebada et al. (2020) studied the anti-SARS-CoV-2 properties of a Jordanian medicinal plant *Crepis sancta*. The phenolic extract of this plant constituted eudesmane sesquiterpenes, (6S,9R)-roseoside, and some methylated flavanols. The *in vitro* antiallergic, anti-inflammatory, and *in silico* main protease inhibition potential of these compounds were evaluated and (6S,9R)-roseoside and methylated flavanols were found effective in both *in vitro* and *in silico* analysis. Chrysosplenetin, which is categorized as methylated flavanol, displayed the most effective outcomes in inhibiting the main protease of SARS-CoV-2, providing a structural scaffold for further investigation of treatments (Ebada et al., 2020).

Puttaswamy et al. (2020) conducted a thorough study of 4704 plant metabolites as anti-SARS-CoV-2 agents. Out of the top-ranked metabolites, 50% consisted of triterpenoids that can bind strongly with the spike-receptor binding domain. More than 32% of compounds with significant interaction with transmembrane serine protease in humans were glycosides of flavonoids or flavonoids. Moreover, more than 16% of flavonol glycosides and anthocyanins showed inhibition against main protease and 13% of flavonol glycosides inhibited RNA-dependent RNA polymerase. Thus, structural scaffolds for effective inhibitors against different SARS-CoV-2 proteins were identified based on the binding affinity, which can indicate the activity of these plant metabolites against SARS-CoV-2 (Puttaswamy et al., 2020).

Liao et al. (2021), in their analysis of Chinese herbal formulations, identified five natural compounds including kaempferol-3-O-gentiobioside, narcissoside, rutin, isoschaftoside, and vicenin-2 as inhibitors of main proteases. Molecular docking and simulation studies reported effective binding of these compounds at the active site of the protease, thus, the structural scaffolds of these compounds showed a positive association with anti-SARS-CoV-2 activity and can be further explored to find more potential candidates (Liao et al., 2021).

Ram et al. (2022) evaluated the composition compounds of AYUSH-64 against the SARS-CoV-2 main protease in an *in silico* approach. Out of the 36 compounds, 35 candidates bound to the active site gave high binding energies compared to the N3 peptide inhibitor reported in the co-crystallized structure. The highest binding energy (-8.4 kcal/mol) compound was akuammicine N-oxide, which showed significant interactions with catalytic residue Cys145 and His164. The N-oxide group of the compound was found to interact with the catalytic Cys145 and thus provide a significant pharmacophore (Ram et al., 2022).

Verma et al. (2021) conducted an *in silico* analysis of Ayurvedic plant metabolites and established that curcumin can effectively bind to ACE-2 host receptors and prevent the interaction of spike protein inhibiting virus entry inside the host cell. Gingerol showed strong binding with RNA-dependent RNA polymerase and spike protein while quercetin binds effectively to the main protease to inhibit its function. These compounds can be further analyzed for their effectiveness and to develop structure-activity relationships (Verma et al., 2021).

Maurya et al. (2020) studied the components of different traditional medicinal plants from Ayurveda for their effectiveness as anti-SARS-CoV-2 compounds and found amarogentin, alpha-amyrin, eufoliatorin, kutkin, caesalpinin, belladonnine, and beta-sitosterol as effective inhibitors of ACE-2 and spike glycoprotein (Maurya et al., 2020).

All these studies provide effective compounds that can target essential proteins of SARS-CoV-2 viral mechanisms and provide natural scaffolds of compounds that can interact with their functional pharmacophores and show activity against the respective targets of SARS-CoV-2. The chemical spaces of these compounds can be further explored and novel effective derivatives can be identified on the basis of SAR approaches. Therefore, the SAR approach has immense potential to accelerate the process of drug discovery and can benefit society by providing natural effective treatment against the COVID-19 disease.

5.7   CONCLUSION

Traditional herbal medicines have gained immense importance due to their exclusive mechanism of targeting multiple pathways, thus providing all-round treatment against diseases. Their lesser side effects, easy availability, and lower cost in turn support their frequent utilization. Viral diseases are becoming highly prevalent, and especially the recent pandemic of COVID-19 has captured the immense attention of the scientific community and has encouraged the discovery of high-potential antivirals. Numerous herbal formulations have been evaluated for their antiviral potential against different viruses including SARS-CoV-2 and their significant bioactive compositions have been explored. Various studies have utilized the structure-activity approach to identify the chemical subgroups or structural components that interact with the potential targets of viral machinery and mediate the antiviral effects. Many studies have utilized the knowledge of effective structural chemical motifs mediating antiviral effects for the derivatization of naturally occurring compounds to develop more potent antiviral compounds compared to the native structures. Hence, this chapter elaborates on the fundamentals of SARs in drug discovery, effective pharmacophores or chemical groups that are capable of mediating different bioactive effects, and their significance. Diverse traditional herbal medicines are effective against different viral infection and their compositions, with emphasis on SAR studies, have been illustrated. Moreover, specific drug targets of SARS-CoV-2 have been discussed followed by effective studies associated with the identification of potential antiviral natural compounds from herbal medicines and their SAR on the basis of specific drug targets displaying antiviral effect.

Eventually, these comprehensive details of SAR studies into potential antiviral compounds of traditional herbal medicines will encourage further research to accelerate the process of antiviral development to prevent future pandemics.

REFERENCES

Ahmad, S., Zahiruddin, S., Parveen, B., Basist, P., Parveen, A., Gautam, G., Parveen, R., & Ahmad, M. (2021). Indian medicinal plants and formulations and their potential against COVID-19–preclinical and clinical research. Frontiers in Pharmacology, 11. https://doi.org/10.3389/fphar.2020.578970

Ali, S., Abuhmed, T., El–Sappagh, S., Muhammad, K., Alonso-Moral, J. M., Confalonieri, R., Guidotti, R., Del Ser, J., Díaz-Rodríguez, N., & Herrera, F. (2023). Explainable artificial intelligence (XAI): What we know and what is left to attain trustworthy artificial intelligence. Information Fusion, 99, 101805. https://doi.org/10.1016/j.inffus.2023.101805

Ali, S. I., Sheikh, W. M., Rather, M. A., Venkatesalu, V., Muzamil, S., & Nabi, S. U. (2021). Medicinal plants: Treasure for antiviral drug discovery. Phytotherapy Research, 35(7), 3447–3483. https://doi.org/10.1002/ptr.7039

Amin, S. A., Banerjee, S., Singh, S., Qureshi, I. A., Gayen, S., & Jha, T. (2021). First structure–activity relationship analysis of SARS-CoV-2 virus main protease (Mpro) inhibitors: An endeavor on COVID-19 drug discovery. Molecular Diversity, 25(3), 1827–1838. https://doi.org/10.1007/s11030-020-10166-3

Atanasov, A. G., Zotchev, S. B., Dirsch, V. M., & Supuran, C. T. (2021). Natural products in drug discovery: Advances and opportunities. Nature Reviews Drug Discovery, 20(3), 200–216. https://doi.org/10.1038/s41573-020-00114-z

Ayipo, Y. O., Ahmad, I., Najib, Y. S., Sheu, S. K., Patel, H., & Mordi, M. N. (2022). Molecular modelling and structure-activity relationship of a natural derivative of o-hydroxybenzoate as a potent inhibitor of dual NSP3 and NSP12 of SARS-CoV-2: In silico study. Journal of Biomolecular Structure & Dynamics, 41(5), 1959–1977. https://doi.org/10.1080/07391102.2022.2026818

Baker, D. H. A., Hassan, E. M., & Gengaihi, S. E. (2023). An overview on medicinal plants used for combating coronavirus: Current potentials and challenges. Journal of Agriculture and Food Research, 13, 100632. https://doi.org/10.1016/j.jafr.2023.100632

Ben-Shabat, S., Yarmolinsky, L., Porat, D., & Dahan, A. (2019). Antiviral effect of phytochemicals from medicinal plants: Applications and drug delivery strategies. Drug Delivery and Translational Research, 10(2), 354–367. https://doi.org/10.1007/s13346-019-00691-6

Bertól, J. W., Rigotto, C., De Pádua, R. M., Kreis, W., Barardi, C. R. M., Braga, F. C., & Simões, C. M. O. (2011). Antiherpes activity of glucoevatromonoside, a cardenolide isolated from a Brazilian cultivar of Digitalis lanata. Antiviral Research, 92(1), 73–80. https://doi.org/10.1016/j.antiviral.2011.06.015

Bertoni, M., Duran-Frigola, M., Badia-I-Mompel, P., Pauls, E., Orozco-Ruiz, M., Guitart-Pla, O., Alcalde, V., DíAz, V. M., Berenguer-Llergo, A., Brun-Heath, I., Villegas, N., De Herreros, A. G., & Aloy, P. (2021). Bioactivity descriptors for uncharacterized chemical compounds. Nature Communications, 12(1). https://doi.org/10.1038/s41467-021-24150-4

Bhatt, P., Sabeena, S., Varma, M., & Arunkumar, G. (2020). Current understanding of the pathogenesis of dengue virus infection. Current Microbiology, 78(1), 17–32. https://doi.org/10.1007/s00284-020-02284-w

Calland, N., Albecka, A., Belouzard, S., Wychowski, C., Duverlie, G., Descamps, V., Hober, D., Dubuisson, J., Rouillé, Y., & Séron, K. (2012). (–)-Epigallocatechin- 3 -gallate is a new inhibitor of hepatitis C virus entry. Hepatology, 55(3), 720–729. https://doi.org/10.1002/hep.24803

Chavda, V. P., Kumar, A., Banerjee, R., & Das, N. M. (2022). Ayurvedic and other herbal remedies for dengue: An update. Clinical Complementary Medicine and Pharmacology, 2(3), 100024. https://doi.org/10.1016/j.ccmp.2022.100024

Cheng, H., Yang, C., Lin, T., Shieh, D., & Lin, C. (2006). Ent-epiafzelechin-(4α→8)-Epiafzelechin extracted from *Cassia javanica* inhibits herpes simplex virus type 2 replication. Journal of Medical Microbiology, 55(2), 201–206. https://doi.org/10.1099/jmm.0.46110-0

Cheng, P., Ng, L., Chiang, L., & Lin, C. (2006). Antiviral effects of saikosaponins on human coronavirus 229e in vitro. Clinical and Experimental Pharmacology and Physiology, 33(7), 612–616. https://doi.org/10.1111/j.1440-1681.2006.04415.x

Cherkasov, A., Muratov, E., Fourches, D., Varnek, A., Baskin, I. I., Cronin, M. T., Dearden, J. C., Gramatica, P., Martin, Y. C., Todeschini, R., Consonni, V., Кузьмин, В. Е., Cramer, R. D., Benigni, R., Yang, C., Rathman, J. F., Terfloth, L., Gasteiger, J., Richard, A. M., & Tropsha, A. (2014). QSAR modeling: Where have you been? Where are you going to? Journal of Medicinal Chemistry, 57(12), 4977–5010. https://doi.org/10.1021/jm4004285

Chia, S., Habchi, J., Michaels, T. C. T., Cohen, S. I., Linse, S., Dobson, C. M., & Knowles, T. P. J. (2018). SAR by kinetics for drug discovery in protein misfolding diseases. Proceedings of the National Academy of Sciences of the United States of America, 115(41), 10245–10250. https://doi.org/10.1073/pnas.1807884115

Cubuk, J., Alston, J. J., Incicco, J. J., Singh, S., Stuchell-Brereton, M. D., Ward, M. D., Zimmerman, M. I., Vithani, N., Griffith, D., Wagoner, J. A., Bowman, G. R., Hall, K. B., Soranno, A., & Holehouse, A. S. (2021). The SARS-CoV-2 nucleocapsid protein is dynamic, disordered, and phase separates with RNA. Nature Communications, 12(1). https://doi.org/10.1038/s41467-021-21953-3

Cumming, J. G., Davis, A. M., Mureşan, S., Haeberlein, M., & Chen, H. (2013). Chemical predictive modelling to improve compound quality. Nature Reviews Drug Discovery, 12(12), 948–962. https://doi.org/10.1038/nrd4128

Đào, T. T., Dang, T. T., Nguyễn, P. H., Kim, E., Thương, P. T., & Oh, W. K. (2012). Xanthones from *Polygala karensium* inhibit neuraminidases from influenza A viruses. Bioorganic & Medicinal Chemistry Letters, 22(11), 3688–3692. https://doi.org/10.1016/j.bmcl.2012.04.028

Đào, T. T., Nguyễn, P. H., Lee, H. S., Kim, E., Park, J., Lim, S. I., & Oh, W. K. (2011). Chalcones as novel influenza A (H1N1) neuraminidase inhibitors from *Glycyrrhiza inflata*. Bioorganic & Medicinal Chemistry Letters, 21(1), 294–298. https://doi.org/10.1016/j.bmcl.2010.11.016

Datta, S., Chatterjee, S., Veer, V., & Chakravarty, R. (2012). Molecular biology of the hepatitis B virus for clinicians. Journal of Clinical and Experimental Hepatology, 2(4), 353–365. https://doi.org/10.1016/j.jceh.2012.10.003

Demeke, C. A., Woldeyohanins, A. E., & Kifle, Z. D. (2021). Herbal medicine use for the management of COVID-19: A review article. *Metabolism Open*, 12, 100141. https://doi.org/10.1016/j.metop.2021.100141

Doytchinova, I. (2022). Drug design—past, present, future. Molecules, 27(5), 1496. https://doi.org/10.3390/molecules27051496

Ebada, S. S., Al-Jawabri, N. A., Youssef, F. S., El-Kashef, D. H., Knedel, T., Albohy, A., Korinek, M., Hwang, T., Chen, B., Lin, G. H., Lin, C. Y., Aldalaien, S. M., Disi, A. M., Janiak, C., & Proksch, P. (2020). Anti-inflammatory, antiallergic and COVID-19 protease inhibitory activities of phytochemicals from the Jordanian hawksbeard: Identification, structure–activity relationships, molecular modeling and impact on its folk medicinal uses. RSC Advances, 10(62), 38128–38141. https://doi.org/10.1039/d0ra04876c

Fanning, L. J. (2010). Retroviruses: Molecular biology, genomics and pathogenesis. *Clinical Infectious Diseases*, 52(2), 280. https://doi.org/10.1093/cid/ciq121

Flores, S. C., & Gerstein, M. (2011). Predicting protein ligand binding motions with the conformation explorer. BMC Bioinformatics, 12(1). https://doi.org/10.1186/1471-2105-12-417

Ganjhu, R. K., Mudgal, P. P., Maity, H., Dowarha, D., Devadiga, S., Nag, S., & Arunkumar, G. (2015). Herbal plants and plant preparations as remedial approach for viral diseases. Virusdisease, 26(4), 225–236. https://doi.org/10.1007/s13337-015-0276-6

Gescher, K., Kühn, J., Hafezi, W., Louis, A., Derksen, A., Deters, A., Lorentzen, E., & Hensel, A. (2011). Inhibition of viral adsorption and penetration by an aqueous extract from *Rhododendron ferrugineum L.* as antiviral principle against herpes simplex virus type-1. Fitoterapia, 82(3), 408–413. https://doi.org/10.1016/j.fitote.2010.11.022

Ghosh, A., Chakraborty, M., Chandra, A., & Alam, M. P. (2021). Structure-activity relationship (SAR) and molecular dynamics study of withaferin-A fragment derivatives as potential therapeutic lead against main protease (Mpro) of SARS-CoV-2. Journal of Molecular Modeling, 27(3). https://doi.org/10.1007/s00894-021-04703-6

Giordano, D., Biancaniello, C., Argenio, M. A., & Facchiano, A. (2022). Drug design by pharmacophore and virtual screening approach. Pharmaceuticals, 15(5), 646. https://doi.org/10.3390/ph15050646

Guha, R. (2013). On exploring structure–activity relationships. Methods in Molecular Biology, 81–94. https://doi.org/10.1007/978-1-62703-342-8_6

Harren, T., Matter, H., Heßler, G., Rarey, M., & Grebner, C. (2022). Interpretation of structure–activity relationships in real-world drug design data sets using explainable artificial intelligence. Journal of Chemical Information and Modeling, 62(3), 447–462. https://doi.org/10.1021/acs.jcim.1c01263

He, W., Han, H., Wang, W., & Gao, B. (2011). Anti-influenza virus effect of aqueous extracts from dandelion. Virology Journal, 8(1). https://doi.org/10.1186/1743-422x-8-538

Huang, Y., Yang, C., Xu, X., Xü, W., & Liu, S. (2020). Structural and functional properties of SARS-CoV-2 spike protein: Potential antivirus drug development for COVID-19. Acta Pharmacologica Sinica, 41(9), 1141–1149. https://doi.org/10.1038/s41401-020-0485-4

Ivanov, J. M., Polshakov, D., Kato-Weinstein, J., Zhou, Q., Li, Y., Granet, R., Garner, L., Deng, Y., Liu, C., Albaiu, D., Wilson, J., & Aultman, C. (2020). Quantitative structure–activity relationship machine learning models and their applications for identifying viral 3CLpro- and RdRp-targeting compounds as potential therapeutics for COVID-19 and related viral infections. ACS Omega, 5(42), 27344–27358. https://doi.org/10.1021/acsomega.0c03682

Jan, J. T., Cheng, T. R., Juang, Y., Hua, H., Wu, Y., Yang, W. B., Cheng, C., Chen, H., Chou, T. H., Shie, J., Cheng, W., Chein, R., Mao, S., Liang, P., Ma, C., Hung, S. C., & Wong, C. H. (2021). Identification of existing pharmaceuticals and herbal medicines as inhibitors of SARS-CoV-2 infection. Proceedings of the National Academy of Sciences of the United States of America, 118(5). https://doi.org/10.1073/pnas.2021579118

Jeong, H. J., Kim, Y. M., Kim, Y. H., Kim, J. Y., Park, J., Park, S., Ryu, Y. B., & Lee, W. S. (2012). Homoisoflavonoids from *Caesalpinia sappan* displaying viral neuraminidases inhibition. Biological & Pharmaceutical Bulletin, 35(5), 786–790. https://doi.org/10.1248/bpb.35.786

Jiang, Y., Yin, W., & Xu, H. (2021). RNA-dependent RNA polymerase: Structure, mechanism, and drug discovery for COVID-19. Biochemical and Biophysical Research Communications, 538, 47–53. https://doi.org/10.1016/j.bbrc.2020.08.116

Jiménez-Luna, J., Grisoni, F., & Schneider, G. (2020). Drug discovery with explainable artificial intelligence. Nature Machine Intelligence, 2(10), 573–584. https://doi.org/10.1038/s42256-020-00236-4

Jin, Z., Du, X., Xu, Y., Deng, Y., Liu, M., Zhao, Y., Zhang, B., Li, X., Zhang, L., Peng, C., Duan, Y., Yu, J., Wang, L., Yang, K., Liu, F., Jiang, R., Yang, X., You, T., Liu, X., ... Yang, H. (2020). Structure of Mpro from SARS-CoV-2 and discovery of its inhibitors. Nature, 582(7811), 289–293. https://doi.org/10.1038/s41586-020-2223-y

Kim, K., Kim, K. H., Kim, H. Y., Cho, H. K., Sakamoto, N., & Cheong, J. (2009). Curcumin inhibits hepatitis C virus replication via suppressing the Akt-SREBP-1 pathway. FEBS Letters, 584(4), 707–712. https://doi.org/10.1016/j.febslet.2009.12.019

Kim, S., Han, L., Yu, B., Hähnke, V., Bolton, E., & Bryant, S. H. (2015). PubChem structure–activity relationship (SAR) clusters. Journal of Cheminformatics, 7(1). https://doi.org/10.1186/s13321-015-0070-x

Klebe, G. (2013). Pharmacophore Hypotheses and Molecular Comparisons. In: Klebe, G. (eds) Drug Design. Springer, Berlin, Heidelberg. https://doi.org/10.1007/978-3-642-17907-5_17

Kumar, S., Sharma, P. P., Shankar, U., Kumar, D., Joshi, S. K., Pena, L., Durvasula, R., Kumar, A., Kempaiah, P., Poonam, & Rathi, B. (2020). Discovery of new hydroxyethylamine analogs against 3CLpro protein target of SARS-CoV-2: Molecular docking, molecular dynamics simulation, and structure–activity relationship studies. Journal of Chemical Information and Modeling, 60(12), 5754–5770. https://doi.org/10.1021/acs.jcim.0c00326

Lau, K. M., Lee, K. M., Koon, C. M., Cheung, C., Lau, C., Ho, H. M. K., Lee, M. Y. H., Au, S. W. N., Chen, C., Lau, C. B. S., Wan, D. C. C., Waye, M. M. Y., Wong, K., Wong, C. K., Lam, C. W. K., Leung, P. C., & Fung, K. (2008). Immunomodulatory and anti-SARS activities of Houttuynia cordata. Journal of Ethnopharmacology, 118(1), 79–85. https://doi.org/10.1016/j.jep.2008.03.018

Lewandowski, W., Lewandowska, H., Golonko, A., Świderski, G., Świsłocka, R., & Kalinowska, M. (2020). Correlations between molecular structure and biological activity in "logical series" of dietary chromone derivatives. PLOS ONE, 15(8), e0229477. https://doi.org/10.1371/journal.pone.0229477

Liao, Q., Chen, Z., Tao, Y., Zhang, B., Wu, X., Yang, L., Wang, Q., & Wang, Z. (2021). An integrated method for optimized identification of effective natural inhibitors against SARS-CoV-2 3CLpro. Scientific Reports, 11(1). https://doi.org/10.1038/s41598-021-02266-3

Li, G., Hilgenfeld, R., Whitley, R. J., & De Clercq, E. (2023). Therapeutic strategies for COVID-19: Progress and lessons learned. Nature Reviews Drug Discovery, 22(6), 449–475. https://doi.org/10.1038/s41573-023-00672-y

Li, H., & Lo, S. (2015). Hepatitis C virus: Virology, diagnosis and treatment. World Journal of Hepatology, 7(10), 1377. https://doi.org/10.4254/wjh.v7.i10.1377

Li, S. Y., Chen, C., Zhang, H., Guo, H. Y., Wang, H., Wang, L., Zhang, X., Hua, S. N., Yu, J., Xiao, P., Li, R. S., & Tan, X. (2005). Identification of natural compounds with antiviral activities against SARS-associated coronavirus. Antiviral Research, 67(1), 18–23. https://doi.org/10.1016/j.antiviral.2005.02.007

Li, T., Liu, L., Wu, H., Chen, S., Zhu, Q., Gao, H., Yu, X., Wang, Y., Su, W., Yao, X., & Peng, T. (2017). Anti-herpes simplex virus type 1 activity of Houttuynoid A, A flavonoid from Houttuynia cordata Thunb. Antiviral Research, 144, 273–280. https://doi.org/10.1016/j.antiviral.2017.06.010

Lin, L., Chen, T. Y., Chung, C. Y., Noyce, R. S., Grindley, T. B., McCormick, C., Lin, T. C., Wang, G. H., Lin, C., & Richardson, C. D. (2011). Hydrolyzable tannins (Chebulagic acid and punicalagin) target viral glycoprotein-glycosaminoglycan interactions to inhibit herpes simplex virus 1 entry and cell-to-cell spread. Journal of Virology, 85(9), 4386–4398. https://doi.org/10.1128/jvi.01492-10

Lin, L., Hsu, W. C., & Lin, C. (2014). Antiviral natural products and herbal medicines. Journal of Traditional and Complementary Medicine, 4(1), 24–35. https://doi.org/10.4103/2225-4110.124335

Lo, Y. C., Rensi, S. E., Torng, W., & Altman, R. B. (2018). Machine learning in chemoinformatics and drug discovery. Drug Discovery Today, 23(8), 1538–1546. https://doi.org/10.1016/j.drudis.2018.05.010

Lubbe, A., Seibert, I., Klimkait, T., & Van Der Kooy, F. (2012). Ethnopharmacology in overdrive: The remarkable anti-HIV activity of Artemisia annua. Journal of Ethnopharmacology, 141(3), 854–859. https://doi.org/10.1016/j.jep.2012.03.024

Luo, L., Zhong, A., Wang, Q., & Zheng, T. (2021). Structure-based pharmacophore modeling, virtual screening, molecular docking, ADMET, and molecular dynamics (MD) simulation of potential inhibitors of PD-L1 from the library of marine natural products. Marine Drugs, 20(1), 29. https://doi.org/10.3390/md20010029

Ma, S., Gao, R., Li, Y., Jiang, J., Gong, N., Li, L., Yang, L., Tang, W., Liu, Y., Qu, J., Lü, H., Li, Y., & Yu, S. (2013). Antiviral spirooliganones A and B with unprecedented skeletons from the roots of Illicium oligandrum. Organic Letters, 15(17), 4450–4453. https://doi.org/10.1021/ol401992s

Maurya, V. K., Kumar, S., Bhatt, M. L. B., & Saxena, S. K. (2020). Antiviral activity of traditional medicinal plants from ayurveda against SARS-CoV-2 infection. Journal of Biomolecular Structure & Dynamics, 40(4), 1719–1735. https://doi.org/10.1080/07391102.2020.1832577

McKinney, J. D. (2000). The practice of structure activity relationships (SAR) in toxicology. Toxicological Sciences, 56(1), 8–17. https://doi.org/10.1093/toxsci/56.1.8

Meuleman, P., Albecka, A., Belouzard, S., Vercauteren, K., Verhoye, L., Wychowski, C., Leroux-Roels, G., Palmer, K. E., & Dubuisson, J. (2011). Griffithsin has antiviral activity against hepatitis C virus. Antimicrobial Agents and Chemotherapy, 55(11), 5159–5167. https://doi.org/10.1128/aac.00633-11

Meyenburg, C., Dolfus, U., Briem, H., & Rarey, M. (2022). Galileo: Three-dimensional searching in large combinatorial fragment spaces on the example of pharmacophores. Journal of Computer-Aided Molecular Design, 37(1), 1–16. https://doi.org/10.1007/s10822-022-00485-y

Morales-Bayuelo, A., Matute, R. A., & Caballero, J. (2015). Understanding the comparative molecular field analysis (CoMFA) in terms of molecular quantum similarity and DFT-based reactivity descriptors. Journal of Molecular Modeling, 21(6). https://doi.org/10.1007/s00894-015-2690-5

Musa, H. H., Musa, T. H., Oderinde, O., Musa, I. H., Shonekan, O. O., Akintunde, T. Y., & Onasanya, A. K. (2022). Traditional herbal medicine: Overview of research indexed in the Scopus database. Advances in Traditional Medicine, 23(4), 1173–1183. https://doi.org/10.1007/s13596-022-00670-2

Najmi, A., Javed, S. A., Bratty, M. A., & Alhazmi, H. A. (2022). Modern approaches in the discovery and development of plant-based natural products and their analogues as potential therapeutic agents. Molecules, 27(2), 349. https://doi.org/10.3390/molecules27020349

Nguyen, T. T. H., Jung, J. H., Kim, M. K., Lim, S., Choi, J. M., Chung, B., Kim, D. W., & Kim, D. (2021). The inhibitory effects of plant derivate polyphenols on the main protease of SARS coronavirus 2 and their structure–activity relationship. Molecules, 26(7), 1924. https://doi.org/10.3390/molecules26071924

Niazi, S. K., & Mariam, Z. (2023). Recent advances in machine-learning-based chemoinformatics: A comprehensive review. International Journal of Molecular Sciences, 24(14), 11488. https://doi.org/10.3390/ijms241411488

Opo, F. A. D. M., Rahman, M. M., Ahammad, F., Ahmed, I., Bhuiyan, M. A., & Asiri, A. M. (2021). Structure based pharmacophore modeling, virtual screening, molecular docking and ADMET approaches for identification of natural anti-cancer agents targeting XIAP protein. Scientific Reports, 11(1). https://doi.org/10.1038/s41598-021-83626-x

Osipiuk, J., Azizi, S., Dvorkin, S., Endres, M., Jedrzejczak, R., Jones, K., Kang, S., Kathayat, R. S., Kim, Y. C., Lisnyak, V. G., Maki, S., Nicolaescu, V., Taylor, C., Tesar, C., Zhang, Y. A., Zhou, Z., Randall, G., Michalska, K., Snyder, S. A., … Joachimiak, A (2021). Structure of papain-like protease from SARS-CoV-2 and its complexes with non-covalent inhibitors. Nature Communications, 12(1). https://doi.org/10.1038/s41467-021-21060-3

Pal, S., Kumar, V., Kundu, B., Bhattacharya, D., Preethy, N., Reddy, M. P., & Talukdar, A. (2019). Ligand-based pharmacophore modeling, virtual screening and molecular docking studies for discovery of potential topoisomerase I inhibitors. Computational and Structural Biotechnology Journal, 17, 291–310. https://doi.org/10.1016/j.csbj.2019.02.006

Payne, S. (2017). Family coronaviridae. Viruses, 149–158. https://doi.org/10.1016/b978-0-12-803109-4.00017-9

Petrera, E., & Coto, C. E. (2009). Therapeutic effect of meliacine, an antiviral derived from Melia azedarach L., in mice genital herpetic infection. Phytotherapy Research, 23(12), 1771–1777. https://doi.org/10.1002/ptr.2850

Polyak, S. J., Oberlies, N. H., Pécheur, E. I., Dahari, H., Ferenci, P., & Pawlotsky, J. (2013). Silymarin for HCV infection. Antiviral Therapy, 18(2), 141–147. https://doi.org/10.3851/imp2402

Puttaswamy, H., Gowtham, H. G., Ojha, M. D., Yadav, A., Choudhir, G., Raguraman, V., Kongkham, B., Selvaraju, K., Shareef, S., Gehlot, P., Ahamed, F., & Chauhan, L. (2020). In silico studies evidenced the role of structurally diverse plant secondary metabolites in reducing SARS-CoV-2 pathogenesis. Scientific Reports, 10(1). https://doi.org/10.1038/s41598-020-77602-0

Ram, T. S., Munikumar, M., Raju, V. N., Devaraj, P., Boiroju, N. K., Hemalatha, R., Prasad, P. V. V., Gundeti, M., Sisodia, B. S., Pawar, S., Prasad, G. P., Chincholikar, M., Goel, S., Mangal, A. K., Gaidhani, S., Srikanth, N., & Dhiman, K. S. (2022). In silico evaluation of the compounds of the ayurvedic drug, AYUSH-64, for the action against the SARS-CoV-2 main protease. Journal of Ayurveda and Integrative Medicine, 13(1), 100413. https://doi.org/10.1016/j.jaim.2021.02.004

Rechtman, M. M., Har-Noy, O., Bar-Yishay, I., Fishman, S., Adamovich, Y., Shaul, Y., Halpern, Z., & Shlomai, A. (2010). Curcumin inhibits hepatitis B virus via down-regulation of the metabolic coactivator PGC-1α. FEBS Letters, 584(11), 2485–2490. https://doi.org/10.1016/j.febslet.2010.04.067

Seidel, T., Wieder, O., Garon, A., & Langer, T. (2020). Applications of the pharmacophore concept in natural product inspired drug design. Molecular Informatics, 39(11). https://doi.org/10.1002/minf.202000059

Shengman, Y., Li, X., Xin, Z., Sun, L., & Shi, J. (2022). Proteomic insights into SARS-CoV-2 infection mechanisms, diagnosis, therapies and prognostic monitoring methods. Frontiers in Immunology, 13. https://doi.org/10.3389/fimmu.2022.923387

Singh, P., & Rawat, P. (2017). Evolving herbal formulations in management of dengue fever. Journal of Ayurveda and Integrative Medicine, 8(3), 207–210. https://doi.org/10.1016/j.jaim.2017.06.005

Singh, R., Singh, A., Kaur, H., Batra, G., Sarma, P., Bhattacharyya, A., Sharma, A., Kumar, S., Upadhyay, S., Tiwari, V., Avti, P., Prakash, A., & Medhi, B. (2021). Promising traditional Indian medicinal plants for the management of novel coronavirus disease: A systematic review. Phytotherapy Research, 35(8), 4456–4484. https://doi.org/10.1002/ptr.7150

Sliwoski, G., Kothiwale, S., Meiler, J., & Lowe, E. W. (2013). Computational methods in drug discovery. Pharmacological Reviews, 66(1), 334–395. https://doi.org/10.1124/pr.112.007336

Spino, C., Dodier, M., & Sotheeswaran, S. (1998). Anti-HIV coumarins from calophyllum seed oil. Bioorganic & Medicinal Chemistry Letters, 8(24), 3475–3478. https://doi.org/10.1016/s0960-894x(98)00628-3

Tamura, S., Yang, G., Yasueda, N., Matsuura, Y., Komoda, Y., & Murakami, N. (2010). Tellimagrandin I, HCV invasion inhibitor from Rosae Rugosae Flos. Bioorganic & Medicinal Chemistry Letters, 20(5), 1598–1600. https://doi.org/10.1016/j.bmcl.2010.01.084

Taubenberger, J. K., & Morens, D. M. (2008). The pathology of influenza virus infections. Annual Review of Pathology-Mechanisms of Disease, 3(1), 499–522. https://doi.org/10.1146/annurev.pathmechdis.3.121806.154316

Temml, V., & Kutil, Z. (2021). Structure-based molecular modeling in SAR analysis and lead optimization. Computational and Structural Biotechnology Journal, 19, 1431–1444. https://doi.org/10.1016/j.csbj.2021.02.018

Tetko, I. V., Aksenova, T., Volkovich, V. V., Kasheva, T. N., Filipov, D. V., Welsh, W. J., Livingstone, D. J., & Villa, A. E. P. (2000). Polynomial neural network for linear and non-linear model selection in quantitative-structure activity relationship studies on the internet. SAR And QSAR in Environmental Research, 11(3–4), 263–280. https://doi.org/10.1080/10629360008033235

Thomas, N., Patil, P., Sharma, A., Kumar, S., Singh, V. K., Alagarasu, K., Parashar, D., & Tapryal, S. (2022). Studies on the antiviral activity of chebulinic acid against dengue and chikungunya viruses and in silico investigation of its mechanism of inhibition. Scientific Reports, 12(1). https://doi.org/10.1038/s41598-022-13923-6

Verma, A. K., Kumar, V., Singh, S., Goswami, B. C., Camps, I., Sekar, A., Yoon, S., & Lee, K. W. (2021). Repurposing potential of ayurvedic medicinal plants derived active principles against SARS-CoV-2 associated target proteins revealed by molecular docking, molecular dynamics and MM-PBSA studies. Biomedicine & Pharmacotherapy, 137, 111356. https://doi.org/10.1016/j.biopha.2021.111356

Vincent, S., Arokiyaraj, S., Saravanan, M., & Dhanraj, M. (2020). Molecular docking studies on the antiviral effects of compounds from Kabasura Kudineer on SARS-COV-2 3CLPro. Frontiers in Molecular Biosciences, 7. https://doi.org/10.3389/fmolb.2020.613401

Wang, L., Geng, C., Ma, Y., Huang, X., Luo, J., Chen, H., & Zhang, X. (2012). Synthesis, biological evaluation and structure–activity relationships of glycyrrhetinic acid derivatives as novel anti-hepatitis B virus agents. Bioorganic & Medicinal Chemistry Letters, 22(10), 3473–3479. https://doi.org/10.1016/j.bmcl.2012.03.081

Wawer, M., Jaramillo, D. E., Dančík, V., Fass, D. M., Haggarty, S. J., Shamji, A. F., Wagner, B. K., Schreiber, S. L., & Clemons, P. A. (2014). Automated structure–activity relationship mining: Connecting chemical structure to biological profiles. J Biomol Screen, 19(5), 738–748. https://doi.org/10.1177/1087057114530783

Wei, X., Huang, T., Yang, Z., Pan, L., Wang, L., & Ding, J. (2024). Quantitative predictive studies of multiple biological activities of TRPV1 modulators. Molecules, 29(2), 295. https://doi.org/10.3390/molecules29020295

Yang, C., Cheng, H., Lin, T., Chiang, L., & Lin, C. (2007a). The in vitro activity of geraniin and 1,3,4,6-tetra-O-galloyl-β-d-glucose isolated from *Phyllanthus urinaria* against herpes simplex virus type 1 and type 2 infection. Journal of Ethnopharmacology, 110(3), 555–558. https://doi.org/10.1016/j.jep.2006.09.039

Yang, C., Cheng, H., Lin, T., Chiang, L., & Lin, C. (2007b). Hippomanin a from acetone extract of *Phyllanthus urinaria* inhibited HSV-2 but not HSV-1 infection in vitro. Phytotherapy Research, 21(12), 1182–1186. https://doi.org/10.1002/ptr.2232

Yap, P., & Gan, C. (2024). Tyrosinase inhibitory peptides: Structure-activity relationship study on peptide chemical properties, terminal preferences and intracellular regulation of melanogenesis signaling pathways. Biochimica Et Biophysica Acta (BBA) - General Subjects, 1868(1), 130503. https://doi.org/10.1016/j.bbagen.2023.130503

Zandi, K., Teoh, B., Sam, S., Wong, P., Mustafa, M. R., & AbuBakar, S. (2012). Novel antiviral activity of baicalein against dengue virus. BMC Complementary and Alternative Medicine, 12(1). https://doi.org/10.1186/1472-6882-12-214

Zhang, D. H., Wu, K. L., Zhang, X., Deng, S. Q., & Peng, B. (2020). In silico screening of Chinese herbal medicines with the potential to directly inhibit 2019 novel coronavirus. Journal of Integrative Medicine, 18(2), 152–158. https://doi.org/10.1016/j.joim.2020.02.005

Zhao, Y., Chen, C. H., Morris-Natschke, S. L., & Lee, K. (2021). Design, synthesis, and structure activity relationship analysis of new betulinic acid derivatives as potent HIV inhibitors. European Journal of Medicinal Chemistry, 215, 113287. https://doi.org/10.1016/j.ejmech.2021.113287

Zheng, W., & Tropsha, A. (1999). Novel variable selection quantitative structure–property relationship approach based on the K-nearest-neighbor principle. Journal of Chemical Information and Computer Sciences, 40(1), 185–194. https://doi.org/10.1021/ci980033m

Zhu, H., Zhou, R., Cao, D., Tang, J., & Li, M. (2023). A pharmacophore-guided deep learning approach for bioactive molecular generation. Nature Communications, 14(1). https://doi.org/10.1038/s41467-023-41454-9

Zhu, S., & Viejo-Borbolla, A. (2021). Pathogenesis and virulence of herpes simplex virus. Virulence, 12(1), 2670–2702. https://doi.org/10.1080/21505594.2021.1982373

6 Molecular Mechanisms of Main Protease Inhibitors from Traditional Herbal Medicines

Arun Bahadur Gurung

6.1 INTRODUCTION

The coronavirus disease 2019 (COVID-19) was a global pandemic which affected millions of people worldwide [1]. The pathogen responsible for COVID-19 is the enveloped beta-coronavirus known as severe acute respiratory syndrome coronavirus 2 (SARS-CoV-2) [2]. The genetic makeup of SARS-CoV-2 is roughly 30 kilobase (kb) in size, and employs the same receptor, angiotensin-converting enzyme 2 (ACE-2), as SARS-CoV [3]. Since ancient times, people have used plants as medicines across the world, including in Asian as well as some African countries [4]. Tribal people have long used these plants, mostly due to their wide availability and affordable price [5]. Plants remain potential therapeutic sources for treating infectious diseases, cancer, diarrhoea, depression, and thrombosis. This offers hope that drug molecules can be derived from plant sources to effectively combat COVID-19. These compounds can bind to viral proteins and enzymes to inhibit viral entry and replication in host cells [6]. The secondary metabolites from medicinal plants are said to be effective against pathogenic microbes [5, 7]. Several preliminary investigations have raised the possibility that bioactive phytochemicals have inhibitory effects against new coronavirus variants [5]. Herbal remedies have assisted in lessening the symptoms of infectious diseases including COVID-19. There is evidence to support the theory that herbal medicines can effectively manage and lower the risk of COVID-19. Herbal medicine has been approved in China for use as an alternate treatment for COVID-19 in conjunction with Western medicines [8]. As traditional treatments did not show effectiveness in suppressing SARS-CoV-2 outbreaks, well-known herbal remedies with antiviral properties were employed as supplemental treatments [9]. Traditional Chinese medicine (TCM) has a great deal of experience managing infectious diseases by promoting body repair, lowering hyperinflammatory states, balancing the immunological response, and enhancing the body's defences against pathogens. According to clinical studies, TCM can be helpful in both treating and preventing COVID-19, offering novel therapeutic options [10]. Another traditional medicinal theory is Ayurveda, which has Indian roots. It places more emphasis on developing mental and physical toughness. Ayurveda offers various therapy alternatives, including steam inhalation, immunomodulators and herbal infusions for respiratory illnesses [11].

Since many medicines are made of plant components or their derivatives, researchers have turned to plant-based medicines because of the lack of a specific therapy against COVID-19 [12]. Even if vaccinations have advanced recently, long-term, multi-stage clinical trials are still necessary to guarantee their safety and true efficacy [12]. Plant-based treatments showed promising efficacy against numerous viruses including SARS-CoV-2 by boosting the immune system [13, 14]. The main protease (M^{pro} or $3CL^{pro}$) enzyme can cleave polyproteins at various sites to produce several functional proteins during virus replication [15]. It is a viable target for the development of broad-spectrum drugs for COVID-19 [14, 16]. Since it does not have counterpart homolog in humans, it is possible to design specific and effective $3CL^{pro}$ inhibitors [17]. In this chapter, we will look at various traditional herbal medicines which can be used in drug discovery and explore how they could be used to identify effective traditional herbal and natural drug components against the main protease enzyme for the treatment of COVID-19.

DOI: 10.1201/9781003452621-6

6.2   HERBAL MEDICINES

6.2.1   TCM FOR COVID-19 TREATMENT

A trend towards plant-based therapeutic approaches has been spurred by the paucity of specific and effective treatments against SARS-CoV-2. The identification of potential anti-COVID-19 herbal remedies is of great interest, as plant-based therapies have demonstrated encouraging antiviral efficacy via strengthening immunity [6]. Treatment and/or prevention of certain outbreaks has been greatly aided by the application of TCM. During the 2003 SARS outbreak, TCM had impressive therapeutic outcomes. The TCM programme has been involved in the analysis and treatment recommendations throughout the COVID-19 pandemic [18]. TCM has demonstrated encouraging effects in lowering the overall death rate, the rate of moderate and/or severe instances, and chronicity. Herbal-based treatment may have direct antiviral, immunomodulatory, and anti-inflammatory benefits when combined with contemporary biomedicine. It may help alleviate hypoxemia and chronic obstructive pulmonary disease [18]. The active component of *Stellaria baicalensis* extracts is baicalin, which continues to be one of the primary TCM herbal components [19]. In addition to cutting down on the amount of time patients must be ventilated in severe cases, another TCM medication called Xuebijing injection has been widely acknowledged to lessen the risk of pneumonia [19]. Furthermore, according to Su et al. *Exocarpium Citri grandis*, a traditional Chinese herb, may be effective in treating and preventing COVID-19 [20].

Combining traditional and modern methods to create potent formulas for a range of ailments is an inventive approach to TCM [3]. For instance, a combination of the liquid fermented broth of *Ganoderma lucidum* and the extract of *Radix Sophorae favescentis* demonstrated effectiveness against the hepatitis B virus. Furthermore, glycyrrhizin, a bioactive component derived from *Lycoris radiate*, showed promise as a treatment for SARS-CoV [21]. TCM frequently uses Shufeng Jiedu capsules (SFJDC) to treat influenza. It is now recommended for COVID-19 treatment. For individuals with mild symptoms of COVID-19, SFJDC remains promising when combined with arbidol [22].

6.2.2   INDIAN TRADITIONAL MEDICINES

Similarly, in India, traditional medicine has been used to treat COVID-19 in addition to modern medicines and vaccines. One of the first components that is still essential to the global healthcare system is traditional Indian medicine [23]. These conventional practices, which include homoeopathy, naturopathy, yoga, Ayurveda, Siddha, and unani, are effective in treating a wide range of illnesses [23]. These traditional methods treat a variety of illnesses by using minerals, plants, and animal products. In South Asia, some 25,000 concoctions and extracts made from herbs have been used in traditional medicine [3]. Recently, a decoction of sunthi (*Zingiber officinale* Roscoe), lavanga (*Syzygium aromaticum*), and maricha (*Piper nigrum*) have been recommended for COVID-19 patients as well as healthy individuals. It promotes humoral and cell-mediated immune responses and reduces airway hypersensitivity [3]. Similarly, it has been shown that the active ingredient in *Curcuma longa* Linn., curcumin, inhibits the release of cytokines, particularly pro-inflammatory cytokines and TNF-α. [24].

The world's traditional medical system, known as Ayurveda, is thought to be safe to employ in treating a variety of infections. Ayurveda is well-equipped with a range of therapeutic modalities for complex harmful illnesses [3]. Experts in Ayurvedic medicine have knowledge of a wide range of microorganisms and the illnesses they might cause. Both Ayurveda and Siddha practices originated in India and are used extensively to treat a variety of illnesses [21]. Identification, isolation, and characterization of the bioactive phytochemicals in medicinal herbs may help fight a variety of illnesses. Viral illnesses have been prevented and treated with Ayurvedic medicine and its extracts. The first form of medicine created by combining spices and plant-based medications is known as "kadha". It is an extract made from spices and herbs from different Indian botanical medications [25]. During the COVID-19 epidemic, the Indian government advised using kadha to boost the immune system and aid in the healing process [25]. A traditional Indian medicine called "Guduchi GhanVati" is typically recommended as an immunomodulatory and antioxidant treatment. Its effectiveness against SARS-CoV-2 infections was recently verified [26]. The Indian

Ayurvedic Pharmacopoeia lists Guduchi GhanVati as a frequent Ayurvedic preparation. It is made from an aqueous *Tinospora cordifolia* extract. It is commonly known as "Guduchi" or "Giloe" [6].

6.3 M^{PRO} ENZYME

The main protease (M^{pro} or 3CL^{pro}) is roughly 34.21 kDa per monomer with three domains–I, II, and III [27] (Figure 6.1A). Domain III is an additional helix domain and it is through its aggregation that 3CL^{pro} dimerizes [28, 29]. The dimeric form of the enzyme acts as a functional unit with

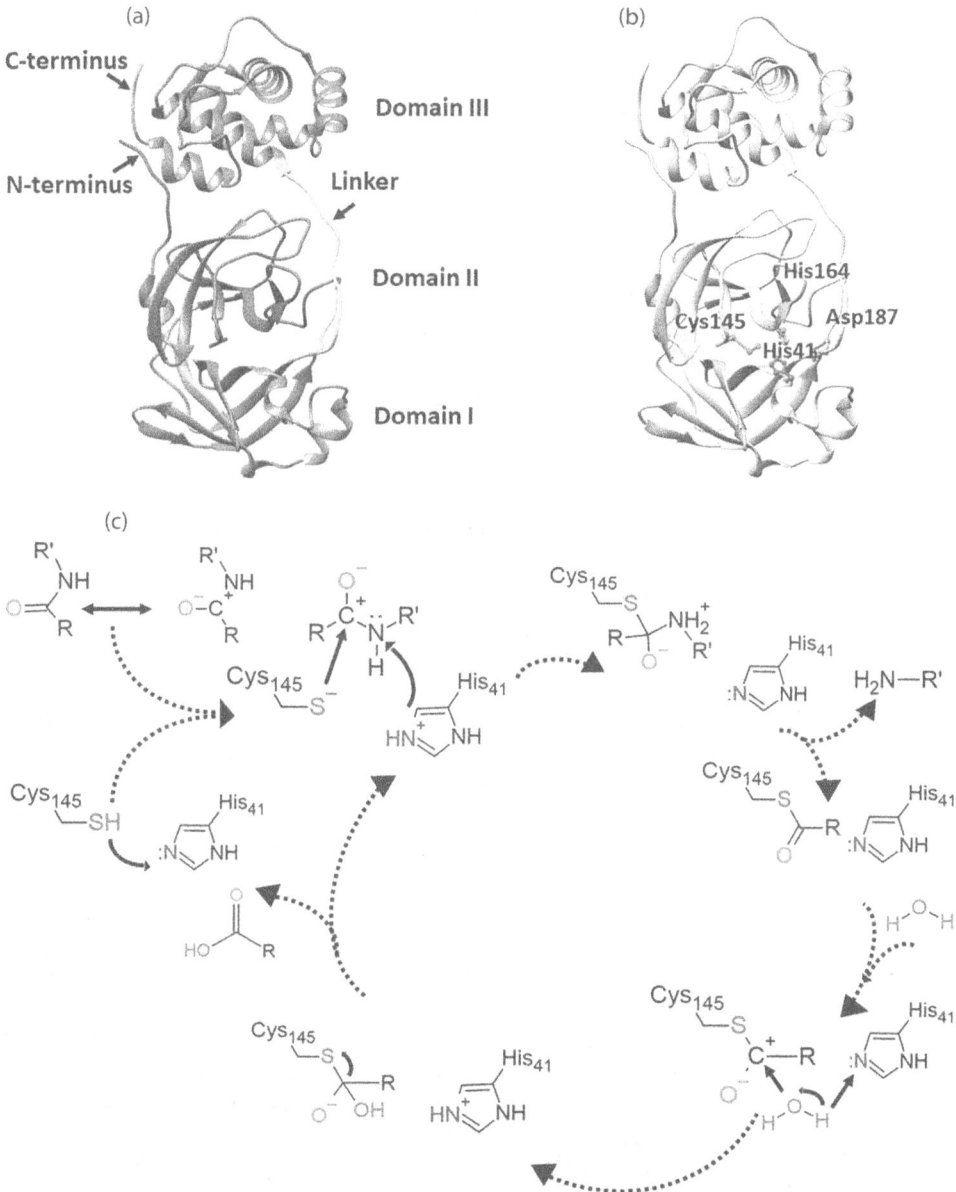

FIGURE 6.1 (A) Three-dimensional structure of SARS-CoV-2 main protease enzyme (PDB ID: 6LU7) with structural domains: domain I (orange), domain II (green), and domain III (pink). (B) Amino acid residues located in the active site pocket (His41 and Cys145 together constitute catalytic dyad). (C) SARS-CoV-2 main protease-mediated catalytic hydrolysis mechanism of amide substrate.

the maximum hydrolytic activity, whereas the monomer is typically less enzymatically active [30, 31]. There are five primary sub-pockets (S1–S5) that make up the 3CL^{pro} catalytic site, which is situated at the junction of domains I and II [32]. Five sub-pockets with critical face residues correspond to five distinct substrate-binding sites. Although P4, P3, and P3′ enhance substrate recognition and binding, P1, P2, and P1′ sites primarily modulate substrate specificity of 3CL^{pro} [30, 33]. In contrast to the catalytic triad of 3-chymotrypsin, Cys145 and His41 constitute the catalytic dyad of 3CL^{pro} [34]. The energetical water required to activate the catalytic dyad is sustained by His164 and Asp187 [35, 36] (Figure 6.1B). Large polyprotein chains are cleaved by 3CL^{pro} at the glutamine residue in the substrate P1 position using the catalytic dyad, where the cysteine thiol acts as the nucleophile [17]. The process by which 3CL^{pro} cleaves polyproteins utilizing a universal nucleophilic-type reaction is illustrated in Figure 6.1C. SARS-CoV-2 3CL^{pro} is a major cysteine protease with 12 cysteine residues, but only three of which are in contact with the solvent and the most significant cysteine is catalytic Cys145 [37]. Cys145 can bind to myricetin when it is oxidized [38]. By creating a Se–S link, ebselen and its derivatives can alter Cys145 [39]. Cys145 can be reacted with peptidomimetic α-acyloxymethylketone warheads via a structure-based selectivity mechanism [40].

6.4 HERBAL MEDICINES AGAINST SARS-CoV-2 M^{PRO}

Herbal medicines contain various phytochemicals such as myricetin, quercetin, epigallocatechin gallate (EGCG), herbacetin, baicalein, etc., which have the potential to inhibit the activity of M^{pro} (Table 6.1). *Withania somnifera* (WS) is one of the most significant herbs in Ayurveda. Its many health advantages have led to its use as Rasayana, a sort of juice, for millennia. WS phytocompounds control detoxification enzymes, boost immunity, and exhibit a wide range of biological actions (including antioxidant, anticancer, and antibacterial properties) [41]. Using *in silico* molecular docking and molecular dynamics simulations, Kandagalla and coworkers identified withasomniferol C (the active component of the Indian traditional medicinal herb WS) as a potential candidate against SARS-CoV-2 3CL^{pro} [42]. In a similar study, Tripathi and coworkers identified withanoside V from Ashwagandha (WS) as a potential SARS-CoV-2 M^{pro} inhibitor [43]. An *in silico* study by Shree and coworkers demonstrated that the active components of Tulsi, Ashwagandha, and Giloy can inhibit the activity of SARS-CoV-2 M^{pro} and this activity was attributed to the presence of phytochemicals such as somniferin, isorientin 4′-O-glucoside 2″-O-p-hydroxybenzoate, withanoside V etc [44]. de Oliveiraa et al. used various theoretical techniques to screen herbal compounds having the potential to inhibit the SARS-CoV-2 M^{pro}. A total of six molecules (physalin B 5,6-epoxide, methyl amentoflavone, withaphysalin C, daphnoline or trilobamine, cepharanoline, and tetrandrine) were identified as potential SARS-CoV-2 M^{pro} inhibitors [45]. Saravanan and coworkers used molecular docking and simulations to screen antiviral compounds from traditional Indian medicinal plants, and their study revealed that amentoflavone, hypericin, and torvoside H can be used as a potential SARS-CoV-2 main protease inhibitor [46]. Singh and coworkers demonstrated that phytochemicals–chebulagic acid, chebulinic acid, and ellagic acid–present in Triphala (a mixture of *Emblica officinalis*, *Terminalia bellirica*, and *Terminalia chebula* fruits) exhibited the highest binding affinity for the main protease of four different coronaviruses including SARS-CoV-2 [47]. Mpiana and coworkers using molecular docking demonstrated that three active components (feralolide, 9-dihydroxyl-2-O-(z)-cinnamoyl-7-methoxy-aloesin, and aloeresin) from the medicinal plant *Aloe vera* showed good binding energy and drug-like properties [48]. The Chinese Pharmacopeia 2015 edition lists Lianhua Qingwen's (LHQW), a TCM with antiviral properties, immunomodulatory and inhibitory effects on virus replication. There are three different dosage forms of LHQW available: pill, granules, and decoction. Thirteen ingredients were used in its formulation [49]. LHQW has potent binding capabilities to ACE-2 and M^{pro} that contribute to its anti-COVID-19 actions [49].

TABLE 6.1

Phytochemicals with Inhibitory Activity Against Coronavirus Main Protease Enzyme

Compound	Class	IC_{50} value (µM)	Reference
Daidzein	Isoflavone	56	[50]
Puerarin	Isoflavone	42 ± 2	[51]
Myricetin	Flavonol	43 ± 1	[52]
Quercetin	Flavonol	93 ± 5	[52]
Quercetagetin	Flavonol	145 ± 6	[52]
Ampelopsin	Flavanonol	128 ± 5	[53]
Naringenin	Flavanone	150 ± 10	[54]
Epigallocatechin gallate (EGCG)	Flavan-3-ol	171 ± 5	[50]
Chlorogenic acid	Hydrocinnamic acid	39.48 ± 5.51	[20]
Caffeic acid	Dihydroxycinnamic acid	197 ± 1	[55]
Ellagic acid	Polyphenol	11.8 ± 5.7	[56]
Gallocatechin gallate (GCG)	Polyphenol	5.774 ± 0.805	[50]
Epicatechin gallate (ECG)	Flavan-3-ol	12.5	[57]
Kaempferol glycosides	Flavonoid	125.00	[58]
Isorhamnetin glycosides	Flavonoid	13.13	[59]
Pectolinarin	Flavonoid	37.7	[60]
Herbacetin	Flavonoid	33.1	[61]
Rhoifolin	Flavonoid	27.4	[61]
Silymarin	Flavonoid	46.88	[46]
Methyl rosmarinate	Polyphenol	21.32	[62]
Baicalein	Flavonoid	0.94 ± 0.20	[20]
Withanoside V	Terpenoid	5.774 ± 0.805	[43]
Ursolic acid	Triterpenoid	12.6	[43]
Curcumin	Terpene	11.9	[63]
Tannic acid	Tannin	13.4	[64]

6.5 CHALLENGES

Plant metabolites have a variety of medicinal uses and they may have synergistic effects that lead to better treatment results. Drug discovery entails several benefits but also several obstacles that must be overcome. The main obstacle is the "druggability" of plant metabolites and the key elements influencing its druggability are its pharmacokinetic parameters. These problems have been solved with the development of innovative drug delivery methods and nanotechnologies. The other issues include (a) the acquisition and verification of plant materials; (b) the use of high-throughput screening bioassays and the scaling-up of bioactive compounds; and (c) the intricacy of the isolation and purification procedures [65]. Furthermore, tests conducted in laboratories that are used to monitor plant metabolites during clinical trials may fail to consider their toxicities. The failure of development efforts at the clinical trial stage is discouraging as it is difficult, time-consuming, and requires a great deal of work to isolate, purify, and test pure plant-derived compounds [66].

6.6 CONCLUSION AND FUTURE PERSPECTIVES

Herbal remedies for COVID-19 management might be complementary, alternative, or independent therapy. Numerous potential metabolites and plant-based herbal formulations produced encouraging results in preclinical research [3]. However, extensive and more rigorous, studies are necessary to assess the safety profiles and efficacy of these herbal formulations [6]. Finding more potent treatment approaches for COVID-19 is urgently needed since the continued pandemic has put global health at considerable risk. The main protease (M^{pro} or $3CL^{pro}$) is one of the most important targets for combating COVID-19. Herbal remedies are known to contain a wide range of chemicals, some of which may interact with different anti-coronavirus targets. Therefore, further research is necessary to fully understand the synergistic effects of various herbal medicine components, which may help explain their anti-COVID-19 properties. Computational modelling of both cell-free and cell-based screening methods has helped to further the hunt for possible medications that target the coronavirus main protease. Future research ought to concentrate on evaluating potential hazardous effects on cells, as even the most promising main protease inhibitors will be useless if they are discovered to have a toxic effect [67]. Evaluating plant-based remedies such as traditional medicines and bioactive compounds to combat the new coronavirus strain may prove to be a huge victory for the skewed public health system.

REFERENCES

1. Elekhnawy E, Negm WA. The potential application of probiotics for the prevention and treatment of COVID-19. Egypt J Med Hum Genet. 2022, **23**(1):1–9.
2. Elekhnawy E, Negm WA, El-Sherbeni SA, Zayed A. Assessment of drugs administered in the Middle East as part of the COVID-19 management protocols. Inflammopharmacology. 2022, **30**(6):1935–1954.
3. Al-Kuraishy HM, Al-Fakhrany OM, Elekhnawy E, Al-Gareeb AI, Alorabi M, De Waard M, Albogami SM, Batiha GE-S. Traditional herbs against COVID-19: back to old weapons to combat the new pandemic. Eur J Med Res. 2022, **27**(1):186.
4. Hoareau L, DaSilva EJ: Medicinal plants: a re-emerging health aid. Electron J Biotechnol. 1999, **2**(2):3–4.
5. Jahan I, Ahmet O. Potentials of plant-based substance to inhabit and probable cure for the COVID-19. Turk J Biol. 2020, **44**(3):228.
6. Alam S, Sarker MMR, Afrin S, Richi FT, Zhao C, Zhou J-R, Mohamed IN. Traditional herbal medicines, bioactive metabolites, and plant products against COVID-19: update on clinical trials and mechanism of actions. Front Pharmacol. 2021, **12**:671498.
7. Semple SJ, Reynolds GD, O'leary MC, Flower RL: Screening of Australian medicinal plants for antiviral activity. J Ethnopharmacol. 1998, **60**(2):163–172.
8. Ang L, Song E, Lee HW, Lee MS. Herbal medicine for the treatment of coronavirus disease 2019 (COVID-19): a systematic review and meta-analysis of randomized controlled trials. J Clin Med. 2020, **9**(5):1583.
9. Panyod S, Ho C-T, Sheen L-Y. Dietary therapy and herbal medicine for COVID-19 prevention: a review and perspective. J Tradit Complement Med. 2020, **10**(4):420–427.
10. Ren J-l, Zhang A-H, Wang X-J. Traditional Chinese medicine for COVID-19 treatment. Pharmacol Res. 2020, **155**:104743.
11. Golechha M. Time to realise the true potential of Ayurveda against COVID-19. Brain Behav Immun. 2020, **87**:130.
12. Silveira D, Prieto-Garcia JM, Boylan F, Estrada O, Fonseca-Bazzo YM, Jamal CM, Magalhães PO, Pereira EO, Tomczyk M, Heinrich M. COVID-19: is there evidence for the use of herbal medicines as adjuvant symptomatic therapy? Front Pharmacol. 2020, **11**:1479.
13. Huang J, Tao G, Liu J, Cai J, Huang Z, Chen J-XJ. Current prevention of COVID-19: natural products and herbal medicine. Front Pharmacol. 2020, **11**:588508.
14. Zhang L, Lin D, Sun X, Curth U, Drosten C, Sauerhering L, Becker S, Rox K, Hilgenfeld R. Crystal structure of SARS-CoV-2 main protease provides a basis for design of improved α-ketoamide inhibitors. Science. 2020, **368**(6489):409–412.
15. Qiao J, Li Y-S, Zeng R, Liu F-L, Luo R-H, Huang C, Wang Y-F, Zhang J, Quan B, Shen C et al. SARS-CoV-2 Mpro inhibitors with antiviral activity in a transgenic mouse model. Science. 2021, **371**(6536):1374–1378.
16. He J, Hu L, Huang X, Wang C, Zhang Z, Wang Y, Zhang D, Ye W. Potential of coronavirus 3C-like protease inhibitors for the development of new anti-SARS-CoV-2 drugs: insights from structures of protease and inhibitors. Int J Antimicrob Agents. 2020, **56**(2):106055.

17. Hu Q, Xiong Y, Zhu GH, Zhang YN, Zhang YW, Huang P, Ge GB. The SARS-CoV-2 main protease (Mpro): structure, function, and emerging therapies for COVID-19. MedComm (2020). 2022, **3**(3):e151.

18. Luo L, Jiang J, Wang C, Fitzgerald M, Hu W, Zhou Y, Zhang H, Chen S. Analysis on herbal medicines utilized for treatment of COVID-19. Acta Pharm Sin B. 2020, **10**(7):1192–1204.

19. Song Z, Xu Y, Bao L, Zhang L, Yu P, Qu Y, Zhu H, Zhao W, Han Y, Qin C. From SARS to MERS, thrusting coronaviruses into the spotlight. Viruses. 2019, **11**(1):59.

20. Su W-W, Wang Y, Li P, Wu H, Zeng X, Shi R, Zheng YY, Li P-L, Peng W. The potential application of the traditional Chinese herb Exocarpium Citri grandis in the prevention and treatment of COVID-19. Tradit Med Res. 2020, **5**(3):160–166.

21. Janiaud P, Axfors C, Schmitt AM, Gloy V, Ebrahimi F, Hepprich M, Smith ER, Haber NA, Khanna N, Moher D et al. Association of convalescent plasma treatment with clinical outcomes in patients with COVID-19: a systematic review and meta-analysis. JAMA. 2021, **325**(12):1185–1195.

22. Chen X, Yin Y-H, Zhang M-Y, Liu J-Y, Li R, Qu Y-Q. Investigating the mechanism of ShuFeng JieDu capsule for the treatment of novel coronavirus pneumonia (COVID-19) based on network pharmacology. Int J Med Sci. 2020, **17**(16):2511.

23. Guo Y-R, Cao Q-D, Hong Z-S, Tan Y-Y, Chen S-D, Jin H-J, Tan K-S, Wang D-Y, Yan Y. The origin, transmission and clinical therapies on coronavirus disease 2019 (COVID-19) outbreak–an update on the status. Mil Med Res. 2020, **7**:1–10.

24. Zeng F, Huang Y, Guo Y, Yin M, Chen X, Xiao L, Deng G. Association of inflammatory markers with the severity of COVID-19: a meta-analysis. Int J Infect Dis. 2020, **96**:467–474.

25. Maurya DK, Sharma D. Evaluation of traditional ayurvedic Kadha for prevention and management of the novel Coronavirus (SARS-CoV-2) using in silico approach. J Biomol Struct Dyn. 2022, **40**(9):3949–3964.

26. Bala PC, Eisenreich BR, Yoo SBM, Hayden BY, Park HS, Zimmermann J. Openmonkeystudio: automated markerless pose estimation in freely moving macaques. Nat Commun. 2020:2020. https://doi.org/10.1101/2020.01.31.928861

27. Denesyuk AI, Permyakov EA, Johnson MS, Permyakov SE, Denessiouk K, Uversky VN. Structural and functional significance of the amino acid differences Val35Thr, Ser46Ala, Asn65Ser, and Ala94Ser in 3C-like proteinases from SARS-CoV-2 and SARS-CoV. Int J Biol Macromol. 2021, **193**:2113–2120.

28. Feng J, Li D, Zhang J, Yin X, Li J. Crystal structure of SARS-CoV 3C-like protease with baicalein. Biochem Biophys Res Commun. 2022, **611**:190–194.

29. Jin Z, Du X, Xu Y, Deng Y, Liu M, Zhao Y, Zhang B, Li X, Zhang L, Peng C et al. Structure of Mpro from SARS-CoV-2 and discovery of its inhibitors. Nature. 2020, **582**(7811):289–293.

30. Noske GD, Nakamura AM, Gawriljuk VO, Fernandes RS, Lima GM, Rosa HVD, Pereira HD, Zeri AC, Nascimento AF, Freire MCLC. A crystallographic snapshot of SARS-CoV-2 main protease maturation process. J Mol Biol. 2021, **433**(18):167118.

31. Pablos I, Machado Y, de Jesus HCR, Mohamud Y, Kappelhoff R, Lindskog C, Vlok M, Bell PA, Butler GS, Grin PM. Mechanistic insights into COVID-19 by global analysis of the SARS-CoV-2 3CLpro substrate degradome. Cell Rep. 2021, **37**(4):109892.

32. Akbulut E. Investigation of changes in protein stability and substrate affinity of 3CL-protease of SARS-CoV-2 caused by mutations. Genet Mol Biol. 2022, **45**:e20210404.

33. Greasley SE, Noell S, Plotnikova O, Ferre R, Liu W, Bolanos B, Fennell K, Nicki J, Craig T, Zhu Y. Structural basis for the in vitro efficacy of nirmatrelvir against SARS-CoV-2 variants. J Biol Chem. 2022, **298**(6):101972.

34. Chen S, Zhang J, Hu T, Chen K, Jiang H, Shen X. Residues on the dimer interface of SARS coronavirus 3C-like protease: dimer stability characterization and enzyme catalytic activity analysis. J Biochem. 2008, **143**(4):525–536.

35. Kneller DW, Phillips G, O'Neill HM, Tan K, Joachimiak A, Coates L, Kovalevsky A. Room-temperature X-ray crystallography reveals the oxidation and reactivity of cysteine residues in SARS-CoV-2 3CL Mpro: insights into enzyme mechanism and drug design. IUCrJ. 2020, **7**(6):1028–1035.

36. Yin J, Niu C, Cherney MM, Zhang J, Huitema C, Eltis LD, Vederas JC, James MNG. A mechanistic view of enzyme inhibition and peptide hydrolysis in the active site of the SARS-CoV 3C-like peptidase. J Mol Biol. 2007, **371**(4):1060–1074.

37. Karges J, Kalaj M, Gembicky M, Cohen SM. ReI tricarbonyl complexes as coordinate covalent inhibitors for the SARS-CoV-2 main cysteine protease. Angew Chem Int Ed Engl. 2021, **60**(19):10716–10723.

38. Kuzikov M, Costanzi E, Reinshagen J, Esposito F, Vangeel L, Wolf M, Ellinger B, Claussen C, Geisslinger G, Corona A et al. Identification of inhibitors of SARS-CoV-2 3CL-pro enzymatic activity using a small molecule in vitro repurposing screen. ACS Pharmacol Transl Sci. 2021, **4**(3):1096–1110.

39. Amporndanai K, Meng X, Shang W, Jin Z, Rogers M, Zhao Y, Rao Z, Liu Z-J, Yang H, Zhang L. Inhibition mechanism of SARS-CoV-2 main protease by ebselen and its derivatives. Nat Commun. 2021, **12**(1):3061.

40. Bai B, Belovodskiy A, Hena M, Kandadai AS, Joyce MA, Saffran HA, Shields JA, Khan MB, Arutyunova E, Lu J. Peptidomimetic α-acyloxymethylketone warheads with six-membered lactam P1 glutamine mimic: SARS-CoV-2 3CL protease inhibition, coronavirus antiviral activity, and in vitro biological stability. J Med Chem. 2021, **65**(4):2905–2925.

41. Singh N, Bhalla M, de Jager P, Gilca M. An overview on ashwagandha: a Rasayana (rejuvenator) of Ayurveda. Afr J Tradit Complement Altern Med. 2011, **8**(5S):208–213.

42. Kandagalla S, Rimac H, Gurushankar K, Novak J, Grishina M, Potemkin V. Withasomniferol C, a new potential SARS-CoV-2 main protease inhibitor from the *Withania somnifera* plant proposed by *in silico* approaches. PeerJ. 2022, **10**:e13374.

43. Tripathi MK, Singh P, Sharma S, Singh TP, Ethayathulla A, Kaur P. Identification of bioactive molecule from Withania somnifera (Ashwagandha) as SARS-CoV-2 main protease inhibitor. J Biomol Struct Dyn. 2021, **39**(15):5668–5681.

44. Shree P, Mishra P, Selvaraj C, Singh SK, Chaube R, Garg N, Tripathi YB. Targeting COVID-19 (SARS-CoV-2) main protease through active phytochemicals of ayurvedic medicinal plants–*Withania somnifera* (Ashwagandha), *Tinospora cordifolia* (Giloy) and *Ocimum sanctum* (Tulsi)–a molecular docking study. J Biomol Struct Dyn. 2022, **40**(1):190–203.

45. de Oliveira OV, Cristina Andreazza Costa M, Marques da Costa R, Giordano Viegas R, Paluch AS, Miguel Castro Ferreira M. Traditional herbal compounds as candidates to inhibit the SARS-CoV-2 main protease: an in silico study. J Biomol Struct Dyn. 2023, **41**(5):1603–1616.

46. Saravanan KM, Zhang H, Senthil R, Vijayakumar KK, Sounderrajan V, Wei Y, Shakila H. Structural basis for the inhibition of SARS-CoV2 main protease by Indian medicinal plant-derived antiviral compounds. J Biomol Struct Dyn. 2022, **40**(5):1970–1978.

47. Singh P, P A, VR H, U.V B, Rafiq M, Rao RP. Screening for potential traditional herbal inhibitors against 3-chymotrypsin-like main protease (3CLpro) from four different coronaviruses: an *in silico* approach. Coronaviruses. 2021, **2**(12):12–27.

48. Mpiana PT, Tshibangu DS, Kilembe JT, Gbolo BZ, Mwanangombo DT, Inkoto CL, Lengbiye EM, Mbadiko CM, Matondo A, Bongo GN. Identification of potential inhibitors of SARS-CoV-2 main protease from Aloe vera compounds: a molecular docking study. Chem Phys Lett. 2020, **754**:137751.

49. Liu L-S, Lei N, Lin Q, Wang W-L, Yan H-W, Duan X-H. The effects and mechanism of Yinqiao powder on upper respiratory tract infection. Int J Biotechnol Wellness Ind. 2015, **4**(2):57.

50. Ghosh R, Chakraborty A, Biswas A, Chowdhuri S. Depicting the inhibitory potential of polyphenols from *Isatis indigotica* root against the main protease of SARS CoV-2 using computational approaches. J Biomol Struct Dyn. 2022, **40**(9):4110–4121.

51. Hu X, Cai X, Song X, Li C, Zhao J, Luo W, Zhang Q, Ekumi IO, He Z. Possible SARS-coronavirus 2 inhibitor revealed by simulated molecular docking to viral main protease and host toll-like receptor. Future Virol. 2020, **15**(6):359–368.

52. Nguyen TTH, Jung J-H, Kim M-K, Lim S, Choi J-M, Chung B, Kim D-W, Kim D. The inhibitory effects of plant derivate polyphenols on the main protease of SARS coronavirus 2 and their structure–activity relationship. Molecules. 2021, **26**(7):1924.

53. Zhu Y, Scholle F, Kisthardt SC, Xie D-Y. Flavonols and dihydroflavonols inhibit the main protease activity of SARS-CoV-2 and the replication of human coronavirus 229E. Virology. 2022, **571**:21–33.

54. Agrawal PK, Agrawal C, Blunden G. Naringenin as a possible candidate against SARS-CoV-2 infection and in the pathogenesis of COVID-19. Nat Prod Commun. 2021, **16**(12):1–16. https://doi.org/10.1177/1934578X211066723

55. Adem Ş, Eyupoglu V, Sarfraz I, Rasul A, Zahoor AF, Ali M, Abdalla M, Ibrahim IM, Elfiky AA. Caffeic acid derivatives (CAFDs) as inhibitors of SARS-CoV-2: CAFDs-based functional foods as a potential alternative approach to combat COVID-19. Phytomedicine. 2021, **85**:153310.

56. Bahun M, Jukić M, Oblak D, Kranjc L, Bajc G, Butala M, Bozovičar K, Bratkovič T, Podlipnik Č, Ulrih NP. Inhibition of the SARS-CoV-2 3CLpro main protease by plant polyphenols. Food Chem. 2022, **373**:131594.

57. Ngwe Tun MM, Luvai E, Nwe KM, Toume K, Mizukami S, Hirayama K, Komatsu K, Morita K. Anti-SARS-CoV-2 activity of various PET-bottled Japanese green teas and tea compounds in vitro. Arch Virol. 2022, **167**(7):1547–1557.

58. Owis AI, El-Hawary MS, El Amir D, Aly OM, Abdelmohsen UR, Kamel MS. Molecular docking reveals the potential of *Salvadora persica* flavonoids to inhibit COVID-19 virus main protease. RSC Adv. 2020, **10**(33):19570–19575.

59. Shahhamzehei N, Abdelfatah S, Efferth T. In silico and in vitro identification of pan-coronaviral main protease inhibitors from a large natural product library. Pharmaceuticals (Basel). 2022, **15**(3):308.
60. Jo S, Kim S, Shin DH, Kim M-S. Inhibition of SARS-CoV 3CL protease by flavonoids. J Enzyme Inhib Med Chem. 2020, **35**(1):145–151.
61. Choudhry N, Zhao X, Xu D, Zanin M, Chen W, Yang Z, Chen J. Chinese therapeutic strategy for fighting COVID-19 and potential small-molecule inhibitors against severe acute respiratory syndrome coronavirus 2 (SARS-CoV-2). J Med Chem. 2020, **63**(22):13205–13227.
62. Ghosh R, Chakraborty A, Biswas A, Chowdhuri S. Evaluation of green tea polyphenols as novel corona virus (SARS CoV-2) main protease (Mpro) inhibitors–an in silico docking and molecular dynamics simulation study. J Biomol Struct Dyn. 2021, **39**(12):4362–4374.
63. Antonopoulou I, Sapountzaki E, Rova U, Christakopoulos P. Inhibition of the main protease of SARS-CoV-2 (Mpro) by repurposing/designing drug-like substances and utilizing nature's toolbox of bioactive compounds. Comput Struct Biotechnol J. 2022, **20**:1306–1344.
64. Wang S-C, Chen Y, Wang Y-C, Wang W-J, Yang C-S, Tsai C-L, Hou M-H, Chen H-F, Shen Y-C, Hung M-C. Tannic acid suppresses SARS-CoV-2 as a dual inhibitor of the viral main protease and the cellular TMPRSS2 protease. Am J Cancer Res. 2020, **10**(12):4538.
65. Bachar SC, Mazumder K, Bachar R, Aktar A, Al Mahtab M. A review of medicinal plants with antiviral activity available in Bangladesh and mechanistic insight into their bioactive metabolites on SARS-CoV-2, HIV and HBV. Front Pharmacol. 2021, **12**:732891.
66. Phu HT, Thuan DT, Nguyen TH, Posadino AM, Eid AH, Pintus G. Herbal medicine for slowing aging and aging-associated conditions: efficacy, mechanisms and safety. Curr Vasc Pharmacol. 2020, **18**(4):369–393.
67. Issa SS, Sokornova SV, Zhidkin RR, Matveeva TV. The main protease of SARS-CoV-2 as a target for phytochemicals against coronavirus. Plants (Basel). 2022, **11**(14):1862.

7 Molecular Mechanisms of Spike Glycoprotein Inhibitors from Traditional Herbal Medicines

*Sucheta Mandal, Anindya Sundar Ray,
Ankita Chakraborty, and Suprabhat Mukherjee*

7.1 INTRODUCTION

By the end of 2019, the world was dealt a lethal blow in the form of a novel strain of coronavirus, severe acute respiratory syndrome-coronavirus-2 (SARS-CoV-2), giving rise to the infamous Coronavirus Disease 2019 (COVID-19) pandemic. On 31 December 2019, Chinese health officials informed the World Health Organization (WHO) that a cluster of pneumonia patients with unknown etiology had been admitted to a hospital in Wuhan after a common exposure to the Wuhan wholesale seafood market where live animals are sold. On 11 January 2020, the first case of casualty was reported from this outbreak. After one month, on 11 February 2020, the WHO named this outbreak COVID-19, and subsequently on 11 March 2020, the COVID-19 outbreak had to be announced as a pandemic, pertaining to the exponentially worsening situation worldwide. According to the WHO, as of 6:59 pm CEST on 6 December 2023, 772,138,818 confirmed cases and 6,985,964 deaths worldwide were reported. Despite these increasing numbers, preventive vaccines as a direct treatment remained elusive. The infectious agent behind the COVID-19 pandemic was the SARS-CoV-2 virus, a single positive-sense-stranded RNA virus, which is a member of the coronavirus family. Historical evidence shows coronaviruses as detrimental to humans with two breakthroughs, i.e. severe acute respiratory syndrome (SARS) and Middle East respiratory syndrome (MERS) caused by SARS-CoV and MERS-CoV, respectively (Table 7.1). The evolutionary tree emphasizes the close relationship of SARS-CoV-2 with the group of SARS- coronaviruses. Out of seven coronaviruses, only SARS-CoV, MERS-CoV, and SARS-CoV-2 have caused fatal zoonotic diseases in humans during the last 18 years; whereas HKU1, NL63, OC43, and 229E are associated with mild symptoms. The virus is responsible for respiratory tract infections ranging from mild (cough-cold) to lethal (pneumonia, lymphopenia, exhausted lymphocytes, and cytokine storm) symptoms.

A very limited number of medicaments like the conventional hydroxychloroquine, remdesivir, etc., were initially available to medical practitioners while attempting to alleviate the symptoms of SARS-CoV-2 infection. Hydroxychloroquine, along with azithromycin and antiretroviral drugs like remdesivir, EIDD-2801, or favipiravir, had been used to reduce the pathogenicity of coronavirus by inhibiting the replication (Caly et al., 2020). Clinical surveys suggest that remdesivir and hydroxychloroquine have adverse effects on the liver, kidney, and heart (Costanzo et al., 2020). Paxlovid, an antiviral oral medication combining two generic drugs, nirmatrelvir and ritonavir, can cause side effects like impaired sense of taste (for example, a metallic taste in the mouth) and diarrhea (Hashemian et al., 2023) On the other hand, antiviral drugs are responsible for anemia, swelling, diarrhea, nausea, hypotension, constipation, and rashes. To overcome such drawbacks, various vaccines have been designed by targeting the morphological protein subunit or the viral RNA. A few companies used inactivated or recombinant viruses for vaccine development. A double-blind study

DOI: 10.1201/9781003452621-7

TABLE 7.1

Comparative Analysis of Various Features of SARS, MERS, and SARS-CoV-2

Features	SARS	MERS	SARS-CoV-2	References
First occurrence	Nov. 16th, 2002 in Foshan, Guangdong, South China	June 2012, in Jeddah, Saudi Arabia	Dec. 07th, 2019 in Wuhan, Hubei, Central China	Roychoudhury et al., 2020
Pathogen	SARS-CoV	MERS-CoV	SARS-CoV-2	
Possible natural reservoir	Bat	Bat	Bat	
Possible intermediate host	Civet cat	Camel	Pangolin/Snake (mostly cited)	
Nucleic acid	RNA	RNA	RNA	
Genome size	27.9 kb	30.1 kb	29.9 kb	
Pathogen linage	B-coronavirus of B lineage	B-coronavirus of C lineage	B-coronavirus of B lineage	
Symptoms	Fever, malaise, myalgia, headache, diarrhea, shivering, cough, and shortness of breath	Cough, fever, and shortness of breath	Fever, cough, and shortness of breath; Gastrointestinal symptoms, including diarrhea	
Key receptor	Angiotensin converting enzyme-2 (ACE2)	Dipeptidyl peptidase 4 (DPP-4)	Angiotensin converting enzyme-2 (ACE2)	
No. of countries infected	17	21	231 (as of 13 September, 2023, 1:28 pm IST)	
Total affected population	8908	2500	772138818 (as of 6 December 2023, 6:59 pm IST)	
Male: Female sex ratio	1.67:1 (as of 11 February 2020)	3·3:1	2·7:1	
Reproductive number (R_0)	1.4–5.5	<1	2.2–2.6	

attempted to observe and analyze the efficacy of a vaccine initiated from mid-2020 to early 2021. COVID vaccines came to market with different trade names viz. Covovax, Nuvaxovid, Spikevax, Comirnaty, Convidecia, Covishield, Covaxin, etc. In India, the vaccines "Covishield" and "Covaxin" were administered in people, but both of them exerted certain side effects like headache, fatigue, myalgia, malaise, pyrexia, chills, arthralgia and headache, fever, body ache, abdominal pain, nausea, vomiting, dizziness, giddiness, tremor, sweating, cold, cough, and injection-site swelling (Mittal et al., 2022). India's Ministry of AYUSH provided some guidelines for COVID-19 treatment, and complete recovery of COVID-19 patients was reported. The interdisciplinary Ayush research and development task force designed and formulated a polyherbal mixture named AYUSH-64 using Yashtimadhu, Ashwagandha, Guduchi, and Pippali. Plants like *Allium sativum, Azadirachta indica,* and *Zingiber officinale* show antiviral effect against adenovirus, *Alloherpesviridae, Arteriviridae, Coronaviridae, Herpesviridae, Poxviridae, Sedoreoviridae, Retroviridae, Togaviridae,* and *Papillomaviridae* (Gbenou et al., 2023). Formulation using these plants is still a novel and unexplored area of research. In this current review, we will explore the use of several herbal medicinal plants and their bioactive compound for treating COVID-19 without any harmful effects.

7.2 SARS-CoV-2 AND COVID-19

The COVID-19 pandemic, caused by the SARS-CoV-2 virus of the *Coronaviridae* family, has taken a extraordinary toll on mankind through innumerable casualties, associated co-morbidities of victims, and socio-economic losses. Wuhan, in the province of Hubei in central China, was identified as the epicenter of this pandemic in 2019 (Das et al., 2022; Patra et al., 2021; Choudhury et al., 2022; Sharma et al., 2020). From that point, about 225 countries, which is almost all of the countries on Earth, were afflicted with the disease, bringing the population down by millions (Choudhury and Mukherjee, 2020).

Phylogenetic analyses reveal this virus to be categorized under the beta-coronavirus family of the 2B group, under the family *Coronaviridae* and the order Nidovirales (Choudhury and Mukherjee, 2020). These taxa also house the names SARS-CoV and MERS-CoV, which had previously caused major damage in Foshan, Guangdong, south China, and Jeddah, Saudi Arabia, respectively (Roychoudhury et al., 2020). SARS-CoV-2 has been reported to be the more communicable member of this family (Das B et al., 2022).

Similar to the other flu viruses, it has an RNA genome encapsulated in a protein sheath, termed a capsid, riddled with elements like the viral spike-glycoprotein, membrane proteins, and envelope surface glycoproteins (Choudhury and Mukherjee, 2020; Mousavizadeh and Ghasemi, 2020). The genome of COVID-19 is an unsegmented, positive-sense, single-stranded RNA of 30 kilobase (kb) (Pal et al., 2020). It codes for four different structural proteins: membrane glycoprotein (M), nucleo-capsid (N), envelope (E), and spike protein (S) (Caputo et al., 2022). Also, some nonstructural proteins (NSPs) such as main protease (Mpro), 3CLpro, PLpro, RNA-dependent RNA polymerase (RdRp), helicase, and nine accessory proteins or open-reading frames (Orfs), namely Orf3a, Orf3b, Orf6, Orf7a, Orf7b, Orf8, Orf9b, Orf9c, and Orf10 are encoded in the genome (Li and De Clerq, 2020; Das B et al., 2022). The genome map of SARS-CoV-2 is depicted in Figure 7.1a.

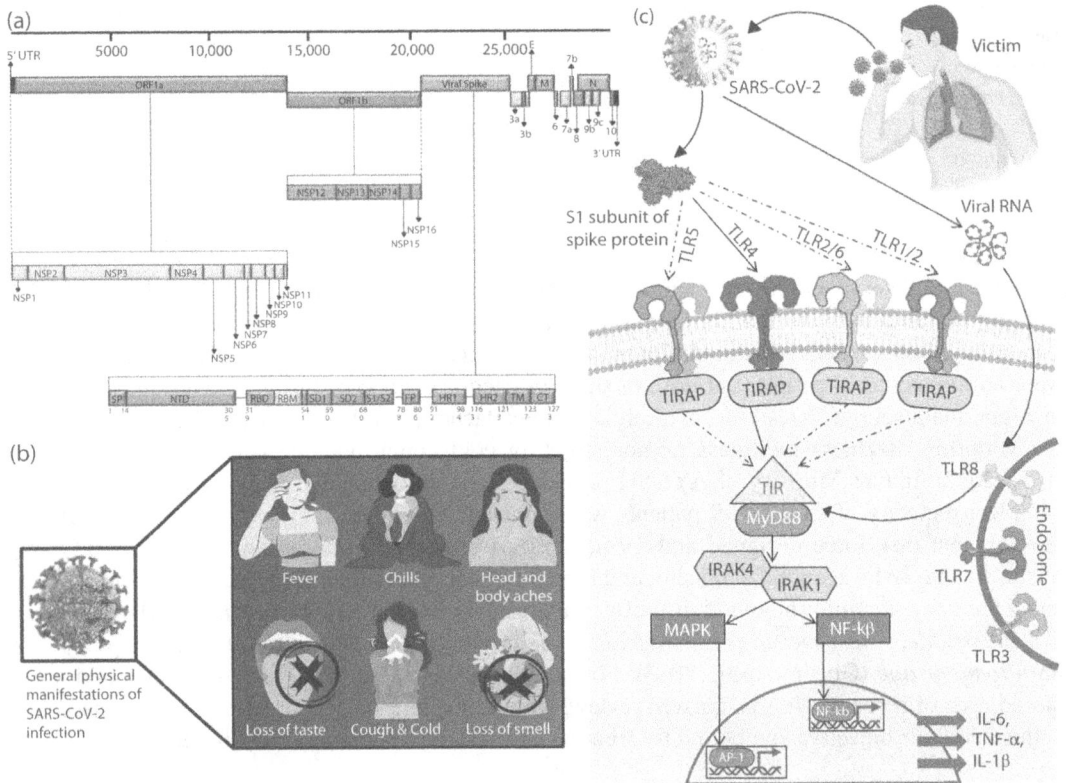

FIGURE 7.1 (a) SARS-CoV-2 genome map elaborating the different regions throughout the genome encoding different proteins. (b) Basic symptoms experienced by a SARS-CoV-2 victim. (c) Overview of the effect of the interaction of viral S-protein subunits with the host toll-like receptors (TLRs) and their consequences.

Typical symptoms of this disease closely resemble the physical manifestations of the common flu, like fever, head and body aches, cough, and chills, although asymptomatic cases have also been widely recorded. Specific symptoms exclusive to this infection include loss of smell and taste (Figure 7.1b) (Borch et al., 2022).

Regarding its pathophysiology, researchers have revealed that SARS-CoV-2 spreads through respiratory droplets mainly, which harness the airway for entry into the host (Choudhury and Mukherjee, 2020). The successful binding of viral spike glycoprotein with the human angiotensin-converting enzyme 2 (ACE2) (present in the human lung cells) is crucially decisive in the initiation of active infection (Wu et al., 2022). ACE2 is an aminopeptidase, and its receptor has been recognized as the prime participant in mediating the entry of the viral RNA into the host cells, along with some other cell surface proteins (Karakus et al., 2020). Post-priming, the spike protein needs to be cleaved at the S1/S2 cleavage site, into its S1 and S2 subunits, and this is duly performed by the serine protease TMPRSS2 and another endosomal protein named Cathepsin L (Bestle et al., 2020; Roshanravan et al., 2020). A study by Hoffman et al. (2020) revealed TMPRSS2 to be more essential in the events of S protein priming as well as the entry of the virus into its host. This S1 subunit of the spike protein comprises an N-terminal domain (NTD), a receptor binding domain (RBD), and a couple of C-terminal domains (CTD) (Li et al., 2022). Viral entry turns out successful only when the S2 subunit binds to the membrane of the host cells taken into target. Again, when the virus particles are endocytosed by the host cell and they proceed into the host phagolysosome, the lysosomal protease named pro-protein convertase, or furin, takes the responsibility of cleaving the S2 subunit at the S2′ cleavage site and releases a fusion peptide, which is hydrophobic in nature, for fusing into the host cell membrane (Wu et al., 2020). Furin is actually a type I transmembrane protein, whose cleavage sites have been reported to be repetitive throughout the S protein of the SARS-CoV-2 virus, which eases the viral fusion in the host membrane on account of increasing the chances of fusion, which consequently enhances the infectivity of the virus (Wu et al., 2020).

Interestingly, the viral pathophysiology showcases high pro-inflammatory responses, evident from the substantially elevated levels of the major pro-inflammatory cytokines like TNF-α, IL-6 G-CSF, IP-10, MCP-1, and MIP-1A in the examined sera of COVID-19 victims (Huang et al., 2020). IL-6 was maximized in casualties compared to the survivors (Gong et al., 2020a,b). Clinical revelations also mention the infiltration of neutrophils in the lung alveoli (Kadhim and Abdullah, 2021). A noteworthy point is that the activation of toll-like receptors (TLRs) is responsible for IL-6 expression. All of these reports collectively indicate the overt involvement of the innate immune response of the host in combating the infection. Again, IL-6 and TNF-α are specifically the products of the TLR4 downstream signaling pathway (Mukherjee et al., 2016). Hence, it was later determined that the viral spike glycoprotein plays a role as a prominent ligand to the human TLR4, which is one of the classical pattern-recognition-receptors (PRRs) in the human body, primarily responsible for triggering of the NF-κβ pathway, resulting in the expression of pro-inflammatory cytokines like IL-6, TNF-α, IL-1β and also plays a role in signaling the activation of inflammasome (Figure 7.1c) (Mukherjee et al., 2016; Sohn et al., 2020). This phenomenon disrupts the inflammatory homeostasis of the host, and a consequent episode of hyper-inflammation with subsequent damage caused to the vital organs, proceeding to a likely death. However, the interaction of viral S protein with other TLRs like TLR1 and TLR6 has also been found (Khanmohammadi and Rezaei, 2021).

Regarding therapeutics, the unfortunate reports of the prevalence of occurring and recurring episodes of the viral infection point toward the fact that the unavailability of properly efficacious chemotherapeutic or immunotherapeutic intervention strategies against this formidable pathogen has lengthened its unwelcome interactions with humans, and the repurposed drugs recommended have only been able to ameliorate the physiological manifestations of the infection to some extent, coming with an array of side effects that turn out to be even more undesirable. Also, the arena of vaccine development needs time to conduct a series of trials and errors to ultimately determine a concluding unitary option (Chakraborty et al., 2023).

7.3 ANTI-SARS-CoV-2 CHEMOTHERAPEUTICS AND THEIR LIMITATIONS

The discovery of idoxuridine against the herpes virus in June 1963 was the gateway to antiviral drug development (De Clercq, 2004). Since then, more than 90 antiviral drugs with different functional groups have been developed as efficacious therapeutics against a series of viral diseases. The pandemic situation of SARS-CoV-2 was treated with hydroxychloroquine or chloroquine, an antimalarial drug, and considered a promising candidate for anti-COVID-19 therapeutics (Sun et al., 2020). Similarly, Caly et al. (2020) reported the use of ivermectin, a US Food and Drug Administration (FDA)-approved antiparasitic drug, for COVID-19 treatment. Angiotensin receptor blockers (ARBs), lopinavir/ritonavir, interferon-alpha (INFα), niclosamide, and arbidol are also considered as treatment options (Li et al., 2020; Lin et al., 2020; Sanchis-Gomar et al., 2020; Xu et al., 2020). Li and co-workers concluded from their study on "therapeutics to control coronavirus and influenza virus" that drugs like remdesivir, teicoplanin, and favipiravir are effective against SARS-CoV, MERS, and influenza (Li et al., 2019). A clinical survey reported patients were also advised to take oseltamivir, azithromycin, vitamin D, and zinc as supplements. Hydroxychloroquine and chloroquine inhibit SARS-CoV-2 by interfering with replication. A virus requires a low pH condition for replication, but both the drugs increase endosomal and lysosomal pH, resulting in low viral count (Fredericksen et al., 2002). Both of these drugs exert various side effects including cardiac problems like conduction problems and cardiomyopathy and neurological and psychiatric dysfunctions. Ferner and Aronson (2020) reported patients treated with hydroxychloroquine and chloroquine faced muscle weakness, diplopia, myasthenia gravis, dyskinesia, irritability, seizures, insomnia, psychosis, aggression, depression, and anxiety.

Remdesivir was used for SARS-CoV-2 treatments for its antiviral properties against single-stranded RNA viruses i.e. Lasagna virus, NIPA virus, Hendra virus, and coronavirus (Beigel et al., 2020). It is an adenosine analog that interferes with functional RNA-dependent RNA polymerase and prevents replication. The side effects most commonly encountered using remdesivir were respiratory failure and organ dysfunction. According to Yang et al. (2021), remdesivir was also responsible for lowering levels of albumin and quantities of erythrocytes and platelets. Gastrointestinal upset, inflammation, or damage to liver cells resulting in low blood pressure, nausea, vomiting, sweating, and chills was observed in patients who received remdesivir injection (Fan et al., 2020).

Azithromycin, an antibacterial drug that inhibits protein synthesis, was used against coronavirus, though COVID-19 is a form of acute respiratory viral disease. Clinical studies suggest that people affected with COVID-19 suffer from pneumonia in parallel, in which case azithromycin is effective. It has also been used for its immunomodulatory effects. Previously, it was used in a clinical trial in patients suffering from influenza, showing promising results (Kakeya et al., 2014). Side effects like headache, loss of appetite, nausea, abdominal pain, diarrhea, and vomiting were common for this drug. Some allergic reactions were also reported in some cases.

Ivermectin, a derivative of macrocyclic lactone with broad-spectrum antiparasitic activity showed *in vitro* activity against SARS-CoV-2 for its anti-inflammatory effect (Yan et al., 2011). Some of the adverse effects associated with ivermectin are diarrhea and nausea. Overdose of the drug brings consequent gastrointestinal problems, hypotension, and neurological problems like confusion, hallucination, decreased consciousness, coma, and, ultimately, death (Barac et al., 2022). Ivermectin interacts with other medications and acts as an anticoagulant in co-morbid patients.

Vitamin D was given to COVID-19 victims along with other drugs. This supplement was found to modulate the immune system (by reducing the synthesis of pro-inflammatory cytokines), improve the lung or respiratory system, and enhance muscle and heart function by maintaining tight junctions (Christakos et al., 2013). However, increased levels of vitamin D in serum were responsible for weakness, fatigue, headaches, loss of appetite, dry mouth, vomiting, and a metallic taste in the mouth of patients. People suffering from atherosclerosis experienced renal failure when supplemented with vitamin D. Zinc also helped in improving the immune apparatus of the body. McPherson et al. (2020) noted the antiviral effect of zinc against influenza and SARS-CoV-2 viruses. But again, prolonged consumption of zinc led to headache, nausea, and vomiting.

Dexamethasone is another drug, approved by the UK National Institutes of Health (NIH), the US National Institute of Health (NIH), the Infectious Diseases Society of America (IDSA), the European Medicines Agency (EMA), and the WHO, for treating SARS-CoV-2 victims in critical need for oxygen. The drug lowered the mortality rate but not without some persisting side effects like gastritis, vomiting, headache, dizziness, insomnia, restlessness, depression, acne, and irregular or absent menstrual cycle (Polderman et al., 2018).

Paxlovid, an antiviral oral medication developed by Pfizer, comprises two generic drugs, nirmatrelvir and ritonavir. It received full FDA approval for adults and emergency use authorization for adolescents and teenagers aged 12 and older who weigh at least 88 pounds. The medication is intended for individuals with mild-to-moderate COVID-19 infection who are at high risk for severe disease. The NIH prioritizes Paxlovid over other treatments for eligible patients.

The medication was authorized in December 2021 for adolescents aged 12 and older and received full approval for adults in May 2023. It is administered as three pills twice daily for 5 days, starting within 5 days of the onset of COVID-19 symptoms. Side effects of this drug include impaired sense of taste, diarrhea, increased blood pressure, and muscle aches. Two of the three pills in each dose contain nirmatrelvir, which prevents the SARS-CoV-2 virus from replicating. The third pill contains ritonavir, which boosts the levels of nirmatrelvir by inhibiting its metabolism in the liver, allowing it to work longer to combat the infection. The FDA's approval of Paxlovid for adults in 2023 was based on Pfizer's submitted scientific evidence, including a study showing an 86% reduction in the risk of COVID-19-related hospitalization or death in patients. According to the latest FDA statement and virology data, Paxlovid retains its activity against currently circulating Omicron variants (Hashemian et al., 2023; Niraj et al., 2022)

The antiviral drug amantadine prevents SARS-CoV-2 entry by blocking the viral purine channel and preventing the release of the virus in the cell cytoplasm (Aranda-Abreu et al., 2020). However, clinical study states that the drug is responsible for neurological, cardiovascular, gastrointestinal, and dermatological problems (Hosenbocus et al., 2013).

All of these drugs were used to try and control the pandemic, but none of them are direct remedies. Consumption of these medications per doctor suggestion provides relief for COVID-19-affected people, but simultaneously many may also die. Each of these drugs has shown its efficiency but they also possess side effects. The side effects of some conventional drugs are summarized in Table 7.2. Sometimes these side effects last for a short period and sometimes they are prolonged. That is

TABLE 7.2

Side Effects Caused by Different COVID-19 Vaccines Available on the Market

Name of Vaccines	Side Effects	References
OxfordAstraZeneca	Injection-site pain, redness/swelling at the injection site, dizziness, imbalance, fatigue, fever, headache, lethargy, nausea, cough, muscle pain, diarrhea, anxiety, insomnia	Aldali et al., 2023
Pfizer-BioNTech	Injection-site pain, redness/swelling at the injection site, fever, headache, lethargy, nausea, muscle pain, anxiety, allergic reaction, small fiber neuropathy, neuromyelitis optica, Bell's palsy, encephalopathy, intracerebral hemorrhage, ischemic stroke	Aldali et al., 2023
Moderna	Injection-site pain, fever, lethargy, nausea, cough, muscle pain, anxiety, herpes zoster, intracerebral hemorrhage, ischemic stroke, epilepsy	Aldali et al., 2023
Sinopharm	Injection-site pain, redness/swelling at the injection site, fever, headache, lethargy, nausea, cough, muscle pain, diarrhea, allergic reaction, multiple sclerosis relapse, neuromyelitis optica	Meo et al., 2023
Sinovac	Injection-site pain, redness/swelling at the injection site, fever, headache, nausea, Bell's palsy	Chohan et al., 2023

why these conventional drugs are not sufficient for combating SARS-CoV-2. Also, increasing drug resistance and adverse side effects of synthetic drugs increase the demand for novel approaches in developing safe antiviral drugs.

7.4 TRADITIONAL MEDICINES AS EFFECTIVE TREATMENT OPTIONS FOR COVID-19

To date, there is no unequivocal vaccine or explicit medication to forestall the novel coronavirus. Scientists are leaving no stone unturned to find an answer. In this situation, herbal therapy can give us hope. Research is still being continued on phyto-therapy, and different parts of traditionally accepted medicinal plants like roots, bark, fruits, flowers, and leaves are being explored for various phyto-compounds with antiviral activities and they are being yoked as novel approaches to combat COVID-19. Innumerable clinical trials have been performed to validate the assumptions regarding the efficacy of phyto-drugs against this disease (Figure 7.2).

"*Immune-boosting*" herbal capsules have been promoted in India. India's Ministry of AYUSH stated "Ayurveda's immunity boosting measures for self-care during the COVID-19 crisis," which incorporated the suggestion to take Chyavanprash, a formulation containing a range of wild plant ingredients, regularly. The Ministry of AYUSH has also received propositions, ideas, and recommendations from practitioners and institutions in the AYUSH Sector regarding possible solutions from naturopathy to restrain the spread of the pandemic.

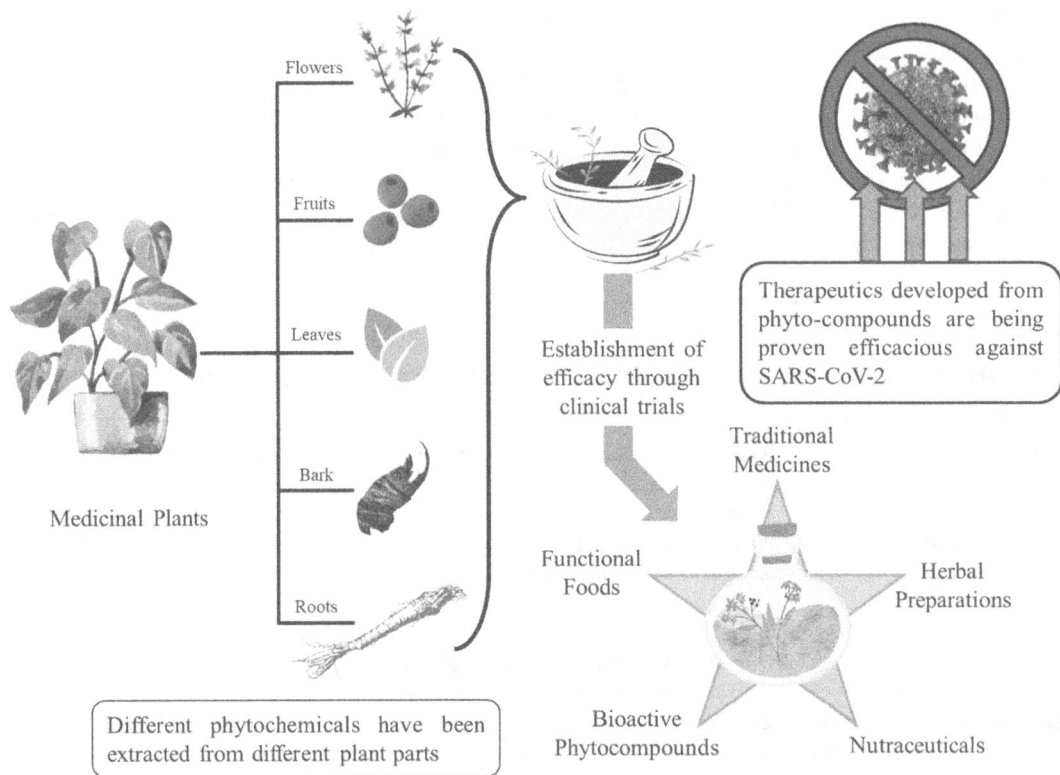

FIGURE 7.2 Researchers are using phyto-therapy against SARS-CoV-2 with different parts of traditionally popular medicinal plants possessing antiviral activities and establishing their anti-SARS-CoV-2 activity through various experimental approaches as well as clinical trials.

The antimalarial drug hydroxychloroquine came to the spotlight after former president of the United States Donald Trump touted it as a "game changer" in the fight against coronavirus, and 29 million doses of hydroxychloroquine were bought by the US from India to combat the pandemic situation.

Previously, in February 2020, the Chinese State Council announced that chloroquine phosphate could be considered as an option for treating COVID-19 patients (Gao et al., 2020). Both hydroxy-chloroquine and chloroquine phosphate are structural analogs of quinine originally extracted from the bark of cinchona trees. Curcumin, the secondary metabolite (a polyphenol) of turmeric (*Curcuma longa*) has the strongest interaction with the protease enzyme of SARS CoV-2 (Ibrahim et al., 2020). Balkrishna and his co-workers (2021) reported that withanone, a steroidal compound, isolated from Ashwagandha (*Withania somnifera*) inhibits the pathogen entry of COVID-19 by disrupting interactions between viral S-protein RBD and host ACE2 receptor. Different *in silico* studies have demonstrated the anti-coronavirus activities of several natural products including scutellarein, silvestrol, tryptanthrin, saikosaponin B2, quercetin, myricetin, caffeic acid, psoralidin, isobavachalcone, and lectins like griffithsin.

Yang et al. (2020) reported that *Astragalus membranaceus, Glycyrrhiza uralensis, Saposhnikoviae divaricata, Rhizoma Atractylodis Macrocephalae, Lonicera japonica* Flos, *Fructus forsythia, Platycodonis radix, Agastache rugosa,* and *Cyrtomium fortunei* are most commonly used Chinese herbs for treating COVID-19 in different provinces that covered northeast, north, central (including Wuhan), south, east, northwest, and southwest China. Systematic review and meta-analysis reports stated herbal drugs, such as *Huoxiang Zhengqi* capsules, *Lianhua qingwen* capsules and granules, and *Radix isatidis* granula, are clinically used for the treatment of novel coronavirus (Fang et al., 2022). Recent clinical studies have elucidated strong activity against upper respiratory tract infections and pharyngotonsillitis of *Andrographis paniculata*, an alternative herbal medicine in co-treatment of COVID-19 (Songvut et al., 2022). Further investigations and isolation of active principles are highly needed for anti-SARS-CoV-2 drug discovery.

Recently, the role of hyaluronan (HA) in acute respiratory distress syndrome (ARDS) and in COVID-19 infection has been brought to light. Doctors are now providing patients with medical-grade hyaluronidase to inhibit the production and accumulation of HA in the lungs.

In this regard, 4-Methylumbelliferone (4-MU), a coumarin derivative (extracted from different members of the family *Apiaceae* or *Umbelliferae*) with potential hyaluronidase activity has shown very effective and promising results. This organic compound can be used by the practitioner to help COVID-19 patients with breathing problems.

7.5   RECENT DEVELOPMENTS AND CHALLENGES

The increasing number of COVID-19 cases created constant pressure on scientists to develop an effective and safe vaccine. According to WHO on 30 March 2023, a total of 382 vaccine products were under clinical development (https://www.who.int/publications/m/item/draft-landscape-of-COVID-19-candidate-vaccines) with 189 and 199 numbers of vaccines in clinical and pre-clinical development, respectively. Vaccines developed under clinical trial are mostly replicating viral vector vaccines, protein subunit vaccines, RNA and DNA vaccines, and virus-like partial vaccines. A vaccine formulated using the inactivated form of the whole virus is the most traditional method of vaccine development. PiCoVacc, BBIBP-CorV, ChAdOx1nCoV19, Ad26.COV2.S, NVX-CoV2373, RBD monomer, S-trimer, GX-19, BNT162b2, ARCoV, and mRNA-1273 are some inactivated vaccines developed by Sinovac, Beijing Institute of Biological Products Ltd., AstraZeneca, Janssen, Novavax, West China Hospital, Sichuan University, Clover Biopharmaceuticals, Genexine Consortium, Pfizer, Walvax Biotech, and Moderna, respectively, using inactivated aluminum hydroxide, DNA, mRNA, protein subunit or non-replicating viral vector (Li et al., 2021).

A cross-sectional study was performed on people in Bangladesh after receiving the COVID-19 vaccine manufactured by Sinopharm, Oxford/AstraZeneca, Moderna, and Pfizer-BioNTech.

Eighty percent of people who received the Pfizer-BioNTech vaccine showed the highest prevalence of side effects followed by Moderna (76.63%), OxfordAstraZeneca (67.72%), Sinopharm (28.23%) and Sinovac (21.05%) vaccines (Table 7.2). A majority of them suffered from injection-site pain (96.51%), fever (94.19%), headache (81.40%), and redness/swelling (Mohsin et al., 2022). Data from the Vaccine Adverse Event Reporting System (VAERS) stated that COVID-19 vaccines have local and systemic neurological complications that occur in different age groups depending on age, sex, and pre-existing immunity. Finsterer (2022) reported that some severe cases appear within 1 day to 1 month after injection. Application of the second dose of the vaccine became more adverse than the first dose (Al Khames Aga et al., 2021). Weakness, numbness, headache, dizziness, imbalance, fatigue, muscle spasms, joint pain, restless leg syndrome, tremors, tinnitus, and herpes zoster are some common side effects of different COVID-19 vaccines. Vaccines based on adenovirus showed the most dangerous neurological complication, cerebral venous sinus thrombosis, in women of childbearing age (Finsterer, 2022). Inactivated virus-based vaccines temporarily increase cytokines that act on blood vessels, muscles, and other tissues which show flu-like syndrome (Zhang et al., 2021).

mRNA-based vaccines are reported for more neurological complications like transverse myelitis, acute diffuse encephalomyelitis (ADEM), Bell's palsy, Guillain-Barré syndrome (GBS), encephalopathy seizures, etc. After the administration of such a vaccine, it releases spike protein, which causes inflammation and, thus, increases glial cell activity and blood-brain barrier permeability. The following events are responsible for the majority of neurological problems (Assiri et al., 2022). Some of the COVID-19 vaccines affect cranial and peripheral nerves, which result in Bell's palsy (Ozonoff et al., 2021), abducens nerve palsy, impaired vision (Santovito and Pinna, 2021), olfactory (Keir et al., 2021), hearing, GBS (Dyer, 2021), small fiber neuropathy (Waheed et al., 2021), Parsonage-Turner syndrome, and herpes zoster. Žorić et al. (2021) reported that Pfizer and AstraZeneca vaccines are responsible for optic nerve inflammation and vision disorders in middle-aged people. Pfizer vaccination leads to facial nerve palsy, vestibulocochlear neuropathy, and glossopharyngeal nerve neuropathy, which results in Ramsay Hunt Syndrome. In addition, some cases were reported with skin blisters in the ear area, which point to Bell's palsy (Woo et al., 2022).

According to the literature review regarding the developed vaccines on the market, it was evident that none of the vaccines can be declared as completely safe or efficacious. But among them, some showed severe side effects for prolonged periods and some for shorter periods and were self-treatable. People with a history of immune-related disease are the most susceptible. Vaccines responsible for cerebral venous sinus thrombosis, transverse myelitis, Bell's palsy, and Ramsay Hunt Syndrome are of more concern. The search for a completely safe drug is still ongoing. Several literature reviews and the history of drug development suggest different medicinal plants used in traditional medicine can provide us with a safe drug for developing a vaccine against coronavirus. It has already been reported by Pan and co-workers (2020) that quercetin was obtained from *Bupleurum chinense* DC. *Cyathula officinalis* (Kuan) can impair the interaction between S protein and ACE2. Molecular docking and network pharmacology studies show isorhamnetin can bind to S protein and ACE2 and inhibit the SARS-CoV-2 spike pseudo-typed virus from entering the ACE2[h]cell (Zhan et al., 2021). Rutin and oroxylin A obtained from *Bupleurum chinense* DC., and *Scutellaria baicalensis* Georgi, respectively, binds to ACE2, suppresses entry of the SARS-CoV-spiked pseudo-typed virus into ACE2 cells amd, thus, inhibits LC3-mediated autophagy of ACE2 cells (Gao et al., 2021; Joshi, 2021). Flavonoids (like irisolidone, luteolin, hesperidin, chrysin, anhydrosafflor yellow B, glabridin, myricitrin, euchrenone, epigallocatechin-3-gallate), terpenoids (like glycyrrhizic acid, uncaric acid, ursolic acid, demethylzeylasteral, maslinic acid, obacunone, andrographolide, atractylenolide III), phenols (like salvianolic acid A, B, C; kobophenol A, neochlorogenic acid, resveratrol), alkaloids (ephedrine, pseudoephedrine, methylephedrine, cryptospirolepine, cryptoquindoline, bicuculline, solanine), and quinones (thymoquinone, emodin, rhein) are reported to bind to the S protein and impair the interaction between S protein and ACE2, which can effectively inhibit the entry of coronavirus (Alazmi and Motwalli, 2020; Basu et al., 2020; Chen et al., 2021; Gangadevi et al., 2021;

Gao et al., 2020; Hu et al., 2021, Niu et al., 2020; Teli et al., 2021; Vardhan and Sahoo, 2020; Xing et al., 2020). Besides this extensive list of bioactive compounds, some glycosides, coumarin, and steroids are also reported for their binding nature to ACE2 to impair the interaction between S-protein and ACE2.

7.6 CONCLUSION AND FUTURE REMARKS

COVID-19 has been the biggest setback to human health in the recent past. Mortality, comorbidity, and socio-economic losses are the major effects humans have suffered so far. Despite the mass campaign of vaccination and administration of antivirals, new SARS-CoV-2 variants are emerge almost every 6 months. Notably, new strains are less susceptible to vaccines and antivirals. Therefore, a search for new antivirals is a major research area. In this regard, traditional medicines, i.e. natural products with antiviral activities, are the hope. As in many diseases, phyto-compounds with their efficacy and non-toxic nature, have been successfully implicated for amelioration as well as complete cure of the ailments. In this chapter, we reviewed the significant findings in projecting traditional herbal medicines in treating and combating SARS-CoV-2 infection, especially in the context of their molecular mechanisms of action at the physiological level. Most of the natural traditional therapeutics have a satisfactory level of efficacy against the virus as well as the ability to improve immunity by sensitizing the activities of the innate immune sensors in recipients. On the other hand, many herbal medicines possess an anti-inflammatory action that is useful in reducing overt immunopathology in severe COVID-19 patients.

ACKNOWLEDGMENTS

We acknowledge all the scientists, doctors, and health professionals who have contributed to combating COVID-19. The use of BioRender.com for drawing the illustrations is gratefully acknowledged.

REFERENCES

Aldali J, Meo SA, Al-Khlaiwi T. Adverse effects of Pfizer (BioNTech), Oxford-AstraZeneca (ChAdOx1 CoV-19), and Moderna COVID-19 vaccines among the adult population in Saudi Arabia: A cross-sectional study. Vaccines (Basel). 2023;11(2):231. doi:10.3390/vaccines11020231

Al Khames Aga QA, Alkhaffaf WH, Hatem TH, et al. Safety of COVID-19 vaccines. J Med Virol. 2021;93(12):6588–6594. doi:10.1002/jmv.27214

Alazmi M, Motwalli O. Molecular basis for drug repurposing to study the interface of the S protein in SARS-CoV-2 and human ACE2 through docking, characterization, and molecular dynamics for natural drug candidates. J Mol Model. 2020;26(12):338. doi:10.1007/s00894-020-04599-8.

Aranda-Abreu GE, Aranda-Martínez JD, Araújo R, Hernández-Aguilar ME, Herrera-Covarrubias D, Rojas-Durán F. Observational study of people infected with SARS-Cov-2, treated with amantadine. Pharmacol Rep. 2020;72(6):1538–1541. doi:10.1007/s43440-020-00168-1.

Assiri SA, Althaqafi RMM, Alswat K, et al. Post COVID-19 vaccination-associated neurological complications. Neuropsychiatr Dis Treat. 2022;18:137–154. doi:10.2147/NDT.S343438.

Balkrishna A, Pokhrel S, Singh H, et al. Withanone from *Withania somnifera* attenuates SARS-CoV-2 RBD and host ACE2 interactions to rescue spike protein induced pathologies in humanized zebrafish model. Drug Des Devel Ther. 2021;15:1111–1133. doi:10.2147/DDDT.S292805.

Barac A, Bartoletti M, Azap O, et al. Inappropriate use of ivermectin during the COVID-19 pandemic: Primum non nocere!. Clin Microbiol Infect. 2022;28(7):908–910. doi:10.1016/j.cmi.2022.03.022.

Basu A, Sarkar A, Maulik U. Molecular docking study of potential phytochemicals and their effects on the complex of SARS-CoV2 spike protein and human ACE2. Sci Rep. 2020;10(1):17699. doi:10.1038/s41598-020-74715-4.

Beigel JH, Tomashek KM, Dodd LE, et al. Remdesivir for the treatment of Covid-19 - Final report. N Engl J Med. 2020;383(19):1813–1826. doi:10.1056/NEJMoa2007764.

Bestle D, Heindl MR, Limburg H, et al. TMPRSS2 and furin are both essential for proteolytic activation of SARS-CoV-2 in human airway cells. Life Sci Alliance. 2020;3(9):e202000786. doi:10.26508/lsa.202000786.

Borch L, Holm M, Knudsen M, Ellermann-Eriksen S, Hagstroem S. Long COVID symptoms and duration in SARS-CoV-2 positive children – A nationwide cohort study. Eur J Pediatr. 2022;181(4):1597–1607. doi:10.1007/s00431-021-04345-z.

Caly L, Druce JD, Catton MG, Jans DA, Wagstaff KM. The FDA-approved drug ivermectin inhibits the replication of SARS-CoV-2 in vitro. Antiviral Res. 2020;178:104787. doi:10.1016/j.antiviral.2020.104787.

Caputo E, Mandrich L. Structural and phylogenetic analysis of SARS-CoV-2 spike glycoprotein from the most widespread variants. Life (Basel). 2022;12(8):1245. doi:10.3390/life12081245.

Chakraborty A, Bayry J, Mukherjee S. (2023). Immunoinformatics Approaches in Designing Vaccines Against COVID-19. In: Reche, P.A. (eds) Computational Vaccine Design. Methods in Molecular Biology (vol 2673). Humana, New York, NY. https://doi.org/10.1007/978-1-0716-3239-0_29

Chen X, Wu Y, Chen C, et al. Identifying potential anti-COVID-19 pharmacological components of traditional Chinese medicine Lianhuaqingwen capsule based on human exposure and ACE2 biochromatography screening. Acta Pharm Sin B. 2021;11(1):222–236. doi:10.1016/j.apsb.2020.10.002.

Chohan HK, Jamal A, Mubeen M, Khan MU, Junaid M, Chohan MK, Imran A, Aslam A, Anwar A, Hashmi AA. The common systemic and local adverse effects of the sinovac COVID-19 vaccine: An observational study from Pakistan. Cureus, 2023;15(5):e38564. https://doi.org/10.7759/cureus.38564.

Choudhury A, Mukherjee S. In silico studies on the comparative characterization of the interactions of SARS-CoV-2 spike glycoprotein with ACE-2 receptor homologs and human TLRs. J Med Virol. 2020;92(10):2105–2113. doi:10.1002/jmv.25987

Choudhury A, Sen Gupta PS, Panda SK, Rana MK, Mukherjee S. Designing AbhiSCoVac - A single potential vaccine for all 'corona culprits': Immunoinformatics and immune simulation approaches. J Mol Liq. 2022;351:118633. doi:10.1016/j.molliq.2022.118633

Christakos S, Hewison M, Gardner DG, et al. Vitamin D: Beyond bone. Ann N Y Acad Sci. 2013;1287(1):45–58. doi:10.1111/nyas.12129.

Costanzo M, De Giglio MAR, Roviello GN. SARS-CoV-2: Recent reports on antiviral therapies based on Lopinavir/Ritonavir, Darunavir/Umifenovir, hydroxychloroquine, remdesivir, favipiravir and other drugs for the treatment of the new coronavirus. Curr Med Chem. 2020;27(27):4536–4541. doi:10.2174/0929867327666200416131117.

Das NC, Chakraborty P, Bayry J, Mukherjee S. In silico analyses on the comparative potential of therapeutic human monoclonal antibodies against newly emerged SARS-CoV-2 variants bearing mutant spike protein. Front Immunol. 2022;12:782506. Published 2022 Jan 10. doi:10.3389/fimmu.2021.782506

Das, B. S., Das, N. C., Swain, S. S., Mukherjee, S., & Bhattacharya, D. Andrographolide Induces anti-SARS-CoV-2 Response Through host-directed mechanism: An in silico study. Future Virology, 2022; 17(9): 651–673. https://doi.org/10.2217/fvl-2021-0171

De Clercq E. Antiviral drugs in current clinical use. J Clin Virol. 2004;30(2):115–133. doi:10.1016/j.jcv.2004.02.009.

Dyer O. Covid-19: Regulators warn that rare Guillain-Barré cases may link to J&J and AstraZeneca vaccines. BMJ. 2021;374:n1786. doi:10.1136/bmj.n1786.

Fan Q, Zhang B, Ma J, Zhang S. Safety profile of the antiviral drug remdesivir: An update. Biomed Pharmacother. 2020;130:110532. doi:10.1016/j.biopha.2020.110532.

Fang L, Zhan Y, Qu L, Sheng H. The clinical efficacy and safety research of Lianhuaqingwen (LHQW) in treatment of COVID-19: A systematic review and meta-analysis. Pharmacol. Res.-Mod. Chin. Med. 2022;3:100092.

Ferner RE, Aronson JK. Chloroquine and hydroxychloroquine in Covid-19. BMJ. 2020;369:m1432. doi:10.1136/bmj.m1432.

Finsterer J. Neurological side effects of SARS-CoV-2 vaccinations. Acta Neurol Scand. 2022;145(1):5–9. doi:10.1111/ane.13550.

Fredericksen BL, Wei BL, Yao J, Luo T, Garcia JV. Inhibition of endosomal/lysosomal degradation increases the infectivity of human immunodeficiency virus. J Virol. 2002;76(22):11440–11446. doi:10.1128/jvi.76.22.11440-11446.2002.

Gangadevi S, Badavath VN, Thakur A, et al. Kobophenol A inhibits binding of host ACE2 receptor with spike RBD domain of SARS-CoV-2, a lead compound for blocking COVID-19. J Phys Chem Lett. 2021;12(7):1793–1802. doi:10.1021/acs.jpclett.0c03119.

Gao J, Ding Y, Wang Y, Liang P, Zhang L, Liu R. Oroxylin A is a severe acute respiratory syndrome coronavirus 2-spiked pseudotyped virus blocker obtained from Radix Scutellariae using angiotensin-converting enzyme II/cell membrane chromatography. Phytother Res. 2021;35(6):3194–3204. doi:10.1002/ptr.7030.

Gao J, Tian Z, Yang X. Breakthrough: Chloroquine phosphate has shown apparent efficacy in treatment of COVID-19 associated pneumonia in clinical studies. Biosci Trends. 2020;14(1):72–73. doi:10.5582/bst.2020.01047.

Gao LQ, Xu J, Chen SD. In silico screening of potential Chinese herbal medicine against COVID-19 by targeting SARS-CoV-2 3CLpro and angiotensin converting enzyme II using molecular docking. Chin J Integr Med. 2020;26(7):527–532. doi:10.1007/s11655-020-3476-x.

Gbenou JD, Toklo PM, Assogba MF, Ahomadegbe MA, Ahoton D, Davo A, ... & Yayi EC. Traditional medicinal plants used in the treatment of viral diseases. Adv Tradit Med. 2023;24:99–131.

Gong J, Dong H, Xia QS, et al. Correlation analysis between disease severity and inflammation-related parameters in patients with COVID-19: A retrospective study. BMC Infect Dis. 2020a;20(1):963. Published 2020 Dec 21. doi:10.1186/s12879-020-05681-5

Gong J, Dong H, Xia SQ et al. Correlation analysis between disease severity and inflammation-related parameters in patients with COVID-19 pneumonia. MedRxiv. 2020b. https://doi.org/10.1101/2020.02.25.20025643 (Epub ahead of print).

Hashemian SMR, Sheida A, Taghizadieh M, et al. Paxlovid (Nirmatrelvir/Ritonavir): A new approach to Covid-19 therapy?. Biomed Pharmacother. 2023;162:114367. doi:10.1016/j.biopha.2023.114367.

Hoffmann M, Kleine-Weber H, Schroeder S, et al. SARS-CoV-2 cell entry depends on ACE2 and TMPRSS2 and is blocked by a clinically proven protease inhibitor. Cell. 2020;181(2):271–280.e278. doi:10.1016/j.cell.2020.02.052.

Hosenbocus S, Chahal R. Amantadine: A review of use in child and adolescent psychiatry. J Can Acad Child Adolesc Psychiatry. 2013;22(1):55–60.

Huang C, Wang Y, Li X, et al. Clinical features of patients infected with 2019 novel coronavirus in Wuhan, China [published correction appears in Lancet. 2020 Feb 15;395(10223):496. doi: 10.1016/S0140-6736(20)30252-X]. Lancet. 2020;395(10223):497–506. doi:10.1016/S0140-6736(20)30183-5

Hu S, Wang J, Zhang Y, et al. Three salvianolic acids inhibit 2019-nCoV spike pseudovirus viropexis by binding to both its RBD and receptor ACE2. J Med Virol. 2021;93(5):3143–3151. doi:10.1002/jmv.26874.

Ibrahim J, Kabiru AY, Abdulrasheed-Adeleke T, Lawal B, Adewuyi AH. Antioxidant and hepatoprotective potentials of curcuminoid isolates from turmeric (Curcuma longa) rhizome on CCl4-induced hepatic damage in Wistar rats. Journal of Taibah University for Science. 2020;14(1):908–915.

Joshi RS, Jagdale SS, Bansode SB, et al. Discovery of potential multi-target-directed ligands by targeting host-specific SARS-CoV-2 structurally conserved main protease. J Biomol Struct Dyn. 2021;39(9):3099–3114. doi:10.1080/07391102.2020.1760137.

Kadhim AS, Abdullah YJ. Serum levels of interleukin-6, ferritin, C-reactive protein, lactate dehydrogenase, D-dimer, and count of lymphocytes and neutrophils in COVID-19 patients: Its correlation to the disease severity. Biomed Biotechnol Res J. 2021;5(1):69.

Kakeya H, Seki M, Izumikawa K, et al. Efficacy of combination therapy with oseltamivir phosphate and azithromycin for influenza: A multicenter, open-label, randomized study. PLoS One. 2014;9(3):e91293. doi:10.1371/journal.pone.0091293.

Karakus U, Pohl MO, Stertz S. Breaking the convention: Sialoglycan variants, coreceptors, and alternative receptors for influenza A virus entry. J Virol. 2020;94(4):e01357–19. doi:10.1128/JVI.01357-19.

Khanmohammadi S, Rezaei N. Role of toll-like receptors in the pathogenesis of COVID-19. J Med Virol. 2021;93(5):2735–2739. doi:10.1002/jmv.26826.

Li G, De Clercq E. Therapeutic options for the 2019 novel coronavirus (2019-nCoV). Nat Rev Drug Discov. 2020;19(3):149–150. doi:10.1038/d41573-020-00016-0.

Li Q, Wang Y, Sun Q, et al. Immune response in COVID-19: what is next? Cell Death Differ. 2022;29(6):1107–1122. doi:10.1038/s41418-022-01015-x

Li CC, Wang XJ, Wang HR. Repurposing host-based therapeutics to control coronavirus and influenza virus. Drug Discov Today. 2019;24(3):726–736. doi:10.1016/j.drudis.2019.01.018.

Li H, Liu SM, Yu XH, Tang SL, Tang CK. Coronavirus disease 2019 (COVID-19): Current status and future perspectives. Int J Antimicrob Agents. 2020;55(5):105951. doi:10.1016/j.ijantimicag.2020.105951.

Li T, Zhang T, Gu Y, Li S, Xia N. Current progress and challenges in the design and development of a successful COVID-19 vaccine. Fundam Res. 2021;1(2):139–150. doi:10.1016/j.fmre.2021.01.011.

Lin S, Shen R, He J, Li X, Guo X. Molecular modeling evaluation of the binding effect of ritonavir, lopinavir and darunavir to severe acute respiratory syndrome coronavirus 2 proteases. bioRxiv; 2020. DOI: 10.1101/2020.01.31.929695.

McPherson SW, Keunen JE, Bird AC, Chew EY, van Kuijk FJ. Investigate Oral zinc as a prophylactic treatment for those at risk for COVID-19. Am J Ophthalmol. 2020;216:A5–A6. doi:10.1016/j.ajo.2020.04.028.

Meo AS, Masood A, Shabbir U, Ali H, Nadeem Z, Meo SA, Alshahrani AN, AlAnazi S, Al-Masri AA, Al-Khlaiwi T. Adverse effects of sinopharm COVID-19 vaccine among vaccinated medical students and health care workers. Vaccines. 2023;11(1):105. https://doi.org/10.3390/vaccines11010105.

Mittal A, Jain B, Varshney AM, Garg G, Sachan D, Singh RB. Adverse effects after first and second dose of Covishield and Covaxin: A longitudinal study. J Family Med Prim Care. 2022;11(11):7339–7345. doi:10.4103/jfmpc.jfmpc_885_22.

Mohsin M, Mahmud S, Uddin Mian A, et al. Side effects of COVID-19 vaccines and perceptions about COVID-19 and its vaccines in Bangladesh: A cross-sectional study. Vaccine X. 2022;12:100207. doi:10.1016/j.jvacx.2022.100207.

Mousavizadeh L, Ghasemi S. Genotype and phenotype of COVID-19: Their roles in pathogenesis. J Microbiol Immunol Infect. 2020;54(2):159–163. doi:10.1016/j.jmii.2020.03.022.

Mukherjee S, Karmakar S, Babu SP. TLR2 and TLR4 mediated host immune responses in major infectious diseases: A review. Braz J Infect Dis. 2016;20(2):193–204. doi:10.1016/j.bjid.2015.10.011.

Niraj N, Mahajan SS, Prakash A, Sarma P, Medhi B. Paxlovid: A promising drug for the challenging treatment of SARS-COV-2 in the pandemic era. Indian J Pharmacol. 2022;54(6):452–458.

Niu W, Wu F, Cui H, Cao W, Chao Y, Wu Z, … & Liang C. Network pharmacology analysis to identify phytochemicals in traditional Chinese medicines that may regulate ACE2 for the treatment of COVID-19. Evid Based Complement Alternat Med. 2020;2020:7493281.

Ozonoff A, Nanishi E, Levy O Bell's palsy and SARS-CoV-2 vaccines. Lancet Infect Dis. 2021;21(4):450–452.

Pal M, Berhanu G, Desalegn C, Kandi V. Severe acute respiratory syndrome coronavirus-2 (SARS-CoV-2): An update. Cureus. 2020;12:e7423.

Pan B, Fang S, Zhang J, Pan Y, Liu H, Wang Y, … & Liu L. Chinese Herbal compounds against SARS-CoV-2: Puerarin and quercetin impair the binding of viral S-protein to ACE2 receptor. Computat Struct Biotechnol J. 2020;18:3518–3527.

Patra R, Chandra Das N, Mukherjee S. Targeting human TLRs to combat COVID-19: A solution?. J Med Virol. 2021;93(2):615–617. doi:10.1002/jmv.26387

Polderman JA, Farhang-Razi V, Van Dieren S, Kranke P, DeVries JH, Hollmann MW, … & Hermanides J. Adverse side effects of dexamethasone in surgical patients. Cochrane Database Syst Rev. 2018;11(11):CD011940.

Roshanravan N, Ghaffari S, Hedayati M. Angiotensin converting enzyme-2 as therapeutic target in COVID-19. Diabetes Metab. Syndr. 2020;14(4):637–639.

Roychoudhury S, Das A, Sengupta P, et al. Viral pandemics of the last four decades: Pathophysiology, health impacts and perspectives. Int J Environ Res Public Health. 2020;17(24):9411. doi:10.3390/ijerph17249411

Sanchis-Gomar F, Lavie CJ, Perez-Quilis C, Henry BM, Lippi G. Angiotensin-Converting Enzyme 2 and Antihypertensives (Angiotensin Receptor Blockers and Angiotensin-Converting Enzyme Inhibitors) in Coronavirus Disease 2019. Mayo Clin Proc. 2020;95(6):1222–1230.

Santovito LS, Pinna G. Acute reduction of visual acuity and visual field after Pfizer-BioNTech COVID-19 vaccine 2nd dose: A case report. Inflamm Res. 2021;70(9):931–933.

Sharma, R., Agarwal, M., Gupta, M., Somendra, S., Saxena, S.K. (2020). Clinical Characteristics and Differential Clinical Diagnosis of Novel Coronavirus Disease 2019 (COVID-19). In: Saxena, S. (eds) Coronavirus Disease 2019 (COVID-19). Medical Virology: From Pathogenesis to Disease Control. Springer, Singapore. https://doi.org/10.1007/978-981-15-4814-7_6

Sohn KM, Lee S-G, Kim HJ et al. COVID-19 patients upregulate toll-like receptor 4-mediated inflammatory signaling that mimics bacterial sepsis. J Korean Med Sci. 2020;35(38):e343.

Songvut P, Suriyo T, Panomvana D, Rangkadilok N, Satayavivad J. A comprehensive review on disposition kinetics and dosage of oral administration of *Andrographis paniculata*, an alternative herbal medicine, in co-treatment of coronavirus disease. Front Pharmacol. 2022;13:952660.

Sun J, Chen Y, Fan X, Wang X, Han Q, Liu Z. Advances in the use of chloroquine and hydroxychloroquine for the treatment of COVID-19. Postgrad Med. 2020;132(7):604–613.

Teli DM, Shah MB, Chhabria MT. In silico screening of natural compounds as potential inhibitors of SARS-CoV-2 main protease and spike RBD: Targets for COVID-19. Front Mol Biosci. 2021;7:599079.

Vardhan S, Sahoo SK. In silico ADMET and molecular docking study on searching potential inhibitors from limonoids and triterpenoids for COVID-19. Comput Biol Med. 2020;124:103936.

Waheed W, Carey ME, Tandan SR, Tandan R. Post COVID-19 vaccine small fiber neuropathy. Muscle Nerve. 2021;64(1):E1–E2.

Woo CJ, Chou OHI, Cheung BMY. Ramsay Hunt syndrome following COVID-19 vaccination. Postgrad Med J. 2022;98(1164):738–739.

Wu C, Zheng M, Yang Y. et al. Furin: A potential therapeutic target for COVID-19. iScience. 2020;23(10):101642.

Wu C, Yin W, Jiang Y, Xu HE. Structure genomics of SARS-CoV-2 and its Omicron variant: Drug design templates for COVID-19. Acta Pharmacol Sin. 2022;43:3021–3033.

Xing Y, Hua Y-R, Shang J, Ge W-H, Liao J. Traditional Chinese medicine network pharmacology study on exploring the mechanism of Xuebijing injection in the treatment of coronavirus disease 2019. Chin J Nat Med. 2020;18(12):941–951.

Xu J, Shi PY, Li H, Zhou J. Broad spectrum antiviral agent niclosamide and its therapeutic potential. ACS Infect Dis. 2020;6(5):909–915.

Yan S, Ci X, Chen NA, Chen C, Li X, Chu X, … & Deng X. Anti-inflammatory effects of ivermectin in mouse model of allergic asthma. Inflamm Res. 2011;60:589–596.

Yang CJ, Wei YJ, Chang HL, Chang PY, Tsai CC, Chen YH, Hsueh PR. Remdesivir use in the coronavirus disease 2019 pandemic: A mini-review. J Microbiol Immunol Infect. 2021;54(1):27–36.

Yang Y, Islam MS, Wang J, Li Y, Chen X. Traditional Chinese medicine in the treatment of patients infected with 2019-new coronavirus (SARS-CoV-2): A review and perspective. Int J Biol Sci. 2020;16(10):1708.

Zhan Y, Ta W, Tang W, Hua R, Wang J, Wang C, Lu W. Potential antiviral activity of isorhamnetin against SARS-CoV-2 spike pseudotyped virus in vitro. Drug Dev Res. 2021;82(8):1124–1130.

Zhang Y, Zeng G, Pan H, Li C, Hu Y, Chu K, … & Zhu F. Safety, tolerability, and immunogenicity of an inactivated SARS-CoV-2 vaccine in healthy adults aged 18–59 years: A randomised, double-blind, placebo-controlled, phase 1/2 clinical trial. Lancet Infect Dis. 2021;21(2):181–192.

Žorić L, Rajović-Mrkić I, Čolak E, Mirić D, Kisić B. Optic neuritis in a patient with seropositive myelin oligodendrocyte glycoprotein antibody during the post-COVID-19 period. Int Med Case Rep J. 2021;14:349–355.

8 Molecular Mechanisms of Envelope Protein Inhibitors from Traditional Herbal Medicines

Abhiruj Navabhatra

8.1 INTRODUCTION

In the context of the COVID-19 pandemic, global interest has converged upon the potential of traditional natural remedies. A notable 90% of a cohort of 214 patients subjected to such treatments exhibited symptomatic improvement. Intriguingly, some herbal interventions have manifested a bifunctional efficacy: they have not only offered prophylactic protection against severe acute respiratory syndrome coronavirus 2 (SARS-CoV-2) for healthy populations but also augmented therapeutic outcomes for individuals with symptoms ranging from moderate to severe. Zhongnan Hospital of Wuhan University has pioneered this therapeutic strategy, diligently integrating traditional medical practices into their established COVID-19 treatment protocols. The selection of herbal interventions is contingent upon the patient's symptomatic severity and disease progression. Several medicinal botanicals, including *Allium sativum*, *Camellia sinensis*, and *Zingiber officinale*, have garnered attention for their purported capability to augment the immune response against SARS-CoV-2. Furthermore, particular phytochemical compounds, such as terpenoids, have evinced promise in their potential to impede viral replication. Alkaloid compounds have been distinctly acknowledged for their potent antiviral properties vis-à-vis the coronavirus (CoV). Collectively, these natural derivatives underscore a compelling trajectory for preventive and therapeutic modalities against the virus. To fully appreciate the therapeutic implications of these treatments, an in-depth understanding of the SARS-CoV-2 viral architecture is imperative. SARS-CoV-2 is taxonomically classified under the beta genus of the *Coronaviridae* family. As delineated by Khan et al., the virus is typified as a single-stranded RNA entity enveloped within a helical nucleocapsid [1]. It comprises 20 distinct proteins, with the predominant ones being spike (S), envelope (E), membrane (M), and nucleocapsid (N). Additionally, it possesses pivotal proteins such as the RNA-dependent RNA polymerase (RdRp). The molecular dynamics of the infection process are of paramount significance. SARS-CoV-2 exhibits tropism in both human and bat cells, predominantly facilitated by the angiotensin-converting enzyme II (ACE2). Leveraging its S protein, the virus establishes affinity with ACE2 receptors, orchestrating its subsequent cellular entry. Following endocytosis, the viral replication machinery is activated. The genomic RNA of the virus orchestrates the synthesis of two primary polyproteins, which undergo subsequent proteolytic cleavage to generate nonstructural proteins. This cascade of events culminates in the host cell fabricating double-membrane vesicles. Upon completion of the viral assembly, virions are extricated to propagate the infection. On a genomic continuum, SARS-CoV-2 exhibits considerable homology with other beta-genus CoVs, notably severe acute respiratory syndrome coronavirus (SARS-CoV) and NL63. However, amidst this genetic congruence, discrepancies in the sequence of the S protein across various strains have been documented. Notwithstanding, the congruency, particularly in the proteinaceous structures, between SARS-CoV-2 and SARS-CoV is undeniable.

DOI: 10.1201/9781003452621-8

The chapter endeavors to expound upon the potential of traditional herbal medicines and phyto-derived products in counteracting COVID-19, with an emphasis on delineating their prospective antiviral mechanisms of action [2–4].

8.2 TRADITIONAL HERBAL MEDICINES AS A TREASURE TROVE

Amidst escalating financial implications and observed adverse effects associated with synthetic pharmaceuticals coupled with a rising incidence of adverse drug reactions, there is a palpable impetus toward identifying inherently safer antiviral agents. Phytochemical constituents derived from herbal sources have garnered increased attention in contemporary research for their capacity to impede viral replication mechanisms. The multifaceted interplay between the SARS-CoV-2 virus and the human host underscores an exigent demand for innovative therapeutic interventions. Lianhuaqingwen, an intricate formulation comprising 11 botanical components and one mineral element, is currently undergoing rigorous scientific scrutiny, particularly within the U.S. context, to ascertain its efficacy in mitigating the ramifications of COVID-19. Preliminary findings, grounded in cell-based studies, lend credence to its antiviral attributes and its capacity to attenuate pro-inflammatory cytokine production–a mechanism of paramount significance given the deleterious cytokine surge observed in acute manifestations of COVID-19. Moreover, it is noteworthy to highlight that, as documented in a Chinese cohort study, approximately 92% of COVID-19-afflicted patients were administered formulations based on Traditional Chinese Medicine (TCM). The phytochemical theaflavin shows notable antiviral potential, but biases in evaluating its efficacy can lead to overextimations or underestimatiions. These discrepancies significantly affect its antiviral effectiveness and associated adverse effects. Historical archives allude to the prophylactic capabilities of certain herbal concoctions against viral pathogens such as SARS and H1N1. While myriad botanical extracts evince antiviral propensities, an exhaustive elucidation of their therapeutic applicability vis-à-vis SARS-CoV-2 remains a research imperative. A systematic evaluation encompassing 200 Chinese botanical extracts identified a subset of six extracts that demonstrated inhibitory effects against SARS-CoV, spotlighting phytochemicals like myricetin and scutellarein as prospective antiviral candidates. Such compounds might pave the way for novel therapeutic or preventive strategies in the battle against SARS-CoV-2 [4, 5].

8.2.1 A GLOBAL HERITAGE OF HEALING

Recent studies have highlighted the potential of natural compounds, derived from various sources, in treating COVID-19. Tylophorine, from the plant *Cynanchum komarovii* Al., has shown promise, especially its optimized form, NK007(S, R), which reduced viral concentrations and lung damage in rat models. Wang et al., enhanced its delivery using nanoparticle formulations. The traditional Chinese medicine, *Venenum bufonis* (ChanSu), derived from toad secretions, has a component, bufadienolide, which demonstrated antiviral capacities against various CoVs, notably telocinobufagin due to its reduced toxicity. In 2008, gallinamide A (GA) was identified in marine cyanobacteria as showing antiviral properties, especially against SARS-CoV-2. Conroy et al.'s research emphasized its potential, especially when combined with other drugs [6]. Cordycepin, derived from fungi, has shown antiviral activities, likely by disrupting viral RNA synthesis. Homofascaplysin A, from marine sponges, has been identified for its potential against various viruses, including SARS-CoV-2. Chirality has been noted to affect the efficacy of compounds, with Wallichin C and D from ferns showing differential effects against SARS-CoV-2 due to chirality differences. Traditional remedies like Qingfei Paidu have shown effectiveness against COVID-19, with compounds from *Glycyrrhiza uralensis* Fisch. being particularly noteworthy. Lastly, glycosylation has been recognized as a process enhancing the therapeutic properties of natural compounds, exemplified by the GA compound in the *Glycyrrhiza* genus [7, 8].

8.2.2 Understanding Pharmacological Properties

The potential therapeutic implications of natural compounds possessing antiviral attributes have attracted significant scholarly attention, particularly concerning their efficacy against CoVs. Some of these bioactive molecules exhibit the capability to impede viral replication or amplify the immune response, thereby mitigating the risk of infections. Critical proteins, namely helicase and the RdRp enzyme, have been identified as prospective therapeutic targets for CoVs. Notably, compounds such as silvestrol, phytochemicals derived from *Isatis indigotica*, and catechins from green tea have manifested promising results in targeting these protein structures. The S glycoprotein is instrumental in facilitating viral entry into host cells. Certain phytochemicals, including those extracted from *Stephania tetrandra* and elderberry, demonstrate potential in inhibiting viral replication and bolstering cellular defense mechanisms. The NLRP3 inflammasome, a key component in mediating inflammatory response, can be modulated by specific flavonoids, including quercetin. The proteases of SARS-CoV, which play a vital role in viral replication, emerge as prospective drug targets, with several natural compounds evincing an affinity to thwart their function. Another protein, TMPRSS2, utilized by SARS-CoV-2 for host cell invasion, has become a central axis of research. Bioactive molecules sourced from milk thistle and wild artichokes may deter essential molecular pathways facilitating viral ingress. Furthermore, certain natural lectins, such as those derived from red algae, showcase pronounced antiviral potential. In essence, the structural intricacies of these natural compounds largely determine their antiviral efficacy, with bioactive molecules like anthocyanins and flavonoids exemplifying potential inhibitory mechanisms. This comprehensive exploration delves deeply into these molecular interactions, offering insights into prospective therapeutic strategies against CoVs [9].

8.2.3 From Traditional Knowledge to Evidence-Based Medicine

Recent advancements in herbal medicine research elucidate that numerous plant species harbor a vast array of medicinal properties, underscored by their complex biochemical compositions. This intricacy poses significant challenges in deciphering their precise therapeutic potential. Many such botanicals have demonstrated antiviral, anti-inflammatory, and immunomodulatory capacities, making them important candidates for consideration amidst the ongoing COVID-19 pandemic. Elevated concentrations of inflammatory markers, including IL-6, erythrocyte sedimentation rate (ESR), and C-reactive protein (CRP), are correlated with severe manifestations of COVID-19, often precipitated by an overwhelming cytokine response. Consequently, botanical preparations with pronounced anti-inflammatory profiles might hold significant therapeutic promise in the overarching strategy against COVID-19. This discourse will delve further into specific botanical interventions tailored for COVID-19. *Vernonia amygdalina*, indigenous to Asia and tropical Africa, is a representative of the *Asteraceae* family, which encompasses approximately 1,000 species within its genus. Historically employed to alleviate ailments such as headaches and fevers, contemporary research accentuates its potential to augment CD4+ cell counts, thereby hinting at its prospective role in antiretroviral therapy for HIV. This plant boasts antimicrobial, antidiabetic, and antioxidant attributes. Nonetheless, apprehensions regarding its potential toxicological profile persist, although animal studies have yet to report fatal outcomes. Investigations into its efficacy against SARS-CoV-2 remain ongoing. *Azadirachta indica*, commonly referred to as neem and a linchpin in traditional medicinal practices, is renowned for ameliorating fever, a salient symptom of COVID-19. Infusions derived from boiled neem leaves are postulated to manifest anti-inflammatory effects. Empirical evidence affirms its antiviral capabilities, notably against the hepatitis C virus. Certain phytochemicals present in neem could potentially interact with structural proteins of COVID-19. Despite its enduring therapeutic legacy, recent studies draw correlations between neem seed oil and potential adverse health effects, thereby underscoring the imperative for rigorous clinical evaluations. *Nigella sativa*, colloquially termed black cumin seed, occupies a prominent position within herbal research paradigms. Extracts derived from this seed exhibit antiviral activities, most markedly against hepatitis C. Animal model investigations reinforce its dual antiviral and immunomodulatory properties. Historically leveraged for ailments like asthma due to its anti-inflammatory potential, prudence is advocated in scenarios entailing prolonged, high-dosage intake. Originating from Malaysia, *Eurycoma longifolia* is celebrated for bolstering male health and might potentiate immune defenses, especially in geriatric populations, through mechanisms involving CD4+ cell proliferation. Demonstrably safe at dosages up to 200 mg daily, the plant exhibits anti-inflammatory attributes, potentially attributable to eurycomalactone. Given its benign safety profile, it emerges as a prospective candidate in mitigating the inflammatory sequelae of COVID-19. *Mentha piperita*, commonly identified as peppermint and with deep-rooted therapeutic antecedents, possesses an essential oil demonstrating efficacy against a spectrum of microbes primarily attributed to constituents such as menthol and menthone. A study conducted in Saudi Arabia drew correlations between peppermint consumption and attenuated hospitalization rates consequent to COVID-19. *A. sativum* (garlic) and *Allium cepa* (onion), foundational elements of Ethiopian traditional medicine, proffer an array of health benefits. Onions have exhibited properties countering the avian flu virus, suggesting their potential relevance in the context of COVID-19. Garlic, enriched with sulfur compounds, might fortify immune defenses against SARS-CoV-2. *Malva sylvestris*, containing compounds such as flavonoids, exerts a suppressive effect on cough reflexes, thus potentially attenuating COVID-19-associated respiratory manifestations. *I. indigotica*, endowed with a rich phytochemical profile, has manifested antiviral properties against pathogens, including hepatitis B. Its potential against COVID-19 merits further exploration. Primarily deployed for dermatological conditions, *Psoralea corylifolia* boasts antimicrobial attributes. Certain phytochemicals therein have evinced promise against SARS-CoV, suggesting a conceivable therapeutic avenue against SARS-CoV-2. Glycyrrhizin, sourced from *Glycyrrhiza glabra*, can impede SARS-coronavirus replication. Structural modifications to this molecule might potentiate its efficacy against SARS-CoV-2, albeit concerns regarding cytotoxicity necessitate caution [10–12].

8.2.4 NAVIGATING THE CHALLENGES AND POTENTIAL APPLICATIONS

Amid the COVID-19 pandemic, the global health sector was transformed due to the virus's rapid mutations, leading to various strains and infection waves. While many treatments and vaccines exist, questions about their effectiveness and side effects persist. Plants have historically been medicinal cornerstones, and the pandemic reignited interest in their antiviral properties. Modern technology allows re-evaluation of existing drugs, and repurposing these may offer quicker solutions than developing new vaccines. Measures like social distancing reduce transmission, but new antiviral agents remain essential. Research increasingly focuses on plant-derived compounds, such as resveratrol and quercetin, for their potential against CoVs. Synthetic drugs often treat symptoms, not causes, leading to significant health issues. Annually, more than 100,000 deaths are linked to synthetic drug issues, causing about 8% of U.S. hospital admissions. In contrast, severe herbal remedy reactions are rare. Medicinal plants typically offer broader therapeutic benefits and reduced toxicity compared to synthetics. As COVID-19 is a recent disease, plant-based treatments demand thorough testing and optimization. Plant medicines, with their affordability and safety, offer promise, especially for resource-limited countries. However, challenges like raw material consistency and understanding their multifaceted effects exist. Both developing and affluent nations can benefit from these treatments, but regulatory changes might be needed. Interdisciplinary research in medicinal plants is crucial for future breakthroughs [13, 14].

8.2.5 EMPOWERING COMMUNITIES AND SUSTAINABLE PRACTICES

The COVID-19 pandemic has significantly impacted global structures, affecting lives, economies, and societal behaviors. International scientists are diligently working toward a vaccine and cure, with a major focus on allopathic treatments, which often have side effects. Despite the more than 160 million COVID-19 cases, there is limited research on the potential of natural medicines. The economic and societal effects of COVID-19, as well as the potential benefits of plant-derived remedies, have been reviewed. Exploration has revealed that these compounds might act as antioxidants, enzyme inhibitors, or block virus receptors, suggesting their potential as alternative treatments. However, the effectiveness of plant-based therapies remains under debate, and there are environmental concerns tied to their extraction. Utilizing agro-industrial waste for phytotherapy is a sustainable solution. Furthermore, merging advancements in nanoscience with herbal medicine is an emerging area that requires more research. Plant-derived compounds offer hope against COVID-19, and incorporating "Green Chemistry" can lead to sustainable healthcare solutions [14–16].

8.3 THE ENVELOPE PROTEINS AND VIRAL LIFE CYCLE

8.3.1 The Role of E Protein of SARS-CoV-2 in Viral Entry

Within the β-coronavirus lineage, there exist three primary CoVs: MERS, SARS-CoV-1, and SARS-CoV-2. Genetically, MERS stands distinct, while SARS-CoV-1 and SARS-CoV-2 exhibit approximately 85% genomic similarity. Inherently, all CoVs possess four fundamental structural components. Among these, the E protein, albeit present in limited quantities within the virus, assumes critical significance. It establishes a unique ion channel within infected cells, precipitating pronounced viral effects. This ion channel is implicated in eliciting robust inflammatory responses in humans. Notably, its absence has been observed to attenuate virus severity in murine models. Following the emergence of SARS-CoV-1, concerted efforts have been directed toward neutralizing this ion channel. Compounds such as amantadine have demonstrated potential therapeutic implications in this regard. Detailed analysis of the E protein's structure unveils a pivotal segment: the PDZ-binding motif (PDZbm). This motif plays an instrumental role in mediating viral interactions with host proteins and is known to augment the virulence of both SARS-CoV and MERS-CoV. It is of paramount importance to note that SARS-CoV-2 exhibits an expanded spectrum of host protein interactions compared to its predecessors, displaying particular affinity toward proteins such as PALS1. In synthesis, despite the evident genetic similarities between SARS-CoV-1 and SARS-CoV-2, the enhanced interactional capacity of the latter with a diverse array of host proteins underscores the potential for a unified therapeutic strategy targeting these CoVs [17–19].

8.3.2 Mutations of SARS-CoV-2 E Protein

In contemporary genomic studies, modifications discerned in SARS-CoV-2 elucidate profound ramifications not only for the diagnostic methodologies pertaining to COVID-19 but also for the strategic formulation of therapeutic interventions and the evolution of vaccine research. Eminent researchers have meticulously cataloged a plethora of changes within the virus's DNA, among which alterations to discrete DNA nucleotides are predominantly observed. Notably, certain mutations, with a distinct emphasis on those located within the E protein, are postulated to potentially attenuate the virus's virulence. Such a premise accentuates the E protein's pivotal role in the sophisticated discourse regarding prospective vaccine developments. Despite the diverse array of mutations exhibited by SARS-CoV-2, the genetic sequence delineating the E protein, encompassing precisely 225 nucleotides, has demonstrated remarkable resilience and stability. Scholarly publications that have undergone rigorous peer-review processes underscore the nuanced modifications in the amino acid sequences of the E protein. Conversely, other protein domains, specifically the S and N proteins, evince a propensity to mutate at an augmented rate. Within the scientific community, there is a prevailing hypothesis suggesting that specific mutations localized within the E protein domain could modulate the virus's pathogenicity by recalibrating its cellular interactions within human hosts. The advent of the Omicron variant has precipitated intense scrutiny within academic domains. Initial investigative endeavors posit that this variant may harbor a genetic modification with potential implications for its pathogenic profile and transmission dynamics. In light of the E protein's quintessential and inherently stable characteristics, assiduous monitoring of its genetic variations remains paramount. Such rigorous scientific undertakings are poised to markedly influence the trajectory of imminent medical and virological research paradigms [20–22].

8.3.3 E PROTEINS AND IMMUNE EVASION

SARS-CoV-2, responsible for severe manifestations of COVID-19, induces a variable cytokine response within the host. This dysregulated immune response may precipitate significant pulmonary impairment, among other systemic effects. A salient feature of this virus is the 2-E component, characterized by two consistent functionalities: its role as an ion channel and its possession of a property termed PBM. These mechanisms are pivotal in cytokine release and inflammasome activation, underscoring the significance of 2-E in elucidating the immune responses elicited by SARS-CoV-2. Upon comparative analysis with other β-coronaviruses, SARS-CoV-1 and SARS-CoV-2 demonstrate shared phenotypic characteristics pertaining to their E-protein ion channel and PBM. This phenotypic overlap posits the potential for analogous immune modulations by these viruses. Previous investigations into SARS-CoV-1 revealed that strains devoid of the E protein exhibit an inability to instigate certain pro-inflammatory cascades. This deficiency results in attenuated cytokine production during murine infections, an observation analogously reported for SARS-CoV-2. In a seminal study, Planès et al. [23] elucidated the capabilities of the 2-E component, identifying its potential to instigate the release of pro-inflammatory cytokines, notably TNF-α and IFN-γ. Given the pronounced elevation of these cytokines in COVID-19-afflicted individuals, their synergistic activity might contribute to exacerbated inflammatory responses, potentially culminating in tissue necrosis and lethal outcomes. New research indicates that 2-E's role in regulating the immune system is more complex than we once thought. It seems to influence cytokine levels by interacting with certain cell-signaling routes. The ion channel functionality of 2-E may underpin its distinctive influence on select immune processes. Complementing this understanding, research led by Yalcinkaya et al. [24] suggests a dynamic interaction of 2-E with immune responses, contingent on the progression of the infection. Initial immunosuppression may be followed by augmented responses in subsequent stages, providing a nuanced perspective on host immune dynamics throughout viral pathogenesis [22, 25, 26].

8.3.4 E PROTEINS IN VIRAL ASSEMBLY

The E protein's pivotal role in modulating infection severity has prompted scholarly discourse regarding the potential of E protein-deficient viruses as vaccine candidates. Preliminary research posits that the ion channels inherent to the E protein may facilitate the translocation of H+ ions. However, Cabrera-Garcia et al. [27] have elucidated certain impediments in authenticating this hypothesis, underscoring the imperative for more sophisticated research methodologies. Analogous to other CoVs, the E protein in SARS-CoV-2 compromises cellular homeostasis, potentially culminating in pronounced health ramifications. Structurally characterized by 75 amino acids and bifurcated into the predominant domains, this protein exhibits multifaceted functionalities within the CoV lifecycle. Its involvement in virion assembly and enhancement of virulence parallels the attributes observed in the E protein of SARS-CoV-1. Intriguingly, humans are endowed with an approximate repertoire of 150 proteins integrated with PDZ domains, which are quintessential for immunological responses. Certain viral strains, notably specific SARS variants, demonstrate an affinity for these proteins, with a predilection for the cellular protein known as PALS1. Such molecular engagements are hypothesized to augment cellular permeability, potentially catalyzing the kinetics of viral dissemination. Within the demographic of geriatric COVID-19 patients harboring coexisting health anomalies, the myriad interactions of the E protein with intracellular proteins warrant meticulous examination. A distinctive amino acid sequence, DLLV, embedded within the protein, is postulated to be instrumental in potentiating the virulence of the virus. The overarching objective of this investigation is to expound upon the structural dynamics underpinning the PALS1-E protein interaction [28, 29].

8.3.5 THE POTENTIAL OF E PROTEIN-TARGETED INTERVENTIONS

In light of the COVID-19 pandemic's onset in 2019, a plethora of scholarly investigations have underscored the pivotal role of the E protein in SARS-CoV-2. This protein is indispensable not only in facilitating the maturation of the virus but also in dictating the trajectory of the disease, exerting influence on cellular architecture and orchestrating alterations in the immune response. Noteworthily, the amino acid sequence of this protein bears a striking resemblance–approximately 95% similarity–to its counterpart in SARS-CoV-1. This observation indicates that extant research on the latter could shed light on the functional dynamics of the E protein in SARS-CoV-2. Presently, a growing contingent of researchers is meticulously analyzing the interface between the E protein and cellular structures, positing it as a viable therapeutic locus. Leveraging state-of-the-art computational methodologies, several molecular entities have been identified as potential antagonists to the E protein, heralding prospective therapeutic avenues for COVID-19. Nonetheless, considering the inherent mutagenic propensity of viruses, exemplified by emergent strains such as Omicron and potential modifications in its E protein, there is an imperative to engineer interventions that engage with conserved viral regions. Further, it is of paramount importance to adopt a forward-thinking stance in pandemic readiness, eschewing a purely reactionary modus operandi [22, 30].

8.4 IDENTIFICATION AND ISOLATION OF E PROTEIN INHIBITORS

8.4.1 HERBAL MEDICINES AS A SOURCE OF ANTIVIRAL COMPOUNDS

Recent investigations in the field of botanical antiviral properties have elucidated the efficacy of myriad specimens. Within this cohort, *Syzygium jambos* has exhibited pronounced efficacy vis-à-vis the HSV1 virus. Furthermore, the pterodontic acid derived from *Laggera pterodonta* has manifested potential in combatting human influenza A. In a similar vein, the rosmarinic acid inherent in *Melissa officinalis* has evidenced notable effectiveness against the enterovirus. A salient focus of these inquiries has centered on the potency of botanical extracts. For instance, extracts derived from *Penthorum chinense* have demonstrated their efficacy against both influenza A and B strains, whereas the diphenanthrenes found in *Bletilla striata* have been underscored for their robust antiviral attributes. Subsequent revelations include the capability of *Oroxylum indicum* to mitigate the effects of the chikungunya virus. Remarkably, the silver nanoparticles sourced from *Andrographis paniculata*, coupled with dehydrojuncusol from *Juncus maritimus*, have both manifested profound antiviral attributes. Delving into the specifics of viral strains, compounds procured from *Kniphofia foliosa* Hochst. offer promising results against HIV-1, and the efficacy of *Paulownia tomentosa* against the dengue virus is notably paramount. The literature has also recorded the antiviral potentials of extracts from *Ocimum tenuiflorum* and *Terminalia chebula* against chikungunya, and the pronounced efficacy of fennel extract in counteracting influenza. Further advancements in this domain include the meticulous extraction of berberine–a compound with putative efficacy against COVID-19–from the root of *Berberis lycium* utilizing the Soxhlet extraction methodology. Empirical studies focused on certain Chinese medicinal plants, notably *Dryopteris crassirhizoma* and *Morus alba*, have illuminated their potent action against the dengue virus. In the persistent quest for efficacious COVID-19 treatments, botanicals such as ashwagandha, Giloy, and Tulsi have been pivotal. Concurrent research has accentuated the antiviral capabilities of specific plant extracts, with the peel of the pomegranate receiving particular attention for its antagonistic action against adenovirus. Additionally, the essential oil extracted from

TABLE 8.1

An Example of Medicinal Plants with Antiretroviral Activity

Plant Name	Major Chemical Constituents	Virus Type
Syzygium jambos	Flavonoids (quercetin, rutin, 5, 4′-dihydroxy, 7-methoxy, 6-methyl-flavone), phenolic compound	HSV1 virus
Laggera pterodonta	Pterodondiol, ilicic acid, artemitin, chrysosplenetin B, 3,5-dihydroxy-3′,4′,6,7-tetramethoxyflavone, chrysosplenol D, 5,6,4′-trihydroxy-3,7-dimethoxyflavone	Influenza A
Melissa officinalis	Volatile compounds (geranial, neral, citronellal and geraniol), triterpenes (ursolic acid and oleanolic acid), phenolic acids (rosmarinic acid, caffeic acid, and chlorogenic acid), and flavonoids (quercetin, rhamnocitrin, and luteolin)	Enterovirus
Penthorum chinense	5-hydroxy-flavanone-7-O-beta-D-glucoside, 2,4,6-trihydroxybenzoic acid, and quercetin	Influenza A and B strains
Kniphofia foliosa	Monomeric anthraquinones (chrysophanol, islandicin, laccaic acid, aloe-emodin, and aloe-emodin acetate)	HIV1
Terminalia chebula	Terminaliate A, gallic acid, methyl gallate, three chebulic acid derivatives, 1,2,6-tri-O-galloyl-β-D-glucopyranose, and arjungenin	Influenza and chikungunya
Berberis lycium	Berberine, berbamine, palmatine, jhelumine, punjabine, sindamine, maleic acid and ascorbic acid, alkaloids, flavonoids, phenols, tannin, terpenoids, resin, and fat	HCV, and COVID19

Salvia officinalis portends potential against the bovine viral diarrhea virus. Amidst the contemporary pandemic milieu, the antiviral characteristics of glycyrrhizin vis-à-vis the CoV have been emphatically articulated, proposing saponins as viable substitutes for synthetic pharmaceuticals [31–34]. As delineated in Table 8.1, botanical specimens have reaffirmed their status as a prolific reservoir of antiviral compounds.

8.4.2 MEDICINAL PLANT FOR COMBATTING SAR CoV

In the quest to identify efficacious interventions for COVID-19, an ideal therapeutic solution should not only neutralize the SARS-CoV-2 virus effectively but also be devoid of adverse side effects, exhibit cytocompatibility, and demonstrate high bioavailability. Historically, medicinal plants have played an integral role in therapeutics, especially in combating myriad diseases, inclusive of respiratory viral infections. Contemporary research has elucidated the potential of numerous phytochemicals and formulations, particularly those grounded in the principles of TCM, recognized for their antiviral efficacies. A salient advantage associated with these botanicals pertains to their dual capacity to modulate inflammatory pathways and augment immunological responses. The World Health Organization (WHO) has ardently endorsed preventive measures, accentuating the imperative of dietary alterations, lifestyle modifications, and the integration of specific medicinal herbs to fortify immunological defenses. Amid the rigorous scientific exploration against COVID-19, phytochemical compounds derived from plants such as *A. sativum* and *G. glabra* have exhibited potential in attenuating CoV replication mechanisms. For instance, phytoconstituents of *Clerodendrum inerme* have been postulated to inhibit the synthesis of viral proteins. Conversely, bioactive compounds from *Strobilanthes cusia* have demonstrated the capability to impede the synthesis of viral RNA and influence other correlated antiviral processes. Noteworthy is the action of botanicals like *Boerhaavia diffusa* and *Punica granatum*, which have evinced inhibitory effects against the ACE2 receptor, a pivotal gateway for SARS-CoV-2 viral entry. Specifically, *P. granatum's* modus operandi appears to be through a competitive inhibition mechanism, while other botanicals may act as nonspecific inhibitors. Subsequent investigations have spotlighted the prospective role of *A. paniculata* in inhibiting proteins intrinsically associated with the pathogenesis of SARS-CoV, and possibly SARS-CoV-2. Furthermore, botanicals such as *Eugenia jambolana* and *Acacia nilotica* have elicited inhibitory activities against HIV proteases, underscoring their potential utility in antiviral therapeutics against CoVs. Empirical analyses have probed the antiviral efficacies of botanicals like *Ocimum sanctum* and *Solanum nigrum* against SARS-CoV-2. It is also pertinent to highlight *Sambucus ebulus*, distinguished for its potential antiviral attributes, which might offer therapeutic benefits in ameliorating COVID-19 symptomatology. An extensive catalog of these therapeutically promising botanicals is delineated in Table 8.2 [33, 35, 36].

TABLE 8.2
Example of Medicinal Plants for Combatting SARS-CoV-2

Plant Name	Major Chemical Constituents	Mechanism
Allium sativum	Organosulfur and lectin	Inhibit viral replication
Glycyrrhiza glabra	Glycyrrhizin, glycyrrhetic acid, isoliquiritin, isoflavones	Inhibit viral replication
Clerodendrum inerme	B-friedoolean-5-ene-3-β-ol, β-sitosterol, stigmasta-5,22, 25-trien-3-β-ol, betulinic acid, and 5-hydroxy-6,7,4'-trimethoxyflavone	Deactivates viral ribosomes
Strobilanthes cusia	Indole alkaloids (IAs), quinolone alkaloids, phenylethanoid glycosides, lignan glycosides, triterpenoids, steroids, amino acids, and flavonoids	Inhibit viral RNA synthesis
Boerhaavia diffusa	Boerhavia acid, isoflavonoids (rotenoids), Punarnavine, sitosterol, boeravinone, palmitic acid, steroids (ecdysteroid), lignan glycosides, and esters of sitosterol	Inhibit ACE2
Punica granatum	Flavonols, flavanols, anthocyanins, proanthocyanidins, ellagitannins, and gallotannins`	Competitive inhibition of ACE2
Andrographis paniculata	Andrographolide, 14-deoxy-11,12-didehydroandrographolide, 14-deoxyandrogra-pholide, 3,14-dideoxyandrographolide, 14-deoxy-11-oxoandro-grapholide, 14-deoxy-12-hydroxyandrographolide, neoandrogra- pholide, andrographiside, and 14-deoxyandrographiside	Suppress NLRP3, IL-1β, and capase-1
Eugenia jambolana	Triterpenes (β-sitosterol, betulinic acid, crategolic acid, and oleanolic acid), phenolic acids (gallic and ellagic acids and their derivatives), and flavonoids (quercetin, myricetin, and myricetin derivatives)	Inhibit HIV protease activity
Acacia nilotica	Ellagic acid, Kaempferol, and Quercetin	Inhibit HIV protease activity
Ocimum sanctum	Methyl eugenol, and β-caryophyllene	Inhibit HIV protease activity
Solanum nigrum	Steroidal saponins, steroidal alkaloids, flavonoids, coumarin, lignin, organic acids, volatile oils, and polysaccharides	Inhibit enveloped viruses
Sambucus ebulus	Monophthalate, palmitic acid, isovaleric acid, α-linoleic acid, phytol, acetic acid, and pentanoic acid	Inhibit enveloped viruses
Torreya nucifera	Luteolin	Inhibit 3CLpro
Broussonetia papyrifera	Papyriflavonol A, 3'-(3-methylbut-2-enyl)-3',4,7-trihydroxyflavane, broussochalcone A,B, broussoflavan A, and 4-Hydroxyisolonchocarpin	Inhibit PLpro

8.4.3 Isolation and Characterization of E Protein Inhibitors

The SARS-CoV E protein exhibits properties reminiscent of ion channels, demonstrating a distinct predilection for sodium ions, with a magnitude of preference approximately ten-fold greater than that for potassium ions. Modulation of a specific amino acid within this protein results in a discernible attenuation of its ion-conduction capacity, thereby accentuating the pivotal role of said amino acid. Moreover, the ablation of the E gene from certain CoV strains exerts pronounced effects on the virus's replicative competence and its cytotoxic potential. These findings indicate that "viroporins," typified by the E protein, represent prospective targets for antiviral therapeutic interventions. In endeavors to modulate the activity of these proteins, several compounds, including adamantanes, HMA, and saccharide analogs, have been delineated. The patch-clamp electrophysiological technique is employed within the scientific community to elucidate the functional attributes of these channels. Notably, whereas a preponderance of ion channels manifests on the cellular periphery, viroporins, inclusive of the E protein, are predominantly intracellular entities. Specifically, their localization is proximal to the cellular nuclei and they evince associations with organelles such as the endoplasmic reticulum and the Golgi apparatus. In a seminal investigation helmed by Breitinger et al., the interaction dynamics of the E protein with a spectrum of inhibitors were meticulously examined. Their empirical findings underscored the efficacy of flavonoids, with EGCG emerging as particularly potent. Additionally, the functional kinetics of the protein were observed to be modulated by environmental pH gradients and evinced homologies with extant viroporins. Noteworthy among the flavonoid cohort, epigallocatechin and quercetin were delineated as formidable inhibitors. Cumulatively, such advancements augur well for the prospective development of novel antiviral pharmacological agents [37].

8.5 MECHANISM OF ACTION OF E PROTEIN INHIBITORS

8.5.1 Review of the Molecular Mechanism of Action of Medicinal Plants Acting on E Proteins

Withania somnifera, valued in Indian medicine for its rich phytochemicals, offers various therapeutic benefits, including antimicrobial and anti-inflammatory effects. Researchers, particularly Alharbi et al., identified four key compounds from the plant with potential enzyme-inhibiting properties. These compounds interacted specifically with the SARS-CoV-2 E protein, particularly the compounds CID 10100411, CID 3035439, CID 101,281,364, and CID 44,423,097, due to their energy barriers and amino-acid interactions. Molecular simulations revealed their stability and potential to inhibit the virus's replication. These compounds hold promise for future SARS-CoV-2 research [38]. Glycyrrhizin, sourced from *G. glabra*, is noted in traditional medicine for its antiviral properties, effective against viruses like HIV and hepatitis B and C. During the SARS CoV-1 outbreak, studies revealed its potential to halt viral spread and prevent cell penetration. This sparked interest in its relevance to SARS-CoV-2. Advanced computational methods have been used to investigate its effects, highlighting its strong interaction with a key SARS-CoV-2 protein and potential to reduce 90% of the virus's virulence. It may also reduce inflammation caused by infections. However, glycyrrhizin is rapidly metabolized in humans, making sustained therapeutic levels challenging. Some suggest molecular modifications might improve its availability, but this may increase its toxicity. Comprehensive clinical trials are needed to fully understand its therapeutic potential and risks [22, 39]. SARS-CoV, part of the *β-coronavirus* genus, has a single-stranded RNA structure. Its envelope protein E, which shares significant sequence similarity with SARS-CoV-2, performs various functions in the virus's life cycle. This protein was studied in a lab using the HEK293 cell line, with a specific targeting sequence ensuring its correct placement. Analytical techniques, including patch-clamp and cell viability assays, showed the protein created electrical currents across the cell membrane. These currents were proton-dependent in low-ion settings but stable in typical conditions. The protein's currents were blocked by inhibitors like amantadine, rimantadine, and hexamethylene amiloride (HMA). Out of ten flavonoids tested, epigallocatechin and quercetin showed strong inhibitory effects, similar to rimantadine. These findings underscore potential directions for antiviral drug development [37]. Research on the E proteins of SARS-CoV and SARS-CoV-2 highlights their multifunctional roles, including in virus assembly, release, and processes like inflammation. Given their similar amino-acid sequences, their assembly mechanisms are believed to be alike, influencing their functions. The E protein, essential in viral replication, forms structures called oligomers and viroporins, affecting the host cell's ionic balance and pH. The E protein's precise role in virus-infected cells is still under study. Initial tests suggest some inhibitors can reduce the SARS-CoV-2 E protein's harmful effects, but their therapeutic potential needs further exploration. Compounds like rutin and doxycycline seem to bind to the E protein, possibly inhibiting viroporin formation [40]. Viroporins are essential in virus replication and release, emphasizing their potential as therapeutic targets. A study on SARS-CoV-2's ORF3a viroporin used cell viability assays and electrophysiological techniques. Introducing ORF3a into HEK293 cells showed its presence on the cell membrane, which was enhanced by adjusting the peptide signal. The study tested various compounds against ORF3a and found established inhibitors, like amantadine and rimantadine, effective. Among ten compounds, kaempferol and curcumin showed inhibition, while others like 6-gingerol did not. The compounds' molecular structures likely influence their effectiveness. Overall, targeting ORF3a viroporin offers a promising approach for antiviral treatments [41]. Recent studies highlight the role of SARS-CoV-2's 2-E protein in creating a cation channel linked to cell death. Through a high-throughput screening of 4376 compounds, 34 showed cell protection, with 15 inhibiting virus replication. Notably, proanthocyanidins bind strongly to the 2-E channel and block its pore, but altering certain residues negates this effect. Given proanthocyanidins' use in cosmetics, they could be potential external disinfectants. This research emphasizes the 2-E channel's therapeutic potential and positions proanthocyanidins as a promising anti-SARS-CoV-2 agent [7]. For a graphical representation elucidating the engagement of medicinal plants with E proteins, readers are directed to Figure 8.1.

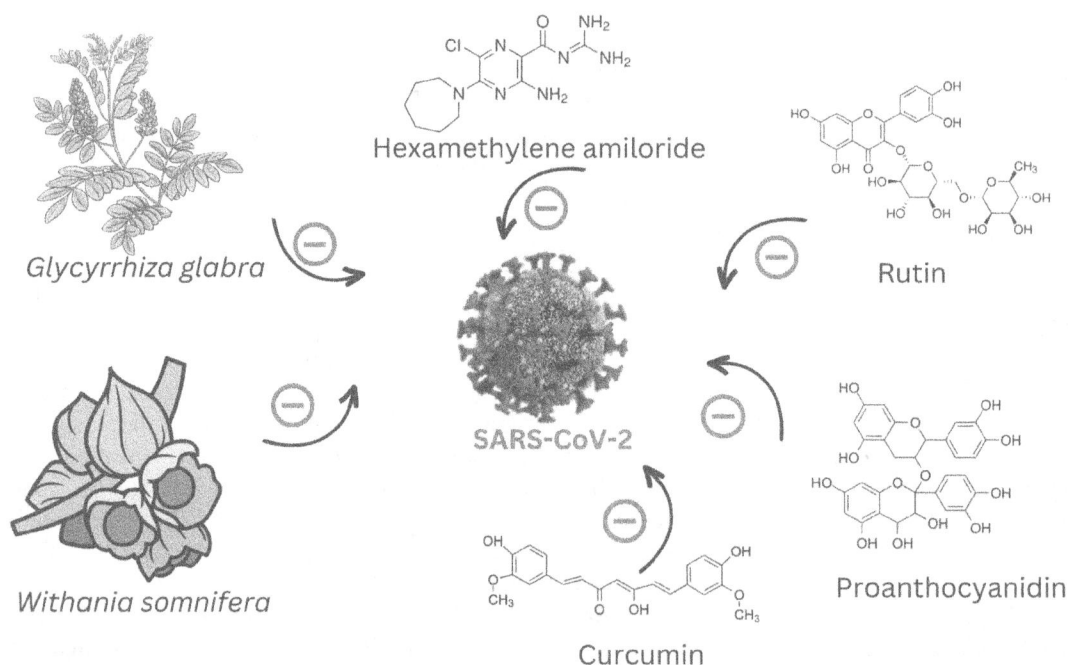

FIGURE 8.1 Graphical representation elucidating the engagement of medicinal plants with E proteins.

8.6 CLINICAL IMPLICATIONS AND CHALLENGES

8.6.1 Clinical Potential of E Protein Inhibitors from Natural Products

The SARS-CoV-2 virus is delineated by the presence of an RNA genome accompanied by four cardinal structural proteins, namely S, N, M, and E. These proteins play an indispensable role in the biosynthesis of mature viral particles. The S protein exhibits an affinity for the human ACE2 enzyme, thus facilitating the viral entry into host cells. Intracellularly, the N protein encapsulates the viral RNA, configuring it into a helical formation, and synergistically collaborates with the M protein in virion assembly. It is imperative to note that the M protein, which is profusely present within the viral architecture, engages in interactions with all other aforementioned structural proteins. Among these, the E protein, albeit the most diminutive in size, is distinguished by its distinct functionalities. Integral organelles implicated in the viral life cycle encompass the endoplasmic reticulum, the Golgi apparatus, and the intermediate compartment interposed between the two. Current scientific literature posits that the E protein possesses the capability to constitute an ion channel, an attribute that may precipitate consequential outcomes such as cellular apoptosis or the elicitation of a robust immune response. Considering these attributes, the E protein has emerged as a focal point in the quest for therapeutic interventions against COVID-19. Subsequently, an array of compounds, spanning herbal formulations, natural derivatives, and time-honored medicinal practices, are under rigorous scrutiny for their potential efficacy in inhibiting the E protein, positioning them as prospective therapeutic candidates for COVID-19. This discourse endeavors to present an exhaustive examination of these potential inhibitors while elucidating their underlying mechanisms of action.

In recent scholarly investigations, certain botanical substances and medicinal formulations, deeply anchored in indigenous knowledge, have exhibited E protein inhibitory attributes *in vitro*, suggesting a promising avenue for the therapeutic management of COVID-19. Nonetheless, empirical examination of these botanical derivatives and medicinal concoctions remains in its early stages in human clinical trials, encompassing both ostensibly healthy participants and those afflicted with COVID-19. Consequently, this section presents select examples of medicinal flora that warrant comprehensive exploration and potential refinement for their applicability in the clinical amelioration of COVID-19 manifestations. An example of the clinical potential of E protein inhibitors from natural products is summarized in Figure 8.2.

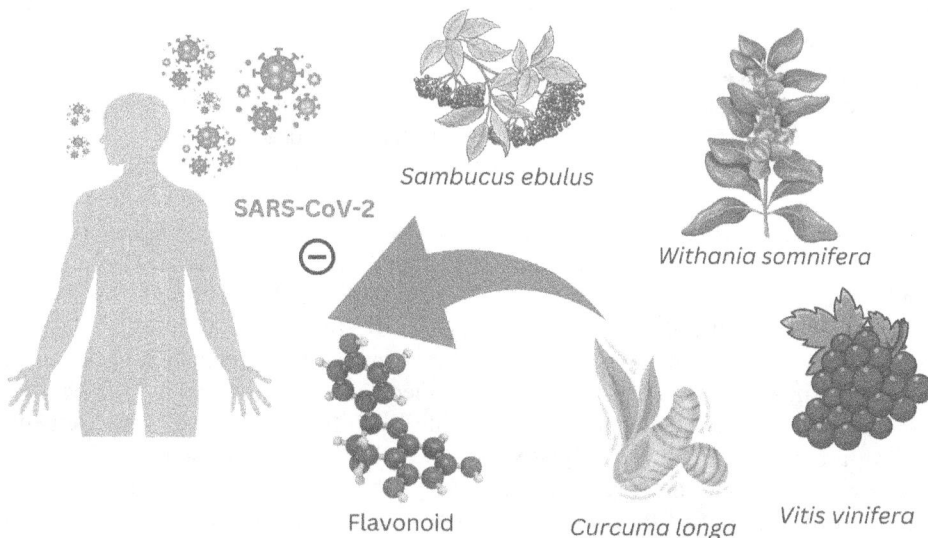

FIGURE 8.2 An example of the clinical potential of envelope protein inhibitors from natural products.

8.6.1.1 Glycyrrhizin

Glycyrrhizin, derived from the botanical *G. glabra*, has garnered significant attention in the realm of traditional medicine owing to its pronounced antiviral attributes. This compound has demonstrated efficacy against an array of formidable viruses, encompassing HIV and hepatitis B and C. Intriguingly, amidst the SARS CoV-1 epidemic, scholarly investigations elucidated glycyrrhizin's capability to impede viral propagation and inhibit its penetration into cellular structures. Such revelations have catalyzed heightened academic discourse, predominantly in the context of SARS-CoV-2. In the academic odyssey to identify efficacious antiviral agents, numerous compounds have undergone rigorous examination employing advanced computational modalities. Leveraging cutting-edge computational methodologies, researchers endeavor to delineate the prospective benefits and limitations inherent to these compounds. Preliminary findings underscore their profound affinity for a pivotal protein present in the SARS-CoV-2 virion, insinuating their prospective antiviral efficacies. Empirical *in vitro* assessments substantiate that these compounds possess the potential to neutralize approximately 90% of the virulence exhibited by the virus. Concurrently, nascent studies indicate that glycyrrhizin may attenuate the endogenous excretion of pro-inflammatory mediators, thereby modulating the intensity of infection-induced symptomatology. However, a salient challenge persists: the rapid metabolic degradation of glycyrrhizin within human physiology, complicates the retention of therapeutic concentrations requisite for efficacious antiviral activity. A cohort of biochemists postulates that nuanced molecular modifications might augment its bioavailability and persistence in systemic circulation. Nevertheless, concomitant with this enhancement in antiviral potency, there is a documented elevation in its cytotoxic potential [22, 39]. In a series of scholarly investigations, the efficacy of *G. glabra* L. in combatting SARS-Cov was meticulously examined. The primary emphasis of these studies was centered on the antiviral potential of glycyrrhizin against CoVs (FFM-1 and FFM-2) derived from patients diagnosed with SARS. The empirical data garnered from these investigations elucidated that glycyrrhizin possesses the capability to inhibit SARS-CoV replication at concentrations deemed noncytotoxic, boasting a selectivity index of 67 and an IC50 measure of 300 mg/mL [42]. In light of these multifaceted revelations, there manifests an imperative for comprehensive clinical evaluations to meticulously discern the therapeutic attributes of glycyrrhizin alongside potential contraindications.

8.6.1.2 *Withania somnifera*

W. somnifera, traditionally recognized in the realm of Indian medicine, is distinguished by its abundant phytochemical constituents and myriad therapeutic benefits. From this botanical source, scholarly investigations have isolated more than 12 alkaloids, 40 withanolides, and a multitude of sitoindosides. In empirical laboratory studies, *W. somnifera* exhibited a spectrum of therapeutic properties, encompassing antimicrobial, anti-inflammatory, antitumor, antistress, neuroprotective, cardioprotective, and antidiabetic modalities. In an in-depth exploration, Alharbi et al. scrutinized four salient compounds derived from *W. somnifera*. They focused on these compounds because of their distinctive charge attributes and marked hydrophobic nature. The ensuing data posited that compounds bearing aromatic configurations from this botanical source might function as prospective enzyme inhibitors. The researchers analyzed the interactions of an array of more than 25 compounds from *W. somnifera* with the active loci of the SARS-CoV-2 E protein. It is worth noting that the compounds CID 10100411, CID 3035439, CID 101,281,364, and CID 44,423,097 emerged as noteworthy due to their optimal energy barriers and consistent affiliations with particular amino acids in the protein structure. Utilizing advanced molecular simulation techniques over an epoch of 30 ns, these compounds manifested commendable stability. Their primary engagements were discerned within the pore region of the SARS-CoV-2 E protein, exhibiting specific affinity to select amino acids. Such molecular binding holds the potential to inhibit the protein's channel functionality, which may, in turn, thwart the virus's replication mechanism. Given this empirical evidence, these compounds from *W. somnifera* warrant consideration for future clinical inquiries pertaining to SARS-CoV-2 [38, 43, 44].

8.6.1.3 *Vitis vinifera*

The grape plant Vitis vinifera is renowned for its beneficial health compounds. While its wines and by-products, including grape remnants, are acknowledged for their potency against a variety of pathogens like bacteria, fungi, and viruses, the properties of grape leaves remain less explored. To address this gap, researchers employed a water-methanol solution to extract compounds from the leaves, subsequently analyzing them via HPLC-MS. This analysis revealed 35 flavonoids, predominantly derivatives of quercetin, with others linked to compounds such as luteolin and kaempferol. The research also delved into the antiviral potential of the leaf extract, especially against viruses like HSV-1 and SARS-CoV-2. Notably, even at minimal concentrations, the extract effectively thwarted the replication of these viruses early in the infection process by targeting particular viral proteins. Such findings are heartening, suggesting that these natural extracts might play a role in antiviral drug development or future vaccine formulations. Given the backdrop of the current pandemic, this breakthrough represents a significant advancement in medicinal research for the treatment of SARS-CoV in a clinical setting [45].

8.6.1.4 *Curcuma longa*

Recent investigations into curcumin, the predominant constituent of *C. longa*, have shown its significant *in vitro* efficacy against myriad viruses, encompassing adenoviruses, dengue, herpes simplex, and notably, coronavirus (SARS-CoV), among others. Mechanistically, curcumin impedes the replication machinery of these viruses, modulates host cell signaling cascades, and precludes viral attachment and subsequent entry by negating the functionality of viral envelopes. The comprehensive scholarly evidence, complemented by empirical observations, underscores the influence of *C. longa* on an array of viruses, thereby rationalizing its prevalent utilization in regions such as Latin America and the Caribbean [46].

8.6.1.5 *Sambucus ebulus*

S. ebulus has been identified for its inhibitory properties against enveloped viruses, thereby positing its potential applicability in combatting a particular virus (the specific virus should be explicitly identified or elaborated upon if not evident from context). While the underlying mechanisms remain to be elucidated, preliminary observations suggest that *S. ebulus* may deter these viruses from penetrating host cells, primarily attributed to the presence of proteins termed lectins that impede viral ingress. Compounds inherent to the plant, such as quercetin 3–0-glucoside and isorhamnetin, have historically exhibited promise in conferring protection against the Ebola virus. Additionally, certain flavonoids intrinsic to the *Sambucus* genus, including diosgenin and yomogenin, have been empirically shown to thwart the hepatitis C virus's entry [33].

8.6.1.6 Flavonoids

Flavonoids, found in foods like fruits and vegetables, possess antioxidant properties and offer health benefits such as anticancer and antiviral effects. Historically recognized in TCM, their therapeutic potential against viruses, especially SARS-CoV-2, has gained recent scientific attention. Notably, quercetin is found to inhibit essential proteins of the CoV and reduce IL-6 levels. Studies on the SARS-CoV showed that its E protein mutation led to milder effects in mice, pointing to the E protein's role in severe inflammatory responses, termed the "cytokine storm." This protein's interactions with a host protein, syntenin, contribute to cytokine imbalance. Current research indicates flavonoids, particularly quercetin and genistein, can inhibit this protein, suggesting potential efficacy against multiple SARS virus strains. Some flavonoids have also shown promise in targeting other viral proteins like the SARS-CoV 3a ion channel [37].

8.6.2 Challenges and Limitations

The SARS-CoV-2 virus, responsible for the ongoing COVID-19 pandemic, presents a profound challenge to global health. As of the current literature review, no therapeutic intervention or vaccine has been definitively validated as efficacious against this disease. The difficulties associated with early detection of the virus compound the challenge, and there remains a dearth of therapeutic modalities. In this context, it becomes imperative for the global scientific community and public health officials to collaborate comprehensively in devising strategies to mitigate the crisis. Consequently, there has been a heightened emphasis on identifying remedies that can be both rapidly developed and disseminated. Natural therapeutics, renowned for their relatively benign side-effect profiles and established roles within the pharmaceutical canon–most notably as antiviral agents–emerge as potential candidates. A plethora of medicinal flora and naturally derived compounds have demonstrated antiviral efficacies against members of the CoV family. Several of these entities impede multiple stages of the viral life cycle, ranging from the initial invasion of host cells to the subsequent processes of replication and virion release. Preliminary evidence, albeit robust, indicates that these botanical entities and their derivatives might be pivotal in the therapeutic landscape of COVID-19. Harnessing the potential of herbal interventions that specifically counteract viral pathogenic pathways could fortify the clinical arsenal. It is noteworthy that certain investigative endeavors have successfully employed ethanol and aqueous extracts in therapeutic modalities targeting COVID-19. However, the employment of isolated phytocompounds may proffer a more targeted therapeutic vector, particularly considering the potential variability in efficacy contingent upon individual metabolic processes. The therapeutic index of these compounds, influenced by factors such as concentration, dosing regimen, and temporal considerations, necessitates meticulous exploration. This discourse seeks to elucidate the prospective utility of natural therapeutics in the context of COVID-19. Nevertheless, a considerable portion of the extant research is nascent, underscoring the need for rigorous *in vivo* studies to delineate the underlying mechanistic intricacies. Further, it is paramount to undertake exhaustive studies elucidating the pharmacokinetic profiles of these botanical agents, encompassing parameters from bioavailability to excretion. Human clinical trials remain integral to authenticate the safety and therapeutic efficacy of these natural interventions. A salient consideration in this paradigm is the potential pharmacodynamic interactions between these botanical agents and established antiviral pharmacologics. For instance, coadministration of ritonavir/lopinavir and *Hypericum perforatum* (commonly known as St. John's Wort) may pose deleterious ramifications. However, certain combinations of pharmaceuticals and botanicals might confer synergistic benefits. There is emerging discourse concerning the potential re-emergence of COVID-19 pathogenicity, accentuating the imperative for therapeutic regimens that preemptively address this contingency and continuously surveil novel viral phenotypes. Strategies augmenting the immunological resilience of afflicted individuals warrant exploration, and myriad botanical entities may offer promise in this domain [35, 47].

8.7   CONCLUSION AND FUTURE PERSPECTIVES

Since the latter part of 2019, the international scholarly community has navigated the multifaceted challenges posed by COVID-19, a phenomenon persisting for a duration surpassing 2 years. In tandem with this global problem, meticulous scientific explorations into the SARS-CoV-2 virus have unveiled the cardinal roles epitomized by the E protein. Integral to these roles is the facilitation of mature virion production and the intricate orchestration of pathogenetic mechanisms, which encompass the induction of cellular perturbations and the initiation of immunological cascades. Despite the prevailing dearth of exhaustive investigations centered on the E protein of SARS-CoV-2, the remarkable homology it maintains with its counterpart in SARS-CoV-1 proffers salient insights ripe for comparative analyses. As academic endeavors forge ahead, elucidating the consequential impacts of the E protein upon cellular substrates, especially with reference to the endoplasmic reticulum, emerges as an imperative. Owing to its centrality in the virus's physiological architecture, the E protein has ascended to prominence as a potential target for pharmacological intercessions. Employing avant-garde computational paradigms, scholars have unearthed an array of compounds, encompassing both naturally occurring constituents and synthetically derived entities, that manifest favorable interactions with the E protein, thus nominating them as viable therapeutic contenders against COVID-19. The evolutionary continuum of the virus unfurls a sophisticated interplay, oscillating between human adaptive measures and its own genomic metamorphoses. Even as medicinal paradigms undergo refinement, the virus showcases its evolutionary dexterity through genomic oscillations. Despite the relative tenacity of the E protein and its sporadic mutational instances, the surfacing of anomalous configurations can precipitate considerable conundrums. The emergence of the Omicron variant, characterized by its heightened transmission kinetics yet modulated virulence, epitomizes this dynamic. The mutation, labeled as T9I, ensconced within its E protein, may indeed be seminal in sculpting its epidemiological contour, accentuating the inherent complexities entwined in curating an efficacious counterstrategy.

In conclusion, the exigency of cultivating a proactive orientation, one that may indeed proffer better solutions compared to post-hoc therapeutic formulations ensuing from emergent mutations, cannot be overstated. A better understanding of the E protein, its integral role in pathogenicity, and the consequences of its genomic variations are essential for the formulation of countermeasures against COVID-19 and potential forthcoming CoV episodes.

REFERENCES

1. Khan, F.I., et al., *Remdesivir Strongly Binds to RNA-Dependent RNA Polymerase, Membrane Protein, and Main Protease of SARS-CoV-2: Indication From Molecular Modeling and Simulations.*
2. Maison, D.P., Deng, Y., and Gerschenson, M., *SARS-CoV-2 and the host-immune response.* Front Immunol, 2023. **14**: p. 1195871.
3. Polatoglu, I., et al., *COVID-19 in early 2023: Structure, replication mechanism, variants of SARS-CoV-2, diagnostic tests, and vaccine & drug development studies.* MedComm (2020), 2023. **4**(2): p. e228.
4. Raman, K., et al., *Role of natural products towards the SARS-CoV-2: A critical review.* Ann Med Surg (Lond), 2022. **80**: p. 104062.
5. Bafandeh, S., et al., *Natural products as a potential source of promising therapeutics for COVID-19 and viral diseases.* Evid Based Complement Alternat Med, 2023. **2023**: p. 5525165.
6. Conroy, T., et al., *Total synthesis, stereochemical assignment, and antimalarial activity of gallinamide A.* Chemistry, 2011. **17**(48): p. 13544–52.
7. Wang, Z., et al., *Bioactive natural products in COVID-19 therapy.* Front Pharmacol, 2022. **13**: p. 926507.
8. Wasilewicz, A., et al., *Identification of natural products inhibiting SARS-CoV-2 by targeting viral proteases: A combined in silico and in vitro approach.* J Nat Prod, 2023. **86**(2): p. 264–275.
9. Kim, C.H., *Anti-SARS-CoV-2 natural products as potentially therapeutic agents.* Front Pharmacol, 2021. **12**: p. 590509.
10. Demeke, C.A., Woldeyohanins, A.E., and Kifle, Z.D., *Herbal medicine use for the management of COVID-19: A review article.* Metabol Open, 2021. **12**: p. 100141.
11. Lin, J.G., Huang, G.J., and Su, Y.C., *Efficacy analysis and research progress of complementary and alternative medicines in the adjuvant treatment of COVID-19.* J Biomed Sci, 2023. **30**(1): p. 30.
12. Sruthi, D., et al., *Curative potential of high-value phytochemicals on COVID-19 infection.* Biochemistry (Mosc), 2023. **88**(1): p. 64–72.
13. Omrani, M., et al., *Potential natural products against respiratory viruses: A perspective to develop anti-COVID-19 medicines.* Front Pharmacol, 2020. **11**: p. 586993.
14. Srivastava, S., et al., *A brief review on medicinal plants-at-arms against COVID-19.* Interdiscip Perspect Infect Dis, 2023. **2023**: p. 7598307.
15. Antonio, A.S., et al., *Efficacy and sustainability of natural products in COVID-19 treatment development: Opportunities and challenges in using agro-industrial waste from Citrus and apple.* Heliyon, 2021. **7**(8): p. e07816.
16. Chauhan, D.S., Yadav, S., and Quraishi, M.A., *Natural products as environmentally safe and green approach to combat Covid-19.* Current Research in Green and Sustainable Chemistry, 2021. **4**.
17. Pezeshkian, W., et al., *Molecular architecture and dynamics of SARS-CoV-2 envelope by integrative modeling.* Structure, 2023. **31**(4): p. 492–503.e7.
18. Santos-Mendoza, T., *The envelope (E) protein of SARS-CoV-2 as a pharmacological target.* Viruses, 2023. **15**(4).
19. Tang, Z., et al., *CD36 mediates SARS-CoV-2-envelope-protein-induced platelet activation and thrombosis.* Nature Communications, 2023. **14**(1).

20. Khetran, S.R., and Mustafa, R., *Mutations of SARS-CoV-2 structural proteins in the Alpha, Beta, Gamma, and Delta variants: Bioinformatics analysis.* JMIR Bioinform Biotech, 2023. **4**: p. e43906.
21. Wang, Y., et al., *Impact of SARS-CoV-2 envelope protein mutations on the pathogenicity of Omicron XBB.* Cell Discov, 2023. **9**(1): p. 80.
22. Zhou, S., et al., *SARS-CoV-2 E protein: Pathogenesis and potential therapeutic development.* Biomed Pharmacother, 2023. **159**: p. 114242.
23. Planes, R., et al., *SARS-CoV-2 envelope (E) protein binds and activates TLR2 pathway: A novel molecular target for COVID-19 interventions.* Viruses, 2022. **14**(5).
24. Yalcinkaya, M., et al., *Modulation of the NLRP3 inflammasome by Sars-CoV-2 envelope protein.* Sci Rep, 2021. **11**(1): p. 24432.
25. Berta, D., et al., *Role of RNA Splicing Mutations in Diffuse Large B Cell Lymphoma.* Int J Gen Med, 2023. **16**: p. 2469–2480.
26. Lu, H., et al., *Potent NKT cell ligands overcome SARS-CoV-2 immune evasion to mitigate viral pathogenesis in mouse models.* PLoS Pathog, 2023. **19**(3): p. e1011240.
27. Cabrera-Garcia, D., et al., *The envelope protein of SARS-CoV-2 increases intra-Golgi pH and forms a cation channel that is regulated by pH.* J Physiol, 2021. **599**(11): p. 2851–2868.
28. Chai, J., et al., *Structural basis for SARS-CoV-2 envelope protein recognition of human cell junction protein PALS1.* Nat Commun, 2021. **12**(1): p. 3433.
29. Dregni, A.J., et al., *The cytoplasmic domain of the SARS-CoV-2 envelope protein assembles into a beta-sheet bundle in lipid bilayers.* J Mol Biol, 2023. **435**(5): p. 167966.
30. Ramirez Salinas, G.L., et al., *In silico screening of drugs that target different forms of E protein for potential treatment of COVID-19.* Pharmaceuticals (Basel), 2023. **16**(2).
31. England, C., et al., *Plants as biofactories for therapeutic proteins and antiviral compounds to combat COVID-19.* Life (Basel), 2023. **13**(3).
32. Hajimonfarednejad, M., et al., *Medicinal plants for viral respiratory diseases: A systematic review on Persian medicine.* Evid Based Complement Alternat Med, 2023. **2023**: p. 1928310.
33. Patel, B., et al., *Therapeutic opportunities of edible antiviral plants for COVID-19.* Mol Cell Biochem, 2021. **476**(6): p. 2345–2364.
34. Sharma, R., et al., *Potential medicinal plants to combat viral infections: A way forward to environmental biotechnology.* Environ Res, 2023. **227**: p. 115725.
35. Abou Baker, D.H., Hassan, E.M., and El Gengaihi, S., *An overview on medicinal plants used for combating coronavirus: Current potentials and challenges.* J Agric Food Res, 2023. **13**: p. 100632.
36. Hossain, A., et al., *Identification of medicinal plant-based phytochemicals as a potential inhibitor for SARS-CoV-2 main protease (M(pro)) using molecular docking and deep learning methods.* Comput Biol Med, 2023. **157**: p. 106785.
37. Breitinger, U., et al., *Inhibition of SARS CoV envelope protein by flavonoids and classical viroporin inhibitors.* Front Microbiol, 2021. **12**: p. 692423.
38. Abdullah Alharbi, R., *Structure insights of SARS-CoV-2 open state envelope protein and inhibiting through active phytochemical of ayurvedic medicinal plants from Withania somnifera.* Saudi J Biol Sci, 2021. **28**(6): p. 3594–3601.
39. Bijelic, K., Hitl, M., and Kladar, N., *Phytochemicals in the prevention and treatment of SARS-CoV-2-clinical evidence.* Antibiotics (Basel), 2022. **11**(11).
40. Cao, Y., et al., *Characterization of the SARS-CoV-2 E protein: Sequence, structure, viroporin, and inhibitors.* Protein Sci, 2021. **30**(6): p. 1114–1130.
41. Fam, M.S., et al., *Channel activity of SARS-CoV-2 viroporin ORF3a inhibited by adamantanes and phenolic plant metabolites.* Sci Rep, 2023. **13**(1): p. 5328.
42. Jamal, Q.M.S., *Antiviral potential of plants against COVID-19 during outbreaks-an update.* Int J Mol Sci, 2022. **23**(21).
43. Das, K., et al., *Inhibition of SARS-CoV2 viral infection with natural antiviral plants constituents: An in-silico approach.* J King Saud Univ Sci, 2023. **35**(3): p. 102534.
44. Ramli, S., et al., *Phytochemicals of Withania somnifera as a future promising drug against SARS-CoV-2: Pharmacological role, molecular mechanism, molecular docking evaluation, and efficient delivery.* Microorganisms, 2023. **11**(4).
45. Zannella, C., et al., *Antiviral activity of Vitis vinifera leaf extract against SARS-CoV-2 and HSV-1.* Viruses, 2021. **13**(7).
46. Husaini, D.C., et al., *Phytotherapies for COVID-19 in Latin America and the Caribbean (LAC): Implications for present and future pandemics.* Arab Gulf Journal of Scientific Research, 2023.
47. Fornari Laurindo, L., et al., *Exploring the impact of herbal therapies on COVID-19 and influenza: Investigating novel delivery mechanisms for emerging interventions.* Biologics, 2023. **3**(3): p. 158–186.

9 Molecular Mechanisms of SARS-CoV-2 Nucleocapsid Protein Inhibitors from Natural Compounds

Adityaraj Patidar, Aditya Khatri, Eeshan Pawar,
Ravindra Verma, and Ashwin Kotnis

9.1 INTRODUCTION

The SARS-CoV-2 virus, which caused the COVID-19 pandemic, first appeared in 2019 and rapidly spread globally, causing widespread disruption to health, economies, and daily routines. Its profound impact has been unparalleled, with far-reaching consequences that extend beyond healthcare. This respiratory illness, presenting a spectrum of symptoms from mild respiratory issues to severe pneumonia, primarily spreads through respiratory droplets. The dynamic nature of the virus has challenged public health responses and prompted unprecedented measures to curb its transmission. The strain on healthcare systems globally has been overwhelming, marked by a surge in hospitalizations and an acute demand for medical resources. The relentless influx of COVID-19 patients has stretched the capacities of hospitals and healthcare professionals to the limit, necessitating innovative approaches and collaborative efforts.

Societal disruptions have reverberated across communities, leading to economic downturns and pervasive fear. Lockdowns, social distancing measures, and travel restrictions have altered the fabric of daily life, creating a new normal characterized by uncertainty and adaptability. The ramifications of these disruptions are felt not only in the immediate health crisis but also in the long-term implications for various sectors. As of November 19, 2023, the global toll of the pandemic stands at 6,939,362 deaths and 697,964,044 confirmed cases. The United States and Brazil have borne the brunt of the mortality, leading to reported deaths. The sheer scale of these numbers underscores the gravity of the situation, prompting introspection on global preparedness for pandemics and the imperative for international cooperation. The quest for effective treatments and vaccines has been a beacon of hope amid the challenges. Scientific communities worldwide have collaborated at an unprecedented pace to develop and distribute vaccines, showcasing the power of collective knowledge and shared commitment to public health.

Despite the progress in vaccination efforts, disparities in vaccine distribution and access persist, emphasizing the need for global solidarity in addressing health crises. The evolving nature of the virus, with new variants emerging, also underscores the importance of ongoing research and surveillance to stay ahead of the curve. Looking ahead, the lessons learned from the COVID-19 pandemic must inform future strategies for pandemic preparedness and response. Strengthening healthcare systems, bolstering international collaboration, and addressing socio-economic inequalities are crucial aspects that demand attention. The COVID-19 pandemic has left an indelible mark on the world, testing the resilience of individuals, communities, and nations. The journey toward recovery is ongoing, and the experiences of this global crisis will undoubtedly shape how societies approach health emergencies in the future.

DOI: 10.1201/9781003452621-9

EVOLUTIONARY LINEAGE OF SARS-CoV-2

Molecular analysis of nucleic acid structure reveals that SARS-CoV-2 is a newly discovered member of the genus *beta-Coronavirus* and subgenus *Sarbecovirus*. This virus has positive sense, single-stranded RNA (ssRNA)genomic features (Mogro et al., 2022).

In the past, many coronaviruses (CoVs) have infected humans and SARS-CoV-2 is seventh in sequence. The six viruses that came before, listed in chronological order, are HCoV-OC43, HCoV-HKU1, HCoV-NL63, HCoV-229E, SARS-CoV, and MERS-CoV (Liu et al., 2021).

SARS-CoV-2 is 96.2% similar to RaTG13 (bat SARS-like coronavirus) in its genetic structure.

SARS-CoV-2 PROTEOME

Turning attention to the SARS-CoV-2 proteome, the virus boasts a genome size of around 30,000 bases. Within its genetic structure are two significant open reading frames (ORF), ORF1a and ORF1b. These ORFs encode four proteins (spike [S], envelope [E], membrane [M], and nucleocapsid [N]) called structural proteins, along with nine accessory proteins. In the CoV's RNA genome, ORFs give rise to two substantial polyproteins, PP1a and PP1ab. About 16 nonstructural proteins (NSPs) are generated when these polyproteins undergo processing facilitated by cysteine proteases. The polyproteins undergo cleavage at the C terminal regions facilitated by a cysteine protease (3CL-PRO) similar to chymotrypsin while a protease (PL-PRO) resembling papain processes the N terminal end. Cleavage of the N terminal portion produces three NSPs that play a pivotal role in the viral replication process by helping in the formation of the replicase-transcriptase complex (Mouffouk et al., 2021; Neuman et al., 2014).

The Nprotein is noteworthy for its significant preservation in the *Coronavirus* genus and its role as one of the most prevalent structural proteins found in cells infected by the virus. Its primary role involves wrapping the viral genome RNA to form a ribonucelocapsid complex, which is long and helical, while it also contributes to assembling the virion by interacting with the genome of the virus and the M protein (Bai et al., 2021).

In addition to its assembly functions, the N protein has a range of other roles, including involvement in viral mRNA transcription and replication, organization of the cytoskeleton, and regulation of the immune response. Also, the N protein plays an important role in RNA silencing by acting as viral suppressor of RNA (VSR)by using its binding activity toward double-stranded RNA (dsRNA) to counter the RNA interference (RNAi)-mediated antiviral response of the host. Given its crucial functions, the N protein becomes an appealing target for developing broad-spectrum inhibitors against the virus.

9.2 STRUCTURE OF NUCLEOCAPSIDPROTEIN

The N proteinof this viruscomprises 419 amino acids and it is generated through the encoding of the ninth ORF. Its primary structure has two distinct and autonomously arranged structural units: the N-terminal domain (NTD) and the C-terminal domain (CTD). These domains are interconnected by a pliable linker region, called the Ser/Arg (SR)-rich central linker (LKR) and the N-arm and C-tail flank the primary structure.

Within the primary structure, the SR-rich zone of the LKR encompasses possible sites for phosphorylation. Previous research has indicated that the N-terminal RNA-binding domain (N-NTD) is accountable for binding with RNA, the N-terminal domain of the C-terminus (N-CTD) is involved in both RNA binding and the formation of dimers, and the intrinsically disordered region (IDR) participates in regulating the crucial RNA-related functions of both the N-NTD and N-CTD, as well as in oligomerization (Figures 9.1 and 9.2).

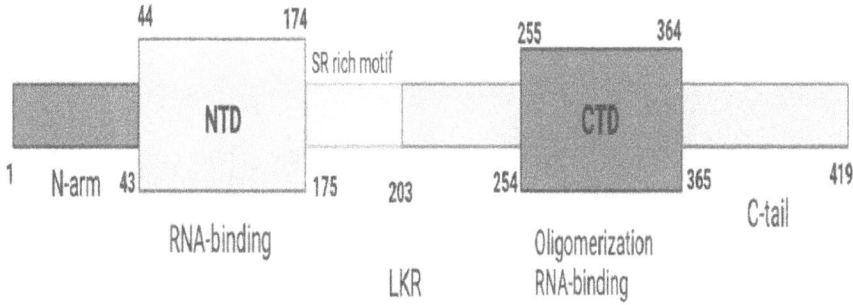

FIGURE 9.1 Schematic representation of domains of nucleocapsid protein of SARS-CoV-2. (Reference: Liu et al., 2021.)

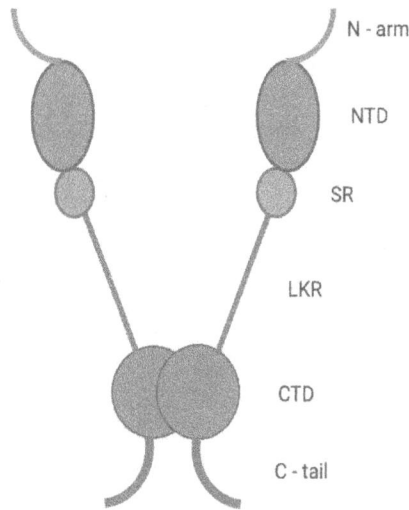

FIGURE 9.2 Schematic representation of nucleocapsid protein dimer of SARS-CoV-2. (Reference: Liu et al., 2021.)

Each N-NTDentity takes on a unique right-handed fist-shaped configuration, consisting of a core subdomain with a quadruple-stranded antiparallel beta-sheet. This core subdomain is nestled between loops and short 3′10 helices and an extended β-hairpin region created by the extension of the β2 and β3 strands from the central core. The surface charge distribution of these protruding loops is notably positive, suggesting a potential site for RNA binding.

On the other hand, the N-CTD creates a condensed homodimer assuming a generally rectangular slab-like configuration. Each protomer takes on a configuration reminiscent of the letter "C," encompassing five α-helices, two β-strands, and two 3′10 helices. The β-hairpin of one protomer is introduced into the opening of the other, resulting in the establishment of a four-stranded, antiparallel β-sheet at the dimeric junction. The β-sheet constitutes a single facet of the slab-shaped dimer, whereas the opposing side is composed of α-helices and loops. The stability of the dimeric configuration is attributed to robust hydrogen bond connections within the two hairpins and hydrophobic interactions between the β-sheet and the α-helices (Bai et al., 2021; Zhou et al., 2020).

9.3 FUNCTIONAL ROLE OF N PROTEIN: PATHOGENESIS AND FUNCTION OF N PROTEIN IN INFECTION

9.3.1 GENE ENCAPSIDATION

The fundamental function of the N protein is to encapsulate the genome of the virus into an RNP particle, which requires its capability to identify and attach to RNA. Inside the N-NTD, The extended β-hairpin (β2′–β3′) is predominantly composed of essential amino acid residues. An in-depth examination of the electrostatic potential on the surface uncovers a positively charged cavity at the intersection of this crucial hairpin and the central structure. This charged pocket acts as a potential site for RNA binding. Both RNA molecules with double-stranded structure (dsRNA) and those with single-stranded structure (ssRNA) exhibit a similar binding pattern within the positively charged gorge positioned amid the fundamental β-hairpin and the nucleus of the N-NTD. This is where arginine residues (R92, R107, and R149) are situated, directly interacting with the RNA.

The existence of an additional RNA-binding domain featuring a positively charged crevice on the helical side of the N-CTD dimer has been confirmed in previous studies. Specifically, the N-CTD of this virus exhibits a positively charged pathway consisting of the amino-acid residues K256, K257, K261, and R262. Additionally, it has been found that both the N-terminal IDR and the LKR exhibit the capability to bind with RNA. These elements enhance the binding allostery and increase binding affinity, allowing the N protein to interact with RNA with strong cooperativity. Collectively, this data emphasizes the collaborative RNA-binding capacities of the NTD, the CTD, and specific unordered sections of the N protein in facilitating the packaging of ribonucleoprotein (RNP) (Bai et al., 2021).

9.3.2 THE N PROTEIN EXPERIENCES LIQUID–LIQUID PHASE SEPARATION TO ELEVATE VIRAL TRANSCRIPTION AND FACILITATE ASSEMBLY

Liquid–liquid phase separation (LLPS) is a prevalent cellular mechanism that arranges biological substances into compartments. In the process of virion assembly, it becomes imperative to generate concentrated compartments containing both proteins and nucleic acids. These compartments isolate cell proteins of host cells, safeguarding the virus against the host immune system and concentrating viral elements to enhance replication efficiency.

Several research teams have demonstrated that the N protein of SARS-CoV-2 can engage in liquid LLPS with different RNAs in laboratory settings. Its conduct is affected by pH, salt concentration, and RNA levels. When the N protein associates with RNA, it strongly experiences LLPS, creating droplets or condensates with liquid-like characteristics.

Additionally, it has been discovered that the phosphorylation of particular segments on the N protein, referred to as the SR region, has the potential to regulate the LLPS triggered by RNA. Phosphorylation in the SR region diminishes these interactions, establishing a more fluidic compartment conducive to viral genome processing.

Recent observations also indicate that the sequence and structure of viral RNA participate in controlling the condensation of the N protein, contingent upon their binding configurations. Sequences fostering LLPS are located at both the 5′ and 3′ extremities of the genome, suggesting their involvement in genome packaging. Conversely, other regions of the genome promote the dissolution of condensates, potentially inhibiting the clumping together of the extensive viral genetic material (Yu et al., 2023).

9.3.3 CONTROL OF THE HOST IMMUNE REACTIONS

Viruses have developed various strategies to replicate in host cells while evading innate immune detection effectively, and contemporary research suggests that the N protein is involved in this mechanism.

Inside the host cells, an inherent antiviral immune defense mechanism, RNAi, exists. This system can break down viral genomes, hindering the replication of the virus. To counteract this response, viruses produce VSRs to thwart the RNAi pathway. The N proteincan oppose RNAi at various phases. In the initial step, it appears that the N protein captures dsRNA within infected cells, potentially obstructing the Dicer's ability to identify and cut viral dsRNA. The N protein also can inhibit RNAi induced by small interfering RNA (siRNA) within cells, demonstrating its counteractive influence in the effector phase. This implies that employing the N protein as a viral suppressor of RNAi is a prevalent immune evasion strategy among CoVs.

Upon encountering infection from various RNA viruses, the RIG-I-like receptor pathway becomes active, prompting the synthesis of IFN-β, a crucial component of the host's innate immune response. When the host is infected with SARS-CoV-2, retinoic acid-inducible gene I (RIG-I) and melanoma differentiation-associated protein (MDA5) identify viral RNA. Subsequently, these proteins move to the mitochondrion and attach to the adaptor protein mitochondrial antiviral signaling (MAVS), creating a signalosome. The signalosome results in the phosphorylation and subsequent movement of interferon regulatory factor (IRF) 3/7 to the nucleus, encouraging the transcription of genes related to type I and III interferons. Moreover, viral infection induces the creation of stress granules, resulting in the association of G3BP1 with RIG-I, consequently enhancing its activation. Multiple research investigations have suggested that the Nprotein hinders the generation of type I interferon prompted by RNA viruses.

After the synthesis of type I interferons, they attach to their receptors and trigger intracellular tyrosine kinases. This results in the phosphorylation and stimulation of the signal transducer and activator of transcription (STAT) 1/2. The Nprotein disrupts the phosphorylation and movement of STAT1/2 to the cell nucleus prompted by type I interferon (Yu et al., 2023).

9.4   SEARCHING FOR POTENTIAL DRUGS USING MOLECULAR DOCKING

Molecular docking is a computer simulation technique utilized in structural biology and drug discovery to predict a ligand's preference and binding affinity, such as a drug or small molecule, to a target protein or biomolecule. The goal of docking is to examine potential binding and interactions between ligands and targets, investigating the strength and mode of binding. This understanding is crucial in deciphering mechanisms and designing new drugs. The process initiates with obtaining the crystal structure of the target protein, in this case, the nucleocapsid phosphoprotein, from a reliable protein database, such as the RCSB Protein Data Bank.

Protein synthesis involves removing crystalline water molecules and bound ligands from the protein structure. Polar hydrogen atoms and Gasteiger charges are incorporated into the protein structure to prepare it for docking simulations. The next step is ligand selection and preparation, where ligands, typically small molecules or drugs with therapeutic potential, are chosen. The chemical structures of these ligands are retrieved or generated, usually from databases like PubChem, and converted to Protein Data Bank (PDB) coordinate file format.

Molecular insertion follows, utilizing a molecular docking tool such as AutoDock Vina to predict a ligand's interaction with a protein's active site. Ligands and receptors (proteins) undergo preprocessing by adding hydrogen atoms, donating charges, and removing unwanted elements. The network around the active site is defined to guide the docking simulation. Docking algorithms are then employed to find different ligand orientations and conformations in the active site, and the docking score is calculated to determine bond affinity. Orientations with the lowest binding strength are selected.

In the analysis and visualization stage, advanced tools like Discovery Studio are employed to examine the intricate interactions between proteins and ligands closely. It's akin to peering through a microscope to explore two-dimensional and three-dimensional interactions at the molecular level, revealing the secrets of specific amino-acid residues engaged in crucial activities like hydrogen bonding and hydrophobic interactions. Absorption Distribution Metabolism Elimination (ADME) prediction serves as a screening process, where compounds are sifted through based on their behavior within the body. Tools like Molinspiration and admetSAR are used to examine how these compounds are absorbed, distributed, metabolized, and excreted.

Moving on to molecular dynamics simulation, compounds enter a virtual world, akin to a simulation video game, using a tool like GROMACS. The compounds undergo a dynamic adventure, and key parameters like hydrogen bond stability, the radius of gyration (Rg), root mean square deviation (RMSD), and potential energy act as plot twists. The simulation runs for 50 virtual seconds, providing a detailed exploration of each compound's behavior. The data is interpreted in the results and discussion phase, examining the strength of the bonds formed, the stability achieved, and the interactions between the ligands and the target protein. The compounds are envisioned to play a crucial role in inhibiting the targeted protein.

9.5   N PROTEIN AS A THERAPEUTIC TARGET OF NATURAL COMPOUNDS

Phytotherapy is one of the alternative methods being investigated by the scientific community to explore and formulate preventive and therapeutic approaches to counter the menace of SARS-CoV-2. The RNA-binding domains at the NTD and CTD of the Nprotein offer promising targets for therapeutic drugs. Naturally occurring compounds, known for their structural diversity and safety, present a valuable source of potential drug candidates. To assess their antiviral potential, *insilico* biology tools were employed to determine their binding effectiveness to the N-NTD and N-CTD of Nprotein.

9.6   HERBAL MEDICINES AND PHYTOCHEMICALS THAT INHIBIT N PROTEIN

In this section, we will investigate the following herbal medicines and their phytochemicals (Table 9.1) that show an affinity for the nucleocapsid protein. We will discuss their sources, action mechanisms, and ongoing clinical trials.

9.6.1   *Nigella sativa*

N. sativa, belonging to the *Ranunculaceae* family, is commonly referred to as black seed/black cumin in English, while it's known in Arabic as Habbatul barakah, and in Hindi and Urdu as Kalonji. When investigating the phytochemical composition of different parts of *N. sativa*, it is evident that the plant contains a wide range of secondary bioactive compounds which are shown in Table 9.1.

TABLE 9.1
List of Herbs Used as Medicines and Their Phytochemical Constituents

Plant Name (Herbs)	Phytochemicals
Nigella sativa	*Terpenoids*: Thymoquinone, p-cymene, trans-anethol, carvacrol, limonene, and dithymoquinone
	Alkaloids: Nigellidine, nigellicimine, nigellicine, Nigellamine, etc.
	Sterols: Sitosterol, stigmasterol, tocopherol, tocotrienol, etc.
	Saponins: Alpha hederin, kalopanaxsaponin, nigella A–D
	Flavanoids: Querceptin, kaempferol, and rutin
	Phenolic acid: Gallic acid, vanillic acid, and trans-cinnamic acid
Cinnamon zeylanicum	Cinnamic acid, eugenol, cinnamaldehyde, and polyphenols
Cinnamomum tamala	Apigenin and catechin
Rheum emodi	Alizarin, aloeemodin, anthraquinone, anthrarufin, anthrone, chryophanol, clicoemodin, dantron, and emodin
Petroselinumcrispum	Apiin, apigenin, luteolin, and querceptin

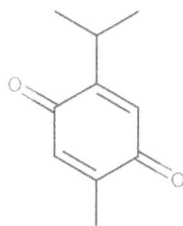

FIGURE 9.3 Structure of thymoquinone. PubChem CID-10281 (https://pubchem.ncbi.nlm.nih.gov/compound/Thymoquinone).

The *N. sativa* seed contains 26–34% fixed oil of which the major fatty acids are linoleic acid (64.6%) and palmitic acid (20.4%). Thymoquinone (2-methyl-5-isopropyl-1,4-benzoquinone) being a monoterpenoid, is a significant volatile constituent (constituting 30–48%) of the *N. sativa* (NS) essential oil, displaying a variety of beneficial bioactivities (Figure 9.3).

Various phytochemicals have shown binding affinity to target proteins, especially aiming at the N-terminal RNA binding domain (NRBD) of the Nprotein and the papain-like protease (PL-PRO) to combat the virus (Rolta et al., 2020).

9.6.2 CINNAMON ZEYLANICUM

Cinnamon, commonly known as Dalchini in India and derived from the bark of various *Cinnamomum* species, has garnered scientific attention not only for its culinary significance but also for potential medicinal characteristics such as its anti-inflammatory, antitumor, anticancer, antidiabetic, antimicrobial, and cardioprotective attributes (Verma & Bagel, 2022; Verma & Bisen, 2022).

Cinnamon boasts a complex chemical composition with its distinctive flavor primarily attributed to cinnamaldehyde while eugenol contributes to its aroma. Cinnamic acid, a precursor to cinnamaldehyde, and polyphenols add antioxidant properties, enhancing potential health benefits. Coumarin, present in varying amounts in different cinnamon types, imparts a sweet scent.

The raw extract derived from cinnamon bark is renowned for alleviating various respiratory conditions, including pneumonia, infectious diseases, and malignant pulmonary edema. Throughout history, formulations based on cinnamon bark have been employed in addressing fever, inflammation, respiratory conditions, gastrointestinal conditions, and pain. Cinnamon essential oil, containing approximately 45–65% cinnamaldehyde and cinnamic acid, exhibits powerful antiviral, antimicrobial, antifungal, anti-atherosclerosis, anticancer, anti-inflammatory, antiulcer, antidiabetic, antihypertensive, antioxidant, cholesterol-reducing, and lipid-reducing effects. Another pivotal element contributing to cinnamon's biological activity is eugenol (Figure 9.4).

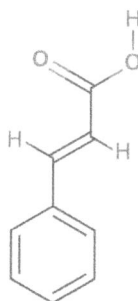

FIGURE 9.4 Structure of cinnamic acid. PubChem CID-444539 (https://pubchem.ncbi.nlm.nih.gov/compound/Cinnamic-Acid).

Recent studies have shown cinnamic acid's efficacy as a therapeutic agent in developing a defense mechanism against this virus through drug-like properties and molecular docking analysis (Husain et al., 2022).

9.6.3 CINNAMOMUM TAMALA

C. tamala, frequently referred to as Tejpatta or Indian bay leaf, is a tree native to India and Nepal. The aromatic leaves of this plant are utilized in culinary practices as a spice. In traditional medicine, *C. tamala* has been ascribed various medicinal properties such as antidiarrheic, anticancer, anti-inflammatory, arthritis fighting, antiparasitic, gastrointestinal and urinary, a ntiparasitic, free radical scavenging, and chemopreventive.

The chemical composition of *C. tamala* is characterized by a spectrum of bioactive compounds in its essential oil derived from the leaves, among them apigenin and catechin, of particular interest. Apigenin, a flavone present in leaves, has been studied for its pharmacological properties, including antioxidants and anti-inflammatory effects. Catechin, a flavonoid with known antioxidant properties, is another significant constituent (Figure 9.5).

Recent studies have shown apigenin and catechin's prophylactic and therapeutic benefits in developing a defense mechanism against SARS-CoV-2 through drug-like properties and molecular docking analysis (Husain et al., 2022).

9.6.4 RHEUM EMODI

R. emodi, scientifically known as *Rheum australe*, belongs to the *Polygonaceae* family. Its roots contain various bioactive compounds, including alizarin, aloeemodin, anthraquinone, anthrarufin, anthrone, chryophanol, clicoemodin, dantron, emodin and many more (Zargar et al., 2011).

Studies have explored the plant's medicinal applications, particularly in gastrointestinal disorders due to its laxative properties. Recent *in silico* studies found that emodin, anthrarufin, alizarine, aloeemodin, and dantron demonstrate advantageous binding affinity at three different sites within the NTD of the Nprotein. This interaction can impede the assembly of virus particles, thereby impeding infection. According to Absorption Distribution Metabolism Elimination Toxicity (ADMET) predictions, these compounds, namely emodin, anthrarufin, alizarine, aloeemodin, and dantron, demonstrate promise as prospective drugs for treating COVID-19 (Rolta et al., 2020) (Figure 9.6).

9.6.5 PETROSELINUM CRISPUM

P. crispum, commonly called parsley, is a culinary herb and a source of potential health benefits. Abundant in vitamins A, C, and K, along with minerals like iron and potassium, parsley is a staple

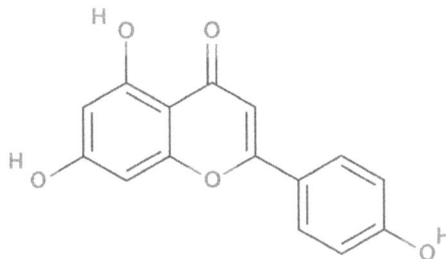

FIGURE 9.5   Structure of apigenin. PubChem CID-5280443 (https://pubchem.ncbi.nlm.nih.gov/compound/Apigenin).

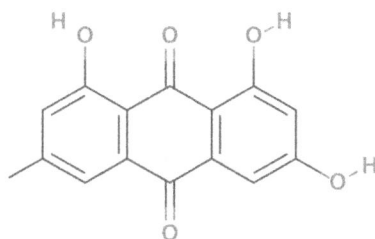

FIGURE 9.6 Structure of emodin. PubChem CID-3220 (https://pubchem.ncbi.nlm.nih.gov/compound/Emodin).

in the culinary traditions of Southeast Asia, India, South America, China, and Mexico. The above-ground components of parsley are employed to alleviate conditions like hemorrhoids and inflammation of the urethra. On the other hand, its roots are employed to aid in passing kidney stones and improve cognitive function and memory (Tang et al., 2015). The herb is acknowledged in folk medicine for its varied uses, encompassing roles as a soothing agent, intestinal tonic, diuretic, antidote, antiurolithiasis, and anti-inflammatory agent. It is employed in the treatment of conditions ranging from dysmenorrhea and amenorrhea to gastrointestinal diseases, hypertension, renal and cardiovascular diseases, metabolic disorders, ear inflammation, runny nose, and various skin-related issues. Furthermore, parsley has demonstrated antiviral activity attributed to flavones like apigenin, luteolin, and flavanol quercetin. Notably, *P. crispum* contains the flavonoid glycoside apiin, studied for its antioxidant properties and role in supporting kidney function and acting as a diuretic (de Menezes Epifanio et al., 2020) (Figure 9.7).

Recent *insilico* studies have shown apiin as a promising preventive and treatment option for SARS-CoV-2, subject to further *in vitro* and *in vivo* exploration (Husain et al., 2022).

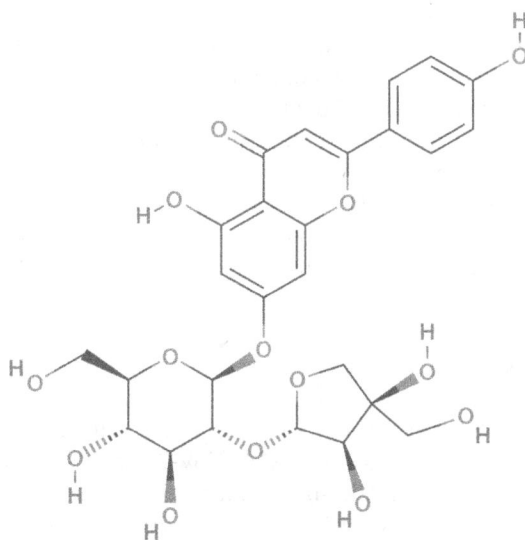

FIGURE 9.7 Structure of Apiin. PubChem CID-5280746 (https://pubchem.ncbi.nlm.nih.gov/compound/Apiin).

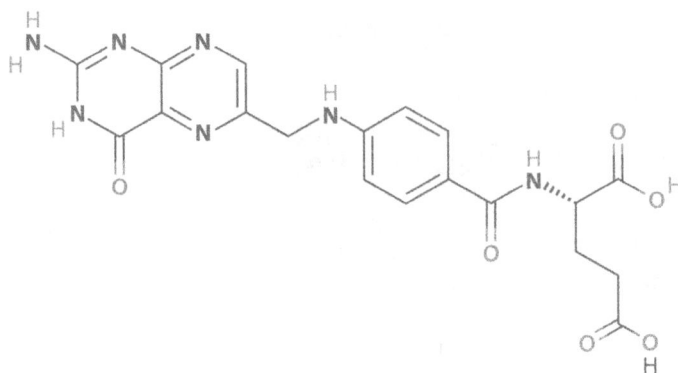

FIGURE 9.8 Structure of folic acid. PubChem CID-135398658 (https://pubchem.ncbi.nlm.nih.gov/compound/Folic-Acid).

9.6.6 FOLIC ACID

During the COVID-19 epidemic in China, many traditional Chinese medicine compositions were developed and widely employed to manage the disease. Many of these formulations demonstrate beneficial clinical outcomes in individuals with COVID-19. Further studies on these traditional Chinese medicines identified new antiviral substances for addressing SARS-CoV-2. One study employing *insilico* technology indicated that folic acid may be an efficacious drug targeting the nucleocapsid protein of SARS-CoV-2 (Chen et al., 2022) (Figure 9.8).

Also, research has shown that folic acid may have broad-spectrum effects against viruses and low toxicity when administered orally (Chen et al., 2022).

Natural compounds derived from other species also showed significant binding to the Nprotein of the virus and can be possibly an antiviral agent. We will now look upon one such species:

9.6.7 *SUBEREA IANTHELLIFORMIS*

Suberea ianthelliformis, commonly known as the "yellow ball sponge," is characterized by its distinctive spherical shape and vibrant yellow coloration. Found in tropical and subtropical waters, particularly in the Caribbean and Western Atlantic regions. *Insilico* binding studies of bromotyrosin derivative compounds, obtained from this sponge, on N proteins of the virus showed S-scores in the range of -5.10 to -7.04 Kcal/mol. Also, these compounds showed significant binding to other proteins such as M protein, S glycoprotein, M glycoprotein, and Main protease compared to native ligands. Sponge-derived compounds include mololipids, fistularin-3, 11-ketofistularin-3, psammaplysene D, anomoian F, and many more such bromotyrosin derivatives (El-Demerdash et al., 2021) (Figure 9.9).

9.7 CONCLUSION

This chapter comprehensively explores the COVID-19 pandemic, covering its origins, global impact, and the molecular intricacies of the SARS-CoV-2 virus, specifically focusing on the N protein. Originating in 2019, the pandemic significantly affected health, economies, and daily life worldwide, leaving a lasting impact. As of November 19, 2023, the toll includes 6,939,362 deaths and 697,964,044 confirmed cases, emphasizing the urgent need for global preparedness and cooperation. As a member of the *beta-Coronavirus* genus, the virus exhibits a near genetic affinity with the bat SARS-like coronavirus (RaTG13).The proteome of the virus, which encompasses structural

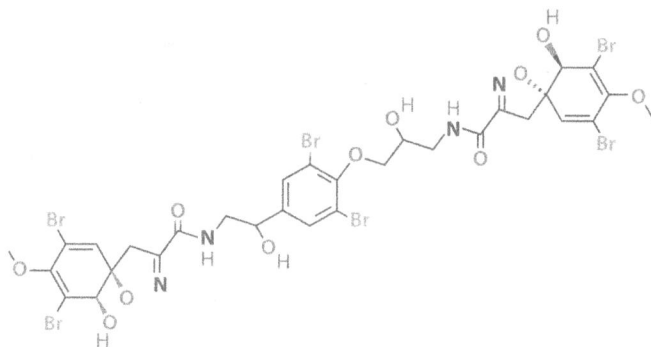

FIGURE 9.9 Structure of Fistularin 3. PubChem CID-10260646 (https://pubchem.ncbi.nlm.nih.gov/compound/10260646).

proteins such as spike, envelope, membrane, and nucleocapsid, is vital for viral assembly, replication, and the modulation of the immune response. The N protein stands out prominently with diverse functions such as gene encapsidation, LLPS, and control of the host immune response. The N protein's structure, with NTD and CTD, contains distinct regions responsible for RNA binding and dimerization. Its role in packaging the viral genome into ribonucleocapsid complexes is vital for the virus's replication process. Furthermore, the N protein experiences liquid–liquid phase transition, creating compartments for efficient viral transcription and assembly, while it simultaneously plays a vital role in modulating the immune response of the host. Due to the importance of the N protein in the viral life cycle, efforts have been undertaken to discover potential pharmaceutical compounds using molecular docking techniques, predicting the binding affinity of various compounds to the N protein, and offering insights into potential therapeutic agents. This involves obtaining the crystal structure of the target protein, ligand selection, molecular insertion, and subsequent analysis and visualization. Moreover, the exploration extends to phytotherapy, investigating the potential of natural compounds in targeting the N protein. Herbal medicines like *N. sativa*, *C.zeylanicum*, *C. tamala*, *R. emodi*, *P. crispum*, and even folic acid have shown promising binding affinities in *insilico* studies. With diverse phytochemical compositions, these compounds show promise as both preventive and therapeutic agents against SARS-CoV-2.

Altogether, the COVID-19 pandemic emphasizes the need for worldwide cooperation, inventive healthcare strategies, and readiness for forthcoming pandemics. Understanding the pathogenesis and identifying potential therapeutic targets for the virus, particularly with insights at the molecular level, are established through knowledge of the N protein. Exploring natural compounds as potential drugs opens avenues for further research and development. As the world strives toward recovery, the lessons learned from this global crisis will undoubtedly shape future strategies for pandemic preparedness and response.

REFERENCES

Bai, Z., Cao, Y., Liu, W., & Li, J. (2021). The SARS-CoV-2 nucleocapsid protein and its role in viral structure, biological functions, and a potential target for drug or vaccine mitigation. *Viruses*, *13*(6), 1115.

Chen, Y.-M., Wei, J.-L., Qin, R.-S., Hou, J.-P., Zang, G.-C., Zhang, G.-Y., & Chen, T.-T. (2022).Folic acid: A potential inhibitor against SARS-CoV-2 nucleocapsid protein, *Pharmaceutical biology*, 60(1), 862–878. DOI: 10.1080/13880209.2022.2063341

de Menezes Epifanio, N. M., Cavalcanti, L. R. I., Dos Santos, K. F., Duarte, P. S. C., Kachlicki, P., Ożarowski, M., … & de Almeida Chaves, D. S. (2020). Chemical characterization and in vivo antioxidant activity of parsley (*Petroselinum crispum*) aqueous extract. *Food & function*, *11*(6), 5346–5356.

El-Demerdash, A., Hassan, A., Abd El-Aziz, T. M., Stockand, J. D., & Arafa, R. K. Marine brominated tyrosine alkaloids as promising inhibitors of SARS-CoV-2. *Molecules,*2021, 26, 6171. https://doi.org/10.3390/molecules26206171

Husain, I., Ahmad, R., Siddiqui, S., Chandra, A., Misra, A., Srivastava, A., Ahamad, T., Khan, M. F., Siddiqi, Z., Trivedi, A., Upadhyay, S., Gupta, A., Srivastava, A. N., Ahmad, B., Mehrotra, S., Kant, S., Mahdi, A. A., & Mahdi, F. (2022). Structural interactions of phytoconstituent(s) from cinnamon, bay leaf, oregano, and parsley with SARS-CoV-2 nucleocapsid protein: A comparative assessment for development of potential antiviral nutraceuticals. *Journal of food biochemistry*, 46(10), e14262. https://doi.org/10.1111/jfbc.14262

Yu H, Guan F, Miller H, Lei J, & Liu C. (2023). The role of SARS-CoV-2 nucleocapsid protein in antiviral immunity and vaccine development. *Emerg Microbes Infect.* Dec;12(1):e2164219. doi: 10.1080/22221751.2022.2164219. PMID: 36583642; PMCID: PMC9980416.

Liu, D. X., Liang, J. Q., & Fung, T. S. (2021). Human coronavirus-229E,-OC43,-NL63, and-HKU1(Coronaviridae). *Encyclopedia of virology*, 428.

Mogro, E. G., Bottero, D., & Lozano, M. J. (2022). Analysis of SARS-CoV-2 synonymous codon usage evolution throughout the COVID-19 pandemic. *Virology*, *568*, 56–71.

Mouffouk, C., Mouffouk, S., Mouffouk, S., Hambaba, L., & Haba, H. (2021). Flavonols as potential antiviral drugs targeting SARS-CoV-2 proteases (3CLpro and PLpro), spike protein, RNA-dependent RNA polymerase (RdRp) and angiotensin-converting enzyme II receptor (ACE2). *European journal of pharmacology*, *891*, 173759.

Neuman, B. W., Chamberlain, P., Bowden, F., & J oseph, J. (2014). Atlas of coronavirus replicase structure. *Virus research*, *194*, 49–66.

Rolta, R., Yadav, R., Salaria, D., Trivedi, S., Imran, M., Sourirajan, A., Baumler, D. J., & Dev, K. (2020): *Insilico* screening of hundred phytocompounds of ten medicinal plants as potential inhibitors of nucleocapsid phosphoprotein of COVID-19: an approach to prevent virus assembly. *Journal of biomolecular structure and dynamics*, DOI: 10.1080/07391102.2020.1804457

Tang, E. L., Rajarajeswaran, J., Fung, S., & Kanthimathi, M. S. (2015). *Petroselinum crispum* has antioxidant properties, protects against DNA damage and inhibits proliferation and migration of cancer cells. *Journal of the science of food and agriculture*, *95*(13), 2763–2771. https://doi.org/10.1002/jsfa.7078

Verma, R., & Bagel, M. P. (2022). Role of functional food additives in regulating the immune response to COVID-19. *Current biotechnology*, *11*(3), 230–239.

Verma, R., & Bisen, P. S. (2022). Cinnamon- An immune modulator food additive to coronavirus. *Journal of food bioactives*, *17*. https://doi.org/10.31665/JFB.2022.17298

Yu, H., Guan, F., Miller, H., Lei, J., & Liu, C. (2023).The role of SARS-CoV-2 nucleocapsid protein in antiviral immunity and vaccine development, *Emerging microbes & infections.*12(1), e2164219. DOI: 10.1080/22221751.2022.2164219

Zargar, B. A., Masoodi, M. H., Ahmed, B., & Ganie, S. A. (2011). Phytoconstituents and therapeutic uses of *Rheum emodi* wall. ex Meissn. *Food chemistry*, *128*(3), 585–589. https://doi.org/10.1016/j.foodchem.2011.03.083

Zhou, R., Zeng, R., von Brunn, A., & Lei, J. (2020). Structural characterization of the C-terminal domain of SARS-CoV-2 nucleocapsid protein. *Molecular biomedicine*, *1*, 1–11.

https://pubchem.ncbi.nlm.nih.gov/compound/Thymoquinone
https://pubchem.ncbi.nlm.nih.gov/compound/Cinnamic-Acid
https://pubchem.ncbi.nlm.nih.gov/compound/Apigenin
https://pubchem.ncbi.nlm.nih.gov/compound/Emodin
https://pubchem.ncbi.nlm.nih.gov/compound/Apiin
https://pubchem.ncbi.nlm.nih.gov/compound/Folic-Acid
https://pubchem.ncbi.nlm.nih.gov/compound/Fistularin-3

10 Molecular Mechanisms of RNA-Dependent RNA Polymerase Inhibition from Traditional Bioactive Medicine

Partha Biswas, Md Imtiaz, Md Moinuddin,
Md Hasibul Hasan, Md Habibur Rahaman,
Labib Shahriar Siam, Sheikh Naem Islam Abhi,
Anwar Parvez, Abdullah Al Noman,
Md Nazmul Hasan, and Shabana Bibi

10.1 INTRODUCTION

RNA and DNA, each coated in a protein sheath, are the genetic materials often found in viruses, the primary agents responsible for infectious diseases [1]. The protein coatings perform a wide range of functions, especially assisting the virus to adhere to the surface of the host cell, infiltrating the host, and using the cellular machinery of the host to replicate [2]. Based on their genetic makeup, viruses are divided into two categories: RNA and DNA viruses. RNA viruses have a notable influence on global healthcare systems. Furthermore, RNA viruses may be divided into three groups—double-stranded (dsRNA), positive-sense single-strand (RNA), and negative-sense single-strand (ssRNA)—according to the kind of genome they contain [3]. Regarding the global morbidity related to viral infections, RNA viruses are mostly responsible. A major threat to public health is the emergence of chronic disorders brought on by recurrent RNA virus infections [4]. Even though well-known viruses like hepatitis C (HCV) and HIV-1, are linked to chronic diseases, research suggests that a number of other RNA viruses, including ones that are emerging like Zika and Ebola, may also cause persistent infections [2, 5]. Additionally, the general population is seriously threatened by pandemics from diseases like avian and swine influenza, as well as the Middle East respiratory syndrome coronavirus (MERS-CoV) and severe acute respiratory syndrome coronavirus (SARS-CoV) [6, 7]. RNA viruses are known to be very common, widely distributed, genetically diverse, and highly susceptible to recombination, all of which contribute to their persistent risk to human health. The advent of severe acute respiratory syndrome coronavirus-2 (SARS-CoV-2), which sparked the coronavirus disease 2019 (COVID-19) pandemic, is a prime example of the harm posed by this virus. In addition to vaccinations, there is a favorable outlook for certain medications to control the ongoing worldwide COVID-19 pandemic propagation [8, 9].

RNA transcription from DNA, protein translation from RNA, and DNA replication are the three main channels by which genetic information flows in cells, according to the fundamental principles of molecular biology [10]. The replication of RNA from RNA, which is aided by enzymes called RNA-dependent RNA polymerases (RdRps), is a newly discovered mechanism that operates outside of this framework. Originally when RNA was the main genetic material, RdRps probably had a major impact on early evolution. They still have important roles in modern biology [11]. Enzymes involved in genome replication, mRNA synthesis, RNA recombination, and other activities are encoded by different RNA viruses [11]. For all major RNA virus groups, RdRps reveal conserved sequence distinctive characteristics [1]. RdRps from positive-strand RNA and dsRNA viruses, reverse transcriptase, and DNA-dependent RNA/DNA polymerases all exhibit structural commonalities in their crystal structures [12]. With fingers, palm, and thumb subdomains vital to polymerization, template binding, nucleoside triphosphate (NTP) access, and other related activities, the RdRp structure of positive-strand RNA viruses is shaped such as a cupped right hand [13]. Another N-terminal subdomain that connects the thumb and finger subdomains functions as the active site for all RdRps. Important communication between the thumb and finger subdomains is facilitated by the limited active site cavity, which supports the active site structural stability. Through its ability to maintain the template RNA and encourage polymerization, the finger subdomain plays a vital role in forming the active site structure. In the meantime, the thumb subdomain's residues help to pack the template RNA, stabilize the initial NTP, and facilitate template translocation by way of large conformational rearrangements [12, 14].

The crucial function of RdRp is integral to the replication of +RNA viruses throughout their viral genome processes. The CoV-2 virus releases its 14 open reading frames (ORFs) into the cytoplasm for transcription and replication as soon as it enters mammalian cells. Two massive replication polyproteins (ppla and pplab) are first produced by ORFs 1a and 1b. These polyproteins are then cleaved by PLpro (nsp3), 3CLpro (nsp5), and exoribonuclease (nsp14) to yield nonstructural proteins (nsps). RdRp (nsp12) and helicase (nsp13) are two crucial proteins involved in controlling viral genomes and coordinating protein production [15, 16]. The potential to impede viral replication across various RNA viruses exists for a wide range of antiviral drugs that target RdRp [17, 18]. The development of antiviral medications to treat RNA virus-related diseases is made possible by the discovery of RdRp as a dependable and potent therapeutic target. Many inhibitors, including favipiravir, galidesivir, sofosbuvir, ribavirin, tenofovir, and remdesivir, have shown efficacy against SARS-CoV-2 [19, 20]. Among the most promising RdRp inhibitors are nucleotide analogs (NAs). Mutagenic nucleosides, mandatory chain terminators, and delayed chain terminators are the three basic types into which these inhibitors are often divided [21, 22]. In order for NAs to effectively compete with the natural substrates of RdRp, host kinases must activate NAs into the pharmacologically active triphosphate form (NTP). By using the energy produced by the hydrolysis of the NTP's pyrophosphate group, this activation mechanism enables them to be integrated into the growing viral RNA [20]. NNIs, which are renowned for their great specificity, work allosterically by preventing polymerase enzymatic activity [23, 24]. Because nsp14 exonuclease has a proofreading role, the utility of nucleoside analog inhibitors (NAIs) as antiviral drugs may be limited. Investigating NNIs is an alternate strategy that often faces less opposition than NAIs. In the clinical treatment of COVID-19, a number of NNIs also exhibit immunomodulatory, cardioprotective, antioxidant, anti-inflammatory, and multifunctional therapeutic properties [25, 26].

Remdesivir is a prodrug of adenosine analog in the form of monophosphoramidate that was first created as a therapy for the Ebola virus outbreak [27]. The active version of this compound, GS-441524, has a broad range of antiviral activity against many viral families, such as filoviruses, paramyxoviruses, and coronaviruses. Remdesivir's monophosphate derivative (RDV-MP) and its active NTP derivative (RDV-TP) are produced via a sequence of biological processes. The active derivative hinders the production of viral RNA by competing with ATP for integration by the viral RdRp. The viral particle avoids the proofreading process of the exoribonuclease enzyme because remdesivir inhibits the action of RdRp, which suppresses RNA transcription [28]. RdRp was intended to be inhibited by the novel synthetic adenosine analog galidesivir [29, 30]. Viral RNA polymerase is given priority over host polymerase by RdRp upon the introduction of galidesivir triphosphate. Then, because of electrostatic interaction, the modified triphosphate residue undergoes an alteration in structure causing RNA to terminate prematurely. During the binding process, galidesivir forms six hydrogen bonds and four hydrophobic interactions with SARS-CoV-2 RdRp [31, 32]. Numerous viral diseases, such as respiratory syncytial virus (RSV), HCV, bunyavirus, herpesvirus, adenovirus, poxvirus, and several viral hemorrhagic fevers, are treated with ribavirin, a guanosine nucleoside homolog [27]. Because it inhibits guanosine production by inhibiting the enzyme inosine monophosphate dehydrogenase, it has antiviral properties. Furthermore, via increasing IFN gene expression and controlling immune responses, ribavirin indirectly affects viral replication [33, 34]. The FPV-RTP-containing RNA primer strand binds to the viral RdRp's active site [31, 35]. According to research, FPV-RTP interacts with Arg533, one of the critical amino acids in SARS-COV-2 RdRp, to influence Asp760–Asp761 and rNTP binding. The binding pocket is changed to an inactive conformation as a consequence of this modification. In the end, deadly mutagenesis caused by FPV-RTP results in RdRp mis-incorporating the antiviral agent into the developing viral RNA [36]. Suramin has garnered renewed interest owing to its strong antiviral characteristics, even though it is a century-old chemical. The viral RdRp has two binding sites when it binds to suramin, according to a new cryogenic electron microscopy (cryo-EM) structural investigation at 2.6 Å [37]. Two of these locations directly obstruct the RNA template strand's ability to attach, while the other one interferes with the RNA primer strand close to the RdRp catalytic active site to effectively block viral RNA replication. Primer binding or elongation may have been obstructed, as shown by the contact with the RNA primer strand close to the catalytic site [34, 38]. The unique chemical groups that define quinoline derivatives enable them to attach to the RdRp active site via a wide range of interactions. Within the RdRp active site, amino-acid residues may form hydrogen bonds with the quinoline nitrogen and other functional groups. Moreover, the quinoline's aromatic ring and the aromatic residues in the active site may interact via π-π stacking [39]. Corilagin seems to target a pocket around the active site of the RdRp palm domain, binding to it directly [40, 41]. For incoming RNA nucleotides to be incorporated into the viral RNA chain, this pocket is essential. According to computational modeling, corilagin forms hydrogen bonding and hydrophobic interactions with important residues of amino acids in this pocket. By interfering with the conformational changes required for nucleotide binding and integration, these interactions hinder viral replication. Interestingly, corilagin binds to RdRp without directly competing with RNA nucleotides for binding [37].

10.2 STRUCTURE AND FUNCTIONS OF RNA DEPENDENT RNA POLYMERASE (PdRp)

Coronaviruses possess an important RNA genome that, along with RdRp, is likely needed for their replication [42, 43]. The SARS-CoV-2 genome contains a single-stranded positive-sense RNA, which shows the ribosomal frameshifting of numerous non-structural replicase genes. Moreover, multiple enzyme functions are encoded in viral replicase-transcriptase polyprotein [44]. RdRp, also called non-structural protein 12 (nsp12), is encoded by whole RNA viruses and a few DNA viruses that initiate the viral genome replication and transcription complex (RTC), including SARS-CoV-2 [45, 46]. The entire RNA viruses, dsRNA, positive-strand RNA, and negative-strand RNA viruses possess RdRps that show three-dimensional structures and multiple sequence alignment of motifs. RNA-positive-stranded viruses and coronaviruses share many RdRp motifs; they make a desirable target for broad-spectrum antivirals [44]. The SARS-CoV-2 RNA genome has 14 ORFs that can be translated and contain two large ORFs, ORF1a and ORF1b. About 67% of the genome comprises 5' ORF1a/1b, which encodes the precursor polyproteins pp1a and pp1ab (via a −1 ribosomal frameshift). These proteins are hydrolyzed to produce 16 non-structural proteins (nsp1 to nsp16) at the 3' end that encodes nine putative accessory factors and four structural proteins [46]. A low amount of polymerase activity is present in the nsp12 alone, but the polymerase activity is significantly increased when the nsp7 and nsp8 co-factors are added, and nsp12 plays a role in the synthesis of RNA virus [46, 47]. Other important subunits like nsp9, nsp10, the helicase nsp13, the exonuclease nsp14, and the methyltransferase nsp16 attach to this complex to produce the RTC [29, 48]. The COVID-19 virus nsp12 structure possesses a distinctive N-terminal extension domain of iridoviruses and C-terminal RdRp that also contains nucleosialyltransferase (NIRAN), which is the element for virus replication [29, 39]. The polymerase domain of RdRp is similar to the 'right hand' shape, which consists of the "fingers," the "thumb," and the "palm" and another β-hairpin that stabilizes the RdRp structure between the "NIRAN" domain and the "palm" subdomain [37, 49]. The NIRAN domain's role is as a catalyst in transferring nucleoside monophosphates that play a function in RNA synthesis, RNA ligation, and transcript capping [29]. The palm domains (residues 582–620 and 680–815), fingers domains (residues 366–581 and 621–679), and thumb domains (residues 816–920) that encircle the RdRp active site, and seven catalytic motifs (motifs A–G) are dispersed throughout the palm (motifs A–E) and fingers (motifs F–G), which facilitates RNA synthesis [36, 43, 45]. The nsp7 and nsp8 subunit join with the thumb subdomain as well as additional interlinkage with the interface domain is formed by the other nsp8 (nsp8.2) subunit, where the finger subdomain is clamped and then stabilizes the conformation of RdRp [37]. To control SARS-CoV-2 replication and transcription, nsp7–nsp16 contribute to forming a sizable RTC. One of these motor proteins, Nsp13 consists of nsp13-1 and nsp13-2 protomers. Nsp13-1 stabilizes the RTC skeleton by interacting to Nsp12 and Nsp8-1. Nsp13-2 unwinds dsRNA and translocates in a 5′→3′ direction. Another subunit nsp9, forms a strong bond with nsp12 (RdRp) NIRAN, which inhibits GTPase activity, entering into nsp12 (Figure 10.1) [50, 51].

The RdRp binds to replication factors and conducts the RNA strand's elongation by adding nucleotides. By introducing them into the newly formed RNA chain, the RdRp employing NAs also catalyzes the termination of RNA elongation [52]. While targeting the viral RdRps, nucleoside analogs are a significant class of antiviral drug candidates that aim to block CoV replication [53, 54]. RNA synthesis is initiated by RdRps via two main mechanisms: *de novo* synthesis (primer-independent) and primer-dependent synthesis. *De novo* has a single nucleotide primer. The mechanism of *de novo* should identify the substrate initiation nucleotide, NTPi, and the template initiation nucleotide, T1. *De novo* initiation is the simplest form, involving the polymerase's active site, a template initiation site (the T1), and the initiation nucleotide (the NTPi), which supplies the 3-hydroxyl needed for

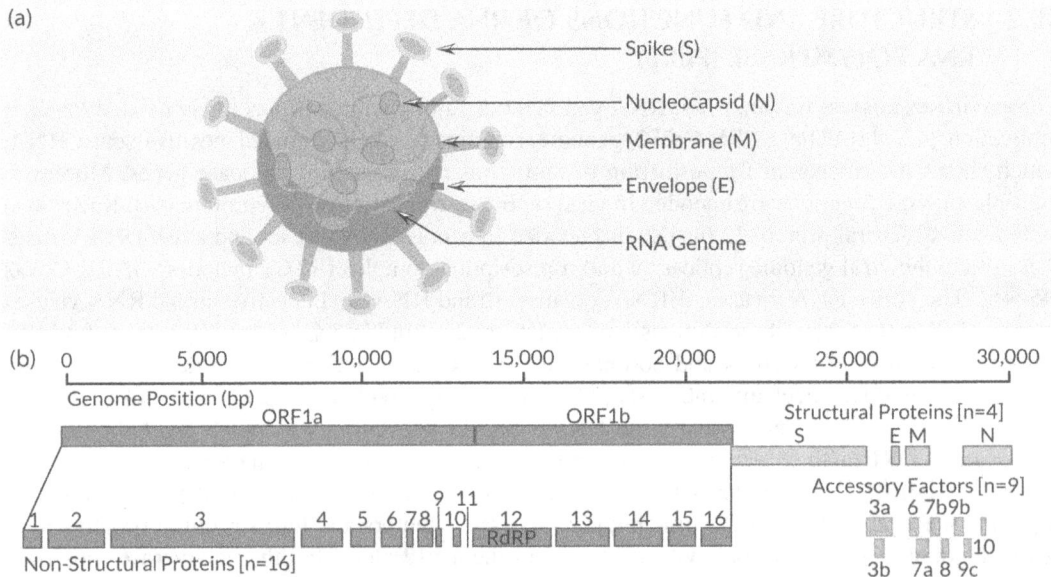

FIGURE 10.1 Illustration of SARS-CoV-2 has four structural proteins, known as the S (Spike), E (envelope), M (membrane), and N (nucleocapsid) proteins and RNA genome (A). The size of the SARS-CoV-2 RNA genome is approximately 30 kb and encodes 16 non-structural proteins (nsp), four structural proteins, and nine putative accessory factors (B).

nucleotidyl transfer to a second NTP. Primer-dependent RNA synthesis in which a protein covalently attached to an oligonucleotide is used to start the synthesis process [55].

10.3 THE SIGNIFICANCE OF RdRp AS A DRUG TARGET

RdRp is engaged in the fundamental role in the life span of the viral RNA virus without processing any homolog. It opens the way for anti-viral drug development and lessens the chances that a protein in human cells may be altered [56, 57]. Focusing on RNA viruses and their supervision is quite difficult because of the inborn viral property of incorporating the genetic material with the host cell genome, which mediates the viruses to go past the human immune system. Then, by taking advantage of host machinery, RNA viruses escalate their genetic content [38]. RdRp functions for the replication of the coronavirus genome that encrypts the fundamental proteins that play a role in the survival of the viruses [58, 59].

SARS-CoV-2 is composed of four structural proteins, S, E, M, and N proteins, which are wound with an RNA genome consisting of viral particles. The large-sized S protein helps interact with the host's pulmonary and bronchial epithelial cells, aiding a coronavirus to reach the epithelial cell. The alveolar epithelial cells contain ACE2, which acts as a receptor for SARS-COV-2. Coronavirus contains positively charged RNA with an ssRNA genome that can be directly translated into proteins inside the host cell where the coronavirus uses ribosomes to replicate the polyproteins (Figure 10.2) [46, 60, 61].

Two enzymes responsible for cleaving the polyproteins to produce 16 non-structural proteins are the main protease (3CLpro) and the papain-like protease [37, 62, 63]. It has been found that, for SARS-CoV-2 replication, the main machinery is in nsp-12, also known as RdRp. Nsp12/RdRp has no catalytic activity itself. As a result, to gain the catalytic ability, it requires the association of nsp7 and nsp8, forming a subcomplex (known as holo-RdRp) to replicate long RNA templates [8, 29, 45]. The three core domains of RdRp, such as fingers, thumb, and palm domains, perform the catalytic function. One of the main functions is that they bind the template entry of nucleoside phosphate and finally go to polymerization [60, 64].

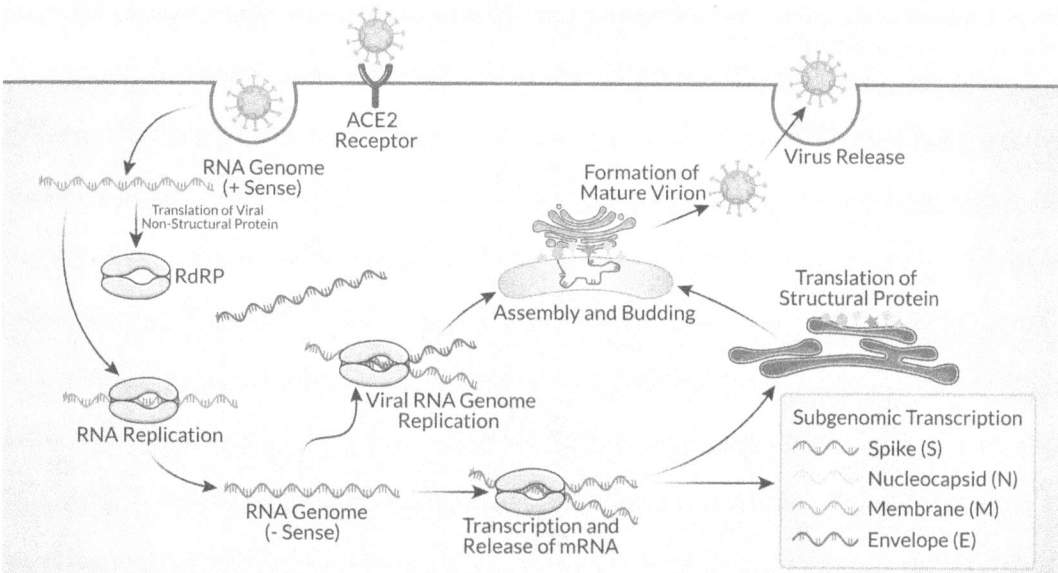

FIGURE 10.2 Presentation of the SARS-CoV-2 viral life cycle in a host cell. The RNA genome encodes 16 nsps. SARS-CoV-2 binds to the ACE2 receptor and then enters the host's cytosol, initiating viral replication.

The two main mechanisms to introduce viral RNA synthesis by RdRp are:

i. Primer-independent (*de novo*) makes genomic RNA by creating a phosphodiester bond with a 3-OH group that is linked to the 5-OH phosphate group of the neighboring nucleotide.
ii. The primer-dependent synthesis forms a new complementary RNA.

Any viral infection can be controlled by blocking the synthesis of viral nucleic acid. In that case, adenine or guanine derivatives, or nucleoside analogs, block viral RNA synthesis (Figure 10.3) [45, 46]. The agents that inhibit the function of RdRp are NAIs.

NAIs are classified into categories like nucleosides, obligate chain terminators, and delayed chain terminators. These NAIs are transformed into the active triphosphate form, which is assisted by the host kinases, leading these to be merged into the nascent viral RNA. For this phenomenon, energy was supplied by hydrolysis and release of the pyrophosphate group of the NTP. After incorporation, NAIs terminate RNA synthesis as a result viral outbreak is inhibited [29, 65].

In other words, after administering into the host, these analogs are converted to an active 5'-triphosphate form that competes with the nucleotides that are already presented in the cell for incorporation into the nascent viral RNA operated by RdRp. The non-natural NTPs that are inserted into the viral genome erroneously by polymerase block the natural pathway for viral particles to generate natural NTPs (Figure 10.3). As a result, viral polymerase lacks natural NTP [45]. SARS-CoV-2, having the presence of a 3-OH group, allows phosphodiester bond formation to the new incoming nucleotide. NAIs act as obligate chain terminators because of the scarcity of 3'-OH, which is needed for RNA chain elongation [8].

Most of the viruses contain polymerase that plays a crucial role not only in viral replication but also in the genetic evolution of the viral RNAs. After adding to the RNA template and incorporating 5'-triphosphate ribonucleases, viral polymerase makes an RNA copy according to the "Watson-Crick" base-pairing rules. This activity can be derived from base-pairing rules and the usual ribose selectivity. As a result, the process becomes an error because of the constitutive (in)fidelity of viral polymerases. It enhances the probability for the polymerase enzyme to undertake the reshaped NAs as substrates. This appears as an essential class of antiviral agents [45, 66]. But there is a drawback: nsp14, which is called exoribonucleases (ExoN), operates the function of proofreading that could lessen the activity of the inhibitors. However, this disadvantage can be overcome by delaying chain termination (Figure 10.4) [8, 37].

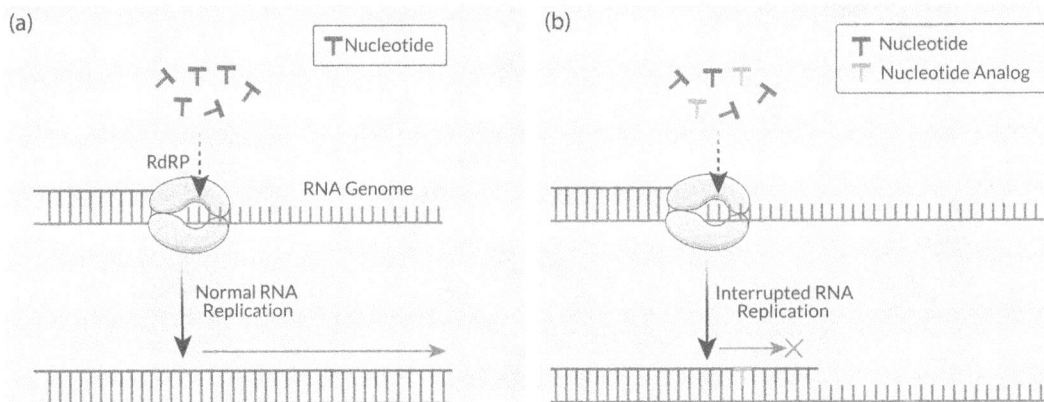

FIGURE 10.3 The normal mechanism of RNA dependent RNA mechanism (A). The introduction of nucleotide analog as an inhibitor that stops RNA elongation (B).

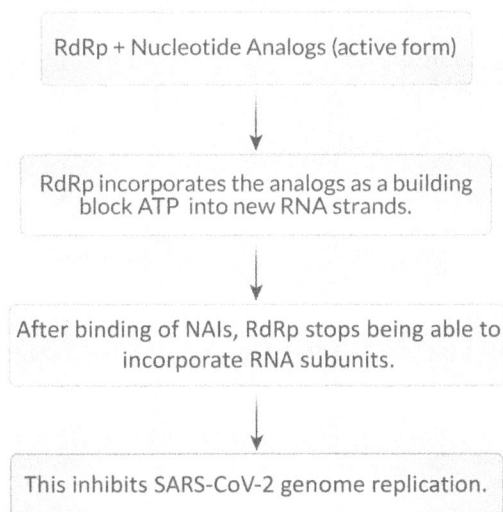

FIGURE 10.4   Mechanism of action of nucleoside analog inhibitors.

10.4   MOLECULAR MECHANISM NAIs COMPOUNDS AGAINST RdRp

Targeting the RNA polymerase of viruses including SARS-CoV, Ebola, influenza, and Zika, nucleoside analogs are useful in treating diseases caused by these infectious agents [29, 67]. Mostly, deoxynucleoside analogs are the medications used to treat these types of diseases. Mutagenic nucleosides, mandatory chain terminators, and delayed chain terminators are the three basic types into which NAIs are commonly divided. Activation of NAIs into the pharmacologically active triphosphate form (NTP) by host kinases is necessary for them to compete with the natural substrates of RdRp [45]. Activation uses the energy created from the hydrolysis of the NTP's pyrophosphate group to enable it to be incorporated into the newly formed viral RNA. When NAIs are integrated, they either stop RNA synthesis from occurring or cause alterations in the RNA that the progenitor produces. To better understand their inhibitory mechanisms, many NAIs have had their structures clarified and their efficacy against SARS-CoV-2 assessed in response to the epidemic (Table 10.1 and Figure 10.5) [9, 68].

10.4.1   REMDESIVIR

Remdesivir (RDV), a monophosphoramidate prodrug of an adenosine analog, exhibits broad-spectrum antiviral action against multiple viral families, such as filoviruses, paramyxoviruses, and coronaviruses (the active form is GS-441524). Remdesivir was first created as an Ebola virus therapy [2]. RDV goes through a series of biochemical reactions that result in the adenine metabolite, the monophosphate derivative (RDV-MP), and ultimately the active NTP (RDV-TP) derivative. The latter is the substrate the viral RdRp contends with ATP for incorporation, which restricts viral RNA synthesis [10]. Remdesivir (RDV) blocks the function of RdRp to suppress RNA transcription, which causes the viral particle to evade its ExoN enzyme proofreading mechanism. This is how Remdesivir (RDV) demonstrates its antiviral activity [69]. Remdesivir had significant therapeutic benefits in non-human primates and reduced the multiplication of the Ebola virus in several human cell lines *in vitro* with an EC50 in the submicromolar range [70]. The RDV demonstrates a potent defense against the transmissible families *Paramyxoviridae*, *Pneumoviridae*, *Filoviridae*,

TABLE 10.1

Chemical Structures of Nucleoside Analog Inhibitor Compounds for RNA Dependent RNA Polymerase

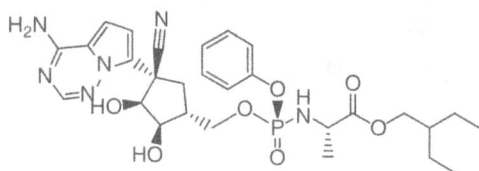

Remdesivir

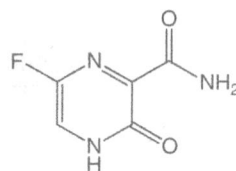

Favipiravir

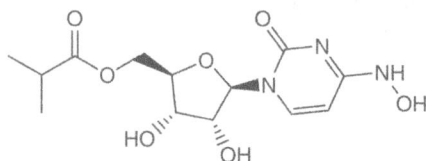

Molnupiravir

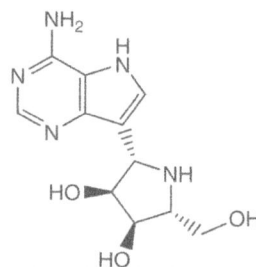

Galidesivir

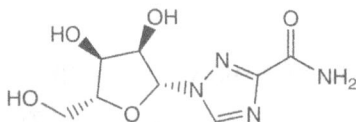

Ribavirin

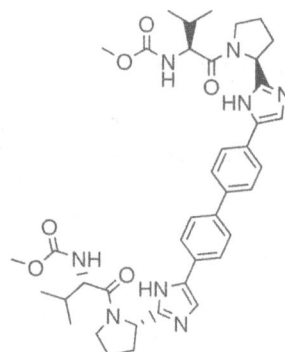

Daclatasvir

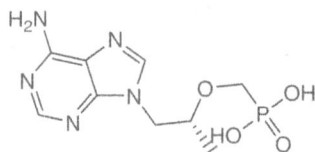

Tenofovir

and *Orthocoronavirinae* [71]. Modulating several dose interventions between 3 mg and 225 mg has been used to assess the pharmacokinetics of RDV in the human body; no toxicity or adverse effects have been observed in the kidney or liver [72]. Remdesivir treatment dramatically lowered the viral load rate in a SARS-CoV-infected mouse's lungs. On Days 4 or 5 postinfection, the virus titer dropped by more than twofold, suggesting that remdesivir ameliorated respiratory symptoms in the lab model. Remdesivir's therapeutic activity against MERS-CoV infection was recently studied in a rhesus macaque model, a non-human primate. Remdesivir was given 3 hours prior to virus inoculation. Emdesivir shielded the lungs against the virus, according to the findings. Nevertheless, when remdesivir was administered to the lab animal 12 hours after the viral inoculation, the same outcomes were seen. Remdesivir is safe to use in humans, according to human trials [45].

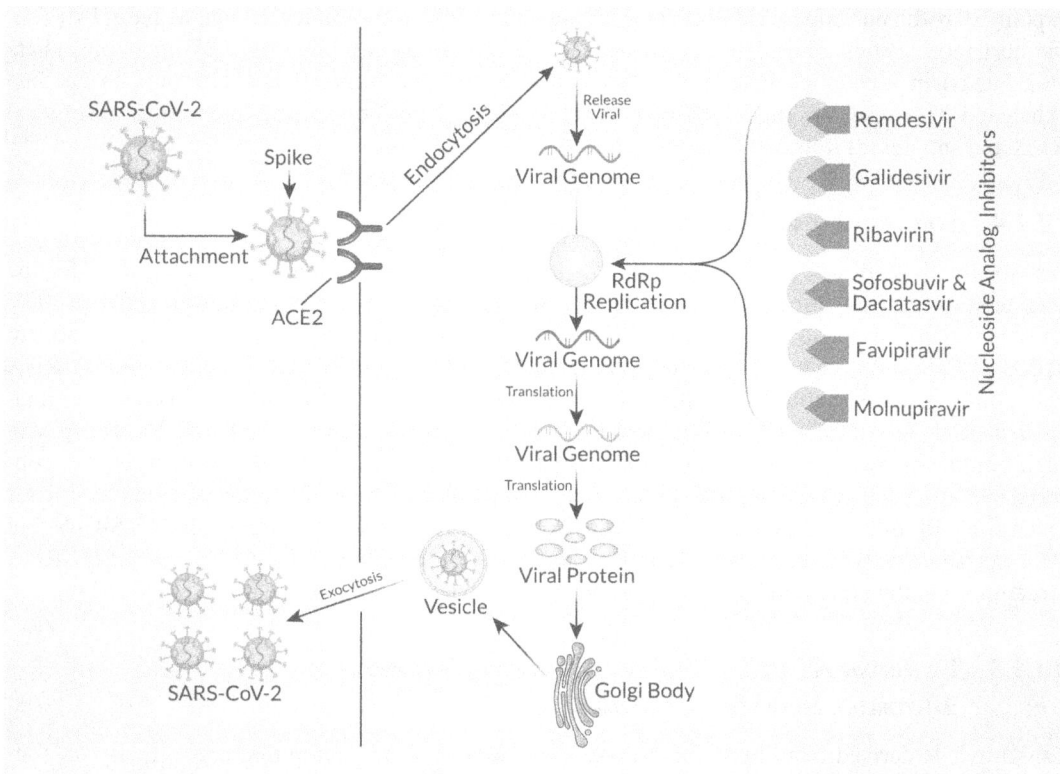

FIGURE 10.5 Molecular mechanism of nucleoside analog inhibitor compounds for the inhibition of RNA dependent RNA polymerase.

10.4.2 GALIDESIVIR

A novel synthetic adenosine analog called galidesivir (BCX4430) was created as an RdRp inhibitor. Incorporating galidesivir triphosphate causes the RdRp to favor viral RNA polymerase over host polymerase. The drug's changed triphosphate residue then suffers a structural change brought on by electrostatic contact [10, 73] that causes RNA to be prematurely terminated [63]. Galidesivir forms six hydrogen bonds and four hydrophobic interactions with SARS-CoV-2 RdRp during this binding process. In addition to inhibiting the replication of numerous RNA viruses, notably Ebola, Zika, Marburg, and yellow fever, galidesivir drastically decreases lung pathology caused by prior infections [71].

10.4.3 RIBAVIRIN

A broad-spectrum antiviral prodrug termed ribavirin (RBV) is used for the treatment of numerous RNA and DNA viruses [71]. RSV, HCV, bunyavirus, herpesvirus, adenovirus, poxvirus, and several viral hemorrhagic fevers are only a few of the viral illnesses that are treated with the aid of the antiviral drug ribavirin, which is a homolog of guanosine nucleoside. The inosine monophosphate dehydrogenase enzyme, which disrupts guanosine synthesis, is what provides ribavirin its antiviral activity. It also indirectly impacts virus replication by enhancing IFN gene expression and controlling immune responses [45]. The synthesis of viral mRNA is halted by ribavirin. To couple with the cytidine triphosphate or uridine triphosphate in the RNA template, RBV is transformed by host kinases into RBV triphosphate (RBV-TP). Its antiviral impact-mediated deadly mutagenesis can be explained by the nucleobase's ambiguous coding nature, which resembles the purine base. IMPDH, the host inosine monophosphate dehydrogenase, is effectively inhibited by RBV monophosphate [10]. Ribavirin monophosphate inhibits IMPDH, causing the intracellular GTP pool to become exhausted. This results in the inhibition of RNA synthesis, which diminishes the replication of viral RNA and suppresses immune function [71].

10.4.4   SOFOSBUVIR AND DACLATASVIR

Sofosbuvir (SOF, β-d-2′-deoxy-2′-α-fluoro-2′-β-C-methyluridine) is a uridine analog used for the treatment of HCV infection [70]. Having the similarity of the replication mechanism between HCV and coronavirus, it is assumed that both SOF and daclatasvir can be possible options in the treatment of SARS-CoV-2 [38]. SOF is a prodrug converted to SOF triphosphate that is inserted into the RNA chain through non-structural protein 5B (NS5B), which acts as a chain terminator in chain terminator in case of HCV [69, 72]. It has been found that SOF strongly attaches to SARS-CoV-2 RdRp with high binding energy, forming seven H-bonds with four RdRp residues and two hydrophobic contacts, and as a result, SOF-triphosphate leads to irreversible blocking of polymerase mediated RNA extension [70]. Both SOF and daclatasvir can show cooperative antiviral effects on SARS-CoV-2 in the respiratory cells. With antiviral activity, daclatasvir unfolds the SARS-CoV-2 secondary RNA structures. On the other hand, SOF triphosphate terminates the polymerase chain reaction [38].

10.4.5   FAVIPIRAVIR (FPV; T-705, AVIGAN, FAVIPIRA; 6-FLUORO-3-HYDROXY-2-PYRAZINE CARBOXAMIDE)

Favipiravir is derived from pyrazine carboxamide, a nucleoside analog that acts directly on conserved active pockets functioning as catalytic substrates for RdRp [45]. It is a guanine analog that competes with purine nucleoside [29]. As it is a prodrug, it must be metabolized in the body, forming active components or drugs through ribosylation and phosphorylation to favipiravir-ribofuranosyl-5′-triphosphate (FPV-RTP). It is identified as a purine nucleotide by RdRp inhibiting viral RNA synthesis [70]. The FPV-RTP is included in the RNA primer strand sticking to the active site of the viral RdRp [65]. Study shows that FPV-RTP interacts with the crucial amino acids of SARS-COV-2 RdRp such as Arg533, which acts on rNTP binding and Asp760-Asp 761 [70]. This transforms the binding pocket into an inactive configuration. Finally, FPV-RTP shows its antiviral effect by lethal mutagenesis that is mis-incorporated into nascent viral RNA by RdRp. As a result, the viral genome containing FPV-RTP will be prone to mutagenesis [38].

10.4.6   MOLNUPIRAVIR

It was formerly known as (EIDD-2801) which is the isopropyl ester prodrug of the ribonucleoside analog β-D-N4-hydroxycytidine (EIDD-1931 or NHC). It effectively blocks the replication of SARS-COV-2 in human respiratory epithelial cells. Taking of EIDD-1931 causes lethal mutagenesis by enticing G to A and C to U interchange [38, 70]. It may be effective in patients having resistance to It may be effective in patients having resistance to Remdesivir (RDV). This explains that has proven the effectiveness of molnupiravir against new variants of the viral genes.

10.5   MOLECULAR MECHANISM NNAI COMPOUNDS AGAINST RdRp

NNIs are highly specific and act allosterically by inhibiting the enzymatic activity of the polymerase. The potential of NAIs as antiviral medications for treatment may be restricted because of the proofreading function of nsp14 exonuclease [60]. Finding NNIs would be a different path to NAIs, which generally exhibit fewer hurdles to resistance. When compared to NAIs, NNIs have a number of additional benefits as well, most notably their increased chemical diversity and perhaps easier administration. Consequently, it may be especially beneficial to find non-nucleoside analog-type medications that target RdRp. Theoretically, the majority of NNIs alter the conformation of RdRp by attaching themselves to allosteric sites on the polymerase's surface, which subsequently influences how the catalytically active region of RdRp binds to substrates [74]. Also, several NNIs have immunomodulatory, cardioprotective, antioxidant, anti-inflammatory, and multifunctional therapeutic effects in the clinical therapy of COVID-19 [46]. The most promising NNIs of SARS-CoV-2 RdRp that have been reported thus far are listed here (Table 10.2 and Figure 10.6).

TABLE 10.2
Chemical Structures of Non-Nucleoside Analog Inhibitor Compounds for RNA Dependent RNA Polymerase

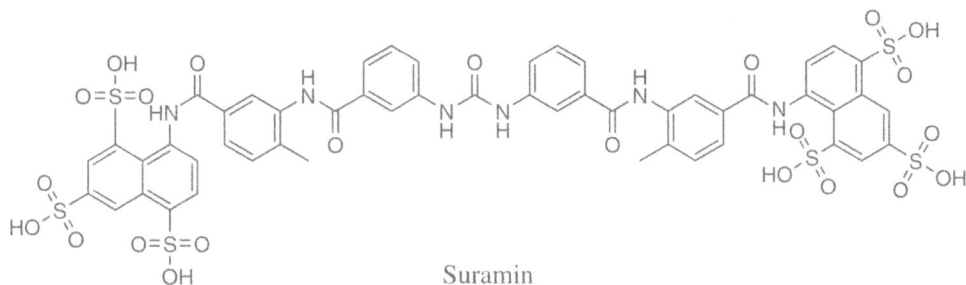

Suramin

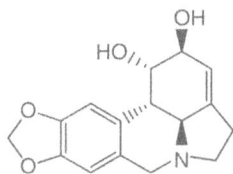

Lycorine

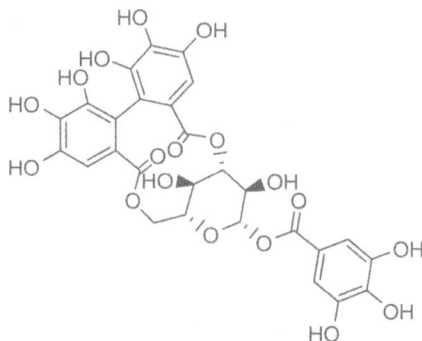

Corilagin

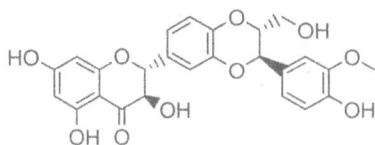

Silibinin

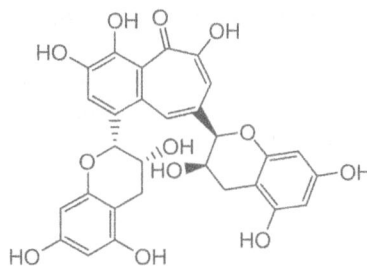

Theaflavin

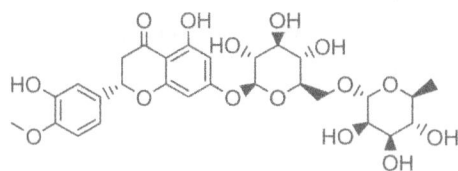

Hesperidin

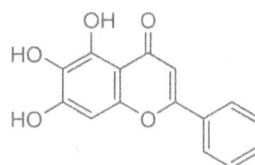

Baicalein

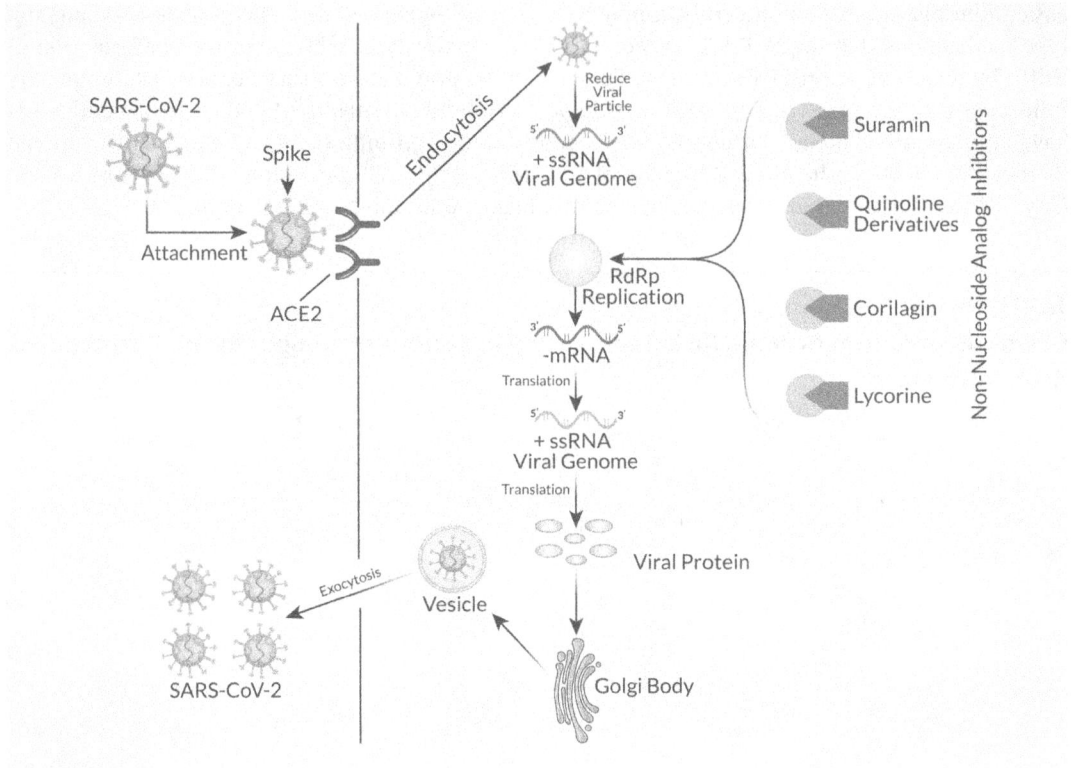

FIGURE 10.6 Molecular mechanism of non-nucleoside analog inhibitor compounds for the inhibition of RNA dependent RNA polymerase.

10.5.1 SURAMIN

Suramin has been around for a century, but its strong antiviral effects are making it popular anew. The viral RdRp bound to suramin has two binding sites, which are revealed by the 2.6 Å cryo-EM structure. One site directly prevents the RNA template strand from binding, while the other site clashes with the RNA primer strand close to the RdRp catalytic active site, preventing viral RNA replication. Near the catalytic site, the other site collides with the RNA primer strand, which may obstruct primer binding or elongation [75]. The binding of suramin may cause conformational changes in the RdRp, hence impacting its overall activity and impeding critical replication process stages. Interactions with RdRp functional domains other than those directly engaged in RNA binding may be responsible for this allosteric impact. Because of its strong negative charge, suramin can interact with a wide range of positively charged molecules, such as RNA and proteins that bind to RNA. Indirect inhibition may result from these interactions if they interfere with RNA processing and transport or interfere with the RdRp complex's ability to assemble [76]. Suramin's efficacy in certain situations may be limited by its size and hydrophobicity, which may prevent it from accessing the RdRp inside infected cells. The structure of suramin is being changed in an effort to increase target selectivity and bioavailability.

10.5.2 QUINOLINE DERIVATIVES

Quinoline derivatives possess specific chemical groups that enable them to bind to the RdRp active site through various interactions. The quinoline nitrogen and other functional groups can form hydrogen bonds with amino-acid residues in the RdRp active site. The aromatic ring of the quinoline can interact with aromatic residues in the active site through π-π stacking interaction. The hydrophobic portions of the quinoline molecule can interact with hydrophobic pockets in the active site. Once bound to the RdRp active site, the quinoline derivative can interfere with the binding and/or polymerization of the RNA substrate in several ways. The bulky quinoline molecule can physically block the RNA substrate from entering the active site. The binding of the quinoline derivative can induce conformational changes in the RdRp that alter the active site geometry, preventing substrate binding or polymerization. The quinoline derivative can compete with the RNA substrate for binding to the same site on the RdRp, effectively preventing the initiation of RNA synthesis. Some quinoline derivatives might also inhibit RdRp activity through other mechanisms, disrupting the interaction between the RdRp and its accessory proteins and inducing the degradation of the RdRp protein. Quinoline derivatives exhibit a diverse array of biological characteristics, including antimicrobial, antifungal, antitumor, and antiviral effects. With therapeutic index (TI) (CC50/EC50) values of 65, 34, and 20, respectively, three quinoline derivatives (I-13e, I-13 h, and I-13i), which have been shown to have significant activity against the influenza A virus RdRp, demonstrate exceptional potency in inhibiting RNA synthesis by targeting SARS-CoV-2 RdRp and relatively low cytotoxicity in a cell-based assay [11, 37]. I-13e is a potent inhibitor of SARS-CoV-2 RdRp that binds to the active site and induces conformational changes that prevent RNA substrate binding. Phenanthroline derivatives are compounds that bind to the RdRp active site and inhibit RNA synthesis through a combination of steric hindrance and competition for binding.

10.5.3 CORILAGIN

It appears that corilagin binds directly to the RdRp palm domain, concentrating on a pocket around the active site. Incoming RNA nucleotides must bind to this pocket for the viral RNA chain to incorporate them. According to computational modeling, corilagin forms hydrophobic and hydrogen-bond interactions with important amino-acid residues in this pocket. Viral replication is essentially stopped by these interactions, which interfere with the conformational modifications required for nucleotide binding and integration. Regarding binding to RdRp, corilagin does not directly compete with RNA nucleotides. Rather, it binds allosterically, changing the structure of the enzyme and impairing its capacity to use the nucleotides correctly. The binding of corilagin stabilizes the RdRp conformation in a manner that prevents the motions required for nucleotide incorporation. The enzyme is essentially locked in an inactive state as a result. Three amino-acid residues that are necessary for the enzymatic activity of the RdRp make up its catalytic triad. The interactions within this triad may be disrupted by corilagin binding, which would further impair RdRp's functionality. Profound antiviral activity of corilagin against SARS-CoV-2 is demonstrated in both cell-free and cell-based tests. It demonstrates resistance to the virus's proofreading ability, enabling it to elude NAIs through a common viral escape route. Studies on corilagin's acute and subchronic toxicity revealed no toxicity. Corilagin was well tolerated *in vivo* as evidenced by the fact that no unfavorable behavioral effects or weight loss were seen in mouse models following oral administration of larger dosages [77, 78].

10.5.4 LYCORINE

An alkaloid found naturally in the *Amaryllidaceae* family, lycorine, has shown promise in the treatment of a number of infections caused by viruses. Lycorine forms hydrogen bonds with the RdRp enzyme to directly interact with it. By preventing the polymerase from engaging and absorbing RNA nucleotides, this binding effectively prevents the enzyme from undergoing critical conformational changes that are necessary for its catalytic activity. According to studies, lycorine blocks incoming nucleotides and the RNA template from entering the RdRp's active site by interacting with particular amino-acid residues. Lycorine may potentially have indirect antiviral effects by interfering with signaling pathways that are important in RdRp gene expression, which would reduce the synthesis of enzymes. It has demonstrated immunomodulatory qualities, which may strengthen the host's defenses against viral infections [79–81].

The benefit of lycorine comes from its non-nucleoside properties. It reduces the possibility of mutations and the development of resistance in viruses because, in contrast to nucleoside analogs, it does not mimic natural nucleotides. However, more investigation is needed to fully comprehend its possible adverse effects and maximize its selectivity for RdRp inhibition due to its wide range of activity targeting different cellular enzymes.

10.5.5   BAICALEIN AND BAICALIN

Scutellaria baicalensis is a traditional herbal plant that contains the naturally occurring flavonoids baicalein and baicalin. It is believed that baicalein and baicalin bind directly to the RdRp active site, where they compete with the viral RNA substrate for binding energy [82, 83]. This efficiently stops viral replication by preventing the polymerase from initiating and elongating RNA chains. Metal ions such as $Mg2+$ are necessary for RdRp activity to catalyze reactions. These metal ions can be chelated by baicalein and baicalin, which stops them from interacting with the enzyme and impairs its function. Strong antioxidant qualities are seen in baicalein and baicalin, such as scavenging free radicals that can harm cellular components and virus RNA. This strengthens cellular resistance to viral infection and lowers viral fitness. Research indicates that by inhibiting pro-inflammatory cytokines and fostering anti-inflammatory pathways, baicalein and baicalin may regulate the immune response. This lessens the harm that inflammation does to tissue and controls the spread of viruses.

Research employing viral RNA templates and purified RdRp has demonstrated that baicalein and baicalin directly and dose-dependently suppress RdRp activity. Baicalein and baicalin have been shown in studies conducted in virus-infected cells to greatly lower viral replication, which further supports their role in RdRp inhibition. More proof of baicalein and baicalin's direct binding mechanism comes from computational simulations that imply they can interact with the RdRp active site.

10.6   CONCLUSIONS AND FUTURE PERSPECTIVES

In vitro and *in vivo* systems have been developed by researchers to find and create inhibitors for the RNA virus RdRp, an enzyme that is essential to viral genome replication. Drug development is still in its infancy, despite the promising results of drugs like remdesivir, galidesivir, ribavirin, sofosbuvir, daclatasvir, favipiravir, molnupiravir, suramin, quinoline derivatives, corilagin, lycorine, baicalein, and baicalin. Considering human bodies might suffer severe and protracted outcomes from RNA virus infection, there is a greater need to develop small-molecule therapeutics in addition to the present anti-RNA virus vaccination strategy. A great deal of information has already been discovered on the precise identification of nucleotide and non-nucleotide analogs, which offers important perspectives for the design of structure-based inhibitors. One prodrug of an adenosine analog in the form of monophosphoramidate is remdesivir, which was first developed as a treatment for the Ebola virus epidemic. Suramin is a molecule that has been around for a century, but because of its strong antiviral properties, it has garnered fresh attention. Finding nucleoside analogs is an interesting path since RNA virus RdRp's proteins have functional and structural similarities with those of other RNA viruses. It is thrilling to think about the potential for future antiviral medicines if new broad-spectrum nucleoside or non-nucleoside compounds are found that may inhibit a variety of RNA viruses, such as HCV, bunyavirus, herpesvirus, adenovirus, and poxvirus, and block RdRp. This method may help develop a more all-encompassing and robust defense against various RNA virus infections.

REFERENCES

1. K. AMQ, "Ninth report of the International Committee on Taxonomy of Viruses." 2012. [Online]. Available: https://cir.nii.ac.jp/crid/1370848662549157251

2. A. H. Bartlett and P. W. Park, "Proteoglycans in Host–Pathogen Interactions: Molecular Mechanisms and Therapeutic Implications," *Expert Rev. Mol. Med.*, vol. 12, p. e5, 2010, doi: 10.1017/S1462399409001367.

3. S. Modrow, D. Falke, U. Truyen and H. Schätzl, "Viruses: Definition, Structure, Classification BT" - *Molecular Virology*, S. Modrow, D. Falke, U. Truyen and H. Schätzl, Eds. Berlin, Heidelberg: Springer Berlin Heidelberg, 2013, pp. 17–30. https://doi.org/10.1007/978-3-642-20718-1_2.

4. R. Carrasco-Hernandez, R. Jácome, Y. López Vidal and S. Ponce de León, "Are RNA Viruses Candidate Agents for the Next Global Pandemic? A Review," *ILAR J.*, vol. 58, no. 3, pp. 343–358, 2017, doi: 10.1093/ilar/ilx026.

5. A. H. Mohsen, P. Easterbrook, C. B. Taylor and S. Norris, "Hepatitis C and HIV-1 Coinfection," *Gut*, vol. 51, no. 4, pp. 601–608, 2002, doi: 10.1136/gut.51.4.601.

6. G. F. Oliver, J. M. Carr and J. R. Smith, "Emerging Infectious Uveitis: Chikungunya, Dengue, Zika and Ebola: A Review," *Clin. Experiment. Ophthalmol.*, vol. 47, no. 3, pp. 372–380, Apr. 2019, https://doi.org/10.1111/ceo.13450.

7. M. Zambon, "Influenza and Other Emerging Respiratory Viruses," *Medicine (Baltimore)*, vol. 42, no. 1, pp. 45–51, 2014, https://doi.org/10.1016/j.mpmed.2013.10.017.

8. K. Van Reeth, "Avian and Swine Influenza Viruses: Our Current Understanding of the Zoonotic Risk," *Vet. Res*, vol. 38, no. 2, pp. 243–260, 2007, doi: 10.1051/vetres:2006062.

9. S. H. Alfaraj, J. A. Al-Tawfiq and Z. A. Memish, "Middle East Respiratory Syndrome Coronavirus (MERS-CoV) Infection during Pregnancy: Report of Two Cases & Review of the Literature," *J. Microbiol. Immunol. Infect.*, vol. 52, no. 3. pp. 501–503, 2019, doi: 10.1016/j.jmii.2018.04.005.

10. C. Basile *et al.*, "Recommendations for the Prevention, Mitigation and Containment of the Emerging SARS-CoV-2 (COVID-19) Pandemic in Haemodialysis Centres," *Nephrol. Dial. Transplant.*, vol. 35, no. 5, pp. 737–741, May 2020, doi: 10.1093/ndt/gfaa069.

11. B. Shu and P. Gong, "Structural Basis of Viral RNA-Dependent RNA Polymerase Catalysis and Translocation," *Proc. Natl. Acad. Sci*, vol. 113, no. 28, pp. E4005–E4014, 2016, doi: 10.1073/pnas.1602591113.

12. S. Bibi, *et al.* "Virtual Screening and Molecular Dynamics Simulation Analysis of Forsythoside A as a Plant-Derived Inhibitor of SARS-CoV-2 3CLpro," *Saudi Pharm. J.*, vol. 30, no. 7, pp. 979–1002, 2022, doi: 10.1016/j.jsps.2022.05.003.

13. M. M. Lai, "Genetic Recombination in RNA Viruses," *Curr. Top. Microbiol. Immunol.*, vol. 176, pp. 21–32, 1992, doi: 10.1007/978-3-642-77011-1_2.

14. K. H. Choi, "Viral Polymerases BT" - *Viral Molecular Machines*, M. G. Rossmann and V. B. Rao, Eds. Boston, MA: Springer US, 2012, pp. 267–304, https://doi.org/10.1007/978-1-4614-0980-9_12.

15. S. Venkataraman, B. V. L. S. Prasad and R. Selvarajan, "RNA Dependent RNA Polymerases: Insights from Structure, Function and Evolution," *Viruses*, vol. 10, no. 2, 2018, doi: 10.3390/v10020076.

16. M. Nur Kabidul Azam *et al.*, "Identification of Antidiabetic Inhibitors from Allophylus villosus and Mycetia sinensis by Targeting α-Glucosidase and PPAR-γ: In-Vitro, in-Vivo, and Computational Evidence," *Saudi Pharm. J.*, vol. 32, no. 1, p. 101884, Jan. 2024, doi: 10.1016/j.jsps.2023.101884.

17. P.-H. Tsai *et al.*, "Genomic Variance of Open Reading Frames (ORFs) and Spike Protein in Severe Acute Respiratory Syndrome Coronavirus 2 (SARS-CoV-2)," *J. Chin. Med. Assoc.*, vol. 83, no. 8, pp. 725–732, Aug. 2020, doi: 10.1097/JCMA.0000000000000387.

18. M. A. R. Mia *et al.*, "The Efficacy of Natural Bioactive Compounds against Prostate Cancer: Molecular Targets and Synergistic Activities," *Phytother. Res.*, vol. 37, no. 12, pp. 5724–5754, Dec. 2023, doi: 10.1002/ptr.8017.

19. P. Leyssen, E. De Clercq and J. Neyts, "Molecular Strategies to Inhibit the Replication of RNA Viruses," *Antiviral Res.*, vol. 78, no. 1, pp. 9–25, 2008, https://doi.org/10.1016/j.antiviral.2008.01.004.

20. N. Nahar *et al.*, "Profiling of Secondary Metabolite and Evaluation of Anti-Diabetic Potency of Crotalaria quinquefolia (L): In-Vitro, In-Vivo, and In-Silico Approaches," *Saudi Pharm. J. SPJ Off. Publ. Saudi Pharm. Soc.*, vol. 32, no. 1, p. 101887, Jan. 2024, doi: 10.1016/j.jsps.2023.101887.

21. S. Khan *et al.*, "A Review on the Interaction of Nucleoside Analogues with SARS-CoV-2 RNA Dependent RNA Polymerase," *Int. J. Biol. Macromol.*, vol. 181, pp. 605–611, 2021, https://doi.org/10.1016/j.ijbiomac.2021.03.112.

22. P. Biswas *et al.*, "Antiviral Phytocompounds Against Animal-to-Human Transmittable SARS-CoV-2 BT" - Ethnopharmacology and Drug Discovery for COVID-19: Anti-SARS-CoV-2 Agents from Herbal Medicines and Natural Products, J.-T. Chen, Ed. Singapore: Springer Nature Singapore, 2023, pp. 189–224. https://doi.org/10.1007/978-981-99-3664-9_7.

23. C. Long, M. E. Romero, D. La Rocco and J. Yu, "Dissecting Nucleotide Selectivity in Viral RNA Polymerases," *Comput. Struct. Biotechnol. J.*, vol. 19, pp. 3339–3348, 2021, https://doi.org/10.1016/j.csbj.2021.06.005.

24. S. Akash *et al.*, "Revolutionizing Anti-Cancer Drug Discovery Against Breast Cancer and Lung Cancer by Modification of Natural Genistein: an Advanced Computational and Drug Design Approach," *Front. Oncol.*, vol. 13, p. 1228865, 2023, doi: 10.3389/fonc.2023.1228865.

25. A. Mishra and A. S. Rathore, "RNA Dependent RNA Polymerase (RdRp) as a Drug Target for SARS-CoV2," *J. Biomol. Struct. Dyn.*, vol. 40, no. 13, pp. 6039–6051, 2022, doi: 10.1080/07391102.2021.1875886.

26. M. A. Sharif *et al.*, "*Phyllanthus emblica* (Amla) Methanolic Extract Regulates Multiple Checkpoints in 15-Lipoxygenase Mediated Inflammopathies: Computational Simulation and In Vitro Evidence," *Saudi Pharm. J.*, vol. 31, no. 8, p. 101681, Aug. 2023, doi: 10.1016/j.jsps.2023.06.014.

27. B. Yousefi, S. Valizadeh, H. Ghaffari, A. Vahedi, M. Karbalaei and M. Eslami, "A Global Treatments for Coronaviruses Including COVID-19," *J. Cell. Physiol*, vol. 235, no. 12, pp. 9133–9142, 2020, doi: 10.1002/jcp.29785.

28. Z. Yuhong *et al.*, "Ribavirin Treatment Up-Regulates Antiviral Gene Expression via the Interferon-Stimulated Response Element in Respiratory Syncytial Virus-Infected Epithelial Cells," *J. Virol.*, vol. 77, no. 10, pp. 5933–5947, May 2003, doi: 10.1128/jvi.77.10.5933-5947.2003.

29. X. Xu *et al.*, "An Update on Inhibitors Targeting RNA-Dependent RNA Polymerase for COVID-19 Treatment: Promises and Challenges," *Biochem. Pharmacol.*, vol. 205, p. 115279, 2022, https://doi.org/10.1016/j.bcp.2022.115279.

30. M. D. H. Rahman *et al.*, "Identification of Therapeutic Leads from *Ficus hispida* Fruit Phytochemicals against Prostate Cancer Using Pharmacoinformatic and Molecular Dynamics Simulation Approach," *bioRxiv*, Jan. 2023, https://doi.org/10.1101/2023.06.04.543596.

31. I. Hadj Hassine, M. Ben M'hadheb and L. Menéndez-Arias, "Lethal Mutagenesis of RNA Viruses and Approved Drugs with Antiviral Mutagenic Activity," *Viruses*, vol. 14, no. 4, 2022. doi: 10.3390/v14040841.

32. M. S. Andalib, P. Biswas, M. R. Sakib, M. N. Hasan, M. H. Rahman and A. Habib, "Identification of Novel MCM2 Inhibitors from *Catharanthus roseus* by Pharmacoinformatics, Molecular Docking and Molecular Dynamics Simulation-Based Evaluation," *Inf. Med. Unlocked*, vol. 39, p. 101251, 2023, https://doi.org/10.1016/j.imu.2023.101251.

33. A. Shannon *et al.*, "Remdesivir and SARS-CoV-2: Structural Requirements at Both nsp12 RdRp and nsp14 Exonuclease Active-Sites," *Antiviral Res.*, vol. 178, p. 104793, 2020, https://doi.org/10.1016/j.antiviral.2020.104793.

34. M. Sohel *et al.*, "Exploring the Anti-Cancer Potential of Dietary Phytochemicals for the Patients with Breast Cancer: A Comprehensive Review," *Cancer Med.*, vol. 12, no. 13, pp. 14556–14583, Jul. 2023, doi: 10.1002/cam4.5984.

35. A. K. M. H. Morshed *et al.*, "Baicalein as Promising Anticancer Agent: A Comprehensive Analysis on Molecular Mechanisms and Therapeutic Perspectives," *Cancers (Basel)*, vol. 15, no. 7, Apr. 2023, doi: 10.3390/cancers15072128.

36. J. A. Brown and I. F. Thorpe, "Dual Allosteric Inhibitors Jointly Modulate Protein Structure and Dynamics in the Hepatitis C Virus Polymerase," *Biochemistry*, vol. 54, no. 26, pp. 4131–4141, 2015, doi: 10.1021/acs.biochem.5b00411.

37. S. A. G. Naidu, G. Mustafa, R. A. Clemens and A. S. Naidu, "Plant-Derived Natural Non-Nucleoside Analog Inhibitors (NNAIs) against RNA-Dependent RNA Polymerase Complex (nsp7/nsp8/nsp12) of SARS-CoV-2," *J. Diet. Suppl.*, vol. 20, no. 2, pp. 254–283, 2023, doi: 10.1080/19390211.2021.2006387.

38. W. Yin *et al.*, "Structural Basis for Inhibition of the SARS-CoV-2 RNA Polymerase by Suramin," *Nat. Struct. Mol. Biol.*, vol. 28, no. 3, pp. 319–325, 2021, doi: 10.1038/s41594-021-00570-0.

39. H. S. Hillen, "Structure and Function of SARS-CoV-2 Polymerase," *Curr. Opin. Virol.*, vol. 48, pp. 82–90, 2021, https://doi.org/10.1016/j.coviro.2021.03.010.

40. A. A. Massoud *et al.*, "Effects of Different Substituents on the Crystal Structures and Antimicrobial Activities of Six Ag(I) Quinoline Compounds," *Inorg. Chem.*, vol. 52, no. 7, pp. 4046–4060, Apr. 2013, doi: 10.1021/ic400081v.

41. P. Biswas *et al.*, "Study of MDM2 as Prognostic Biomarker in Brain-LGG Cancer and Bioactive Phytochemicals Inhibit the p53-MDM2 Pathway: A Computational Drug Development Approach," *Molecules*, vol. 28, no. 7, Mar. 2023, doi: 10.3390/molecules28072977.

42. M. Dejmek *et al.*, "Non-Nucleotide RNA-Dependent RNA Polymerase Inhibitor That Blocks SARS-CoV-2 Replication," *Viruses*, vol. 13, no. 8. 2021. doi: 10.3390/v13081585.

43. A. Jabin *et al.*, "Target-Specificity of Different Amyrin Subunits in Impeding HCV Influx Mechanism Inside the Human Cells Considering the Quantum Tunnel Profiles and Molecular Strings of the CD81 Receptor: A Combined In Silico and In Vivo Study," *Silico Pharmacol.*, vol. 11, no. 1, p. 8, 2023, doi: 10.1007/s40203-023-00144-6.

44. C. C. Posthuma, A. J. W. te Velthuis and E. J. Snijder, "Nidovirus RNA Polymerases: Complex Enzymes Handling Exceptional RNA Genomes," *Virus Res.*, vol. 234, pp. 58–73, 2017, https://doi.org/10.1016/j.virusres.2017.01.023.

45. I. Vicenti, M. Zazzi and F. Saladini, "SARS-CoV-2 RNA-Dependent RNA Polymerase as a Therapeutic Target for COVID-19," *Expert Opin. Ther. Pat*, vol. 31, no. 4, pp. 325–337, 2021, doi: 10.1080/13543776.2021.1880568.

46. W. Zhu, C. Z. Chen, K. Gorshkov, M. Xu, D. C. Lo and W. Zheng, "RNA-Dependent RNA Polymerase as a Target for COVID-19 Drug Discovery," *SLAS Discov. Adv. Sci. Drug Discov*, vol. 25, no. 10, pp. 1141–1151, 2020, doi: 10.1177/2472555220942123.

47. R. N. Kirchdoerfer and A. B. Ward, "Structure of the SARS-CoV nsp12 Polymerase Bound to nsp7 and nsp8 Co-factors," *Nat. Commun.*, vol. 10, no. 1, p. 2342, 2019, doi: 10.1038/s41467-019-10280-3.

48. M. D. H. Rahman *et al.*, "An In-Silico Identification of Potential Flavonoids against Kidney Fibrosis Targeting TGFβR-1," *Life (Basel, Switzerland)*, vol. 12, no. 11, Nov. 2022, doi: 10.3390/life12111764.

49. R. Singh, M. Khalid, N. Batra, P. Biswas, L. Singh and R. Bhatti, "Exploring the Anticonvulsant Activity of Aqueous Extracts of *Ficus benjamina* L. Figs in Experimentally Induced Convulsions," *J. Chem*, vol. 2023, p. 6298366, 2023, doi: 10.1155/2023/6298366.

50. Q. Peng *et al.*, "Structural and Biochemical Characterization of the nsp12-nsp7-nsp8 Core Polymerase Complex from SARS-CoV-2," *Cell Rep.*, vol. 31, no. 11, p. 107774, 2020, doi: 10.1016/j.celrep.2020.107774.

51. A. Abdullah *et al.*, "Molecular Dynamics Simulation and Pharmacoinformatic Integrated Analysis of Bioactive Phytochemicals from *Azadirachta indica* (Neem) to Treat Diabetes Mellitus," *J. Chem.*, vol. 2023, p. 4170703, 2023, doi: 10.1155/2023/4170703.

52. C. Wu, W. Yin, Y. Jiang and H. E. Xu, "Structure Genomics of SARS-CoV-2 and Its Omicron Variant: Drug Design Templates for COVID-19," *Acta Pharmacol. Sin.*, vol. 43, no. 12, pp. 3021–3033, 2022, doi: 10.1038/s41401-021-00851-w.

53. L. Yan *et al.*, "Cryo-EM Structure of an Extended SARS-CoV-2 Replication and Transcription Complex Reveals an Intermediate State in Cap Synthesis," *Cell*, vol. 184, no. 1, pp. 184–193.e10, 2021, doi: 10.1016/j.cell.2020.11.016.

54. S. Aziz *et al.*, "A Review on Influence of Biochar Amendment on Soil Processes and Environmental Remediation," *Biotechnol. Genet. Eng. Rev.*, pp. 1–35, Feb. 2023, doi: 10.1080/02648725.2022.2122288.

55. M. Yusuf *et al.*, "SARS-CoV-2 Spike Protein and RNA Dependent RNA Polymerase as Targets for Drug and Vaccine Development: A Review," *Biosaf. Heal.*, vol. 03, no. 05, pp. 249–263, Oct. 2021, doi: 10.1016/j.bsheal.2021.07.003.

56. M. A. Rohaim, R. F. El Naggar, E. Clayton and M. Munir, "Structural and Functional Insights into Non-Structural Proteins of Coronaviruses," *Microb. Pathog.*, vol. 150, p. 104641, 2021, https://doi.org/10.1016/j.micpath.2020.104641.

57. S. K. Baral *et al.*, "A Comprehensive Discussion in Vaginal Cancer Based on Mechanisms, Treatments, Risk Factors and Prevention," *Front. Oncol.*, vol. 12, p. 883805, 2022, doi: 10.3389/fonc.2022.883805.

58. Y. Jiang, W. Yin and H. E. Xu, "RNA-Dependent RNA Polymerase: Structure, Mechanism, and Drug Discovery for COVID-19," *Biochem. Biophys. Res. Commun.*, vol. 538, pp. 47–53, 2021, https://doi.org/10.1016/j.bbrc.2020.08.116.

59. G. Les, K. Min-Ju and K. C. Cheng, "Requirements for De Novo Initiation of RNA Synthesis by Recombinant Flaviviral RNA-Dependent RNA Polymerases," *J. Virol*, vol. 76, no. 24, pp. 12526–12536, 2002, doi: 10.1128/jvi.76.24.12526-12536.2002.

60. S. Pathania, R. K. Rawal and P. K. Singh, "RdRp (RNA-Dependent RNA Polymerase): A Key Target Providing Anti-Virals for the Management of Various Viral Diseases," *J. Mol. Struct.*, vol. 1250, p. 131756, 2022, https://doi.org/10.1016/j.molstruc.2021.131756.

61. J. Huang, W. Song, H. Huang and Q. Sun, "Pharmacological Therapeutics Targeting RNA-Dependent RNA Polymerase, Proteinase and Spike Protein: From Mechanistic Studies to Clinical Trials for COVID-19," *Journal of Clinical Medicine*, vol. 9, no. 4, 2020. doi: 10.3390/jcm9041131.

62. C. Gil *et al.*, "COVID-19: Drug Targets and Potential Treatments," *J. Med. Chem.*, vol. 63, no. 21, pp. 12359–12386, Nov. 2020, doi: 10.1021/acs.jmedchem.0c00606.

63. P. Biswas *et al.*, "A Comprehensive Analysis and Anti-Cancer Activities of Quercetin in ROS-Mediated Cancer and Cancer Stem Cells," *Int. J. Mol. Sci.*, vol. 23, no. 19, Oct. 2022, doi: 10.3390/ijms231911746.

64. P. Biswas *et al.*, "Evaluation of Melongosides as Potential Inhibitors of NS2B-NS3 Activator-Protease of Dengue Virus (Serotype 2) by Using Molecular Docking and Dynamics Simulation Approach," *J. Trop. Med.*, vol. 2022, p. 7111786, 2022, doi: 10.1155/2022/7111786.

65. D. Dey *et al.*, "Natural Flavonoids Effectively Block the CD81 Receptor of Hepatocytes and Inhibit HCV Infection: A Computational Drug Development Approach," *Mol. Divers.*, vol. 27, no. 3, pp. 1309–1322, Jun. 2023, doi: 10.1007/s11030-022-10491-9.

66. D. Dey *et al.*, "Investigating the Anticancer Potential of Salvicine as a Modulator of Topoisomerase II and ROS Signaling Cascade," *Front. Oncol.*, vol. 12, p. 899009, 2022, doi: 10.3389/fonc.2022.899009.

67. N. Ferdausi *et al.*, "Point-Specific Interactions of Isovitexin with the Neighboring Amino Acid Residues of the hACE2 Receptor as a Targeted Therapeutic Agent in Suppressing the SARS-CoV-2 Influx Mechanism," *J. Adv. Vet. Anim. Res.*, vol. 9, no. 2, pp. 230–240, Jun. 2022, doi: 10.5455/javar.2022.i588.

68. M. Munshi *et al.*, "Novel Compounds from Endophytic Fungi of Ceriops decandra Inhibit Breast Cancer Cell Growth through Estrogen Receptor Alpha in In-Silico Study," *Inf. Med. Unlocked*, vol. 32, p. 101046, 2022, https://doi.org/10.1016/j.imu.2022.101046.

69. D. Tian *et al.*, "An Update Review of Emerging Small-Molecule Therapeutic Options for COVID-19," *Biomed. Pharmacother.*, vol. 137, p. 111313, 2021, https://doi.org/10.1016/j.biopha.2021.111313.

70. H. Abolhassani, G. Bashiri, M. Montazeri, H. Kouchakzadeh, S. A. Shojaosadati and S. E. R. Siadat, "Ongoing Clinical Trials and the Potential Therapeutics for COVID-19 Treatment BT" - *COVID-19: Science to Social Impact*, M. Rahmandoust and S.-O. Ranaei-Siadat, Eds. Singapore: Springer Singapore, 2021, pp. 27–89. https://doi.org/10.1007/978-981-16-3108-5_2.

71. R. A. Mizrahi, K. J. Phelps, A. Y. Ching and P. A. Beal, "Nucleoside Analog Studies Indicate Mechanistic Differences between RNA-Editing Adenosine Deaminases," *Nucleic Acids Res.*, vol. 40, no. 19, pp. 9825–9835, 2012, doi: 10.1093/nar/gks752.

72. M. Chien *et al.*, "Nucleotide Analogues as Inhibitors of SARS-CoV-2 Polymerase, a Key Drug Target for COVID-19," *J. Proteome Res.*, vol. 19, no. 11, pp. 4690–4697, Nov. 2020, doi: 10.1021/acs.jproteome.0c00392.

73. M. T. Sarker *et al.*, "Identification of Blood-Based Inflammatory Biomarkers for the Early-Stage Detection of Acute Myocardial Infarction," *Netw. Model. Anal. Heal. Informatics Bioinforma.*, vol. 11, no. 1, p. 28, 2022, doi: 10.1007/s13721-022-00371-5.

74. M. Machitani, M. Yasukawa, J. Nakashima, Y. Furuichi and K. Masutomi, "RNA-Dependent RNA Polymerase, RdRP, a Promising Therapeutic Target for Cancer and Potentially COVID-19," *Cancer Sci.*, vol. 111, no. 11, pp. 3976–3984, Nov. 2020, https://doi.org/10.1111/cas.14618.

75. Y. Ma *et al.*, "Structural Basis and Functional Analysis of the SARS Coronavirus nsp14–nsp10 Complex," *Proc. Natl. Acad. Sci.*, vol. 112, no. 30, pp. 9436–9441, Jul. 2015, doi: 10.1073/pnas.1508686112.

76. M. J. Sofia, W. Chang, P. A. Furman, R. T. Mosley and B. S. Ross, "Nucleoside, Nucleotide, and Non-Nucleoside Inhibitors of Hepatitis C Virus NS5B RNA-Dependent RNA-Polymerase," *J. Med. Chem.*, vol. 55, no. 6, pp. 2481–2531, 2012, doi: 10.1021/jm201384j.

77. X. Zheng *et al.*, "Discovery of Benzoazepinequinoline (BAQ) Derivatives as Novel, Potent, Orally Bioavailable Respiratory Syncytial Virus Fusion Inhibitors," *J. Med. Chem.*, vol. 61, no. 22, pp. 10228–10241, Nov. 2018, doi: 10.1021/acs.jmedchem.8b01394.

78. G. Zhang *et al.*, "Design, Synthesis and In Vitro Anti-Influenza A Virus Evaluation of Novel Quinazoline Derivatives Containing S-Acetamide and NH-Acetamide Moieties at C-4," *Eur. J. Med. Chem.*, vol. 206, p. 112706, 2020, https://doi.org/10.1016/j.ejmech.2020.112706.

79. Z. Wan, D. Hu, P. Li, D. Xie and X. Gan, "Synthesis, Antiviral Bioactivity of Novel 4-Thioquinazoline Derivatives Containing Chalcone Moiety," *Molecules*, vol. 20, no. 7, pp. 11861–11874, 2015, doi: 10.3390/molecules200711861.

80. J. Zhao *et al.*, "Quinoline and Quinazoline Derivatives Inhibit Viral RNA Synthesis by SARS-CoV-2 RdRp," *ACS Infect. Dis.*, vol. 7, no. 6, pp. 1535–1544, Jun. 2021, doi: 10.1021/acsinfecdis.1c00083.

81. B. U. Reddy *et al.*, "A Natural Small Molecule Inhibitor Corilagin Blocks HCV Replication and Modulates Oxidative Stress to Reduce Liver Damage," *Antiviral Res.*, vol. 150, pp. 47–59, 2018, https://doi.org/10.1016/j.antiviral.2017.12.004.

82. H. Xiao *et al.*, "Lycorine and Organ Protection: Review of Its Potential Effects and Molecular Mechanisms," *Phytomedicine*, vol. 104, p. 154266, 2022, https://doi.org/10.1016/j.phymed.2022.154266.

83. S. Bibi, *et al.* "Impact of Traditional Plants and Their Secondary Metabolites in the Discovery of COVID-19 Treatment," Current Pharmaceutical Design 27.9, 2021, 1123–1143. doi: 10.2174/138161282 6666201118103416.

11 Protein Targets for Anti-SARS-CoV-2 Activity of Traditional Herbal Medicine

Maria Shah, Shabana Bibi, Faryal Jahan,
Sadia Aziz, Amina Shamas, and Dipta Dey

11.1 INTRODUCTION

In recent years, newly emerging viral infections (VIs) have grown to pose a serious global threat to public health. Global public health is seriously threatened by the new coronavirus (CoV)-caused by a VI. The World Health Organization (WHO) was first notified of the novel CoV outbreak on December 31, 2019. The WHO classified this virus as a novel CoV on January 12, 2020, calling it "2019-nCoV" (World Health Organization, 2020). The pathogen responsible for the illness was quickly found to be a novel CoV and presented an identity containing virus sequences that were about 89.1% similar to severe acute respiratory syndrome coronavirus (SARS-CoV) (Kong et al., 2020). Hence, the novel virus was called severe acute respiratory syndrome coronavirus-2 (SARS-CoV-2) (Gorbalenya et al., 2020).

CoVs are encapsulated viruses through a positive sense of single-stranded RNA. CoVs have the biggest genomes amongst RNA viruses, presenting a length range (26–32 kilobase [kb]). Three groups of CoVs have been identified based on genetic and antigenic standards: alpha-CoVs, beta-CoVs, and gamma-CoVs. Although CoVs mostly infect birds and mammals through enzootic infections, they have also been found in recent decades to infect humans (Jayameena, Sivakumari, Ashok, & Rajesh, 2018; Schoeman & Fielding, 2019). According to the literature, the 2003 SARS epidemic and the more recent outbreak of respiratory diseases in the Middle East have verified the fatality of CoVs once they enter any human skin barrier and infect the patient.

A shift toward herbal medicine-based treatment options has been encouraged by the paucity of targeted and potent therapies against SARS-CoV-2 (Rahman et al., 2022). A major factor is possibly several medicines made from botanical substances or their bioactive constituents. Many people are interested in finding possible anti-COVID-19 herbal remedies (HRs) because they have shown intriguing abilities to fight different viruses by enhancing immunity (Bibi et al., 2021). Additionally, HRs have been effective in reducing the symptoms of VIs like SARS-CoV-2. Research suggests that using HRs to manage and lower the risk of COVID-19 may be beneficial (Ahmed et al., 2023). In conjunction with contemporary medicine, HRs have been used as a substitute treatment for COVID-19 and have produced a number of recommendations regarding HRs (Moher, Liberati, Tetzlaff, Altman, & PRISMA Group*, 2009). The efficacy of many plant-derived treatment options as anti-VI agents has been tested for medicinal purposes and shouldn't be undervalued. Many HRs with anti-VI properties are effectively being used as an alternative treatment to suppress SARS-CoV-2 (Ang, Song, Lee, & Lee, 2020). Several medicinal plant isolates are recommended and offer promising results in the treatment and prevention of COVID-19, according to recent studies. Additionally, Table 11.1 provides an overview of some of the common natural medicinal plants used to treat COVID-19 patients (Demeke, Woldeyohanins, & Kifle, 2021).

DOI: 10.1201/9781003452621-11

TABLE 11.1

Summarized Information of Plant Extracts Signifies in Handling of SARS-CoV-2 Infection

Plant Extract Name	Active Ingredient	Molecular Activity	References
Malva sylvestris L.	Polysaccharides	Potential antitussive agent	(Nosalova et al., 2005)
Aloe barbadensis Mill.	HF1Z (polysaccharide)	Potential antitussive agent	(Nosalova et al., 2005)
Salvia officinalis L.	Polysaccharide	Strong antitussive and emollient properties	(Nosalova et al., 2005)
Cynara scolymus L.	Cynaroside	Suppression of ACE	(Hoever et al., 2005; Khan & Kumar, 2019)
Erigeron abajoensis Cronquist	Flavone (Scutellarin)	Suppression of ACE	(Utomo, Ikawati, & Meiyanto, 2020)
Hibiscus sabdariffa L.	Anthocyanins	Suppression of ACE	(Nwachukwu, Aneke, Obika, & Nwachukwu, 2015)
Hancornia speciosa Gomes	Chlorogenic acid	Suppression of ACE	(Santos et al., 2016)
Isatis indigotica	Aloe emodin, hesperetin, indigo, sinigrin, 3-indolone, 2,2-Di (3-indolyl), and phaitanthrin D	Suppression of COVID-19 protein (3CL pro)	(Lin et al., 2005)
Alnus japonica (**Thunb.**) **Steud.**	Diarylheptanoid (hirsutenone)	Suppression of COVID-19 protein (PL pro)	(Park et al., 2012)
Paulownia tomentosa **Steud.**	Geranylated flavonoids	Suppression of COVID-19 protein (PL pro)	(Cho et al., 2013)
Torreya Nucifera (L.) **Siebold & Zucc.**	Genuine flavones (apigenin) and amentoflavone	Suppression of COVID-19 protein (3CL pro)	(Ryu et al., 2010)
Citrus **spp.**	Aspirin and hesperetin Neohesperidin and rhoifolin	Suppression of COVID-19 protein (3CL pro)	(Adem, Eyupoglu, Sarfraz, Rasul, & Ali, 2020)
Psoralea corylifolia L.	Corylifol A, psoralidin, isobavachalcone, 40-Omethylbavachalcone, neobavaisoflavone, and bevacizin	Suppression of COVID-19 protein (PL pro)	(Kim et al., 2014)
Artemisia capillaries **Thunb.**	Capillarisin (flavone)	Reduce TNF-α, IL-1β, IL-1α, and IL-6 while using antipyretics	(Han et al., 2013)
Glycyrrhiza glabra L.	Arabinogalactan macromolecule	The protective effects of citric acid on mucous membranes and spasmolytic activity in Guinea pigs	(Saha et al., 2011)

11.2 TRADITIONAL CHINESE MEDICINE (TCM) FOR COVID-19

In the People's Republic of China, the most frequently used medicine is known as Traditional Chinese Medicine (TCM) and has been used to treat the wellbeing of COVID-19 patients as an alternative medicine. Chinese researchers conducted systematic research on COVID-19 patients and checked their acceptance rate of TCM with no significant symptoms and hence proved a significant treatment for SARS-CoV-2 (Pan et al., 2023).

Literature reported that the bioactive component of *Sanctellaria baicalensis* extracts contains both baicalin, a beneficial TCM herbal constituent, and hesperetin, is a bioactive ingredient found in the peel of tangerine (Song et al., 2019). Both are effective substances that have contents that reduce the symptoms linked with COVID-19. Another TCM drug, xuebijing injection, is well known to reduce the risk of pneumonia transmission as well as regulation of respiration rate among patients with respiratory infections. Moreover, a researcher (Su et al., 2020) suggested *Exocarpium Citri grandis*, a TCM, is effective in treating and preventing COVID-19.

More than 100 natural TCM compositions have been long established in China to protect lives during outbreaks of illnesses. Following the Wuhan outbreak, these ancient achievements of adopting TCM to combat epidemics were exploited and extensively implemented in all Chinese hospitals. The Chinese government's effective fight against COVID-19 is now attributed to TCM (Lu & XS, 2020).

11.3 TRADITIONAL HERBAL MEDICINE FOR COVID-19

Similar to this, traditional medicine (TM) is used in Asian nations to treat COVID-19 in addition to modern treatments and vaccines. One of the earliest components of the global healthcare system, traditional Indian medicine is essential (Bibi et al., 2022). These conventional practices, which include homeopathy, naturopathy, yoga, ayurveda, unani, and siddha, are effective in treating a wide range of illnesses. These TM-dependent methods treat a variety of illnesses by using minerals, plants, and animal yields. In TM, about 25,000 herb-based preparations and extracts have been used in South Asia (Guo et al., 2020).

11.3.1 BINDING OF ANTI-SARS-CoV-2 POTENTIAL PROTEINS WITH *MALVA SYLVESTRIS* EXTRACTS

Malva sylvestris (wood mallow) is famous for lung ailments, effective for sore throats and dry coughs (Brasil & Sanitária, 2011; Silveira et al., 2020). Flavonoids make up the chemical components of *M. sylvestris*. The main components of mucilage are galacturonic acids, along with glucose, glucuronic, trehalose, galactose, and rhamnose (Paul, 2016).

Preclinical data showed that preparations of wood mallow plants have been researched for fever and lung diseases. Mucilage and rhamnogalacturonan, two of its isolated constituents, were examined for their antitussive qualities in cats. Both medications suppress bronchitis and manage the wheezing response, particularly in the laryngopharyngeal region (Nosalova et al., 2005). In recent literature, it's reported that *in vivo* models presented its pain-relaxants and anti-inflammatory activity (Seddighfar, Mirghazanfari, & Dadpay, 2020). The wood mallow plant has long been famous for fever and cough cures, and might help with COVID-19 symptoms by reducing irritation of the respiratory tract. A recent study elaborated the investigation on the molecular binding of components from wood mallow plants with the proteins in SARS-CoV-2. NADPH (PDB ID: 2CDU) macromolecular opposition was estimated against SAR-COV-2 infections. The binding contacts of the isolated compounds with the NADPH macromolecule have been investigated to understand the selectivity and structural analysis of the separated medicines. By introducing discrete pharmaceuticals into the binding standards of the NADPH macromolecule, molecular docking results have been achieved. Three isolates presented good binding contacts and energies as −5.09, −4.68, and −5.72 Kcal/mol (Irfan et al., 2021). The outcomes of the docking simulation highlighted that the extracted phytochemical has anti-SARS-CoV-2 effects. This shows a remarkable result to highlight the effects of explored drugs on SARS-CoV-2. This plant's extract mainly contained a maximum of total phenol constituents and total flavonoid constituents (Irfan et al., 2021).

The isolates of wood mallow plants have presented excellent activities as an antioxidant and anti-VI drugs when bound with the 6LU7 target protein, currently encouraging for drug development for COVID-19. Molecular docking-based calculated energies are demonstrated in Figure 11.1 with respect to each isolate studied by Irfan et al., 2021. This discovery paves the way for additional investigation into drug extracts and their probable uses in the handling and deterrence of VIs (Irfan et al., 2021).

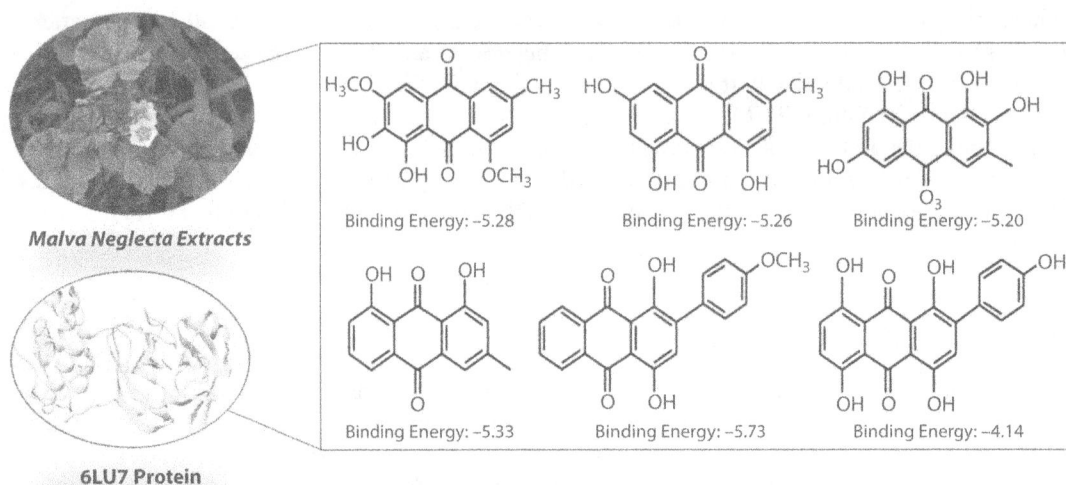

Malva Neglecta Extracts

6LU7 Protein

FIGURE 11.1 Molecular docking estimations of isolated compounds with the 6LU7 COVID-19 protein (Irfan et al., 2021).

11.3.2 BINDING OF ANTI-SARS-CoV-2 POTENTIAL PROTEINS WITH *AZADIRACHTA INDICA* (NEEM) EXTRACTS

The medical indications of VIs are elevated body temperature and neem's plant extracts could minimize it. In the past, boiling and consuming *A. indica,* neem plant leaves, was used to treat COVID-19-related fever. These leaves have presented remarkable anti-inflammatory qualities in animal studies (Paterson et al., 2020). The extracts from the leaves of *A. indica* and their biochemical constituents, flavonoids and polysaccharides, have been proven to possess antiviral properties against the hepatitis C virus in animal models, studied by computational drug designing (Lage et al., 2014). Particular to coronaviral disease, a macromolecular binding investigation revealed that neem-derived isolates such as: cycloartenol, nimbolin, and nimocin, could be good binders for the COVID-19 envelope (E) macromolecule, membrane (M) molecule, and other glycoproteins as potential suppressing agents (Borkotoky & Banerjee, 2021). Its leaf extracts positively impact immunoregulation to increase immunity in animal models (Bizer & Seaborne, 2004). Neem-derived seed extract was administered intravenously and increased the assembly of IFN-γ in mice given the Brucella Rev-1 vaccine after the inoculation procedure (Asal & Al Zubaidy, 2016). Pregnant women should be careful about neem-derived seed extracts because animal models have shown that they do not support pregnancy (Atay, Gercel-Taylor, & Taylor, 2011). Antihuman chorionic gonadotropin effects have been documented in clinical investigations (Talwar et al., 1995). According to studies, consuming neem-leaves boiled water for medicinal purposes is very important as an anti-VI agent (Lim, Teh, & Tan, 2021). In traditional use, nothing is known about the toxicity profile of *A. indica* plant leaf; however, the vegetable oil extract has been reported to cause renal damage and clinical acidosis in the body system in an elderly person in India; hence, Neem oil must be used conscientiously (Mishra & Dave, 2013).

For the M protein, binding studies with neem isolates showed two promising interacting cavities: Nimocin's and Nimbolin A's within the transmembrane conserved region, with assistance from 20 significant amino acids at the C-terminal. Hence the C-terminal tail comprised Lys 199, Gly 201, Tyr 203, and Arg 204, which are important for the suppression of action of the M protein of MERS-CoV in the trans-Golgi network (TGN). A post-simulation Molecular Mechanics Poisson-Boltzmann

Surface Area (MM-PBSA) investigation highlighted Gly 202 and Tyr 204 within the active cavity of the C-terminal end of M-protein interactions (Perrier et al., 2019).

Another study presented that beta-Amyrin, 24-Methylenecycloartan-3-one, and phytosterol were involved in viral genomic assembly by binding with active residues of the C-terminal region ((Arndt, Larson, & Hogue, 2010). Therefore, the following plant isolates identified active binding with crucial regions of the M protein involved in TGN localization. The novel drugs inhibiting the E and M proteins of SARS-CoV-2 using computer-assisted docking and simulations studies were identified. Nimocin and Cycloartanol could interact with both coronaviral proteins and affect the propagation and pathophysiology of CoV (Sarkar et al., 2020) The virus assembly is achieved by the strongest bonds among the M protein and other nucleocapsid (N) proteins (Arndt et al., 2010). Another study by Kuo, Koetzner, & Masters (2016) also revealed the mechanism by which genome assembly is attained.

The E protein is a homopentameric, short, integral membrane protein of 76–109 amino acids (Pervushin et al., 2009) M and E proteins are essential and adequate to form a constituent part of other beta-CoV (Mortola & Roy, 2004); hence, they are considered to play a key role in the assembly of viral components. The assembly of the virus depends on the binding contacts of the M protein with the other main structural proteins, including N, spike (S), and E. It is believed that the involvement of the ion-channel protein E in membrane bending and contact is essential to the development of virus budding (Schoeman & Fielding, 2019). Prior investigation indicates that the ion-channel function of E is necessary for the spread of viruses; therefore, inhibitors that attack the E protein can aid in the prevention of virus propagation (Borkotoky & Banerjee, 2021). It has also been documented that the flu-related viral medication "amantadine" can bind to ion channels made by the E protein, although the impact was quite mild. The binding contacts between the top five chemicals occurred within the pentamer-formed core channel, according to our docking study with E (Torres et al., 2007).

In one study (Borkotoky & Banerjee, 2021), strong bonds were observed between the significant derivatives in terms of binding energy and the transmembrane and C-terminal domains of the E protein. In the E protein pentamer, nearly every chemical exhibited hydrophobic contact with residues from each of the five monomeric chains. Nimbolin A's 16 hydrophobic contacts presented an interacting score of −11.2 kcal/mol. With interacting scores of −11.0 kcal/mol, −10.9 kcal/mol, −10.8 kcal/mol, and −10.5 kcal/mol, respectively, the other bioactive isolates–nimocin, 7-deacetyl-7-benzoylgedunin, 24 methylenecycloartanol, and cycloeucalenone–interacted with the E protein pentamer, exhibiting at least 10 hydrophobic contacts in each instance.

11.3.3 Binding of Anti-SARS-CoV-2 Potential Proteins with *Nigella sativa* Extracts

Black cumin seeds (*Nigella sativa*) were among the natural remedies with the greatest amount of useful information available. In hepatitis C patients, ethanol-based extracts of *N. sativa* seeds were shown to combat viruses by lowering the propagation of viral, alpha-fetoprotein and improving hepatic activity indices (Abdel-Moneim, Morsy, Mahmoud, Abo-Seif, & Zanaty, 2013). Its seed oil has anti-VI and immuno-regulatory properties for cytomegalovirus, suppressing infections. Alleviating the CD3 and CD4 levels, causing a rise in the release of IFN-γ in macrophages and Natural Killer T-cells, can improve the immunological response (Salem et al., 2017) through a decreased expression of the gene of several leukocyte transient receptors protein. It is highly effective for respiratory disorders like asthma. In clinical tests, long-term (3 months) use of 3 g/day black cumin seed ingestion did not result in substantial adverse effects on liver or renal function (Bamosa, Kaatabi, Lebdaa, Elq, & Al-Sultanb, 2010). To find the compounds' binding affinities with SARS-CoV-2-related macromolecular structure, such as proteases (6LU7 and 6Y2E), main peptidase (2GTB), angiotensin-converting enzyme 2 (ACE2), and heat shock protein A5, the molecular docking of isolated compounds from *N sativa* and various anti-VI medicines was done in a study (Bouchentouf & Missoum, 2020). Nigelledine, an alkaloid of black cumin seeds, bound with

6LU7 presented optimal scores. When docked with 2GTB active sites, the saponin α-hederin from *N. sativa* exhibited a higher energy score compared to favipiravir, hydroxychloroquine, and chloroquine. The major essential oil ingredient of black cumin seeds, thymoquinone, displayed interacting scores with 6LU7, ACE2, and heat shock protein A5, with scores smaller than hydroxychloroquine in 6LU7 and ACE2. Furthermore, a saponin isolated from *N. sativa*, hederagenin, displayed a lower binding score in 6LU7 and 6Y2E when docked with ACE2, GRP78, 6LU7, and 6Y2E active sites than saquinavir (Rajapaksa, Perera, Nisansala, Perera, & Dissanayake, 2020; Sampangi-Ramaiah, Vishwakarma, & Shaanker, 2020). With SARS-CoV-2 6LU7, endoribonucleoase, ADP-ribose-1-phosphatase, RNA-dependent RNA polymerase, the interface domain of the SARS-CoV-2 S protein, and human ACE2, thymohydroquinone exhibited modest docking energy (Silva, Figueiredo, Byler, & Setzer, 2020). Nigellidine had a strong affinity for SARS-CoV-2 enzymes and proteins, including S-glycoprotein, nucleocapsid, 6LU7, N-terminus protease, and nonstructural protein. High binding energy was demonstrated by nigellidine for human receptors, inflammatory signaling molecules, and other proteins, including human IL1R (1ITB), TNFR1 (1NCF), and TNFR2 (3ALQ) (Maiti, Banerjee, Nazmeen, Kanwar, & Das, 2020) (Table 11.2).

TABLE 11.2

A Summary of Effects of *Nigella sativa* Compounds on Severe Acute Respiratory Syndrome-Coronavirus Disease 2 (SARS-CoV-2) Targets (Koshak & Koshak, 2020)

Bioactive Compounds	Protein IDs	Control	Molecular Activities	Reference
Thymoquinone	6LU7	–	Presented moderate interacting potential toward 6LU7 macromolecule.	(Barakat, Shoman, Dina, & Alfarouk, 2010)
Nigellidine, α-hederin	6LU7, 2GTB	Chloroquine	The highest binding affinities with 6LU7 and 2GTB were found with nigellidine and α-hederin.	(Ulasli et al., 2014)
		HCQ	Nigellidine outperformed favipiravir and HCQ.	
		Favipiravir	α-hederin is superior to activity than favipiravir, HCQ, and chloroquine.	
Hederagenin	6LU7, 6Y2E	Saquinavir	Drug bound to 6LU7 with a high affinity, but less so than saquinavir, and 6Y2E nearly as well as saquinavir.	(Oyero et al., 2016)
Nigellidine	6LU7, NSP2, 6VSB, QHD43415_3, QHD43423, IL1R, TNFR1, & TNFR2	–	Presented strong interacting potential as an anti-inflammatory agent.	(Dorra, El-Berrawy, Sallam, & Mahmoud, 2019)
Hederagenin	ACE2, GRP78	–	Presented highest binding affinity selected macromolecule.	(Barakat, El Wakeel, & Hagag, 2013)
Thymoquinone	6LU7, ACE2	HCQ	Compared to HCQ, thymoquinone's binding affinity to 6LU7 and ACE2 1R42 was moderate.	(Onifade, Jewell, & Adedeji, 2013)
Thymoquinone	HSPA5	–	Presented moderate affinity with selected macromolecule.	(Onifade, Jewell, Ajadi, Rahamon, & Ogunrin, 2013)
Thymohydroquinone	6LU7, Nsp15/NendoU, ADRP, RdRp, rS, and ACE2	–	Exhibited a moderate affinity activity with multiple molecular targets of SARS-CoV-2.	(Romano & Tatonetti, 2019)

11.3.4 BINDING OF ANTI-SARS-COV-2 POTENTIAL PROTEINS WITH *MENTHA PIPERITA* EXTRACTS

The most recognized ancient HR for a variety of infections is peppermint (*M. piperita*). Dry peppermint has been cultivated for hundreds of years and is also known as an ancient Egyptian, Greek, and Chinese traditional drug and noted for its significance. *M. piperita* has strong antibacterial and antifungal properties (Nayak, Kumar, Gupta, & Joshi, 2020). A researcher from Saudi Arabia highlighted that 78% of nonhospitalized patients using peppermint, and 22% of hospitalized patients not using the *M. piperita* supplement, obsereved against the CoV disease (Aldwihi et al., 2021). Three substances presenting strong binding scores for blocking the SARS-CoV-2 coding PDB-5R7Y structure were found in mint leaves by the study team (Gervasoni et al., 2020). Among them are isorhoifolin, rutin, and hesperidin. Furthermore, the ChemPLP values of rutin and hesperidin in particular were lower than arbidol, whereas values of chloroquine, darunavir, lopinavir, and remdesivir were higher. According to the results, rutin and hesperidin have a more pronounced direct inhibitory effect on SARS-CoV-2 protease enzyme receptors than other bioactive derivatives (Bellavite & Donzelli, 2020).

11.3.5 BINDING OF ANTI-SARS-COV-2 POTENTIAL PROTEINS WITH *ALLIUM SATIVUM* AND *ALLIUM CEPA* EXTRACTS

It has been noted that the phytochemicals found in onions and garlic interfere with virus propagation by inhibiting the formation of proteins and genome in virus (Castrillo & Carrasco, 1987) (Zandi et al., 2011). Garlic has strong anti-VI qualities. Garlic extract, for instance, has been shown in studies to reduce influenza A and B virus infection. Garlic also works wonders against a variety of viruses, including rotavirus, rhinovirus, CMV, HIV, herpes simplex virus 1 and 2. Garlic was found to considerably reduce the incidence of the common cold virus in another investigation (Josling, 2001). Organosulfur components such allicin, diallyl trisulfide, and ajoene are the major molecules that impart anti-VI capabilities to garlic (Hughes, Murray, North, & Lawson, 1989; Weber et al., 1992). These substances were found to have an impact on the oxidative stress response mechanism (Hall, Troupin, Londono-Renteria, & Colpitts, 2017). By preventing the virus from attaching itself to the cell, onion isolates have been shown to help reduce the spread of the New Castle Disease virus (Harazem, El Rahman, & El-Kenawy, 2019). Thus, onions and garlic are plants that may be utilized as a substitute therapy for VIs and to stop the onset of serious illnesses.

Proteases are responsible for the replication of numerous viruses and, following proteolytic cleavage, generate functional proteins. Proteases are, therefore, thought to be effective and appropriate protein targets for the creation and advancement of antiviral pharmacological treatments (Paraskevis et al., 2020). Protease has been successfully structured and repositioned in PDB (PDB ID: 6LU7) and is considered a possible target protein to inhibit the infection of the virus by inhibiting its replication. The binding contacts between viral protease (MPro) and efficacious isolates of garlic (*A. sativum*) and (*A. cepa*) onion have been studied and also compared with drugs in use for the treatment of COVID-19 like hydroxychloroquine. To identify the best suitable TM against the COVID-19 infection pathway, these isolated entities were selected based on binding energies via the molecular docking approach. From these two plants, 23 effective compounds were selected and out of 23 isolates, seven isolates were found as the most potent inhibitors against the COVID-19 target protein 6LU7. The result from molecular docking exhibits that *alliin*, a garlic extract, presented properties for the handling of CoV proteases (Pandey, Khan, Kumar, Srivastava, & Jha, 2021).

11.3.6 BINDING OF ANTI-SARS-COV-2 POTENTIAL PROTEINS WITH *EURYCOMA LONGIFOLIA* EXTRACTS

The popular herb *E. longifolia*, which has long been used to improve men's health, is widely available in Malaysia (Rehman, Choe, & Yoo, 2016). It is one of the few natural products on the market with safety data and established standardization. The recommended amount of 200 mg/day of the standardized aqueous extract of *E. longifolia* has been shown to positively induce immunity in the elderly population by raising the number of CD4+ cells, but no direct anti-VI effects have been documented with clinical data (George et al., 2016).

Additionally, there is preclinical evidence supporting *E. longifolia's* anti-inflammatory qualities. Among the robust NF-κB inhibitory actions of the strong anti-inflammatory bioactive substances derived from *E. longifolia* are eurycomalactone, 1-4-1-5β-dihydroklaieanone, and 1-3-2-1 dehydroeurycomanone (Tran et al., 2014). It has also been observed that several phenolic components derived from *E. longifolia* roots greatly reduce the production of IL-6 in lipopolysaccharide-stimulated RAW macrophages (Ruan et al., 2019). There is ample data regarding the safety profile of *E. longifolia*. Additionally, additional research on the plant's possible anti-inflammatory properties may be found in relation to COVID-19 preventive and treatment strategies (Choudhary, Bommu, Ming, & Zulkawi, 2012).

11.4   CONCLUSION

There are several proteins associated with the VIs propagation, such as S proteins, E proteins, M proteins, and N proteins. These are the important structural proteins of COVID-19 studied frequently in recent years with respect to the binding of bioactive natural compounds for the development of novel VI drugs. Traditional plants discussed in this chapter are important for the extraction of multiple therapeutic compounds, and their isolates have shown binding contacts with protein targets of SARS-CoV-2, causing hindrance in its propagation. It is considered that they may have a valuable role in SARS-CoV-2 anticipation and management. More advanced computational and artificial intelligence-based drug design applications are encouraged to be used for plant-based database screening and the development of anti-VI medicine.

REFERENCES

Abdel-Moneim, A., Morsy, B. M., Mahmoud, A. M., Abo-Seif, M. A., & Zanaty, M. I. (2013). Beneficial therapeutic effects of *Nigella sativa* and/or *Zingiber officinale* in HCV patients in Egypt. *EXCLI Journal*, *12*, 943.
Adem, S., Eyupoglu, V., Sarfraz, I., Rasul, A., & Ali, M. (2020). Identification of potent COVID-19 main protease (Mpro) inhibitors from natural polyphenols: an in silico strategy unveils a hope against CORONA.
Ahmed, S. I., Jamil, S., Ismatullah, H., Hussain, R., Bibi, S., Khandaker, M. U., … Emran, T. B. (2023). A comprehensive perspective of traditional Arabic or Islamic medicinal plants as an adjuvant therapy against COVID-19. *Saudi Journal of Biological Sciences*, 103561.
Aldwihi, L. A., Khan, S. I., Alamri, F. F., AlRuthia, Y., Alqahtani, F., Fantoukh, O. I., … Almohammed, O. A. (2021). Patients' behavior regarding dietary or herbal supplements before and during COVID-19 in Saudi Arabia. *International Journal of Environmental Research and Public Health*, *18*(10), 5086.
Ang, L., Song, E., Lee, H. W., & Lee, M. S. (2020). Herbal medicine for the treatment of coronavirus disease 2019 (COVID-19): A systematic review and meta-analysis of randomized controlled trials. *Journal of Clinical Medicine*, *9*(5), 1583.
Arndt, A. L., Larson, B. J., & Hogue, B. G. (2010). A conserved domain in the coronavirus membrane protein tail is important for virus assembly. *Journal of virology*, *84*(21), 11418–11428.
Asal, S. N., & Al Zubaidy, I. A. (2016). Seroprevelance study of *Toxoplasma gondii* in horses and camels animal in Wasit province. *The Iraqi Journal of Veterinary Medicine*, *40*(1), 177–182.
Atay, S., Gercel-Taylor, C., & Taylor, D. D. (2011). Human trophoblast-derived exosomal fibronectin induces pro-inflammatory IL-1β production by macrophages. *American Journal of Reproductive Immunology*, *66*(4), 259–269.
Bamosa, A. O., Kaatabi, H., Lebdaa, F. M., Elq, A.-M., & Al-Sultanb, A. (2010). Effect of *Nigella sativa* seeds on the glycemic control of patients with type 2 diabetes mellitus. *Indian J Physiol Pharmacol*, *54*(4), 344–354.

Barakat, A. B., Shoman, S. A., Dina, N., & Alfarouk, O. R. (2010). Antiviral activity and mode of action of *Dianthus caryophyllus* L. and *Lupinus termes* L. seed extracts against in vitro herpes simplex and hepatitis A viruses infection. *J Microbiol Antimicrob*, 2(3), 23–29.

Barakat, E. M. F., El Wakeel, L. M., & Hagag, R. S. (2013). Effects of *Nigella sativa* on outcome of hepatitis C in Egypt. *World Journal of Gastroenterology: WJG*, 19(16), 2529.

Bellavite, P., & Donzelli, A. (2020). Hesperidin and SARS-CoV-2: New light on the healthy function of citrus fruits. *Antioxidants*, 9(8), 742.

Bibi, S., Khan, M. S., El-Kafrawy, S. A., Alandijany, T. A., El-Daly, M. M., Yousafi, Q., ... Azhar, E. I. (2022). Virtual screening and molecular dynamics simulation analysis of Forsythoside A as a plant-derived inhibitor of SARS-CoV-2 3CLpro. *Saudi Pharmaceutical Journal*, 30(7), 979–1002.

Bibi, S., Sarfraz, A., Mustafa, G., Ahmad, Z., Zeb, M. A., Wang, Y.-B., ... Yu, H. (2021). Impact of traditional plants and their secondary metabolites in the discovery of COVID-19 treatment. *Current Pharmaceutical Design*, 27(9), 1123–1143.

Bizer, C., & Seaborne, A. (2004). Proceedings of the 3rd International Semantic Web Conference (ISWC2004).

Borkotoky, S., & Banerjee, M. (2021). A computational prediction of SARS-CoV-2 structural protein inhibitors from *Azadirachta indica* (Neem). *Journal of Biomolecular Structure and Dynamics*, 39(11), 4111–4121.

Bouchentouf, S., & Missoum, N. (2020). Identification of Compounds from *Nigella sativa* as New Potential Inhibitors of 2019 Novel Coronasvirus (Covid-19): Molecular Docking Study.

Brasil, & Sanitária, A. N. d. V. (2011). Formulário de Fitoterápicos. da Farmacopéia Brasileira: Anvisa Brasília, DF.

Castrillo, J. L., & Carrasco, L. (1987). Action of 3-methylquercetin on poliovirus RNA replication. *Journal of Virology*, 61(10), 3319–3321.

Cho, J. K., Curtis-Long, M. J., Lee, K. H., Kim, D. W., Ryu, H. W., Yuk, H. J., & Park, K. H. (2013). Geranylated flavonoids displaying SARS-CoV papain-like protease inhibition from the fruits of Paulownia tomentosa. *Bioorganic & Medicinal Chemistry*, 21(11), 3051–3057.

Choudhary, Y. K., Bommu, P., Ming, Y. K., & Zulkawi, N. B. (2012). Acute, sub-acute, and subchronic 90-days toxicity of *Eurycoma longifolia* aqueous extract (Physta) in wistar rats. *International Journal of Pharmacy and Pharmaceutical Sciences*, 4(3).

Demeke, C. A., Woldeyohanins, A. E., & Kifle, Z. D. (2021). Herbal medicine use for the management of COVID-19: A review article. *Metabolism Open*, 12, 100141.

Dorra, N., El-Berrawy, M., Sallam, S., & Mahmoud, R. (2019). Evaluation of antiviral and antioxidant activity of selected herbal extracts. *Journal of High Institute of Public Health*, 49(1), 36–40.

George, A., Suzuki, N., Abas, A. B., Mohri, K., Utsuyama, M., Hirokawa, K., & Takara, T. (2016). Immunomodulation in middle-aged humans via the ingestion of Physta® standardized root water extract of *Eurycoma longifolia* Jack—a randomized, double-blind, placebo-controlled, parallel study. *Phytotherapy Research*, 30(4), 627–635.

Gervasoni, S., Vistoli, G., Talarico, C., Manelfi, C., Beccari, A. R., Studer, G., ... Pedretti, A. (2020). A comprehensive mapping of the druggable cavities within the SARS-CoV-2 therapeutically relevant proteins by combining pocket and docking searches as implemented in pockets 2.0. *International Journal of Molecular Sciences*, 21(14), 5152.

Gorbalenya, A. E., Baker, S. C., Baric, R. S., de Groot, R. J., Drosten, C., Gulyaeva, A. A., ... Neuman, B. W. (2020). Severe acute respiratory syndrome-related coronavirus: The species and its viruses–a statement of the Coronavirus Study Group. *BioRxiv*.

Guo, Y.-R., Cao, Q.-D., Hong, Z.-S., Tan, Y.-Y., Chen, S.-D., Jin, H.-J., ... Yan, Y. (2020). The origin, transmission and clinical therapies on coronavirus disease 2019 (COVID-19) outbreak–an update on the status. *Military Medical Research*, 7, 1–10.

Hall, A., Troupin, A., Londono-Renteria, B., & Colpitts, T. M. (2017). Garlic organosulfur compounds reduce inflammation and oxidative stress during dengue virus infection. *Viruses*, 9(7), 159.

Han, S., Lee, J. H., Kim, C., Nam, D., Chung, W.-S., Lee, S.-G., ... Ahn, K. S. (2013). Capillarisin inhibits iNOS, COX-2 expression, and proinflammatory cytokines in LPS-induced RAW 264.7 macrophages via the suppression of ERK, JNK, and NF-κB activation. *Immunopharmacology and Immunotoxicology*, 35(1), 34–42.

Harazem, R., El Rahman, S. A., & El-Kenawy, A. (2019). Evaluation of antiviral activity of *Allium cepa* and *Allium sativum* extracts against Newcastle disease virus. *Alexandria Journal for Veterinary Sciences*, 61(1).

Hoever, G., Baltina, L., Michaelis, M., Kondratenko, R., Baltina, L., Tolstikov, G. A., … Cinatl, J. (2005). Antiviral activity of glycyrrhizic acid derivatives against SARS– coronavirus. *Journal of Medicinal Chemistry*, *48*(4), 1256–1259.

Hughes, B., Murray, B., North, J., & Lawson, L. (1989). Antiviral constituents from Allium sativum. *Planta Medica*, *55*(01), 114–114.

Irfan, A., Imran, M., Khalid, M., Ullah, M. S., Khalid, N., Assiri, M. A., … Hussein, M. (2021). Phenolic and flavonoid contents in *Malva sylvestris* and exploration of active drugs as antioxidant and anti-COVID19 by quantum chemical and molecular docking studies. *Journal of Saudi Chemical Society*, *25*(8), 101277.

Irfan, A., Imran, M., Khalid, N., Hussain, R., Basra, M. A. R., Khaliq, T., … Qayyum, M. A. (2021). Isolation of phytochemicals from Malva neglecta Wallr and their quantum chemical, molecular docking exploration as active drugs against COVID-19. *Journal of Saudi Chemical Society*, *25*(12), 101358.

Jayameena, P., Sivakumari, K., Ashok, K., & Rajesh, S. (2018). In silico molecular docking studies of Rutin compound against apoptotic proteins (Tumor Necrosis Factor, Caspase-3, NF-Kappa-B, P53, Collagenase, Nitric oxide synthase and Cytochrome C). *Journal of Cancer Research and Treatment*, *6*(2), 28–33.

Josling, P. (2001). Preventing the common cold with a garlic supplement: A double-blind, placebo-controlled survey. *Advances in Therapy*, *18*, 189–193.

Khan, M. Y., & Kumar, V. (2019). Mechanism & inhibition kinetics of bioassay-guided fractions of Indian medicinal plants and foods as ACE inhibitors. *Journal of Traditional and Complementary Medicine*, *9*(1), 73–84.

Kim, D. W., Seo, K. H., Curtis-Long, M. J., Oh, K. Y., Oh, J.-W., Cho, J. K., … Park, K. H. (2014). Phenolic phytochemical displaying SARS-CoV papain-like protease inhibition from the seeds of Psoralea corylifolia. *Journal of Enzyme Inhibition and Medicinal Chemistry*, *29*(1), 59–63.

Kong, R., Yang, G., Xue, R., Liu, M., Wang, F., Hu, J., … Chang, S. (2020). COVID-19 docking server: A meta server for docking small molecules, peptides and antibodies against potential targets of COVID-19. *Bioinformatics*, *36*(20), 5109–5111.

Koshak, A. E., & Koshak, E. A. (2020). *Nigella sativa* L as a potential phytotherapy for coronavirus disease 2019: A mini review of in silico studies. *Current Therapeutic Research*, *93*, 100602.

Kuo, L., Koetzner, C. A., & Masters, P. S. (2016). A key role for the carboxy-terminal tail of the murine coronavirus nucleocapsid protein in coordination of genome packaging. *Virology*, *494*, 100–107.

Lage, G. A., Medeiros, F. d. S., Furtado, W. d. L., Takahashi, J. A., Filho, J. D. d. S., & Pimenta (2014). The first report on flavonoid isolation from Annona crassiflora Mart. *Natural Product Research*, *28*(11), 808–811.

Lim, X., Teh, B., & Tan, T. (2021). Medicinal plants in COVID-19: Potential and limitations. *Front Pharmacol*, *2021*(12), 1–8.

Lin, C.-W., Tsai, F.-J., Tsai, C.-H., Lai, C.-C., Wan, L., Ho, T.-Y., … Chao, P.-D. L. (2005). Anti-SARS coronavirus 3C-like protease effects of Isatis indigotica root and plant-derived phenolic compounds. *Antiviral Research*, *68*(1), 36–42.

Lu, Z., & XS, L. (2020). Qingfei Paidu decoction demonstrates the anti-epidemic effects and self-confidence of traditional Chinese medicine. *Journal of Traditional Chinese Medicine*, *61*(10), 833–834.

Maiti, S., Banerjee, A., Nazmeen, A., Kanwar, M., & Das, S. (2020). Active-site Molecular docking of Nigellidine to nucleocapsid/Nsp2/Nsp3/MPro of COVID-19 and to human IL1R and TNFR1/2 may stop viral-growth/cytokine-flood, and the drug source *Nigella sativa* (black cumin) seeds show potent antioxidant role in experimental rats.

Mishra, A., & Dave, N. (2013). Neem oil poisoning: Case report of an adult with toxic encephalopathy. *Indian Journal of Critical Care Medicine: Peer-Reviewed, Official Publication of Indian Society of Critical Care Medicine*, *17*(5), 321.

Moher, D., Liberati, A., Tetzlaff, J., Altman, D. G., & PRISMA Group*. (2009). Preferred reporting items for systematic reviews and meta-analyses: The PRISMA statement. *Annals of Internal Medicine*, *151*(4), 264–269.

Nayak, P., Kumar, T., Gupta, A., & Joshi, N. (2020). Peppermint a medicinal herb and treasure of health: A review. *Journal of Pharmacognosy and Phytochemistry*, *9*(3), 1519–1528.

Nosalova, G., Sutovska, M., Mokry, J., Kardosova, A., Capek, P., & Khan, M. (2005). Efficacy of herbal substances according to cough reflex. *Minerva Biotecnologica*, *17*(3), 141.

Nwachukwu, D. C., Aneke, E. I., Obika, L. F., & Nwachukwu, N. Z. (2015). Effects of aqueous extract of Hibiscus sabdariffa on the renin-angiotensin-aldosterone system of Nigerians with mild to moderate essential hypertension: A comparative study with lisinopril. *Indian Journal of Pharmacology*, *47*(5), 540.

Onifade, A., Jewell, A., Ajadi, T., Rahamon, S., & Ogunrin, O. (2013). Effectiveness of a herbal remedy in six HIV patients in Nigeria. *Journal of Herbal Medicine, 3*(3), 99–103.

Onifade, A. A., Jewell, A. P., & Adedeji, W. A. (2013). *Nigella sativa* concoction induced sustained seroreversion in HIV patient. *African Journal of Traditional, Complementary and Alternative Medicines, 10*(5), 332–335.

Oyero, O. G., Toyama, M., Mitsuhiro, N., Onifade, A. A., Hidaka, A., Okamoto, M., & Baba, M. (2016). Selective inhibition of hepatitis C virus replication by Alpha-zam, a *Nigella sativa* seed formulation. *African Journal of Traditional, Complementary and Alternative Medicines, 13*(6), 144–148.

Pan, B., Yin, H.-w., Yu, Y., Xiang, X., Yu, C., Yan, X.-J., … Hong, J. (2023). Acceptance and attitude towards the traditional Chinese medicine among asymptomatic COVID-19 patients in Shanghai Fangcang hospital. *BMC Complementary Medicine and Therapies, 23*(1), 97.

Pandey, P., Khan, F., Kumar, A., Srivastava, A., & Jha, N. K. (2021). Screening of potent inhibitors against 2019 novel coronavirus (Covid-19) from *Allium sativum* and *Allium cepa*: An in silico approach. *Biointerface Research in Applied Chemistry, 11*(1), 7981–7993.

Paraskevis, D., Kostaki, E. G., Magiorkinis, G., Panayiotakopoulos, G., Sourvinos, G., & Tsiodras, S. (2020). Full-genome evolutionary analysis of the novel corona virus (2019-nCoV) rejects the hypothesis of emergence as a result of a recent recombination event. *Infection, Genetics and Evolution, 79*, 104212.

Park, J.-Y., Jeong, H. J., Kim, J. H., Kim, Y. M., Park, S.-J., Kim, D., … Ryu, Y. B. (2012). Diarylheptanoids from *Alnus japonica* inhibit papain-like protease of severe acute respiratory syndrome coronavirus. *Biological and Pharmaceutical Bulletin, 35*(11), 2036–2042.

Paterson, R. W., Brown, R. L., Benjamin, L., Nortley, R., Wiethoff, S., Bharucha, T., … Zambreanu, L. (2020). The emerging spectrum of COVID-19 neurology: Clinical, radiological and laboratory findings. *Brain, 143*(10), 3104–3120.

Paul, D. (2016). A review on biological activities of common Mallow (*Malva sylvestris* L.). *Journal of Life Science, 4*, 1–5.

Perrier, A., Bonnin, A., Desmarets, L., Danneels, A., Goffard, A., Rouillé, Y., Dubuisson, J., & Belouzard, S. (2019). The C-terminal domain of the MERS coronavirus M protein contains a trans-Golgi network localization signal. *Journal of Biological Chemistry, 294*(39), 14406–14421.

Pervushin, K., Tan, E., Parthasarathy, K., Lin, X., Jiang, F. L., Yu, D.,. . . Torres, J. (2009). Structure and inhibition of the SARS coronavirus envelope protein ion channel. *PLoS Pathogens, 5*(7), e1000511.

Rahman, M. M., Bibi, S., Rahaman, M. S., Rahman, F., Islam, F., Khan, M. S., … Maeesa, S. K. (2022). Natural therapeutics and nutraceuticals for lung diseases: Traditional significance, phytochemistry, and pharmacology. *Biomedicine & Pharmacotherapy, 150*, 113041.

Rajapaksa, R., Perera, B. T., Nisansala, M., Perera, W., & Dissanayake, K. (2020). Potential of inhibiting the receptor binding mechanism of sarscov-2 using phytochemical extracts of medicinal herb; moleculer docking study.

Rehman, S. U., Choe, K., & Yoo, H. H. (2016). Review on a traditional herbal medicine, *Eurycoma longifolia* Jack (Tongkat Ali): Its traditional uses, chemistry, evidence-based pharmacology and toxicology. *Molecules, 21*(3), 331.

Romano, J. D., & Tatonetti, N. P. (2019). Informatics and computational methods in natural product drug discovery: A review and perspectives. *Frontiers in Genetics, 10*, 368.

Ruan, J., Li, Z., Zhang, Y., Chen, Y., Liu, M., Han, L., … Wang, T. (2019). Bioactive constituents from the roots of *Eurycoma longifolia*. *Molecules, 24*(17), 3157.

Ryu, Y. B., Jeong, H. J., Kim, J. H., Kim, Y. M., Park, J.-Y., Kim, D., … Park, K. H. (2010). Biflavonoids from *Torreya nucifera* displaying SARS-CoV 3CLpro inhibition. *Bioorganic & Medicinal Chemistry, 18*(22), 7940–7947.

Saha, S., Nosál'ová, G., Ghosh, D., Flešková, D., Capek, P., & Ray, B. (2011). Structural features and in vivo antitussive activity of the water extracted polymer from *Glycyrrhiza glabra*. *International Journal of Biological Macromolecules, 48*(4), 634–638.

Salem, A. M., Bamosa, A. O., Qutub, H. O., Gupta, R. K., Badar, A., Elnour, A., & Afzal, M. N. (2017). Effect of *Nigella sativa* supplementation on lung function and inflammatory mediators in partly controlled asthma: A randomized controlled trial. *Annals of Saudi Medicine, 37*(1), 64–71.

Sampangi-Ramaiah, M. H., Vishwakarma, R., & Shaanker, R. U. (2020). Molecular docking analysis of selected natural products from plants for inhibition of SARS-CoV-2 main protease. *Current Science, 118*(7), 1087–1092.

Santos, U. P., Campos, J. F., Torquato, H. F. V., Paredes-Gamero, E. J., Carollo, C. A., Estevinho, L. M., … Dos Santos, E. L. (2016). Antioxidant, antimicrobial and cytotoxic properties as well as the phenolic content of the extract from *Hancornia speciosa* Gomes. *PloS One, 11*(12), e0167531.

Sarkar, C., Mondal, M., Torequl Islam, M., Martorell, M., Docea, A. O., Maroyi, A.,. . . Calina, D. (2020). Potential therapeutic options for COVID-19: current status, challenges, and future perspectives. *Frontiers in Pharmacology*, *11*, 572870.

Schoeman, D., & Fielding, B. C. (2019). Coronavirus envelope protein: Current knowledge. *Virology Journal*, *16*(1), 1–22.

Seddighfar, M., Mirghazanfari, S. M., & Dadpay, M. (2020). Analgesic and anti-inflammatory properties of hydroalcoholic extracts of *Malva sylvestris*, *Carum carvi* or *Medicago sativa*, and their combination in a rat model. *Journal of Integrative Medicine*, *18*(2), 181–188.

Silva, J. K. R. D., Figueiredo, P. L. B., Byler, K. G., & Setzer, W. N. (2020). Essential oils as antiviral agents. Potential of essential oils to treat SARS-CoV-2 infection: An *in-silico* investigation. *International Journal of Molecular Science*, *21*(10), 3426.

Silveira, D., Prieto-Garcia, J. M., Boylan, F., Estrada, O., Fonseca-Bazzo, Y. M., Jamal, C. M., ... Heinrich, M. (2020). COVID-19: Is there evidence for the use of herbal medicines as adjuvant symptomatic therapy? *Frontiers in Pharmacology*, *11*, 581840.

Song, Z., Xu, Y., Bao, L., Zhang, L., Yu, P., Qu, Y., ... Qin, C. (2019). From SARS to MERS, thrusting coronaviruses into the spotlight. *Viruses*, *11*(1), 59.

Su, W.-W., Wang, Y., Li, P., Wu, H., Zeng, X., Shi, R., ... Su, W. (2020). The potential application of the traditional Chinese herb *Exocarpium Citri grandis* in the prevention and treatment of COVID-19. *Traditional Medicine Research*, *5*(3), 160–166.

Talwar, G., Pal, R., Singh, O., Garg, S., Taluja, V., Upadhyay, S., ... Sehgal, S. (1995). Safety of intrauterine administration of purified neem seed oil (Praneem Vilci) in women & effect of its co-administration with the heterospecies dimer birth control vaccine on antibody response to human chorionic gonadotropin. *The Indian Journal of Medical Research*, *102*, 66–70.

Torres, J., Maheswari, U., Parthasarathy, K., Ng, L., Liu, D. X., & Gong, X. (2007). Conductance and amantadine binding of a pore formed by a lysine-flanked transmembrane domain of SARS coronavirus envelope protein. *Protein Science, 16*(9), 2065–2071.

Tran, T. V. A., Malainer, C., Schwaiger, S., Atanasov, A. G., Heiss, E. H., Dirsch, V. M., & Stuppner, H. (2014). NF-κB inhibitors from *Eurycoma longifolia*. *Journal of Natural Products*, *77*(3), 483–488.

Ulasli, M., Gurses, S. A., Bayraktar, R., Yumrutas, O., Oztuzcu, S., Igci, M., ... Arslan, A. (2014). The effects of *Nigella sativa* (Ns), *Anthemis hyalina* (Ah) and *Citrus sinensis* (Cs) extracts on the replication of coronavirus and the expression of TRP genes family. *Molecular Biology Reports*, *41*, 1703–1711.

Utomo, R. Y., Ikawati, M., & Meiyanto, E. (2020). Revealing the potency of citrus and galangal constituents to halt SARS-CoV-2 infection.

Weber, N. D., Andersen, D. O., North, J. A., Murray, B. K., Lawson, L. D., & Hughes, B. G. (1992). In vitro virucidal effects of *Allium sativum* (garlic) extract and compounds. *Planta Medica*, *58*(05), 417–423.

World Health Organization. (2020). Clinical management of severe acute respiratory infection when novel coronavirus (2019-nCoV) infection is suspected: interim guidance, 28 January 2020: World Health Organization.

Zandi, K., Teoh, B.-T., Sam, S.-S., Wong, P.-F., Mustafa, M. R., & AbuBakar, S. (2011). Antiviral activity of four types of bioflavonoid against dengue virus type-2. *Virology Journal*, *8*(1), 1–11.

12 Herbal Medicines for COVID-19

The Impact of Flavonoids on Protein Targets

Panneerselvam Saamisha, Karuppannan Swathi,
Shanmugam Velayuthaprabhu, Kruthi Ashok Kumar,
Swetha M. Menon, Gunna Sureshbabu Suruthi,
Ramesh Nivedha, and Arumugam Vijaya Anand

12.1 INTRODUCTION

The global impacts of COVID-19 and the recent advent of highly virulent mutant strains of SARS-CoV-2 have prompted a greater investigation into therapeutic medicines (Bhowmik *et al.*, 2021; Abidharini *et al.*, 2023). Within the realm of various secondary metabolites in therapeutic research, flavonoids and their subclasses stand as a focal point that has proven antiviral properties. Flavonoids, a type of phenolic phytochemicals that have a diverse group of more than 8,000 physiologically active compounds, demonstrate other biological properties like antioxidant and antiphlogistic effects (Boro *et al.*, 2023; Mutha *et al.*, 2021; Pandey & Rizvi, 2009; Mehta *et al.*, 2020) and also play a significant part in culinary and therapeutic applications (Colunga Biancatelli *et al.*, 2020). As portrayed in Table 12.1 and Table 12.2, flavonoids are produced by various plants in reaction to numerous environmental stimuli and are documented in various forms, such as aglycones, glycosides, and methyl derivatives in different plant tissues (Zhu *et al.*, 2022). The classification of anthocyanins, flavones, flavonols, flavanones, flavans, isoflavones, biflavones, and other compounds is based on the level of pyrene epoxidation and the methods of substitution (Wang *et al.*, 2020). They offer potential as vitamins, medicines, and food additives due to their low toxicity and high safety profile (Kowalczyk *et al.*, 2021). Achieving optimal bioavailability and solubility of flavonoids is vital to fully exploit their therapeutic potential. Significantly, their remarkable antiviral effects compared to synthetic medicines assert the importance of these enrichments in medicinal utilization (Boro *et al.*, 2023; Latha *et al.*, 2024).

In the context of coronavirus, structural and sequence similarities between the glycoproteins S of SARS-CoV-2 and SARS-CoV (Krych & Gebicka, 2013) indicate the potential for cross-immunity. Within the intricate landscape of viral replication, this virus employs three crucial proteases such as 3CLpro, PLpro, and Mpro and each serves distinct functions. Specifically, 3CLpro is vital for SARS-CoV, PLpro facilitates SARS-CoV-2 replication, and Mpro recreates an essential role in maturing proteins in SARS-CoV-2 (Jain *et al.*, 2021; Jo *et al.*, 2020). Referring to this, flavonoids demonstrate their antiviral effects through manifold mechanisms, such as impeding viral neuraminidase, protease, and DNA/RNA polymerases while modifying various viral proteins (Boro *et al.*, 2023). These flavonoid actions strategically target membrane glycoprotein (M), spike protein (S1 and S2), envelope (E), nucleocapsid protein (N), and non-structural proteins (nsps) like nsp12 RdRp, nsp3 (PLpro) and nsp5 (Mpro), providing a nuanced approach within the intricate cellular milieu of SARS-CoV-2 (Agrawal *et al.*, 2021; Ahmed *et al.*, 2021; Meng *et al.*, 2023).

DOI: 10.1201/9781003452621-12

TABLE 12.1

Classes of Flavonoids and Their Major Sources

Flavonoid Subgroup	Members	Food Sources
Chalcones	Phloretin, xanthohumol, isoliquiritigenin, velutin	Strawberries, apples (Park *et al.*, 2016)
Flavones	Apigenin, luteolin, hispidulin, wogonin, oroxylin, scutellarin, rhamnocitrin, baicalein, chrysin, morusin, tangeretin, pectolinarigenin	Chamomile, mint, celery, parsley, tomatoes, fruit skin, red wine, medicinal plants (Wang *et al.*, 2020)
Flavonols	Kaempferol, quercetin, fisetin, myricetin, morin, rutin	Onion, kale, grapes, berries, lettuce, tea, red wine, olive oil (Zhu *et al.*, 2022)
Isoflavonoids	Genistein, glycitein, daidzein	Soya (Güler *et al.*, 2021)
Flavanols	Catechin, epicatechin, epigallocatechin-3-gallate	Green tea, cacao beans, peaches, bananas, blueberries, pears, medicinal plants (Mutha *et al.*, 2021)
Flavanones	Hesperidin, naringenin, naringin, taxifolin, eriodictyol	Oranges, oregano, grapes lemons (Mahmoud *et al.*, 2012)
Anthocyanidins	Cyanidin, delphinidin, apigenidin, malvidin	Black-/cran-/rasp-/straw-/blueberries, grapes, cherries, blackcurrants, nuts (Wang *et al.*, 2020)

TABLE 12.2

Important Flavonoids, Dietary Source, and their Biological Effects

Flavonoid	Food Sources	Biological Effects
Hesperetin	Lemon, peppermint, sweet oranges	Anti-inflammatory, antioxidant, anti-allergic, antiviral (Mahmoud *et al.*, 2012)
Rutin	Buckwheat, unpeeled apple, apricots, green tea	Antimicrobial, anti-inflammatory, antiarthritic, antiulcer (Rizzuti *et al.*, 2021)
Luteolin	Green pepper, chamomile tea	Anti-cancer, anti-allergic, anti-inflammatory, antiviral (Wang *et al.*, 2020)
Morin	Indian guava, almond, fig, white mulberry	Anti-angiogenic, antioxidant, neuroprotective (Pandey & Rizvi, 2009)
Naringin	Grapefruit, cherries, tomatoes	Anti-inflammatory, antioxidant, antiviral, antineoplastic (Mutha *et al.*, 2021)
Quercetin	Fennel, pepper, onions, berries	Antihypertensive, antiobesity, anti-inflammatory, antiviral (Gasmi *et al.*, 2022; Saakre *et al.*, 2021)
Apigenin	Chamomile tea, spinach, parsley	Antioxidant, anticancer, anti-inflammatory, antiviral (Chu *et al.*, 2000)
Galangin	Propolis, lesser galangal leaves	Antiviral, anti-inflammatory, antibacterial (Godinho *et al.*, 2021)
Chrysin	Honey, passion flowers, carrots	Anticancer, anti-inflammatory, neuroprotective (Kumar *et al.*, 2022)
Fisetin	Grapes, acacia leaves, strawberry fruits	Antiviral, immunosuppressive, anticancer, anti-inflammatory (Liu *et al.*, 2008)

12.2   FLAVONOIDS WITH ANTIVIRAL POTENTIAL AND MECHANISM

Each flavonoid, akin to a molecular sentinel, exhibits a unique strategy, influencing critical phases of the viral life cycle. The nuanced mechanisms through which flavonoids orchestrate their inhibitory effects on SARS-CoV-2 Mpro are multiple, suggesting specific flavonoids such as isorhamnetin, oroxylin A, and EGCG specifically target and prevent the entrance of SARS-CoV-2

(Bhowmik *et al.*, 2021). Naringenin, hesperidin, quercetin, cyanidin, kaempferol, catechin, luteolin-7-glucoside, genistein, apigenin-7-glucoside, EGC, and rutin specifically target Mpro, a crucial player in the replication of viruses (Camini *et al.*, 2017; Leal *et al.*, 2021). Chrysin shows a strong affinity with Mpro in MERS-CoV, SARS-CoV, and SARS-CoV-2 (Alzaabi *et al.*, 2021).

Quercetin, herbacetin, and isobavachalcone demonstrate the ability to inhibit crucial enzymes like PLpro and 3CLpro in SARS and MERS-CoV (Solnier & Fladerer, 2021). Compounds such as hesperetin and daidzein from *Isatis indigotica* emerge as promising contenders, showcasing their potential in inhibiting SARS-CoV 3CLpro (Lin *et al.*, 2005). Additionally, sciadopytiscin, bilobetin, amentoflavone, and ginkgetin from *Torreya nucifera* (Ryu *et al.*, 2010) exhibit the highest inhibitory activity against 3CLpro, employing a non-competitive mechanism. Nelfinavir and lopinavir emerge as potential treatment options, while curcumin, kaempferol, luteolin-7-glucoside, quercetin, apigenin-7-glucoside, naringenin, catechin, and ECG are identified as having the highest potential as COVID-19 Mpro inhibitors (Khaerunnisa *et al.*, 2020).

Further, *in vitro* efficacy against SARS-CoV proteases is found in alkylated chalcones from *Angelika keiskei*, with isobavachalcone (Park *et al.*, 2016) particularly inhibiting SARS-CoV PLpro via non-competitive pathways. Helichrysetin and herbacetin exhibit putative inhibitory actions against MERS-CoV 3CLpro, emphasizing the significance of specific hydroxyl groups in binding to viral proteases. Prenylated flavonol derivatives found in the fruits of *Paulownia tomentosa* (Cho *et al.*, 2013) and *Broussonetia papyrifera* (Park *et al.*, 2017) possess substantial resistance against SARS-CoV PLpro. Baicalin and baicalein, the main components of Shuanghuanglian, are recognized as the first non-covalent, non-peptidomimetic antagonists of SARS-CoV-2 3CLpro (Su *et al.*, 2020). Theaflavins (Chen *et al.*, 2005) and procyanidins found in black tea demonstrate inhibitory properties on SARS-CoV 3CLpro, suggesting their potential for decreasing or eliminating coronavirus infection.

The 3a protein of coronaviruses, including SARS-CoV-2, forms a cation-selective channel, potentially influencing ion transport in infected cells (Schwarz *et al.*, 2014). The flavonol kaempferol's glycoside derivate, juglanin, inhibits the SARS-CoV 3a-mediated current, suggesting potential antiviral effects on the cation-selective channel crucial for virus release (Chandra Manivannan *et al.*, 2022). Other glycosyl flavonoids, such as pectolinarin and baicalin, exhibit potent antagonistic effects on key targets governing viral replication, including S glycoprotein, SARS-CoV-2 PLpro, RdRp, and SARS-CoV-2 3CLpro (Walls *et al.*, 2020). Rutin, in its glycoside forms (Rizzuti *et al.*, 2021), preserves important pharmacological characteristics, while quercetin inhibits SARS-CoV-2 3CLpro. Flavonoids such as quercitrin, myricetin 3-O-rutinoside, and rutin, with their strong affinity and interaction suggestive of inhibitory potential, emerge as promising therapeutic options in contradiction of SARS-CoV-2 3CLpro (Khaerunnisa *et al.*, 2020). A docking analysis identifies scutellarin as a promising candidate to target 3CLpro. From *Siparuna cristata*, 3,7,30,40-Tetra-O-methyl-quercetin and flavonols 3,7-di-O-methyl-kaempferol efficiently impede the reproduction of SARS-CoV-2. EGCG's inhibition of HCoV-OC43 and HCoV-229E's 3CLpro impacts SARS-CoV-2.

In vitro tests on myricetin products, such as 7-O-methyl myricetin, reveal strong anti-SARS-CoV-2 3CLpro effects. Myricetin and its byproducts show higher inhibitory effects compared to derivatives of flavanonol ampelopsin (DHM). In the absence of B-ring 3′-OH, taxifolin is predicted to have a more powerful inhibitory effect than myricetin (Kaul *et al.*, 2021). Amentoflavone, isolated from *T. nucifera*, is documented to be a more powerful inhibitor of SARS-CoV 3CLpro than luteolin and apigenin, corresponding flavones, and biflavonoid derivatives with different methoxy groups (Zhu *et al.*, 2022). Additionally, endo-lysosomal TPCs, which are essential for coronavirus biology, may be inhibited by naringenin, which might impede SARS-CoV-2 replication. Combining TPC2 silencing with naringenin therapy substantially minimizes SARS-CoV-2 replication, indicating that TPC2 can be useful as an anti-coronavirus intervention via modulating TPC channels (Hoffmann *et al.*, 2020).

Other impeding mechanisms of flavonoids include efficacious inhibition of pseudoviruses that are carried out by the flavonoid compounds such as baicalin, herbacetin, pectolinin, rutin, EGCG,

GCG, luteolin, oroxylin A, scutellarein, kaempferol, quercetin, and myricetin. Inhibition of S glycoprotein by glycosyl flavonoids, quercetin, baicalin, naringin, pelargonidin, hesperidin, aglycone hesperetin, isoquercitrin (Godinho *et al.*, 2021). Inhibition of SARS-CoV-2 RdRp by baicalein and baicalin from *Scutellaria baicalensis* that exhibit strong anti-SARS-CoV-2 RdRp actions by inhibiting RdRp via binding to NiRAN and palm subdomains. The nsP15 of SARS-CoV-2 is inhibited by an extract having EGCG and ECG, whereas the green tea catechin GCG disrupts the SARS-CoV-2 nucleocapsid protein (Kaul *et al.*, 2021).

The multifaceted antiviral mechanisms of flavonoids extend to targeting key entry points and intracellular processes in SARS-CoV-2 infection. Compounds like hesperetin, myricetin, linebacker, caflanone, naringenin, baicalin, and chrysin exhibit a robust affinity for ACE2 and/or S protein, effectively preventing viral entry. Additionally, neohesperidin, myricitrin, quercitrin, naringin, icariin, and silybin demonstrate a high level of affinity to TMPRSS2, a crucial factor for viral entry (Colunga Biancatelli *et al.*, 2020). The inhibitory properties of quercetin on endocytosis, interference with viral genome transcription, and facilitation of eukaryotic translation initiation factor 4G cleavage (Chen *et al.*, 2005) contribute to its comprehensive antiviral role. Moreover, hesperetin and amentoflavone act as inhibitors for viral proteases and helicase/ATPase enzymes (Colunga Biancatelli *et al.*, 2020). The innovative use of nano drones coated with flavonoids further enhances the potential for targeted drug delivery to the lungs for treating COVID-19 (Jannat *et al.*, 2021).

12.3   SYNERGISTIC EFFECT OF FLAVONOIDS AS AN ANTI-COVID-19 AGENT

Herbal remedies with traditional medical roots offer alternative treatments for a range of illnesses including viral infections. The common plants with proven antiviral activity include *Avicennia marina*, *Andrographis paniculata*, *Azadirachta indica*, *Moringa oleifera*, and *Catharanthus roseus* (Liskova *et al.*, 2021; Meyyazhagan *et al.*, 2022) and mostly documented in south India (Bharathi *et al.*, 2023). Basic culinary ingredients that have antiviral properties, which are effective against SARS-CoV, include black cumin seeds, garlic, mushrooms, cinnamon, black pepper, drumsticks, and honey (Meyyazhagan *et al.*, 2022). The potential phytochemicals derived from cloves, including eugenol and casuarictin, act as Mpro antagonists against SARS-CoV-2 (Chandra Manivannan *et al.*, 2022). Flavanols, cyanidin-3-O-glucoside, and hypericin exhibit significant inhibition of PLpro deubiquitinase, offering a nuanced understanding of their antiviral mechanisms (Kaul *et al.*, 2021). Though medicinal plants possessing active flavonoids have been revealed to have antiviral action, the crucial factors for ensuring safety and effective therapy are the dosage and/or appropriate concentration.

The ongoing pursuit of effective treatments against COVID-19 involves investigating the antiviral properties of various compounds and exploring synergistic nutritional approaches. An intake of flavonoids at 250 to 500 mg/day is indicated for potential antiviral effects (Peluso & Palmery, 2015). Nutritional interventions, encompassing a balanced diet and essential vitamins (A, B6, B12, C, D, and E) (Alzaabi *et al.*, 2021; Gasmi *et al.*, 2022; Kumar *et al.*, 2022), along with trace elements (selenium, copper, magnesium), and omega-3 polyunsaturated fatty acids, show a vital role in bolstering immunity (Shanmugam *et al.*, 2020). Additionally, vitamin K emerges as a noteworthy nutraceutical, playing a part in the reduction of COVID-19 mortality rates (Ramya *et al.*, 2023). These findings underscore the importance of a balanced nutritional approach in fortifying the body's defense mechanisms against viral threats.

Moreover, hesperidin and naringin, with favorable properties like water solubility and resistance to crossing the blood-brain barrier, defy Lipinski's rule of five guidelines predicting oral drug bioavailability. These compounds, showing stability in molecular dynamics simulations, interact with SARS-CoV-2 target proteins. Notably, naringin exhibits the highest dynamic stability (Kumar *et al.*, 2022). Many flavonoid formulations show promising effects such as *Psoralea corylifolia* seeds, containing flavonoids with modest IC_{50} values against SARS-CoV PLpro, which present a potential avenue for targeted antiviral intervention (Chandra Manivannan *et al.*, 2022).

Flavonoids like baicalin, pectolinarin, quercetin, and isoquercitrin exhibit inhibitory influences on crucial viral proteases, implying their capacity to impede viral replication (Khaerunnisa *et al.,* 2020). Furthermore, the development of inhalable zein-chitosan nanoparticles incorporating curcumin and kaempferol introduces a novel therapeutic option, specifically designed to wield antiviral and anti-inflammatory effects on the respiratory system (Kuchi Bhotla *et al.,* 2021a). The suggested dosage concentrations for flavonoids in an antiviral medication provide a practical guideline, with a progressive rise from 0.5 g/day to a total of 3 g/day for a 70 kg individual (Ninfali *et al.,* 2020).

12.4 FLAVONOIDS IN MODULATING SARS-CoV-2 REPLICATION AND ACE2 INTERACTION

In laboratory studies using Vero cells, flavonoids found in *S. baicalensis*, specifically baicalein and baicalin, have shown significant effectiveness in inhibiting the activity of the RdRp of the SARS-CoV-2 virus, with IC_{50} values of 4.5 ± 0.2 µM and 9.0 ± 0.08 µM, respectively. Moreover, baicalein, one of these flavonoids, has demonstrated a broad-spectrum antiviral effect by inhibiting oxygen consumption in Vero E6 cells, displaying an IC_{50} of 10 µM (Huang *et al.,* 2020). Another flavonoid, GCG, derived from *S. baicalensis*, has exhibited disruption of the N protein of the virus (Kaul *et al.,* 2021). These findings highlight the diverse and potent antiviral properties of flavonoids present in *S. baicalensis*, suggesting their potential as candidate molecules in combating viral infections.

The relevance of these flavonoids to COVID-19 dosage lies in their specific features that contribute to their antioxidant effectiveness. The targeted interaction with ACE2 residues, such as Gln325/Glu329 and Asp38/Gln42, suggests a potential role in modulating the cellular environment associated with COVID-19 infection. The planarity, especially the arrangement of hydroxyl groups at critical positions, impacts electron shifting, influencing dissociation constants for efficient binding to target molecules. This intricate binding mechanism might be crucial in affecting processes related to viral entry or replication. The identification of potential inhibitory sites on the hACE2 receptor further highlights their possible relevance in the context of COVID-19, as targeting these sites could have implications for the virus-host interaction. Understanding these molecular interactions can contribute to determining appropriate dosage concentrations for potential therapeutic effects against COVID-19. For instance, hesperetin demonstrates inhibitory effects on the SARS-CoV-2 S1 S glycoprotein and ACE2 (Godinho *et al.,* 2021). Furthermore, quercetin exhibits notable antiviral potential compared to isoquercitrin, EGCG, and epicatechin, displaying efficacy in inhibiting HCoV-229E replication in Huh-7 cells (Mutha *et al.,* 2021).

12.5 FLAVONOIDS IN OTHER VIRAL INFECTIONS

Flavonoids have been demonstrated for their notable benefits as opposed to SARS-CoV-2 infection. However, with varying antiviral activity of flavonoids opposed to influenza, coronaviruses, HIV, and HSV include baicalin, luteolin, and kaempferol. These activities of polyphenolic secondary metabolites such as flavonoids further demonstrate their importance as antiviral agents other than SARS-CoV-2. More precisely, catechin and its derivatives (epicatechin, EGC, EGCG) show efficacy against influenza virus, HIV, herpesviruses, HBV and HCV, Zika virus, and CHIKV via mechanisms such as CD4 binding and inhibition of gene expression in viruses (Boro *et al.,* 2023; Chu *et al.,* 2000). Flavonoids such as apigenin, baicalin, quercetin, and rutin each have strong antiviral effects via distinct mechanisms. Apigenin inhibits the HCV via reducing mature miRNA122. Substances like demethylated gardenin A, robinetin, chrysin, acacetin, and apigenin demonstrate inhibitory actions against HIV-1 proteinase and unique transcription inhibition mechanisms that obstruct HIV-1 activation (Mehta *et al.,* 2020; Walls *et al.,* 2020).

Silymarin plays a part in developing antiviral drugs at various stages of infection through immune-mediated mechanisms and is in phase II studies for HCV (Lalani & Poh, 2020). Quercetin in synergy with naringenin and pinocembrin, inhibits the canine distemper virus, and a combination of quercetin and quercitrin demonstrates increased inhibition against Dengue virus (DENV-2) with lower cell cytotoxicity, indicating synergistic antiviral behavior (Gasmi *et al.*, 2022). Likewise, baicalin prevents DENV-2 from adhering to and fusing with HIV-1. By preventing glycoprotein-mediated cell-cell fusion, rutin suppresses HIV-1, whereas quercetin targets HCV and HSV through enzyme inactivation (Lalani & Poh, 2020).

Among the dominant antiviral phytochemicals, quercetin notably acts by binding to the hemagglutinin (HA2 subunit). This process prevents influenza virus strains, such as H1N1 and H3N2, from entering the body. Quercetin exhibits post-attachment antiviral action in a mouse model of rhinovirus-induced COPD, which goes beyond entrance inhibition reduces inflammation, and inhibits viral replication (Gasmi *et al.*, 2022).

12.6 MITIGATING ROLE OF FLAVONOIDS AGAINST COVID-19 PATHOGENESIS

Comorbidities associated with COVID-19, including diabetes due to altered glucose metabolism, ketoacidosis and immune disturbances (Arumugam *et al.*, 2020), obesity, renal diseases, dyslipidemia, cardiovascular disease, and respiratory disease, amplify susceptibility. Notably, unbalanced gut flora influences pulmonary functions, and reciprocally, pulmonary flora affects the gastrointestinal tract. The extensive use of antibiotics and corticosteroids in COVID-19 treatment leads to dysbiosis, opening the gate for cascading influences such as fungal infections, especially due to *Candida* and *Aspergillus* (Kuchi Bhotla *et al.*, 2021b), *Clostridioides difficile* infection, and inflammatory bowel disease (Fitero *et al.*, 2022).

Pregnant women with severe COVID-19, especially in the third trimester, who have an elevated body mass index, are at increased risk for perinatal complications, morbidity, and mortality with long-term effects on the newborn (Meng *et al.*, 2023). The relationship between COVID-19 pathogenesis and treatment complexities involves several interconnected factors. Mucormycosis, caused by Mucorales, becomes a concern in individuals with weakened immune systems, where corticosteroids, while potentially reducing mortality in severe COVID-19 cases, increase susceptibility to such fungal infections by affecting the host's iron levels (Pushparaj *et al.*, 2022).

Flavonoids could be employed to relieve or even eradicate the post-COVID-19 symptoms related to post-infection inflammation (Bardelčíková *et al.*, 2022; Table 12.3). Flavonoids have antioxidant properties that may help to regulate viral pathogenesis by reducing cytokine storms and oxidative stress. Platelets may enhance SARS-CoV-2 entrance, stimulate degranulation, produce inflammatory cytokines, and perhaps contribute to COVID-19 pathogenesis through surface receptors such as toll-like receptors (TLRs) and CD13 (Kuchi Bhotla *et al.*, 2020). Quercetin, the only clinically studied flavonoid in COVID-19 patients, offers a potential therapeutic avenue by influencing D-dimer levels and addressing thrombotic complications (Quintal Martínez & Segura Campos, 2023). Baicalin, by converting Ang II to Ang VII, upregulates the PI3K/AKT/eNOS pathway through ACE2 and Mas receptor modulation.

This pathway, activated by baicalin, reduces oxidative stress, protecting endothelial cells from Ang II-induced dysfunction and stress, especially in cardiovascular diseases such as COVID-19 co-morbid condition, it aids in maintaining a homeostasis condition (Arumugam *et al.*, 2020). This suggests a potential role in countering endothelial dysfunction, a crucial aspect of COVID-19 pathogenesis (Wei *et al.*, 2015). Various flavonoids like quercetin, chrysin, apigenin, baicalein, hesperidin, and naringin, along with luteolin, exhibit promising preventive antidiabetic effects (Meng *et al.*, 2023). Likewise, bayberry fruit extract with abundant flavonoids has demonstrated significant

TABLE 12.3

Effectiveness of Flavonoids on the Post COVID Consequences

Post COVID-19 Complications	Clinical Manifestation	Mechanism of Action of Flavonoids	Flavonoid
Pulmonary post-COVID-19 syndrome	Interstitial lung disease	Inhibition of TGF-β1 and MMP-9 expression	Apigenin, quercetin, naringenin, grape seed extract, epigallocatechin (Dejani *et al.*, 2021)
Cardiovascular post-COVID-19 syndrome	Hypertension	ACE inhibitors	Quercetin, procyanidin, elphinidin-3-oglucoside, luteolin, kaempferol, isosakuranetin, hesperidin, naringenin, epicatechin, EGCG, genistein, dihydrokaempferide, daidzein, betuletol (Arumugam *et al.*, 2020)
	Heart injury	Restrain the release of tissue-damaging proteases, reduction in the coronary perfusion pressure induced by angiotensin II, increase PGI2 release	Procyanidins (Chen *et al.*, 2005)
	Inflammatory cardiomyopathy	COX-2, antioxidant effect, inhibition of pro-inflammatory cytokine and inflammatory enzymes PLA2	Kaempferol, myricetin, quercetin, catechin, hesperidin, naringenin (Yamaguchi *et al.*, 2021)
	Blood clots	Inhibition of COX1, decrease of TXA2, interaction with GPIIb/IIIa receptors	Epicatechin, rutin, genistein, fisetin, quercetin, equol, kaempferol, morin, daidzein tangeretin, naringin, naringenin, myricetin (Quintal Martínez & Segura Campo, 2023)
Neurological post-COVID-19 syndrome	Cognitive dysfunction-memory disorders, chronic fatigue syndrome, sleep disorders	Attenuation of oxidative stress, decrease of TNFα levels	EGC, curcumin, naringin, thymoquinone (Mendonca & Soliman, 2020)
	Anxiety depression	Modulators of ion channels GIRK and Herg, positive modulators of GABAA	Hesperidin, diosmin, neohesperidin, naringin, baicalein, chrysin, astilbin, gossypin, luteolin, linarin, kaempferitrin, vitexin, nobiletin (Bardelčíková *et al.*, 2022)
Nephrological post-COVID-19 syndrome/kidney failure	Renal fibrosis, kidney tubular injury, hematuria, proteinuria	Inhibition of TGFβ/SMAD3 and JAK2/STAT3	Pectolinarigenin, rutin, quercetin (Bardelčíková *et al.*, 2022)
Gastrointestinal post-COVID-19 syndrome	Liver injury elevated AST, ALT	Decrease of ALT, AST, ALP, CAT, GPx, GSH Reduction of NF-κB and TNF-α, MDA Increase of SOD	Morin, silymarin, quercetin, hydroxyethylrutoside, ternatin, glycone (rutin), gossypin (Fitero *et al.*, 2022)
	Diabetes mellitus type I, acute pancreatitis, pancreas injury	Increase in SOD and CAT activity and GSH content, hepatic, AST, LDH, and CK-MB, serum insulin level, a decrease in glucose level, and muscle glycogen content	Hesperidin, naringin, chrysin apigenin, quercetin, baicalein (Ahmed *et al.*, 2021)

antidiabetic effects (Zhang *et al.*, 2015). These flavonoids may mitigate COVID-19 pathogenesis by addressing diabetes-related vulnerabilities.

12.7 FLAVONOIDS AS IMMUNOMODULATORS DURING SARS-CoV-2 INFECTION

Flavonoids exhibit notable immunomodulatory effects, recreating a vital role in controlling the immune system. These phytochemicals showcase a diverse range of activities that contribute to immune regulation, influencing various immune cells and their related pathways. The pathophysiological cascade of SARS-CoV-2 infection encompasses viral ingress into host cells via the ACE2 receptor, predominantly mediated by TLR-4. Within the pulmonary cellular milieu, this activates an immune response illustrated by the activation of macrophages, monocytes, cytokines, and adaptive B and T cell pathways. Viral-induced pyroptosis ensues, leading to viral egress and inflammatory manifestations, including the release of IL, IL-1β, a pivotal inflammatory cytokine (Kowalczyk *et al.*, 2021). Compounds like liquiritigenin, liquiritin, and liquiritin apioside from *Glycyrrhiza uralensis* exhibit notable effectiveness in easing pulmonary inflammation. This is accomplished by diminishing the infiltration of inflammatory cells and particularly by curbing the release of pivotal inflammatory mediators, TNF-α and IL-1β. These findings underscore the promising potential of these licorice flavonoids in alleviating lung inflammation associated with COVID-19 (Godinho *et al.*, 2021).

Flavonoids exhibit diverse anti-inflammatory and immunomodulatory effects with potential relevance to COVID-19. Naringenin and hesperetin both inhibit the ERK and NF-κB pathways related to immune response, while naringenin additionally reduces TNF-α, iNOS, and COX-2 (inflammatory molecules). Quercetin influences aryl hydrocarbon receptor binding, which inhibits EGFR and NF-κB pathways, regulates Th1/Th2 balance, and impairs T-cell activation, regulating immune responses and maintaining immune homeostasis. Flavonoids like fisetin, luteolin, chyrosin, apigenin, and caflanone exert intricate immunomodulatory effects by selectively targeting various immune molecules and pathways. Fisetin demonstrates a broad spectrum of actions, including the suppression of MPC1 levels that impact cellular metabolism and mitochondrial function,

TNF-α leads to decreased inflammation and attenuation of the immune response, CCL5 regulates the immune cells contributing to inflammation, ERK1/2 pathways modulate the immune cell proliferation, survival, and differentiation, PKC-δ recreates a vital role in immune regulation by modulating signaling pathways that influence cell activation, proliferation, and inflammatory responses, and COX-2 mediates anti-inflammatory effects. Chyrosin acts as a PPAR-γ agonist, inhibiting NF-κB, COX-2, and MPO, contributing to the host defense against pathogens through oxidative, TNF-α, and iNOS (nitrosative responses) while activating macrophages and inhibiting nitric oxide. Apigenin selectively inhibits ICAM1, CCL5, and IL-6, contributing to reduced inflammation. Luteolin enhances CD4+, CD25+ Tregs, pivotal in immune tolerance, while suppressing other immune cells. Caflanone inhibits both 5-lipoxygenase and microsomal prostaglandin E synthase 1, modulating leukotriene and prostaglandin synthesis associated with inflammatory responses (Alzaabi *et al.*, 2021).

In individuals afflicted with severe viral infections, the BALF reveals the presence of FCN1+ macrophages originating from inflammatory monocytes. This particular condition coincides with an elevated count of peripheral blood CD14+ and CD16+ inflammatory monocytes, contributing to the onset of a cytokine storm (Godinho *et al.*, 2021). The activation of antiviral mechanisms crucially relies on type I IFN. However, the SARS-CoV-2 ORF6 protein induces immunological dysregulation by impeding type I IFN production and signaling. This disruption contributes to hyperinflammation, pulmonary damage, coagulopathy, and multiorgan failure (Dejani *et al.*, 2021). Glycosyl flavonoids, including scutellarin and baicalin, exert an influence on signaling pathways and inflammatory mediators, specifically TLRs and NLRP3 inflammasomes. It has been proven that the antiviral and immunomodulatory effects of these flavonoids opposed to COVID-19 infection, operating through interactions with TLRs or NLRP3 inflammasomes (Godinho *et al.*, 2021).

The inhibitory consequences of hesperetin on the TLR4 and NF-κB signaling pathway reduce neuroinflammation and neuronal death. Resveratrol exhibits anti-inflammatory properties in individuals with KD by reducing the synthesis of proinflammatory cytokines and adhesion molecules in the endothelial cells of the coronary artery (Huang *et al.*, 2017). Luteolin reduces inflammation in the heart and protects cardiomyocytes from oxidative stress and inflammation caused by high glucose levels in diabetic rats (Dejani *et al.*, 2021). Flavonoids, specifically wogonin, pinostrobin chalcone, cardamomin, chrysin, and pinocembrin, have been demonstrated to prevent angiogenesis caused by VEGF (Tian *et al.*, 2014). Further, green tea extract inhibits Ang II-induced cardiomyocyte hypertrophy (Zakaryan *et al.*, 2017).

Flavonoids have the ability to induce anti-inflammatory responses, activate natural killer cells, and affect macrophage profiles. Flavonoids also have the ability to activate natural killer cells. The synthetic flavonoid LFG-500 suppresses the activation of NF-κB, decreases cytokines in lung tissues *in vitro* and interacts with the JNK, MAPK and p38 pathways. Luteolin, pinocembrin, and oroxylin-A modify pro-inflammatory cytokines and signaling pathways to facilitate lipopolysaccharide-induced acute respiratory distress syndrome both *in vivo* and *in vitro*. By influencing pro-inflammatory genes, flavonoids suppress the activity of enzymes including iNOS, COX, and lipoxygenase, control the formation of cytokines like IL-4 and IL-13, and reduce inflammation (Santana *et al.*, 2021).

Naringenin is known to have positive effects on COVID-19 treatment when combined with other compounds such as wogonin, kaempferol, luteolin, isorhamnetin, and quercetin. These compounds function by limiting cytokine storm (Mehta *et al.*, 2020) and protecting target organs through the inhibition of viral adsorption and replication, regulation of inflammatory mediators, and exhibit anti-inflammatory and immune-regulatory characteristics. There are reports of antiviral effects that continue to exist even after the consumption of medication has been discontinued. The quercetin protein is essential in facilitating the cleavage of eukaryotic translation initiation factor 4G, resulting in a reduction in the viral protein translation. Additionally, it enhances the host's immune reaction to viruses present in the mitochondria (Alzaabi *et al.*, 2021).

Quercetin also strengthens innate immune pathways by promoting the synthesis of 2′,5′-oligoadenylate synthetase, which amplifies TNF's suppressive impact against viruses (Khazeei Tabari *et al.*, 2021). Flavonoids protect host cells from viral infection-induced damage by lowering complement activation, decreasing inflammatory cell adherence to the endothelium, and resulting in a decreased inflammatory response (Wang *et al.*, 2020). Rutin, administered at 80 mg/kg, significantly reduced paw edema in both acute and chronic phases, completely suppressing it on days 21 and 30 post-inflammation. This implies a possible role in managing the inflammatory aspects of COVID-19. Comparatively, related flavonoids quercetin and hesperidin also demonstrate significant albeit smaller anti-inflammatory effects (Wei *et al.*, 2015).

Pharmacologically activating Nrf2 has shown promise in regulating genes related to respiratory virus infectivity and immune response, providing antiviral benefits. Activating Nrf2 has the possibility to inhibit viral replication, alleviate symptoms, and enhance the capacity of the immune system against cytokine storm in COVID-19. Heme oxygenase 1 is a phase II antioxidant that aids in preventing or delaying oxidative cell damage and is one of the enzymes that Nrf2 regulates. Innate defense against viral infections has been proven to benefit from Nrf2 activation, which is also known to influence natural killer cell priming.

Two particular flavonoids stand out for their strong antiviral action in addition to their function as Nrf2 inducers: EGCG and thymoquinone. Numerous studies have been performed on EGCG, which is present in green tea. Thymoquinone, a derivative of black cumin (*Nigella sativa*), shows antiviral effectiveness against different kinds of viruses. Thymoquinone also suppresses neuro-inflammation and the formation of pro-inflammatory cytokines by activating Nrf2 and blocking NF-κB-dependent pathways, according to studies. Thymoquinone treatment may have the possibility to prevent the COVID-19 condition or mitigate the rigor of the condition. This is because corona-viruses can invade the nervous system and cause oxidative stress, and thymoquinone can counteract these effects (Mendonca & Soliman, 2020). The majority of phytochemicals, including flavonoids, can act as reverse transcriptase inhibitors (Anand *et al.*, 2021).

12.8 FLAVONOIDS IN ACE RECEPTOR

The access of SARS-CoV-2 into host cells is reduced by the interaction between its S1 and S2 and the ACE2 receptor. The RBD of the S1 subunit has a high association with the ACE2 receptor, which contains a zinc atom and acts as a dipeptidyl carboxypeptidase (Balasuriya & Rupasinghe, 2011). Notably, this affinity is 10–20 times more than that observed in the case of SARS-CoV (Alzaabi *et al.*, 2021).

Following the engagement of ACE2, the S2 subunit undergoes conformational changes, enabling fusion with the cell membrane of the host. In type II pneumocytes, the cellular protease TMPRSS2 recreates a critical part in priming the virus for membrane fusion, a pivotal step that facilitates the discharge of genetic material and initiates the viral replication process. Following the engagement of ACE2, the S2 subunit undergoes conformational changes, enabling fusion with the host cell membrane. Quercetin in synergy with naringenin and pinocembrin, inhibits the canine distemper virus, and a combination of QR and quercitrin demonstrates increased inhibition against DENV-2 with lower cell cytotoxicity, indicating synergistic antiviral behavior (Gasmi *et al.*, 2022). Baicalin prevents DENV-2 from adhering to and fusing with HIV-1. By preventing glycoprotein-mediated cell–cell fusion, rutin suppresses HIV-1, whereas quercetin targets HCV and HSV through enzyme inactivation. Catechin and its derivatives (epicatechin, ECG, EGC, and EGCG) show efficacy against influenza virus, HIV, herpesviruses, HBV, HCV, Zika virus, and CHIKV via mechanisms such as CD4 binding and viral gene expression inhibition. Varying antiviral activity against influenza, coronaviruses, HIV, and HSV include baicalin, luteolin, and kaempferol.

Type II pneumocytes in the lungs play a critical role in COVID-19 as they express TMPRSS2, a protease facilitating viral entry, and ACE2, a receptor for SARS-CoV-2, contributing to the virus's cellular infection and pathogenesis. This priming involves TMPRSS2 cleaving the S protein, essential for effective viral entry (Russo *et al.*, 2020). The infection of these cells can lead to respiratory difficulties, highlighting their significance in the context of COVID-19 (Alzaabi *et al.*, 2021).

Simultaneously, ACE2 serves as a key player in counterbalancing the effects of ACE and Ang II. By binding with SARS-CoV-2, ACE2 converts Ang II to Ang (1–7) (South *et al.*, 2020). Superior binding capabilities against ACE2 receptors have been determined using six RBD amino acids, which is important for determining the host tropism of viruses similar to SARS-CoV. Proteases found in humans notably do not cleave the functional group by hybrid analogs, a finding that needs more investigation (Poochi *et al.*, 2020).

However, the stimulation of the RAAS at this juncture involves oxidative stress, triggering a cytokine storm, escalating viral load, and leading to acute hypoxemia. Furthermore, the viral RdRp, crucial for viral replication, exacerbates tissue hypoxia. This intricate interplay between viral entry, ACE2 function, RAAS activation, and RdRp activity contributes to the pathogenesis of severe COVID-19, creating a cascade of events that ultimately leads to respiratory distress and potentially severe outcomes to tissue hypoxia (Oostra *et al.*, 2007). Viral-ACE2 binding causes signal reorganization that affects biological functions and enhances cell cycle arrest (Hoffmann *et al.*, 2020).

Flavonoids, including hesperetin, myricetin, linebacker, and caflanone, demonstrate a robust ability to impede viral entry by strongly binding to crucial elements of the SARS-CoV-2 virus, such as the S protein, helicase, and ACE2 receptor (Ngwa *et al.*, 2020). Certain flavonoids, including kaempferol, leutolin, quercetin, herbacetin, and isobavachalcone, show promise in inhibiting various coronavirus targets, despite challenges such as limited bioavailability and stability. These flavonoids can interfere with crucial components like proteases, helicases, and S proteins that bind to the ACE2 receptor on the host cell membranes. This interaction is vital for the virus to enter target cells (Solnier & Fladerer, 2021).

Notably, naringenin exhibits remarkably lower binding energy to ACE2, exhibiting a particularly strong affinity. In comparative analyses, baicalin surpasses abacavir and hydroxychloroquine in their affinity for the S protein, suggesting potential antiviral efficacy against other illnesses (Pandey & Rizvi, 2009). Chrysin, another flavonoid, disrupts the interface between ACE2 and the S protein. Furthermore, flavonoids like icariin, myricitrin, quercitrin, naringin, neohesperidin, and silybin (Alzaabi *et al.*, 2021) display strong binding affinity to TMPRSS2, a key factor for viral entry (Chikhale *et al.*, 2021). Furthermore, flavonoids, particularly flavan-3-ols and anthocyanins, showcase effective ACE inhibitory properties *in vitro*. Catechins and their polymers emerge as the most potent ACE inhibitors, while isoflavones exhibit intermediate inhibition. The mechanism of action of flavonoids is demonstrated by their typical competitive inhibition of ACE (Güler *et al.*, 2021). Flavonoids found in citrus fruits can modulate immunity by targeting ACE2 (Cheng *et al.*, 2022). In conclusion, these outcomes collectively underscore the multifaceted antiviral possibility of flavonoids against SARS-CoV-2, offering a promising avenue for therapeutic development.

12.9 FLAVONOIDS AS A MODULATOR OF CYTOKINES

Recent evidence documented the impact of flavonoids on cytokine modulation and their importance in the control of viral infection, particularly in COVID-19. The *in vitro* study stated that cytokine production was inhibited by flavonoids like luteolin, apigenin, fisetin, scutellarein, and fustin (Nguyen *et al.*, 2012). Upon detection of SARS-CoV-2, certain individuals undergo a pronounced immunological reaction known as a cytokine storm, which lasts for a duration of 5 to 7 days following infection. Critically ill patients show significant increases in G-CSF, cytokines, and other biomarkers (D-dimer, ferritin and C-reactive protein) (Agrawal *et al.*, 2021; Zhang *et al.*, 2015).

These cytokines attract immune cells to the site of infection, resulting in lymphopenia and an increase in the neutrophil-to-lymphocyte ratio. Further, flavonoids regulate the cytokine-mediated formation of cytotoxic CD8+ T cells and antigen-specific B cells by helper CD4+ T cells. Patients with COVID-19 show variable lymphopenia, increased white blood cell counts, and increased neutrophils. IL-6 levels correlate with disease severity, with higher levels in severely ill patients. Flavonoids have shown several anti-inflammatory effects directed at specific cytokines related to the development of COVID-19. The flavonoids such as naringenin reduce IL-6, hesperetin inhibits IL-1β and IL-6, fisetin inhibits IL-6 and IL-8, chrysin inhibits IL-8 and IL-1β, and apixenin inhibits IL-6 (Alzaabi *et al.*, 2021). Numerous flavonoids such as baicalein, diosmin, genistein, biocanin A, silymarin, catechin, theaflavin, liquiritigenin, eriodictyol, taxifolin, pinocembrin, myricetin, casticin, galangin, kaempferol, fisetin, apigenin, chrysin, wogonin, velutin, and formononetin also have similar inhibitory effects on cytokine substrates (Godinho *et al.*, 2021). The synthetic flavonoid LFG-500 exerts a protective effect *in vivo* by inhibiting the cytokines in lung tissue (Dejani *et al.*, 2021, Godinho *et al.*, 2021). *Nerium oleander* chlorogenic acids and kaempferol showed positive binding abilities for IL-1β and IL-6. Kaempferol, in particular, has emerged as a drug candidate to combat inflammation and pain due to its strong IL-6 binding ability (Karmakar *et al.*, 2019).

Quercetin, which is understood to be a powerful inhibitor of NLRP3 inflammasome-mediated IL-1β production, acts at multiple sites in a relevant pathway (Anand *et al.*, 2021). Non-survivors had increased IL-6 compared to survivors during the COVID-19 infection, which shows their role in the critical conditions. The elevated concentrations of cytokines are observed in individuals with COVID-19. Patients with severe disease have higher expression of inflammatory cytokines, which may be biomarkers of disease progression. Specific flavonoid interventions show promising results in modulating cytokine levels. Epicatechin, found in tea, represses the secretion of IL-6, IL-8, and TNF-α. Quercetin, rutin, and the glycoside naringenin inhibit IL-6, IL-1β, and TNF-α in cell lines and animal models. Luteolin reduces IL-6 and TNF-α in a mouse model of lung injury. Microcospaniculate and apigenin C-glycoside (ACG) reduce pulmonary edema and microvascular permeability by reducing LPS-induced TNF-α, IL-1β, and IL-6 expression. The resistant system-mediated backhanded influence of SARS-CoV-2 on the cardiovascular framework gives rise to an arrangement of results, counting a cytokine storm, insecurity of coronary plaques, and extreme myocardial harm characterized by increased levels of IL-1, interferon-gamma, and different cytokines. In occurrences of hypertension-associated COVID-19, the dysregulation of CD4 and CD8 cells leads to a safe lopsidedness, activating an intemperate discharge of various pro-inflammatory cytokines. Flavonoids, particularly kaempferol, luteolin, and quercetin, demonstrated significant inhibitory effects on cytokine expression. Flavonoids demonstrate favorable multitarget effects as opposed to SARS-CoV-2, making them appropriate for usage in both present and anticipated forthcoming epidemics (Agrawal *et al.*, 2020; Gour *et al.*, 2021).

12.10   CONCLUSION

In conclusion, flavonoids have beneficial effects based on protein targets associated with anti-inflammatory, immunomodulatory, cytokine regulation, and antiviral activities. This suggests that flavonoids can be a promising treatment strategy as an alternative to conventional medicines for COVID-19 and other deadly viral contagions.

ACKNOWLEDGMENT

This work was supported by MHRD-Rashtriya Uchchatar Shiksha Abhiyan (RUSA) under the BEICH 2.0 (BU/RUSA/BEICH/2019/299-39) scheme and Bhrathiar University, Coimbatore, India.

DATA AVAILABILITY

The authors confirm that the data supporting the findings of this study are available within the book/chapter.

LIST OF ABBREVIATIONS

3CLpro	3C-Like Protease
ACE2	Angiotensin-Converting Enzyme 2
ACG	Apigenin C-Glycoside
AKT	Protein Kinase B
ALP	Alkaline phosphatase
ALT	Alanine aminotransferase
Ang II	Angiotensin II
Ang VII	Angiotensin VII
ARDS	Acute Respiratory Distress Syndrome
AST	Aspartate aminotransferase
BALF	Bronchoalveolar Lavage Fluid
CAT	Catalase
CCL5	C-C Motif Chemokine Ligand 5

CD14+ and CD16+	Cluster of Differentiation 14 and 16
CHIKV	Chikungunya Virus
COPD	Chronic Obstructive Pulmonary Disease
COVID-19	Coronavirus Disease 2019
COX-2	Cyclooxygenase-2
DENV-2	Dengue Virus-2
DHM	Dihydromyricetin
DNA/RNA	Deoxyribonucleic Acid/Ribonucleic Acid
E	Envelope protein
EC50	Half-Maximal Effective Concentration
ECG	Epicatechin Gallate
EGC	Epigallocatechin
EGCG	Epigallocatechin Gallate
EGFR	Epidermal Growth Factor Receptor
eNOS	Endothelial Nitric Oxide Synthase
ERK	Extracellular Signal-Regulated Kinase
ERK1/2	Extracellular Signal-Regulated Kinases 1 and 2
FCN1+	Ficolin-1 Positive
GCG	Gallocatechin Gallate
G-CSF	Granulocyte Colony-Stimulating Factor
GPx	Glutathione peroxidase
GSH	Glutathione
H1N1	Influenza A Virus Subtype H1N1
H3N2	Influenza A Virus Subtype H3N2
hACE2	Human Angiotensin-Converting Enzyme 2
HBV	Hepatitis B Virus
HCoV-229E	Human Coronavirus 229E
HCoV-OC43	Human Coronavirus OC43
HCV	Hepatitis C Virus
HIV	Human Immunodeficiency Virus
HSV	Herpes Simplex Virus
IC50	Half-maximal inhibitory concentration
ICAM1	Intercellular Adhesion Molecule 1
IFN	Interferon
IL	Interleukin
iNOS	Inducible Nitric Oxide Synthase
JNK	c-Jun N-terminal Kinase
KD	Kawasaki-like Disease
LFG-500	Synthesized flavonoid, substituted with a piperazine and a benzyl group
LPS	Lipopolysaccharide
M	Membrane protein
MAPK	Mitogen-Activated Protein Kinase
Mas receptor	Mas Proto-Oncogene Receptor Tyrosine Kinase
MDA	malondialdehyde
MERS-CoV	Middle East Respiratory Syndrome Coronavirus
MPC1	Mitochondrial Pyruvate Carrier 1
MPO	Myeloperoxidase
Mpro	Main Protease
N	Nucleocapsid Protein
NF-κB	Nuclear Factor-кappa B
NiRAN	Nidovirus RdRp-Associated Nucleotidyltransferase

NLRP3	NLR Family Pyrin Domain-Containing 3
Nrf 2	Nuclear Factor Erythroid 2-Related Factor 2
Nsps	Nonstructural proteins
nsP15	Nonstructural Protein 15
ORF6	Open Reading Frame 6
PI3K	Phosphoinositide 3-kinase
PKC-δ	Protein Kinase C-delta
PLpro	Papain-Like Protease
PPAR-γ	Peroxisome Proliferator-Activated Receptor Gamma
QR	Quercetin
RAAS	Renin-Angiotensin-Aldosterone System
RBD	Receptor-Binding Domain
RdRp	RNA-dependent RNA polymerase
S	Spike protein
S1 and S2	Subunits 1 and 2 of the Spike protein
SARS	Severe Acute Respiratory Syndrome
SARS-CoV	Severe Acute Respiratory Syndrome Coronavirus
SARS-CoV-2	Severe Acute Respiratory Syndrome Coronavirus 2
SOD	Superoxide dismutase
Th1/Th2	T-helper 1/T-helper 2
TLRs	Toll-like receptors
TLR-4	Toll-Like Receptor 4
TMPRSS2	Transmembrane Protease Serine 2
TNF-α	Tumor Necrosis Factor-α
TPCs	Two-Pore Channels
Tregs	Regulatory T cells
VEGF	Vascular Endothelial Growth Factor

REFERENCES

Abidharini, J. D., Souparnika, B. R., Elizabeth, J., Vishalini, G., Nihala, S., Shanmugam, V., & Anand, A. V. (2023). Herbal Formulations in Fighting Against the SARS-CoV-2 Infection. In Jen-Tsung Chen (Ed.), *Ethnopharmacology and Drug Discovery for COVID-19: Anti-SARS-CoV-2 Agents from Herbal Medicines and Natural Products* (pp. 85–113). Springer. https://doi.org/10.1007/978-981-99-3664-9_4.

Agrawal, P. K., Agrawal, C., & Blunden, G. (2020). Quercetin: Antiviral significance and possible COVID-19 integrative considerations. *Natural Product Communications*, *15*(12). https://doi.org/10.1177/1934578X20976293.

Agrawal, P. K., Agrawal, C., & Blunden, G. (2021). Naringenin as a possible candidate against SARS-CoV-2 infection and in the pathogenesis of COVID-19. *Natural Product Communications*, *16*(12). https://doi.org/10.1177/1934578X211066723.

Ahmed, M. H., Hassan, A., & Molnár, J. (2021). The role of micronutrients to support immunity for COVID-19 prevention. *Revista Brasileira de Farmacognosia*, *31*(4):361–374.

Alzaabi, M. M., Hamdy, R., Ashmawy, N. S., Hamoda, A. M., Alkhayat, F., Khademi, N. N., …, & Soliman, S. S. (2021). Flavonoids are promising safe therapy against COVID-19. *Phytochemistry Reviews,* 21(1):291–312.

Anand, A. V., Balamuralikrishnan, B., Kaviya, M., Bharathi, K., Parithathvi, A., Arun, M., … & Dhama, K. (2021). Medicinal plants, phytochemicals, and herbs to combat viral pathogens including SARS-CoV-2. *Molecules*, 26(6):1775.

Arumugam, V. A., Thangavelu, S., Fathah, Z., Ravindran, P., Sanjeev, A. M. A., Babu, S., …, & Tiwari, R. (2020). COVID-19 and the world with co-morbidities of heart disease, hypertension and diabetes. *Journal of Pure and Applied Microbiology*, *14*(3):1623–1638.

Balasuriya, B. W. N., & Rupasinghe, H. P. V. (2011). Plant flavonoids as angiotensin converting enzyme inhibitors in regulation of hypertension. *Functional Foods in Health and Disease*, *1*(5):172–188.

Bardelčíková, A., Miroššay, A., Šoltýs, J., & Mojžiš, J. (2022). Therapeutic and prophylactic effect of flavonoids in post-COVID-19 therapy. *Phytotherapy Research*, *36*(5):2042–2060.

Bharathi, K., Sivasangar Latha, A., Jananisri, A., Bavyataa, V., Rajan, B., Balamuralikrishnan, B., …, & Anand, A. V. (2023). Antiviral Properties of South Indian Plants against SARS-CoV-2. In Jen-Tsung Chen (Ed.) *Ethnopharmacology and Drug Discovery for COVID-19: Anti-SARS-CoV-2 Agents from Herbal Medicines and Natural Products* pp. 447–478). Springer. DOI:10.1007/978-981-99-3664-9_17

Bhowmik, D., Nandi, R., Prakash, A., & Kumar, D. (2021). Evaluation of flavonoids as 2019-nCoV cell entry inhibitor through molecular docking and pharmacological analysis. *Heliyon*, *7*(3):e06515.

Boro, A., Reji Souparnika, B., Atchaya, A., Ramya, S., Senthilkumar, N., Velayuthaprabhu, S., Rengarajan, R. L., & Vijaya Anand, A. (2023). Antiviral Activities of the Flavonoids Compound against COVID-19 and Other Viruses Causing Ailments. In: Chen, JT. (eds). *Bioactive Compounds against SARS-CoV-2*. CRC Press.

Camini, F. C., da Silva Caetano, C. C., Almeida, L. T., & de Brito Magalhães, C. L. (2017). Implications of oxidative stress on viral pathogenesis. *Archives of Virology*, *162*:907–917.

Chandra Manivannan, A., Malaisamy, A., Eswaran, M., Meyyazhagan, A., Arumugam, V. A., Rengasamy, K. R. R., …, & Liu, W.-C. (2022). Evaluation of clove phytochemicals as potential antiviral drug candidates targeting SARS-CoV-2 main protease: Computational docking, molecular dynamics simulation, and pharmacokinetic profiling. *Frontiers in Molecular Biosciences*, *9*:918101.

Chen, C.-N., Lin, C. P. C., Huang, K.-K., Chen, W.-C., Hsieh, H.-P., Liang, P.-H., & Hsu, J. T.-A. (2005). Inhibition of SARS-CoV 3C-like protease activity by theaflavin-3, 3'-digallate (TF3). *Evidence-Based Complementary and Alternative Medicine*, *2*:209–215.

Cheng, L., Zheng, W., Li, M., Huang, J., Bao, S., Xu, Q., & Ma, Z. (2022). Citrus fruits are rich in flavonoids for immunoregulation and potential targeting ACE2. *Natural Products and Bioprospecting*, *12*(1):4.

Chikhale, R. V., Gupta, V. K., & Eldesoky, G. E. et al. (2021) Identification of potential anti-TMPRSS2 natural products through homology modelling, virtual screening and molecular dynamics simulation studies. *Journal of Biomolecular Structure & Dynamics*. *39*(17), 6660–6675.

Cho, J. K., Curtis-Long, M. J., Lee, K. H., Kim, D. W., Ryu, H. W., Yuk, H. J., & Park, K. H. (2013). Geranylated flavonoids displaying SARS-CoV papain-like protease inhibition from the fruits of *Paulownia tomentosa*. *Bioorganic & Medicinal Chemistry*, *21*(11):3051–3057.

Chu, Y., Chang, C., & Hsu, H. (2000). Flavonoid content of several vegetables and their antioxidant activity. *Journal of the Science of Food and Agriculture*, *80*(5):561–566.

Colunga Biancatelli, R. M. L., Berrill, M., Catravas, J. D., & Marik, P. E. (2020). Quercetin and vitamin C: An experimental, synergistic therapy for the prevention and treatment of SARS-CoV-2 related disease (COVID-19). *Frontiers in Immunology*, *11*:1451.

Dejani, N. N., Elshabrawy, H. A., Bezerra Filho, C. D S., & de Sousa, D. P. (2021). Anticoronavirus and immunomodulatory phenolic compounds: Opportunities and pharmacotherapeutic perspectives. *Biomolecules*, *11*(8):1254.

Fitero, A., Bungau, S. G., Tit, D. M., Endres, L., Khan, S. A., Bungau, A. F., …, & Tarce, A. G. (2022). Comorbidities, associated diseases, and risk assessment in COVID-19—A systematic review. *International Journal of Clinical Practice*, *2022*:1571826.

Gasmi, A., Mujawdiya, P. K., Lysiuk, R., Shanaida, M., Peana, M., Gasmi Benahmed, A., …, & Bjørklund, G. (2022). Quercetin in the prevention and treatment of coronavirus infections: A focus on SARS-CoV-2. *Pharmaceuticals*, *15*(9):1049.

Godinho, P. I. C., Soengas, R. G., & Silva, V. L. M. (2021). Therapeutic potential of glycosyl flavonoids as anti-coronaviral agents. *Pharmaceuticals*, *14*(6):546.

Gour, A., Manhas, D., Bag, S., Gorain, B., & Nandi, U. (2021). Flavonoids as potential phytotherapeutics to combat cytokine storm in SARS-CoV-2. *Phytotherapy Research*, *35*(8):4258–4283.

Güler, H. I., Şal, F. A., Can, Z., Kara, Y., Yildiz, O., Beldüz, A. O., …, & Kolayli, S. (2021). Targeting CoV-2 spike RBD and ACE2 interaction with flavonoids of *Anatolian propolis* by *in silico* and *in vitro* studies in terms of possible COVID-19 therapeutics. *Turkish Journal of Biology*, *45*(7):530–548.

Hoffmann, M., Kleine-Weber, H., Schroeder, S., Krüger, N., Herrler, T., Erichsen, S., …, & Nitsche, A. (2020). SARS-CoV-2 cell entry depends on ACE2 and TMPRSS2 and is blocked by a clinically proven protease inhibitor. *Cell*, *181*(2):271–280.

Huang, F.-C., Kuo, H.-C., Huang, Y.-H., Yu, H.-R., Li, S.-C., & Kuo, H.-C. (2017). Anti-inflammatory effect of resveratrol in human coronary arterial endothelial cells via induction of autophagy: Implication for the treatment of Kawasaki disease. *BMC Pharmacology and Toxicology*, *18*:1–8.

Huang, S., Liu, Y., Zhang, Y., Zhang, R., Zhu, C., Fan, L., …, & Shi, Y. (2020). Baicalein inhibits SARS-CoV-2/VSV replication with interfering mitochondrial oxidative phosphorylation in a mPTP dependent manner. *Signal Transduction and Targeted Therapy*, *5*(1):266.

Jain, A. S., Sushma, P., Dharmashekar, C., Beelagi, M. S., Prasad, S. K., Shivamallu, C., ..., & Prasad, K. S. (2021). *In silico* evaluation of flavonoids as effective antiviral agents on the spike glycoprotein of SARS-CoV-2. *Saudi Journal of Biological Sciences*, 28(1):1040–1051.

Jannat, K., Paul, A. K., Bondhon, T. A., Hasan, A., Nawaz, M., Jahan, R., ..., & Pereira, M. de L. (2021). Nanotechnology applications of flavonoids for viral diseases. *Pharmaceutics*, 13(11):1895.

Jo, S., Kim, S., Kim, D. Y., Kim, M.-S., & Shin, D. H. (2020). Flavonoids with inhibitory activity against SARS-CoV-2 3CLpro. *Journal of Enzyme Inhibition and Medicinal Chemistry*, 35(1):1539–1544.

Karmakar, B., Talukdar, P., & Talapatra, S. N. (2019). An *in silico* study for two anti-inflammatory flavonoids of Nerium oleander on proinflammatory receptors. *Research Journal of Life Sciences, Bioinformatics, Pharmaceuticals and Chemical Sciences*, 5(1):582–596.

Kaul, R., Paul, P., Kumar, S., Büsselberg, D., Dwivedi, V. D., & Chaari, A. (2021). Promising antiviral activities of natural flavonoids against SARS-CoV-2 targets: Systematic review. *International Journal of Molecular Sciences*, 22(20):11069.

Khaerunnisa, S., Kurniawan, H., Awaluddin, R., Suhartati, S., & Soetjipto, S. (2020). Potential inhibitor of COVID-19 main protease (Mpro) from several medicinal plant compounds by molecular docking study. *Preprints*, 2020:2020030226.

Khazeei Tabari, M. A., Iranpanah, A., Bahramsoltani, R., & Rahimi, R. (2021). Flavonoids as promising antiviral agents against SARS-CoV-2 infection: A mechanistic review. *Molecules*, 26(13):3900.

Kowalczyk, M., Golonko, A., Świsłocka, R., Kalinowska, M., Parcheta, M., Swiergiel, A., & Lewandowski, W. (2021). Drug design strategies for the treatment of viral disease. Plant phenolic compounds and their derivatives. *Frontiers in Pharmacology*, 12:709104.

Krych, J., & Gebicka, L. (2013). Catalase is inhibited by flavonoids. *International Journal of Biological Macromolecules*, 58:148–153.

Kuchi Bhotla, H., Balasubramanian, B., Arumugam, V. A., Pushparaj, K., Easwaran, M., Baskaran, R., ..., & Meyyazhagan, A. (2021a). Insinuating cocktailed components in biocompatible-nanoparticles could act as an impressive neo-adjuvant strategy to combat COVID-19. *Natural Resources for Human Health*, 1(1):3–7.

Kuchi Bhotla, H., Balasubramanian, B., Meyyazhagan, A., Pushparaj, K., Easwaran, M., Pappusamy, M., ..., & Di Renzo, G. C. (2021b). Opportunistic mycoses in COVID-19 patients/survivors: Epidemic inside a pandemic. *Journal of Infection and Public Health*, 14(11):1720–1726.

Kuchi Bhotla, H., Kaul, T., Balasubramanian, B., Easwaran, M., Arumugam, V. A., Pappusamy, M., ..., & Meyyazhagan, A. (2020). Platelets to surrogate lung inflammation in COVID-19 patients. *Medical Hypotheses*, 143:110098.

Kumar, S., Paul, P., Yadav, P., Kaul, R., Maitra, S. S., Jha, S. K., & Chaari, A. (2022). A multi-targeted approach to identify potential flavonoids against three targets in the SARS-CoV-2 life cycle. *Computers in Biology and Medicine*, 142:105231.

Kumar, U., Zoha, R., Kodali, M. V. R. M., Smriti, K., Patil, V., Gadicherla, S., & Singh, A. (2022). Role of dietary flavonoids in preventing COVID-19 infection and other infectious diseases: A mini review. *European Journal of General Dentistry*, 11(03):158–165.

Lalani, S., & Poh, C. L. (2020). Flavonoids as antiviral agents for Enterovirus A71 (EV-A71). *Viruses*, 12(2):184.

Latha, A. S., Anita, P. P., Sidhic, N., James, E., Kaviya, M., Bharathi, K., ..., & Anand, A. V. (2024). Phytonutrients and Secondary Metabolites to Cease SARS-CoV-2 Loop. In: Chen, JT. (eds). *Bioactive Compounds against SARS-CoV-2* (pp. 111–124). CRC Press. https://doi.org/10.1201/9781003323884

Leal, C. M., Leitão, S. G., Sausset, R., Mendonça, S. C., Nascimento, P. H., de Araujo R. Cheohen, C. F., ..., & Leitão, G. G. (2021). Flavonoids from *Siparuna cristata* as potential inhibitors of SARS-CoV-2 replication. *Revista Brasileira de Farmacognosia*, 31(5):658–666.

Lin, C. W., Tsai, F. J., Tsai, C. H., Lai, C. C., Wan, L., Ho, T. Y., ..., & Chao, P. D. L. (2005). Anti-SARS coronavirus 3C-like protease effects of *Isatis indigotica* root and plant-derived phenolic compounds. *Antiviral Research*, 68:36–42.

Liskova, A., Samec, M., Koklesova, L., Samuel, S. M., Zhai, K., Al-Ishaq, R. K., ..., & Kubatka, P. (2021). Flavonoids against the SARS-CoV-2 induced inflammatory storm. *Biomedicine & Pharmacotherapy*, 138:111430.

Liu, A. L., Wang, H. D., Lee, S. M., Wang, Y. T., & Du, G. H. (2008). Structure-activity relationship of flavonoids as influenza virus neuraminidase inhibitors and their *in vitro* anti-viral activities. *Bioorganic & Medicinal Chemistry*, 16(15):7141–7147.

Mahmoud, A. M., Ashour, M. B., Abdel-Moneim, A., & Ahmed, O. M. (2012). Hesperidin and naringin attenuate hyperglycemia-mediated oxidative stress and proinflammatory cytokine production in high fat fed/streptozotocin-induced type 2 diabetic rats. *Journal of Diabetes and its Complications*, 26(6), 483–490.

Mehta, P., McAuley, D. F., Brown, M., Sanchez, E., Tattersall, R. S., & Manson, J. J. (2020). COVID-19: Consider cytokine storm syndromes and immunosuppression. *The Lancet*, *395*(10229):1033–1034.

Mendonca, P., & Soliman, K. F. (2020). Flavonoids activation of the transcription factor Nrf2 as a hypothesis approach for the prevention and modulation of SARS-CoV-2 infection severity. *Antioxidants*, *9*(8):659.

Meng, J. R., Liu, J., Fu, L., Shu, T., Yang, L., Zhang, X., ..., & Bai, L. P. (2023). Anti-entry activity of natural flavonoids against SARS-CoV-2 by targeting spike RBD. *Viruses*, *15*(1):160.

Meyyazhagan, A., Pushparaj, K., Balasubramanian, B., Kuchi Bhotla, H., Pappusamy, M., Arumugam, V. A., Easwaran, M., Pottail, L., Mani, P., Tsibizova, V., & Di Renzo, G. C. (2022). COVID-19 in pregnant women and children: Insights on clinical manifestations, complexities, and pathogenesis. *International Journal of Gynaecology and Obstetrics*, *156*(2):216–224.

Mutha, R. E., Tatiya, A. U., & Surana, S. J. (2021). Flavonoids as natural phenolic compounds and their role in therapeutics: An overview. *Future Journal of Pharmaceutical Science*, *7*:25.

Nguyen, T. T. H., Woo, H. J., Kang, H. K., Nguyen, V. D., Kim, Y. M., Kim, D. W., & Kim, D. (2012). Flavonoid-mediated inhibition of SARS coronavirus 3C-like protease expressed in *Pichia pastoris*. *Biotechnology Letters*, *34*:831–838.

Ngwa, W., Kumar, R., Thompson, D., Lyerly, W., Moore, R., Reid, T. E., Lowe, H., & Toyang, N. (2020). Potential of flavonoid-inspired phytomedicines against COVID-19. *Molecules (Basel, Switzerland)*, *25*(11):2707.

Ninfali, P., Antonelli, A., Magnani, M., & Scarpa, E. S. (2020). Antiviral properties of flavonoids and delivery strategies. *Nutrients*, *12*(9):2534.

Oostra, M., Te Lintelo, E. G., Deijs, M., Verheije, M. H., Rottier, P. J. M., & De Haan, C. A. M. (2007). Localization and membrane topology of coronavirus non-structural protein 4: Involvement of the early secretory pathway in replication. *Journal of Virology*, *81*(22):12323–12336.

Pandey, K. B., & Rizvi, S. I. (2009). Plant polyphenols as dietary antioxidants in human health and disease. *Oxidative Medicine and Cellular Longevity*, *2*(5):270–278.

Park, J. Y., Ko, J. A., Kim, D. W., Kim, Y. M., Kwon, H. J., Jeong, H. J., ..., & Ryu, Y. B. (2016). Chalcones isolated from *Angelica keiskei* inhibit cysteine proteases of SARS-CoV. *Journal of Enzyme Inhibition and Medicinal Chemistry*, *31*:23–30.

Park, J. Y., Yuk, H. J., Ryu, H. W., Lim, S. H., Kim, K. S., Park, K. H., ..., & Lee, W. S. (2017). Evaluation of polyphenols from *Broussonetia papyrifera* as coronavirus protease inhibitors. *Journal of Enzyme Inhibition and Medicinal Chemistry*, *32*:504–515.

Peluso, I., & Palmery, M. (2015). Flavonoids at the pharma-nutrition interface: Is a therapeutic index in demand? *Biomedicine & Pharmacotherapy*, *71*:102–107.https://doi.org/10.1016/j.biopha.2015.02.028

Poochi, S. P., Easwaran, M., Balasubramanian, B., Anbuselvam, M., Meyyazhagan, A., Park, S., ..., & Keshavarao, S. (2020). Employing bioactive compounds derived from *Ipomoea obscura* (L.) to evaluate potential inhibitor for SARS-CoV-2 main protease and ACE2 protein. *Food Frontiers*, *1*(2):168–179.

Pushparaj, K., Bhotla, H. K., Arumugam, V. A., Pappusamy, M., Easwaran, M., Liu, W.-C., ..., & Balasubramanian, B. (2022). Mucormycosis (black fungus) ensuing COVID-19 and comorbidity meets-magnifying global pandemic grieve and catastrophe begins. *Science of The Total Environment*, *805*:150355.

Quintal Martínez, J. P., & Segura Campos, M. R. (2023). Flavonoids as a therapeutical option for the treatment of thrombotic complications associated with COVID-19. *Phytotherapy Research*, *37*(3):1092–1114.

Ramya, S., Boro, A., Geofferina, I. P., Jananisri, A., Vishalini, G., Balamuralikrishna, B., ..., & Anand, A. V. (2023). The Potential Contribution of Vitamin K as a Nutraceutical to Scale Down the Mortality Rate of COVID-19. In: Chen, JT. (eds). Bioactive Compounds against SARS-CoV-2 (pp. 140–159). CRC Press.

Rizzuti, B., Grande, F., Conforti, F., Jimenez-Alesanco, A., Ceballos-Laita, L., Ortega-Alarcon, D., ..., & Velazquez-Campoy, A. (2021). Rutin is a low micromolar inhibitor of SARS-CoV-2 main protease 3CLpro: Implications for drug design of quercetin analogs. *Biomedicines*, *9(4)*:375.

Russo, M., Moccia, S., Spagnuolo, C., Tedesco, I., & Russo, G. L. (2020). Roles of flavonoids against coronavirus infection. *Chemico-Biological Interactions*, *328*:109211.

Ryu, Y. B., Jeong, H. J., Kim, J. H., Kim, Y. M., Park, J. Y., Kim, D., ..., & Lee, W. S. (2010). Biflavonoids from Torreya nucifera displaying SARS-CoV 3CLpro inhibition. *Bioorganic & Medicinal Chemistry*, *18*(22):7940–7947.

Saakre, M., Mathew, D., & Ravisankar, V. (2021). Perspectives on plant flavonoid quercetin-based drugs for novel SARS-CoV-2. *Beni-Suef University Journal of Basic and Applied Sciences*, *10*(1):1–13.

Santana, F. P. R., Thevenard, F., Gomes, K. S., Taguchi, L., Câmara, N. O. S., Stilhano, R. S., ..., & Lago, J. H. G. (2021). New perspectives on natural flavonoids on COVID-19-induced lung injuries. *Phytotherapy Research*, *35*(9):4988–5006.

Schwarz, S., Sauter, D., Wang, K., Zhang, R., Sun, B., Karioti, A., ..., & Schwarz, W. (2014). Kaempferol derivatives as antiviral drugs against the 3a channel protein of coronavirus. *Planta Medica*, *80*(02/03):177–182.

Shanmugam, R., Thangavelu, S., Fathah, Z., Yatoo, M. I., Tiwari, R., Pandey, M. K., Dhama, J., Chandra, R., Malik, Y., Dhama, K., Sha, R., & Chaicumpa, W. Shanmugam, V., & Arumugam, A. V. (2020). SARS-CoV-2/COVID-19 pandemic-an update. *Journal of Experimental Biology and Agricultural Sciences*, *8*(1):S219–S245.

Solnier, J., & Fladerer, J. P. (2021). Flavonoids: A complementary approach to conventional therapy of COVID-19? *Phytochemistry Reviews*, *20*(4):773–795. https://doi.org/10.1007/s11101-020-09720-6

South, A. M., Brady, T. M., & Flynn, J. T. (2020). ACE2 (Angiotensin-converting enzyme 2), COVID-19, and ACE inhibitor and Ang II (Angiotensin II) receptor blocker use during the pandemic: The pediatric perspective. *Hypertension*, *76*(1):16–22.

Su, H., Yao, S., Zhao, W., Li, M., Liu, J., Shang, W. ..., & Xu, Y. (2020). Discovery of baicalin and baicalein as novel, natural product inhibitors of SARS-CoV-2 3CL protease *in vitro*. *BioRxiv*, 2020–04.

Tian, S. S., Jiang, F. S., Zhang, K., Zhu, X. X., Jin, B., Lu, J. J., & Ding, Z. S. (2014). Flavonoids from the leaves of *Carya cathayensis* Sarg. inhibit vascular endothelial growth factor-induced angiogenesis. *Fitoterapia*, *92*:34–40.

Walls, A. C., Park, Y. J., Tortorici, M. A., Wall, A., McGuire, A. T., & Veesler, D. (2020). Structure, function, and antigenicity of the SARS-CoV-2 spike glycoprotein. *Cell*, *181*(2):281–292.

Wang, L., Song, J., Liu, A., Xiao, B., Li, S., Wen, Z., ..., & Du, G. (2020). Research progress of the antiviral bioactivities of natural flavonoids. *Natural Products and Bioprospecting*, *10*:271–283.

Wei, X., Zhu, X., Hu, N., Zhang, X., Sun, T., Xu, J., & Bian, X. (2015). Baicalin attenuates angiotensin II-induced endothelial dysfunction. *Biochemical and Biophysical Research Communications*, *465*(1):101–107.

Yamaguchi, T., Hoshizaki, M., Minato, T. *et al.* ACE2-like carboxypeptidase B38-CAP protects from SARS-CoV-2-induced lung injury. *Nat Commun* **12**, 6791 (2021). https://doi.org/10.1038/s41467-021-27097-8

Zakaryan, H., Arabyan, E., Oo, A., & Zandi, K. (2017). Flavonoids: Promising natural compounds against viral infections. *Archives of Virology*, *162*:2539–2551.

Zhang, X., Huang, H., Zhao, X., Lv, Q., Sun, C., Li, X., & Chen, K. (2015). Effects of flavonoids-rich Chinese bayberry (Myrica rubra Sieb. et Zucc.) pulp extracts on glucose consumption in human HepG2 cells. *Journal of Functional Foods*, *14*:144–153.

Zhu, Y., Scholle, F., Kisthardt, S. C., & Xie, D.-Y. (2022). Flavonols and dihydroflavonols inhibit the main protease activity of SARS-CoV-2 and the replication of human coronavirus 229E. *Virology*, *571*:21–33.

13 Herbal Medicines for COVID-19

The Impact of Alkaloids on Protein Targets

Bancha Yingngam

13.1 INTRODUCTION

The outbreak of novel coronavirus disease 2019 (COVID-19), caused by severe acute respiratory syndrome coronavirus 2 (SARS-CoV-2), emerged in late 2019 and rapidly evolved into a global health crisis (Alkafaas et al., 2023). The genome of SARS-CoV-2 shares approximately 82% similarity with the genome of the earlier known SARS-CoV, the virus associated with the SARS-CoV-2. Characterized by symptoms ranging from mild respiratory issues to SARS, COVID-19 has posed unprecedented challenges to healthcare systems worldwide. The illness caused by COVID-19 is transmitted through contact with infected individuals and by inhalation of the virus. The time from exposure to the onset of symptoms ranges between 2 and 14 days (Raman et al., 2023). The high transmission rate of the virus, along with its ability to mutate, resulting in various strains, has further complicated the global response (Adzdzakiy et al., 2024). These mutations occasionally lead to strains with increased transmissibility or resistance to existing vaccines and treatments, thereby intensifying the need for versatile and effective therapeutic strategies (Xiao et al., 2024). The impact of COVID-19 extends beyond immediate health implications, affecting economics, social structures, and daily life; therefore, identifying effective treatments is urgently needed (Raghav et al., 2023). SARS-CoV-2 infection is notably contagious but generally has a low mortality rate, ranging from 1.0% to 3.5%. However, this rate is more common among elderly individuals who have additional health conditions. Approximately 15–20% of those infected with the virus experience severe pneumonia, and 5–10% of these cases are severe enough to necessitate intensive care services (Sayed et al., 2023). As reported by the World Health Organization (WHO) Coronavirus (COVID-19) Dashboard (https://data.who.int/dashboards/covid19/cases?n=c), the global number of confirmed COVID-19 cases surpassed 773,119,173 as of December 31, 2023, with approximately 6,990,067 deaths. Additionally, as of December 24, 2023, a total of 13.59 billion vaccine doses had been administered. Despite the mitigation of illness severity through existing vaccines and antiviral treatments, the risk of infection remains, particularly with the emergence of new virus variants (Guruprasad et al., 2023; Wilkinson et al., 2023). The effectiveness of these vaccines is still limited, and some antiviral drugs used for treating SARS-CoV-2 have been linked to viral mutations, suggesting that these measures alone may not be sufficient to completely stop the virus's spread (Dhawan et al., 2023; Lan et al., 2024; Zhang, 2023). Despite these circumstances, it appears increasingly likely that SARS-CoV-2 will continue to circulate globally.

In light of these challenges, herbal medicine has surfaced as a considerable area of interest. Historically, herbal remedies have been employed to manage and treat various viral infections by leveraging the wide range of pharmacological properties found in plants (Yoon, 2023). These traditional practices, which are deeply rooted in cultural medicine systems worldwide, offer a rich

DOI: 10.1201/9781003452621-13

repository of knowledge and resources that can be utilized for developing novel treatments (You et al., 2023). The advantage of herbal medicines lies not only in their pharmacological potential but also in their general accessibility and affordability, making them a viable option for large-scale use, especially in resource-limited settings (Houeze et al., 2023). In the context of COVID-19, the exploration of herbal medicine is particularly promising. The focus is on identifying and harnessing compounds with antiviral properties that can act either independently or in conjunction with existing therapies (Lu et al., 2023). This approach involves not only finding immediate remedies but also expanding the arsenal of long-term strategies to combat not only COVID-19 but also future viral outbreaks (Song et al., 2023). The potential of herbal medicine in this scenario lies in its ability to offer new mechanisms of action against the virus, providing hope for more effective and inclusive therapeutic solutions.

One of the significant challenges in combating COVID-19 is the inherent ability of the SARS-CoV-2 virus to mutate and adapt rapidly. These mutations can potentially diminish the effectiveness of existing antiviral drugs and vaccines, which were developed based on earlier strains of the virus (Hillary & Ceasar, 2023). As the virus evolves, it can acquire genetic alterations that may enable it to evade the immune response triggered by vaccines or become resistant to current therapeutic agents (Alquraan et al., 2023). This dynamic nature of the virus highlights the need for alternative and more adaptive therapeutic strategies. Herbal medicines, with their specific bioactive compounds, present a promising frontier in this regard (Shahali et al., 2024). Approximately 20–30% of plants originating from tropical and temperate areas are recognized for possessing antiviral properties (Mohanty et al., 2023). Compounds derived from plants, such as flavonoids, phenolics, alkaloids, lignans, terpenoids, and tannins, exhibit antiviral qualities and are especially effective against SARS-CoV-2 (Pradeep Kumar, 2024). In particular, alkaloids have garnered increased amounts of attention. These are a diverse group of naturally occurring compounds that are predominantly found in plants and known for their potential to inhibit various viral mechanisms (Yonamine et al., 2023). The unique properties of these compounds, including their structural diversity and biological activity, make them candidates for antiviral drug development. This approach is particularly relevant for identifying novel mechanisms of action for coronaviruses.

This chapter aims to:

1. Review alkaloid-rich herbs and identify and discuss various herbs known for their significant alkaloid content and antiviral properties. This chapter involves exploring the historical and current use of these herbs in traditional and modern medicine, providing a comprehensive understanding of their role in antiviral therapy.
2. Elucidate how these alkaloids interact with key protein targets of the SARS-CoV-2 virus, such as the spike protein, the main protease (Mpro), and RNA-dependent RNA polymerase (RdRp), we investigated these interactions. This study explored the molecular interactions and binding mechanisms involved, shedding light on the potential pathways through which alkaloids can inhibit viral replication and propagation.
3. Evaluate the efficacy and safety of alkaloid-based therapies. This assessment will cover not only the therapeutic potential of these compounds but also their pharmacokinetics, toxicity, and possible side effects, which are crucial factors in developing safe and effective treatments.
4. Explore future research directions involving identifying gaps in the current body of knowledge and proposing future studies for the development of alkaloid-based antiviral therapies. This forward-looking perspective is vital for not only addressing the ongoing pandemic but also preparing for future viral outbreaks.

The remainder of this chapter is organized as follows: Section 13.2 provides a detailed exploration of the nature and significance of alkaloids found in various herbs. Section 13.3 focuses on identifying and understanding the specific proteins in the virus that are potential targets for alkaloid interactions. Section 13.4 examines the biochemical dynamics between alkaloids and these viral proteins, elucidating the mechanisms through which these natural compounds could inhibit or alter the virus lifecycle. In Section 13.5, the author presents a discussion of the underlying challenges, prospective pathways for future research, and potential integration of alkaloid-based treatments in clinical settings. Finally, Section 13.6 summarizes the key insights and discusses forward-looking perspectives on the role of alkaloids in antiviral therapy, especially for COVID-19.

13.2   ALKALOIDS IN HERBAL MEDICINE

Alkaloids, a large group of naturally occurring organic compounds, are predominantly found in plants, with more than 20,000 types identified across various species and in some fungi and animals (Phukan et al., 2023). These compounds, mostly containing basic nitrogen atoms, serve as defense mechanisms for organisms, deterring herbivores or pathogens through their potent effects (Muhammad et al., 2024). The presence and concentration of specific alkaloids vary significantly among different plant parts (such as leaves, roots, or bark) and are influenced by growing conditions (Faisal et al., 2023). Characterized by their diverse chemical properties and molecular configurations, alkaloids are generally basic (alkaline), though some are neutral or weakly acidic and capable of forming ammonium salts when reacting with acids. Their complex structures often include one or more carbon atom rings, frequently with heterocyclic rings, leading to a variety of biological activities (Halder & Jha, 2023). While typically more soluble in organic solvents, their solubility can vary, with some alkaloids being water soluble, particularly as salts. Many alkaloids are optically active due to the chiral centers in their molecules and are capable of rotating the plane of polarized light (Zhang et al., 2023). These compounds can undergo a range of chemical reactions, including the formation of salts with acids and the participation of oxidation, reduction, and substitution reactions. Pure alkaloids usually appear as colorless crystals at room temperature, but impurities can impart color, and they often have a characteristic bitter taste. The classification of alkaloids based on chemical structure is complex due to the diversity of these compounds. Each class of alkaloids has distinct structural characteristics and biological activities. The classification is not always clear, as some alkaloids may fit into multiple categories or may not fit neatly into any of the previously mentioned groups. However, these compounds are generally categorized based on the structure of their core ring system or the precursor amino acids from which they are biosynthetically derived.

Figure 13.1 illustrates the various basic alkaloid skeletons commonly found in natural products, each of which is distinct in its molecular framework and biological significance. Starting with the simple five-membered ring structure of pyrrole, known for its presence in complex natural compounds, the chemical structure progresses to pyrrolidine, a saturated analog of pyrrole that is integral to several bioactive alkaloids. Pyridine, with its characteristic nitrogen-containing six-membered ring, is foundational in many alkaloids, including nicotine. Piperidine, a saturated counterpart of pyridine, is frequently observed in alkaloid structures. Quinoline and isoquinoline, both with fused benzene and nitrogen-containing rings, are key skeletons in antimalarial and analgesic compounds, respectively. Pyrrolizidine reappears, emphasizing its prevalence among natural alkaloids. Quinolizidine, a fusion of quinoline and pyrrolidine structures, features lupin alkaloids. Indole, renowned for its role in tryptophan-derived alkaloids, and indolizidine, a hybrid of indole and pyrrolidine, are pivotal for accessing numerous pharmacologically active compounds. Tropane, a nitrogen-containing seven-membered ring, is a core structure of several powerful alkaloids, such as atropine. Imidazole, which has a distinctive five-membered ring containing two nitrogen atoms, is a key component of histamine and other biologically important compounds. Purine, a bicyclic structure, forms the backbone of essential biomolecules such as DNA and caffeine. Finally, aporphine, which has a tetracyclic system, is a significant plant-derived alkaloid. Each structure represents a fundamental framework upon which numerous natural alkaloids are built, often through modifications such as methylation,

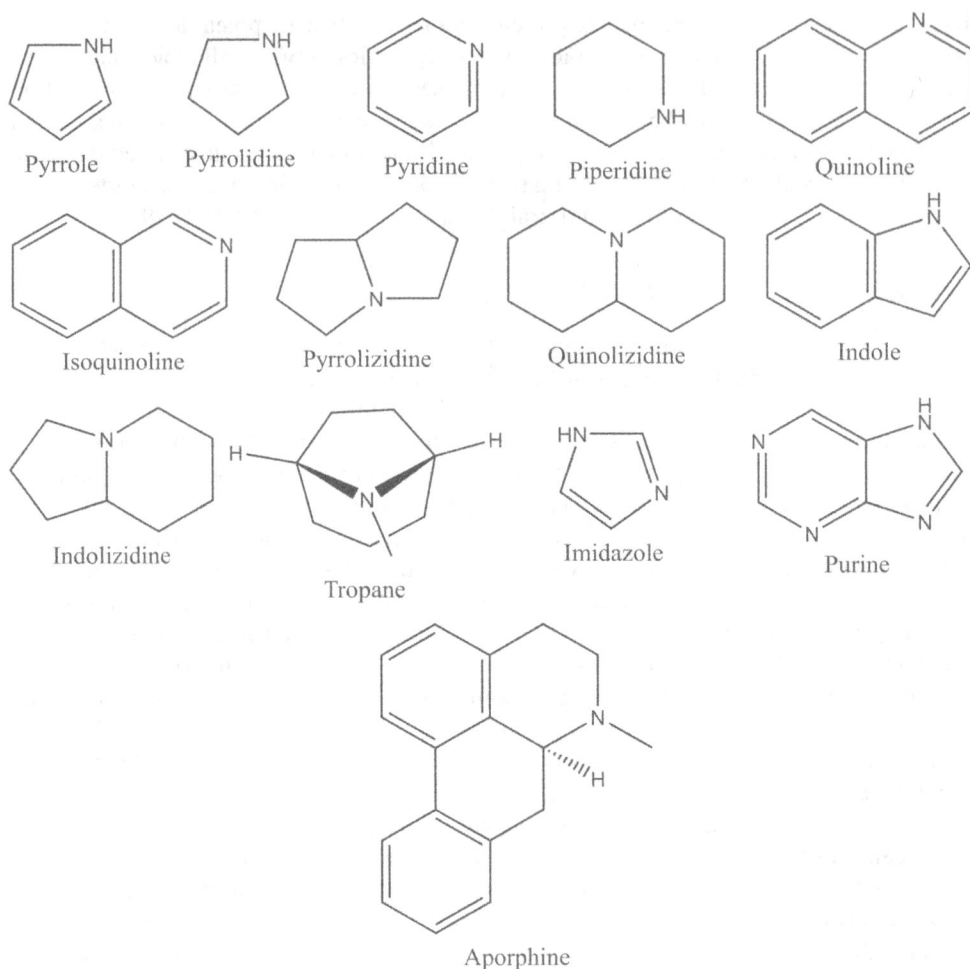

FIGURE 13.1 Selected basic alkaloid skeletal structures in natural products. (Generated with ChemDraw 20.1.1.)

hydroxylation, or ring fusion. These skeletal structures not only are crucial for understanding natural product chemistry but also play a pivotal role in medicinal chemistry, as many alkaloids are key in drug development and synthesis. Alkaloids are typically classified based on their amino-acid precursors, and some of the main classes include the following sections.

13.2.1 TRUE ALKALOIDS

True alkaloids are a major class of alkaloids that are primarily derived from amino acids and characterized by the presence of nitrogen in a heterocyclic ring (Powder-George, 2024). This class of compounds can be further divided based on the specific amino acid from which they originate and the structure of the heterocyclic ring. The following is a detailed look at each subclass.

13.2.1.1 Pyridine-piperidine Alkaloids

Pyridine-piperidine alkaloids, characterized by the fusion of pyridine and piperidine rings, represent a unique class of natural compounds with significant pharmacological properties. The pyridine moiety, a six-membered ring containing one nitrogen atom, contributes to the alkalinity of alkaloids and facilitates interactions with biological targets, particularly proteins. This structural component often plays a pivotal role in binding receptors and enzymes, influencing pharmacokinetics and pharmacodynamics. On the other hand, the piperidine ring, a saturated six-membered ring with one nitrogen atom, imparts rigidity to the molecule and affects its spatial orientation, which is critical for specific protein-alkaloid interactions.

13.2.1.2 Tropane Alkaloids

Tropane alkaloids are characterized by a bicyclic structure consisting of a seven-membered tropane ring system that includes a nitrogen atom. This structure is foundational to compounds such as atropine and scopolamine, which are derived from plants in the *Solanaceae* family. The tropane ring imparts a rigid framework to these molecules, contributing to their specific interactions with biological targets, including muscarinic acetylcholine receptors, which are pivotal in various physiological processes. The ability of these peptides to cross the blood–brain barrier due to their lipophilic nature also plays a significant role in their pharmacological effects. The stereochemistry of tropane alkaloids is a key factor in determining their binding affinity and activity at protein targets, influencing their therapeutic and toxicological profiles. Additionally, modifications of the tropane scaffold, such as esterification or hydroxylation, further diversify the biological activities of the materials and increase their potential as therapeutic agents.

13.2.1.3 Isoquinoline Alkaloids

Isoquinoline alkaloids, characterized by their heterocyclic isoquinoline core–a benzene ring fused to a pyridine ring–exhibit a broad spectrum of biological activities relevant to COVID-19 treatment. This structural framework is foundational for a variety of pharmacologically active compounds, such as morphine and codeine, derived primarily from the *Papaveraceae* family. The isoquinoline nucleus provides an essential platform for various structural modifications, including methylation, hydroxylation, and glycosylation, which significantly influence the interaction of these modifications with biological targets. These modifications can alter alkaloid solubility, distribution, and affinity for specific proteins, including those involved in the inflammatory response and immune modulation, key aspects of the pathogenesis of COVID-19 (Verma et al., 2021). The diverse array of isoquinoline alkaloids, which have complex molecular architectures and varied pharmacological effects, underscores the importance of understanding their chemistry in the pursuit of novel therapeutic strategies against COVID-19.

13.2.1.4 Indole Alkaloids

Indole alkaloids are characterized by their indole core, a fused structure of a benzene ring and a pyrrole ring; these compounds serve as versatile backbones for myriad natural compounds with significant pharmacological activity. This class includes notable examples, such as reserpine and yohimbine, which are derived from various plant families, including *Apocynaceae* and *Rubiaceae*. The indole nucleus is pivotal in these compounds because it can interact with an array of protein targets, including neurotransmitter receptors and transporters. This interaction is crucial in the modulation of neurological and physiological processes and may be relevant in the symptomatology and treatment of COVID-19 (Ang et al., 2023). The chemical diversity of indole alkaloids is further enhanced by various structural modifications, such as methylation, hydroxylation, and glycosylation. These modifications not only affect the pharmacokinetics and pharmacodynamics of these alkaloids but also their specificity and affinity for different protein targets (Mohseni et al., 2022).

13.2.1.5 Purine Alkaloids

Purine alkaloids, primarily caffeine and theophylline, are characterized by their purine base, a bicyclic structure composed of a pyrimidine ring fused to an imidazole ring. This structure forms the core of important biomolecules such as adenosine triphosphate (ATP) and DNA, underscoring the fundamental role of purines in biological systems. Purine alkaloids, particularly those found in common beverages such as coffee and tea, are of interest due to their stimulatory effect on the central nervous system and potential impact on immune function. The biochemical properties of purine alkaloids stem from their ability to mimic and interfere with the action of adenosine, a regulatory molecule, by binding to its receptors. This antagonism can modulate various physiological processes, including inflammation, bronchodilation, and neural activity, which may be relevant to the symptoms and complications of COVID-19. The lipophilicity of these alkaloids enhances their ability to cross biological membranes, influencing their absorption and distribution within the body. Moreover, structural variations, such as methylation, in purine alkaloids alter their potency and selectivity toward different adenosine receptor subtypes, thereby modulating their pharmacological effects.

13.2.2 PROTOALKALOIDS

Protoalkaloids represent a unique subclass of alkaloids characterized by the presence of nitrogen in a heterocyclic ring, similar to true alkaloids; however, distinctively, they are not directly derived from amino acids. One of the most well-known examples of protoalkaloids is mescaline. Mescaline is found in certain cacti, such as the peyote cactus, and is famous for its psychoactive and hallucinogenic properties. Unlike true alkaloids, which typically originate from amino acids such as lysine or tryptophan, protoalkaloids are synthesized through alternative biosynthetic pathways. This distinction in origin results in a diverse range of structures and functions, setting protoalkaloids apart from the more traditional alkaloid compounds. Despite their differing origins, the presence of nitrogen in a heterocyclic ring is a common structural feature that they share with true alkaloids.

13.2.3 PSEUDOALKALOIDS

Pseudoalkaloids, unlike typical alkaloids, are not primarily derived from amino acids; instead, they are synthesized from other precursors, such as terpenoids, steroids, or fatty acids. This distinct origin imparts unique structural and pharmacological properties to pseudoalkaloids. For example, solanine and chaconine, steroid-derived pseudoalkaloids found in *Solanaceae* family plants, have shown potential as anti-inflammatory agents (Lee et al., 2015). The structural diversity of pseudoalkaloids, which are often more complex than traditional alkaloids, allows for a wide range of biological activities, making them intriguing candidates for therapeutic exploration. The mechanisms of action of these enzymes can vary significantly, from interacting with membrane lipids to modulating specific enzyme activities, reflecting their diverse biosynthetic origins.

13.2.4 POLYAMINE ALKALOIDS

Polyamine alkaloids, characterized by multiple amine groups, stand out due to their linear or cyclic structures composed of two or more amino groups linked by alkyl chains. This structure is exemplified in compounds such as spermine and spermidine, which are derived from ornithine or arginine. The presence of multiple amine groups in these alkaloids contributes to their strong interaction with various biological macromolecules, particularly nucleic acids and proteins, influencing cellular processes such as gene expression, enzyme activity, and ion channel function. The biosynthesis of polyamine alkaloids involves enzymatic decarboxylation, amination, and methylation, processes that result in a wide variety of compounds with diverse biological activities.

13.2.5 PEPTIDE AND CYCLOPEPTIDE ALKALOIDS

Peptide and cyclopeptide alkaloids are characterized by their incorporation of amino-acid residues into linear (peptide) or cyclic (cyclopeptide) structures, often linked to a nitrogen atom within a heterocyclic ring. This unique combination of peptide bonds with alkaloid structures results in a diverse array of bioactive molecules, exemplified by compounds such as ergotamine and cycloviolacin.

The cyclic nature of cyclopeptide alkaloids often confers stability and specificity in their biological interactions, making them potent agents for modulating physiological processes. The presence of multiple chiral centers in these molecules also introduces a complex process that significantly affects their pharmacokinetics and pharmacodynamics. Additionally, the variability in the amino-acid composition and sequence of these alkaloids allows for a high degree of diversity in their structure and function (Ebob et al., 2021).

13.2.6 Alkaloid Amides

Alkaloid amides are characterized by the presence of an amide functional group, which is typically formed by the conjugation of an alkaloid moiety with a carboxylic acid derivative. This class of compounds includes capsaicin and piperine, which are derived from plants such as *Capsicum* spp. (chili peppers) and *Piper nigrum* (black pepper), respectively. The amide linkages in these molecules impart specific physicochemical properties, influencing their solubility, stability, and ability to interact with biological targets. Additionally, the lipophilic nature of many alkaloid amides enhances their ability to permeate cell membranes, potentially influencing their interaction with intracellular targets (Wansri et al., 2022).

13.3 TARGET PROTEINS OF SARS-CoV-2

SARS-CoV-2 is an enveloped virus with a positive single-stranded RNA genome that affects both humans and animals (Cupertino et al., 2023). Its genome encodes a variety of proteins, notably the 3-chymotrypsin-like cysteine protease (CCP or 3CLpro), also known as Mpro, the papain-like protease (PLpro), and RdRp (Ghosh et al., 2023; Liu et al., 2023). The virus also produces helicases, spike glycoproteins, and other accessory proteins. Among known viruses, coronaviruses, including SARS-CoV-2, which encode both structural and accessory proteins as well as the viral polyproteins pp1a and pp1ab, possess the largest RNA genomes. These polyproteins are processed by papain-like protease (PLP) and 3C-like protease (3CLpro), processes essential for viral replication and transcription (Brier et al., 2023).

In the SARS-CoV-2 virus, which is responsible for COVID-19, several key proteins play crucial roles in the life cycle and interaction with the host. Figure 13.2 categorizes the key proteins of the SARS-CoV-2 virus based on their specific roles and functions in the life cycle. The proteins are divided into three main categories: viral entry and assembly proteins, viral replication and transcription proteins, and immune response modulation and viral processing proteins. In the first category, proteins such as the spike (S) protein, envelope (E) protein, membrane (M) protein, and nucleocapsid (N) protein are highlighted for their roles in facilitating virus entry into host cells and contributing to virus assembly and structure. The second category included proteins essential for viral RNA synthesis and replication, namely, RdRp, 3CLpro (or Mpro), Nsp12, and Nsp13. The final category included proteins such as PLpro, Nsp1, Nsp3, and Nsp15, which are involved in modulating the host immune response and processing viral components.

Briefly, the S protein is central to the virus's ability to enter host cells by binding to the angiotensin-converting enzyme 2 (ACE2) receptor, a critical step in initiating infection. Once inside, the virus relies on structural proteins such as the E, M, and N proteins for the assembly, shape, and packing of the viral RNA. The virus replication and transcription machinery includes vital components such as RdRp and 3CLpro (or Mpro) and several nonstructural proteins (Nsps), such as Nsp12, which controls RdRp activity, and Nsp13, a helicase involved in RNA unwinding. Additionally, the virus employs mechanisms to evade and modulate the host immune response involving proteins such as PLpro, Nsp1, and Nsp15, which are crucial for processing viral polyproteins, inhibiting host gene expression, and assisting in RNA processing and immune evasion (Liu et al., 2023). These proteins collectively facilitate virus replication, assembly, and interaction with host defenses, making them key targets for therapeutic intervention and vaccine development against COVID-19.

Viral entry and assembly
- Spike (S) protein: Facilitates entry into host cells by binding to the ACE2 receptor.
- Envelope (E) protein: Involved in virus assembly and release.
- Membrane (M) protein: Plays a role in virus assembly, shape, and budding.
- Nucleocapsid (N) protein: Binds to viral RNA, crucial in replication and modulating cellular responses.

Viral replication and transcription
- RNA-dependent RNA polymerase (RdRp): Essential for viral RNA synthesis.
- 3-chymotrypsin-like protease (3CLpro or Mpro): Processes the polyproteins translated from viral RNA.
- Nsp12 (Nonstructural protein 12): Houses the RdRp activity.
- Nsp13 (Nonstructural protein 13): Helicase that unwinds RNA.

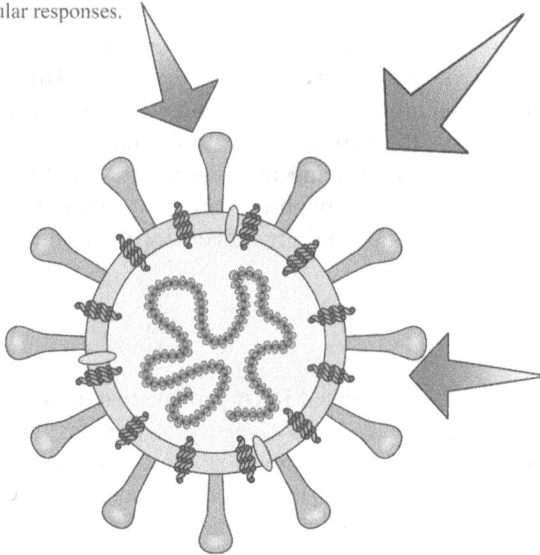

Immune Response Modulation and Viral Processing:
- Papain-like protease (PLpro): Processes viral polyproteins and modulates host immune responses.
- Nsp1 (Nonstructural protein 1): Inhibits host gene expression and immune response.
- Nsp3 (Nonstructural protein 3): Contains PLpro, involved in replication.
- Nsp15 (Nonstructural protein 15): Endoribonuclease for RNA processing and immune evasion.

FIGURE 13.2   Functional grouping of target proteins in the SARS-CoV-2 life cycle.

In this section, key target proteins in the SARS-CoV-2 life cycle, including the S protein, Mpro, and RdRp, are discussed in detail, emphasizing the structural and functional characteristics crucial for developing effective therapeutic strategies.

13.3.1   CORONAVIRUS SPIKE GLYCOPROTEIN

The S protein of SARS-CoV-2 is a multifunctional glycoprotein critical for the virus's interaction with host cells. The proteins are structurally large, ranging from 1200 to 1400 amino acids in length and 180–200 kDa in length (Muhammed et al., 2021). It comprises two primary segments: the S1 and S2 regions. The S1 region is key for initial host interaction and contains a receptor-binding domain (RBD) that specifically attaches to ACE2 receptors on human cells, facilitating viral adhesion. The S2 region plays a vital role in merging the viral envelope with the host cell membrane, a process essential for viral entry. The S protein is characterized by extensive glycosylation involving both *N*-linked and *O*-linked glycosylation. This glycosylation not only is structural but also functional, influencing the protein's behavior and interaction with host cells (Harries et al., 2024). Furthermore, the protein undergoes critical proteolytic cleavage at specific sites, namely, the S1-S2 boundary and the S2′ site. This cleavage, performed by host cell proteases, is pivotal for activating the S protein for membrane fusion, essentially priming it to facilitate virus entry into the host cell (Choi et al., 2024). A remarkable feature of the S protein is its ability to undergo dramatic conformational changes. Initially, in a metastable prefusion state, the protein transforms to a postfusion conformation as part of the fusion process. This transformation involves refolding of the heptad repeats in the S2 region into a six-helix bundle, bringing the viral and host cell membranes into close proximity. This change is complex, triggered by receptor binding and proteolytic cleavage and is critical for the fusion of the viral envelope with the host cell membrane, allowing the viral genome to enter the host cell. The S protein, prominently exposed on the virus surface, is a major target for the host immune response. It elicits the production of neutralizing antibodies, which are key in immune defense against the virus (Tarigan et al., 2023). However, its glycosylation can act as a 'shield', masking epitopes and thus evading immune detection to some extent. Despite this shielding, the immune system often successfully develops antibodies against the S protein, particularly those targeting the RBD (Zhang et al., 2024).

13.3.2 MAIN PROTEASE

Mpro of SARS-CoV-2, also referred to as 3-chymotrypsin-like protease (3CLpro) (33.8 kDa cysteine protease with a nonclassical Cys-His dyad [Cys145-His41]), is a vital enzyme in the lifecycle of the virus (Yoosefian et al., 2023). Mpro plays a crucial role in the life cycle of coronaviruses. It is essential for both viral replication and transcription, especially because of the lack of similar enzymes in humans (T. T. Le et al., 2023). This protease comprises 306 amino acids and shares structural and sequential similarities with the Mpro of SARS-CoV. It functions as a dimer, and mutations that hinder its inherent dimerization significantly diminish its enzymatic activity (Xu et al., 2023). Each monomer of the protease contains three domains (I–III), with domains I and II creating the catalytic binding site, while domain III facilitates the dimerization of the enzyme (Figure 13.3). For Mpro to

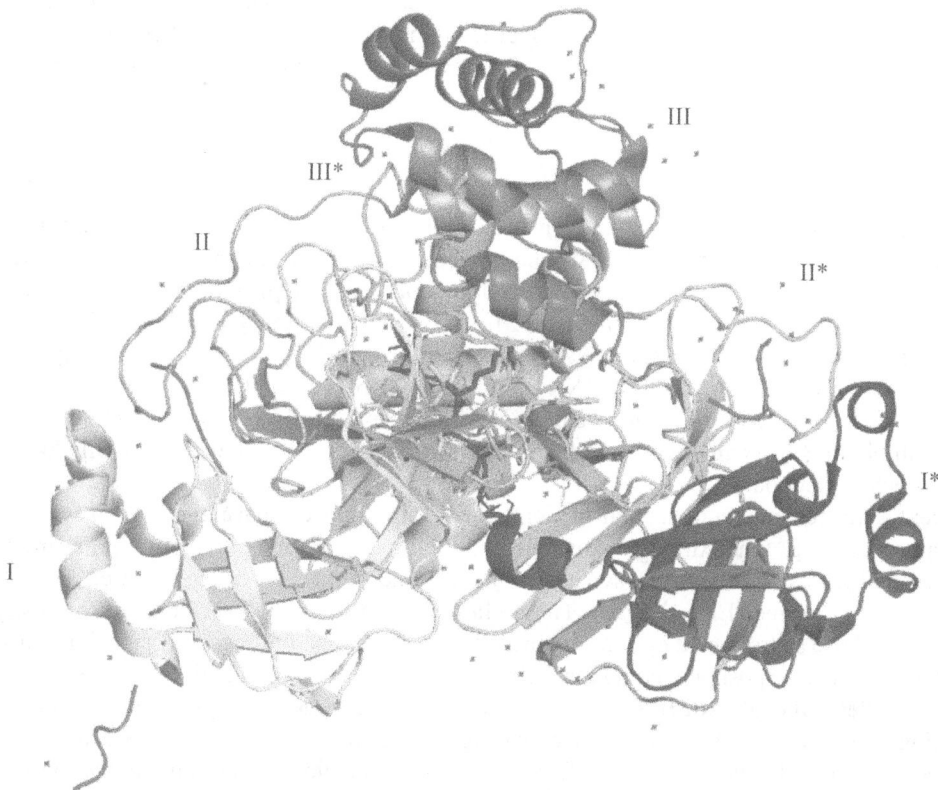

FIGURE 13.3 Three-dimensional crystal structure of the free SARS-CoV-2 main protease (PDB ID: 1UK4) obtained from the Protein Data Bank and generated using PyMol software version 2.5.7 (From Schrodinger, LLC). The figure includes roman numerals I, II, III, some marked with asterisks, indicating specific subunits or important regions of the protein structure that are discussed in the text.

be functional, it must possess a conserved catalytic dyad comprising HIS-41-CYS-145 at its active site. Consequently, these catalytic residues undergo alteration.

As a cysteine hydrolase, it is crucial for the processing of viral polyproteins, a step essential for viral replication. Mpro cleaves polyproteins at specific sites, leading to the production of functional viral proteins necessary for the virus to replicate and assemble. Mpro acts on the large protein 1ab (replicase 1ab, approximately 790 kDa), segmenting it at 11 distinct locations. Most of these cleavage sites are recognized by the sequence Leu-Gln↓(Ser/Ala/Gly), with the ↓ symbol denoting the point of cleavage (Goyal & Goyal, 2020). Without the activity of Mpro, the virus cannot generate the proteins needed to replicate and infect other cells, rendering it a critical target for therapeutic intervention. Given its central role in the viral lifecycle, Mpro has been recognized as a primary target for developing treatments against COVID-19. Inhibiting this protease can effectively halt the replication of the virus, thereby limiting the spread of infection within the host. Researchers and pharmaceutical companies have focused on designing and developing Mpro inhibitors as a strategy to treat COVID-19. This involves finding compounds that can bind to the protease, inhibiting its function and thereby curbing the virus's ability to replicate. The development of effective inhibitors of Mpro poses several challenges, primarily due to the need for high specificity and potency. It is crucial that these inhibitors selectively target Mpro without affecting similar human enzymes to minimize potential side effects. This is because, since there are no reported human proteases with similar cleavage specificities, inhibitors targeting Mpro are unlikely to be toxic (Goyal & Goyal, 2020). Advances in understanding the structural features of Mpro have facilitated the development of inhibitors. Various approaches, including peptidomimetic and nonpeptidomimetic compounds, as well as natural substances, are being explored to discover efficacious Mpro inhibitors. These efforts are crucial in the ongoing fight against COVID-19, providing a promising direction for therapeutic development.

13.3.3 RDRP

The RdRp of SARS-CoV-2, identified as nsp12, is pivotal in the virus's life cycle and is primarily responsible for replicating its RNA genome. The RdRp complex in SARS-CoV-2, which is composed of nsps 12, 7, and 8, is crucial for RNA replication. Nsp7 and nsp8 serve as auxiliary enhancers of polymerase activity, while nsp12 functions as the primary RdRp, marking it as a key component in the viral replication process (Pokharkar et al., 2023). As an enzyme, RdRp catalyzes the synthesis of viral RNA from an RNA template, a process central to the replication and proliferation of the virus within the host. Due to its critical role, RdRp is a prime target for antiviral drugs. Inhibiting RdRp can effectively disrupt the viral replication process, potentially halting the progression of infection.

Recent advancements in molecular modeling and drug discovery have highlighted several clinically approved drugs that exhibit promising results in binding to and potentially inhibiting RdRp. Compounds such as sofosbuvir, ribavirin, galidesivir, and remdesivir have shown effectiveness in binding to this enzyme (Elfiky, 2020). The ability of these drugs to interact with RdRp suggests their potential use as treatment options against COVID-19. These findings are crucial because they pave the way for repurposing existing antiviral drugs and accelerating the development of effective treatments (Bekheit et al., 2023). A focus on RdRp as a drug target has significant implications for managing and treating COVID-19. Understanding the binding affinity of various drugs for RdRp can lead to the development of more effective therapies. Ongoing research is directed toward not only testing existing drugs but also developing novel inhibitors that can selectively and effectively bind to RdRp. Continuous exploration of the structure and functionality of this enzyme is vital for developing therapeutic strategies against SARS-CoV-2 and potentially other RNA viruses in the future (Gangadharan et al., 2022).

13.4   INTERACTIONS OF ALKALOIDS WITH PROTEIN TARGETS

In this section, the author explains the molecular interactions between alkaloids and various protein targets of SARS-CoV-2. Alkaloids have garnered attention for their potential therapeutic actions against this virus (Aziz et al., 2023). Key molecular targets of these alkaloids include the S glycoprotein, which is critical for virus entry into host cells; the coronavirus 3CLpro; papain-like protease (PLpro), which are both essential for viral replication; the RdRp, which is crucial for viral RNA synthesis; and the N protein, which is important for viral assembly. Additionally, the author explored how alkaloids interact with other vital viral components, such as helicases, which are enzymes that are instrumental in viral replication and transcription. Understanding these interactions is pivotal for developing novel therapeutic strategies against COVID-19.

13.4.1   Targeting The Coronavirus Spike Glycoprotein

The intricate involvement of specific human receptors and enzymes in the invasion of cells by SARS-CoV-2 opens new possibilities for treatment strategies. Central to this process is the trimeric glycoprotein known as the S protein, which extends from the surface of the virus, playing a key role in facilitating its entry into host cells. Its primary function is the recognition and binding of ACE2 to host cells. As demonstrated in Figure 13.4, the interaction between the virus S protein and human ACE2 triggers this invasive sequence. This binding event destabilizes the trimer structure, setting the stage for the fusion of the S protein with the cell membrane. A crucial phase in this process is the sequential cleavage of the S protein, initially by the furin enzyme at the S1/S2 subunit, followed by the transmembrane protease serine 2 (TMPRSS2) at the S2 subunit (Palshetkar et al., 2024).

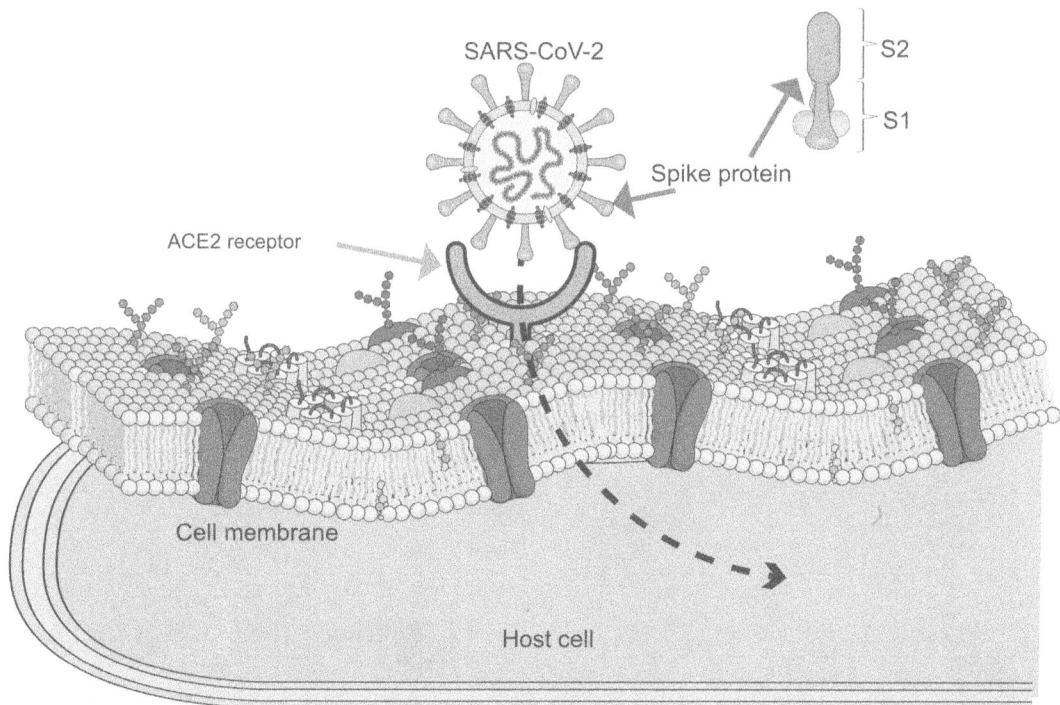

FIGURE 13.4 Schematic representation of the severe acute respiratory syndrome coronavirus 2 (SARS-CoV-2) structure. This diagram illustrates the virus's spike proteins and their ability to bind to the angiotensin-converting enzyme 2 (ACE2) receptor on the host cell surface, a key mechanism in the virus infection process.

In particular, TMPRSS2 plays a pivotal role in activating the spike protein of SARS-CoV-2, which allows the virus to penetrate human cells by cleaving the S protein, thus facilitating the merging of viral and cellular membranes (Rizka Nurcahyaningtyas et al., 2023). Therefore, disrupting the interaction between TMPRSS2 and the spike protein has emerged as a compelling approach in the development of novel COVID-19 therapies.

Several alkaloids have emerged as potential inhibitors of this protein. The mechanism of action of alkaloids on the spike protein involves several key steps, contributing to their potential as antiviral agents against COVID-19.

13.4.1.1 Binding to the Spike Protein

The initial step in the mechanism of action is the binding of alkaloids to the S protein. This protein is responsible for the recognition and binding of the virus to host cell receptors, particularly the ACE2 receptor. Certain alkaloids have been found to bind directly to the S protein, which can interfere with its ability to engage with the ACE2 receptor. This binding is typically characterized by noncovalent interactions, such as hydrogen bonding (Emirik, 2022). Computational docking studies play a vital role in identifying these potential binding sites, allowing researchers to predict how and where these alkaloids might interact with the S protein. These studies often involve simulating the molecular dynamics of the interaction to understand how the binding could alter the protein's function. Conformational changes and inhibition of viral entry (Junqueira Ribeiro, 2024). Upon binding to the S protein, alkaloids can induce conformational changes that hinder the protein's ability to undergo the necessary structural rearrangements for virus–cell fusion. The S protein must transition from a prefusion to a postfusion conformation to facilitate viral entry into host cells. Alkaloids can stabilize the S protein in a configuration that is not conducive to this transition, effectively preventing the virus from entering the cell (Ma et al., 2021). A previous study conducted by Gyebi et al. (2022) investigated the interaction of drug-like alkaloids with the SARS-CoV-2 S glycoprotein, ACE2, and TMPRSS2. These authors focused on how 97 alkaloids can inhibit the S protein, thereby preventing the virus from entering human cells. This study identified specific alkaloids with high binding affinities, suggesting their potential for altering the ability of viruses to enter host cells. Key alkaloids, including cryptospirolepine, 10-hydroxyusambarensine, and cryptoquindoline (Figure 13.5), demonstrated strong binding patterns and interactions with target proteins, indicating their potential as inhibitors of coronavirus entry.

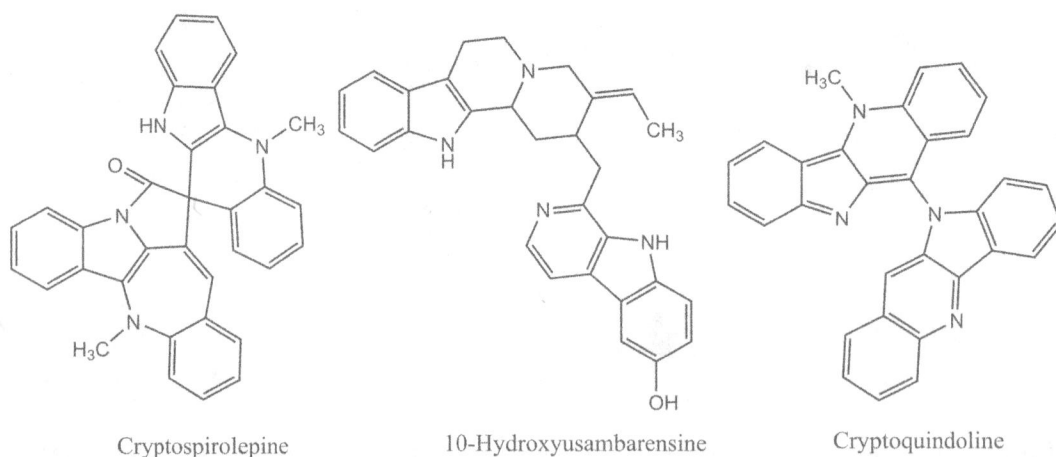

Cryptospirolepine   10-Hydroxyusambarensine   Cryptoquindoline

FIGURE 13.5 Chemical structures of cryptospirolepine, 10-hydroxyusambarensine, and cryptoquindoline. (Created with ChemDraw 20.1.1.)

13.4.1.2 Conformational Changes and Inhibition of Viral Entry

The S protein exists in two main conformations: prefusion and postfusion. The transition from the prefusion to the postfusion state is essential for the virus to fuse with the host cell membrane, a critical step in the infection process (Bejoy et al., 2023). Alkaloids can inhibit this process by binding to the S protein and stabilizing it in the prefusion conformation. This stabilization prevents the structural rearrangements necessary for the protein to mediate fusion with the host cell membrane, thus preventing the virus from entering the cell. The molecular basis of this stabilization involves altering the dynamics of the protein's subunits or interfering with regions critical for conformational change (Sugano et al., 2023).

Research on bioactive compounds from Indonesian marine invertebrates has revealed promising results in the search for inhibitors of TMPRSS2 and the SARS-CoV-2 Omicron variant S protein. Utilizing computational methods such as molecular docking and dynamic simulation, scientists have identified the significant potential of acanthomanzamine C, a compound found in *Acanthostrongylophora ingens*. This alkaloid has shown exceptional effectiveness in targeting TMPRSS2 and the S protein, surpassing the efficacy of other reference ligands used against TMPRSS2. Notably, acanthomanzamine C demonstrated superior binding energies. Variations observed in molecular dynamics simulations after 50 nanoseconds indicated a stable binding affinity to both TMPRSS2 and the S protein. Furthermore, *in silico* absorption, distribution, metabolism, excretion, and toxicity (ADMET) analysis suggested that acanthomanzamine C could be feasibly administered orally (Rizka Nurcahyaningtyas et al., 2023). These findings hold significant promise for advancing the development of treatments for SARS-CoV-2 infections.

In addition, calcium (Ca^{2+}) signaling plays a vital role in this process of virus entry. In a study conducted by Huang et al. (2021), berbamine, a bis-benzylisoquinoline alkaloid, was shown to exhibit anti-SARS-CoV-2 properties by affecting the TRPML-mediated endolysosomal transport of ACE2. Berbamine influences Ca^{2+} signaling both *in vitro* and *in vivo*. It was observed that berbamine significantly reduces the capacity of Gly-Phe β-naphthylamide (GPN) to induce Ca^{2+} release from lysosomes, indicating its potential to inhibit lysosomal Ca^{2+} channels. Given that lysosomal Ca^{2+} channels are involved in various endolysosomal trafficking events, it is plausible that berbamine might obstruct these channels, thereby disrupting ACE2 trafficking and ultimately impeding viral entry. In cell studies using Vero-E6 cells, berbamine notably reduced the viral yield, as measured by a qRT–PCR assay, with an effective concentration (EC_{50}) of approximately 2.4 μM.

13.4.1.3 Disruption of the Spike Protein-ACE2 Interaction

The RBD of the S protein is the primary site for interaction with the ACE2 receptor in host cells. Alkaloids that can bind to the RBD can potentially disrupt this critical interaction. By attaching to the RBD, these compounds can prevent the S protein from effectively binding to the ACE2 receptor, thus inhibiting the initial step of viral entry (Junqueira Ribeiro, 2024). This mechanism is particularly important because it targets a critical early stage in the virus's life cycle. The effectiveness of this disruption depends on the affinity of the alkaloid for the RBD and its ability to compete with the ACE2 receptor for binding sites (Ahmad et al., 2023).

For example, He et al. (2021) identified a group of bis-benzylisoquinoline alkaloids, including hernandezine, as effective broad-spectrum inhibitors of coronavirus entry through a process of compound screening. These alkaloids have been shown to be effective at safeguarding various cell lines from infection by SARS-CoV-2 and other coronaviruses *in vitro*. The cell lines tested included 293T-ACE2 cells (which are HEK 293T cells overexpressing human ACE2), A549 cells, and Calu-3 cells. The mechanism of action for these alkaloids involves blocking host calcium channels, which in turn inhibits the fusion of the viral and endosomal membranes facilitated by Ca^{2+}. This process effectively hinders the ability of the virus to enter host cells.

Fangchinoline, which is extracted from the plant *Stephania tetrandra*, has been reported to be an inhibitor of coronavirus entry and is effective against both SARS-CoV-2 and MERS-CoV in various *in vitro* assays. This bisbenzylisoquinoline alkaloid exhibited strong antiviral effects on SARS-CoV-2, particularly at a concentration of 10 µM. In therapeutic applications, fangchinoline has been shown to reduce viral loads in the lungs and alleviate airway inflammation in hACE2 transgenic (Tg) mice and Syrian hamsters infected with SARS-CoV-2. Notably, a synthetic derivative of fangchinoline, named MK-04-003, demonstrated even greater efficacy in inhibiting SARS-CoV-2 and its variants, including B.1.617.2 and BA.5, in mouse models (Sadhu et al., 2023). These findings collectively suggest that fangchinoline is a broad-spectrum inhibitor of beta coronaviruses and has potential as a treatment option for emerging coronavirus diseases.

13.4.2 Targeting Mpro

The role of alkaloids in targeting SARS-CoV-2 has garnered significant interest, especially concerning their interaction with Mpro of the virus. Mpro is a critical enzyme that is significantly involved in facilitating viral replication and transcription in coronaviruses, which is particularly notable due to the lack of similar enzymes in humans (T. T. T. T. Le et al., 2023). As a result, targeting Mpro for inhibition is a viable antiviral treatment. Numerous studies have explored the ability of various alkaloids to inhibit Mpro. For example, aristolactam glycosides, a type of aristolactone, were extracted from the roots of *Asarum heterotropoides* var. *mandshuricum* F. Maekawa, which belongs to the *Aristolochiaceae* family. These compounds were then analyzed through molecular docking with the SARS-CoV-2 Mpro structure (Protein Data Bank [PDB] ID: 6LU7). These findings indicate that these alkaloids hold potential for the development of novel anti-SARS-CoV-2 medications (T. T. T. T. Le et al., 2023). In addition to terrestrial plants, research using an *in silico* approach has been conducted on imidazole-based alkaloids derived from marine sponges (specifically the *Leucetta* and *Clathrina* species). Compounds such as Naamidine I, Naamidine E, and Pyronaamidine have shown potential as antiviral agents targeting SAR-CoV-2 Mpro, as indicated by their interaction with the protein (PDB ID: 6 W63) (Solo & Arockia doss, 2023). Furthermore, thorough computational screening combined with studies on structure–activity relationships has identified marine batzelladine alkaloids, especially batzelladines H and I, as promising candidates for targeting the Mpro of SARS-CoV-2. Studies on structure–activity relationships highlight the significant impact of fused tricyclic guanidinic structures, their level of unsaturation, the placement of the N–OH group and the length of the side chain that acts as a connector between two active sites. These findings reveal key structural and pharmacophoric characteristics essential for effective protein–ligand interactions (Elgohary et al., 2022).

Sayed et al. (2023) utilized docking and molecular dynamics simulations to target Mpro of SARS-CoV-2 *in vitro*. They constructed a library of 65 mostly axially chiral naphthylisoquinoline alkaloids and their analogs, characterized by diverse molecular architectures and structural analogs, to evaluate their activity against SARS-CoV-2. Despite the common oversight of axial chirality in natural biaryls, these compounds can bind to protein targets in an atroposelective manner. Integrating docking results with steered molecular dynamics simulations, researchers have identified korupensamine A as a particularly effective inhibitor of SARS-CoV-2 Mpro, significantly more so than the reference covalent inhibitor GC376 (IC$_{50}$ values of 2.52 $\pm$ 0.14 µM and 0.88 $\pm$ 0.15 µM, respectively). Korupensamine A also substantially reduced viral growth *in vitro* by five orders of magnitude (IC$_{50}$ = 4.23 $\pm$ 1.31 µM). The authors further explored the binding pathway and mode of interaction of korupensamine A within the protease's active site using Gaussian accelerated molecular dynamics simulations, which corroborated the docking pose of korupensamine A inside the enzyme's active site. This study identified naphthylisoquinoline alkaloids as a promising new class of potential anti-COVID-19 agents.

However, computational studies, while valuable for initial screening and hypothesis generation, are not conclusive. These findings provide insights based on theoretical models and predictions of how molecules might behave. To validate these predictions, empirical studies are necessary. This involves synthesizing alkaloids, testing them *in vitro* (in cell culture or enzymatic assays), and eventually *in vivo* (in animal models) to confirm their efficacy and safety. These steps are critical for determining whether these alkaloids can be effective and safe for use in humans.

13.4.3 TARGETING RDRP

The RdRp of SARS-CoV-2, also known as nsp12, plays a critical role in the virus replication cycle, making it a significant target for antiviral drug development. RdRp is considered a promising target for antiviral drugs due to its lack of homology with host-cell components, which suggests minimal toxicity to human cells. Furthermore, RdRp is notable for its absence of proofreading activity (Rehman et al., 2023). The mechanisms through which alkaloids interact with and potentially inhibit RdRp are complex, highlighting their potential as antiviral agents in the fight against COVID-19 (Bekheit et al., 2023).

The primary mechanism by which alkaloids exert their inhibitory effect on RdRp is through direct binding to its active site. RdRp is responsible for the replication of viral RNA, and its active site is a critical region where new RNA strands are synthesized. Several alkaloids that bind to this active site have been identified through computational modeling and empirical studies. This binding can interfere with the enzyme's ability to catalyze the RNA synthesis process. Binding interactions often involve hydrogen bonds, hydrophobic interactions, and ionic interactions, which are essential for the specificity and stability of the alkaloid-RdRp complex (Bekheit et al., 2023). Upon binding to RdRp, alkaloids can induce conformational changes in the enzyme's structure. These changes can alter the spatial arrangement of critical amino acids in the active site or other important regions, thereby reducing the catalytic efficiency of the enzyme. Conformational changes might also result in the distortion of the overall structure of the enzyme, further inhibiting its function. These modifications in the enzyme structure can impede the replication of viral RNA, effectively reducing viral proliferation (Ananda et al., 2024).

In a study conducted by Borquaye et al. (2020), 13 alkaloids isolated from the African herb *Cryptolepis sanguinolenta* were analyzed through *in silico* studies for their potential anti-SARS-CoV-2 properties, specifically targeting RdRP (PDB ID: 6M71) and the main protease (Mpro, PDB ID: 6LU7). This study revealed that alkaloids such as cryptomisrine, cryptospirolepine, crypto-quindoline, and biscryptolepine demonstrated promising affinity for the RdRp enzyme, suggesting their potential as inhibitors of both proteins. Additionally, the study evaluated the drug-like properties and potential toxicity of these alkaloids. The quantitative structure-activity relationship (QSAR) and ADMET predictions indicated that most alkaloids conformed to Lipinski's rules for drug likeness. Cryptomisrine exhibited the highest number of hydrogen bond donors, while crypto-heptine was the most effective at accepting hydrogen bonds. The alkaloids were predominantly hydrophobic, with varying solubility levels; hydroxycryptolepine was noted for its complete solubility. These compounds were able to cross the blood–brain barrier and generally exhibited high gastrointestinal absorption. Compared to standard drugs such as lopinavir, these alkaloids exhibit lower solubility but greater permeability and metabolism. The distribution volumes and predicted short half-lives of these compounds indicated good bioavailability. Drug-likeness scores ranged from −1 to −0.2, supporting the potential of these compounds as drug candidates. The LD50 values, which indicate toxicity levels, were generally high, suggesting low toxicity, except for that of cryptospirolepine. Notably, cryptoquindoline had a high probability of mutagenicity, while crypto-spirolepine and biscryptolepine (Figure 13.6) were not identified as potential mutagens or carcinogens. Prior research on *C. sanguinolenta* extracts and alkaloids, the sources of these compounds, has investigated their anticancer potential. These studies have shown that plant extracts are safe at concentrations up to 500 mg/kg but may become toxic at very high concentrations. Further research using animal models to determine tolerable dosages could provide deeper insights into the toxicity, mutagenicity, and carcinogenicity of these compounds.

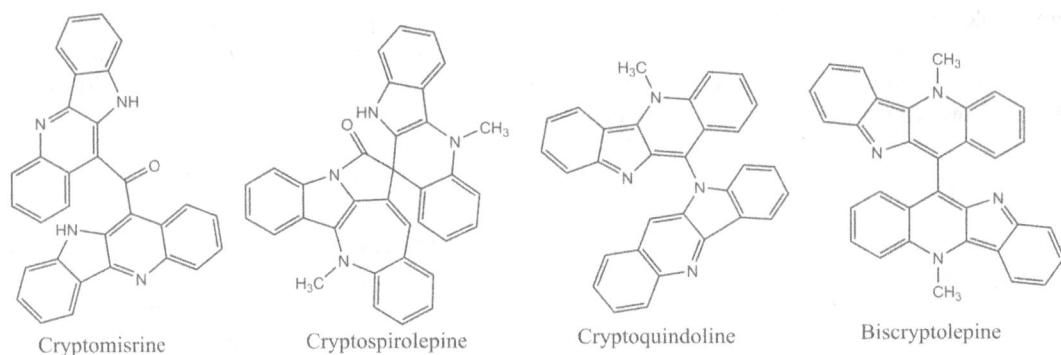

FIGURE 13.6 Chemical structures of cryptomisrine, cryptospirolepine, cryptoquindoline, and biscryptolepine.

13.4.4 Targeting Nucleocapsid Protein

The N protein of SARS-CoV-2 plays a crucial role in the virus replication cycle. It is primarily involved in packaging viral RNA into a ribonucleoprotein complex, which is essential for virus assembly and release. The N protein also has roles in viral genome replication, modulating the host cell's response to infection and evasion of the host's immune response (Yu et al., 2023). Some alkaloids offer a unique approach to targeting N proteins. The binding of specific alkaloids to the N protein could interfere with its ability to form a ribonucleoprotein complex, thereby hampering virus assembly and replication (Wink, 2020). Like *Sophora flavescens*, matrine, an alkaloid found in plants, has shown potential against SARS-CoV-2. While its primary action is noted against other viral components, there is growing interest in exploring its effects on the N protein. By binding to the N protein, matrine can disrupt the viral RNA packaging process, limiting the ability of the virus to replicate and spread (He et al., 2023). Another alkaloid, harmine, known for its presence in plants such as the Syrian Rue, has been studied for its antiviral properties. Research into its interaction with the N protein could reveal the mechanisms by which it impedes the virus's replication process, particularly in the later stages of the virus life cycle (Rani et al., 2022). Targeting N proteins with alkaloids represents a new frontier in COVID-19 treatment strategies. Specific binding and interference with N-protein function could lead to a reduction in viral replication and spread. This approach could be especially valuable in developing treatments that are less susceptible to resistance, as the N protein is less prone to mutations than other viral proteins, such as the S protein.

13.4.5 Other Protein Targets

Alkaloids have shown promising activity against a range of viral targets, extending beyond the well-known S protein, 3CLpro, PLpro, RdRp, and N protein of SARS-CoV-2. The keys among these proteins are protein kinase C (PKC) and the SARS-CoV helicase, both of which are crucial due to their roles in ATP activity. PKC, which is vital for regulating cellular functions such as signal transduction and cell proliferation, has become a significant target in the context of viral infections. Alkaloids can modulate PKC activity, thus influencing both viral replication and the host-cell response. Berberine, for instance, is known to inhibit PKC activity in specific cellular settings, potentially disrupting the replication mechanism of SARS-CoV-2, as PKC is integral to intracellular signaling pathways exploited by viruses for replication and assembly (Figure 13.7) (Gildenhuys et al., 2023). The SARS-CoV helicase protein, which is essential for viral replication due to its role

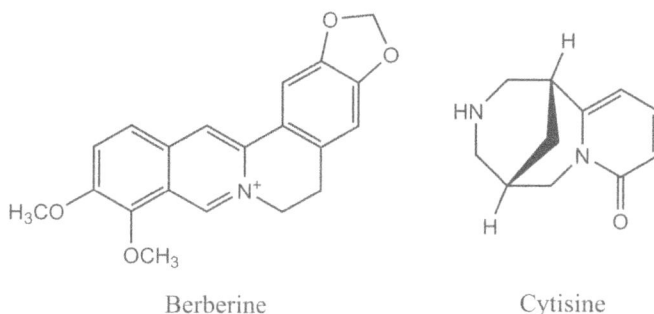

FIGURE 13.7   Chemical structures of berberine and cytisine. (Created with ChemDraw 20.1.1.) (Gildenhuys et al., 2023; deAndrés-Galiana et al., 2022.)

in unwinding RNA structures, is another critical target. Alkaloids can inhibit the ATPase activity of this protein, which is fundamental for its helicase function. Cytisine has been shown to impede the ATPase activity of the SARS-CoV helicase, directly affecting the virus's RNA replication and subsequently its life cycle. This approach of targeting ATP-dependent activities of key proteins can effectively hinder virus replication and proliferation within the host (deAndrés-Galiana et al., 2022). The exploration of the ability of alkaloids to target proteins such as PKC and SARS-CoV helicase underscores the expansive potential of these natural compounds in antiviral therapy. Their unique mechanisms, particularly in influencing ATP activity and altering enzyme functions, complement the action of direct-acting antivirals. Ongoing research into the diverse structures of alkaloids could lead to the discovery of novel inhibitors effective against various viral components. Furthermore, integrating alkaloid-based treatments with other antiviral medications may increase the efficacy of these treatments and reduce the chances of drug resistance. Such a comprehensive approach is vital for the development of more effective COVID-19 treatments and could also be beneficial for tackling other viral infections.

13.4.6   CASE STUDY OF CEPHARANTHINE

The development of drugs, especially for urgent health crises such as COVID-19, is a complex and time-consuming process that can be expedited through collaborative efforts. This is exemplified in the case of cepharanthine, an alkaloid isolated from the medicinal plant *Schizothrix* sp., which has been clinically used since 1951 for various ailments (Yang & Wang, 2023). Notably, cepharanthine has demonstrated a broad spectrum of antiviral effects *in vitro*, including activity against SARS-CoV (IC_{50} value = 9.50 μM), HIV-1 (IC_{50} value = 0.026 μM), and HCoV-OC43 (IC_{50} value = 0.83 μM), showing promising inhibitory concentrations against these viruses (Kim et al., 2019; Okamoto et al., 1998; Zhang et al., 2005).

Figure 13.8 illustrates the chemical structure of cepharanthine. Cepharanthine, a bisbenzylisoquinoline alkaloid, has a complex and unique chemical profile that significantly influences its biological activity and therapeutic potential. Its molecular structure consists of two isoquinoline units linked through a central carbon, forming a rigid and planar bisbenzylisoquinoline framework connected by multiple aromatic rings. These rings, coupled with various functional groups, including methoxy groups and a secondary amine, endow cepharanthine with distinct chemical properties. The lipophilic nature of the molecule, attributed to these aromatic structures and methoxy groups, results in low water solubility but facilitates easier crossing of cell membranes, impacting pharmacokinetic behaviors such as absorption and distribution. Additionally, the stereochemistry of

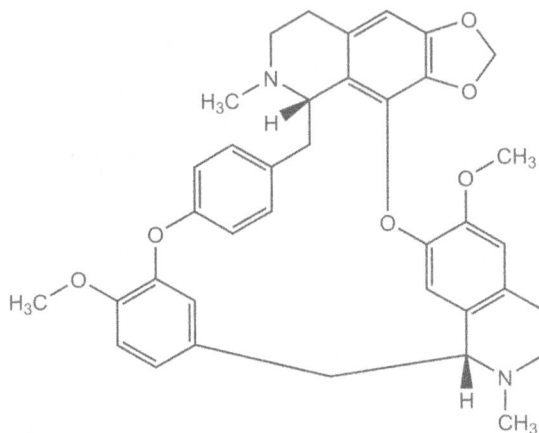

FIGURE 13.8 Chemical structure of cepharanthine. (Created with ChemDraw 20.1.1.)

cepharanthine, which is marked by chiral centers, plays a critical role in its biological interactions, particularly by influencing its efficacy and specificity in binding to biological targets, including receptors and enzymes. The stability of cepharanthine under physiological conditions is a key factor in its effectiveness as a therapeutic agent, as it retains structural integrity and functional capacity within the body. Moreover, the reactivity of cepharanthine, influenced by its aromatic and methoxy groups, is essential for interactions at the molecular level, including with proteins and nucleic acids, which is pivotal for its antiviral action against pathogens such as SARS-CoV-2.

In early 2020, through the use of a pangolin coronavirus as a model, research by Fan et al. (2020) in Beijing revealed that cepharanthine had significant anti-SARS-CoV-2 effects on Vero E6 cells, and low toxicity was observed. Subsequent studies by Ohashi et al. (2020) in Japan confirmed its potent inhibition of SARS-CoV-2 replication in specialized cell lines. One of the key interactions of cepharanthine is with the SARS-CoV-2 S protein, where it blocks attachment to the ACE2 receptor. Research from the United States by White et al. (2020) highlighted cepharanthine's mechanism of action, targeting the SARS-CoV-2 nsp13 (helicase), a vital protein for viral replication. Additionally, Li et al. (2021) in Beijing demonstrated that cepharanthine can counteract SARS-CoV-2 infection by influencing the heat shock response and hypoxia pathways.

These various studies provide strong evidence for the therapeutic potential of cepharanthine for treating COVID-19. Research groups worldwide, including He et al. (2021) in Chongqing, China and Drayman et al. (2021) in the United States, have shown its efficacy in inhibiting SARS-CoV-2 in various cell lines, often outperforming known antivirals such as remdesivir. Notably, simulations by Hijikata et al. (2022) in Japan identified the diphenyl ether moiety of cepharanthine as the anti-SARS-CoV-2 pharmacophore. Zhang et al. (2022) in Beijing, China further demonstrated its effectiveness against the SARS-CoV-2 B.1.351 variant, with significant results in reducing both the viral load and lung injury in a mouse model. Moreover, the synergistic effects observed with the combination of cepharanthine and other compounds against the B.1.351 variant underscore its potential.

These promising outcomes led to clinical trials, such as a phase II trial conducted in 2022 (ClinicalTrials.gov ID: NCT05398705, Status: Completed) (www.ClinicalTrials.gov) at the School of Medicine, Shanghai Jiao Tong University, China. This trial assessed the efficacy and safety of cepharanthine for nonhospitalized COVID-19 patients. A double-blind, three-arm study investigated the effects of orally administered high or low doses of cepharanthine compared with a placebo in nonhospitalized asymptomatic or mildly symptomatic adult participants with COVID-19. The results indicated a favorable safety profile for cepharanthine and the potential for shortening the disease course in patients with asymptomatic or mild symptoms. While further large-scale, double-blind trials are necessary to confirm these findings, progress in cepharanthine development underscores the rapid advancement possible in natural product-based drug discovery, particularly when supported by collaboration among various scientific and medical disciplines.

13.5 CHALLENGES, FUTURE DIRECTIONS AND CLINICAL IMPLICATIONS

In this section, the associated challenges, potential future research directions, and clinical implications of using alkaloids as therapeutic agents for COVID-19 treatment will be explored, as summarized in Figure 13.9 and detailed in the subsections. The focus will be on identifying gaps in current research, suggesting areas for future studies, and discussing how the findings could be translated into clinical practice.

13.5.1 EXPANDING THE RANGE OF ALKALOIDS STUDIED

Despite significant advancements in understanding the therapeutic potential of certain alkaloids against COVID-19, a vast array of these compounds has not been thoroughly investigated. The current focus has been primarily on a limited set of alkaloids, often those already known for their medicinal properties in other contexts. For example, quinine, an alkaloid derived from the bark of the cinchona tree, has been extensively studied for its antimalarial properties and has drawn

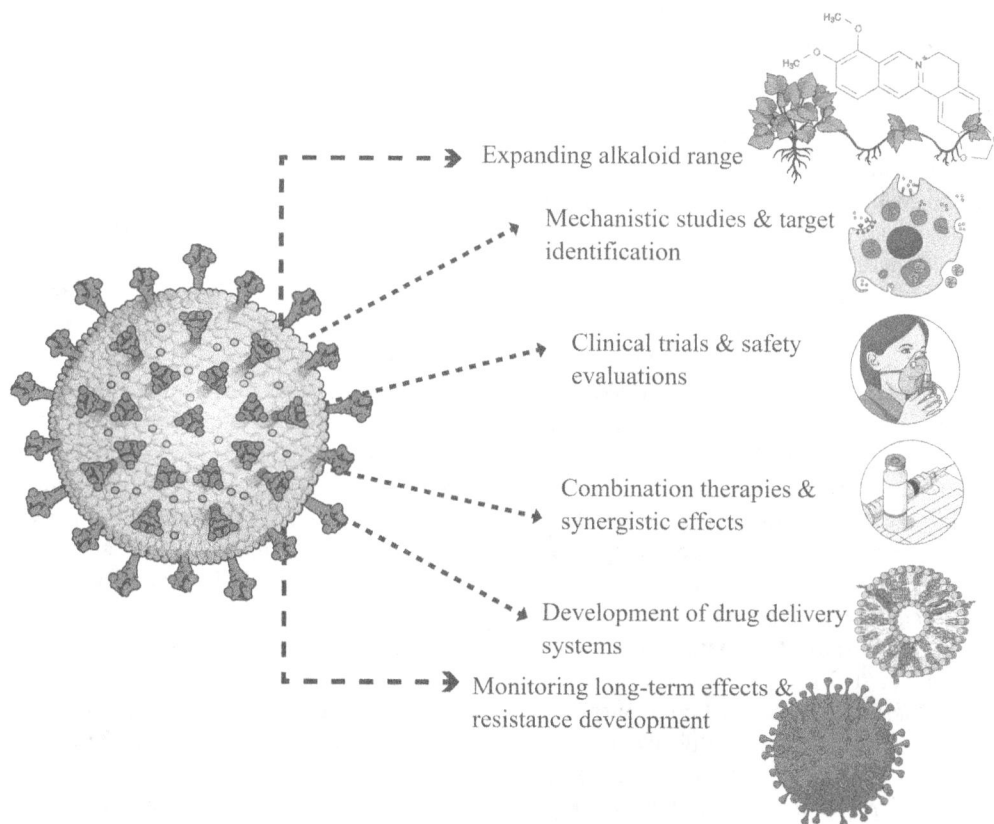

FIGURE 13.9   Overview of future directions and clinical implications of alkaloids in COVID-19 treatment.

attention for its potential application in COVID-19 treatment (Große et al., 2021). Similarly, berberine, found in plants such as *Berberis* sp., is known for its anti-inflammatory and antiviral effects and has been considered for COVID-19 therapy (Oner et al., 2023). However, these compounds represent only a fraction of the alkaloids found in nature. The plant kingdom is rich in thousands of alkaloids, many of which are found in less-studied species (Phukan et al., 2023). These lesser-known alkaloids could possess unique structures and mechanisms of action that make them effective against SARS-CoV-2. Furthermore, the development of synthesized alkaloid analogs offers another promising avenue. Synthetic analogs can be designed to enhance the efficacy, specificity, and safety profiles of natural alkaloids. For instance, modifications in the molecular structure could increase the affinity of alkaloids for specific viral proteins of SARS-CoV-2, potentially leading to more effective inhibition of the virus. The synthesis of new compounds also allows for the exploration of chemical space not covered by natural alkaloids, potentially leading to the discovery of novel mechanisms of antiviral action (N-T-H. Le et al., 2023).

13.5.2   Mechanistic Studies And Target Identification

Mechanistic studies and target identification are essential in developing alkaloid-based therapies for COVID-19. The authors focused on how alkaloids interact with SARS-CoV-2 proteins to understand their mechanism of action, guiding the development of more effective treatments (T. T. T. Le et al., 2023). Due to their structural diversity, alkaloids can interact with various viral and host proteins. For example, some inhibit SARS-CoV-2 entry by binding to its S protein, while others target viral replication processes such as RdRp or Mpro (Sayed et al., 2023). Initial studies on hydroxychloroquine, which is thought to inhibit viral entry and replication, highlighted this approach, although its efficacy and safety have not been fully elucidated (Infante et al., 2021). These studies also explored how alkaloids affect the host immune response, such as by modulating cytokine release to prevent severe COVID-19-related inflammatory responses (Bhar et al., 2023). Identifying targets and understanding alkaloid interactions require biochemical, virological, and computational methods. Biochemical assays screen for alkaloids that interact with specific proteins, virological studies assess their impact on viral replication, and computational approaches such as molecular docking simulate alkaloid binding to targets.

For instance, Kakimoto et al. (2024) investigated the antiviral effects of extracts from two types of *Ephedra* plants, *Ephedra sinica* and *Ephedra przewalskii*, on various strains of the SARS-CoV-2 virus, including those causing COVID-19. Traditionally used in East Asian medicine, *E. sinica* is known for treating common colds and flu but has side effects due to ephedrine alkaloids. In contrast, *E. przewalskii*, which is used in Uyghur and Mongolian traditional medicine, is free from these alkaloids, making it a safer alternative. The present research involved infecting VeroE6/TMPRSS2 cells with different SARS-CoV-2 strains (conventional, delta and omicron variants) and treating them with extracts from both plants. The study measured the viral infectious titer using the 50% tissue culture infectious dose (TCID$_{50}$) method and calculated the half-maximal IC$_{50}$ to compare the antiviral efficacy of the two extracts. Additionally, the effectiveness of *E. przewalskii* extract was compared with that of remdesivir against a conventional virus strain. The results showed that the *E. przewalskii* extract was significantly more effective than the *E. sinica* extract at inhibiting the virus, and its antiviral efficacy was 2.7 to 10.8 times greater across various strains. Compared to that of remdesivir, the efficacy of the *E. przewalskii* extract was lower but still notable. Furthermore, while *E. sinica* exhibited minimal virus inactivation, *E. przewalskii* exhibited substantial viral inactivation. This study suggested that *E. przewalskii* extract has greater antiviral activity against SAR-CoV-2 than does *E. sinica* and could be used for treating viral infections, including COVID-19. However, it is important to note that the antiviral efficacy of *E. przewalskii*, though promising, is not as high as that of remdesivir. Understanding these interactions helps in identifying which stage of the viral life cycle is most vulnerable to intervention by alkaloid-based treatments.

13.5.3 CLINICAL TRIALS AND SAFETY EVALUATIONS

Translating *in vitro* and *in vivo* findings into effective clinical treatments for COVID-19 involves rigorous clinical trials and safety evaluations, which are crucial for assessing the safety, efficacy, and optimal dosing of alkaloid-based treatments in humans. Clinical trials progress through phases, from initial safety assessments (Phase I) to larger-scale efficacy and side effect evaluations (Phases II and III), each designed to answer specific research questions while prioritizing participant safety. Safety is paramount due to the potentially toxic effects of alkaloids at certain doses or in specific patient contexts (Bhar et al., 2023). For example, while alkaloids such as quinidine have therapeutic effects, they also pose risks such as cardiotoxicity. In patients with COVID-19, determining a safe dosage range is essential, especially given the variability in patient responses. Efficacy evaluation is also critical for determining whether treatment reduces COVID-19 severity or the viral load. Trials such as those for hydroxychloroquine, which showed limited efficacy despite initial promise, exemplify this phenomenon (Infante et al., 2021). Understanding the pharmacokinetics (how the drug is processed in the body) and pharmacodynamics (the drug's biological effects and action mechanisms) is crucial for optimal dosing and predicting responses in diverse patient populations. This includes considering how comorbid conditions such as liver or kidney disease might affect drug efficacy and safety.

13.5.4 COMBINATION THERAPIES AND SYNERGISTIC EFFECTS

Combining alkaloids with other antiviral agents is a promising strategy for treating COVID-19, leveraging the potential synergistic effects of multiple drugs via different mechanisms. Synergy occurs when the combined effect of drugs exceeds their individual effects, possibly enhancing the efficacy of the virus combination, reducing disease severity, and mitigating drug resistance risks (Valipour et al., 2023). This approach mirrors successful HIV/AIDS treatments, where antiretroviral therapy involves the use of a combination of drugs targeting different HIV life cycle stages, improving outcomes, and reducing drug-resistant virus strains. Similarly, for COVID-19, pairing an alkaloid that inhibits virus entry into cells with a drug that suppresses viral replication could lead to more effective treatments. This not only enhances efficacy but also reduces the likelihood of the virus developing resistance, as multiple simultaneous mutations are required to overcome combined drug effects, which is a less likely scenario. Combination therapies can also potentially reduce side effects and toxicity. Using multiple drugs at lower doses may achieve therapeutic effects with fewer adverse effects than when used alone; this approach is important for diverse COVID-19 patient populations, including those with underlying health conditions more susceptible to drug complications (de Loera, 2022).

13.5.5 DEVELOPMENT OF DRUG DELIVERY SYSTEMS

Innovative drug delivery systems are crucial for enhancing the bioavailability and efficacy of alkaloids for COVID-19 treatment. These systems transport pharmaceutical compounds in the body to achieve the desired therapeutic effect safely and effectively, ensuring drug release at the right site, time, and amount, thus maximizing efficacy and minimizing side effects. Nanoparticles are groundbreaking materials, especially for accessing alkaloids because of solubility or stability issues (El-Fakharany et al., 2023). Encapsulating alkaloids in nanoparticles improves absorption, allowing for lower dosages and fewer side effects. Nanoparticles can be designed to target specific cells or tissues, enhancing treatment efficacy by focusing on the therapeutic effect where needed. Targeted delivery mechanisms deliver drugs directly to the infection site, increasing the local drug concentration and reducing systemic exposure. Inhalable alkaloid formulations, for instance, can directly target the lungs, the primary infection site in COVID-19, enhancing antiviral activity while reducing systemic toxicity (Bhattacharjee et al., 2023). Controlled-release systems allow gradual drug release over time and maintain steady therapeutic levels, which are crucial for chronic conditions or diseases requiring constant drug concentrations (Maheshwari et al., 2023). In patients with COVID-19, controlled release could provide sustained antiviral activity without frequent dosing. These advanced systems also improve treatment practicality, increase bioavailability, reduce dosing frequency, and improve patient adherence. Minimizing side effects through targeted or controlled release significantly improves patient quality of life (Maheshwari et al., 2023).

13.5.6 MONITORING LONG-TERM EFFECTS AND RESISTANCE DEVELOPMENT

Monitoring the long-term effects of alkaloid-based COVID-19 therapies is crucial for ensuring patient safety and treatment effectiveness. While short-term effects are observed in clinical trials, long-term impacts, emerging weeks to years later, require ongoing surveillance. This approach is vital for identifying adverse effects that are not initially evident and understanding the full impact over time. The key to this monitoring is detecting delayed adverse effects, such as potential liver or heart toxicity, through regular medical checks and tests. Additionally, identifying the development of resistance in viruses such as SARS-CoV-2 is important, as these viruses can mutate and reduce treatment efficacy. Continuous monitoring of the viral response helps adjust treatment protocols in a timely manner. Moreover, resistance development affects both individual treatment and public health strategies, especially for fast-spreading viruses such as COVID-19. Tracking treatment outcomes and viral mutations demands collaboration between healthcare providers, researchers, and public health authorities. Finally, long-term monitoring offers valuable data for future research, aiding in the development of more effective, safer treatments and strategies to manage resistance. This involves exploring new alkaloids or combinations to enhance efficacy and reduce side effects (Abookleesh et al., 2022).

13.6   CONCLUSION

This chapter explored the crucial role that alkaloids play in targeting proteins associated with COVID-19. The complex interactions of these proteins with viral components, especially the SARS-CoV-2 S protein, Mpro, and RdRp, highlight a promising direction for therapeutic development. The efficacy of alkaloids in modulating immune responses and inhibiting viral replication offers a glimmer of optimism in the ongoing battle against this pandemic. However, translating these promising laboratory discoveries into clinical applications is fraught with challenges. The complexity of alkaloid biosynthesis, combined with the intricacies of interactions between human and viral proteins, calls for a careful and methodical approach. The inconsistency in the efficacy of alkaloids and the risk of adverse reactions highlight the necessity for thorough clinical testing. Moreover, the rapid mutation rate of SARS-CoV-2 necessitates a dynamic and adaptable approach for therapeutic development. Future efforts should concentrate on employing computational models and extensive screening processes to expedite the identification of effective alkaloid-based treatments. The collaborative combination of scientists is imperative to fully utilize the potential of alkaloids in combating COVID-19. Investigating the combined effects of alkaloids and existing antiviral treatments could lead to more comprehensive and effective treatment options. In conclusion, continued research in this domain is vital not only for addressing COVID-19 but also for helping the medical field better respond to emerging viral threats.

REFERENCES

Abookleesh, F. L., Al-Anzi, B. S., & Ullah, A. (2022). Potential antiviral action of alkaloids. *Molecules*, *27*(3), 903. https://doi.org/10.3390/molecules27030903

Adzdzakiy, M. M., Sutarno, S., Asyifa, I. Z., Sativa, A. R., Fiqri, A. R. A., Fibriani, A., … Saputra, S. (2024). SARS-CoV-2 genetic variation and bacterial communities of naso-oropharyngeal samples in middle-aged and elderly COVID-19 patients in West Java, Indonesia. *Journal of Taibah University Medical Sciences*, *19*(1), 70–81. https://doi.org/10.1016/j.jtumed.2023.09.001

Ahmad, A., Makhmutova, Z., Cao, W., Majaz, S., Amin, A., & Xie, Y. (2023). Androgen receptor, a possible anti-infective therapy target and a potent immune respondent in SARS-CoV-2 spike binding: a computational approach. *Expert Review of Anti-Infective Therapy*, 21(3), 317–327. https://doi.org/10.1080/147 87210.2023.2179035

Alkafaas, S. S., Abdallah, A. M., Hussien, A. M., Bedair, H., Abdo, M., Ghosh, S., … Hessien, M. (2023). A study on the effect of natural products against the transmission of B.1.1.529 Omicron. *Virology Journal*, 20(1), 191. https://doi.org/10.1186/s12985-023-02160-6

Alquraan, L., Alzoubi, K. H., & Rababa'h, S. Y. (2023). Mutations of SARS-CoV-2 and their impact on disease diagnosis and severity. *Informatics in Medicine Unlocked*, 39, 101256. https://doi.org/10.1016/j.imu.2023.101256

Ananda, A. N., Triawanti, T., Setiawan, B., Makati, A. C., Putri, J. A., & Raharjo, S. J. (2024). In silico study of the flavonoid compound of Sauropus androgynus leaves ON RNA-Dependent RNA polymerase (RdRp) SARS-CoV-2. *Aspects of Molecular Medicine*, 3, 100032. https://doi.org/10.1016/j.amolm.2023.100032

Ang, D., Kendall, R., & Atamian, H. S. (2023). Virtual and in vitro screening of natural products identifies indole and benzene derivatives as inhibitors of SARS-CoV-2 main protease (Mpro). *Biology*, 12(4), 519. https://doi.org/10.3390/biology12040519

Aziz, S., Waqas, M., Mohanta, T. K., Halim, S. A., Iqbal, A., Ali, A., … Al-Harrasi, A. (2023). Identifying nonnucleoside inhibitors of RNA-dependent RNA-polymerase of SARS-CoV-2 through per-residue energy decomposition-based pharmacophore modeling, molecular docking, and molecular dynamics simulation. *Journal of Infection and Public Health*, 16(4), 501–519. https://doi.org/10.1016/j.jiph.2023.02.009

Bejoy, J., Williams, C. I., Cole, H. J., Manzoor, S., Davoodi, P., Battaile, J. I., … Suzuki, Y. J. (2023). Effects of spike proteins on angiotensin converting enzyme 2 (ACE2). *Archives of Biochemistry and Biophysics*, 748, 109769. https://doi.org/10.1016/j.abb.2023.109769

Bekheit, M. S., Panda, S. S., & Girgis, A. S. (2023). Potential RNA-dependent RNA polymerase (RdRp) inhibitors as prospective drug candidates for SARS-CoV-2. *European Journal of Medicinal Chemistry*, 252, 115292. https://doi.org/10.1016/j.ejmech.2023.115292

Bhar, A., Jain, A., & Das, S. (2023). Natural therapeutics against SARS CoV-2: the potentiality and challenges. *Vegetos*, 36(2), 322–331. https://doi.org/10.1007/s42535-022-00401-7

Bhattacharjee, A., Thomas, S., & Palit, P. (2023). Chapter 19 - Nebulizer spray delivery of phytopharmaceutical nanosuspension via oral and nasal route: a challenging approach to fight against COVID-19. In S. Thomas, N. Kalarikkal & A. R. Abraham (Eds.), *Applications of Multifunctional Nanomaterials* (pp. 437–457). Elsevier. https://doi.org/10.1016/B978-0-12-820557-0.00017-5

Borquaye, L. S., Gasu, E. N., Ampomah, G. B., Kyei, L. K., Amarh, M. A., Mensah, C. N., … Aboagye, C. I. (2020). Alkaloids from *Cryptolepis sanguinolenta* as potential inhibitors of SARS-CoV-2 viral proteins: an in silico study. *BioMed Research International*, 2020, 5324560. https://doi.org/10.1155/2020/5324560

Brier, L., Hassan, H., Hanoulle, X., Landry, V., Moschidi, D., Desmarets, L., … Charton, J. (2023). Novel dithiocarbamates selectively inhibit 3CL protease of SARS-CoV-2 and other coronaviruses. *European Journal of Medicinal Chemistry*, 250, 115186. https://doi.org/10.1016/j.ejmech.2023.115186

Choi, A., Kots, E. D., Singleton, D. T., Weinstein, H., & Whittaker, G. R. (2024). Analysis of the molecular determinants for Furin cleavage of the spike protein S1/S2 site in defined strains of the prototype coronavirus murine hepatitis virus (MHV). *Virus Research*, 340, 199283. https://doi.org/10.1016/j.virusres.2023.199283

Cupertino, M. C., Freitas, A. N. D., Meira, G. S. B., Silva, P. A. M. d., Pires, S. d. S., Cosendey, T. d. A., … Siqueira-Batista, R. (2023). COVID-19 and one health: potential role of human and animals in SARS-CoV-2 life cycle. *Science in One Health*, 2, 100017. https://doi.org/10.1016/j.soh.2023.100017

de Loera, D. (2022). Chapter 18 - the role of traditional medicine in the fight against SARS-CoV-2. In S. Rosales-Mendoza, M. Comas-Garcia & O. Gonzalez-Ortega (Eds.), *Biomedical Innovations to Combat COVID-19* (pp. 339–385). Academic Press. https://doi.org/10.1016/B978-0-323-90248-9.00019-X

deAndrés-Galiana, E. J., Fernández-Martínez, J. L., Álvarez-Machancoses, Ó., Bea, G., Galmarini, C. M., & Kloczkowski, A. (2022). Analysis of transcriptomic responses to SARS-CoV-2 reveals plausible defective pathways responsible for increased susceptibility to infection and complications and helps to develop fast-track repositioning of drugs against COVID-19. *Computers in Biology and Medicine*, 149, 106029. https://doi.org/10.1016/j.compbiomed.2022.106029

Dhawan, M., Saied, A. A., & Sharma, M. (2023). Virus-like particles (VLPs)-based vaccines against COVID-19: where do we stand amid the ongoing evolution of SARS-CoV-2? *Health Sciences Review*, 9, 100127. https://doi.org/10.1016/j.hsr.2023.100127

Drayman, N., DeMarco, J. K., Jones, K. A., Azizi, S.-A., Froggatt, H. M., Tan, K., ... Dvorkin, S. (2021). Masitinib is a broad coronavirus 3CL inhibitor that blocks replication of SARS-CoV-2. *Science*, *373*(6557), 931–936. https://doi.org/10.1126/science.abg5827

Ebob, O. T., Babiaka, S. B., & Ntie-Kang, F. (2021). Natural products as potential lead compounds for drug discovery against SARS-CoV-2. *Natural Products and Bioprospecting*, *11*(6), 611–628. https://doi.org/10.1007/s13659-021-00317-w

El-Fakharany, E. M., El-Gendi, H., El-Maradny, Y. A., Abu-Serie, M. M., Abdel-Wahhab, K. G., Shabana, M. E., & Ashry, M. (2023). Inhibitory effect of lactoferrin-coated zinc nanoparticles on SARS-CoV-2 replication and entry along with improvement of lung fibrosis induced in adult male albino rats. *International Journal of Biological Macromolecules*, *245*, 125552. https://doi.org/10.1016/j.ijbiomac.2023.125552

Elfiky, A. A. (2020). Ribavirin, Remdesivir, Sofosbuvir, Galidesivir, and Tenofovir against SARS-CoV-2 RNA dependent RNA polymerase (RdRp): A molecular docking study. *Life Sciences*, *253*, 117592. https://doi.org/10.1016/j.lfs.2020.117592

Elgohary, A. M., Elfiky, A. A., Pereira, F., Abd El-Aziz, T. Sobeh, M., Arafa, M., El-Demerdash R. K. (2022). Investigating the structure-activity relationship of marine polycyclic batzelladine alkaloids as promising inhibitors for SARS-CoV-2 main protease (Mpro). *Computers in Biology and Medicine*, *147*, 105738. https://doi.org/10.1016/j.compbiomed.2022.105738

Emirik, M. (2022). Potential therapeutic effect of turmeric contents against SARS-CoV-2 compared with experimental COVID-19 therapies: in silico study. *Journal of Biomolecular Structure and Dynamics*, *40*(5), 2024–2037. https://doi.org/10.1080/07391102.2020.1835719

Faisal, S., Badshah, S. L., Kubra, B., Emwas, A.-H., & Jaremko, M. (2023). Alkaloids as potential antivirals. A comprehensive review. *Natural Products and Bioprospecting*, *13*(1), 4. https://doi.org/10.1007/s13659-022-00366-9

Fan, H. H., Wang, L. Q., Liu, W. L., An, X. P., Liu, Z. D., He, X. Q., ... Tong, Y. G. (2020). Repurposing of clinically approved drugs for treatment of coronavirus disease 2019 in a 2019-novel coronavirus-related coronavirus model. *Chinese Medical Journal*, *133*(9), 1051–1056. https://doi.org/10.1097/CM9.0000000000000797

Gangadharan, S., Ambrose, J. M., Rajajagadeesan, A., Kullappan, M., Patil, S., Gandhamaneni, S. H., ... Surapaneni, K. M. (2022). Repurposing of potential antiviral drugs against RNA-dependent RNA polymerase of SARS-CoV-2 by computational approach. *Journal of Infection and Public Health*, *15*(11), 1180–1191. https://doi.org/10.1016/j.jiph.2022.09.007

Ghosh, A. K., Shahabi, D., Imhoff, M. E. C., Kovela, S., Sharma, A., Hattori, S.-I., ... Mesecar, A. D. (2023). SARS-CoV-2 papain-like protease (PLpro) inhibitory and antiviral activity of small molecule derivatives for drug leads. *Bioorganic & Medicinal Chemistry Letters*, *96*, 129489. https://doi.org/10.1016/j.bmcl.2023.129489

Gildenhuys, S., Steenkamp, P. A., Sekele, L., Mosebi, S., & James, C. C. (2023). An analysis in identifying the mechanism of action of the alkaloids found in *Berberis vulgaris* on the treatment and prevention of SARS-CoV-2. *Natural Product Communications*, *18*(12), 1934578X231212835. https://doi.org/10.1177/1934578X231212835

Goyal, B., & Goyal, D. (2020). Targeting the dimerization of the main protease of coronaviruses: A potential broad-spectrum therapeutic strategy. *ACS Combinatorial Science*, *22*(6), 297–305. https://doi.org/10.1021/acscombsci.0c00058

Große, M., Ruetalo, N., Layer, M., Hu, D., Businger, R., Rheber, S., ... Schubert, U. (2021). Quinine inhibits infection of human cell lines with SARS-CoV-2. *Viruses*, *13*(4), 647. https://doi.org/10.3390/v13040647

Guruprasad, L., Naresh, G. K. R. S., & Boggarapu, G. (2023). Taking stock of the mutations in human SARS-CoV-2 spike proteins: from early days to nearly the end of COVID-19 pandemic. *Current Research in Structural Biology*, *6*, 100107. https://doi.org/10.1016/j.crstbi.2023.100107

Gyebi, G. A., Adegunloye, A. P., Ibrahim, I. M., Ogunyemi, O. M., Afolabi, S. O., & Ogunro, O. B. (2022). Prevention of SARS-CoV-2 cell entry: insight from in silico interaction of drug-like alkaloids with spike glycoprotein, human ACE2, and TMPRSS2. *Journal of Biomolecular Structure and Dynamics*, *40*(5), 2121–2145. https://doi.org/10.1080/07391102.2020.1835726

Halder, M., & Jha, S. (2023). Medicinal plants and bioactive phytochemical diversity: a fountainhead of potential drugs against human diseases. In S. Jha & M. Halder (Eds.), *Medicinal Plants: Biodiversity, Biotechnology and Conservation* (pp. 39–93). Springer Nature Singapore. https://doi.org/10.1007/978-981-19-9936-9_2

Harries, M., Jaeger, V. K., Rodiah, I., Hassenstein, M. J., Ortmann, J., Dreier, M., ... Lange, B. (2024). Bridging the gap - estimation of 2022/2023 SARS-CoV-2 healthcare burden in Germany based on multidimensional data from a rapid epidemic panel. *International Journal of Infectious Diseases*, *139*, 50–58. https://doi.org/10.1016/j.ijid.2023.11.014

He, C.-L., Huang, L.-Y., Wang, K., Gu, C.-J., Hu, J., Zhang, G.-J., ... Huang, A.-L. (2021). Identification of bis-benzylisoquinoline alkaloids as SARS-CoV-2 entry inhibitors from a library of natural products. *Signal Transduction and Targeted Therapy*, *6*(1), 131. https://doi.org/10.1038/s41392-021-00531-5

He, Z., Yuan, J., Zhang, Y., Li, R., Mo, M., Wang, Y., & Ti, H. (2023). Recent advances toward natural plants as potential inhibitors of SARS-CoV-2 targets. *Pharmaceutical Biology*, *61*(1), 1186–1210. https://doi.org/10.1080/13880209.2023.2241518

Hijikata, A., Shionyu-Mitsuyama, C., Nakae, S., Shionyu, M., Ota, M., Kanaya, S., ... Shirai, T. (2022). Evaluating cepharanthine analogs as natural drugs against SARS-CoV-2. *FEBS Open Biology*, *12*(1), 285–294. https://doi.org/10.1002/2211-5463.13337

Hillary, V. E., & Ceasar, S. A. (2023). An update on COVID-19: SARS-CoV-2 variants, antiviral drugs, and vaccines. *Heliyon*, *9*(3), e13952. https://doi.org/10.1016/j.heliyon.2023.e13952

Houeze, E. A., Wang, Y., Zhou, Q., Zhang, H., & Wang, X. (2023). Comparison study of Beninese and Chinese herbal medicines in treating COVID-19. *Journal of Ethnopharmacology*, *308*, 116172. https://doi.org/10.1016/j.jep.2023.116172

Huang, L., Yuen, T. T., Ye, Z., Liu, S., Zhang, G., Chu, H., & Yue, J. (2021). Berbamine inhibits SARS-CoV-2 infection by compromising TRPMLs-mediated endolysosomal trafficking of ACE2. *Signal Transduction and Targeted Therapy*, *6*(1), 168. https://doi.org/10.1038/s41392-021-00584-6

Infante, M., Ricordi, C., Alejandro, R., Caprio, M., & Fabbri, A. (2021). Hydroxychloroquine in the COVID-19 pandemic era: in pursuit of a rational use for prophylaxis of SARS-CoV-2 infection. *Expert Review of Anti-Infective Therapy*, *19*(1), 5–16. https://doi.org/10.1080/14787210.2020.1799785

Junqueira Ribeiro, M. M. (2024). Computer-aided drug discovery approaches in the identification of natural products against SARS-CoV-2: a review. *Current Computer-Aided Drug Design*, *20*(4), 313–324. https://doi.org/10.2174/1573409919666230329090403

Kakimoto, M., Nomura, T., Nazmul, T., Yamamoto, A., Sasaki, H., Higashiura, A., ... Sakaguchi, T. (2024). In vitro anti-severe acute respiratory syndrome coronavirus 2 effect of Ephedra przewalskii Stapf extract. *Journal of Ethnopharmacology*, *319*, 117341. https://doi.org/10.1016/j.jep.2023.117341

Kim, D. E., Min, J. S., Jang, M. S., Lee, J. Y., Shin, Y. S., Park, C. M., ... Kwon, S. (2019). Natural bis-benzylisoquinoline alkaloids-tetrandrine, fangchinoline, and cepharanthine, inhibit human coronavirus oc43 infection of mrc-5 human lung cells. *Biomolecules*, *9*(11), 696. https://doi.org/10.3390/biom9110696

Lan, Q., Yan, Y., Zhang, G., Xia, S., Zhou, J., Lu, L., & Jiang, S. (2024). Clinical development of antivirals against SARS-CoV-2 and its variants. *Current Research in Microbial Sciences*, *6*, 100208. https://doi.org/10.1016/j.crmicr.2023.100208

Le, N-T-H., De Jonghe, S., Erven, K., Neyts, J., Pannecouque, C., Vermeyen, T., ... Tuenter, E. (2023). A new alkaloid from *Pancratium maritimum* - structure elucidation using computer-assisted structure elucidation (CASE) and evaluation of cytotoxicity and anti-SARS-CoV-2 activity. *Phytochemistry Letters*, *58*, 1–7. https://doi.org/10.1016/j.phytol.2023.09.006

Le, T. T., Cao, T. Q., Ha, M. T., Han, K.-H., Kim, Y.-B., Kim, J. A., & Min, B. S. (2023). Structural characterization and SARS-CoV-2 inhibitory effects of alkaloids from the roots of *Asarum heterotropoides var. mandshuricum* (Aristolochiaceae). *Phytochemistry Letters*, *56*, 57–66. https://doi.org/10.1016/j.phytol.2023.06.005

Lee, K.-G., Lee, S.-G., Lee, H.-H., Lee, H. J., Shin, J.-S., Kim, N.-J., ... Lee, K.-T. (2015). α-Chaconine isolated from a *Solanum tuberosum* L. cv Jayoung suppresses lipopolysaccharide-induced pro-inflammatory mediators via AP-1 inactivation in RAW 264.7 macrophages and protects mice from endotoxin shock. *Chemico-Biological Interactions*, *235*, 85–94. https://doi.org/10.1016/j.cbi.2015.04.015

Li, S., Liu, W., Chen, Y., Wang, L., An, W., An, X., ... Lu, C. (2021). Transcriptome analysis of cepharanthine against a SARS-CoV-2-related coronavirus. *Briefings in Bioinformatics*, *22*(2), 1378–1386. https://doi.org/10.1093/bib/bbaa387

Liu, W., Wang, J., Wang, S., Yue, K., Hu, Y., Liu, X., ... Xu, X. (2023). Discovery of new noncovalent and covalent inhibitors targeting SARS-CoV-2 papain-like protease and main protease. *Bioorganic Chemistry*, *140*, 106830. https://doi.org/10.1016/j.bioorg.2023.106830

Lu, C.-l., Yang, L.-Q., Liu, X.-H., Jin, X.-Y., Wang, F.-X., Friedemann, T., ... Liu, J.-P. (2023). Chinese herbal medicine Shufeng Jiedu Capsule for patients with mild to moderate coronavirus disease 2019 (COVID-19): protocol for a randomized, blinded, placebo control trial. *European Journal of Integrative Medicine*, *62*, 102286. https://doi.org/10.1016/j.eujim.2023.102286

Ma, L.-L., Liu, H.-M., Liu, X.-M., Yuan, X.-Y., Xu, C., Wang, F., ... Zhang, D.-K. (2021). Screening S protein–ACE2 blockers from natural products: strategies and advances in the discovery of potential inhibitors of COVID-19. *European Journal of Medicinal Chemistry*, *226*, 113857. https://doi.org/10.1016/j.ejmech.2021.113857

Maheshwari, M., Patel, B., & Acharya, N. (2023). Phytomolecules and novel drug delivery approach for COVID-19. In R. Shegokar & Y. Pathak (Eds.), *Viral Drug Delivery Systems: Advances in Treatment of Infectious Diseases* (pp. 375–405). Springer International Publishing. https://doi.org/10.1007/978-3-031-20537-8_17

Mohanty, S. S., Sahoo, C. R., Paidesetty, S. K., & Padhy, R. N. (2023). Role of phytocompounds as the potential anti-viral agent: an overview. *Naunyn-Schmiedeberg's Archives of Pharmacology*, *396*(10), 2311–2329. https://doi.org/10.1007/s00210-023-02517-2

Mohseni, M., Bahrami, H., Farajmand, B., Hosseini, F. S., Amanlou, M., & Salehabadi, H. (2022). Indole alkaloids as potential candidates against COVID-19: an in silico study. *Journal of Molecular Modeling*, *28*(6), 144. https://doi.org/10.1007/s00894-022-05137-4

Muhammad, M., Basit, A., Wahab, A., Li, W.-J., Shah, S. T., & Mohamed, H. I. (2024). Chapter 24 - Response mechanism of plant stresses to secondary metabolites production. In K. A. Abd-Elsalam & H. I. Mohamed (Eds.), *Fungal Secondary Metabolites* (pp. 469–492). Elsevier. https://doi.org/10.1016/B978-0-323-95241-5.00012-5

Muhammed, Y., Yusuf Nadabo, A., Pius, M., Sani, B., Usman, J., Anka Garba, N., ... Sambo, M. (2021). SARS-CoV-2 spike protein and RNA dependent RNA polymerase as targets for drug and vaccine development: a review. *Biosafety and Health*, *3*(5), 249–263. https://doi.org/10.1016/j.bsheal.2021.07.003

Ohashi, H., Watashi, K., Saso, W., Shionoya, K., Iwanami, S., Hirokawa, T., ... Kim, K. S. (2020). Multidrug treatment with nelfinavir and cepharanthine against COVID-19. *BioRxiv*, 2020. https://doi.org/10.1101/2020.04.14.039925

Okamoto, M., Ono, M., & Baba, M. (1998). Potent inhibition of HIV type 1 replication by an antiinflammatory alkaloid, cepharanthine, in chronically infected monocytic cells. *AIDS Research and Human Retroviruses*, *14*(14), 1239–1245. https://doi.org/10.1089/aid.1998.14.1239

Oner, E., Al-Khafaji, K., Mezher, M. H., Demirhan, I., Suhail Wadi, J., Belge Kurutas, E., ... Choowongkomon, K. (2023). Investigation of berberine and its derivatives in SARS-CoV-2 main protease structure by molecular docking, PROTOX-II and ADMET methods: in machine learning and in silico study. *Journal of Biomolecular Structure and Dynamics*, *41*(19), 9366–9381. https://doi.org/10.1080/07391102.2022.2142848

Palshetkar, A. D., Rasal, A. U., Murugan, A., & Desai, N. D. (2024). Natural product-derived phytochemicals as potential inhibitors of angiotensin converting enzyme 2 (ACE2): promising drug candidates for COVID-19. *Current Drug Therapy*, *19*(1), 13–19. https://doi.org/10.2174/157488551866623050210315

Phukan, M. M., Sangma, S. R., Kalita, D., Bora, P., Das, P. P., Manoj, K., ... Meenakshi Sundaram, K. (2023). Chapter 8 - alkaloids and terpenoids: synthesis, classification, isolation and purification, reactions, and applications. In C. Verma & D. K. Verma (Eds.), *Handbook of Biomolecules* (pp. 177–213). Elsevier. https://doi.org/10.1016/B978-0-323-91684-4.00017-7

Pokharkar, O., Anumolu, H., Zyryanov, G. V., & Tsurkan, M. V. (2023). Natural products from red algal genus Laurencia as potential inhibitors of RdRp and nsp15 enzymes of SARS-CoV-2: an in silico perspective. *Microbiology Research*, *14*(3), 1020–1048. https://doi.org/10.3390/microbiolres14030069

Powder-George, Y. L. (2024). Chapter 8 - alkaloids. In S. B. McCreath & Y. N. Clement (Eds.), *Pharmacognosy (Second Edition)* (pp. 167–209). Academic Press. https://doi.org/10.1016/B978-0-443-18657-8.00005-0

Pradeep Kumar, R. (2024). Need and possibilities of phytocompounds against SARS-CoV-2: recent advances in COVID-19 therapy [Short survey]. *Current Traditional Medicine*, *10*(1), 144–160, e210223213863. https://doi.org/10.2174/2215083809666230221151814

Raghav, P. K., Mann, Z., Ahluwalia, S. K., & Rajalingam, R. (2023). Potential treatments of COVID-19: drug repurposing and therapeutic interventions. *Journal of Pharmacological Sciences*, *152*(1), 1–21. https://doi.org/10.1016/j.jphs.2023.02.004

Raman, K., Rajagopal, K., Swaminathan, G., Jupudi, S., Dhama, K., Barua, R., ... Khandaker, M. U. (2023). A critical review on the potency of phytoconstituents in the management of COVID-19. *Journal of Pure and Applied Microbiology*, *17*(3), 1320–1340. https://doi.org/10.22207/JPAM.17.3.38

Rani, J., Bhargav, A., Khan, F. I., Ramachandran, S., Lai, D., & Bajpai, U. (2022). In silico prediction of natural compounds as potential multitarget inhibitors of structural proteins of SARS-CoV-2. *Journal of Biomolecular Structure and Dynamics*, *40*(22), 12118–12134. https://doi.org/10.1080/07391102.2021.1968497

Rehman, H. M., Sajjad, M., Ali, M. A., Gul, R., Naveed, M., Aslam, M. S., ... Amin, A. (2023). Identification of RdRp inhibitors against SARS-CoV-2 through E-pharmacophore-based virtual screening, molecular docking and MD simulations approaches. *International Journal of Biological Macromolecules*, *237*, 124169. https://doi.org/10.1016/j.ijbiomac.2023.124169

Rizka Nurcahyaningtyas, H., Irene, A., Tri Wibowo, J., Yunovilsa Putra, M., & Yanuar, A. (2023). Identification of potential Indonesian marine invertebrate bioactive compounds as TMPRSS2 and SARS-CoV-2 Omicron spike protein inhibitors through computational screening. *Arabian Journal of Chemistry*, *16*(9), 104984. https://doi.org/10.1016/j.arabjc.2023.104984

Sadhu, S., Dandotiya, J., Dalal, R., Khatri, R., Mykytyn, A. Z., Batra, A., … Awasthi, A. (2023). Fangchinoline inhibits SARS-CoV-2 and MERS-CoV entry. *Antiviral Research*, *220*, 105743. https://doi.org/10.1016/j.antiviral.2023.105743

Sayed, A. M., Ibrahim, A. H., Tajuddeen, N., Seibel, J., Bodem, J., Geiger, N., … Abdelmohsen, U. R. (2023). Korupensamine A, but not its atropisomer, korupensamine B, inhibits SARS-CoV-2 in vitro by targeting its main protease (Mpro). *European Journal of Medicinal Chemistry*, *251*, 115226. https://doi.org/10.1016/j.ejmech.2023.115226

Shahali, A., Azar, Z. J., & Ardeshir, R. A. (2024). A comprehensive review on potentially therapeutic agents against COVID-19 from natural sources. *Current Traditional Medicine*, *10*(1), 91–101, e030223213428. https://doi.org/10.2174/2215083809666230203142343

Solo, P., & Arockia doss, M. (2023). Imidazole-based alkaloids from marine sponges (*Leucetta and Clathrina*) as potential inhibitors targeting SARS-CoV-2 main protease: an in silico approach. *Polycyclic Aromatic Compounds*. https://doi.org/10.1080/10406638.2023.2182796

Song, J.-B., Zhao, L.-Q., Wen, H.-P., & Li, Y.-P. (2023). Herbal combinations against COVID-19: a network pharmacology, molecular docking and dynamics study. *Journal of Integrative Medicine*, *21*(6), 593–604. https://doi.org/10.1016/j.joim.2023.09.001

Sugano, A., Murakami, J., Kataguchi, H., Ohta, M., Someya, Y., Kimura, S., & Takaoka, Y. (2023). In silico binding affinity of the spike protein with ACE2 and the relative evolutionary distance of S gene may be potential factors rapidly obtained for the initial risk of SARS-CoV-2. *Microbial Risk Analysis*, *25*, 100278. https://doi.org/10.1016/j.mran.2023.100278

Tarigan, S., Dharmayanti, N. I., Sugiartanti, D., Putri, R., AndrianiNuradji, H., … Djufri, F. (2023). Characterization of two linear epitopes SARS CoV-2 spike protein formulated in tandem repeat. *PLoS One*, *18*(1), e0280627. https://doi.org/10.1371/journal.pone.0280627

Valipour, M., Hosseini, A., Di Sotto, A., & Irannejad, H. (2023). Dual action anti-inflammatory/antiviral isoquinoline alkaloids as potent naturally occurring anti-SARS-CoV-2 agents: a combined pharmacological and medicinal chemistry perspective. *Phytotherapy Research*. https://doi.org/10.1002/ptr.7833

Verma, V. A., Saundane, A. R., Meti, R. S., & Vennapu, D. R. (2021). Synthesis of novel indolo[3,2-c]isoquinoline derivatives bearing pyrimidine, piperazine rings and their biological evaluation and docking studies against COVID-19 virus main protease. *Journal of Molecular Structure*, *1229*, 129829. https://doi.org/10.1016/j.molstruc.2020.129829

Wansri, R., Lin, A. C. K., Pengon, J., Kamchonwongpaisan, S., Srimongkolpithak, N., Rattanajak, R., … Hengphasatporn, K. (2022). Semisynthesis of N-aryl amide analogs of piperine from *Piper nigrum* and evaluation of their antitrypanosomal, antimalarial, and anti-SARS-CoV-2 main protease activities. *Molecules*, *27*(9), 2841. https://doi.org/10.3390/molecules27092841

White, M. A., Lin, W., & Cheng, X. (2020). Discovery of COVID-19 inhibitors targeting the SARS-CoV-2 Nsp13 helicase. *Journal of Physical Chemistry Letters*, *11*(21), 9144–9151. https://doi.org/10.1021/acs.jpclett.0c02421

Wilkinson, B., Patel, K. S., Smith, K., Walker, R., Wang, C., Greene, A. M., … Chen, R. T. (2023). A Brighton collaboration standardized template with key considerations for a benefit/risk assessment for the Novavax COVID-19 Vaccine (NVX-CoV2373), a recombinant spike protein Vaccine with Matrix-M adjuvant to prevent disease caused by SARS-CoV-2 viruses. *Vaccine*, *41*(45), 6762–6773. https://doi.org/10.1016/j.vaccine.2023.07.040

Wink, M. (2020). Potential of DNA intercalating alkaloids and other plant secondary metabolites against SARS-CoV-2 causing COVID-19. *Diversity*, *12*(5), 175. https://doi.org/10.3390/d12050175

Xiao, B., Xin, Z., & Wang, L. (2024). COVID-19's influence on life history strategy: insights from cross-temporal meta-analysis and experimental research. *Personality and Individual Differences*, *219*, 112505. https://doi.org/10.1016/j.paid.2023.112505

Xu, L., Chen, R., Liu, J., Patterson, T. A., & Hong, H. (2023). Analyzing 3D structures of the SARS-CoV-2 main protease reveals structural features of ligand binding for COVID-19 drug discovery. *Drug Discovery Today*, *28*(10), 103727. https://doi.org/10.1016/j.drudis.2023.103727

Yang, L., & Wang, Z. (2023). Bench-to-bedside: innovation of small molecule anti-SARS-CoV-2 drugs in China. *European Journal of Medicinal Chemistry*, *257*, 115503. https://doi.org/10.1016/j.ejmech.2023.115503

Yonamine, D. K., Narciso dos Reis, V. E., Feu, A. E., de Souza Borges, W., Cardoso, C. L., & Dinamarco, T. M. (2023). Ligand fishing approach to explore Amaryllidaceae alkaloids as potential antiviral candidates targeting SARS-CoV-2 Nsp4. *Journal of Pharmaceutical and Biomedical Analysis*, 115935. https://doi.org/10.1016/j.jpba.2023.115935

Yoon, H.-C. (2023). Herbal medicine use in Republic of Korea to alleviate side effects of COVID-19 vaccines: a cross-sectional study. *Journal of Integrative Medicine*, *21*(4), 361–368. https://doi.org/10.1016/j.joim.2023.06.002

Yoosefian, M., Dashti, R., Mahani, M., Montazer, L., & Mir, A. (2023). A suitable drug structure for interaction with SARS-CoV-2 main protease between boceprevir, masitinib and rupintrivir; a molecular dynamics study. *Arabian Journal of Chemistry*, *16*(9), 105051. https://doi.org/10.1016/j.arabjc.2023.105051

You, L.-Z., Dai, Q.-Q., Zhong, X.-Y., Yu, D.-D., Cui, H.-R., Kong, Y.-F., … Shang, H.-C. (2023). Clinical evidence of three traditional Chinese medicine drugs and three herbal formulas for COVID-19: a systematic review and meta-analysis of the Chinese population. *Journal of Integrative Medicine*, *21*(5), 441–454. https://doi.org/10.1016/j.joim.2023.08.001

Yu, H., Guan, F., Miller, H., Lei, J., & Liu, C. (2023). The role of SARS-CoV-2 nucleocapsid protein in antiviral immunity and vaccine development. *Emerging Microbes & Infections*, *12*(1), e2164219. https://doi.org/10.1080/22221751.2022.2164219

Zhang, C. H., Wang, Y. F., Liu, X. J., Lu, J. H., Qian, C. W., Wan, Z. Y., … Qi, S. Y. (2005). Antiviral activity of cepharanthine against severe acute respiratory syndrome coronavirus in vitro. *Chinese Medical Journal*, *118*(6), 493–496.

Zhang, L. (2023). Biomedical equipment, vaccine and drug in the prevention, diagnosis and treatment of COVID-19. *Heliyon*, *9*(7), e18089. https://doi.org/10.1016/j.heliyon.2023.e18089

Zhang, L., Zhang, S., Liao, L., Tang, H., Wang, W., Yin, F., … Zhao, Y. (2023). Discovery, structure, and mechanism of the (R, S)-norcoclaurine synthase for the chiral synthesis of benzylisoquinoline alkaloids. *ACS Catalysis*, *13*(22), 15164–15174. https://doi.org/10.1021/acscatal.3c03296

Zhang, S., Huang, W., Ren, L., Ju, X., Gong, M., Rao, J., … Wang, J. (2022). Comparison of viral RNA–host protein interactomes across pathogenic RNA viruses informs rapid antiviral drug discovery for SARS-CoV-2. *Cell Research*, *32*(1), 9–23. https://doi.org/10.1038/s41422-021-00581-y

Zhang, Y., Wang, D., Xiang, Q., Hu, X., Zhang, Y., Wu, L., … Li, J. (2024). A potent neutralizing nanobody targeting a unique epitope on the receptor-binding domain of SARS-CoV-2 spike protein. *Virology*, *589*, 109925. https://doi.org/10.1016/j.virol.2023.109925

14 Herbal Medicines for COVID-19

The Impact of Terpenoids on Protein Targets

Abhiruj Navabhatra

14.1 INTRODUCTION

Herbal medicine has gained increased recognition and research support, especially during the coronavirus disease 2019 (COVID-19) pandemic, where its effectiveness has become a significant research focus. Derived mainly from plants, these remedies have shown promise in improving various health conditions, including COVID-19 symptoms. A meta-analysis and various studies, such as those by Demeke et al. [1], have highlighted the reduction of symptoms such as fever and cough. Many have used these herbal methods as preventive and therapeutic measures, with *Echinacea purpurea* and curcumin from turmeric being notable for their antiviral and anti-inflammatory properties. Other herbs, such as *Cinchona* sp. and xanthorrhizol, along with traditional Persian medicine and Lianhuaqingwen capsules, have also been identified for their potential benefits. In places such as Indonesia, herbal practices are deeply rooted and widely practiced, as seen with their significant turmeric production. Safety assessments, including clinical trials, report minimal severe side effects. However, the complexity of botanical extracts, with their bioactive compounds, requires careful research to ensure effective and safe use. Consequently, patients should be cautious and seek medical advice before using herbal treatments for COVID-19. Healthcare professionals need to increase their knowledge of herbal medicines to guide patients effectively, as many currently lack a thorough understanding of these agents [2]. Terpenes are organic compounds that can be classified into multiple groups based on their chemical structures, including monoterpenes (e.g., limonene), sesquiterpenes (e.g., elemene), diterpenes (e.g., camphene), and polyterpenes (Figure 14.1). These compounds are soluble in organic solvents and have unique configurations that facilitate specific chemical reactions, which are crucial for their separation and identification. Terpenes have medical significance, particularly in developing new cancer treatments, with elemene showing promise owing to its minimal side effects and effectiveness against various cancers. Beyond medicine, terpenes are used in products ranging from adhesives to pesticides and are important in industries such as agriculture and chemistry. Understanding their biosynthesis, heavily influenced by microbial interactions and the production of precursor molecules such as isopentenyl diphosphate (IPP) and its congener, dimethylallyl diphosphate (DMAPP), is key to controlling terpene content. This involves manipulating the Daqu fermentation process and the enzymes involved [3]. This research chapter will delve into the role of terpenoids against viral infections, including their potential against COVID-19, by examining their interactions with vital viral proteins and their impact on the viral life cycle. It will address the challenges of incorporating terpenoid-based treatments into contemporary medicine and their broader implications for antiviral research and global health.

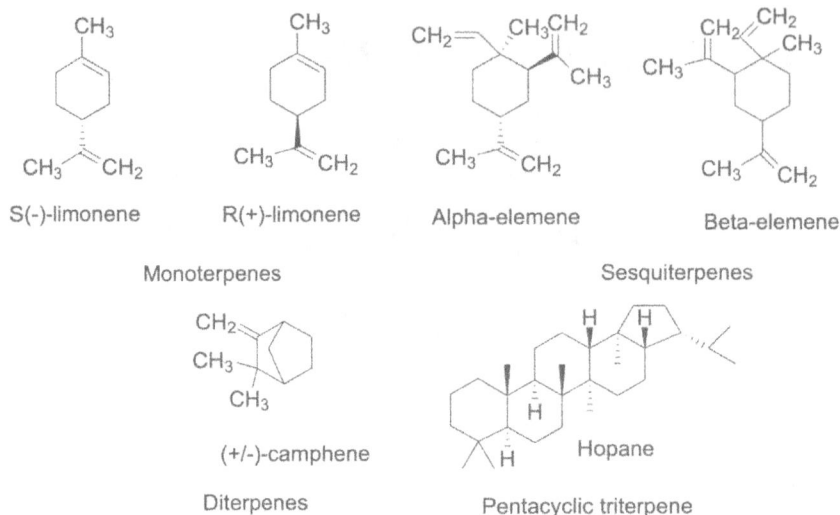

FIGURE 14.1 An example of the chemical structures of terpenoids.

14.2 UNVEILING VIRAL PROTEIN TARGETS

14.2.1 OVERVIEW OF SARS-CoV-2

The COVID-19 pandemic, caused by the severe acute respiratory syndrome coronavirus 2 (SARS-CoV-2) virus related to the 2002–2004 severe acute respiratory syndrome (SARS) virus, has led to significant scientific engagement, as we face over 6 million deaths and widespread impact over 3 years. Extensive genomic analysis has provided insight into the virus's evolution into new strains with varying characteristics, aiding in the development of containment strategies and medical responses. Challenges persist, including new variants, increased transmission, vaccine hesitancy, and nonadherence to health measures [4]. Misidentification of symptoms and reluctance toward testing, along with the questionable accuracy of self-tests, underscore the need for better diagnostics. More than 50 vaccines have been approved, but the virus's rapid mutations, such as the Omicron variant, challenge their long-term effectiveness and demand ongoing vaccine evaluation. The urgency for precise diagnostics and effective, safe antivirals is paramount, alongside concerns about increased antibiotic use and resulting resistance. The pandemic highlights the necessity for better preparedness for future outbreaks, including rapid sharing of viral data, improved medical supply chains, and agile vaccine technologies such as mRNA [5].

14.2.2 KEY PROTEIN TARGETS IN SARS-CoV-2 AND THEIR ROLE IN THE VIRUS LIFE CYCLE AND DISEASE PROGRESSION

14.2.2.1 Spike (S) Protein

The spike (S) protein of coronaviruses, essential for infection, is a glycosylated trimeric class I fusion protein with three main domains–S1, S2, and S2′. S1 binds to the angiotensin-converting enzyme 2 (ACE2) receptor in human cells, S2 fuses the virus with the cell membrane, and S2′ assists this fusion. The S1 domain's receptor binding domain (RBD) is crucial for this binding, with certain amino acids playing a key role in binding efficiency. The RBD can switch between "down" (closed) and "up" (open and receptive) configurations, affecting the virus's ability to bind to receptors. The RBD of SARS-CoV-2 is variable, affecting its binding and disease severity [6].

Structural differences in the S proteins of different coronaviruses could inspire new treatments. Antibodies targeting the S protein may neutralize the virus, and mRNA vaccines work by inducing cells to produce parts of the S protein, boosting immunity. The virus enters cells via RBD attachment to ACE2, followed by membrane fusion or endocytosis. The S protein is activated by cleavage through cellular proteases such as TMPRSS2 for direct fusion or furin and cathepsin B/L for endocytosis, leading to viral RNA release inside host cells [7].

14.2.2.2 Membrane (M) Protein

The membrane (M) protein of coronaviruses is a transmembrane glycoprotein with three transmembrane domains and 222 amino acids, which are crucial for the virus's structure and assembly. It interacts with other viral proteins (envelope [E], nucleoprotein [N], and S) to encapsulate RNA and helps maintain cellular balance. It also stimulates the production of neutralizing antibodies, especially in SARS. While similar across various coronaviruses, such as Bat coronaviruses (BAT-CoV), SARS-CoV, and SARS-CoV-2, there are subtle differences compared to Middle East respiratory syndrome coronavirus (MERS-CoV) [8]. Although critical for virus shape and assembly, the exact functions of the M protein have been partly mysterious. Recent cryo-electron microscopy has revealed that the M protein can take on two shapes and form dimers resembling a mushroom with a tri-helical structure and additional domains. It is capable of forming complex structures aided by a "hinge" region that allows structural flexibility. The M protein is structurally similar to the SARS-CoV-2 open reading frame 3a protein (ORF3a), ORF3a, and is key in bringing together the nucleocapsid protein and viral RNA, with its electropositive domain playing a significant role. These insights into the structure and function of the M protein could lead to new targeted treatments [9, 10].

14.2.2.3 Nucleoprotein (N)

The nucleoprotein (N) of coronaviruses has a complex structure with three main parts: an RNA-binding domain at the N-terminus, a serine/arginine-rich central linker, and a C-terminal domain that enables dimerization. This protein is essential for packing viral RNA into the nucleocapsid and releasing the viral genome into the ER-to-Golgi intermediate compartment (ERGIC). The N-terminal domain, particularly a segment of approximately 140 amino acids, is structured for optimal RNA binding, making it a target for COVID-19 treatments. N proteins play diverse roles, including in viral RNA transcription and replication, cell cycle regulation, cytoskeleton reconfiguration, and triggering host cell apoptosis, making them potential drug targets. Similarities in the N proteins of different coronaviruses, such as BAT-CoV, SARS-CoV, and SARS-CoV-2, suggest that treatments targeting these proteins could be broadly effective. The N protein also assists in translating the viral genome, forming the replication-transcription complex, and creating replication organelles linked to the endoplasmic reticulum (ER). Viral RNAs are replicated within double-membrane vesicles and serve as templates for new RNA genomes and subgenomic RNAs, which are essential for protein synthesis. Virions are then released from the host cell through exocytosis or lysosomal transport. The N protein is key in managing these steps, manipulating cellular functions, inducing immune responses, and using the host's mechanisms to incorporate RNA into new virus particles. Overall, the N protein is vital in the SARS-CoV-2 life cycle, from cell entry to virion assembly and replication [11].

14.2.2.4 Replicase Polyprotein

While the significance of structural proteins in relation to SARS-CoV-2 has been duly noted in prior literature, it is imperative to underscore the critical role of replicase proteins in the virus's pathogenic progression. A predominant fraction of the replicase genome is dedicated to the ORF1ab gene, which gives rise to two principal polyproteins, denoted pp1a and pp1ab. The replicase polyprotein is delineated by three salient domains: the macro domain, the papain-like domain (PLpro), and the main protease (Mpro). Collectively, these domains are indispensable in the degradation of host RNA and the subsequent replication of viral RNA. Following their synthesis, these polyproteins are meticulously processed into 16 nonstructural proteins (NSPs) via the enzymatic activity of proteolytic agents. Notably, the Mpro and PLpro enzymes specifically cleave the C-terminal and N-terminal ends of the aforementioned polyproteins (PPs), respectively. Embedded within ORF 1ab is a specialized region, the RNA-dependent RNA polymerase (RdRp), which assumes the cardinal responsibility of transcribing and replicating viral RNA. A comprehensive analysis of the extant literature delineates a pronounced sequence congruence in 3C-like protease (3CLpro) across BAT-CoV, SARS-CoV, and SARS-CoV-2. This discernible similarity markedly contrasts with MERS-CoV, which remains devoid of such sequence parallelism [6].

14.2.2.5 Main Protease (Mpro or 3CLpro) and Papain-Like Protease (PLpro)

14.2.2.5.1 Mpro or 3CLpro

Within the context of SARS-CoV-2's biological mechanisms, the main protease, Mpro, emerges as a salient entity, rendering it a formidable target for therapeutic interventions aimed at mitigating COVID-19. This enzyme, integral to the virus's functionality, synergizes with a plethora of viral components, orchestrating the replication and transcription of the RNA intrinsic to the virus. The pivotal role of Mpro in the intricate processing of polyproteins–indispensable for viral assembly –has subsequently elevated its stature within the realm of coronavirus pharmacological research. From a structural perspective, the Mpro of SARS-CoV-2 manifests as a protein intricately composed of 306 amino acid residues. This assembly delineates into three distinct domains: domain I (residues 8–101), domain II (residues 102–184), and domain III (residues 201–303). Bridging the spatial gap between domains II and III is a loop encapsulating 15 residues, distinguished by its Cys-His catalytic dyad. Intriguingly, the substrate-binding locus of Mpro, reminiscent of its counterparts in other coronaviruses, is strategically positioned at the interstice between domains I and II. A comparative analysis underscores a profound sequence congruence between the Mpro of SARS-CoV and that of SARS-CoV-2, evidenced by an approximate sequence alignment of 96%. In its operational capacity, Mpro meticulously cleaves polyproteins, culminating in the genesis of NSPs. These NSPs are instrumental in the formation of the replicase-transcriptase complex, an essential cog in the virus lifecycle. The enzymatic specificity of Mpro is accentuated by its selective targeting of fewer than 11 discrete cleavage sites present on the replicase 1ab polyprotein. Contemporary investigations have illuminated the crystal structure of SARS-CoV-2 Mpro, offering a promising avenue for bespoke drug design against COVID-19. The strategic emphasis on the conserved residues within Mpro might indeed catalyze the evolution of efficacious therapeutic modalities against SARS-CoV-2 [12].

14.2.2.5.2 PLpro

The SARS-CoV-2 genome is approximately 30 kilobases long, with at least 12 ORFs. The last third encodes four key structural proteins (S, E, M, and N) and several accessory proteins, while the first two-thirds encode two large replicase polyproteins, pp1a and pp1ab, which are cleaved into 16 NSPs by virus-specific enzymes. These proteins form a replication complex near double-membrane vesicles on the ER. The protease PLpro, which is essential for processing polyproteins and conserved across various viruses, is present in only one form in SARS-CoV-2, unlike some coronaviruses, which have two isoforms. The PLpro of SARS-CoV-2 cuts at specific points between NSPs, which are critical for viral RNA synthesis and lifecycle maintenance. Enzymatic studies show that the S2 subsite of PLpro has a strong preference for glycine, the S4 subsite for hydrophobic amino acids, and the S3 subsite has broad specificity, highlighting potential targets for drug development [13, 14].

14.2.2.6 Envelope (E) Protein

Both SARS-CoV-1 and SARS-CoV-2 E proteins undergo N-glycosylation at the same residue and cysteine palmitoylation, affecting intracellular movement and interactions. These proteins form ion channels, primarily in the Golgi and ER, crucial for viral architecture and ion transport, influencing virus-host interactions and inflammation. Viruses without these ion channels show reduced virulence, underscoring the role of the E protein in pathogenicity. Efforts to target the E protein to reduce the severity of infection have led to live attenuated vaccine research, showing promising immune responses. A unique motif in the E protein interacts with host proteins –post synaptic density protein (PSD95), Drosophila Disc large tumor suppressor (Dlg1), and zonula occludens-1 protein (zo-1), collectively known as PDZ proteins. This interaction affects virus spread and lung tissue integrity. Despite similarities, SARS-CoV-2 might interact with more host proteins due to variations near this motif, potentially altering virulence [15]. The SARS-CoV-2 lifecycle involves its genome directing the creation of double-membrane vesicles from the ER, leading to membrane networks where genomic replication occurs. Structural proteins N, S, M, and E transit through the ER and ERGIC, with N and M proteins driving virion assembly and release. Although a minor component, the E protein is integral to the coronavirus lifecycle and virion assembly. Research is ongoing to understand the full role and potential of the E protein as a drug target for both SARS-CoV-1 and SARS-CoV-2, given their shared conserved domains [16].

14.3 TERPENOIDS: NATURE'S MEDICINAL TREASURE TROVE

14.3.1 Brief Overview of Terpenoids

Plants have evolved complex secondary metabolic processes to adapt to their environments, playing key roles in interactions and defense against threats such as insects and pathogens. Terpenoids, a major group of these metabolites, are central to plant defense and ecological interactions. Understanding the biosynthesis and regulation of terpenoids is crucial for insights into plant defense strategies. Terpenes, the basic hydrocarbons, are vital for plant health and ecological balance, while terpenoids, their oxygenated derivatives, feature various functional groups and undergo structural changes such as oxidation and modification of methyl groups [17]. More than 80,000 terpenoids have been identified as key volatile organic compounds in plants, mainly synthesized through processes similar to isoprene formation. There are two primary pathways: the methylerythritol-phosphate (MEP) pathway in plastids for monoterpenes and diterpenes and the mevalonic acid (MVA) pathway for sesquiterpenes. Due to their functional diversity, terpenoids are of significant interest in botanical and biochemical research. Studies show that some terpenes, such as carvacrol, carvone, eugenol, geraniol, and thymol, are effective antimicrobials against bacteria such as *Staphylococcus aureus*, working by disrupting cells and hindering biological functions such as protein and DNA synthesis. Monoterpenoids, in particular, have antibacterial effects, curbing microbial growth and activity. These phytochemicals, also found in aromatic and medicinal plants, offer eco-friendly alternatives to traditional pesticides and could enhance flavors in food, highlighting their potential in disease resistance and food processing industries [18].

14.3.2 Medicinal Properties and Therapeutic Potential of Terpenoids

The full potential of terpenoids in medicine has yet to be confirmed, but combining "omics" and molecular network pharmacology could unlock their mechanisms and structure-activity relationships, accelerating their development into drugs. Currently, approximately 50,000 terpenoids are known, with plants as their main source, and some have attracted commercial interest for their health and nutritional benefits. These compounds are promising for advancing pharmaceuticals and creating innovative drug delivery systems, potentially leading to their central role in future therapies.

TABLE 14.1

Medicinal Plants that Contain Terpenoids and Their Pharmacological Effects

Plant Name	Terpenoids	Pharmacologic(al) Properties	References
Stevia rebaudiana	β-pinene, α-terpinolene, β-elemene	Ameliorates insulin resistance	[21]
Artemisia annua	Artemisinin	Inhibition in ACE2 overexpressing in human lung cells	[22]
Cannabis indica	β-myrcene	Anxiolytic, antioxidant, anti-inflammatory, and analgesic	[23]
Citrus × latifolia	D-limonene	Enhances therapeutic effect of tamoxifen in breast cancer cells	[24]
Panax quinquefolius	Ginsenoside	Inhibits ACE2 and TMPRSS2	[25]
Salvia miltiorrhiza	Tanshinone IIA	Inhibits ARS-CoV-2 nsp14 protein	[26]
Stevia rebaudiana	β-pinene, α-terpinolene, β-elemene	Ameliorates insulin resistance	[21]
Glycyrrhiza species	Glycyrrhizin	Binds to the ACE2 enzyme and blocks the entry of SARS-CoV-2, inhibits the viral replication process	[27]

Terpenoids' applications in wellness and cosmetics also suggest significant economic opportunities. As green chemistry advances, terpenes and essential oils (EOs) are fostering collaborations across science, industry, and academia. Terpenoids offer eco-friendly alternatives to chemical pesticides, which have led to resistance and health concerns. Compounds such as carvacrol, limonene, and linalool show promise as bioagents against pests. Moreover, terpenes are recognized for their antimicrobial effects, with plants such as pine, cumin, and rosemary being rich sources. In malaria research, terpenes such as β-myrcene and limonene are being studied as cost-effective treatments, with artemisinin from *Artemisia annua* already a breakthrough antimalarial. In cancer, dietary terpenes such as D-limonene and perillyl alcohol show potential against various cancers. Antiviral research on terpenoids is emerging, with some showing effects against viruses such as herpes simplex. For metabolic diseases such as type 2 diabetes, compounds from plants such as stevia are being explored for their therapeutic effects, and terpenes are being investigated for their anti-inflammatory and analgesic properties. Cardiovascular health could benefit from terpenoids such as tanshinone IIA from *Salvia miltiorrhiza*, while tuberculosis might be addressed by isosteviol derivatives. The study of terpenoids is expanding, indicating their versatile potential in various health domains [19, 20]. Table 14.1 provides a sample listing of various medicinal plants that contain terpenoids, along with a description of their health benefits.

14.4   MECHANISMS OF ACTION: TERPENOIDS AND VIRAL

14.4.1   Terpenoids in Common Medicinal Plants and Their Potential Role Against COVID-19

The COVID-19 pandemic has shifted global priorities to finding effective treatments against the virus and its variants. The SARS-CoV-2 main protease enzyme, Mpro, is vital for viral replication and is a key target for drug development. Research has focused on plant-derived terpenes, with studies on terpenoids from *Xylocarpus moluccensis* showing that they can inhibit Mpro. Out of 67 terpenoids analyzed, a few, including angolensic acid methyl ester and moluccensin V, have shown high binding affinity to Mpro, with strong docking scores and stable interactions, suggesting their potential as drug candidates. These terpenoids also meet drug-likeness and absorption, distribution, metabolism, and excretion-toxicity (ADMET) criteria, indicating their promise as therapies warranting further *in vitro* testing [28]. Ursolic acid, a compound with anti-inflammatory, antibacterial, and anticancer properties, has shown potential to affect the SARS-CoV-2 Mpro enzyme, which is key to the virus's life cycle. Supported by molecular simulations, ursolic acid demonstrated a high affinity for the enzyme, indicating possible therapeutic use. Similarly, a terpenoid from the marine sponge *Cacospongia mycofijiensis* outperformed some peptide inhibitors, with stable hydrogen bond interactions suggesting effectiveness against COVID-19. These findings highlight the therapeutic promise of natural compounds in combatting COVID-19 [29]. A study found that seven tanshinones from *S. miltiorrhiza* can act as noncompetitive inhibitors against SARS-CoV proteases, suggesting their use in antiviral drug development. Tanshinone IIA also shows effectiveness against HIV-1 and viral myocarditis, while cryptotanshinone is active against influenza A. Furthermore, quinone-methide triterpenes from *Tripterygium regelii* and terpenes from *Chamaecyparis obtusa*, *Juniperus formosana*, and *Cryptomeria japonica* have been identified as inhibitors of SARS-CoV in cell assays. Glycyrrhizin from the *Glycyrrhiza genus* has been confirmed to inhibit SARS-CoV adsorption, penetration, and replication. Despite their chemical similarities, the diversity in their molecular structures and functional groups makes it difficult to determine a clear structure-activity relationship. However, these compounds have shown significant antiviral effects and safety, highlighting their potential as leads for anticoronavirus drugs [30]. In a detailed study, researchers evaluated eight diterpenoid compounds from *Torreya nucifera* for their ability to inhibit a protease linked to SARS-CoV (or SARS-CoV-2). These compounds, including the most effective compound, ferruginol (IC$_{50}$ of 49.6 μM), showed varying degrees of inhibitory action, with IC$_{50}$ values between 49.6 and 283.5 μM. Computational analysis further revealed two terpenoid compounds (22-hydroxyhopan-3-one and 6-oxoisoiguesterin) from African plants and two alkaloids as strong inhibitors of the 3CLpro protein essential for SARS-CoV-2 replication, highlighting their therapeutic potential [31]. Andrographolide, an antibiotic diterpenoid from *Andrographis paniculata* of the *Acanthaceae* family, shows promise against inflammation, cancer, obesity, and diabetes. Sa-Ngiamsuntorn et al. [32] found that it inhibits SARS-CoV-2 replication in lung cells in a dose-dependent way, with effectiveness comparable to remdesivir, having an IC$_{50}$ of 0.034 μmol/L. It disrupts multiple stages of the virus's life cycle, especially later stages. Enmozhi et al. [33] confirmed andrographolide's safety, noting low toxicity. Its strong affinity for the virus's S protein suggests that it may block viral entry into cells, highlighting its potential as an antiviral agent. An example of medicinal plants containing terpenoids known for their anti-SARS-CoV-2 properties is listed in Figure 14.2.

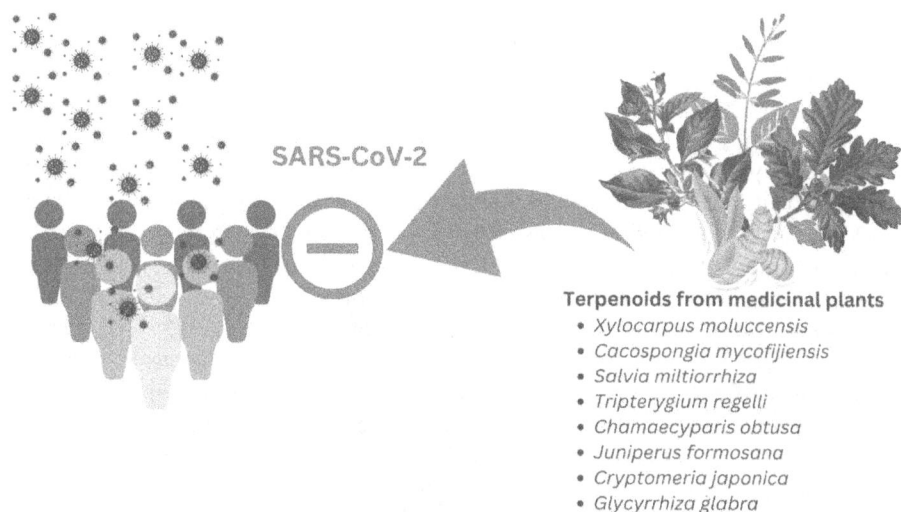

Terpenoids from medicinal plants
- *Xylocarpus moluccensis*
- *Cacospongia mycofijiensis*
- *Salvia miltiorrhiza*
- *Tripterygium regelli*
- *Chamaecyparis obtusa*
- *Juniperus formosana*
- *Cryptomeria japonica*
- *Glycyrrhiza glabra*

FIGURE 14.2 A list of medicinal plants containing terpenoids known for their anti-SARS-CoV-2 properties.

14.4.2   MOLECULAR MECHANISMS: HOW TERPENOIDS INTERACT WITH KEY SARS-CoV-2 PROTEINS

With more than 80,000 variants, terpenes have shown a range of pharmacological effects but are understudied as treatments for diseases such as SARS-CoV-2. This research focuses on their role as therapeutic agents using data from studies on various human coronaviruses, molecular docking, and analysis of mechanistic pathways. A key study examined 14 limonoids and terpenoids and found that deacetylnomilin, ichangin, nomilin, and β-amyrin interacted significantly with SARS-CoV-2's Mpro, affecting viral replication. In particular, deacetylnomilin and ichangin disrupted the replication process by interacting with the enzyme's active site. However, these compounds showed less stable interactions with the S protein, raising questions about their effectiveness in blocking virus attachment to the ACE2 receptor. Preliminary *in silico* studies suggest that these triterpenoids could be strong inhibitors of the Mpro enzyme, warranting further *in vitro* and *in vivo* testing [34]. In modern drug discovery, thoroughly screening antiviral marine terpenoids for potential leads against SARS-CoV-2 is crucial. Using computational tools, compounds such as brevione F and liouvillosides A and B have shown promise due to their strong affinity to SARS-CoV-2 enzymes, surpassing current antivirals, with stachyflin emerging as a notable candidate pending further study [35]. Research has highlighted the antiviral effects of plant-derived terpenes on various coronaviruses, such as the effectiveness of friedelane-type triterpenoids from *Euphorbia neriifolia* L. against HCoV-229E. Terpenoids from various plants have demonstrated antiviral capabilities against viruses such as influenza and herpes simplex, with saikosaponins showing a broad antiviral range. Tanshinones from *S. miltiorrhiza* and quinone-methide triterpenes from *T. regelii* have potential as coronavirus inhibitors. Terpenes from *C. obtusa* and *J. formosana* have shown effectiveness against SARS-CoV in cell studies. Finally, glycyrrhizin from Glycyrrhiza spp. has been recognized for its antiviral capacity against viruses, including SARS-CoV, and its ability to affect viral processes, making it a significant candidate for antiviral therapy [30]. This study examined the antiviral potential of NT-VRL-1, a terpene formulation, against human coronavirus (HCoV-229E). Additionally, using MOE 09 software, the research investigated the effects of 19 tannins on SARS-CoV-2, focusing on their interactions with the virus's main protease (3CLpro) at the Cys145 and His41 residues. Compounds such as pedunculagin, tercatan, and castalin demonstrated strong interactions with the binding sites and catalytic dyad of SARS-CoV-2, suggesting that they could block the virus's S protein from binding to the ACE2 receptor and, thus, prevent infection. These findings indicate that plant-based chemicals and their derivatives could serve as effective treatments against human coronaviruses, paving the way for future research [36]. Ursolic acid is a chemical compound extracted from the ethanol extracts of plants, notably *Mimusops elengi* (Spanish cherry), *Ilex paraguariensis* (Yerba mate), and *Glechoma hederacea* (ground ivy). It is known for a variety of health benefits, including reducing inflammation, combating bacteria, and providing antioxidant, antidiabetic, and anticancer effects. During the COVID-19 pandemic, research has identified ursolic acid as an effective inhibitor of a key coronavirus enzyme, SARS-CoV-2 Mpro. Further molecular studies have confirmed this, showing that ursolic acid can consistently bind to certain protease enzymes—crucial in the virus's life cycle—during simulations. Specifically, it has a strong tendency to attach to the PLpro, an enzyme important for virus replication, by forming a hydrogen bond with a component known as Asp 108 and interacting with other components, including Ala 107, Pro 248, and Tyr 264, which may disrupt the enzyme's function [37]. In a comprehensive study, 125 molecules from licorice (*Glycyrrhiza uralensis* Fisch.) were analyzed using computer simulations to understand their interaction with the SARS-CoV-2 S protein's RBD, which is crucial for the virus's entry into cells. Two molecules, licorice-saponin A3 and glycyrrhetinic acid, were effective in targeting different viral parts—nsp7 and the RBD, respectively. This study is the first to confirm the antiviral properties of these molecules against SARS-CoV-2. Their effectiveness, despite sharing similar structures but targeting different viral components, underscores the complex nature of herbal medicine and its role in the effective treatment of COVID-19 [38]. Columbin and ganodermanontriol, compounds

TABLE 14.2

Terpenoids Known for Their anti-SARS-CoV-2 Properties and Their Mechanisms of Action

Terpenoids	Mechanism Related Against SARS-CoV-2	References
Angolensic acid methyl ester and moluccensin V	Inhibition of mPro of SARS-CoV-2	[28]
Brevione F, and stachyflin	Inhibition of mPro of SARS-CoV-2	[35]
Celastrol, pristimerin, tingenone, and iguesterin	Reduction of SARS-CoV replication by competitive inhibition of SARS-CoV 3CLpro protease activity	[30]
Deacetylnomilin, ichangin, nomilin, and β-amyrin	Inhibition of mPro of SARS-CoV-2	[34]
Glycyrrhizin	Binds to the ACE2 enzyme and blocks the entry of SARS-CoV-2, inhibits the viral replication process	[27]
Liouvilloside A and B	Inhibition of SARS-CoV-2–RdRp	[35]
Pinusolidic acid, forskolin, and cedrane-3-β,12-diol	Inhibition of SARS CoV-induced infection/apoptosis and prolonged cellular survival after virus infection	[30]
Saikosaponin A, B2, C, and D	Inhibition of HCoV-229E activity	[30]
Tanshinones	Inhibition of SARS-CoV 3CLpro and PLpro	[30]

isolated from *Tinospora cordifolia* and *Ganoderma lucidum*, respectively, have demonstrated a robust affinity for the TMPRSS2 protein, as evidenced by their substantial binding energies of −8.1 and −8.2 kcal/mol. This indicates their potential to impede viral entry into host cells, thereby establishing them as promising agents in the prophylaxis of viral infections [39]. A recent computer-based (*in silico*) study investigated the effectiveness of certain antiviral compounds, known as diterpenoids, against the main protease (Mpro) of SARS-CoV-2. The study discovered that diterpenoids from specific groups–abietane, jatrophane, segetane, pepluane, and paraliane–bind more effectively to the virus's main protease than GC373, a well-known inhibitor of this enzyme. Additionally, using tools such as SwissADME for pharmacokinetic analysis and molecular dynamics simulation, it was suggested that some of these diterpenoids have characteristics similar to drugs. This research points to the promise of these plant-derived compounds, suggesting that they should be further studied in clinical trials as potential treatments for COVID-19 [40].

Examples of terpenoids with anti-SARS-CoV-2 properties are listed in Table 14.2.

14.4.3 REVIEW OF RELEVANT RESEARCH STUDIES

Research is honing in on the antiviral potential of medicinal plants against coronaviruses, focusing on EOs and cannabinoids such as cannabidiol (CBD) and tetrahydrocannabinol (THC). Terpenes in plants such as *Origanum acutidens* and *Cannabis sativa* are being studied for their ability to hinder viral replication and entry into cells while also offering anti-inflammatory benefits. Studies indicate that the interaction of CBD with the body's endocannabinoid system could alleviate symptoms of viral infections such as SARS-CoV-2. Promising results from molecular and cellular studies suggest that compounds in these plants may serve as a basis for new treatments, with certain CBD formulations showing effectiveness in reducing viral load in three-dimensional tissue models. However, more *in vivo* and preclinical studies are needed to confirm these preliminary findings and fully understand how these natural substances could contribute to treating COVID-19 [41]. Mangrove ecosystems, thriving in saline coastal areas, are biologically rich and cover a significant portion of the world's coastlines, with potential environmental and medical benefits. These biomes are rich in bioactive compounds such as saponins, alkaloids, and terpenoids, particularly from the mangrove species *X. moluccensis*. Scientists are interested in the species' terpenoids for their ability to block the Mpro protein, which is crucial for SARS-CoV-2 replication. Using computational methods, researchers have identified several compounds from *X. moluccensis*, such as angolensic acid methyl ester and moluccensin V, as promising for inhibiting the virus's Mpro protein. These findings are supported by virtual screenings, molecular dynamics simulations, and pharmacokinetic evaluations, although these compounds show limited blood–brain barrier permeability. They demonstrate good gastrointestinal (GI) absorption and a nongenotoxic profile, positioning them as potential anti-COVID-19 drug candidates [28]. The study utilizes computational methods to prescreen compounds for drug development, thus saving time and resources. Notably, terpenoids such as limonin and glycyrrhizin demonstrate promise in their ability to bind to SARS-CoV-2 proteins, suggesting potential as COVID-19 treatments. Although they do not significantly disrupt the interaction between the S protein and ACE2 complex, some compounds, including deacetylnomilin and β-amyrin, effectively bind to the virus's main protease, potentially inhibiting viral replication. These findings highlight the therapeutic potential of terpenoids beyond their direct antiviral activity, including reducing inflammation and enhancing the effectiveness of other drugs. Despite these encouraging results, further studies are essential to confirm the efficacy of these natural compounds against COVID-19. This research lays the groundwork for future strategies using natural products to fight the disease [34]. In the search for COVID-19 treatments, researchers have focused on proteins crucial to viral entry, specifically targeting the early stages of SARS-CoV-2 infection. A study analyzed 106 terpenoids from African medicinal plants for their ability to bind to human proteins essential for viral entry, such as human angiotensin-converting enzyme 2 (hACE2), transmembrane protease, serine 2 (TMPRSS2), and coronavirus S proteins. Through molecular docking analysis, the study evaluated terpenoids for absorption, distribution, metabolism, excretion, and toxicity. Eight terpenoids showed strong binding affinities to these proteins, particularly 24-methylene cycloartenol and isoiguesterin to hACE2, abietane diterpenes to TMPRSS2, and 3-benzoylhosloppone and cucurbitacin to different parts of the S protein. These interactions, which are stable in simulated environments, suggest that these terpenoids are promising candidates for drug development. The results, underlining the potential of these compounds to block SARS-CoV-2 cellular entry, call for further *in vitro* and *in vivo* studies to develop effective COVID-19 therapies [42]. The Chaga mushroom (*Inonotus obliquus*), from the *Hymenochaetaceae* family, has attracted research interest for its health benefits, including nutritional, antioxidant, and therapeutic potential. Studies on its mycochemicals, especially triterpenoids, have shown promise in treating various diseases, such as cancer and chronic conditions. Chaga has demonstrated antiviral effects against RNA viruses, including HIV and hepatitis C, and its compounds, such as polyphenols and polysaccharides, contribute to its medicinal properties. Notably, certain terpenoids from Chaga have shown the ability to inhibit SARS-CoV and SARS-CoV-2 by binding to the virus's S protein, potentially preventing it from entering and replicating in host cells. Molecular docking studies revealed that terpenoids, particularly betulinic acid and inonotusane C, might effectively neutralize SARS-CoV-2. These findings highlight the need for more in-depth research, including *in vitro* studies, to fully understand Chaga's potential as a therapeutic against COVID-19 [43]. The study employed a computer simulation to assess how 23 terpenoid compounds might attach themselves to key parts of the SARS-CoV-2 virus–the main protease, which is essential for the virus to replicate, and the S RBD, which allows the virus to attach to and enter human cells. The simulation identified certain amino acids in these proteins that are vital for successful attachment. Furthermore, the study predicted the behavior of these compounds within the human body and their potential effectiveness using advanced computer-based evaluations. The initial results are promising, indicating that these compounds from plants might combat SARS-CoV-2 effectively. However, actual laboratory experiments are necessary to confirm whether these terpenoids could be developed into COVID-19 treatments [44]. An exhaustive computational analysis was undertaken to evaluate 13 polyphenolic and terpenoid constituents derived from *Rosmarinus officinalis* L. for their potential inhibitory action on the main protease of the SARS-CoV-2 pathogen, the etiological agent responsible for the COVID-19 pandemic. The investigation employed sophisticated *in silico* methodologies, notably molecular docking and simulation techniques, to ascertain the therapeutic viability of these compounds. Notably, apigenin, betulinic acid, luteolin, carnosol, and rosmarinic acid manifested promising drug-like attributes and antiviral efficacies that were analogous to established antiviral agents such as remdesivir and favipiravir. In particular, luteolin and betulinic acid were distinguished as compounds of interest. While the outcomes of this computational inquiry are indeed promising, they necessitate validation through empirical *in vivo* and *in vitro* trials to establish their medicinal efficacy. The findings provided by this study may be instrumental in guiding the synthesis and optimization of novel therapeutic agents against SARS-CoV-2 infection [45].

14.5 CASE STUDIES AND CLINICAL POTENTIAL OF MEDICINAL PLANTS AND THEIR TERPENOIDS AGAINST COVID-19

14.5.1 CURCUMA LONGA

Curcumin derived from turmeric binds effectively to mutated S proteins, potentially weakening the virus's attachment to human receptors and destabilizing the virus-receptor complex. Simulations show that curcumin stably binds to the Omicron variant S protein, suggesting its potential as a treatment for this strain of SARS-CoV-2 [46]. A pilot open-label, randomized controlled trial was conducted at Mayo Hospital, King Edward Medical University in Lahore, Pakistan. In this randomized trial, patients with mild to moderate COVID-19 received standard care, with half also receiving CQC for two weeks. The CQC group exhibited faster virus clearance, better symptom relief, and reduced inflammation, without adverse effects, suggesting that CQC could enhance standard treatment and hasten recovery. These promising results require confirmation from larger trials [47].

14.5.2 GLYCYRRHIZA GLABRA

Glycyrrhiza glabra, the licorice plant, is recognized for fighting various viruses, including influenza, hepatitis B and C, herpes, and respiratory syncytial virus. Its antiviral properties are mainly due to triterpenoid saponins, especially glycyrrhizin, which binds to ACE2 receptors and might block viral entry, supporting immune defense against SARS-CoV-2 and reducing inflammation. However, glycyrrhizin's quick metabolism and clearance from the body make maintaining effective levels difficult. Researchers are modifying its structure to enhance bioavailability and antiviral efficacy, although this increases toxicity risks. This underscores the need for clinical trials to fully evaluate and optimize glycyrrhizin's therapeutic use against viruses [48]. Controlled trials suggest that traditional Chinese medicine (TCM) can effectively treat COVID-19 symptoms and expedite recovery. Jinhua Qinggan granules have been shown to lower viral levels and alleviate pneumonia symptoms, while Lianhuaqingwen (LHQW) granules have helped reduce fatigue, fever, and cough. Integrating TCM with Western medicine has resulted in better recovery outcomes and less severe symptoms, such as intense illness and muscle pain. Studies indicate that LHQW capsules may interact with ACE2 receptors and have minor side effects, Shufeng Jiedu® (SFJD) capsules have a high effectiveness rate and reduce viral load, and Ma Xing Shi Gan Decoction (MXSG) and Qingfei Paidu decoction (QFPD) preparations can decrease fever and cough. Glycyrrhiza-containing TCM treatments notably reduce mortality and improve outcomes in severe cases. Additionally, D-Reglis® and glycyrrhizin capsules have been effective in relieving symptoms and lowering respiratory support needs. A large study involving 2000 participants revealed that a Glycyrrhiza-based anti-inflammatory formulation could mitigate COVID-19's inflammatory effects, with licorice potentially offering postrecovery benefits [27].

14.5.3 ARTEMISIA ANNUA

A. annua, a botanical entity renowned for its antimalarial properties, has garnered significant attention within the scientific community due to its potential therapeutic application against *Plasmodium falciparum*, the etiological agent of malaria. The exploration of *A. annua* extracts, specifically the bioactive constituent artemisinin, has intensified, prompted by the exigencies of the COVID-19 pandemic and the quest for efficacious treatments. Phytochemical analyses have revealed a diverse array of terpenoids present in *A. annua*, with sesquiterpenes predominating, followed by triterpenes, diterpenes, and monoterpenes. Of particular interest are compounds such as germacrene-D and trans-β-farnesene, whose concentrations have been observed to increase notably [49]. Artemisinin's low toxicity profile stands out among antiviral agents, offering a safer alternative for treating COVID-19, particularly for patients with other infections. Currently, nine global intervention trials are exploring its effectiveness against COVID-19, with six registered in the United States and three in China, although one is suspended. A forthcoming phase II trial aims to assess *A. annua* as an early treatment to improve outcomes for high-risk COVID-19 patients but has not started recruiting. There is a clear need for more clinical research to define clear therapeutic protocols for Artemisia-based treatments, especially for outpatients with comorbidities such as heart diseases. Artemisia's antiviral potential, high zinc content, and empirically backed safety call for rigorous clinical trials to establish its use in COVID-19 treatment regimens, alone or with other drugs. Strict adherence to health protocols and medical supervision are crucial when combining Artemisia with other treatments to understand drug interactions and impacts on the body's inflammatory response and the hACE2 receptor, which is essential for SARS-CoV-2 entry [50].

14.6 CLINICAL IMPLICATIONS AND CHALLENGES

14.6.1 Potential Benefits of Using Terpenoids in COVID-19 Treatment

Researchers are exploring new compounds to combat SARS-CoV-2, with cell culture techniques advancing the production of potential antiviral agents such as terpenoids, of which over 100,000 types exist in plants. Current clinical trials on these compounds are limited. A phase II trial is examining a treatment combining curcumin (a terpenoid), quercetin (a flavonoid), and vitamin D in early COVID-19 symptoms to reduce infection severity. More than 100 adults confirmed to be infected by reverse transcription polymerase chain reaction (RT–PCR) and symptomatic were enrolled. They received a dosage of 42 mg curcumin, 65 mg quercetin, and 90 IU vitamin D in softgel capsules four times daily for 2 weeks. The study aims to measure the treatment's impact on symptoms and viral load [47, 51]. *In silico* studies highlight the antiviral promise of terpenoids from ayurvedic plants, particularly ursolic acid from Tulsi and withanoside V from ashwagandha. Other Indian plants, including curcumin from turmeric, have shown potential as COVID-19 interventions. Monoterpenes such as limonene from *Citrus limon* and glycyrrhizic acid from *G. glabra* have shown effects on SARS-CoV-2 proteins. Although glycyrrhizin is approved by the US Food and Drug Administration, its fast metabolism and increased toxicity with structural modification are concerns. Additionally, artemisinin and its derivative artesunate from *A. annua* have known antiviral effects and are used traditionally in Asia and Africa, with potential against SARS-CoV-2 [48]. EOs and compounds from plants such as black pepper and *A. paniculata* also show promise against the virus. Despite the potential of plant-derived compounds and advanced drug discovery methods, their efficacy must be validated through extensive clinical trials due to the limited scope of initial studies. The natural safety profiles of these phytochemicals offer advantages over synthetic drugs in the fight against SARS-CoV infections [52].

14.6.2 Challenges in the Development of Terpenoid-Based Drugs

The pharmaceutical industry must accelerate the synthesis and strategic reserve of small-molecule pharmacotherapeutics, which are characterized by their diminutive molecular size and favorable pharmacokinetic profiles, in anticipation of the sustained influence of SARS-CoV-2 and the potential emergence of novel viral pathogens. Phytochemicals, particularly terpenoids, have exhibited substantial antiviral efficacy. The integration of cutting-edge technological advancements in drug discovery, referred to as innovative screening platforms, holds the promise of catalyzing the formulation of an extensive repertoire of novel therapeutic agents. These agents would be tailored to confront the challenges posed by extant coronaviruses, including SARS-CoV-2 [53]. Terpenoids, which are essential for plant defense mechanisms, confer resistance to herbivores, pathogens, and diseases. The synthesis of these secondary metabolites generally occurs via biosynthetic pathways. Secondary metabolites are crucial in mediating plant-environment interactions involved in the wild range of microorganisms and animals. Multiple studies have reported that terpenoids plausibly exhibit anticancer, anti-inflammatory, and antimicrobial activities. In addition, terpenoids have been suggested to promote immune and neuronal functions. Notwithstanding their therapeutic promise, the elucidation of the complete spectrum of biological activities of numerous terpenoids warrants rigorous empirical inquiry. These phytochemicals, profuse in medicinal flora, present expansive prospects for medicinal development and deployment. Beyond traditional plant extraction methodologies, terpenoids can be synthesized through avant-garde techniques such as metabolic engineering, synthetic biology, and biotransformation processes, which ameliorate the limitations associated with natural reserves and permit scalable production for pharmaceutical industries. Conventional extraction from botanical sources often results in suboptimal yields and purities of terpenes, exemplified by elemene and limonene, thereby rendering the process economically and technically challenging. Conversely, microbial biosynthesis emerges as a vanguard approach, offering augmented purity and regulated production, which potentially broadens the application of terpenes in diverse industrial realms, as proposed by Fan et al. [3]. Deciphering the molecular mechanisms underpinning terpenoid action is paramount to the advancement of pharmacognosy. Progress in this domain is propelled by omics technologies and the application of molecular network pharmacology, which facilitates the stratification of terpenoids with pronounced bioactivity, serving as novel therapeutic agents or scaffolds for structural enhancement. Nevertheless, terpenoids are not devoid of limitations, notably issues pertaining to solubility and chemical stability that compromise their therapeutic performance and bioavailability. The advent of sophisticated drug delivery systems is poised to surmount these impediments, integrating cutting-edge medical technology with natural product pharmacology to potentiate therapeutic outcomes, enable precision diagnostics, and foster responsive treatment modalities, as anticipated by Tang et al. [54]. However, drug discovery from natural matrices remains fraught with complexities. The bioefficacy of plant-derived metabolites upon human physiology is subject to variability, attributable to concentration-dependent receptor interactions, yielding a gamut of pharmacological or toxicological consequences. Certain metabolites, classified as panassay interference compounds (PAINS) and invalid metabolic panaceas (IMPS), have the propensity to confound bioassays and manifest substandard bioavailability, thereby curtailing their clinical applicability. The use of natural substances in treating illnesses faces a major obstacle: these substances often have poor solubility and low bioavailability, meaning that human bodies have difficulty dissolving and absorbing them. This problem significantly impedes the assessment of their effectiveness in clinical trials. To avoid unnecessary expenditures on such costly trials, it is crucial to address bioavailability issues early on. Bioavailability is critical in determining how much and how fast a drug reaches its site of action in the body. It depends on the drug's bioaccessibility, its release and absorption after consumption, and bioactivity, its ability to produce a biological effect. Assessments of bioavailability involve *in vitro* tests simulating human digestion or *in vivo* tests in living organisms. For instance, Caco-2 cells are used to mimic the intestinal barrier in laboratories. After these tests, animal studies confirm bioavailability before human trials. For phytotherapeutics such as terpenes, bioavailability is complex. These compounds must be released from plant material and transformed within the body into a usable form, often coupling with dietary fats to aid absorption. Unlike direct effects from intravenous routes, terpenes in EOs are typically administered via skin ingestion or inhalation. Skin absorption is influenced by various factors, including the application area and duration. Once ingested, the intestines primarily absorb terpenes, while inhalation allows pulmonary uptake. After absorption, terpenes are metabolized mainly in the liver and eliminated through breath, urine, or feces, preventing bioaccumulation. For example, camphor is converted to hydroxycamphor, and metabolized menthol is excreted as glucuronides. Research has shown that EOs can enhance transdermal drug delivery, such as galangal oil with flurbiprofen and menthol with ligustrazine hydrochloride. However, when taken orally, the absorption of terpenes varies due to factors such as lipophilicity and the GI tract's pH levels, affecting the absorption and efficacy of P-glycoprotein substrate drugs, with terpenes typically absorbing best in the duodenum [17]. In addition, monoterpenes are known for their cytotoxicity in research, while terpenes such as β-caryophyllene offer therapeutic effects, including anti-inflammation and cell protection. Some terpenes, such as α-terpineol and terpinolene, are notably toxic. Terpene toxicity, measured by IC50/LC50 or LD50 values, varies with dosage and time and can lead to cellular and mitochondrial damage, particularly affecting the liver, kidneys, lungs, and nervous system. Therefore, it is vital to thoroughly evaluate terpenes, commonly used in food and medicine, for any potential health risks [55]. We need to find more effective delivery methods for these natural treatments, particularly for plant-derived compounds such as phenolics and terpenes, which have recognized therapeutic effects. Nanotechnology may provide an innovative solution by creating tiny carriers that can transport these natural medicines more effectively into the body. Additionally, taking these natural treatments in small, regular doses might be a viable approach to preventing viruses from entering cells, which could slow or stop the progression of infections. Additionally, the intricacies of disease pathogenesis and remediation entail multifaceted interactions within the body's biological systems, delineating a complex network where these phytochemicals exert their effects. An in-depth comprehension of these dynamisms is indispensable for harnessing the therapeutic potential of terpenoids and congeneric natural entities [51].

14.7 CONCLUSION AND FUTURE PERSPECTIVES

In the current global landscape, the persistent resurgence of COVID-19 represents a formidable challenge to public health infrastructures worldwide. The scholarly community is increasingly attentive to the potential of phytotherapeutics, encompassing ethnomedicinal practices, bioactive phytochemicals, and nutraceuticals, which promise substantial progress in the ongoing struggle against this pandemic. The discovery and characterization of diminutive molecules with the capability to lower the mortality associated with SARS-CoV-2 are recognized as a critical scientific imperative. Pharmaceutical agents, such as chloroquine and hydroxychloroquine, while employed therapeutically, exhibit suboptimal attributes, notably due to accessibility constraints and documented adverse effects. In contrast, natural products, particularly plant extracts and terpenoids, have demonstrated considerable antiviral efficacy, notably in the inhibition of enzymatic processes critical to viral replication. Computational modeling methodologies have facilitated the identification of a suite of promising antiviral candidates, which now necessitate rigorous empirical validation via *in vitro* and *in vivo* studies to verify their therapeutic potential. Historically, the pharmacological constituents of medicinal flora have been integral to the amelioration of a multitude of pathologies, thereby warranting their exploration as prospective agents in the armamentarium against COVID-19. These botanical interventions may be administered as monotherapies or as adjuncts to allopathic treatment regimens. Preliminary investigations have posited the antiviral competence of specific terpenoids, prompting the initiation of clinical trials to evaluate the efficacy of such phytoconstituents. Although these agents may not act prophylactically, their immunomodulatory capabilities could be conducive to enhanced patient prognoses. Nonetheless, the adoption of plant-based therapies mandates a comprehensive assessment of their safety and therapeutic efficacy. The international scientific community acknowledges the impediments associated with the therapeutic deployment of natural products and terpenoids. A lacuna persists in sequential research that could consolidate the nascent discoveries of phytomedicinal treatments. Moreover, the prospect of pharmacodynamic interactions between these natural entities and conventional antiviral pharmacotherapies presents a tangible risk. For example, the concomitant application of certain herbal extracts may attenuate the clinical effectiveness of antiretroviral agents such as ritonavir/lopinavir. It is, however, noteworthy that not all drug-phytoconstituent interactions are deleterious; synergistic effects are a viable possibility. Additionally, the research agenda must incorporate considerations of viral reactivation and its subsequent health ramifications. There is a pronounced necessity for investigative pursuits aimed at optimizing detoxification processes and fortifying the immunological responses of individuals affected by COVID-19. The resolution of this pandemic is predicated on the synergistic cooperation of global citizens, the research community, medical practitioners, and policy-makers. The integration of phytocompounds and terpenoids into the therapeutic landscape offers not only a homage to traditional medicinal wisdom but also a potential reduction in the socioeconomic impact borne by individuals grappling with COVID-19, potentially attenuating the mortality incidence in the enduring tenure of this global health crisis.

REFERENCES

1. Demeke, C.A., A.E. Woldeyohanins, and Z.D. Kifle, *Herbal medicine use for the management of COVID-19: A review article.* Metabol Open, 2021. **12**: p. 100141.
2. Komariah, M., et al., *The efficacy of herbs as complementary and alternative therapy in recovery and clinical outcome among people with COVID-19: A systematic review, meta-analysis, and meta-regression.* Ther Clin Risk Manag, 2023. **19**: p. 611–627.
3. Fan, M., et al., *Application of terpenoid compounds in food and pharmaceutical products.* Fermentation, 2023. **9**(2): p. 119.
4. Markov, P.V., et al., *The evolution of SARS-CoV-2.* Nat Rev Microbiol, 2023. **21**(6): p. 361–379.
5. Polatoglu, I., et al., *COVID-19 in early 2023: Structure, replication mechanism, variants of SARS-CoV-2, diagnostic tests, and vaccine & drug development studies.* MedComm (2020), 2023. **4**(2): p. e228.
6. Shamsi, A., et al., *Potential drug targets of SARS-CoV-2: From genomics to therapeutics.* Int J Biol Macromol, 2021. **177**: p. 1–9.
7. Metzdorf, K., et al., *TMPRSS2 is essential for SARS-CoV-2 beta and omicron infection.* Viruses, 2023. **15**(2): p. 271.
8. Mohammed, M.E.A., *SARS-CoV-2 proteins: Are they useful as targets for COVID-19 drugs and vaccines?* Curr Mol Med, 2022. **22**(1): p. 50–66.
9. Liu, L., et al., *SARS-CoV-2 ORF3a sensitizes cells to ferroptosis via Keap1-NRF2 axis.* Redox Biol, 2023. **63**: p. 102752.
10. Zhang, J., et al., *Understanding the role of SARS-CoV-2 ORF3a in viral pathogenesis and COVID-19.* Front Microbiol, 2022. **13**: p. 854567.
11. Wu, W., et al., *The SARS-CoV-2 nucleocapsid protein: Its role in the viral life cycle, structure and functions, and use as a potential target in the development of vaccines and diagnostics.* Virol J, 2023. **20**(1): p. 6.
12. Hu, Q., et al., *The SARS-CoV-2 main protease (M(pro)): Structure, function, and emerging therapies for COVID-19.* MedComm (2020), 2022. **3**(3): p. e151.
13. Han, H., et al., *A covalent inhibitor targeting the papain-like protease from SARS-CoV-2 inhibits viral replication.* RSC Adv, 2023. **13**(16): p. 10636–10641.
14. Jiang, H., P. Yang, and J. Zhang, *Potential inhibitors targeting papain-like protease of SARS-CoV-2: Two birds with one stone.* Front Chem, 2022. **10**: p. 822785.
15. Santos-Mendoza, T., *The envelope (E) protein of SARS-CoV-2 as a pharmacological target.* Viruses, 2023. **15**(4): p. 1000.
16. Zhou, S., et al., *SARS-CoV-2 E protein: Pathogenesis and potential therapeutic development.* Biomed Pharmacother, 2023. **159**: p. 114242.
17. Masyita, A., et al., *Terpenes and terpenoids as main bioactive compounds of essential oils, their roles in human health and potential application as natural food preservatives.* Food Chem X, 2022. **13**: p. 100217.
18. Li, C., et al., *Advances in the biosynthesis of terpenoids and their ecological functions in plant resistance.* Int J Mol Sci, 2023. **24**(14): p. 11561.
19. Dash, D.K., et al., *Revisiting the Medicinal Value of Terpenes and Terpenoids*, in *Revisiting Plant Biostimulants*, V. Meena, Editor. 2022, IntechOpen Limited: 5 Princes Gate Court, London, SW7 2QJ, United Kingdom.

20. Singh, L.J., D.A. Challam, and B.D. Senjam, *Medicinal plants as a sources of terpenoids and their impact on Central Nervous System disorders: A review.* J Phytopharmacol, 2023. **12**(2): p. 104–110.

21. Han, J.Y., M. Park, and H.J. Lee, *Stevia (Stevia rebaudiana) extract ameliorates insulin resistance by regulating mitochondrial function and oxidative stress in the skeletal muscle of db/db mice.* BMC Complement Med Ther, 2023. **23**(1): p. 264.

22. Nair, M.S., et al., *SARS-CoV-2 omicron variants are susceptible in vitro to Artemisia annua hot water extracts.* J Ethnopharmacol, 2023. **308**: p. 116291.

23. Surendran, S., et al., *Myrcene-what are the potential health benefits of this flavoring and aroma agent?* Front Nutr, 2021. **8**: p. 699666.

24. Mandal, D., B.R. Sahu, and T. Parija, *Combination of tamoxifen and D-limonene enhances therapeutic efficacy in breast cancer cells.* Med Oncol, 2023. **40**(8): p. 216.

25. Boopathi, V., et al., *In silico and in vitro inhibition of host-based viral entry targets and cytokine storm in COVID-19 by ginsenoside compound K.* Heliyon, 2023. **9**(9): p. e19341.

26. De, A., et al., *Exploring the pharmacological aspects of natural phytochemicals against SARS-CoV-2 Nsp14 through an in silico approach.* In Silico Pharmacol, 2023. **11**(1): p. 12.

27. Banerjee, S., et al., *Glycyrrhizin as a promising kryptonite against SARS-CoV-2: Clinical, experimental, and theoretical evidence.* J Mol Struct, 2023. **1275**: p. 134642.

28. Lokhande, K.B., et al., *Terpenoid phytocompounds from mangrove plant Xylocarpus moluccensis as possible inhibitors against SARS-CoV-2: In silico strategy.* Comput Biol Chem, 2023. **106**: p. 107912.

29. Sepay, N., et al., *Anti-COVID-19 terpenoid from marine sources: A docking, ADMET and molecular dynamics study.* J Mol Struct, 2021. **1228**: p. 129433.

30. Diniz, L.R.L., et al., *Bioactive terpenes and their derivatives as potential SARS-CoV-2 proteases inhibitors from molecular modeling studies.* Biomolecules, 2021. **11**(1): p. 74.

31. Sruthi, D., et al., *Curative potential of high-value phytochemicals on COVID-19 infection.* Biochemistry (Mosc), 2023. **88**(1): p. 64–72.

32. Sa-Ngiamsuntorn, K., et al., *Anti-SARS-CoV-2 activity of Andrographis paniculata extract and its major component andrographolide in human lung epithelial cells and cytotoxicity evaluation in major organ cell representatives.* J Nat Prod, 2021. **84**(4): p. 1261–1270.

33. Enmozhi, S.K., et al., *Andrographolide as a potential inhibitor of SARS-CoV-2 main protease: An in silico approach.* J Biomol Struct Dyn, 2021. **39**(9): p. 3092–3098.

34. Giofre, S.V., et al., *Interaction of selected terpenoids with two SARS-CoV-2 key therapeutic targets: An in silico study through molecular docking and dynamics simulations.* Comput Biol Med, 2021. **134**: p. 104538.

35. Sahoo, A., et al., *Potential of marine terpenoids against SARS-CoV-2: An in Silico Drug Development Approach.* Biomedicines, 2021. **9**(11): 1505.

36. Al-Harrasi, A., et al., *Targeting natural products against SARS-CoV-2.* Environ Sci Pollut Res Int, 2022. **29**(28): p. 42404–42432.

37. Jaber, S.A., *The effect of a small polyphenolic and terpenoids phytochemical constituent on curing and preventing of Covid-19 infections.* Pharmacia, 2023. **70**(3): p. 665–672.

38. Yi, Y., et al., *Natural triterpenoids from licorice potently inhibit SARS-CoV-2 infection.* J Adv Res, 2022. **36**: p. 201–210.

39. Pooja, M., et al., *Unraveling high-affinity binding compounds toward transmembrane protease serine 2 enzyme in treating SARS-CoV-2 infection using molecular modeling and docking studies.* Eur J Pharmacol, 2021. **890**: p. 173688.

40. Abdelrady, Y.A., et al., *In silico assessment of diterpenes as potential inhibitors of SARS-CoV-2 main protease.* Future Virology, 2023. **18**(5): p. 295–308.

41. Santos, S., et al., *Cannabidiol and terpene formulation reducing SARS-CoV-2 infectivity tackling a therapeutic strategy.* Front Immunol, 2022. **13**: p. 841459.

42. Gyebi, G.A., et al., *SARS-CoV-2 host cell entry: An in silico investigation of potential inhibitory roles of terpenoids.* J Genet Eng Biotechnol, 2021. **19**(1): p. 113.

43. Basal, W.T., A. Elfiky, and J. Eida, *Chaga medicinal mushroom inonotus obliquus (Agaricomycetes) terpenoids may interfere with SARS-CoV-2 spike protein recognition of the host cell: A molecular docking study.* Int J Med Mushrooms, 2021. **23**(3): p. 1–14.

44. Hadni, H., et al., *Identification of terpenoids as potential inhibitors of SARS-CoV-2 (main protease) and spike (RBD) via computer-aided drug design.* J Biomol Struct Dyn, 2023: p. 1–14.

45. Patel, U., et al., *Bioprospecting phytochemicals of Rosmarinus officinalis L. for targeting SARS-CoV-2 main protease (M(pro)): A computational study.* J Mol Model, 2023. **29**(5): p. 161.

46. Nag, A., et al., *Curcumin inhibits spike protein of new SARS-CoV-2 variant of concern (VOC) Omicron, an in silico study.* Comput Biol Med, 2022. **146**: p. 105552.
47. Khan, A., et al., *Oral co-supplementation of curcumin, quercetin, and vitamin D3 as an adjuvant therapy for mild to moderate symptoms of COVID-19-results from a pilot open-label, randomized controlled trial.* Front Pharmacol, 2022. **13**: p. 898062.
48. Bijelic, K., M. Hitl, and N. Kladar, *Phytochemicals in the prevention and treatment of SARS-CoV-2-clinical evidence.* Antibiotics (Basel), 2022. **11**(11): p. 1614
49. Sankhuan, D., et al., *Variation in terpenoids in leaves of Artemisia annua grown under different LED spectra resulting in diverse antimalarial activities against Plasmodium falciparum.* BMC Plant Biol, 2022. **22**(1): p. 128.
50. Orege, J.I., et al., *Artemisia and artemisia-based products for COVID-19 management: Current state and future perspective.* Adv Tradit Med, 2021. **23**(1): p. 85–96.
51. Yilmaz Aydin, D. and S. GÜRÜ, *Potential of natural therapeutics against SARS-CoV-2: Phenolic compounds and terpenes.* Namık Kemal Tıp Dergisi, 2022. **10**(2): p. 119–128.
52. Issa, S.S., et al., *The main protease of SARS-CoV-2 as a target for phytochemicals against coronavirus.* Plants (Basel), 2022. **11**(14): p. 1862.
53. Nguyen, H.T., et al., *The potential of ameliorating COVID-19 and sequelae from Andrographis paniculata via bioinformatics.* Bioinform Biol Insights, 2023. **17**: p. 1–14.
54. Tang, P., et al., *Challenges and opportunities for improving the druggability of natural product: Why need drug delivery system?* Biomed Pharmacother, 2023. **164**: p. 114955.
55. Agus, H.H., *Terpene Toxicity and Oxidative Stress*, in *Toxicology Oxidative Stress and Dietary Antioxidants*. 2021, Elsevier Inc.: Patel, V.B. p. 33–42.

15 Anti-Inflammatory Activity of Herbal Medicines and the Potential for Combating COVID-19

*Mani Manoj, Arumugam Sembiyaa,
Jayakumar Dharshini, Balasubramaniam Rashmika,
Chenniappan Anchana Devi, Natarajan Pushpa,
Jen-Tsung Chen, and Arumugam Vijaya Anand*

15.1 INTRODUCTION

The coronavirus or COVID-19 spread worldwide in December 2019 and became dangerous to the global population. Because of its rapid transmission, it affected large populations of all ethnicities. Even though many allopathy medicines and vaccines provide immunity to the disease, herbal medicine has shown more promising results without any side effects. Traditional Chinese Medicine (TCM), which uses only natural products, has shown some impressive results against COVID-19 over other traditional medicine (Luo *et al.*, 2020; Shah *et al.*, 2021). COVID-19 is also known as severe acute respiratory syndrome coronavirus 2 (SARS-CoV-2), which originated in Wuhan, China. It is a positive sense RNA virus that has a diameter of 60 to 140 nm with a spike protein on its surface that gives a crown-like appearance when visualized under an electron microscope (Singhal, 2020). Recently, the exploration of herbal medicine's anti-inflammatory properties has acquired noteworthy attention, particularly in the context of combating COVID-19. This chapter delves into the potential of herbal remedies to modulate inflammatory responses, shedding light on their role in managing the immune system and potentially mitigating the severity of COVID-19 symptoms. The combination of traditional medicinal plants and modern research has given rise to herbal drugs that are composed of one or more medicinal plants. Shufeng Jiedu capsules are one of the herbal drugs which shown more promising results than any other herbal medicines. It has shown effective treatment for upper respiratory tract infections and other kind of respiratory diseases (Huang *et al.*, 2020; Ling, 2020; Lu *et al.*, 2021). Exploring the intersection between traditional herbal medicine and its anti-inflammatory attributes offers a nuanced perspective on potential therapeutic avenues against COVID-19 (Ang *et al.*, 2020).

15.2 OVERVIEW OF HERBAL MEDICINE AND ITS TRADITIONAL USE IN INFLAMMATION

Inflammation is a biological process that occurs mammals due to exposure to pathogenic microorganisms, toxins, chemicals and physical damage (Verma, 2016). These cells try to trap pathogens and other agents to heal injured tissues. Inflammation is classified as either acute or chronic inflammation (Barnes, 2002). Acute inflammation, or a short-term inflammation, is resolved within minutes, hours or days. The clinical signs include vasodilation, oedema, tissue damage and fever. Chronic inflammation, an inappropriate immune response, leads to prolonged inflammatory response and results in allergic and hypersensitivity (Kumar *et al.*, 2013). Acute inflammation is the first line of defence,

DOI: 10.1201/9781003452621-15

is non-specific, and is the initial reaction of the body to risk factors such as trauma, infections, etc. This type of inflammation leads to the accumulation of fluid in the plasma, intravascular activation of platelets, and polymorpho-nuclear neutrophils as provocative cells (Singh *et al.*, 2008). If the risk factors are not dismissed, then acute inflammation will turn into chronic inflammation. It is related to the presence of lymphocytes, macrophages, blood cell proliferation, tissue necrosis and fibrosis (Sen *et al.*, 2010).

In ancient times, ancestors used herbal medicine to cure a large variety of diseases. Plants synthesize a phytochemical compound as secondary metabolites, which can be used as an effective treatment for chronic disease (Verma and Singh, 2008). These therapeutic plants are a primary source for the production of modern medicines, food supplements and pharmaceutical products. Herbs are the natural products which stimulate our body's immune system. Attention to herbal medicines has increased in recent years due to bacterial resistance, a decrease in the efficacy of some drugs and an increase in harmful side effects of synthetic drugs (Khumalo *et al.*, 2022). Medicinal plants are a comprehensive variety of biologically active components for treating various diseases and are used by many countries. Herbal medicines and healthcare preparation are described in various ancient texts such as the Vedas and the Bible. Natural products with anti-inflammatory properties are used to treat conditions such as fever, pain, migraine and arthritis (Vikrant and Arya, 2011).

The World Health Organization declared a global pandemic in March 2020, due to the global spread of Coronavirus 2019 (COVID-19), which has caused at least 3 million deaths. The causative factor for COVID-19 is SARS-CoV-2 (Arumugam *et al.*, 2020; Bhotla *et al.*, 2020; Shanmugam *et al.*, 2020; Bhotla *et al.*, 2021; Harwansh and Bahadur, 2022; Meyyazhagan *et al.*, 2022; Pushparaj *et al.*, 2022). It is an enclosed β-coronavirus that has non-segmented positive sense RNA. It has genetic information of 30 kilobase (kb). SARS-CoV-2 induces disease in the respiratory tract of humans by binding to the receptor of angiotensin-converting enzyme 2 (ACE2), an encoded protein that catalyzes the separation of angiotensin 1 to angiotensin 1–9 and angiotensin II into the vasodilator angiotensin 1–7; (Guo *et al.*, 2020). The encoded protein acts as a functional receptor for the human SARS-CoV and SARS-CoV-2. SARS-CoV-2 in extreme cases, initiates the inflammatory immune response and release of proinflammatory cytokines, which causes multi-organ dysfunction. There is no efficient remedy for COVID-19 infection (Zhang and Holmes, 2020). Many types of vaccines are now known to inhibit the COVID-19 pandemic. People, particularly from African and Asian countries such as India, China and Japan, have used herbs as medications to relieve symptoms of numerous diseases. Herbs are low cost and they could help in developing a drug with anti-COVID-19 efficacy (Poochi *et al.*, 2020; Bhotla *et al.*, 2021; Chandra Manivannan *et al.*, 2022; Al-Kuraishy *et al.*, 2022; Abidharini *et al.*, 2023; Anand *et al.*, 2021; Bharathi *et al.*, 2023; Boro *et al.*, 2023; Kaviya *et al.*, 2023; Latha *et al.*, 2023; Ramya *et al.*, 2023).

15.3   THE MECHANISMS OF INFLAMMATION AND IMMUNE RESPONSE

An immune response is activated depending on the presence of a pathogen, whether it is inside or outside of the host cell and the location of the response. Pathogen recognition is associated with the interaction of foreign organisms with the host cell. These recognition molecules are mostly secreted recognition molecules, soluble receptors or membrane-bound receptors (Wolfe *et al.*, 2007). Ligand, which binds to the recognition molecules, includes pathogens and products secreted by pathogens. Ligand binding triggers an intracellular or extracellular cascade of events, which identifies and destroys the pathogen. This type of immune response involves a complex system of cells that leads to engulfing the pathogen or recognising and killing the pathogen, as well as the secretion of soluble proteins that involves the labelling and destruction of foreign organisms (Owen *et al.*, 2013).

15.3.1   THE IMMUNE RESPONSE TOWARDS COVID-19

People become affected by SARS-CoV-2 when exposed to microdroplets emitted from the breath of an infected person, or by contact with viral particles on contaminated objects (Guan *et al.*, 2020).

Once COVID-19 enters a healthy person's lungs, it reaches the bronchioles and alveolar spaces, targeting the bronchial epithelium and type II ACE2 pneumocytes of the alveolar epithelium (Wu *et al.*, 2020). SARS-CoV-2 infection induces separation of the basal membrane and inhibits the expression of ACE2 and autophagy. Angiotensin II binds the AT1aR receptor, eventually causing acute lung injury (Shi *et al.*, 2020).

15.3.1.1 Innate Immune Response to COVID-19

Innate immunity serves as a first line of defence and is essential for immunity to viruses. SARS-CoV-2 predominantly infects airways, vascular endothelial cells, alveolar epithelial cells and macrophages. SARS-CoV activates different innate recognition and reaction pathways (Paces *et al.*,2020). SARS-CoV-2, a single-strand RNA virus, first attaches to the pattern recognition receptors (PRRs) via cytosolic retinoic acid-inducible gene I (RIG-I)-like receptors and extracellular and endosomal toll-like receptors (TLRs) (Diao *et al.*, 2020). Upon PRR binding, cascades of downstream signalling trigger secretion of cytokines such as type 1 and type III interferons are important antiviral defences. Other cytokines including interleukin (IL)-1, IL-6, IL-8 and proinflammatory tumour necrosis factor-alpha (TNF-α) are released (Vabret *et al.*, 2020). Chemokines attract more immune responses such as monocytes, polymorphonuclear leukocytes, dendritic cells (DC) and natural killer (NK) cells, as well as produce IP-10, MIG and MCP-1, which are competent in inducing lymphocytes that will recognize the viral antigens presented by DC (Xu *et al.*, 2020). SARS-CoV-2 has a better ability to replicate in pulmonary tissue, evade the antiviral effects of IFN-1 (type 1 Interferon) and IFN-III, activate the innate immune response and cause the formation of cytokines needed for the hiring of adaptive immune response (Garcia, 2020).

15.3.1.2 Adaptive Immune Response to COVID-19

15.3.1.2.1 Cell-Mediated Immunity

Research related to adaptive immune response is limited in COVID-19. The humoral and cellular responses are identified and it induces a Th1 type immune reaction. In patients with COVID-19, the count of CD8+ is reported to be decreased, and in extreme cases, regulatory T cells and memory CD4+ counts are reduced in lymph nodes (Paces *et al.*, 2020). Once the SARS-CoV-2 virus is inside the tissue cells, such as respiratory epithelial cells, it's presented through major histocompatibility complex class 1 (MHC Class-I) proteins to CD8+ cells. CD8+ cells are activated and start to show clonal expansion and development of the virus-specific effector and memory T cells (Jansen *et al.*, 2019). The virus-infected tissue cells are lysed by CD8+ cytotoxic cells. For a short duration of time, whole virus and virus particles are identified by antigen-presenting cells, which are primarily DC, macrophages and viral peptides, are presented through MHC class-II proteins to CD4+ cells (Siu *et al.*, 2009).

Flow cytometric analysis stated that the percentage of CD4+ naïve T cells increased and memory helper T cells decreased in peripheral blood. In severe cases, the percentage of cytotoxic T cells also decreased (Palomares *et al.*, 2017). There was no notable difference in activated total T cells and activated cytotoxic T cells. These individuals also presented with lower levels of regulatory T cells (Azkur *et al.*, 2020). For these instances, the drug targeting lymphocyte proliferation or inhibiting apoptosis such as IL-2 and IL-7 or programmed cell death protein 1 inhibitors could control lymphopenia or restore lymphocyte counts. The percentage of lymphocytes is suggested as a biomarker for severity and recovery (Ng *et al.*, 2016). At the first time point, days 10–12 after symptoms, lymphocyte percentage higher than 20% was categorized as mild to moderate and patients should recover quickly. Patients who had fewer than 20% lymphocytes at the second time point, days 17–19 following the onset of symptoms, were classified as being in recovery. Individuals who had lymphocyte percentages of 5–20% are still considered vulnerable. According to Tan *et al.* (2020), a patient who has a lymphocyte percentage below 5% is considered critically unwell and has a high risk of death.

15.3.1.2.2 Humoral Mediated Immunity

IgM (Immunoglobulin M) antibody levels rise in COVID-19 cases during the acute phase, while IgG antibody levels rise later. The use of a serum infusion from a recuperated patient has been demonstrated to improve lung function and reduce inflammation, oxygenation and viral load; it may also prevent or limit SARS-CoV-2 (To *et al.,* 2020). IgG or IgM antibody levels against the SARS-CoV-2 nucleoprotein (NP) or receptor binding domain (RBD) rise ten days after the onset of symptoms. A normal antibody response is anticipated to include a rise in IgM, which is later followed by the production of IgG (Okba *et al.,* 2020). IgA specific to SARS-CoV may contribute to the mice's protection against SARS vaccination. According to Tamura *et al.* (1990), the establishment of mucosal immunity through IgA may be crucial for averting SARS-CoV infections.

15.4 HERBS WITH ANTI-INFLAMMATORY PROPERTIES

The absence of accurate and effective therapies and treatments for SARS-CoV has triggered some scientists to move to plant-based treatments. This is because abundant drugs are either from plant materials or derivatives of their herbal bioactive components (Al-Kuraishy *et al.,* 2022). In the specialization of medicine associated with herbal investigation, there may be great medical importance in using a single plant species as it may comprise a wide range of bioactive compounds. These phytochemical constituents may act either by themselves or in grouping with other components to produce the desired pharmacological products. The useful impacts of herbs for medicinal purposes come from their biologically active secondary metabolites which include triterpenes, glycosides, steroids and alkaloids (Vaou *et al.,* 2021). At present, the major concern for research in medicine is the innovation and development of anti-inflammatory and antiviral agents. While exhibiting the antiviral properties directly, herbal drugs which have anti-inflammatory properties also play a prominent part in COVID-19 treatment (Ghildiyal *et al.,* 2020).

A genus called *Prunella*, which is a part of herbaceous annual plants, exists in the family of *Labiatae*. Within this genus, *Prunella vulgaris* is the most widely researched. Prunella has exhibited advanced antiviral, antiphlogistic, antibacterial, anti-oxidative, immunoregulatory and antitumorous effects. It is an herbaceous plant which is generally recognized for its self-healing and healing effects. It has also been tested and reported to produce several biological properties which include antimicrobial, anti-inflammation, and anticancerous properties (Bai *et al.,* 2016). *Prunellae spica* comprises several bioactive components in its chemical composition, which includes flavonoids, triterpenes, carbohydrates and phenolic compounds. These have been linked to various protective effects like antiphlogistic, anticancerous, immunosuppressive, neuroprotective activity and anti- human immunodeficiency virus (HIV) activity (Shahzad *et al.,* 2020).

Garlic (*Allium sativum*) is an herbaceous plant which contains aromatic properties, that has been broadly used worldwide. Despite garlic being used commonly, this specific plant contains great medicinal values, as garlic has produced antiviral, antibacterial, antitumor and even antifungal effects. More than 200 chemical constituents present in garlic can shield human health from numerous infections (Tesfaye, 2021). As a medicinal agent, Ginger *(Zingiber officinalis)* has shown great promise. Ginger is a most frequently used spice that is abundant in polysaccharides, terpenes, organic acids, lipids, phenolic compounds and raw fibres. It has been stated that the benefits for health from ginger are primarily due to the phenolic components it possesses like shogao and gingerols. Much research has been conducted supporting that ginger can possess a variety of activities for biological processes like antiphlogistic, antioxidant, antimicrobial, cardiovascular protective, anticancer and antiviral properties (Mao *et al.,* 2019).

Evidence from clinical and basic research indicates that species like *Radix isatidis*, *Rhizoma polygoni*, *Radix bupleuri*, *Cuspidati*, *Herba verbenae*, *Herba patriniae*, *Rhizoma phragmitis* and *Radix glycyrrhizae*. They can show antiphlogistic and immunomodulatory functions by themselves or in grouping with other chemical therapeutics. Remarkably, it has been better established that the general medications for patients with COVID-19 are significant in improving the immune response of the host against viral RNA infection. Hence, it can also be treated as a choice to boost the anti-phlogistic properties and immunity of the host against the contamination of novel coronavirus (Tao *et al.*, 2020).

Currently, as vaccines, there are no efficient treatments against COVID-19. A traditional Chinese herb called *Artemisia annua* can be a better option (Law *et al.*, 2020). Species like *Radix scutellariae, Rhizoma coptidis, Fructus gardenia* and *Cortex phellodendri* have been used for the cure of sepsis for more than 1,700 years. It is broadly used as an efficient medicine in China for treatment against COVID-19 (Qin *et al.*, 2021). In Figure 15.1, traditional medicine and its traditional uses are illustrated. Cinnamon and *Coptis chinensis* is a well-known recorded prescription for the treatment of insomnia, which has been used for centuries. Recently, researchers found that it is capable of reducing the activity of acetylcholinesterase (AChE) and increasing the activity of choline acetyltransferase (ChAT) when AChE is elevated, the cholinergic pathway is activated and cognitive function is improved. Its role in COVID-19 infection and inflammation which has been explored was noteworthy (Qin *et al.*, 2021).

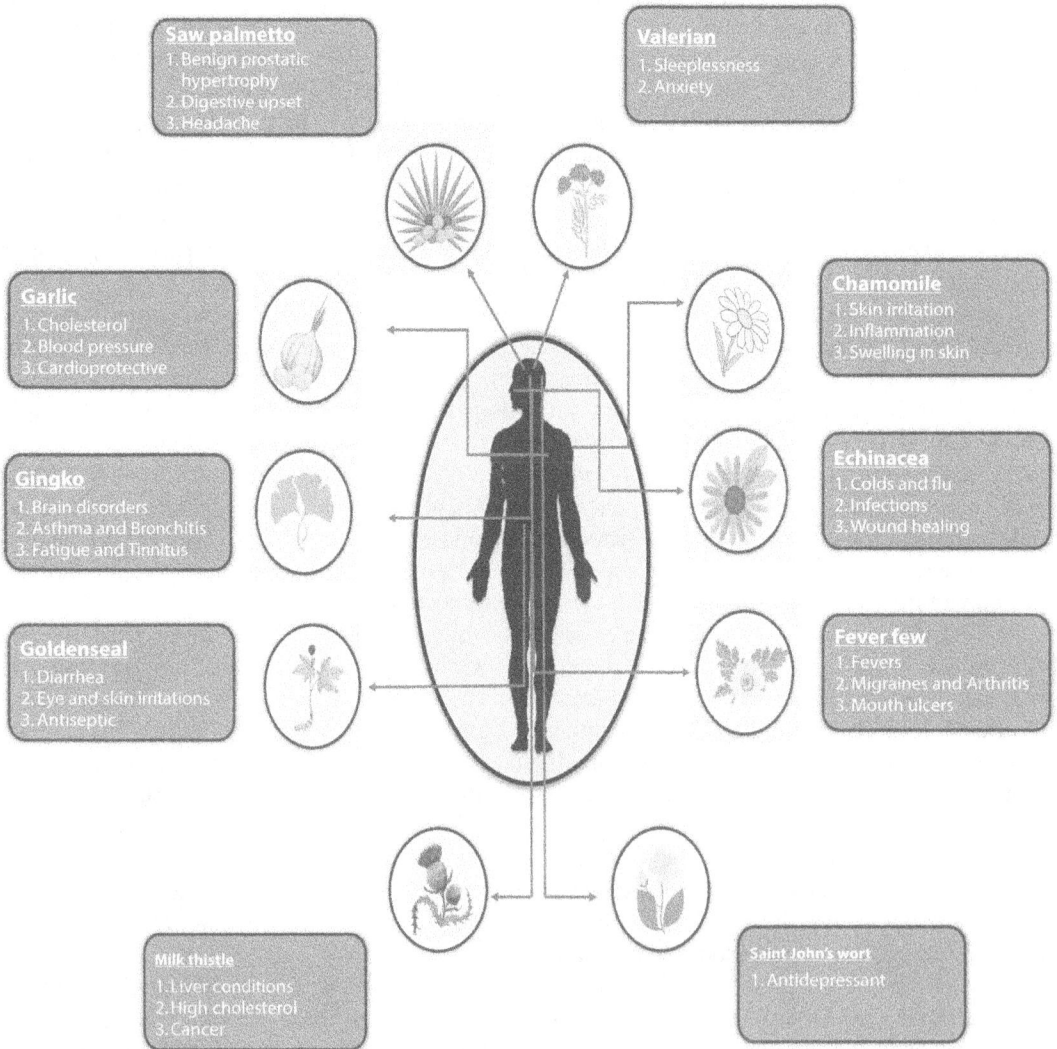

Saw palmetto
1. Benign prostatic hypertrophy
2. Digestive upset
3. Headache

Valerian
1. Sleeplessness
2. Anxiety

Garlic
1. Cholesterol
2. Blood pressure
3. Cardioprotective

Chamomile
1. Skin irritation
2. Inflammation
3. Swelling in skin

Gingko
1. Brain disorders
2. Asthma and Bronchitis
3. Fatigue and Tinnitus

Echinacea
1. Colds and flu
2. Infections
3. Wound healing

Goldenseal
1. Diarrhea
2. Eye and skin irritations
3. Antiseptic

Fever few
1. Fevers
2. Migraines and Arthritis
3. Mouth ulcers

Milk thistle
1. Liver conditions
2. High cholesterol
3. Cancer

Saint John's wort
1. Antidepressant

FIGURE 15.1   Traditional herbs and its traditional uses.

15.5 HERBAL COMPOUNDS AND THEIR EFFECTS ON INFLAMMATION

Anti-inflammatory herbs have proved to be beneficial by combating inflammation that can lead to severe deformity in biological systems. Medicinal herbs contain bioactive compounds that can prevent abominable inflammatory processes and they can also possess anti-inflammatory activity (Yatoo *et al.*, 2018). Therapeutic herbs have been the basis of a wide variety of biologically active constituents for many years and have been used significantly as crude material as well as for treating various diseases. The usage of herbal medicine is becoming more favoured due to the toxicity and adverse reactions of conventional medicines (Kumar *et al.*, 2013). Medicinal herbs consist of phytoconstituents that can avert adverse inflammatory processes and also possess anti-inflammatory activity. Steroids, phenolics, glycosides, alkaloids, flavonoids, polysaccharides, cannabinoids, terpenoids and fatty acids are some of the common phytoconstituents that exist in medicinal plants (Yatoo *et al.*, 2018).

Prunella is an herbaceous plant that is usually well-known for its self-healing as well as healing properties. It is known to have triterpenoids along with phenolic acids, saponins, sterols and associated glycosides, organic acids, volatile oil, flavonoids and saccharides. The diverse pharmacological study investigating Prunella has verified the superior antibacterial, antiviral, antiphlogistic, immunoregulatory, anti-oxidative and antitumour effects it contains (Shahzad *et al.*, 2020). Garlic is known to have numerous sulfur-containing phytoconstituents such as ajoenes, alliin, flavonoids and vinyldithiins, including quercetin. *A. sativum* is a sweet-smelling herbaceous plant which is used by all over the world as a food as well as traditional medicine to care for a variety of diseases. It has several biological properties including antihypertensive, anti-oxidant, anticarcinogenic, antibacterial, antifungal, renoprotective, antidiabetic and anti-atherosclerotic within indigenous medicines (El-Saber Batiha *et al.*, 2020).

P. spica belongs to *Labiatae* family of *P. vulgaris* which is dried spica and it is broadly used in the pharmaceutical and health fields (Wu *et al.*, 2023). It contains various plant-derived compounds by its chemical combination including flavonoids, triterpenes, carbohydrates and phenolic compounds. These components have been related to various protective impacts such as antiphlogistic, immunosuppressive, anticancer, anti-HIV action and neuroprotective activity (Shahzad *et al.*, 2020). Ginger is rich in active components, such as phenolic terpene and phenolic compounds. In ginger, there are also many additional phenolic compounds present, including zingerone, quercetin, 6-dehydrogingerdione and gingerenone-A. Additionally, there are numerous terpene constituents in ginger, including α-curcumene, β-bisabolene, zingiberene, β-sesquiphellandrene and α-farnesene, which are regarded as the main constituent of ginger essential oils (Mao *et al.*, 2019). Paradols, shogaols and gingerols in ginger are extensive phenolic compounds. The main therapeutic actions of ginger are antitumorigenic, anti-emetic actions, immuno-modulatory, anti-apoptotic, anti-inflammatory, antilipidemic and antihyperglycemic. It is a powerful antioxidant substance which may also diminish or avoid the formation of free radicals. Ginger is an herbal medicine with little to no adverse consequences or side effects (Ali *et al.*, 2008).

Forsythiae fructus is a Chinese medicine due to its detoxifying and heat-clearing effects. It is broadly known as a single herb in addition to a compound mixture in Asia. Present pharmacology has proved that *F. fructus* acquires various medicinal effects, both *in vivo* and *in vitro*, such as antiviral, antibacterial and anti-inflammatory activities (Dong *et al.*, 2017). More than 200 chemical constituents were isolated and identified in this plant, together with cyclohexane ethanol derivatives, phenylethanoid glycosides, lignans, terpenes, organic acids, flavonoids, alkaloids and others (Li *et al.*, 2022). *F. fructus* is potent with anti-inflammatory capacities, and is also commonly used to treat chronic and acute inflammation. *F. fructus*'s anti-inflammatory activity is ranked in the top 5 among 81 TCM tested (Wang *et al.*, 2018). Banlangen (*Radix isatidis*) is a dried root belonging to the *Isatis indigotica* Fort. plant. Organic acids, alkaloids and phenylpropanoids are the three major chemical fractions in *Radix isatidis* (Ping *et al.*, 2016). It has been confirmed that it has excellent antioxidant and anti-inflammation activities (Xiao *et al.*, 2014). Earlier research showed that Banlangen has several pharmacological activities such as antiviral, antimicrobial, anticancer and anti-inflammatory effects. When investigated *in vitro*, *Radix isatidis* was found to have antiphlogistic and anti-oxidant activities (Tong *et al.*, 2020).

Herba patriniae has been utilized in China for thousands of years as a classic herbal therapy with detoxifying and heat-clearing properties. There are 233 compounds to be identified in this herbal plant, which include saponins, triterpenoid organic acids, flavonoids, volatiles and iridoids. *Patrinia scabiosifolia* Fisch. is rich in volatiles and triterpenoid saponins though *Patrinia villosa* Juss. contains an increased number of flavonoids. These two sources of the *Herba patriniae* species gave related pharmacological impacts including antiphlogistic, anticancer, antimicrobial, antioxidant, hypnotic and sedative effects (Gong *et al.*, 2021). *Glycyrrhizae radix* is considered one of the most admired and generally used as an herbal medicine and plays a major role in conventional Chinese medicine. Glycyrrhiza consists of several secondary metabolites including various types of flavonoids, triterpene saponins, polysaccharides, coumarins and other phenolic compounds (Li *et al.*, 2020). The *Leguminosae* family of this species has been widely used in various TCM formulas. They possess an extensive range of immunological effects including antiviral, anti-inflammatory, antioxidant and anti-obesity activities (Jiang *et al.*, 2018). The properties of conventional medicine are illustrated in Figure 15.2.

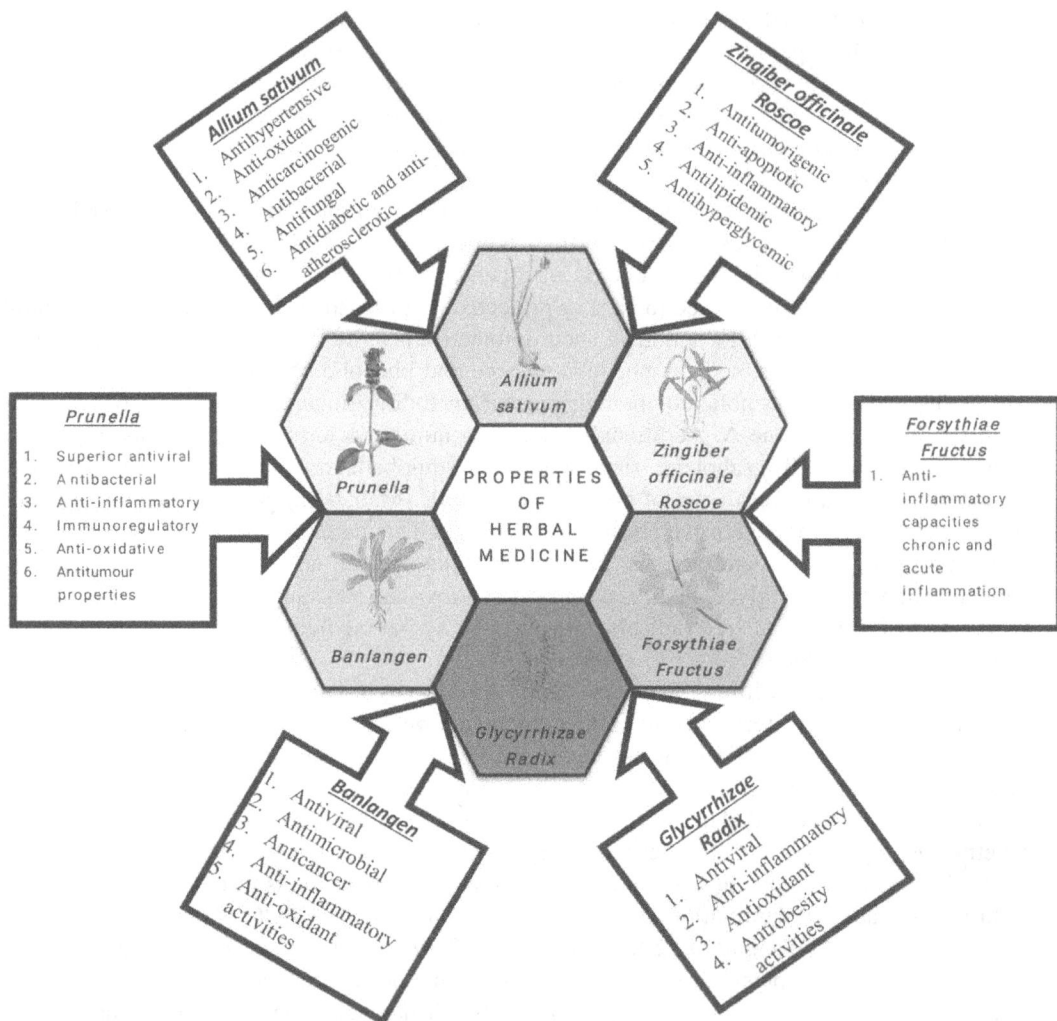

FIGURE 15.2   Properties of herbal medicinal plants.

15.6 HERBAL MEDICINE AND COVID-19: POTENTIAL MECHANISMS OF ACTION

COVID-19 is a recent infectious disease caused by the acute respiratory disorder, SARS-CoV-2 called coronavirus 2 which pertains to the family of coronavirus (Nugraha *et al.*, 2020). The coronavirus has become a serious hazard in a short time and affected the global economic system. The genome of SARS-COV-2 is 70% similar to SARS-COV. To prevent the clinical symptoms of the virus, herbal medicine is used as an alternate treatment that has a wide variety of chemicals that have potential antiviral activity. Herbal components like andrographolide, catechins, hesperidin, scutellarein, biorobin, shikonin, silvestrol, tryptanthrin, myricetin, vitexin quercetin, caffeic acid, luteolin, psoralidin, etc., show prospective inhibitory effect against SARS-CoV-2 (Harwansh and Bahadur, 2022). TCM shows promising results in treating this disease. Numerous herbs can be given in a combined form to get effective results in treating the disease (Chao *et al.*, 2017). However, it is a complex task to find large amounts of an efficient combination herbal pair along with numerous herbs in addition to identifying its prospective mechanisms. A combination of herbs from TCM is broadly used for the healing of diseases. To prevent and treat infections, TCM played a significant part in the avoidance and treatment of transmissible diseases like influenza A (H1N1) in 2009 and severe acute respiratory syndrome in 2003 (Xia *et al.*, 2021).

Herbal remedies also lessened the effects of infectious disorders like SARS-CoV. Evidence supports the theory that making herbal medicine more accessible could help with administration and lessen the risk of COVID-19. The use of phytomedicine in China has been approved by the National Health Commission as an alternative drug to COVID-19 in conjunction with Western medicine (Alam *et al.*, 2021). The flavonoid quercetin is a common compound that usually occurs in vegetables and fruits (Lakhanpal and Rai, 2007). Besides this dietary property, quercetin possesses several biological properties, such as its abilities to combat inflammation, viruses, oxidants, allergies, mood deterioration and cancer, similar to vasoprotective medicines. Preceding studies demonstrate that quercetin has antiviral activity against several viruses, including influenza A virus (IAV), SARS-CoV, enterovirus (EV-71), hepatitis C Virus (HCV), etc. (Al-Doori *et al.*, 2021). In the SARS virus, quercetin exhibits a relatively soaring inhibition rate and half maximal inhibitory concentration (IC_{50}) values of 73 µM and 82% correspondingly, as opposed to SARS-CoV 3CL^{pro} in *Pichia pastoris* fungus (Nguyen *et al.*, 2012).

ToujieQuwen (TJQW) is a TCM that could exert anti-inflammatory, antiviral and antioxidant effects, collectively modulating the immune responses. Therefore, TJQW plays an essential protective aspect in the lungs, persuading potential biotargets before COVID-19. TJQW effects caspase 3 (CASP3), formyl-peptide receptor-2 (FPR2), BCL-2-like protein 1 (BCL2L1), estrogen receptor 1 (ESR1), B2 bradykinin receptor (BDKRB2), ACE and myeloperoxidase (MPO) (Huang *et al.*, 2020). These are identified as potential therapeutic as well as core pharmacological targets, which are reported to affect pulmonary inflammation. TJQW includes 16 components of TCM (Sun *et al.*, 2021). The name of the compounds includes *Forsythia suspense* dried fruit; dried roots of *Bupleurum chinense* DC; Vahl; edible tulip of the *Liliaceae* family, *Tulipa gesneriana* L; *Lonicera japonica* Thunb from the *Caprifoliaceae* family; the root of *Pseudostellaria heterophylla;* leaf of *Isatis indigotica* Fortune ex Lind; dried root of the medicinal plant *Scutellariae baicalensis* Georgi; *Artemisia apiacea* Hance from the *Compositae* family; desiccated roots of *Peucedanum praeruptorum* Dunn; the used shell of *Cryptotympana pustulata* Fabricius; *Fritillaria cirrhosa* D. Don, from the *Liliaceae* family; *Fritillaria thunbergii* Miq., also from *Liliaceae* family; unripe fruit of *Prunus mume;* desiccated sclerotium of *Wolfiporia cocos; Astragalus propinquus* Schischkin, from the *Fabaceae* family; and desiccated root of *Scrophularia ningpoensis* Hemsl (Huang *et al.*, 2020). The bioactive compounds present in TJQW, including oroxylin A, quercetin, umbelliprenin, luteolin, kaempferol, stigmasterol and praeruptorin E, have vital antiviral and anti-inflammatory properties and also have potential to be effective against SARS-CoV-2 infection (Alam *et al.*, 2021).

Emodin, an anthraquinone from *Reynoutria multiflora* (Thunb.), Moldenke and *Rheum officinale* Baill contain antibacterial and antiphlogistic effects. Emodin decreases the contagion of the spike (S) protein pseudo-typed retrovirus to Vero E6 cells and prevents the S protein from binding to ACE2. Emodin blocks the interaction between ACE2 in a concentration-dependent manner among IC_{50} of 200 µM and S protein, suggesting it might be a latent medicinal agent used for SARS therapy (Huang *et al.*, 2020). *Curcuma xanthorrhiza* Roxb or Java turmeric is an herbal plant which is commonly utilised in Southeast Asian countries (Oon *et al.*, 2015). Java turmeric has anti-inflammatory, antimicrobial, antihyperglycemic, antioxidant, antihypertensive, nephroprotective and antiplatelet properties and is an additional agent for systemic lupus erythematosus and has anticancer properties (Nugraha *et al.*, 2020). The occurrence of xanthorrhizol is specific and it can differentiate this plant from *Curcuma longa* (Sayuti and Rusita, 2023). The investigation concerning xanthorrhizol's active constituents indicated that facilitating this compound might decrease inflammatory genes in the muscle, liver and adipose tissue in patients with diabetes mellitus. The xanthorrhizol compound constitutes an immunosuppressant that could be utilised as a remedy for COVID-19 with its ability to reduce pro-inflammatory cytokines (Banerjee *et al.*, 2023). The benefit of xanthorrhizol might reduce the pro-inflammatory reaction in a patient with COVID-19 with or without cytokine release syndrome. Utilizing xanthorrhizol meant for therapy and avoidance of COVID-19 still needs more assessment, particularly in clinical practice (Akram *et al.*, 2023).

Fever is COVID-19's primary clinical symptom. To reduce the fever, neem (*Azadirachta indica*) plants contain valuable outcomes (Paterson *et al.*, 2020). Neem leaves are conventionally boiled and taken orally for managing fever related to COVID-19; using reported antiphlogistic properties in animal experiments. Neem leaf extract and its metabolic constituents contain polysaccharides and flavonoids and have been shown to have direct antiviral actions against many diseases, along with HCV (Lage *et al.*, 2014). This was supported by both the *in silico* docking investigations and animal studies. Specifically to SARS-CoV-2, the molecular docking research has shown to facilitate neem-derived constituents including cycloartenol, nimocin and nimbolin, which bind to the SARS-CoV-2 glycoproteins, membrane (M) and envelope (E) proteins, which inhibits its functions. Neem leaves have favourable immunoregulatory effects, boosting immune reactions in animal studies (Demeke *et al.*, 2021). Natural products and medicinal plants are considered hopeful alternatives to treat or prevent several diseases. In Figure 15.3, herbal plants used in COVID-19 precluding and with therapy are illustrated. Since the beginning of the COVID-19 pandemic in December 2019, a variety of classic herbal medications have been used, resulting in optimistic health consequences among COVID-19 patients, mostly in China (Benarba and Pandiella, 2020).

15.7 INVESTIGATIONS AND CLINICAL TRIALS ON HERBAL MEDICINE FOR COVID-19

Herbal medicines and conventional Chinese medicines can be efficiently used as a protective treatment against COVID-19 and after unbeaten clinical trials, those latent treatments can be encouraged by countries all over the world (Nile and Kai, 2021). Worldwide, researchers and scientists have worked extremely hard to create COVID-19 vaccination and preventive therapy medications. Preclinical investigations have confirmed the potential of several interventions against SARS-CoV-2, but large-scale trials are still needed to validate their effectiveness and control the spread and progression of COVID-19 in the community. As a result, there is a critical need to grow effectual therapies to prevent and treat as well as to raise the mortality rate from COVID-19 (Lythgoe and Middleton, 2020). Furthermore, the outcomes of current research have verified that the combination of TCM and conventional medicine for COVID-19 can enhance the elimination of numerous clinical manifestations and lessen the negative effects of medications used to treat the illness (Setayesh *et al.*, 2022).

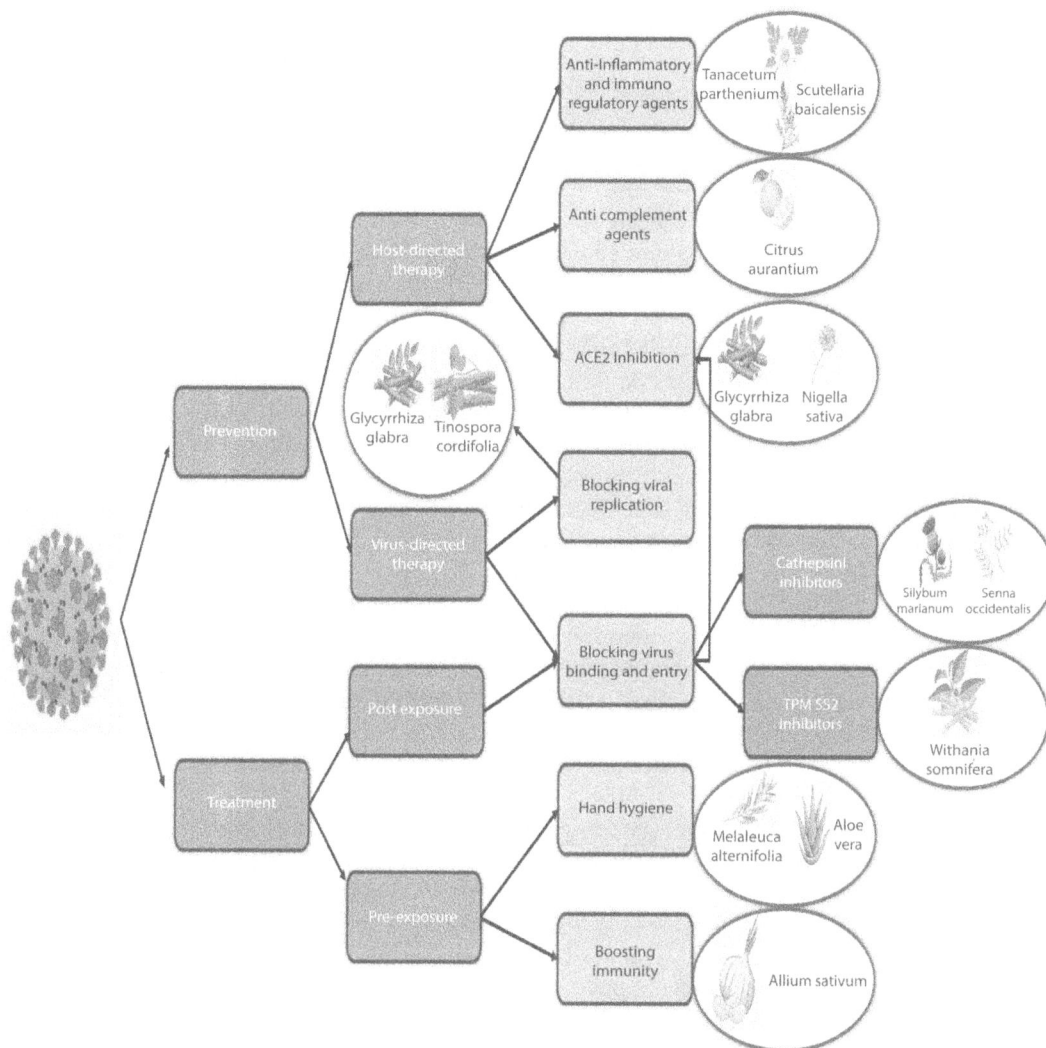

FIGURE 15.3 Herbal plants use in COVID-19 prevention and treatment.

The medicinal properties of TCM in treating COVID-19 have lately been examined by preliminary clinical trials (Han, 2020). It was described that 139 clinical studies together with 47 methods including TCM intervention have been certified by either the American Clinical Trial Registry or the Chinese Clinical Trial Registry. A preliminary randomized controlled trial with a total of 52 patients was conducted in a Wuhan, China hospital at the height of the outbreak (Luo *et al.*, 2020). The TCM therapy exhibited decreases of body temperature, symptom relief, shortened the standard hospital visit, a drop in the incidence of the intermediate stage moving to the rigorous stage and improved comeback rates, however, a more cases are needed to explain the outcomes. It was described that 33 of the 86 COVID-19 instances (together with 31 critical cases and 65 severe cases) have been cured with complementary therapies by reassuring signs and averting the decline to a rigorous stage (Lv *et al.*, 2020). The Chinese herbal formulation Lian Hua Qing Wen puree considerably decreased the manifestation of cough, fever, shortness of breath and fatigue (Xiang *et al.*, 2021).

15.8   SAFETY AND EFFICACY OF HERBAL REMEDIES

In the study of herbal medicine, it is usual to analyse various pharmacological effects of one plant. In herbal medicine, it is evident that a single plant can exhibit a varied series of phytoconstituents (Gribner *et al.*, 2023). By showing antiviral effects directly, plants with medicinal properties can exhibit pleiotropic roles with testified anti-inflammatory properties for the administration of COVID-19, as the increase of provocative markers like C-reactive protein (CRP), IL-6, and the erythrocyte sedimentation rate have been associated with severe disease with poor outcomes for patients with COVID-19 (Gilzad-Kohan and Jamali, 2020). As a plant containing several phyto-chemicals with the capacity to produce multi-modal mechanisms of action, acetone, methanol, and ethanol extracts also indicated that, for anti-inflammatory properties for animals in the laboratory, to focus on the variation in levels of inflammatory mediators and cytokines. Although this plant has been shown to have strong effects on immunological and inflammatory responses, its toxicity profile is yet unknown (Lim *et al.*, 2021). With drugs that have antiviral, anti-inflammatory, and immune-regulating qualities, TCM is effective in preventing and controlling COVID-19. This can lower hospitalisation rates and hasten the resolution of symptoms (Kang *et al.*, 2022).

P. vulgaris L. is an herbal Chinese medicine which has high efficacy against several diseases and has substantial anti-inflammatory, antiviral, immunomodulatory effects and anti-oxidative properties. Initial studies have tested for inhibition of *P. vulgaris* on Ebola, HCV and HIV infections (Yang *et al.*, 2023). The spike of *Prunella vlgaris*, which is also called *P. spica*, is frequently employed for the treatment and prevention of several diseases in TCM. These components have been connected to various defensive aspects like anti-inflammatory, anti-cancer, neuroprotective, anti-HIV activity, and immunosuppressive activity. The results of the experiment confirmed the major antiviral capability that Prunella has exhibited against viral infections like herpes simplex virus and a study has been demonstrated for testing the efficacy of viral infections like COVID-19 (Shahzad *et al.*, 2020).

Though there are some medicinal properties in *A. sativum*, there is no proof of its efficacy for COVID-19. Severe effects caused by it include allergy, nausea, light-headedness, coagulation and modification in platelet function, hypotension, diaphoresis and a burning sensation in the gastrointestinal tract have been linked to the usage of garlic (Omokhua-Uyi and Van Staden, 2021). Some health benefits have been associated with *Z. officinalis*, but no proof of its capability to treat COVID-19. A few side effects associated with ginger include mouth irritation, diarrhoea and heartburn. Though no reports on its contraindication have been stated, there were a few concerns that anticoagulants were formed due to their interconnection with ginger (Jafarzadeh *et al.*, 2021). Ceylon cinnamon ethanolic extracts have anti-inflammatory properties and antagonize TLR2 and TLR4 activation in a dose-dependent manner with the least impact on cell culture viability. We discovered a synergy between the antiphlogistic properties of the various compounds: compared to the pure active components of cinnamon, the complex combination exhibited increased effectiveness (Lucas *et al.*, 2021).

As a potent nuclear factor erythroid 2-related factor 2 (NRF2) inducer, cinnamaldehyde helps to detoxify reactive oxygen species (ROS)/reactive nitrogen species (RNS). Additionally, it can inhibit metastasis using mitigation and angiogenesis of the phosphoinositide 3-kinase (PI3K)/protein kinase B (Akt) pathway. A specific repression of angiogenesis along with vascular endothelial growth factor receptor 2 kinase was revealed for a water-based extract of Ceylon cinnamon. An extract from Ceylon cinnamon was shown to reduce the atherogenic risk and shield the aorta from dexamethasone-induced atherosclerosis in model organisms (Cuevas *et al.*, 2015).

15.9   TRADITIONAL HERBAL MEDICINE AND MODERN MEDICINE

15.9.1   TRADITIONAL HERBAL MEDICINE FOR COMBATING COVID-19

COVID-19 was declared a pandemic as new cases of infected pneumonia increased worldwide. TCM combined with modern medicine has shown the greatest positive result in combating COVID-19

(Zhang *et al.*, 2020). The infection caused by SARS-CoV-2 is related to ACE2 (Wang *et al.*, 2020). The S protein of COVID-19 binds to the ACE2 receptor and drives the virus to reproduce and attack new cells. The clinical symptoms include fever, cough, fatigue, diarrhoea and, in severe cases, multi-organ dysfunction (Yu-Jie *et al.*, 2020). A literature analysis found that TCM has anti-inflammatory, antiviral and immunoregulation functions. The Office of the State Administration of TCM has selected QingfeiPaidu decoction as a general prescription for the treatment and prevention of COVID-19 (Cui and He, 2023). This prescription is based on Shegan Mahuang decoction, Xiaochaihu Tang, Mahuang decoction, Wuling powder and Ju-Zhi-Jiang decoction and is suitable for patients with mild symptoms as well as critical and convalescent patients (Liu *et al.*, 2020). Through the TCM analysis combined with clinical observation, QingfeiPaidu decoction could play an antiviral effect and shorten patient treatment time (Shen *et al.*, 2020). The huashibaidu prescription could monitor a *qi* of the lung and is recommended for severe pneumonia (Zhao *et al.*, 2020). Convalescent patients often have a *qi* deficiency. Shengmai San is a classical prescription for the treatment of *qi* and *fin* deficiency, it could also increase the recovery time and prevent the occurrence of sequelae (Zheng *et al.*, 2022).

15.9.2 Modern Medicine for Combating COVID-19

For COVID-19 patients, the commonly used physical life support systems are ventilators, artificial liver support systems (ALSS), blood purification systems and extracorporeal membrane oxygenation (ECMO) (Wu *et al.*, 2021). Disease progression can be categorized into three stages– early infection, pulmonary phase with severe hyperinflammation and the renin-angiotensin system-mediated bradykinin storm (RBS) phase (Farah Yusuf Mohamud *et al.*, 2020). During the pulmonary phase, the antiviral drug remdesivir is used for the treatment and dexamethasone, an anti-inflammatory drug, is combined with oxygen for the treatment of the pulmonary phase to hyperinflammation and RBS phase of COVID-19. The efficacy of remdesivir in some clinical trials has failed to show any clinical benefits (Spinner *et al.*, 2020).

Beigel *et al.* (2020) in their research suggested that remdesivir was superior to a placebo that shortens the recovery time of hospitalized adults with COVID-19. Evidence of remdesivir reducing respiratory tract infections has also been shown (Beigel *et al.*, 2020). Monoclonal antibodies and convalescent plasma may be useful for severely ill patients as treatment of passive antibodies. The new hypothesis states that RBS developed a wide range of symptoms caused by COVID-19 infection. Existing US Food and Drug Administration (FDA)-approved pharmaceuticals used for treating RBS including vitamin D may help reverse and treat COVID-19 (Garvin *et al.*, 2020).

15.10 FUTURE DIRECTIONS AND RESEARCH OPPORTUNITIES

COVID-19 is a dangerous risk to worldwide health. Few medicines have shown efficacies in treatment clinically against SARC-CoV-2 and its complications. Drugs with different combinations and various labelled uses are presently being used as supportive treatments (Asif *et al.*, 2020). Though the recent studies in drug repositioning initiate current development, there is no selective or potential treatment and drug approved for COVID-19. Hence, to combat the effects of COVID-19, more research clearly must be conducted. Enhanced investigations are exhibited based on clinical trials on these compounds (Wondmkun and Mohammed, 2020). The overuse of these medicinal herbs to elucidate those beneficial components is our perception of the capability for treating COVID-19. The accurate mode of action for fulfilling clinical requirements needs to be completely understood. The presence of several products derived from plants with their antiviral abilities through various mechanisms of action would be a potential medication for COVID-19. Testing of several herbal extracts on SARS-CoV-2 from different countries, together with the usage of medicines to compete with this virus, which would ultimately enhance the recovery process, is also greatly recommended (Salama *et al.*, 2021).

Collectively with their favourable ability to avert herb-based preparations, resistance development, should ideally form a core constituent of exploration for therapy of COVID-19 (Attah *et al.*, 2021). While the consequence of herb-based preparations ultimately depends on the constituent of herbs interacting with organic macromolecules, their use further depends significantly on pharmacokinetics (Fugh-Berman, 2000). As active antiviral drugs and defensive vaccines are not accessible for the remedy of numerous viruses, eradicating this viral infection appears difficult. A natural product serves as a great basis of biodiversity for a rising pioneering antiviral, with further structure-activity relations and effective and therapeutic medical approaches to combat viral infections (Boukhatem and Setzer, 2020). Numerous aspects are to be considered in the antiviral action of utilising and developing therapeutic herbs, such as the extraction technique used (Mukhtar *et al.*, 2008). Further research should also examine the probability of combining this treatment with alternative natural constituents or with usual medicines, as a multiple-target solution might help reduce the infection probability of drug-resistant viruses. We hope that organic remedies, including pure oil compounds, essential oils obtained from medicinal plants and sweet-smelling herbs, will carry over to play a significant role and contribute to the growth and improvement of anti-coronavirus drugs (Basavegowda and Baek, 2022).

15.11 CONCLUSION

The chapter delves into the exploration of herbal medicinal plants as a valuable reservoir of natural compounds with potential anti-inflammatory properties. The focal point revolves around the anti-inflammatory activity exhibited by these medicinal plants, underscoring their pivotal role in modulating the immune response and potentially alleviating the inflammatory processes. Notably, the chapter places specific emphasis on the relevance of herbal medicinal plants within the context of COVID-19. It delves into the exploration of their anti-inflammatory properties as a strategic avenue for managing the inflammatory responses intricately linked with the virus.

A key theme within the chapter is the recognition of natural compounds inherent in herbal medicinal plants as valuable resources for the development of anti-inflammatory agents. This perspective posits that such compounds may offer a more holistic and potentially safer approach to addressing inflammation when compared to synthetic alternatives. Furthermore, the chapter explores the potential of herbal remedies in immunomodulation, elucidating their ability to balance the immune system's reaction and mitigate excessive inflammation, particularly observed in extreme cases of COVID-19. Positioning the use of herbal medicinal plants as a complementary approach to conventional medical interventions, the chapter advocates for a synergistic integration of traditional and modern medicine in the management of inflammation associated with COVID-19. In doing so, it presents a nuanced perspective that recognizes the potential benefits derived from combining the strengths of both approaches.

ACKNOWLEDGEMENT

All the authors are thankful to their respective Universities and Institutions for their valuable support. The corresponding author is thankful to Skill Development Centre, BEICH, RUSA2.0 (SDC Project order No. BU/RUSA/SDC/BEICH 2020/15; 28.12.2020), Bharathiar University for providing financial support.

DATA AVAILABILITY

The authors confirm that the data supporting the findings of this study are available within the book/chapter.

REFERENCES

Abidharini, J. D., Souparnika, B. R., Elizabeth, J., Vishalini, G., Nihala, S., Shanmugam, V., … Anand, A. V. Herbal formulations in fighting against the SARS-CoV-2 infection. In *Ethnopharmacology and Drug Discovery for COVID-19: Anti-SARS-CoV-2 Agents from Herbal Medicines and Natural Products*. Singapore: Springer Nature Singapore. 2023, pp. 85–113.

Akram, M., Chishti, M. A., Laila, U., Zainab, R., Iftikhar, M., Talha, M. & Parmar, P. Herbal Remedies with potential COVID-19 activities. *International Archives of Integrated Medicine*. 2023, *10*(9), 27–36.

Alam, S., Sarker, M. M. R., Afrin, S., Richi, F. T., Zhao, C., Zhou, J. R. & Mohamed, I. N. Traditional herbal medicines, bioactive metabolites, and plant products against COVID-19: Update on clinical trials and mechanism of actions. *Frontiers in Pharmacology*. 2021, *12*, 671498.

Al-Doori, A., Ahmed, D., Kadhom, M. & Yousif, E. Herbal medicine as an alternative method to treat and prevent COVID-19. *Baghdad Journal of Biochemistry and Applied Biological Sciences*. 2021, *2*(01), 1–20.

Ali, B. H., Blunden, G., Tanira, M. O.& Nemmar, A. Some phytochemical, pharmacological and toxicological properties of ginger (*Zingiber officinale* Roscoe*):* A review of recent research. *Food and Chemical Toxicology*. 2008, *46*(2), 409–420.

Al-Kuraishy, H. M., Al-Fakhrany, O. M., Elekhnawy, E., Al-Gareeb, A. I., Alorabi, M., De Waard, M. & Batiha, G. E. S. Traditional herbs against COVID-19: Back to old weapons to combat the new pandemic. *European Journal of Medical Research*. 2022, *27*(1), 186.

Anand, A. V., Balamuralikrishnan, B., Kaviya, M., Bharathi, K., Parithathvi, A., Arun, M., … Dhama, K. Medicinal plants, phytochemicals, and herbs to combat viral pathogens including SARS-CoV-2. *Molecules*. 2021, *26*(6), 1775.

Ang, L., Lee, H. W., Choi, J. Y., Zhang, J. & Lee, M. S. Herbal medicine and pattern identification for treating COVID-19: A rapid review of guidelines. *Integrative Medicine Research*. 2020, *9*(2), 100407.

Arumugam, V. A., Thangavelu, S., Fathah, Z., Ravindran, P., Sanjeev, A. M. A., Babu, S., … Tiwari, R. COVID-19 and the world with co-morbidities of heart disease, hypertension and diabetes. *Journal of Pure and Applied Microbiology*. 2020, *14*(3), 1623–1638.

Asif, M., Saleem, M., Saadullah, M., Yaseen, H. S., & Al Zarzour, R. COVID-19 and therapy with essential oils having antiviral, anti-inflammatory, and immunomodulatory properties. *Inflammopharmacology*. 2020, *28*, 1153–1161.

Attah, A. F., Fagbemi, A. A., Olubiyi, O., Dada-Adegbola, H., Oluwadotun, A., Elujoba, A. & Babalola, C. P. Therapeutic potentials of antiviral plants used in traditional African medicine with COVID-19 in focus: A Nigerian perspective. *Frontiers in Pharmacology*. 2021, *12*, 596855.

Azkur, A. K., Akdis, M., Azkur, D., Sokolowska, M., van de Veen, W., Brüggen, M. C. & Akdis, C. A. Immune response to SARS-CoV-2 and mechanisms of immunopathological changes in COVID-19. *Allergy*. 2020, *75*(7), 1564–1581.

Bai, Y., Xia, B., Xie, W., Zhou, Y., Xie, J., Li, H. & Li, C. Phytochemistry and pharmacological activities of the genus *Prunella*. *Food Chemistry*. 2016, *204*, 483–496.

Banerjee, A., Somasundaram, I., Das, D., Jain Manoj, S., Banu, H., Mitta Suresh, P. & Pathak, S. Functional foods: A promising strategy for restoring gut microbiota diversity impacted by SARS-CoV-2 variants. *Nutrients*. 2023, *15*(11), 2631.

Barnes, P. M. Complementary and alternative medicine use among adults, United States. *Advance Data*. 2002, *343*, 1–19.

Basavegowda, N. & Baek, K. H. Combination strategies of different antimicrobials: An efficient and alternative tool for pathogen inactivation. *Biomedicines*. 2022, *10*(9), 2219.

Beigel, J. H., Tomashek, K. M. & Dodd, L. E. Remdesivir for the treatment of Covid-19-preliminary report. *New England Journal of Medicine*. 2020, *383*(10), 994.

Benarba, B. & Pandiella, A. Medicinal plants as sources of active molecules against COVID-19. *Frontiers in Pharmacology*. 2020, *11*, 1189.

Bharathi, K., Sivasangar Latha, A., Jananisri, A., Bavyataa, V., Rajan, B., Balamuralikrishnan, B., … Anand, A. V. Antiviral properties of South Indian plants against SARS-CoV-2. In *Ethnopharmacology and Drug Discovery for COVID-19: Anti-SARS-CoV-2 Agents from Herbal Medicines and Natural Products*. Springer. 2023, pp. 447–478.

Bhotla, H. K., Balasubramanian, B., Arumugam, V. A., Pushpraj, K., Easwaran, M., Baskaran, R., … Meyyazhagan, A. Insinuating cocktailed components in biocompatible-nanoparticles could act as an impressive neo-adjuvant strategy to combat COVID-19. *Natural Resources for Human Health*. 2021, *1*(1), 3–7.

Bhotla, H. K., Balasubramanian, B., Meyyazhagan, A., Pushparaj, K., Easwaran, M., Pappusamy, M., ... Alfalih, A. M. Opportunistic mycoses in COVID-19 patients/survivors: Epidemic inside a pandemic. *Journal of Infection and Public Health*. 2021, *14*(11), 1720–1726.

Bhotla, H. K., Kaul, T., Balasubramanian, B., Easwaran, M., Arumugam, V. A., Pappusamy, M., ... Meyyazhagan, A. Platelets to surrogate lung inflammation in COVID-19 patients. *Medical Hypotheses*. 2020, *143*, 110098.

Boro, A., Souparnika, B. R., Atchaya, S., Ramya, S., Senthilkumar, N., Velayuthaprabhu, S., Rengarajan R. L., & Vijaya Anand, A. Antiviral activities of the flavonoids compound against COVID-19 and other viruses causing ailments. In Chen, JT. (eds). *Bioactive Compound against SARS-CoV-2*. CRC Press. 2023, pp. 68–81.

Boukhatem, M. N. & Setzer, W. N. Aromatic herbs, medicinal plant-derived essential oils, and phytochemical extracts as potential therapies for coronaviruses: Future perspectives. *Plants*. 2020, *9*(6), 800.

Chandra Manivannan, A., Malaisamy, A., Eswaran, M., Meyyazhagan, A., Arumugam, V. A., Rengasamy, K. R. R., ... Liu, W.-C. Evaluation of clove phytochemicals as potential antiviral drug candidates targeting SARS-CoV-2 main protease: Computational Docking, molecular dynamics simulation, and pharmacokinetic profiling. *Frontiers in Molecular Biosciences*. 2022, *9*, 918101.

Chao, J., Dai, Y., Verpoorte, R., Lam, W., Cheng, Y. C., Pao, L. & Chen, S. Major achievements of evidence-based traditional Chinese medicine in treating major diseases. *Biochemical Pharmacology*. 2017, *139*, 94–104.

Cuevas, S., Yang, Y., Konkalmatt, P., Asico, L. D., Feranil, J., Jones, J. & Jose, P. A. Role of nuclear factor erythroid 2-related factor 2 in the oxidative stress-dependent hypertension associated with the depletion of dj-1. *Hypertension*. 2015, *65*(6), 1251–1257.

Cui, Y. & He, W. Study on the law of TCM medication for mild and medium-sized coronavirus pneumonia based on an ancient doctor case cloud platform. *MEDS Chinese Medicine*. 2023, *5*(9), 1–9.

Demeke, C. A., Woldeyohanins, A. E. & Kifle, Z. D. Herbal medicine uses for the management of COVID-19: A review article. *Metabolism Open*. 2021, *12*, 100141.

Diao, B., Wang, C., Tan, Y., Chen, X., Liu, Y., Ning, L., Chen, L., Li, M., Liu, Y. & Wang, G. online may 1, 2020. Reduction and functional exhaustion of T cells in patients with coronavirus disease 2019 (COVID-19). *Frontiers in Immunology*. 2020, *11*, 287.

Dong, Z., Lu, X., Tong, X., Dong, Y., Tang, L. & Liu, M. *Forsythiae Fructus*: A review on its phytochemistry, quality control, pharmacology and pharmacokinetics. *Molecules*. 2017, *22*(9), 1466.

El-Saber Batiha, G., Magdy Beshbishy, A., G. Wasef, L., Elewa, Y. H., A. Al-Sagan, A., Abd El-Hack, M. E. & Prasad Devkota, H. Chemical constituents and pharmacological activities of garlic (*Allium Sativum L.*): A review. *Nutrients*. 2020, *12*(3), 872.

Farah Yusuf Mohamud, M., Garad Mohamed, Y., Mohamed Ali, A. & Ali Adam, B. Loss of taste and smell are common clinical characteristics of patients with COVID-19 in Somalia: A retrospective double Centre study. *Infection and Drug Resistance*. 2020, *13*, 2631–2635.

Fugh-Berman, A. Herb-drug interactions. *The Lancet*. 2000, *355*(9198), 134–138.

García, L. F. Immune response, inflammation, and the clinical spectrum of COVID-19. *Frontiers in Immunology*. 2020, *11*, 1441.

Garvin, M. R., Alvarez, C., Miller, J. I., Prates, E. T., Walker, A. M., Amos, B. K. & Jacobson, D. A mechanistic model and therapeutic interventions for COVID-19 involving a RAS-mediated bradykinin storm. *Elife*. 2020, *9*, e59177.

Ghildiyal, R., Prakash, V., Chaudhary, V. K., Gupta, V. & Gabrani, R. Phytochemicals as antiviral agents: Recent updates. Plant-derived bioactives: Production, properties and therapeutic applications. *Plant-derived Bioactives*. 2020, 279–295.

Gilzad-Kohan, H. & Jamali, F. Anti-inflammatory properties of drugs used to control COVID-19 and their effects on the renin-angiotensin system and angiotensin-converting enzyme-2. *Journal of Pharmacy & Pharmaceutical Sciences*. 2020, *23*, 259–277.

Gong, L., Zou, W., Zheng, K., Shi, B. & Liu, M. The *Herba Patriniae* (*Caprifoliaceae*): A review on traditional uses, phytochemistry, pharmacology and quality control. *Journal of Ethnopharmacology*. 2021, *265*, 113264.

Gribner, C., Moura, P. F., Rigoni, A. A. R., Clemente, M., Rattmann, Y. D. & Gomes, E. C. Herbal medicines from the industry in the unified health system: Challenge faced by medical professionals. *Brazilian Journal of Pharmaceutical Sciences*. 2023, *58*, e18701.

Guan, W. J., Ni, Z. Y., Hu, Y., Liang, W. H., Ou, C. Q., He, J. X., Liu, L., Shan, H., Lei, C., Hui, D. S. C., Du, B., Li, L., Zeng, G., Yuen, K., Chen, R., Tang, C., Wang, T., Chen, P., Xiang, J. & Zhong, N. Clinical characteristics of coronavirus disease 2019 in China. *New England Journal of Medicine*. 2020, *382*(18), 1708–1720.

Guo, Y. R., Cao, Q. D., Hong, Z. S., Tan, Y. Y., Chen, S. D., Jin, H. J. & Yan, Y. The origin, transmission and clinical therapies on coronavirus disease 2019 (COVID-19) outbreak-an update on the status. *Military Medical Research*. 2020, 7, 1–10.

Han, Y. Y. Application of integrative medicine protocols in treatment of coronavirus disease 2019. *Chinese Traditional and Herbal Drugs*. 2020, 24, 878–882.

Harwansh, R. K. & Bahadur, S. Herbal medicines to fight against COVID19: New battle with an old weapon. *Current Pharmaceutical Biotechnology*. 2022, 23(2):235–260.

Huang, F., Li, Y., Leung, E. L. H., Liu, X., Liu, K., Wang, Q. & Luo, L. A review of therapeutic agents and Chinese herbal medicines against SARS-COV-2 (COVID-19). *Pharmacological Research*. 2020, 158, 104929.

Huang, J., Tao, G., Liu, J., Cai, J., Huang, Z. & Chen, J. X. Current prevention of COVID-19: Natural products and herbal medicine. *Frontiers in Pharmacology*. 2020, 11, 588508.

Huang, Y., Zheng, W. J., Ni, Y. S., Li, M. S., Chen, J. K., Liu, X. H. & Li, J. Q. Therapeutic mechanism of ToujieQuwen granules in COVID-19 based on network pharmacology. *BioData Mining*. 2020, 13, 1–21.

Jafarzadeh, A., Jafarzadeh, S. & Nemati, M. Therapeutic potential of ginger against COVID-19: Is there enough evidence? *Journal of Traditional Chinese Medical Sciences*. 2021, 8(4), 267–279.

Jansen, J. M., Gerlach, T., Elbahesh, H., Rimmelzwaan, G. F. & Saletti, G. Influenza virus-specific CD4+ and CD8+ T cell-mediated immunity induced by infection and vaccination. *Journal of Clinical Virology*. 2019, 119, 44–52.

Jiang, Y. X., Dai, Y. Y., Pan, Y. F., Wu, X. M., Yang, Y., Bian, K. & Zhang, D. D. Total flavonoids from *Radix Glycyrrhiza* exert anti-inflammatory and antitumorigenic effects by inactivating INOS signaling pathways. *Evidence-Based Complementary and Alternative Medicine*. 2018, 2018, 6714282.

Kang, X., Jin, D., Jiang, L., Zhang, Y., Zhang, Y., An, X. & Lian, F. Efficacy and mechanisms of traditional Chinese medicine for COVID-19: A systematic review. *Chinese Medicine*. 2022, 17(1), 30.

Kaviya, M., Geofferina, I. P., Poornima, P., Rajan, A. P., Balamuralikrishnan, B., Arun, M., … & Anand, A. V. Dietary plants, spices, and fruits in curbing SARS-CoV-2 virulence. In *Ethnopharmacology and Drug Discovery for COVID-19: Anti-SARS-CoV-2 Agents from Herbal Medicines and Natural Products*. Singapore: Springer Nature Singapore. 2023, pp. 265–316.

Khumalo, G. P., Van Wyk, B. E., Feng, Y. & Cock, I. E. A review of the traditional use of southern African medicinal plants for the treatment of inflammation and inflammatory pain. *Journal of Ethnopharmacology*. 2022, 283, 114436.

Kumar, S., Bajwa, B. S., Kuldeep, S. & Kalia, A. N. Anti-inflammatory activity of herbal plants: A review. *International Journal of Advances in Pharmacy, Biology and Chemistry*. 2013, 2(2), 272–281.

Lage, G. A., Medeiros, F. D. S., Furtado, W. D. L., Takahashi, J. A., Filho, J. D. D. S. & Pimenta, L. P. S. The first report on flavonoid isolation from *Annona Crassiflora* Mart. *Natural Product Research*. 2014, 28(11), 808–811.

Lakhanpal, P. & Rai, D. K. Quercetin: A versatile flavonoid. *Internet Journal of Medical Update*. 2007, 2(2), 22–37.

Latha, A. S., Anita, P. P., Sidhic, N., James, E., Kaviya, M., Bharathi, K., … & Anand, A. V. Phytonutrients and secondary metabolites to cease SARS-CoV-2 loop. In *Bioactive Compounds Against SARS-CoV-2*. CRC Press. 2023, pp. 111–124.

Law, S., Leung, A. W. & Xu, C. Is the traditional Chinese herb "*Artemisia Annua*" possible to fight against COVID-19? *Integrative Medicine Research*. 2020, 9(3), 100474.

Li, C., Sun, C. Z., Yang, Y. H., Ma, N., Wang, Y. J., Zhang, F. X. & Pei, Y. H. A novel strategy by integrating chemical profiling, molecular networking, chemical isolation, and activity evaluation to target isolation of potential anti-ACE2 candidates in *Forsythiae Fructus*. *Phytomedicine*. 2022, 96, 153888.

Li, F., Liu, B., Li, T., Wu, Q., Xu, Z., Gu, Y. & Lei, H. Review of constituents and biological activities of triterpene saponins from *Glycyrrhizae Radix* et *Rhizoma* and its solubilization characteristics. *Molecules*. 2020, 25(17), 3904.

Lim, X. Y., Teh, B. P. & Tan, T. Y. C. Medicinal plants in COVID-19: Potential and limitations. *Frontiers in Pharmacology*. 2021, 12, 611408.

Ling, C. Q. Traditional Chinese medicine is a resource for drug discovery against 2019 novel coronavirus (SARS-CoV-2). *Journal of Integrative Medicine*. 2020, 18(2), 87.

Liu, Y., Liu, L., Cao, M. M., Wen, J., Zhou, S., Jiang, H. & Yang, Y. H. Literature analysis of Chinese patent medicine treatment in the observation period of clinical management of corona virus disease 2019 (trial 6th edition). *Clinical Medicine Journal*. 2020, 18(2), 62–66.

Lu, X. I. A., Yujing, S. H. I., Jie, S. U., Friedemann, T., Zhenggang, T. A. O., Lu, Y. & Schröder, S. ShufengJiedu, a promising herbal therapy for moderate COVID-19: Antiviral and anti-inflammatory properties, pathways of bioactive compounds, and a clinical real-world pragmatic study. *Phytomedicine.* 2021, *85*, 153390.

Lucas, K., Fröhlich-Nowoisky, J., Oppitz, N. & Ackermann, M. Cinnamon and Hop extracts as potential immunomodulators for severe COVID-19 cases. *Frontiers in Plant Science.* 2021, *12*, 589783.

Luo, L., Jiang, J., Wang, C., Fitzgerald, M., Hu, W., Zhou, Y. & Chen, S. Analysis on herbal medicines utilized for treatment of COVID-19. *Acta Pharmaceutica Sinica B.* 2020, *10*(7), 1192–1204.

Lv, R., Wang, W. & Li, X. Novel coronavirus pneumonia suspected cases treated with Lianhua Qingwen Decoction: A clinical observation of 63 cases. *Journal of Traditional Chinese Medicine.* 2020, *61*(2020), 695–659.

Lythgoe, M. P. & Middleton, P. Ongoing clinical trials for the management of the COVID-19 pandemic. *Trends in Pharmacological Sciences.* 2020, *41*(6), 363–382.

Mao, Q. Q., Xu, X. Y., Cao, S. Y., Gan, R. Y., Corke, H., Beta, T. & Li, H. B. Bioactive compounds and bioactivities of ginger (*Zingiber officinale* Roscoe). *Foods.* 2019, *8*(6), 185.

Meyyazhagan, A., Pushparaj, K., Balasubramanian, B., Kuchi Bhotla, H., Pappusamy, M., Arumugam, V. A., … Di Renzo, G. C. COVID-19 in pregnant women and children: Insights on clinical manifestations, complexities, and pathogenesis. *International Journal of Gynecology & Obstetrics.* 2022, *156*(2), 216–224.

Mukhtar, M., Arshad, M., Ahmad, M., Pomerantz, R. J., Wigdahl, B. & Parveen, Z. Antiviral potentials of medicinal plants. *Virus Research.* 2008, *131*(2), 111–120.

Ng, O. W., Chia, A., Tan, A. T., Jadi, R. S., Leong, H. N., Bertoletti, A. & Tan, Y. J. Memory T cell responses targeting the SARS coronavirus persist up to 11 years postinfection. *Vaccine.* 2016, *34*(17), 2008–2014.

Nguyen, T. T., Woo, H. J., Kang, H. K., Nguyen, V. D., Kim, Y. M. & Kim, D. W. Flavonoid mediated inhibition of SARS coronavirus 3C-like protease expressed in Pichia pastoris. *Biotechnology Letters.* 2012, *34*, 831–838.

Nile, S. H. & Kai, G. Recent clinical trials on natural products and traditional Chinese medicine combating the COVID-19. *Indian Journal of Microbiology.* 2021, *61*, 10–15.

Nugraha, R. V., Ridwansyah, H., Ghozali, M., Khairani, A. F.& Atik, N. Traditional herbal medicine candidates as complementary treatments for COVID-19: A review of their mechanisms, pros and cons. *Evidence-Based Complementary and Alternative Medicine.* 2020, *2020, p. 2560645.*

Okba, N. M. A., Müller, M. A., Li, W., Wang, C., GeurtsvanKessel, C. H., Corman, V. M., Lamers, M. M., Sikkema, R. S., de Bruin, E., Chandler, F. D., Yazdanpanah, Y., Le Hingrat, Q., Descamps, D., Houhou-Fidouh, N., Reusken, C. B. E. M., Bosch, B., Drosten, C., Koopmans, M. P. G., & Haagmans, B. L. Severe acute respiratory syndrome coronavirus-2 specific antibody responses in coronavirus disease 2019 patients. *Emerging Infectious Diseases.* 2020, *26*(7), 1478–1488.

Omokhua-Uyi, A. G. & Van Staden, J. Natural product remedies for COVID-19: A focus on safety. *South African Journal of Botany.* 2021, *139*, 386–398.

Oon, S. F., Nallappan, M., Tee, T. T., Shohaimi, S., Kassim, N. K., Sa'ariwijaya, M. S. F. & Cheah, Y. H. Xanthorrhizol: A review of its pharmacological activities and anticancer properties. *Cancer Cell International.* 2015, *15*, 1–15.

Owen, J. A., Punt, J., Stranford, S. A., Jones, P. P. & Kuby, J. *Kuby Immunology* (7th ed.). W.H. Freeman. 2013, pp. 11–23.

Paces, J., Strizova, Z., Daniel, S. M. R. Z. & Cerny, J. COVID-19 and the immune system. *Physiological Research.* 2020, *69*(3), 379.

Palomares, O., Akdis, M., Martín-Fontecha, M. & Akdis, C. A. Mechanisms of immune regulation in allergic diseases: The role of regulatory T and B cells. *Immunological Reviews.* 2017, *278*(1), 219–236.

Paterson, R. W., Brown, R. L., Benjamin, L., Nortley, R., Wiethoff, S., Bharucha, T. & Zandi, M. S. The emerging spectrum of COVID-19 neurology: Clinical, radiological and laboratory findings. *Brain.* 2020, *143*(10), 3104–3120.

Ping, X., Weiyang, Y., Jianwei, C. & Xiang, L. Antiviral activities against influenza virus (FM1) of bioactive fractions and representative compounds extracted from *Banlangen* (*Radix Isatidis*). *Journal of Traditional Chinese Medicine.* 2016, *36*(3), 369–376.

Poochi, S. P., Easwaran, M., Balasubramanian, B., Anbuselvam, M., Meyyazhagan, A., Park, S., … Kaul, T. Employing bioactive compounds derived from Ipomoea obscura (L.) to evaluate potential inhibitor for SARS-CoV-2 main protease and ACE2 protein. *Food Frontiers.* 2020, *1*(2), 168–179.

Pushpraj, K., Bhotla, H. K., Arumugam, V. A., Pappusamy, M., Easwaran, M., Liu, W. C., … Balasubramanian, B. Mucormycosis (black fungus) ensuing COVID-19 and comorbidity meets-magnifying global pandemic grieve and catastrophe begins. *Science of the Total Environment.* 2022, *805*, 150355.

Qin, Z., Xiang, K., Su, D. F., Sun, Y. & Liu, X. Activation of the cholinergic anti-inflammatory pathway as a novel therapeutic strategy for COVID-19. *Frontiers in Immunology*. 2021, *11*, 3870.

Ramya, S., Boro, A., Geofferina, I. P., Jananisri, A., Vishalini, G., Balamuralikrishna, B., ... & Anand, A. V. The potential contribution of vitamin K as a nutraceutical to scale down the mortality rate of COVID-19. In *Bioactive Compounds Against SARS-CoV-2*. CRC Press. 2023, pp. 140–159.

Salama, M., Rashed, S. A., Fayez, A., Hassanein, S. S., Sharaby, M. R., Tawfik, N. M. & Adel, M. Medicinal plant-derived compounds as potential phytotherapy forCOVID-19: Future perspectives. *Journal of Pharmacognosy and Phytotherapy*. 2021, *13*(3), 68–81.

Sayuti, N. A. & Rusita, Y. D. Traditional medicinal herbs for healthiness and fitness during the Covid-19 pandemic in Indonesia: Literature review. *Journal of Advanced Pharmacy Education & Research*. 2023, *13*(1), 81.

Sen, S., Chakraborty, R., De, B., Ganesh, T., Raghavendra, H. G. & Debnath, S. Analgesic and anti-inflammatory herbs: A potential source of modern medicine. *International Journal of Pharmaceutical Sciences and Research*. 2010, *1*(11), 32.

Setayesh, M., Karimi, M., Zargaran, A., Abousaidi, H., Shahesmaeili, A., Amiri, F. & Hasheminasab, F. S. Efficacy of a Persian herbal medicine compound on coronavirus disease 2019 (COVID-19): A randomized clinical trial. *Integrative Medicine Research*. 2022, *11*(3), 100869.

Shah, M. A., Rasul, A., Yousaf, R., Haris, M., Faheem, H. I., Hamid, A. & Batiha, G. E. S. Combination of natural antivirals and potent immune invigorators: A natural remedy to combat COVID-19. *Phytotherapy Research*. 2021, *35*(12), 6530–6551.

Shahzad, F., Anderson, D. & Najafzadeh, M. The antiviral, anti-inflammatory effects of natural medicinal herbs and mushrooms and SARS-CoV-2 infection. *Nutrients*. 2020, *12*(9), 2573.

Shanmugam, R., Thangavelu, S., Fathah, Z., Yatoo, M. I., Tiwari, R., Pandey, M. K., ... & Arumugam, V. A. SARS-CoV-2/COVID-19 pandemic-an update. *Journal of Experimental Biology and Agricultural Sciences*. 2020, *8(1),* S219–S245.

Shen, A. M., Zhang, W., Wu, Z., Wang, W. L. & Hua, J. J. TCM theory analysis of Qing-Fei-Pai-Du-Tang in treating COVID-19. *Liaoning Journal of Traditional Chinese Medicine*. 2020, *47*, 106–108.

Shi, Y., Wang, Y., Shao, C., Huang, J., Gan, J., Huang, X., Bucci, E., Piacentini, M., Ippolito, G. & Melino, G. COVID-19 infection: The perspectives on immune responses. *Cell Death and Differentiation*. 2020, *27*(5), 1451–1454.

Singh, A., Malhotra, S. & Subban, R. Anti-inflammatory and analgesic agents from Indian medicinal plants. *International Journal of Integrative Biology*. 2008, *3*, 57–72.

Singhal, T. A review of coronavirus disease-2019 (COVID-19). *The Indian Journal of Pediatrics*. 2020, *87*(4), 281–286.

Siu, K. L., Kok, K. H., Ng, M. J., Poon, V. K. M., Yuen, K. Y., Zheng, B. J. & Jin, D. Y. Severe acute respiratory syndrome coronavirus M protein inhibits type I interferon production by impeding the formation of TRAF3. Tank. *Journal of Biological Chemistry*. 2009, *284*(24), 16202–16209.

Spinner, C. D., Gottlieb, R. L., Criner, G. J., Arribas López, J. R., Cattelan, A. M., Soriano Viladomiu, A., Ogbuagu, O., Malhotra, P., Mullane, K. M., Castagna, A., Chai, L. Y. A., Roestenberg, M., Tsang, O. T. Y., Bernasconi, E., Le Turnier, P., Chang, S. C., SenGupta, D., Hyland, R. H., Osinusi, A. O., GS-US-540-5774 Investigators. Effect of Remdesivir vs standard care on clinical status at 11 days in patients with moderate COVID-19: A randomized clinical trial. *JAMA*. 2020, *324*(11), 1048–1057.

Sun, P., Yan, D., Li, B., Tang, L., Xu, L. & Wang, F. ToujieQuwen granule used with conventional western therapy for coronavirus disease 2019: A protocol for systematic review and meta-analysis. *Medicine*. 2021, *100*(24), e26370.

Tamura, S., Funato, H., Hirabayashi, Y., Kikuta, K., Suzuki, Y., Nagamine, T., Aizawa, C., Nakagawa, M. & Kurata, T. Functional role of respiratory tract haemagglutinin-specific IgA antibodies in protection against influenza. *Vaccine*. 1990, *8*(5), 479–485.

Tan, L., Wang, Q., Zhang, D., Ding, J., Huang, Q., Tang, Y. Q., Wang, Q. & Miao, H. Lymphopenia predicts disease severity of COVID-19: A descriptive and predictive study. *Signal Transduction and Targeted Therapy*. 2020, *5*(1), 33.

Tao, Z., Zhang, L., Friedemann, T., Yang, G., Li, J., Wen, Y. & Shen, A. Systematic analyses on the potential immune and anti-inflammatory mechanisms of ShufengJiedu Capsule against Severe Acute Respiratory Syndrome Coronavirus 2 (SARS-CoV-2)-caused pneumonia. *Journal of Functional Foods*. 2020, *75*, 104243.

Tesfaye, A. Revealing the therapeutic uses of garlic (*Allium Sativum*) and its potential for drug discovery. *The Scientific World Journal*. 2021, *2021*, 8817288.

To, K. K., Tsang, O. T., Leung, W. S., Tam, A. R., Wu, T. C., Lung, D. C., Yip, C. C., Cai, J. P., Chan, J. M., Chik, T. S., Lau, D. P., Choi, C. Y., Chen, L. L., Chan, W. M., Chan, K. H., Ip, J. D., Ng, A. C., Poon, R. W., Luo, C. T. & Yuen, K. Y. Temporal profiles of viral load in posterior oropharyngeal saliva samples and serum antibody responses during infection by SARS-CoV- 2: An observational cohort study. *Lancet Infectious Diseases*. 2020, *20*(5), 565–574.

Tong, C., Chen, Z., Liu, F., Qiao, Y., Chen, T. & Wang, X. Antiviral activities of *Radix Isatidis* polysaccharide against pseudorabies virus in swine testicle cells. *BMC Complementary Medicine and Therapies*. 2020, *20*, 1–5. 57

Vabret, N., Britton, G. J., Gruber, C., Hegde, S., Kim, J., Kuksin, M. & Laserson, U. Immunology of COVID-19: Current state of the science. *Immunity*. 2020, *52*(6), 910–941.

Vaou, N., Stavropoulou, E., Voidarou, C., Tsigalou, C. & Bezirtzoglou, E. Towards advances in medicinal plant antimicrobial activity: A review study on challenges and future perspectives. *Microorganisms*. 2021, *9*(10), 2041.

Verma, S. & Singh, S. P. Current and future status of herbal medicines. *Veterinary World*. 2008, *2*(2), 347–350.

Verma, S. Medicinal plants with anti-inflammatory activity. *The Journal of Phytopharmacology*. 2016, *5*(4), 157–159.

Vikrant, A. & Arya, M. L. A review on anti-inflammatory plant barks. *International Journal of PharmTech Research*. 2011, *3*(2), 899–908.

Wang, S. X., Wang, Y., Lu, Y. B., Li, J. Y., Song, Y. J., Nyamgerelt, M. & Wang, X. X. Diagnosis and treatment of novel coronavirus pneumonia based on the theory of traditional Chinese medicine. *Journal of Integrative Medicine*. 2020, *18*(4), 275–283.

Wang, Z., Xia, Q., Liu, X., Liu, W., Huang, W., Mei, X. & Ma, Z. Phytochemistry, pharmacology, quality control and future research of *Forsythia Suspensa* (Thunb.) Vahl: A review. *Journal of Ethnopharmacology*. 2018, *210*, 318–339.

Wolfe, N. D., Dunavan, C. P. & Diamond, J. Origins of major human infectious diseases. *Nature*. 2007, *447*(7142), 279–283.

Wondmkun, Y. T. & Mohammed, O. A. A review on novel drug targets and future directions for COVID-19 treatment. *Biologics: Targets and Therapy*. 2020, *14*, 77–82.

Wu, F., Zhao, S., Yu, B., Chen, Y. M., Wang, W., Song, Z. G. & Zhang, Y. Z. A new coronavirus associated with human respiratory disease in China. *Nature*. 2020, *579*(7798), 265–269.

Wu, J., Sun, B., Hou, L., Guan, F., Wang, L., Cheng, P. & Lam, W. Prospective: Evolution of Chinese medicine to treat COVID-19 patients in China. *Frontiers in Pharmacology*. 2021, *11*, 615287.

Wu, S. R., Feng, W. H., Chen, K. M., Guan, L. J., Chen, L. M., Wang, Z. M. & Song, Z. H. Chemical composition and antioxidant activity of different parts of *Prunella Vulgaris* by UPLC-Q-TOF-MS/MS and UPLC. *Zhongguo Zhong yao za zhi= ZhongguoZhongyaoZazhi= China Journal of Chinese Materia Medica*. 2023, *48*(17), 4569–4588.

Xia, S., Zhong, Z., Gao, B., Vong, C. T., Lin, X., Cai, J. & Li, C. The important herbal pair for the treatment of COVID-19 and its possible mechanisms. *Chinese Medicine*. 2021, *16*(1), 1–16.

Xiang, M. F., Jin, C. T., Sun, L. H., Zhang, Z. H., Yao, J. J. & Li, L. C. Efficacy and potential mechanisms of Chinese herbal compounds in coronavirus disease 2019: Advances of laboratory and clinical studies. *Chinese Medicine*. 2021, *16*(1), 1–13.

Xiao, P., Huang, H., Chen, J. & Li, X. In vitro antioxidant and anti-inflammatory activities of *Radix Isatidis* extract and bioaccessibility of six bioactive compounds after simulated gastro-intestinal digestion. *Journal of Ethnopharmacology*. 2014, *157*, 55–61.

Xu, Z., Shi, L., Wang, Y., Zhang, J., Huang, L., Zhang, C., Liu, S., Zhao, P., Liu, H., Zhu, L., Tai, Y., Bai, C., Gao, T., Song, J., Xia, P., Dong, J., Zhao, J. & Wang, F. S. Pathological findings of COVID-19 associated with acute respiratory distress syndrome. *Lancet Respiratory Medicine*. 2020, *8*(4), 420–422.

Yang, X. L., Wang, C. X., Wang, J. X., Wu, S. M., Yong, Q., Li, K. & Yang, J. R. In silico evidence implicating novel mechanisms of Prunella vulgaris L. as a potential botanical drug against COVID-19-associated acute kidney injury. *Frontiers in Pharmacology*. 2023, *14*, 1188086.

Yatoo, M., Gopalakrishnan, A., Saxena, A., Parray, O. R., Tufani, N. A., Chakraborty, S. & Iqbal, H. Anti-inflammatory drugs and herbs with special emphasis on herbal medicines for countering inflammatory diseases and disorders-a review. *Recent Patents on Inflammation & Allergy Drug Discovery*. 2018, *12*(1), 39–58.

Yu-Jie, D. A. I., Shi-Yao, W., Shuai-Shuai, G., Jin-Cheng, L., Fang, L. I. & Jun-Ping, K. Recent advances of traditional Chinese medicine on the prevention and treatment of COVID-19. *Chinese Journal of Natural Medicines*. 2020, *18*(12), 881–889.

Zhang, L., Yu, J., Zhou, Y., Shen, M. & Sun, L. Becoming a faithful defender: Traditional Chinese medicine against coronavirus disease 2019 (COVID-19). *The American Journal of Chinese Medicine*. 2020, *48*(04), 763–777.

Zhang, Y. Z. & Holmes, E. C. A genomic perspective on the origin and emergence of SARS-CoV-2. *Cell*. 2020, *181*(2), 223–227.

Zhao, Y. S., Hou, X. S., Gao, Z. H. & Wang, T. Research on medication for severe type of COVID-19 based on Huashi Baidu prescription. *Chinese Archives of Traditional Chinese Medicine*. 2020, *38*(6), 14–17.

Zheng, S., Xue, T., Wang, B., Guo, H. & Liu, Q. Application of network pharmacology in the study of the mechanism of action of traditional Chinese medicine in the treatment of COVID-19. *Frontiers in Pharmacology*. 2022, *13*, 926901.

16 Traditional Herbal Medicines against the SARS-CoV-2 Induced Inflammatory Storm

The Mode of Action

Muhammad Furqan Shah, Shabana Bibi,
Rashid Hussain, Faryal Jahan, Swastika Maitra,
Athanasius Alexiou, and Asim Mehmood

16.1 INTRODUCTION

In the contemporary scientific milieu, there is an unparalleled emphasis on dissecting the intricacies of the COVID-19 pandemic–a malaise of significant respiratory perturbations not just in humans but also in select mammals and avian species. Following an incubation period that spans approximately 5–7 days, the clinical nuances of COVID-19 become palpable, characterized by a spectrum of symptoms from fever and fatigue to more pronounced respiratory challenges and lymphopenia (Mahdy et al., 2020). While a considerable proportion of the infected populace manifests asymptomatic to moderate clinical presentations, there is an undeniable subset that spirals into advanced complications, including multi-organ dysfunction and acute respiratory distress syndrome (ARDS). Certain demographic and clinical strata, notably the geriatric population and those with pre-existing comorbidities such as chronic kidney disorders, cardiovascular pathologies, hypertension, and metabolic diseases, exhibit an accentuated vulnerability to severe disease trajectories (Hasöksüz et al., 2020). The gamut of complications attributable to severe SARS-CoV-2 infection is vast, ranging from hemostatic anomalies and hepatic challenges to heightened predispositions to opportunistic infections (Wu et al., 2020).

The human immune architecture, an intricate nexus of coordinated mechanisms, is intrinsically designed to offer a bulwark against an array of pathogenic challenges, encompassing toxins, foreign antigens, and neoplasms. However, aberrations within this sanctum can trigger discordant responses, often exacerbating the very pathologies they aim to thwart (Schultze & Aschenbrenner, 2021). Inflammation, while fundamental to host defense, is envisaged as an acute, ephemeral response. This cascade is initiated with macrophage mobilization and culminates in the secretion of pivotal cytokines like interleukin (IL)-1, tumor necrosis factor-alpha (TNF-α), and IL-6, which modulate the functional dynamics of endothelial cells and fibroblasts, thereby influencing vascular permeability (Wang & Perlman, 2022).

Cytokines, quintessential glycoprotein modulators, are pivotal in the orchestration of immune dynamics. Depending on their cellular provenance and function, they either act as harbingers of impending immunological threats or arbiters of inflammatory resolution (Sierawska et al., 2022; Zanza et al., 2022). A seminal facet of this cascade involves the interaction and migration of cells, a process facilitated by cytokine-induced adhesion molecules, crucial for effective tissue infiltration and immunological defense (Bernitsa et al., 2022). Chronicity in inflammatory landscapes emerges when antigens persistently engage the immune system or when acute

DOI: 10.1201/9781003452621-16

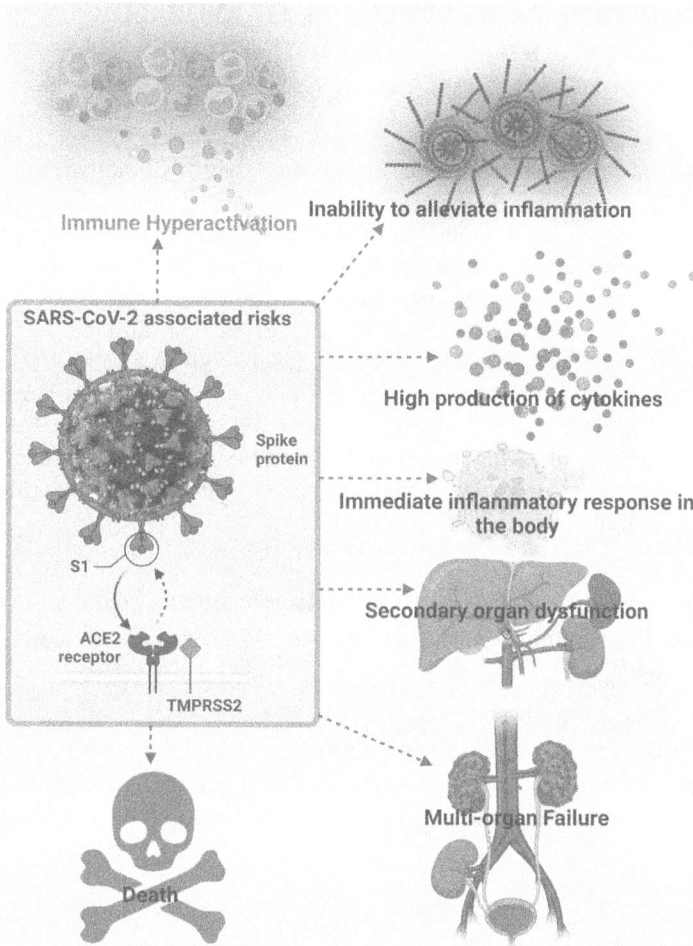

FIGURE 16.1 Immunological features of the COVID-19-associated cytokine storm. (Created with BioRender.com.)

inflammatory feedback loops remain unchecked. This protracted state induces a milieu replete with immune cells, sustained cytokine production, and enhanced antimicrobial activity (Tirelli et al., 2023). Figure 16.1 shows the immunological features of COVID-19-associated cytokine storm.

An inadvertent culmination of unchecked inflammation is the phenomenon termed "cytokine storm"–a cataclysmic, overwhelming surge of pro-inflammatory cytokines that jeopardizes vascular robustness and organ integrity (Montazersaheb et al., 2022). It is paramount to cognize the dual-faceted nature of inflammation–protective when in equilibrium, yet potentially deleterious when in disarray. Such knowledge is instrumental in safeguarding physiological integrity (Li et al., 2022). In subsequent sections, we will embark on an analytical voyage, meticulously unraveling the mechanistic intricacies underpinning SARS-CoV-2's engagement with the host immune system, with an emphasis on cytokine dynamics, cellular repercussions, and potential therapeutic avenues. By comprehensively deconstructing the enigma of the cytokine storm and pinpointing efficacious intervention strategies, our objective remains to ameliorate clinical outcomes and diminish the overarching morbidity and mortality associated with COVID-19.

16.2 IMMUNOMODULATORY POTENTIAL OF AYURVEDIC MEDICINAL PLANTS IN COUNTERACTING VIRAL PATHOGENS: A FOCUS ON SARS-CoV-2

In the multifaceted realm of therapeutic interventions against viral pathologies, botanical remedies have historically played a salient role, especially within traditional medicinal paradigms. India, with its rich tapestry of Ayurvedic knowledge, has underscored the significance of a cadre of medicinal plants, extolled not only for their immunomodulatory properties but also their potential to attenuate the proliferation of pathogens. Renowned Ayurvedic practitioners have curated a repository of these herbs, such as Neem, Amalaki, Kutki, Guduchi, Drumstick, Tulsi, Ashwagandha, cinnamon, black pepper, ginger, turmeric, licorice, aloe, Harsingar, Satavar, almond, and broccoli, emphasizing their profound impact on immune dynamics (Mahaboob Ali et al., 2022). Lauded as the vanguard of Ayurvedic therapeutics, these plants not only augment host immunity but also potentially deter infectious agents. In the contemporary scenario of emergent viral threats, epitomized by SARS-CoV-2, leveraging these botanic assets to fortify immune defenses could serve as a pragmatic adjunct to conventional treatment modalities, thereby attenuating vulnerability to infections (Rastogi et al., 2022) [Table 16.1].

16.2.1 NEEM (*AZADIRACHTA INDICA*): POTENTIAL IMMUNOMODULATORY AND ANTIVIRAL PROPERTIES (BOTANICAL CLASSIFICATION: *MELIACEAE* FAMILY)

The Neem plant, a member of the *Meliaceae* family, possesses an array of phytochemicals that modulate several biological pathways. Compounds such as nimbidin, nimbidol, and quercetin have been identified as potential therapeutic agents (Eze et al., 2022). Nimbin ($C_{30}H_{36}O_9$), a primary bitter constituent derived from Neem oil, not only enhances immunity but also offers defense against various pathogens. It does so by promoting the phagocytic function of macrophages, a mechanism

---

TABLE 16.1

A Tabulated Delineation of the Aforementioned Herbs and Their Primary Medicinal Plants

Herb	Primary Medicinal Attribute
Neem	Immunomodulation, antiviral
Amalaki	Immune booster, antioxidant
Kutki	Liver protection, immune enhancement
Guduchi	Adaptogen, immunostimulant
Drumstick	Nutrient-rich, antimicrobial
Tulsi	Immunostimulant, antioxidant
Ashwagandha	Adaptogen, stress alleviation
Cinnamon	Antimicrobial, blood sugar regulation
Black pepper	Bioavailability enhancer, anti-inflammatory
Ginger	Anti-inflammatory, gastrointestinal health
Turmeric	Anti-inflammatory, antioxidant
Licorice	Antiviral, respiratory health
Aloe	Skin health, digestive aid
Harsingar	Anti-inflammatory, analgesic
Satavar	Female reproductive health, adaptogen
Almond	Nutrient-rich, brain health
Broccoli	Nutrient-rich, detoxification agent

TABLE 16.2

Key Active Constituents of Neem and Their Functional Roles

Constituent	Function
Nimbin	Enhances immunity and fights infections
Nimbidol	Immunomodulatory property
Quercetin	Anti-inflammatory and antioxidant properties
Gedunin	Antimalarial and antifungal properties

that involves stimulating cytokine production, thus augmenting the body's immune response. In the context of COVID-19, the potential of Neem to modulate cytokine production could influence the body's response to SARS-CoV-2 infections (Thakur et al., 2022). Notably, recent molecular analyses have identified Neem leaf phytochemicals that bind effectively to the COVID-19 main protease, suggesting a potential strategy for virus inhibition (Pramanik et al., 2023) [Table 16.2].

16.2.2 AMALAKI (*EMBLICA OFFICINALIS*): A POWERHOUSE OF IMMUNITY (BOTANICAL CLASSIFICATION: *PHYLLANTHACEAE* FAMILY)

Amalaki, part of the *Phyllanthaceae* lineage, is revered for its capability to modulate immune response. Rich in vitamin C and flavonoids like gallic acid and kaempferol, Amalaki augments immune response by boosting IL-2 and natural killer (NK) cell production, which are pivotal in the body's defense mechanisms against pathogens like SARS-CoV-2 (Das, 2022). Considering the COVID-19 pandemic, vitamin C's role in immunity suggests that regular consumption of Amalaki might offer resilience against respiratory challenges, including those induced by the virus. Specific compounds from Amalaki, such as phyllaemblicin B, have shown inhibitory effects against viruses like coxsackie, hinting at a broader spectrum of antiviral efficacy (Bharathi et al., 2023) [Table 16.3].

16.2.3 KUTKI (*PICRORHIZA KURROA*): HARNESSING IMMUNE STRENGTH (BOTANICAL CLASSIFICATION: *SCROPHULARIACEAE* FAMILY)

A renowned member of the *Scrophulariaceae* family, Kutki is noted for its profound immunomodulatory traits. Picroside, one of its main phytochemicals, amplifies the immune response by enhancing phagocytic action and fostering both humoral and cell-mediated immunity. These actions could potentially bolster the body's defenses against viral incursions, including SARS-CoV-2 (Thakur et al., 2022). Moreover, recent explorations have uncovered compounds in Kutki that might inhibit viral protein R, shedding light on another potential therapeutic avenue in the context of viral infections like COVID-19 (Jafari et al., 2022) [Table 16.4].

TABLE 16.3

Principal Components of Amalaki and Their Potential Benefits

Constituent	Benefit
Vitamin C	Boosts immunity
Gallic acid	Antioxidant property
Kaempferol	Anti-inflammatory and cardioprotective
Ellagic acid	Antioxidant and anticancer properties

TABLE 16.4
Prominent Phytochemicals of Kutki and Their Healthy Impacts

Constituent	Impact
Picroside	Enhances immune system
Vanillic acid	Antioxidant and anti-inflammatory
Kutkin	Hepatoprotective property
Androsin	Antibacterial and antioxidant

16.2.4 TULSI (*OCIMUM SANCTUM*): SACRED HERB WITH IMMUNE MODULATION (BOTANICAL CLASSIFICATION: *LAMIACEAE* FAMILY)

Tulsi, part of the *Lamiaceae* family, is native to the Indian subcontinent. Revered as a sacred plant in Hindu culture, it has been used for millennia in traditional Ayurvedic medicine. Tulsi leaves are a rich repository of several bioactive compounds that offer numerous health benefits [Table 16.5]:

- **Eugenol:** A primary compound with antioxidant and anti-inflammatory properties.
- **Ursolic acid:** Demonstrated anticancer, anti-inflammatory, and antioxidant activities.
- **Apigenin:** A flavonoid known for its anti-inflammatory, antioxidant, and antiviral properties.
- **Rosmarinic acid:** An antioxidant that is also found in rosemary and has anti-inflammatory benefits.

16.2.4.1 Mechanisms of Action
- **Immunomodulation:** Tulsi enhances the body's immune response by elevating the production of crucial immune cells and cytokines, such as NK cells and interferon-gamma (IFN-γ) (Mondal et al., 2011). This bolstered immune response can be instrumental in combating various pathogens, including viruses.
- **Antioxidant activity:** The abundant antioxidants in Tulsi neutralize free radicals, preventing oxidative stress and associated damage to cells and tissues. Oxidative stress is believed to exacerbate inflammatory conditions and diseases, suggesting Tulsi's potential protective role.
- **Anti-inflammatory properties:** Chronic inflammation is a hallmark of various diseases, including respiratory ailments. Tulsi's potent anti-inflammatory compounds like eugenol and ursolic acid help in modulating inflammatory responses in the body.

TABLE 16.5
Essential Components of Tulsi and Their Therapeutic Roles

Constituent	Role
Ursolic acid	Anti-inflammatory and antioxidative
Apigenin	Anti-inflammatory, antioxidant, and anticancer
1,8-Cineole	Antimicrobial
Eugenol	Analgesic and anti-inflammatory

16.2.4.2 Relevance to Respiratory Diseases and COVID-19

Given the respiratory nature of COVID-19 and its associated inflammatory lung damage includes:

- **Antiviral potential:** While direct evidence of Tulsi's effectiveness against SARS-CoV-2 is limited, extracts from the plant have shown antiviral capabilities against other viruses, such as H9N2. This suggests the possibility of a broader spectrum of antiviral efficacy (Hasan et al., 2023).
- **Anti-inflammatory response:** In severe cases of COVID-19, an overwhelming inflammatory response known as a "cytokine storm" can occur. Given Tulsi's anti-inflammatory properties, it might play a supportive role in mitigating excessive inflammation.
- **Lung protection:** Oxidative stress has been identified as a potential exacerbating factor in COVID-19. With its antioxidant properties, Tulsi may offer some protection against the oxidative damage that can occur in the lungs of infected individuals.

Tulsi has an established place in traditional medicine systems for its multifaceted health benefits. Its immunomodulatory, antioxidant, and anti-inflammatory actions highlight its potential therapeutic value. While preliminary research indicates potential benefits in the context of viral infections, rigorous scientific investigations are essential to elucidate its specific role concerning COVID-19.

16.2.5 GINGER (*ZINGIBER OFFICINALE*): SPICING UP IMMUNITY (BOTANICAL CLASSIFICATION: *ZINGIBERACEAE* FAMILY)

Originating in Southeast Asia, ginger belongs to the *Zingiberaceae* family and is renowned for its rhizomes, which are commonly used in culinary and medicinal preparations. The bioactive components present in ginger impart its characteristic flavor and medicinal properties [Table 16.6]. These compounds include:

- **Gingerol:** The main bioactive component responsible for the majority of its medicinal properties.
- **Shogaol:** Produced when ginger is dried, it's considered to have even stronger anti-inflammatory and antioxidant effects than gingerol.
- **Zingiberene:**Contributes to the aromatic scent of ginger and has a role in its therapeutic effects.
- **Paradols:** Display antioxidant properties.

16.2.5.1 Mechanisms of Action
- **Anti-inflammatory activity:** Ginger has been known to inhibit the synthesis of pro-inflammatory cytokines, reducing inflammation at the cellular level. Gingerol and shogaol are primarily responsible for this anti-inflammatory action.

TABLE 16.6
Pivotal Active Ingredients of Ginger and Their Immunomodulatory Actions

Constituent	Action
Gingerol	Enhances both humoral and cell-mediated immunity
Zingiberene	Immunostimulatory
β-Bisabolene	Anti-inflammatory
Shogaol	Anti-inflammatory and analgesic

- **Antioxidant activity:** Ginger's potent antioxidants neutralize free radicals, limiting oxidative stress and cellular damage. This plays a role in chronic disease prevention and general health promotion.
- **Gastrointestinal health:** Ginger has a long-standing reputation as a remedy for gastrointestinal issues, including nausea and indigestion. It may act by accelerating gastric emptying.

16.2.5.2   Relevance to Respiratory Diseases and COVID-19

Considering the respiratory afflictions associated with COVID-19 and the related inflammatory responses include:

- **Antiviral potential:** Though direct evidence on ginger's activity against SARS-CoV-2 is limited, its general antiviral properties, such as against the rhinovirus, indicate potential broad-spectrum antiviral efficacy.
- **Anti-inflammatory response:** The "cytokine storm" associated with severe cases of COVID-19 cause's excessive inflammation. Ginger's known anti-inflammatory properties suggest it might have a supportive role in tempering this response.
- **Immune boosting:** Compounds like gingerol have demonstrated the ability to stimulate immune responses, which could potentially enhance the body's defense against viral infections.
- **Lung protection:** Given the anti-inflammatory and antioxidant properties of ginger, it might offer some protection against the oxidative and inflammatory damage occurring in the lungs of COVID-19 patients.

Ginger, with its array of bioactive compounds, offers multifaceted health benefits, from anti-inflammatory to antioxidant actions. These attributes, combined with its long history in traditional medicine, make it an area of interest in the context of viral infections, including COVID-19. However, while its supportive role in health is undeniable, it's crucial to remember that ginger supplements or extracts should be consumed in moderation and should not replace mainstream medical treatments. As always, it's essential to consult with healthcare professionals for therapeutic decisions.

16.2.6   TURMERIC

Turmeric, another member of the *Zingiberaceae* family, is a highly prevalent and extensively cultivated plant with diverse uses worldwide. The rhizome of turmeric is valued for its medicinal properties, primarily as an immune stimulant, blood purifier, and general tonic (Jagetia & Aggarwal, 2007). These beneficial effects can be attributed to the existence of active ingredients, including zingiberene, curcumin, alpha, and beta-turmerone. Curcumin ($C_{21}H_{20}O_6$) is a crucial complex known for its ability to activate various immune cells such as B cells, T cells, neutrophils, macrophages, dendritic cells, and NK cells. It also enhances antibody responses, particularly at lower doses (Pan et al., 2017). The immunomodulatory action of turmeric was evaluated through the assessment of nitric oxide-mediated making in RAW264.7 macrophages. The results demonstrated better immune reactions, tissue healing, and improved lifetime with an enhanced immune system. Recently, in a non-randomized study involving seven cancer patients with endometrial cancer, the immunomodulatory effects of curcumin phytosome were examined by assessing T cell activation, revealing important immune-stimulating activity (Tuyaerts et al., 2019). The use of turmeric was employed in the context of COVID due to its broad range of properties against viruses. Scientific evidence from the past has shown that curcumin exhibits antiviral activity against various viruses, including influenza and hepatitis, as well as evolving viruses like Zika and chikungunya. Furthermore, clinical reports have indicated that curcumin prevents herpes simplex virus 2 (HSV-2), human papillomavirus, and human immunodeficiency virus (Mounce et al., 2017).

16.2.7 ALMOND

Almond, a member of the *Rosaceae* family, normally known as "badam," is widely obtainable in numerous countries. The seed of the almond is highly efficient in treating pneumonia and chronic coughs, providing soothing relief to the throat. It is enriched with vitamin E, which enhances immunity. The seed primarily consists of amino acids, oil, and vitamin E (Peter et al., 2021). The isolated components from the oil include diolein and triolein. Notably, almonds are found to have high levels of cytokines, such as interferon-alpha (IFN-α), interleukins (IL-12), and IFN-γ. These cytokines exert a crucial effect in improving the immune investigation of peripheral blood mononuclear cells against viral infections (Levy et al., 2020). Additionally, almonds have been observed to significantly reduce the duplication of HSV-2. The extract derived from almonds stimulates the high-level production of cytokines, including IFN-α, interleukins, IFN-γ, and TNF-α (Helali et al., 2021). This immune-boosting effect enhances the immune response of blood mononuclear cells against viral infections.

16.2.8 BROCCOLI

Broccoli is a member of the *Brassicaceae* family, usually cultivated in many countries, and its flower possesses numerous health-promoting properties. The flower of broccoli is rich in glucosinolates, flavonoids, isothiocyanates, vitamins, essential oils, minerals, polypeptides, and phenolic compounds (Bousquet et al., 2021). One notable compound found in broccoli is sulforaphane, an isothiocyanate that acts as an immune promoter. Sulforaphane enhances the peritoneal macrophages and reduces the raised production of TNF-α by lipopolysaccharide-stimulated macrophages (Mohammed, 2023). The immune action of pectin taken out from broccoli (*Brassica oleracea* var. italica) has been investigated both *in vivo* and *in vitro*. These studies focused on the activation of macrophages, which have a crucial role in immunity (Hamid et al., 2021).

16.2.9 HARSINGAR PLANT

The Harsingar plant, belonging to *Oleaceae* family, is renowned in India for its therapeutic properties. The leaves of Harsingar have demonstrated strong immunomodulatory activity. They contain tertiary and quaternary alkaloids that contribute to the immune-boosting effects (Panda et al., 2017). The primary constituents found in leaves include nyctanthic acid, nicotiflorin, oleanolic acid, lupeol, mannitol, friedeline, fructose, and glucose. On the other hand, the flowers consist of nyctanthin, d-mannitol, carotenoids, and β-monogentiobioside ester of α-crocetin (Agarwal et al., 2020). The elements from leaves have been shown to improve the productivity of antibodies and increase white blood cell (WBC) counts, thus promoting immunity. In another study, the aqueous extract of plants from the flowers was investigated, which has beneficial effects such as splenocyte proliferation and cytokine induction (Nasirzadeh et al., 2021). Due to its antiviral potential, it acts as a strong antiviral agent against coronavirus. An *in vitro* study verified its antiviral action against encephalomyocarditis and Semliki Forest viruses, when the ethanolic extract of the plant was tested (Renda & Gökkaya.Şöhretoğlu, 2022). Additionally, the acetone, ethanol, and water extracts of *Nyctanthes arbor-tristis* (Harsingar) were assessed for their antiviral activity against enterovirus 71 in a research study, where they were found to exhibit potent antiviral properties (Amin et al., 2022).

16.2.10 BLACK PEPPER

Black pepper from the *Piperaceae* family is a widely used spice worldwide. Black pepper is renowned for its effectiveness in treating throat infections, coughs, and colds (Soy et al., 2020). The therapeutic characteristics of black pepper are attributed to the existence of piperic acid, piperine, piperlonguminine, piperolein B, pellitorine, piperettinep, iperamide, and (-)-kusunokinin, among others (Kunnumakkara et al., 2021). Piperine, specifically, exhibits potent immune modulatory effects. It enhances the process of phagocytosis, promotes the propagation of macrophage cells, and increases WBC count, thus boosting the immune system. Recently, black pepper has been used in the cure of COVID-19 patients due to its prevailing properties against viral infections (Palai et al., 2020). Earlier research has also demonstrated the strong activity of black pepper against viruses. Extract from black pepper exhibited greater activity against the hepatitis B virus act by reducing the hepatitis B virus antigen and secretion of hepatitis B virus surface antigen (Congestion & Ginger, 2023).

Another study investigated the antiviral potential of chloroform and ethanolic extracts from the seeds of *Piper longum* against human parainfluenza virus and the vesicular stomatitis virus. The results indicated potent antiviral activity, probably attributed to the existence of piperidine (Tagde et al., 2021). Overall, black pepper and its constituents have shown immunostimulant properties and powerful antiviral activity against various viruses, making it a valuable natural resource for therapeutic applications.

16.3 BIOACTIVE COMPOUNDS IN VIRUS-HOST DYNAMICS, A MECHANISTIC INSIGHT AND THERAPEUTIC PROMISE

The realm of bioactive compounds, complex in their molecular intricacy, stands as a beacon in the vast expanse of interactions between viruses and their host organisms (Table 16.7). These compounds, often derived from natural or synthesized sources, serve as tools that can strategically target, modulate, or even disrupt specific pathways pivotal to the life cycle and pathogenesis of viruses within their hosts. Such molecular interventions can influence viral entry, replication, assembly, or release, thus holding the potential to dictate the course of viral infections. Understanding the precise

TABLE 16.7

Summary of Herbal Plants, Their Main Product, and Mode of Action

Names	Family	Biologically Active Compound	Immunological Storm	Protective Activities	References
Neem	*Meliaceae*	Nimbin	Produces IFN-γ, and TNF-α	Mitogenic response of spleenocytes	(Kunnumakkara et al., 2021)
Amalaki	*Phyllanthaceae*	Rich vitamin C content	Promote natural killer cell activity, produce IL-2	Targets the reverse transcriptase enzyme of the virus	(Sai Ram et al., 2002)
Kutki	*Scrophulariaceae*	Picroside	Increase phagocytosis and humoral immunity	Protects against liver and respiratory disorders	(Striata, 2022)
Tulsi	*Lamiaceae*	Ursolic acid	Production of IL-4, IFN-γ, NK cells, and T-helper cells	Enhances phagocytosis	(Mediratta et al., 2002)
Ginger	*Zingiberaceae*	Terpene compounds	Production of IL-6, IL-1β, and TNF-α	Treats throat infections, dry coughs, gastrointestinal infections	(Grzanna et al., 2005)
Turmeric	*Zingiberaceae*	Curcumin	Macrophages, T cells, neutrophils, B cells, natural killer cells	Purifies blood, stimulates immunity	(Jagetia & Aggarwal, 2007)

(Continued)

TABLE 16.7 *(Continued)*
Summary of Herbal Plants, Their Main Product, and Mode of Action

Names	Family	Biologically Active Compound	Immunological Storm	Protective Activities	References
Almond	*Rosaceae*	Vitamin E	INF-α, IL-12, INF-γ	Treats chronic cough and pneumonia, providing soothing relief to the throat	(Liskova et al., 2021)
Broccoli	*Brassicaceae*	Sulforaphane	Immune booster	Enhances the phagocytic activity of peritoneal macrophages	(Bousquet et al., 2021)
Harsingar	*Oleaceae*	Alkaloids	Production of circulating antibodies	Increases total white blood cell counts	(Panda et al., 2017)
Black pepper	*Piperaceae*	Piperine	Increases the white blood cell count	Effective in treating throat infections, cough, and cold	(Palai et al., 2020)

mechanistic roles these bioactive entities play is crucial. For instance, some may thwart viral entry by binding to receptors on the host cell's surface, while others might inhibit vital enzymes required for the virus's replication machinery. Yet, others might bolster the host's innate defense mechanisms, enabling a more robust resistance against viral invasions. As the field of virology continues to evolve, so does our comprehension of these compounds. Harnessing this knowledge equips us with a more formidable arsenal against both existing and emerging viral threats. By channeling resources into rigorous research and development efforts centered on these bioactive compounds, we not only deepen our understanding of virus–host dynamics but also unveil a plethora of promising therapeutic avenues ripe for exploration and application.

16.3.1 PLANT-DERIVED COMPOUNDS: PIONEERING DEFENSES AGAINST SARS-CoV-2 ENTRY

The realm of botanically sourced compounds offers a robust suite of molecular tools equipped to deter viral entry into host cells. In navigating the intricacies of SARS-CoV-2, entry-inhibiting compounds can be systematically categorized into two foundational groups: the primary group encompasses molecules targeting the TMPRSS2 and ACE2 receptors, while the secondary group pertains to agents obstructing the intricate dance of viral–cellular receptor engagement and the subsequent fusion process (Hudson B. et al., 2021). From the primary group, gingerol, a prized extract from *Z. officinale*, stands out as an effective agent, demonstrating a remarkable ability to inhibit coronavirus amplification by modulating the host's TMPRSS2 receptor interactions. Concurrently, flavonoids have emerged as molecular sentinels, adept at neutralizing the ACE2 receptor. Their dual functionality encompasses both viral replication interruption and entry blockade. Traditional Chinese Medicine (TCM) constituents, notably epigallocatechin gallate (EGCG), baicalin, and an array of kaempferol derivatives, further bolster this defense, each showcasing potent antiviral attributes (Rastogi et al., 2022).

Delving into the secondary group, hesperetin, a class of flavonoids, has been identified as a potential juggernaut against ACE2, blocking SARS-CoV-2's entry by selectively engaging with viral elements like helicase and the spike (S) protein situated on ACE2 receptors. Advanced bioinformatics analysis has spotlighted the prowess of molecules such as baicalin, hesperidin, and kaempferol

3-O-rutinoside. These compounds diminish virus–cell adhesion dynamics, effectively curtailing SARS-CoV-2's invasion (Bhowmik et al., 2017). Compounds from the *Cinnamomi cortex*, including butanol and procyanidins, present another line of defense at the RNA transcriptional level, while the revered bioactive molecule from *Camellia sinensis*, EGCG, disrupts the viral progression pathways (Kale & Keskin, 2023).

Emodin, an anthraquinone derivative within plants like rhubarb and *Aloe vera*, delineates a fascinating molecular narrative. By binding to the viral S proteins, it effectively mitigates the virulence of SARS-CoV, curbing its optimal activity patterns. This molecule's repertoire further extends to obstructing the synthesis of non structural protein (Nsp), a linchpin for replication, and impeding SARS-CoV's interactions with ACE2 receptors (de Oliveira et al., 2020). Parallel findings have unveiled terpenoids, specifically oleanane-type saikosaponins from *Bupleurum* spp., as formidable barriers to viral ingress into host cells (Yapasert et al., 2021). In synthesis, this plethora of plant-derived molecular instruments, characterized by their distinct mechanisms and diverse botanical origins, offer a pioneering foundation for devising robust strategies against SARS-CoV-2.

16.3.2   THERAPEUTIC POTENTIAL OF PROTEASE INHIBITORS IN SARS-CoV AND SARS-CoV-2 INTERVENTION

Proteases represent pivotal biochemical entities in the life cycle of both SARS-CoV and SARS-CoV-2, orchestrating key processes such as the priming of S proteins and facilitation of viral replication. As a consequence, protease inhibitors have emerged as significant therapeutic candidates in the quest to mitigate the impact of COVID-19 (Khan et al., 2021). In this therapeutic landscape, compounds such as curcumin and eugenol have garnered attention due to their pronounced affinity towards the main protease (Mpro) and S proteins of SARS-CoV. Specifically, studies have highlighted the potential of curcumin to obstruct the activity of SARS-CoV 3-chymotrypsin-like proteases (3CLpro). Similarly, an array of phenolic derivatives predominant in tea–including 3-isotheaflavin-3-gallate, tannic acid, and theaflavin-3,3-gallate–has shown inhibition capabilities against SARS-CoV 3CLpro, with respective IC_{50} values aligning at 3 μM, 7 μM, and 9.5 μM (Jain et al., 2022). In a cellular context, the efficacy of sinigrin, an active constituent found in edible plants such as broccoli, Brussels sprouts, and black mustard seeds, has also been underscored with empirical data, suggesting its role in disrupting the cleavage activities of 3CLpro (IC_{50} at 752 μM) (Yang et al., 2022).

Delving deeper, compounds encompassing polyphenols, *Scutellaria baicalensis*-derived polysaccharides, and various polyglycans have been identified to modulate immunity via their pronounced antioxidant and antiviral attributes. Within the flavonoid family, compounds such as baicalin and scutellarein, originating from the *Scutellaria* species, manifest inhibitory actions against the SARS-CoV Nsp13 helicase. Furthermore, myricetin's efficacy, underscored by an IC_{50} value of 2.71 μM against the virus, is noteworthy, especially in the context of inhibiting Nsp13 *in vitro* by perturbing the ATPase functionality inherent to SARS-CoV (Pei et al., 2022). In essence, this burgeoning field of study, encapsulating diverse botanical derivatives, underpins the significant promise and potential of protease inhibitors in offering robust therapeutic interventions against SARS-CoV and its successor, SARS-CoV-2.

16.3.3   HARNESSING POLYPHENOL COMPOUNDS, AN ADVANCEMENT IN CURBING CORONAVIRUS REPLICATION

In the strategic landscape of combating coronavirus diseases, leveraging inhibitors of viral replication emerges as a cornerstone. A deeper probe into natural derivatives uncovers a constellation of polyphenolic compounds possessing intriguing antiviral potentials. Compounds originating from sources like *Melia azedarach* and the well-regarded green tea have evinced capabilities to counter RNA-dependent proteases, pivotal in the coronavirus RNA replication mechanism. Additionally, the beneficial arsenal of tea extracts might extend beyond conventional health advantages, possibly modulating virus morphogenesis and release. The prowess of stilbene derivatives, especially resveratrol, offers a fresh perspective. Documented studies indicate its potential in substantially mitigating MERS-CoV-induced cytotoxicity and substantially curtailing infectious MERS-CoV replication. The presence of such derivatives in botanical species, including the *Vaccinium* genus and *Vitis vinifera* L. grape, adds another layer of interest. It's posited that these botanicals might function as barriers to cellular viral entry via intricate cellular pathways.

Flavonoids, including catechin, quercetin, naringenin, and hesperetin, perennially celebrated for their dietary significance, are predominantly abundant in various fruits and vegetables. Pertinent research has accentuated naringenin's efficacy in moderating SARS-CoV-2 replication, especially in specific cellular environments. One cannot overlook the therapeutic attributes of the *Pelargonium sidoides* extract (EPS 7630), renowned for its polyphenolic richness. Recognized in numerous countries as a bronchitis remedy, its potency at specific concentrations demonstrates interference with the replication of a broad spectrum of viruses, spanning human coronaviruses to diverse respiratory viruses. The antiviral foray extends to essential oils of *Laurus nobilis* and *Salvia officinalis*, both delineating potential inhibitory action against SARS-CoV replication. Echoing similar antiviral attributes, the oils from *Thuja orientalis* and the compound aescin, sourced from the horse chestnut tree, both present promising replication inhibitory profiles. To encapsulate, the exploration and optimal harnessing of these polyphenolic entities could be instrumental in devising innovative therapeutic strategies, fortifying our approach in the enduring battle against coronavirus afflictions (Bousquet et al., 2021).

16.3.4 Implications of Virucidal Compounds in Viral Defense: A Focus on *Echinacea purpurea*

The targeted inactivation of viral entities emerges as a pivotal strategy in the armory against viral afflictions. Among the vanguard of these potential virucidal agents stands *Echinacea purpurea*. Specific extracts from this plant, exemplified by the commercially sourced Echinaforce, have shown significant prowess in thwarting the infectivity of 229E in respiratory epithelial frameworks. Diverse studies have showcased its potency, with noted IC50 values oscillating between 3.2 µg/mL to 9 ± 3 µg/mL (as observed in studies by Signer et al. [2020]).

The underlying mechanism suggests a multifaceted intervention by *E. purpurea*. The myriad constituents within the extract appear to engage irreversibly with the viral docking platforms, precluding the pathogen's invasive strategies. Parallel observations have elucidated Echinaforce's capacity to deter MERS-CoV; at a concentration of 10 µg/mL, there was a staggering 99.9% reduction in viral infectivity, underscoring its potent antiviral attributes (Nicolussi et al., 2022). A sophisticated therapeutic amalgamation is now under consideration–melding the viricidal prowess of *E. purpurea* with the immunological reinforcements offered by vitamin C, vitamin D, and zinc. Such a coalition aims to fortify the body's defenses against SARS-CoV-2. An analytical review opines that the synchronized action of echinacea extracts alongside vitamin D, vitamin C, and zinc can be instrumental in impeding the onset and progression of respiratory ailments. Notably, there's a potential for diminishing the symptomatic duration and attenuating the symptom severity (as extrapolated from findings by Grant et al. (2020).

16.3.5 Phytochemical Strategies in Addressing SARS-CoV-2-Induced Cytokine Storms

The SARS-CoV-2 infection dynamics underscore the intricate relationship between viral loads and the clinical severity of patients. The cytokine storm, characterized by an unchecked surge in cytokines, notably interleukin-1β and TNFα, poses a critical therapeutic challenge. Phytochemicals, derived from the botanical realm, emerge as potential interventions in this context (Figure 16.2).

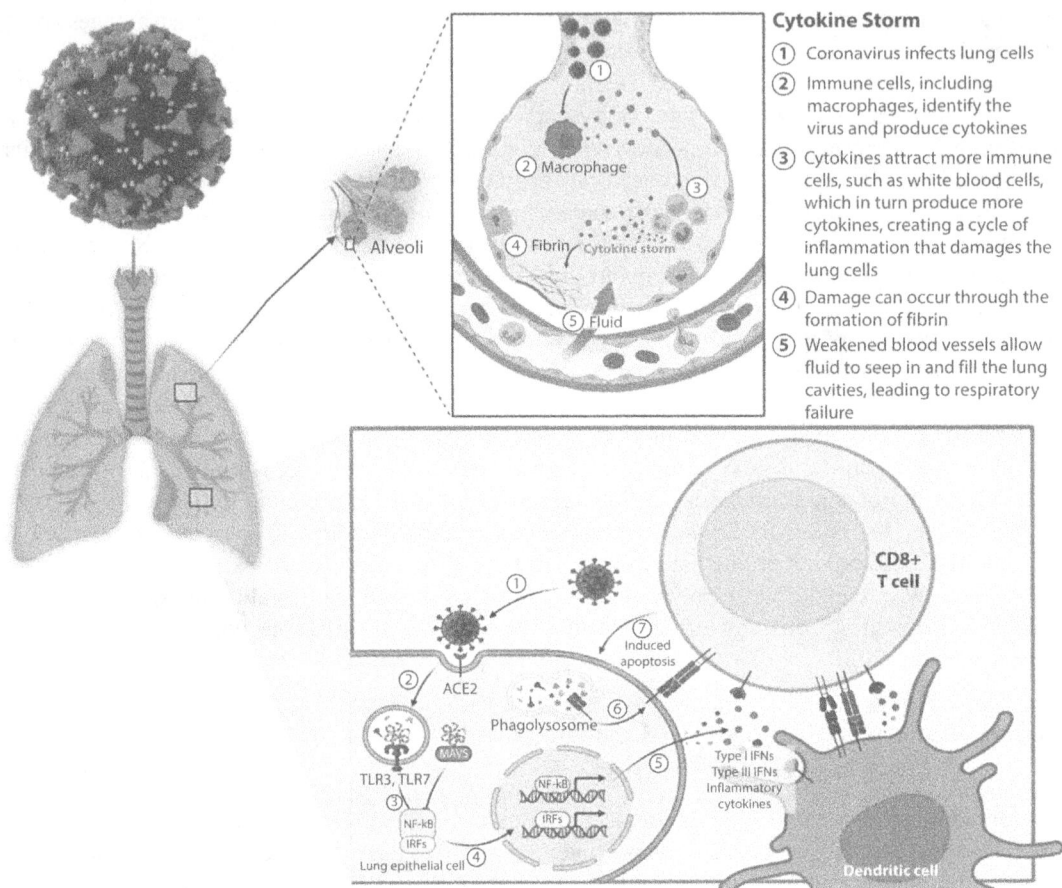

FIGURE 16.2 Mechanism of action of SARS-COV-2 infection leading to a cytokine storm. (Created with BioRender.com.)

16.3.5.1 Tomentin E from *Paulownia Tomentosa*

Derived from *Paulownia tomentosa*, the geranylated flavonoid, Tomentin E, has shown noteworthy antiviral and anti-inflammatory capabilities. Studies by Rabaan et al. (2021) suggest that this compound not only has an inhibitory effect on SARS-CoV but also modulates key inflammatory markers such as TNFα and IL-1β, offering a dual mechanism to counteract the complications of severe COVID-19.

16.3.5.2 *Lonicera japonica* and its Therapeutic Constituents

The multifaceted phytochemical profile of *Lonicera japonica*, encompassing bioactive entities like luteoloside, chlorogenic acid, and quercetin, has been explored for diverse therapeutic potentials. Gao et al. (2023) have elucidated the capacity of these compounds to counteract bacterial and viral pathogens, all while providing antioxidative and anti-inflammatory benefits, especially targeting key inflammatory molecules like TNF-β and IL-1β (Gao et al., 2023).

16.3.5.3 The Promise of Anthocyanins from Red Fruits

Anthocyanins, predominantly found in vibrant fruits such as blackberries and species of *Vaccinium*, have generated interest for their potential anti-inflammatory properties. In studies by Giovinazzo et al. (2020), specific metabolites, notably protocatechuic acid, showcased a propensity to modulate nitric oxide (NO) production and TNF-α levels in cell-based models. Complementing this, gallic acid demonstrated potential in suppressing inflammatory markers, with a focus on ICAM-1, MCP-1, and VCAM-1. Nevertheless, it's pertinent to recognize the potential discrepancies between the concentrations utilized in experimental setups and those achieving physiological relevance (Giovinazzo et al., 2020).

16.3.5.4 Echinacea's Immunomodulatory Potential

Echinacea's role in mitigating the immunoinflammatory responses seen in COVID-19 has been emphasized, particularly its suggested interactions with cannabinoid-2 (CB2) receptors and the activation of PPARγ, as noted by (Vallée, 2022). In summation, as we endeavor to expand the therapeutic repertoire against COVID-19, the botanical realm offers intriguing possibilities. The linchpin for translating these findings remains robust methodological validation, discernment in dosing paradigms, and an integrated approach to potential synergistic interactions with established therapeutic regimens.

16.4 ADVANCING THERAPEUTIC PARADIGMS: HERBAL MEDICINES-INSPIRED PROTACs AS A STRATEGY AGAINST SARS-CoV-2

Emerging research avenues underscore the profound implications of integrating the inherent properties of phytochemicals into the realm of proteolysis targeting chimeras (PROTACs) for countering SARS-CoV-2. As we embark on deciphering this potential synergy, a holistic understanding of the mechanistic intricacies and precision targeting becomes paramount. The potential for leveraging phytochemicals as a foundation for creating PROTACs for combating SARS-CoV-2 presents an enticing direction in drug research (Mukerjee & Ghosh, 2023). To delve into this, it's pivotal to understand the underlying mechanism of PROTACs and the potential target proteins in the context of COVID-19. The incorporation of phytochemicals into PROTACs poses an avant-garde strategy that fuses traditional therapeutic wisdom with modern drug design. The envisioned approach promises not just viral inhibition but a systematic degradation of its critical proteins. As we look to the future, realizing this potential necessitates comprehensive research matrices–spanning molecular docking studies, advanced *in vitro* assays, and rigorous clinical evaluations–to forge a new vanguard in antiviral therapeutics against SARS-CoV-2.

16.5 CONCLUSION AND FUTURE DIRECTION

In synthesizing our knowledge, it becomes evident that delving into traditional herbal pharmacopeia, including remedies like ginger, offers an integrative and forward-thinking strategy for counteracting the inflammatory cascade precipitated by SARS-CoV-2. The diversity of bioactive constituents within botanical sources has illustrated a multifaceted inhibitory capacity on the virus, encompassing interventions at key junctures such as viral entry, protease activity disruption, replication suppression, as well as direct viricidal dynamics, and immune modulation (Rani et al., 2022). Remarkably, specific herbal candidates have not only demonstrated viral-containment properties but also displayed capabilities in rejuvenating immune functions and restoring a balanced immune milieu. This synergy illuminates the dual potential of botanical remedies: curtailing viral propagation whilst tempering the host's hyperinflammatory response.

However, to transition from ethnobotanical anecdotes to clinical mainstays, stringent evidentiary scientific evaluations conclusively affirm their safety, define their therapeutic index, and calibrate appropriate dosing regimens. Molecular mechanistic elucidation forms the bedrock of such evaluations, aiming to detail the nuanced interactions of these botanical compounds with viral and host systems, and the potential for advantageous synergism when coupled with contemporary therapeutic agents (Bibi et al., 2022). Furthermore, the challenge of ensuring consistency in herbal therapeutic matrices underscores the importance of evolved pharmacokinetic research, rigorous safety auditing, and unwavering adherence to quality assurance paradigms. Such meticulous endeavors seek to ensure that these herbal interventions retain their therapeutic promise while minimizing unforeseen adverse outcomes.

As we envision the future trajectory of this domain, the emphasis should be anchored in fostering interdisciplinary dialogues, curating expansive knowledge repositories, and catalyzing global alliances. The inception of specialized educational frameworks and skill enhancement programs could seamlessly meld ancestral botanical wisdom with contemporary scientific methodologies, fostering evidence-centric applications. Sustainability should be the cornerstone when exploiting these botanical repositories, ensuring their perennial availability. By weaving together rigorous scientific rigor with international collaborations and sustainable practices, the vast potential harbored within traditional herbal medicines stands poised to offer novel adjunctive strategies, fortifying our global arsenal against the multifaceted challenges of COVID-19.

REFERENCES

Agarwal, S., Darbar, S., Saha, S., & Deb, T. (2020). Sanative effect of a low-cost Novel Green Formulation â€"I M-SSS20 to minimize the inflammatory and cytokine storm against Respiratory Diseases. *Innoriginal: International Journal of Sciences*, 1–11.

Amin, M. N., Bahoosh, S. R., Eftekhari, M., & Hosseinzadeh, L. (2022). Herbal sources of magnesium as a promising multifaceted intervention for the management of COVID-19. *Natural Product Communications*, 17(8). https://doi.org/10.1177/1934578X221116235

Bernitsa, S., Dayan, R., Stephanou, A., Tzvetanova, I. D., & Patrikios, I. S. (2022). Natural biomolecules and derivatives as anticancer immunomodulatory agents. *Frontiers in Immunology*, 13. https://doi.org/10.3389/FIMMU.2022.1070367

Bharathi, K., Sivasangar Latha, A., Jananisri, A., Bavyataa, V., Rajan, B., Balamuralikrishnan, B., Arasu, M. V., Al-Dhabi, N. A., Catharine, B., & Anand, A. V. (2023). Antiviral Properties of South Indian Plants Against SARS-CoV-2. In *Ethnopharmacology and Drug Discovery for COVID-19: Anti-SARS-CoV-2 Agents from Herbal Medicines and Natural Products*, 447–478. https://doi.org/10.1007/978-981-99-3664-9_17

Bhowmik, D., Nandi, R., Prakash, A., & Kumar, D. (2021). Evaluation of flavonoids as 2019-nCoV cell entry inhibitor through molecular docking and pharmacological analysis. *Heliyon*, 7(3). https://doi.org/10.1016/j.heliyon.2021.e06515

Bibi, S., Khan, M. S., El-Kafrawy, S. A., Alandijany, T. A., El-Daly, M. M., Yousafi, Q., … Azhar, E. I. (2022). Virtual screening and molecular dynamics simulation analysis of forsythoside A as a plant-derived inhibitor of SARS-CoV-2 3CLpro. *Saudi Pharmaceutical Journal*, 30(7), 979–1002.

Bousquet, J., Le Moing, V., Blain, H., Czarlewski, W., Zuberbier, T., de la Torre, R., Pizarro Lozano, N., Reynes, J., Bedbrook, A., Cristol, J. P., Cruz, A. A., Fiocchi, A., Haahtela, T., Iaccarino, G., Klimek, L., Kuna, P., Melén, E., Mullol, J., Samolinski, B., … Anto, J. M. (2021). Efficacy of broccoli and glucoraphanin in COVID-19: From hypothesis to proof-of-concept with three experimental clinical cases. *World Allergy Organization Journal*, 14(1), 100498. https://doi.org/10.1016/j.waojou.2020.100498

Das, K. (2022). Herbal plants as immunity modulators against COVID-19: A primary preventive measure during home quarantine. *Journal of Herbal Medicine*, 32, 100501. https://doi.org/10.1016/J.HERMED.2021.100501

de Oliveira, P. G., Termini, L., Durigon, E. L., Lepique, A. P., Sposito, A. C., & Boccardo, E. (2020). Diacerein: A potential multi-target therapeutic drug for COVID-19. *Medical Hypotheses*, 144, 109920. https://doi.org/10.1016/J.MEHY.2020.109920

Eze, M. O., Ejike, C. E. C. C., Ifeonu, P., Udeinya, I. J., Udenigwe, C. C., & Uzoegwu, P. N. (2022). Anti-COVID-19 potential of *Azadirachta indica* (Neem) leaf extract. *Scientific African*, 16, e01184. https://doi.org/10.1016/J.SCIAF.2022.E01184

Gao, B., Zhu, L., Liu, Z., Li, Y., He, X., Wu, X., Pehrsson, P., Sun, J., Xie, Z., Slavin, M., & Yu, L. L. (2023). Chemical composition of honeysuckle (*Lonicerae japonicae*) extracts and their potential in inhibiting the SARS-CoV-2 spike protein and ACE2 binding, suppressing ACE2, and scavenging radicals. *Journal of Agricultural and Food Chemistry*. https://doi.org/10.1021/ACS.JAFC.3C00584

Giovinazzo, G., Gerardi, C., Uberti-Foppa, C., & Lopalco, L. (2020). Can natural polyphenols help in reducing cytokine storm in COVID-19 patients? *Molecules*, 25(24), 5888. https://doi.org/10.3390/MOLECULES25245888

Grant, W. B., Lahore, H., McDonnell, S. L., Baggerly, C. A., French, C. B., Aliano, J. L., & Bhattoa, H. P. (2020). Evidence that vitamin D supplementation could reduce risk of influenza and COVID-19 infections and deaths. *Nutrients, 12*(4), 988.

Grzanna, R., Lindmark, L., & Frondoza, C. G. (2005). Ginger - An herbal medicinal product with broad anti-inflammatory actions. *Journal of Medicinal Food, 8*(2), 125–132. https://doi.org/10.1089/jmf.2005.8.125

Hamid, H., Thakur, A., & Thakur, N. S. (2021). Role of functional food components in COVID-19 pandemic: A review. *Annals of Phytomedicine: An International Journal, 10*(Sp-Issue1 (COVID-19): S4-S12), 240–250. https://doi.org/10.21276/ap.covid19.2021.10.1.22

Hasan, M. R., Alotaibi, B. S., Althafar, Z. M., Mujamammi, A. H., Hasan, M. R., Saud Alotaibi, B., Althafar, Z. M., Mujamammi, A. H., & Jameela, J. (2023). An update on the therapeutic anticancer potential of *Ocimum sanctum L*.: "Elixir of life." *Molecules, 28*(3), 1193. https://doi.org/10.3390/MOLECULES28031193

Hasöksüz, M., Kiliç, S., & Saraç, F. (2020). Coronaviruses and SARS-COV-2. *Turkish Journal of Medical Sciences, 50*(9), 549–556. https://doi.org/10.3906/sag-2004-127

Helali, A., Hammadi, M. W., Houalef, M., & Benchachou, K. (2021). Natural substances and coronavirus: Review and potential for the inhibition of SARS-CoV-2. *Current Perspectives on Medicinal and Aromatic Plants (CUPMAP), 4*, 128–150. https://doi.org/10.38093/cupmap.1029572

Hudson B., S., Kolte, V., Khan, A., & Sharma, G. (2021). Dynamic tracking of variant frequencies depicts the evolution of mutation sites amongst SARS-CoV-2 genomes from India. *Journal of Medical Virology, 93*(4), 2534–2537. https://doi.org/10.1002/JMV.26756

Jafari, S., Dadmehr, M., Sharifi, Y., Manshouri, S., Kamali, M., Vahidi Emami, Z., Rasizadeh, R., & Seif, F. (2022). The potential effects of *Scrophularia Striata Boiss* on COVID-19. *Immunoregulation, 4*(2), 69–72. https://doi.org/10.32598/Immunoregulation.4.2.3

Jagetia, G. C., & Aggarwal, B. B. (2007). "Spicing Up" of the immune system by curcumin. *Journal of Clinical Immunology, 27*(1), 19–35. https://doi.org/10.1007/s10875-006-9066-7.

Jain, M., Anand, A., & Shah, A. (2022). Exploring the potential role of theaflavin-3,3′-digallate in inhibiting various stages of SARS-CoV-2 life cycle: An in-silico approach. *Chemistry Africa, 5*(4), 883. https://doi.org/10.1007/S42250-022-00376-7

Kale, O., & Keskin, G. (2023). Investigation of supplement products preferred by healthcare professionals during COVID-19 pandemic process. *Journal of Contemporary Medicine, 13*(1), 107–113. https://doi.org/10.16899/JCM.1175032

Khan, M. A., Mahmud, S., Alam, A. S. M. R. U., Rahman, M. E., Ahmed, F., & Rahmatullah, M. (2021). Comparative molecular investigation of the potential inhibitors against SARS-CoV-2 main protease: A molecular docking study. *Journal of Biomolecular Structure and Dynamics*, 1–7. https://doi.org/10.1080/07391102.2020.1796813

Kunnumakkara, A. B., Rana, V., Parama, D., Banik, K., Girisa, S., Henamayee, S., Thakur, K. K., Dutta, U., Garodia, P., Gupta, S. C., & Aggarwal, B. B. (2021). COVID-19, cytokines, inflammation, and spices: How are they related? *Life Sciences, 284*, 119201. https://doi.org/10.1016/j.lfs.2021.119201

Levy, E., Delvin, E., Marcil, V., & Spahis, S. (2020). Can phytotherapy with polyphenols serve as a powerful approach for the prevention and therapy tool of novel coronavirus disease 2019 (COVID-19)? *American Journal of Physiology - Endocrinology and Metabolism, 319*(4), E689–E708. https://doi.org/10.1152/ajpendo.00298.2020

Li, B., Deng, A., Li, K., Hu, Y., Li, Z., Shi, Y., Xiong, Q., Liu, Z., Guo, Q., Zou, L., Zhang, H., Zhang, M., Ouyang, F., Su, J., Su, W., Xu, J., Lin, H., Sun, J., Peng, J., … Lu, J. (2022). Viral infection and transmission in a large, well-traced outbreak caused by the SARS-CoV-2 Delta variant. *Nature Communications, 13*(1), 1–9. https://doi.org/10.1038/s41467-022-28089-y

Liskova, A., Samec, M., Koklesova, L., Samuel, S. M., Zhai, K., Al-Ishaq, R. K., Abotaleb, M., Nosal, V., Kajo, K., Ashrafizadeh, M., Zarrabi, A., Brockmueller, A., Shakibaei, M., Sabaka, P., Mozos, I., Ullrich, D., Prosecky, R., La Rocca, G., Caprnda, M., … Kubatka, P. (2021). Flavonoids against the SARS-CoV-2 induced inflammatory storm. *Biomedicine and Pharmacotherapy, 138*. https://doi.org/10.1016/j.biopha.2021.111430

Mahaboob Ali, A. A., Bugarcic, A., Naumovski, N., & Ghildyal, R. (2022). Ayurvedic formulations: Potential COVID-19 therapeutics? *Phytomedicine Plus, 2*(3), 100286. https://doi.org/10.1016/J.PHYPLU.2022.100286

Mahdy, M. A. A., Younis, W., & Ewaida, Z. (2020). An Overview of SARS-CoV-2 and Animal Infection. In *Frontiers in Veterinary Science* (Vol. 7). Frontiers Media S.A. https://doi.org/10.3389/fvets.2020.596391

Mediratta, P. K., Sharma, K. K., & Singh, S. (2002). Evaluation of immunomodulatory potential of *Ocimum sanctum* seed oil and its possible mechanism of action. *Journal of Ethnopharmacology*, *80*(1), 15–20. https://doi.org/10.1016/S0378-8741(01)00373-7

Mohammed, M. A. (2023). Fighting cytokine storm and immunomodulatory deficiency: By using natural products therapy up to now. *Frontiers in Pharmacology*, *14*(April). https://doi.org/10.3389/fphar.2023.1111329

Mondal, S., Varma, S., Deepak Bamola, V., Narayan Naik, S., Ranjan Mirdha, B., Mohan Padhi, M., Mehta, N., & Chandra Mahapatra, S. (2011). Double-blinded randomized controlled trial for immunomodulatory effects of Tulsi (*Ocimum sanctum Linn.*) leaf extract on healthy volunteers. *Elsevier*, *136*, 452–456. https://doi.org/10.1016/j.jep.2011.05.012

Montazersaheb, S., Hosseiniyan Khatibi, S. M., Hejazi, M. S., Tarhriz, V., Farjami, A., Ghasemian Sorbeni, F., Farahzadi, R., & Ghasemnejad, T. (2022). COVID-19 infection: An overview on cytokine storm and related interventions. *Virology Journal*, *19*(1), 1–15. https://doi.org/10.1186/S12985-022-01814-1

Mounce, B. C., Cesaro, T., Carrau, L., Vallet, T., & Vignuzzi, M. (2017). Curcumin inhibits Zika and chikungunya virus infection by inhibiting cell binding. *Antiviral Research*, *142*, 148–157. https://doi.org/10.1016/j.antiviral.2017.03.014

Mukerjee, N., & Ghosh, A. (2023). Revolutionizing viral disease treatment: PROTACs therapy could be the ultimate weapon of the future. *Journal of Medical Virology*, *95*(8), e28981.

Nasirzadeh, A., Bazeli, J., Hajavi, J., Yavarmanesh, N., … Islam, S. M. S. (2021). Inhibiting IL-6 during cytokine storm in COVID-19: Potential role of natural products. https://doi.org/10.20944/preprints202106.0131.v1

Nicolussi, S., Ardjomand-Woelkart, K., Stange, R., Gancitano, G., Klein, P., & Ogal, M. (2022). Echinacea as a potential force against coronavirus infections? A mini-review of randomized controlled trials in adults and children. *Microorganisms*, *10*(2), 211. https://doi.org/10.3390/microorganisms10020211

Palai, S., Dehuri, M., & Patra, R. (2020). Spices boosting immunity in COVID-19. *Annals of Phytomedicine: An International Journal*, *9*(2). https://doi.org/10.21276/ap.2020.9.2.7

Pan, M. H., Wu, J. C., Ho, C. T., & Badmaev, V. (2017). Effects of water extract of *Curcuma longa* (L.) roots on immunity and telomerase function. *Journal of Complementary and Integrative Medicine*, *14*(3). https://doi.org/10.1515/jcim-2015-0107

Panda, S. K., Padhi, L., Leyssen, P., Liu, M., Neyts, J., & Luyten, W. (2017). Antimicrobial, anthelmintic, and antiviral activity of plants traditionally used for treating infectious disease in the Similipal Biosphere Reserve, Odisha, India. *Frontiers in Pharmacology*, *8*(OCT), 1–15. https://doi.org/10.3389/fphar.2017.00658

Pei, T., Yan, M., Huang, Y., Wei, Y., Martin, C., & Zhao, Q. (2022). Specific flavonoids and their biosynthetic pathway in *Scutellaria baicalensis*. *Frontiers in Plant Science*, *13*, 866282. https://doi.org/10.3389/fpls.2022.866282

Peter, A. E., Sandeep, B. V., Rao, B. G., & Kalpana, V. L. (2021). Calming the storm: Natural immunosuppressants as adjuvants to target the cytokine storm in COVID-19. *Frontiers in Pharmacology*, *11*(December 2019), 1–22. https://doi.org/10.3389/fphar.2020.583777

Pramanik, P., Tewari, S., Mukherjee, P., Banka, R., & Sen, S. (2023). Indian traditional plants: Medicinal properties and human health. *Journal of Survey in Fisheries Sciences*, *10*(1S), 6170–6179. https://sifisheriessciences.com/journal/index.php/journal/article/view/2122

Rabaan, A. A., Al-Ahmed, S. H., Garout, M. A., Al-Qaaneh, A. M., Sule, A. A., Tirupathi, R., … & Dhama, K. (2021). Diverse immunological factors influencing pathogenesis in patients with COVID-19: a review on viral dissemination, immunotherapeutic options to counter cytokine storm and inflammatory responses. *Pathogens*, *10*(5), 565.

Rani, I., Kalsi, A., Kaur, G., Sharma, P., Gupta, S., Gautam, R. K., … & Emran, T. B. (2022). Modern drug discovery applications for the identification of novel candidates for COVID-19 infections. *Annals of Medicine and Surgery*, *80*, 104125.

Rastogi, S., Pandey, D. N., & Singh, R. H. (2022). COVID-19 pandemic: A pragmatic plan for ayurveda intervention. *Journal of Ayurveda and Integrative Medicine*, *13*(1), 100312. https://doi.org/10.1016/J.JAIM.2020.04.002

Renda, G., & Gökkaya.Şöhretoğlu, D. (2022). Immunomodulatory properties of triterpenes. *Phytochemistry Reviews*, *21*(2), 537–563. https://doi.org/10.1007/s11101-021-09785-x

Sai Ram, M., Neetu, D., Yogesh, B., Anju, B., Dipti, P., Pauline, T., Sharma, S. K., Sarada, S. K. S., Ilavazhagan, G., Kumar, D., & Selvamurthy, W. (2002). Cyto-protective and immunomodulating properties of Amla (*Emblica officinalis*) on lymphocytes: An in-vitro study. *Journal of Ethnopharmacology*, *81*(1), 5–10. https://doi.org/10.1016/S0378-8741(01)00421-4

Schultze, J. L., & Aschenbrenner, A. C. (2021). COVID-19 and the human innate immune system. *Cell.Com*, *184*, 1671–1692. https://doi.org/10.1016/j.cell.2021.02.029

Sierawska, O., Małkowska, P., Taskin, C., Hrynkiewicz, R., Mertowska, P., Grywalska, E., Korzeniowski, T., Torres, K., Surowiecka, A., Niedźwiedzka-Rystwej, P., & Strużyna, J. (2022). Innate immune system response to burn damage—Focus on cytokine alteration. *International Journal of Molecular Sciences*, *23*(2), 716. https://doi.org/10.3390/IJMS23020716

Signer, J., Jonsdottir, H. R., Albrich, W. C., Strasser, M., Züst, R., Ryter, S., Ackermann-Gäumann, R., Lenz, N., Siegrist, D., Suter, A., Schoop, R., & Engler, O. B. (2020). In vitro virucidal activity of Echinaforce®, an Echinacea purpurea preparation, against coronaviruses, including common cold coronavirus 229E and SARS-CoV-2. *Virology Journal*, *17*(1). https://doi.org/10.1186/S12985-020-01401-2

Soy, M., Keser, G., Atagündüz, P., Tabak, F., Atagündüz, I., & Kayhan, S. (2020). Cytokine storm in COVID-19: Pathogenesis and overview of anti-inflammatory agents used in treatment. *Clinical Rheumatology*, *39*(7), 2085–2094. https://doi.org/10.1007/s10067-020-05190-5

Tagde, P., Tagde, S., Tagde, P., Bhattacharya, T., Monzur, S. M., Rahman, M. H., Otrisal, P., Behl, T., Hassan, S. S. U., Abdel-Daim, M. M., Aleya, L., & Bungau, S. (2021). Nutraceuticals and herbs in reducing the risk and improving the treatment of covid-19 by targeting sars-cov-2. *Biomedicines*, *9*(9). https://doi.org/10.3390/biomedicines9091266

Thakur, A., Gautam, S., Kaushal, K., Kumari, A., Bhatt, K., & Jasta, S. (2022). A review on therapeutic potential of Indian medicinal plants against COVID-19 pandemic. *Annals of Phytomedicine*. https://doi.org/10.54085/ap.covid19.2022.11.3.6

Tirelli, C., De Amici, M., Albrici, C., Mira, S., Nalesso, G., Re, B., Corsico, A. G., Mondoni, M., & Centanni, S. (2023). Exploring the role of immune system and inflammatory cytokines in SARS-CoV-2 induced lung disease: A narrative review. *Biology*, *12*(2), 177. https://doi.org/10.3390/BIOLOGY12020177

Tuyaerts, S., Rombauts, K., Everaert, T., Van Nuffel, A. M. T., & Amant, F. (2019). A phase 2 study to assess the immunomodulatory capacity of a lecithin-based delivery system of curcumin in endometrial cancer. *Frontiers in Nutrition*, *5*(January). https://doi.org/10.3389/fnut.2018.00138

Usman, M. R. M., Jain, B. V., Patil, A. L., & Jain, P. S. (2023). Chest congestion and infusions of ginger, honey, tulsi, black pepper and other home remedies as OTC dispensed pharma products. *Journal of Clinical Otorhinolaryngology, Head, and Neck Surgery*, 1281–9.

Vallée, A. (2022). Cannabidiol and SARS-CoV-2 infection. *Frontiers in Immunology*, *13*, 870787. https://doi.org/10.3389/fimmu.2022.870787

Wang, Y., & Perlman, S. (2022). COVID-19: Inflammatory profile. *Annual Review of Medicine*, *73*, 65–80. https://doi.org/10.1146/ANNUREV-MED-042220-012417

Wu, C., Chen, X., Cai, Y., Xia, J., Zhou, X., Xu, S., Huang, H., Zhang, L., Zhou, X., Du, C., Zhang, Y., Song, J., Wang, S., Chao, Y., Yang, Z., Xu, J., Zhou, X., Chen, D., Xiong, W., … Song, Y. (2020). Risk factors associated with acute respiratory distress syndrome and death in patients with coronavirus disease 2019 pneumonia in Wuhan, China. *JAMA Internal Medicine*, *180*(7), 934–943. https://doi.org/10.1001/JAMAINTERNMED.2020.0994

Yang, F., Jiang, X. L., Tariq, A., Sadia, S., Ahmed, Z., Sardans, J., Aleem, M., Ullah, R., & Bussmann, R. W. (2022). Potential medicinal plants involved in inhibiting 3CLpro activity: A practical alternate approach to combating COVID-19. *Journal of Integrative Medicine*, *20*(6), 488–496. https://doi.org/10.1016/J.JOIM.2022.08.001

Yapasert, R., Khaw-On, P., Banjerdpongchai, R., Sabatier, J.-M., & Rapposelli, S. (2021). Coronavirus infection-associated cell death signaling and potential therapeutic targets. *Molecules*, *26*(24), 7459. https://doi.org/10.3390/MOLECULES26247459

Zanza, C., Romenskaya, T., Manetti, A. C., Franceschi, F., La Russa, R., Bertozzi, G., Maiese, A., Savioli, G., Volonnino, G., & Longhitano, Y. (2022). Cytokine storm in COVID-19: Immunopathogenesis and therapy. *Medicina*, *58*(2), 144. https://doi.org/10.3390/MEDICINA58020144

17 Curcumin and COVID-19

Therapeutic Scopes and Limitations

Prem Rajak and Abhratanu Ganguly

17.1 INTRODUCTION

The entire world witnessed the recent coronavirus disease 19 (COVID-19) pandemic spread by severe acute respiratory syndrome coronavirus 2 (SARS-CoV-2). The disease has shown rapid spread and colonized almost every corner of the globe. The disease was initially reported to cause upper respiratory tract infection, however, with the advancement of the disease, several critical ailments like cardiovascular injuries, intense inflammation, and mortality have been frequently observed. In many cases, the patients are asymptomatic without any severe health issues. Several treatment measures have been adopted to ameliorate the disease, nevertheless, none are fully effective and still require additional treatment procedures to feel relief from the disease.

The causative agent SARS-CoV-2 is a retrovirus containing ss-RNA as the genetic material. The surface of the virus bears multiple glycoproteins that help the virions to invade the host cell. The spike protein of the virus is a trimeric glycoprotein with a polybasic furin cleavage site. The spike protein contains a receptor binding domain (RBD) that attaches to the ACE-2 receptor (Wrapp et al., 2020). TMPRSS2 and furin cleave the spike protein at the S1/S2 interface of protein, leading to the fusion of viral and host cell membranes. This enables the subcellular internalization of viral contents, including the genetic materials. The subcellular invasion of the virus is mediated by the membrane-bound S2 subunit that harbors heptapeptide repeat sequences 1 and 2 (HR1 and HR2), a fusion peptide, a transmembrane region, and cytosolic domain (Xia et al., 2020). Later, the virus hijacks the host's replication transcription machinery to synthesize subsets of proteins essential for subcellular replication of the virus.

ACE-2 is widely expressed in the heart, liver, kidney, gut, and vascular endothelium. This receptor is an integral component of the renin-angiotensin-aldosterone (RAAS) system. It converts angiotensin I (Ang I) to Ang 1–9 and Ang II to Ang 1–7 (Gheblawi et al., 2020). Ang 1–7 has several health-promoting impacts on physiology. It exerts antioxidant, anti-inflammatory, antithrombotic, antifibrotic, and hypertrophic impacts. Blocking of ACE-2 results in concentration up-regulation of Ang II. Ang II, when found in abundance, interacts with the ATR1 receptor and favors proinflammatory cascades leading to oxidative stress, vasoconstriction, hypertrophy, and tissue injuries (Kuba et al., 2005). Therefore, inhibiting ACE-2 is crucial to maintain a healthy physiological status.

Several drugs have been formulated and developed to cure the disease against the nonstructural proteins of SARS-CoV-2. These drugs interfere with the viral proteins to impede virus replication within cells. However, the high rate of mutation in the genetic material of the virus and the frequent emergence of variants of concern posed a serious threat to global health. Many of such drugs may not be effective against the variants of concern because of mutations in spike proteins. Moreover, many chemical anti-SARS-CoV-2 drugs fail to provide 100% security to patients against reinfection by new variants. Hence, continuous drug repurposing is essential to target the newly emerged variants of concern effectively.

Drug repurposing is an excellent method to identify potential natural drugs to combat any disease. Approaches like molecular docking, molecular simulation, binding affinities, and determination of intermolecular interactions could be helpful in screening antiviral agents from a hub of natural compounds. Studies involving these approaches have suggested the efficacy of remdesivir against major proteins of SARS-CoV-2, such as RNA-dependent RNA polymerase, main protease (Mpro), exonuclease, and helicase (Beck et al., 2020). Madecassic acid has been reported to bear binding affinities for

DOI: 10.1201/9781003452621-17

a number of nonstructural proteins (Rajak et al., 2023). In addition, the inhibitory potential of epicat-echin gallate, a secondary metabolite of tea, has been shown against a subset of nonstructural proteins of SARS-CoV-2 (Rajak and Ganguly, 2023a; Rajak and Ganguly, 2023b). Interestingly, computational studies have indicated that madecassic acid might also help modulate entry and replication of SARS-CoV-2 (Ganguly et al., 2023). Curcumin has binding affinities for several critical proteins of SARS-CoV-2. Therefore, curcumin is a promising drug candidate for in vitro and in vivo investigations.

Curcumin (1,7-bis[4-hydroxy-3-methoxyphenyl]-1,6-heptadiene-3,5-dione), or diferuloylmethane, is an important constituent of turmeric, which belongs to the rhizomatous herbaceous perennial plant *Curcuma longa* of the ginger family. The phytocompound has both cultural and medicinal values. It is traditionally used in many countries for its antioxidant, antimutagenic, anti-inflammatory, anticancer, and antimicrobial properties (Hewlings and Kalman, 2017). Several studies have demonstrated the subcellular impacts of curcumin. Curcumin boosts the antioxidant status to reverse oxidative injuries. Moreover, it can directly neutralize reactive oxygen species (ROS) and reactive nitrogen species (RNS) at subcellular moiety (Ak and Gülçin, 2008). Curcumin impairs the functions of lipoxygenase/cyclooxygenase and xanthine hydrogenase/oxidase to reduce free radical production. Curcumin blocks the activation of NF-κβ in response to various inflammatory stimuli. The compound is also helpful in combating metabolic syndrome by suppressing adipogenesis, increasing blood pressure, inflammation, and oxidative stress, and improving insulin sensitivity (Hewlings and Kalman, 2017).

Curcumin can be administered by various routes, including oral ingestion, subcutaneous, intraperitoneal, intravenous, topical, and nasal therapy (Prasad and Tyagi, 2015). Nevertheless, studies indicate that curcumin has limited absorption, biodistribution, metabolism, and bioavailability despite of its considerable pharmacological importance (Dei Cas and Ghidoni, 2019). Following ingestion, curcumin undergoes hepatic metabolism via phase I and II biotransformation reactions, also facilitated by gut flora (Dei Cas and Ghidoni, 2019). During the phase I reaction, alcohol dehydrogenase reduces the double bonds of curcumin, resulting in the formation of dihydro-curcumin, tetrahydro-curcumin, hexahydro-curcumin, and octahydro-curcumin (Nelson et al., 2017). In the later stage of phase II reaction, curcumin and its phase I metabolites undergo fast conjugation with glucuronic acid by uridine 5′-diphospho-glucuronosyltransferases (UGTs) and sulfate by sulfotransferases (SULTs) at the phenol position (Dei Cas and Ghidoni, 2019).

Therefore, curcumin has several health benefits both at the subcellular and organismal levels. Several recent studies have unraveled the anti-SARS-CoV-2 potential of curcumin. Hence, this chapter aims to present the studies that have reported the promising potential of curcumin in combating COVID-19. In addition, the existing limitations of the therapeutic efficiency of the compound will also be elucidated.

17.2 ANTI-VIRAL ACTIVITY AGAINST SARS-CoV-2

Several *in silico* investigations have demonstrated the direct antiviral activity of curcumin against SARS-CoV-2. The phytocompound has been reported to target multiple proteins that are critical for replication and assembly of the virions.

The spike glycoprotein of the virus binds with the ACE-2 receptor to facilitate the subcellular internalization of the virions. Therefore, both ACE-2 and the viral spike protein are potential drug targets. Curcumin has been shown to interact with spike protein and ACE-2 through the hydrogen bonds. Interestingly, the compound displayed better binding affinities for the target proteins than nafamostat, hydroxychloroquine, and captopril (Maurya et al., 2020).

TMPRSS2 is an essential protease that catalyzes the cleavage of spike protein at the S1/S2 interface to facilitate viral-host cell membrane fusion. This enables the viral contents to enter the host cytoplasm. A recent study has documented that curcumin binds with TMPRSS2 through one stable hydrogen bond. The intermolecular binding was further strengthened by four hydrophobic interactions between curcumin and TMPRSS2 (Motohashi et al., 2020). The results were in line with a previous finding where curcumin treatment reduced TMPRSS2 in cancer cells (Zhang et al., 2007). Therefore, curcumin can block the activities of TMPRSS2 to suppress the subcellular entry of the viral particles.

Mpro is inevitable for the subcellular maturation of various viral proteins responsible for the generation of new virions. Catalytic activities of Mpro are essential for the maturation of helicase and RNA-dependent RNA polymerase. An *in silico* study has revealed that curcumin can interact with active amino acid residues of the protease with high binding affinities. Interestingly, curcumin showed better binding affinities than other tested drugs, including remdesivir and entecavir. Binding was comparable to N3 control (Huynh et al., 2020). Therefore, targeting Mpro with curcumin could impair the maturation of nonstructural proteins, impeding subcellular replication of SARS-CoV-2.

SARS-CoV-2 can invade the host cell through the endosomal route. Entry through the endosomal route requires an acidic pH maintained by endosomal proteases cathepsin S and cathepsin L. An ion channel, namely, the vacuolar ATPase pump, also plays a critical role in reducing endosomal pH. Studies have indicated that curcumin can control pH by inhibiting the expression of vacuolar ATPase pumps (Vishvakarma et al., 2011). Another study involving SARS-CoV-2 pseudo-virions demonstrated that curcumin can bind with ACE-2 and TMPRSS2 in both cell-free and cell-based assays and impair their activities, which finally affect the subcellular entry of the virus. Moreover, the compound was found to moderately increase endosomal/lysosomal pH to impede subcellular entry of the virions (Goc et al., 2021).

Therefore, both the *in silico* and *in vitro* studies have reinforced the multi-target efficacy of curcumin against the entry factors of SARS-CoV-2.

17.3   INHIBITION OF VIRAL ATTACHMENT AND REPLICATION

Curcumin shows a strong affinity for ACE-2 receptors and can potentially impede SARS-CoV-2 invasion (Zahedipour et al., 2020). Works have shown that derivatives of curcumin, when administered at a relatively modest dosage, reduce the amount of Ang II and, thus, increase the expression of the ACE-2 protein (Pang et al., 2015). Furthermore, researchers have investigated the effectiveness of curcumin in impeding viral entry through an *in silico* simulation study. Curcumin has been reported to exhibit a strong binding affinity toward the RBD of S-protein and human ACE-2 (Jena et al., 2021). Subsequently, it was discovered that the enol and keto forms of curcumin interact with the receptor binding motif (RBM) of the spike glycoprotein at crucial residues (Q493, N501, Y505, Y489, and Q498) (Shanmugarajan et al., 2020). Furthermore, curcumin may inhibit ADAM17 (Borah et al. 2016). Curcumin has been found to exhibit a high binding affinity (19.86 kJ/mol) with TMPRSS2 (Motohashi et al., 2020). Recent studies have indicated that curcumin can impede the translation of viral proteins by binding to the active sites of 3CLPro/chymotrypsin-like protease (NSP3) and PLPro/papain-like protease (NSP5) (Das et al., 2021; Laksmiani et al., 2020).

17.4   ANTI-INFLAMMATORY POTENTIAL OF CURCUMIN

One of the major features of COVID-19 is the exacerbated cytokine storm. It leads to more intense inflammatory outcomes in patient and could initiate multi-organ injuries. The implications of NFkβ are central to SARS-CoV-2-mediated cytokine storms. NFkβ can stimulate a subset of TNF-α, IL-1β, IL-6, and IL-18. SARS-CoV-2 can activate the NFkβ pathway through multiple cascades. The viral envelope and nucleocapsid proteins can directly activate the NFkβ pathway. The virus also activates the Ang II/AT1R axis, which in turn stimulates NFkβ activation (Crowley and Rudemiller, 2017). Ang II phosphorylates the p65 subunit of NFkβ, leading to its activation and subsequent expression of proinflammatory cytokines such as IL-1β, IL-6, TNF-α, and IL-18 (Ruiz-Ortega et al., 2001). Moreover, disintegrin and metalloprotease 17 (ADAM17), activated by Ang II/AT1R axis process membrane form of mIL-6Ra to its soluble state, i.e., sIL-6Ra. sIL-6Ra activates STAT-3 via gp130. STAT-3 further induces the NFkβ pathway for the release of proinflammatory cytokines (Murakami et al., 2019). Interestingly curcumin inhibits STAT-3-mediated activation of NFkβ and impairs the positive feedback loop between activation of NFkβ and release of proinflammatory cytokines (Yadav et al., 2015).

NFkβ is kept inactive by the Ikβ. Phosphorylation of Ikβ releases the NFkβ from the complex and promotes the translocation of NFkβ to the nucleus for the expression of proinflammatory cytokines. Curcumin inhibits the phosphorylation of Ikβ, therefore, preventing the translocation of NFkβ to the nucleus and reducing the subsequent expression of proinflammatory cytokines (Wang et al., 2018).

NFkβ stimulates NLRP-3-mediated inflammasome formation and the release of IL-1β and IL-18. COVID-19 patients have shown increased NLRP-3, IL-1β, and IL-18 in their serum (Rodrigues et al., 2021). Curcumin-mediated suppression in the activation of NFkβ could impede the formation of inflammasome and release of inflammatory cytokines in patients, thereby reducing disease severity. The impact of nano-curcumin on inflammatory mediators has been studied. Patients with SARS-CoV-2 infection contain greater levels of IL-16 and IL-1β in their serum, suggesting potential contribution to disease severity.

Interestingly, nano-curcumin has been found to reduce these inflammatory cytokines in SARS-CoV-2-infected patients (Valizadeh et al., 2020). In another investigation, nano-curcumin was demonstrated to lower IL-17, IL-21, IL-23, and Granulocyte-macrophage colony-stimulating factor (GM-CSF)levels in SARS-CoV-2-infected patients (Tahmasebi et al., 2021).

Therefore, multiple studies have indicated that curcumin can reduce the COVID-19-mediated cytokine storm by suppressing the activation of NF-kβ and associated signaling cascades in patients (Figure 17.1).

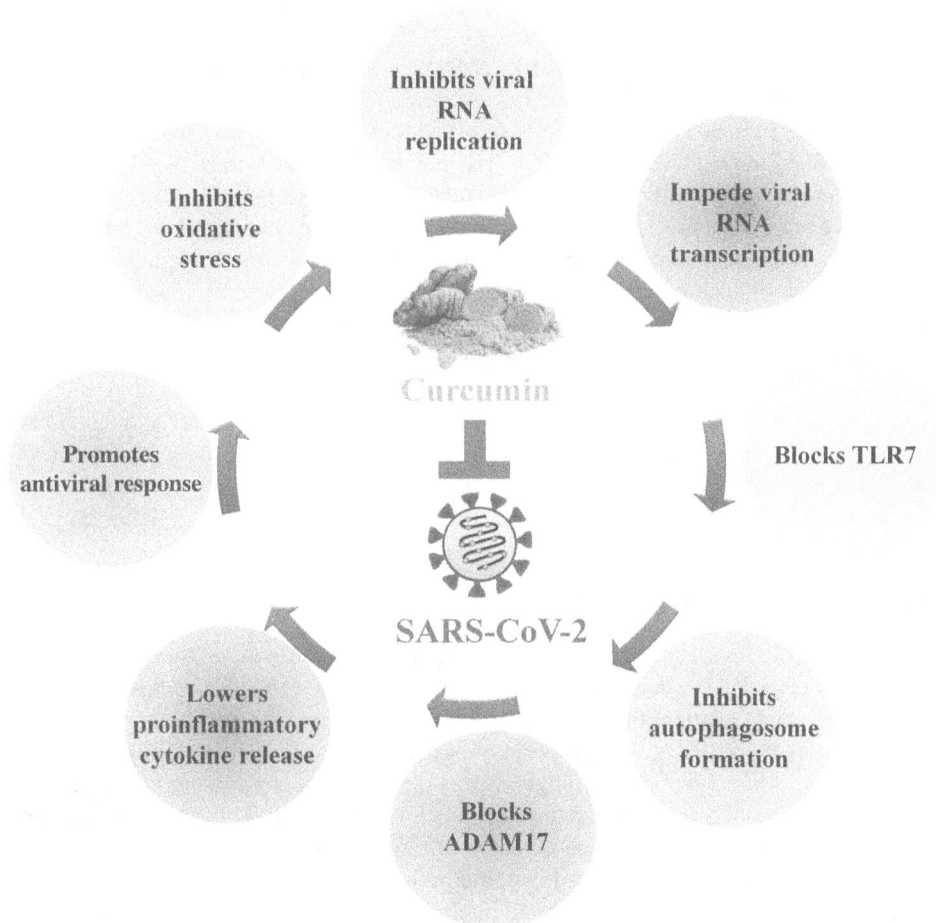

FIGURE 17.1   Anti-SARS-CoV-2 activities of curcumin.

17.5   CARDIOPROTECTIVE POTENTIAL OF CURCUMIN

The exacerbated cytokine storm during COVID-19 imposes mild to severe injuries to various organs. The inflammatory cytokines arrive at the organs and damage tissues' normal architectural organization. The cardiovascular system is sensitive to intense inflammation and oxidative damage that could be triggered upon SARS-CoV-2 invasion. Proinflammatory cytokines damage the cardiac tissue linings and damage the pericardial membrane alongside fine blood capillaries. Cardiovascular tissue injuries release troponin I, a striking marker of cardiac injuries. In addition, thrombus is another risk factor for cardiac damage upon intense inflammation and cell lysis induced by viral infection.

Patients suffering from COVID-19 have been reported with cardiac issues. Patients with COVID-19 have increased levels of highly sensitive cardiac Troponin I with myocardial injuries (Huang et al., 2020). Notably, cardiac issues in some patients were more prominent compared to classic COVID-19 symptoms, such as fever and cough (Deng et al., 2020). In some instances, underlying cardiovascular disease was linked to a higher mortality rate in COVID-19 patients. Studies have documented that patients with underlying cardiovascular diseases and increased levels of troponin T had a mortality rate of 69.4% (Guo et al., 2020).

In COVID-19 patients, cardiac ailments mainly include heart failure, acute myocardial injury, infarction, arrhythmia, and acute coronary syndrome (Lang et al., 2020). Histopathological analysis of cardiac tissues in COVID-19 patients revealed necrosis, myocardial infarction, inflammation, and neutrophilic infiltration (Rapkiewicz et al., 2020). Therefore, cardiac tissue is one of the major targets of COVID-19.

Studies have also reported cardiac issues like chest pain and palpitations in post-COVID-19 individuals (Patell et al., 2020). Thrombosis events are documented in many patients, further increasing the post-COVID-19 health concerns. In a sepsis model, curcumin was protected from cardiac injuries and improved the survival parameters. The phytocompound also reduced the levels of troponin I and the products of lipids (Yang et al., 2013). Oxidative stress can severely injure cardiac tissues. Oxidative stress favors the activation of matrix metalloproteinase, degrading the extracellular matrix. This causes thinning of the ventricular wall and progressive expansion of infarction and dilation of the cardiac chamber (Wang et al., 2012). In a study, curcumin was found to prevent the metalloproteinase-mediated degradation of the extracellular matrix and promote the synthesis of collagen and the formation of myofibroblasts. This resulted in reduced thinning of the ventricular wall and improved functioning of cardiac tissues (Wang et al., 2012).

It is essential to note that no direct report is currently available that claims benefits of curcumin in preventing cardiac injuries due to COVID-19. However, studies involving curcumin in other aspects suggest that this compound could also be effective in reducing the cardiac injuries escalated during SARS-CoV-2 infection.

17.6   PULMONARY PROTECTIVE POTENTIAL OF CURCUMIN

SARS-CoV-2 primarily targets alveolar type II cells (ATII) cells. Infection induces necrosis and the subsequent spread of virions to neighboring pulmonary cells. The infection causes the development of multi-nucleated giant cells and the appearance of fibrin-rich hyaline membranes. Such impact of the virus on pulmonary cells leads to the onset of acute respiratory distress syndrome (ARDS). In a study, lung injury induced by benzo (a) pyrene was attenuated by treatment with curcumin. This phytocompound reduced the levels of inflammatory cytokines such as TNF-α, IL-6, and C-reactive protein in serum and prevented the apoptotic death of pulmonary tissues (Almatroodi et al., 2020).

Mechanical lung ventilation is required in some patients with COVID-19. However, inadequate mechanical ventilation could result in lung damage. Studies have shown that ventilator-induced lung injury is linked to the induction of inflammatory mediators and oxidative stress. Curcumin attenuates lung injury by inhibiting the activation of NFkβ and oxidative stress (Wang et al., 2018).

COVID-19 patients were detected with NETosis in pulmonary tissues (Middleton et al., 2020). NETosis causes cell damage and is an indicator of inflammatory attack. Inflammatory mediator IL-17A induces the release of CXCL-1, which aggravates neutrophil aggregation. Curcumin has been found to decrease the levels of IL-17A and neutrophil infiltration in pulmonary tissues (Chai et al., 2020). Therefore, controlling inflammation and neutrophils with curcumin could help reduce pulmonary damage in patients with COVID-19.

Regulatory T cells (Tregs) are important subsets of T cells that regulate the immune microenvironment through their anti-inflammatory and anti-apoptotic functions. Curcumin was shown to induce CD4+ T cells to differentiate into the Tregs. This differentiation is regulated by IL-10, which acts as an anti-inflammatory cytokine (Chai et al., 2020). IL-10 impacts macrophage reprogramming by modulating the mTORC1 complex. The M2 macrophage exerts anti-inflammatory and tissue repair in lungs during sepsis (Ip et al., 2017). Patients with COVID-19 manifested an increased population of M1 macrophages with aggravated pulmonary inflammation while the control individuals showed a reduced population of M1 macrophages and an increased population of M2 macrophages (Liao et al., 2020). Tests with curcumin showed a reduction in M1 macrophages while increasing M2 macrophages, therefore participating in macrophage polarization (Chai et al., 2020).

In another study, curcumin was observed to lower oxidative injuries in the lungs which were triggered by cyclophosphamide. Here, curcumin restored redox balance and minimized lipid peroxidation (Ashry et al., 2013). Curcumin has also been demonstrated to lessen LPS-induced lung injury by reducing pulmonary edema. In another investigation with acute lung injury and ARDS, curcumin protected pulmonary tissues by reducing the levels of inflammatory cytokines (Xu et al., 2013).

Direct investigations showing the impact of curcumin on lungs injured by COVID-19 are still lacking. Nonetheless, the pulmonary protective potential of curcumin attracts investigations involving COVID-19.

17.7   CURCUMIN AND BLOOD CIRCULATION

Infection by SARS-CoV-2 was linked to abnormal blood clotting. A higher incidence of thromboembolism and hyperplatelet activation was demonstrated in COVID-19 patients. Elevated levels of D-dimers are the characteristic feature of thromboembolism and platelet activation (Zhou et al., 2020). SARS-CoV-2 can directly interact with the ACE-2 and TMPRSS2 receptors expressed on the platelet cell surface to favor platelet activation (Zhang et al., 2020).

Platelets express P selectin and glycoprotein VI and interact with subendothelial collagen for adhesion. Curcumin reduces the expression of P selectin and glycoprotein VI, preventing platelet adhesion to the endothelial and subendothelial lining (Zhang et al., 2008). Therefore, curcumin can potentially subvert platelet activation to prevent thromboembolism during COVID-19.

COVID-19 is associated with neutrophil traps (NET), culminating in cell death (NETosis), which can escalate inflammation and injuries. Products of NETosis could also obstruct the blood vessels in the lungs. *In vivo* studies have shown that curcumin can lessen NETosis and reduce neutrophil infiltration in lung tissues induced by LPS (Antoine et al., 2013). Platelets bind neutrophils through their P selectins and favor neutrophil infiltration. Curcumin-mediated reduction in P selectin, therefore, helps in the prevention of NETosis and intense inflammation.

Increased Ang II induces tissue factor expression, which triggers a procoagulant response. Tissue factor binds with factor VIIa and such binding fuels coagulation cascades. This results in thrombin formation and protofibril deposition (Sathler, 2020). Curcumin can potentially inhibit tissue factor expression induced by TNF-α, thrombin, and LPS (Pendurthi et al., 1997). Platelets secret thromboxane A2 (TXA2) that attracts other platelets to promote coagulation. Curcumin impedes TXA2 release by platelets and, therefore, prevents platelet aggregation and blood coagulation (Shah et al., 1999).

TABLE 17.1

Studies Demonstrating the Anti-SARS-CoV-2 Potential of Curcumin

Type of Study	Model Animal/ Cell Line	Effective Concentrations of Curcumin	Pharmaceutical Impacts	References
In vitro study	Vero E6 cells	10 μg/mL	Reduction in proinflammatory cytokines (IL-1β, IL-6, and IL-8)	Marín-Palma et al., 2021
In vitro study	Vero E6	IC50 = 1.659–8.828 μM	Enhanced COX-1/2 properties	Srour et al., 2021
In vitro study	Vero E6 and human Calu-3 cells	125-1 μg /mL	Reduced SARS-CoV-2 RNA levels in cell culture supernatants	Bormann et al., 2021
In vitro study	Vero E6 cells	10 μM	SARS-CoV-2 replication (reduction >99%)	Zupin et al 2022
Computational study	Not applicable	Not applicable	Binding interaction with the viral spike protein and the cognate host cell receptor ACE-2	Jena et al., 2021
In silico study	Not applicable	Not applicable	Binding interactions with spike glycoproteins, nucleocapsid phosphoprotein, membrane glycoprotein along with nsp10	Suravajhala et al., 2020
In vitro study	Vero E6 cells A549 cells	EC50 of 13.63 μM EC50 of 4.57 μM	Anti-SARS-CoV-2 activity Inhibited cytopathic effect	Mohd Abd Razak et al., 2023
In silico study	Not applicable	Not applicable	Having the potential for destabilizing the structural integrity of SARS-CoV-2 receptor proteins	Srivastava and Singh, 2020

In an *in vivo* model of intravascular coagulation, curcumin administration reduced the circulating TNF-α and peripheral platelets and decreased plasma fibrinogen (Chen et al., 2007). In another study, administration of curcumin for 15 days at 10 mg significantly reduced the circulating plasma fibrinogen levels (Ramirez et al., 2000).

Therefore, thrombotic and coagulative issues in COVID-19 are detrimental to health. Curcumin has the potential to mitigate the circulation related issues in COVID-19 (Table 17.1).

17.8   LIMITATIONS

Although several studies have shown good results for curcumin, its substantial limitations prevent its widespread clinical usage in treatment of viral infections. Curcumin has low bioavailability. For instance, in a study, after oral administration of a 12 g dose of curcumin extract, the concentration of curcumin in the bloodstream was measured to be only 57.6 ng/mL (Lao et al., 2006). The oral bioavailability of curcumin is often limited because it is not well absorbed by the small intestine. Additionally, curcumin undergoes substantial metabolism in the liver, involving both reduction and conjugation processes. Finally, curcumin is eliminated from the body through the gall bladder. Curcumin interacts with the enterocyte proteins further limiting its bioavailability (Heger et al., 2014). Curcumin exhibits significant phase I and II biotransformation. The liver, the intestine, and gut bacteria are identified as the major sites where curcumin undergoes metabolism. A reductase enzyme subsequently reduces the double bonds of curcumin in enterocytes and hepatocytes. Phase II metabolism is highly active in both the intestinal and hepatic cytosol, acting on both curcumin and its phase I metabolites. Curcumin is also metabolized by intestinal bacteria, including *Escherichia coli* and *Blautia* sp., through an alternate pathway (Hassaninasab et al., 2011).

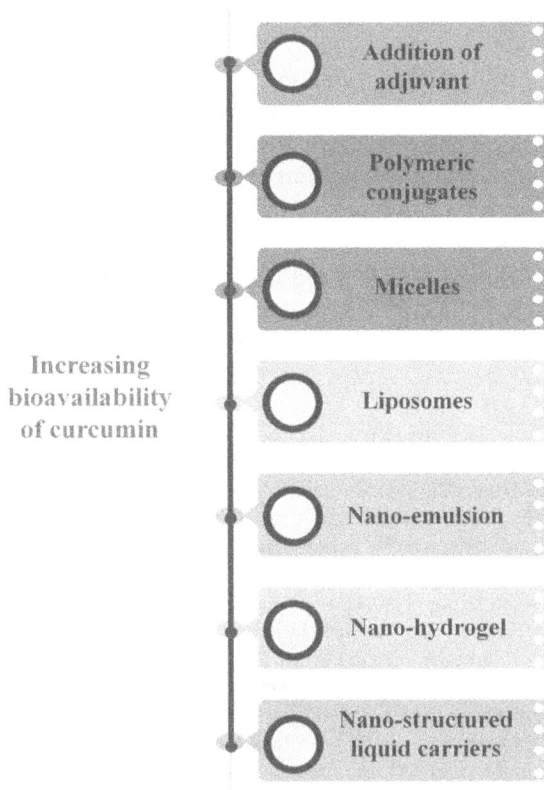

FIGURE 17.2 Potential measures that can be employed to enhance bioavailability of curcumin.

Such active metabolism of curcumin and its derivatives promote its rapid excretion out of the body. This further reduces the bioavailability of the compound. However, curcumin can be combined with other compounds such as lecithin, piperine, hydrophilic nanoparticles, and lipid-based formulation to enhance bioavailability.

Another issue with curcumin is its toxicity or less therapeutic responsiveness. Storka et al. have documented the harmful impacts of curcumin in lipid-based formulations on red blood cells (Storka et al., 2015). In another investigation, certain individuals exhibited hematological side effects following the intravenous infusion of liposomal curcumin (Greil et al., 2018) (Figure 17.2).

17.9 CONCLUSION AND OUTLOOK

COVID-19 has fueled the most recent global pandemic. It rapidly spread to every corner of the world. The causative agent of the disease is the beta coronavirus, i.e., the SARS-CoV-2 that contains double-stranded RNA as genetic material and harbors various structural and nonstructural proteins. Many of the structural proteins including spike, membrane, envelope, and nucleocapsid proteins are essential for viral invasion. Nonstructural proteins are mainly involved in the replication of the virions within the host cells. Various drugs target structural and nonstructural proteins of the virus to subvert the viral life cycle. However, many of such drugs may pose toxicities to the patients. Curcumin is well tolerable even at higher doses and can potentially mitigate various health issues fueled by SARS-CoV-2. Curcumin has potential to control cytokine storm and hyperinflammation-induced health complications. Curcumin has cardioprotective effects. It can reduce thrombotic events in blood vessels. Curcumin could also help maintain pulmonary functions. Thus, curcumin can potentially mitigate the health issues triggered by COVID-19. However, investigations involving the direct impact of curcumin on patients with COVID-19 are warranted.

Vaccines formulated against RNA viruses mainly target the spike protein. However, a high mutation rate alters the spike protein, which might reduce effectiveness of the vaccines. Thus, only vaccination may not provide complete protection from COVID-19. Multi-level treatment is required to mitigate the infection and related issues. Curcumin, with other drugs, may provide enhanced protection from the disease. It can also enhance the efficacy of conventional drugs, as curcumin can potentially ameliorate the morbidities related to various systems of the body. The water solubility of curcumin makes it readily execrable, and hence, the possibility of toxicity is less with this compound.

REFERENCES

Ak, Tuba, and Ilhami Gülçin. "Antioxidant and radical scavenging properties of curcumin." *Chemico-biological interactions* 174, no. 1 (2008): 27–37.

Almatroodi, Saleh A., Faris Alrumaihi, Mohammed A. Alsahli, Mazen Fahad Alhommrani, Arif Khan, and Arshad Husain Rahmani. "Curcumin, an active constituent of turmeric spice: Implication in the prevention of lung injury induced by benzo (a) pyrene (BaP) in rats." *Molecules* 25, no. 3 (2020): 724.

Antoine, Francis, Jean-Christophe Simard, and Denis Girard. "Curcumin inhibits agent-induced human neutrophil functions in vitro and lipopolysaccharide-induced neutrophilic infiltration in vivo." *International immunopharmacology* 17, no. 4 (2013): 1101–1107.

Ashry, Nora A., Nariman M. Gameil, and Ghada M. Suddek. "Modulation of cyclophosphamide-induced early lung injury by allicin." *Pharmaceutical biology* 51, no. 6 (2013): 806–811.

Beck, Bo Ram, Bonggun Shin, Yoonjung Choi, Sungsoo Park, and Keunsoo Kang. "Predicting commercially available antiviral drugs that may act on the novel coronavirus (SARS-CoV-2) through a drug-target interaction deep learning model." *Computational and structural biotechnology journal* 18 (2020): 784–790.

Borah PK, Chakraborty S, Jha AN, Rajkhowa S, Duary RK. "In silico approaches and proportional odds model towards identifying selective ADAM17 inhibitors from anti-inflammatory natural molecules." *The Journal of Molecular Graphics and Modelling* 70, (2016): 129–139.

Bormann, Maren, Mira Alt, Leonie Schipper, Lukas van de Sand, Vu Thuy Khanh Le-Trilling, Lydia Rink, Natalie Heinen et al. "Turmeric root and its bioactive ingredient curcumin effectively neutralize SARS-CoV-2 in vitro." *Viruses* 13, no. 10 (2021): 1914.

Chai, Yu-Sen, Yan-Qing Chen, Shi-Hui Lin, Ke Xie, Chuan-Jiang Wang, Yuan-Zheng Yang, and Fang Xu. "Curcumin regulates the differentiation of naïve CD4+ T cells and activates IL-10 immune modulation against acute lung injury in mice." *Biomedicine & pharmacotherapy* 125 (2020): 109946.

Chen, Hsiang-Wen, Hung-Tien Kuo, Chee-Yin Chai, Jian-Liang Ou, and Rei-Cheng Yang. "Pretreatment of curcumin attenuates coagulopathy and renal injury in LPS-induced endotoxemia." *Journal of endotoxin research* 13, no. 1 (2007): 15–23.

Crowley, Steven D., and Nathan P. Rudemiller. "Immunologic effects of the renin-angiotensin system." *Journal of the American society of nephrology: JASN* 28, no. 5 (2017): 1350.

Das, Sourav, Sharat Sarmah, Sona Lyndem, and Atanu Singha Roy. "An investigation into the identification of potential inhibitors of SARS-CoV-2 main protease using molecular docking study." *Journal of biomolecular structure and dynamics* 39, no. 9 (2021): 3347–3357.

Dei Cas, Michele, and Riccardo Ghidoni. "Dietary curcumin: Correlation between bioavailability and health potential." *Nutrients* 11, no. 9 (2019): 2147.

Deng, Qing, Bo Hu, Yao Zhang, Hao Wang, Xiaoyang Zhou, Wei Hu, Yuting Cheng, Jie Yan, Haiqin Ping, and Qing Zhou. "Suspected myocardial injury in patients with COVID-19: Evidence from front-line clinical observation in Wuhan, China." *International journal of cardiology* 311 (2020): 116–121.

Ganguly, Abhratanu, Mandi, Moutushi, Dutta, Anik, Rajak, Prem. "In Silico Analysis Reveals the Inhibitory Potential of Madecassic Acid against Entry Factors of SARS-CoV-2." *ACS applied bio materials* 6, (2023): 652–662.

Gheblawi, Mahmoud, Kaiming Wang, Anissa Viveiros, Quynh Nguyen, Jiu-Chang Zhong, Anthony J. Turner, Mohan K. Raizada, Maria B. Grant, and Gavin Y. Oudit. "Angiotensin-converting enzyme 2: SARS-CoV-2 receptor and regulator of the renin-angiotensin system: Celebrating the 20th anniversary of the discovery of ACE2." *Circulation research* 126, no. 10 (2020): 1456–1474.

Greil, Richard, Sigrun Greil-Ressler, Lukas Weiss, Charlotte Schönlieb, Teresa Magnes, Bianca Radl, Gordon T. Bolger, Brigitta Vcelar, and Peter P. Sordillo. "A phase 1 dose-escalation study on the safety, tolerability and activity of liposomal curcumin (Lipocurc™) in patients with locally advanced or metastatic cancer." *Cancer chemotherapy and pharmacology* 82 (2018): 695–706.

Goc, Anna, Sumera, Waldemar, Rath, Matthias, Niedzwiecki, Aleksandra. "Phenolic compounds disrupt spike-mediated receptor-binding and entry of SARS-CoV-2 pseudo-virions." *PLoS One* 16 (2021): e0253489.

Guo, Tao, Yongzhen Fan, Ming Chen, Xiaoyan Wu, Lin Zhang, Tao He, Hairong Wang, Jing Wan, Xinghuan Wang, and Zhibing Lu. "Cardiovascular implications of fatal outcomes of patients with coronavirus disease 2019 (COVID-19)." *JAMA cardiology* 5, no. 7 (2020): 811–818.

Hassaninasab, Azam, Yoshiteru Hashimoto, Kaori Tomita-Yokotani, and Michihiko Kobayashi. "Discovery of the curcumin metabolic pathway involving a unique enzyme in an intestinal microorganism." *Proceedings of the national academy of sciences* 108, no. 16 (2011): 6615–6620.

Heger, Michal, Rowan F. van Golen, Mans Broekgaarden, and Martin C. Michel. "The molecular basis for the pharmacokinetics and pharmacodynamics of curcumin and its metabolites in relation to cancer." *Pharmacological reviews* 66, no. 1 (2014): 222–307.

Hewlings, Susan J., and Douglas S. Kalman. "Curcumin: A review of its effects on human health." *Foods* 6, no. 10 (2017): 92.

Huang, Yuan, Chan Yang, Xin-Feng Xu, Wei Xu, and Shu-Wen Liu. "Structural and functional properties of SARS-CoV-2 spike protein: Potential antivirus drug development for COVID-19." *Acta pharmacologica sinica* 41, no. 9 (2020): 1141–1149.

Huynh, Tien, Haoran Wang, and Binquan Luan. "In silico exploration of the molecular mechanism of clinically oriented drugs for possibly inhibiting SARS-CoV-2's main protease." *The journal of physical chemistry letters* 11, no. 11 (2020): 4413–4420.

Ip, WK Eddie, Namiko Hoshi, Dror S. Shouval, Scott Snapper, and Ruslan Medzhitov. "Anti-inflammatory effect of IL-10 mediated by metabolic reprogramming of macrophages." *Science* 356, no. 6337 (2017): 513–519.

Jena, Atala B., Namrata Kanungo, Vinayak Nayak, G. B. N. Chainy, and Jagneshwar Dandapat. "Catechin and curcumin interact with S protein of SARS-CoV2 and ACE2 of human cell membrane: Insights from computational studies." *Scientific reports* 11, no. 1 (2021): 2043.

Kuba, Keiji, Yumiko Imai, Shuan Rao, Hong Gao, Feng Guo, Bin Guan, Yi Huan et al. "A crucial role of angiotensin converting enzyme 2 (ACE2) in SARS coronavirus–induced lung injury." *Nature medicine* 11, no. 8 (2005): 875–879.

Laksmiani, Ni Putu Linda, Luh Putu Febryana Larasanty, Anak Agung Gde Jaya Santika, Putu Agus Andika Prayoga, Anak Agung Intan Kharisma Dewi, and Ni Putu Ayu Kristiara Dewi. "Active compounds activity from the medicinal plants against SARS-CoV-2 using in silico assay." *Biomedical and pharmacology journal* 13, no. 2 (2020): 873–881.

Lang, Joshua P., Xiaowen Wang, Filipe A. Moura, Hasan K. Siddiqi, David A. Morrow, and Erin A. Bohula. "A current review of COVID-19 for the cardiovascular specialist." *American heart journal* 226 (2020): 29–44.

Lao, Christopher D., Mack T. Ruffin, Daniel Normolle, Dennis D. Heath, Sandra I. Murray, Joanne M. Bailey, Martha E. Boggs, James Crowell, Cheryl L. Rock, and Dean E. Brenner. "Dose escalation of a curcuminoid formulation." *BMC complementary and alternative medicine* 6, no. 1 (2006): 1–4.

Liao, Mingfeng, Yang Liu, Jing Yuan, Yanling Wen, Gang Xu, Juanjuan Zhao, Lin Cheng et al. "Single-cell landscape of bronchoalveolar immune cells in patients with COVID-19." *Nature medicine* 26, no. 6 (2020): 842–844.

Marín-Palma, Damariz, Jorge H. Tabares-Guevara, María I. Zapata-Cardona, Lizdany Flórez-Álvarez, Lina M. Yepes, Maria T. Rugeles, Wildeman Zapata-Builes, Juan C. Hernandez, and Natalia A. Taborda. "Curcumin inhibits in vitro SARS-CoV-2 infection in Vero E6 cells through multiple antiviral mechanisms." *Molecules* 26, no. 22 (2021): 6900.

Maurya, Vimal K., Swatantra Kumar, Anil K. Prasad, Madan LB Bhatt, and Shailendra K. Saxena. "Structure-based drug designing for potential antiviral activity of selected natural products from ayurveda against SARS-CoV-2 spike glycoprotein and its cellular receptor." *Virusdisease* 31 (2020): 179–193.

Middleton, Elizabeth A., Xue-Yan He, Frederik Denorme, Robert A. Campbell, David Ng, Steven P. Salvatore, Maria Mostyka et al. "Neutrophil extracellular traps contribute to immunothrombosis in COVID-19 acute respiratory distress syndrome." *Blood, the journal of the American society of hematology* 136, no. 10 (2020): 1169–1179.

Mohd Abd Razak, Mohd Ridzuan, Nur Hana Md Jelas, Amirrudin Muhammad, Noorsofiana Padlan, Muhammad Nor Farhan Sa'at, Muhammad Afif Azizan, Siti Nur Zawani Rosli et al. "In vitro anti-SARS-CoV-2 activities of curcumin and selected phenolic compounds." *Natural product communications* 18, no. 9 (2023): 1934578X231188861.

Motohashi, Noboru, Anuradha Vanam, and Rao Gollapudi. "In silico study of curcumin and folic acid as potent inhibitors of human transmembrane protease serine 2 in the treatment of COVID-19." *INNOSC theranostics and pharmacological sciences* 3, no. 2 (2020): 935.

Murakami, Masaaki, Daisuke Kamimura, and Toshio Hirano. "Pleiotropy and specificity: Insights from the interleukin 6 family of cytokines." *Immunity* 50, no. 4 (2019): 812–831.

Nelson, Kathryn M., Jayme L. Dahlin, Jonathan Bisson, James Graham, Guido F. Pauli, and Michael A. Walters. "The essential medicinal chemistry of curcumin: Miniperspective." *Journal of medicinal chemistry* 60, no. 5 (2017): 1620–1637.

Pang, Xue-Fen, Li-Hui Zhang, Feng Bai, Ning-Ping Wang, Ron E. Garner, Robert J. McKallip, and Zhi-Qing Zhao. "Attenuation of myocardial fibrosis with curcumin is mediated by modulating expression of angiotensin II AT1/AT2 receptors and ACE2 in rats." *Drug design, development and therapy* 9, (2015): 6043–6054.

Patell, Rushad, Thomas Bogue, Anita Koshy, Poorva Bindal, Mwanasha Merrill, William C. Aird, Kenneth A. Bauer, and Jeffrey I. Zwicker. "Postdischarge thrombosis and hemorrhage in patients with COVID-19." *Blood, the journal of the American society of hematology* 136, no. 11 (2020): 1342–1346.

Pendurthi, Usha R., J. Todd Williams, and L. Vijaya Mohan Rao. "Inhibition of tissue factor gene activation in cultured endothelial cells by curcumin: Suppression of activation of transcription factors Egr-1, AP-1, and NF-κB." *Arteriosclerosis, thrombosis, and vascular biology* 17, no. 12 (1997): 3406–3413.

Prasad, Sahdeo, and Amit K. Tyagi. "Curcumin and its analogues: A potential natural compound against HIV infection and AIDS." *Food & function* 6, no. 11 (2015): 3412–3419.

Rajak, Prem, Abhratanu Ganguly, Sukhendu Dey, Anik Dutta, and Moutushi Mandi. "In silico study unravels binding potential of madecassic acid against non-structural proteins of SARS-CoV-2." *Pharmacological research-modern Chinese medicine* 9 (2023): 100320.

Rajak, Prem, Ganguly, Abhratanu. "Computational study unravels inhibitory potential of epicatechin gallate against inflammatory and pyroptosis-associated mediators in COVID-19." *MedComm – Future Medicine* 2, (2023a):e52.

Rajak, Prem, Ganguly, Abhratanu. "In silico study unfolds inhibitory potential of epicatechin gallate against SARS-CoV-2 entry and replication within the host cell." *Mechanobiology in Medicine* 1, (2023b): 100015.

Ramirez, Boscá A., Alfonso Soler, Miguel A. Carrión-Gutiérrez, Mira D. Pamies, Zapata J. Pardo, Joaquín Diaz-Alperi, August Bernd, Almagro E. Quintanilla, and Jaime Miquel. "An hydroalcoholic extract of *Curcuma longa* lowers the abnormally high values of human-plasma fibrinogen." *Mechanisms of ageing and development* 114, no. 3 (2000): 207–210.

Rapkiewicz, Amy V., Xingchen Mai, Steven E. Carsons, Stefania Pittaluga, David E. Kleiner, Jeffrey S. Berger, Sarun Thomas et al. "Megakaryocytes and platelet-fibrin thrombi characterize multi-organ thrombosis at autopsy in COVID-19: A case series." *EClinicalMedicine* 24 (2020): 100434.

Rodrigues TS, de Sá KSG, Ishimoto AY, Becerra A, Oliveira S, Almeida L, Gonçalves AV, et al. "Inflammasomes are activated in response to SARS-CoV-2 infection and are associated with COVID-19 severity in patients. *Journal of Experimental Medicine* 2018, (2021):e20201707.

Ruiz-Ortega, Marta, Oscar Lorenzo, Yusuke Suzuki, Mónica Rupérez, and Jesús Egido. "Proinflammatory actions of angiotensins." *Current opinion in nephrology and hypertension* 10, no. 3 (2001): 321–329.

Sathler, Plinio C. "Hemostatic abnormalities in COVID-19: A guided review." *Anais da academia Brasileira de Ciencias* 92 (2020): e20200834.

Shah, Bukhtiar H., Zafar Nawaz, Shamim A. Pertani, Asad Roomi, Hammad Mahmood, Sheikh A. Saeed, and Anwar H. Gilani. "Inhibitory effect of curcumin, a food spice from turmeric, on platelet-activating factor-and arachidonic acid-mediated platelet aggregation through inhibition of thromboxane formation and Ca2+ signaling." *Biochemical pharmacology* 58, no. 7 (1999): 1167–1172.

Shanmugarajan, Dhivya, P. Prabitha, B. R. Prashantha Kumar, and B. Suresh. "Curcumin to inhibit binding of spike glycoprotein to ACE2 receptors: Computational modelling, simulations, and ADMET studies to explore curcuminoids against novel SARS-CoV-2 targets." *RSC advances* 10, no. 52 (2020): 31385–31399.

Srivastava, Akhileshwar, and Divya Singh. "Destabilizing the structural integrity of SARS-CoV2 receptor proteins by curcumin along with hydroxychloroquine: An Insilco approach for a combination therapy." (2020).

Srour, Aladdin M., Siva S. Panda, Ahmed Mostafa, Walid Fayad, May A. El-Manawaty, Ahmed AF Soliman, Yassmin Moatasim et al. "Synthesis of aspirin-curcumin mimic conjugates of potential antitumor and anti-SARS-CoV-2 properties." *Bioorganic chemistry* 117 (2021): 105466.

Storka, Angela, Brigitta Vcelar, Uros Klickovic, Ghazaleh Gouya, Stefan Weisshaar, Stefan Aschauer, Gordon Bolger, Lawrence Helson, and M. Woltz. "Safety, tolerability and pharmacokinetics of liposomal curcumin (Lipocurc™) in healthy humans." *International journal of clinical pharmacology and therapeutics* 53, no. 1 (2015): 54–65.

Suravajhala R, Parashar A, Choudhir G, Kumar A, Malik B, Nagaraj VA, Padmanaban G, Polavarapu R, Suravajhala P, Kishor PBK. "Molecular docking and dynamics studies of curcumin with COVID-19 proteins". *Netw Model Anal Health Inform Bioinform* 10 (2021): 44.

Tahmasebi, Safa, Mohamed A. El-Esawi, Zaid Hameed Mahmoud, Anton Timoshin, Hamed Valizadeh, Leila Roshangar, Mojtaba Varshoch et al. "Immunomodulatory effects of nanocurcumin on Th17 cell responses in mild and severe COVID-19 patients." *Journal of cellular physiology* 236, no. 7 (2021): 5325–5338.

Valizadeh, Hamed, Samaneh Abdolmohammadi-Vahid, Svetlana Danshina, Mehmet Ziya Gencer, Ali Ammari, Armin Sadeghi, Leila Roshangar et al. "Nano-curcumin therapy, a promising method in modulating inflammatory cytokines in COVID-19 patients." *International immunopharmacology* 89 (2020): 107088.

Vishvakarma, Naveen Kumar, Anjani Kumar, and Sukh Mahendra Singh. "Role of curcumin-dependent modulation of tumor microenvironment of a murine T cell lymphoma in altered regulation of tumor cell survival." *Toxicology and applied pharmacology* 252, no. 3 (2011): 298–306.

Wang, Xun, Xiaojing An, Xiaocen Wang, Chen Bao, Jing Li, Dong Yang, and Chunxue Bai. "Curcumin ameliorated ventilator-induced lung injury in rats." *Biomedicine & pharmacotherapy* 98 (2018): 754–761.

Wang, Yiqing, Qichun Tang, Peibei Duan, and Lihua Yang. "Curcumin as a therapeutic agent for blocking NF-κB activation in ulcerative colitis." *Immunopharmacology and immunotoxicology* 40, no. 6 (2018): 476–482.

Wang, Ning-Ping, Zhang-Feng Wang, Stephanie Tootle, Tiji Philip, and Zhi-Qing Zhao. "Curcumin promotes cardiac repair and ameliorates cardiac dysfunction following myocardial infarction." *British journal of pharmacology* 167, no. 7 (2012): 1550–1562.

Wrapp, Daniel, Nianshuang Wang, Kizzmekia S. Corbett, Jory A. Goldsmith, Ching-Lin Hsieh, Olubukola Abiona, Barney S. Graham, and Jason S. McLellan. "Cryo-EM structure of the 2019-nCoV spike in the prefusion conformation." *Science* 367, no. 6483 (2020): 1260–1263.

Xia, Shuai, Meiqin Liu, Chao Wang, Wei Xu, Qiaoshuai Lan, Siliang Feng, Feifei Qi et al. "Inhibition of SARS-CoV-2 (previously 2019-nCoV) infection by a highly potent pan-coronavirus fusion inhibitor targeting its spike protein that harbors a high capacity to mediate membrane fusion." *Cell research* 30, no. 4 (2020): 343–355.

Xu, Fang, Shi-Hui Lin, Yuan-Zheng Yang, Rui Guo, Ju Cao, and Qiong Liu. "The effect of curcumin on sepsis-induced acute lung injury in a rat model through the inhibition of the TGF-β1/SMAD3 pathway." *International immunopharmacology* 16, no. 1 (2013): 1–6.

Yadav, Renu, Babban Jee, and Sudhir Kumar Awasthi. "Curcumin suppresses the production of pro-inflammatory cytokine interleukin-18 in lipopolysaccharide stimulated murine macrophage-like cells." *Indian journal of clinical biochemistry* 30 (2015): 109–112.

Yang, Cheng, Keng Wu, Shang-Hai Li, and Qiong You. "Protective effect of curcumin against cardiac dysfunction in sepsis rats." *Pharmaceutical biology* 51, no. 4 (2013): 482–487.

Zahedipour, Fatemeh, Seyede Atefe Hosseini, Thozhukat Sathyapalan, Muhammed Majeed, Tannaz Jamialahmadi, Khalid Al-Rasadi, Maciej Banach, and Amirhossein Sahebkar. "Potential effects of curcumin in the treatment of COVID-19 infection." *Phytotherapy research* 34, no. 11 (2020): 2911–2920.

Zhang, Li, Zhen-Lun Gu, Zheng-Hong Qin, and Zhong-Qin Liang. "Effect of curcumin on the adhesion of platelets to brain microvascular endothelial cells in vitro 1." *Acta pharmacologica sinica* 29, no. 7 (2008): 800–807.

Zhang, Hui-Na, Chun-xiao Yu, Peng-Ju Zhang, Wei-Wen Chen, An-Li Jiang, Feng Kong, Jing-Ti Deng, Jian-Ye Zhang, and Charles YF Young. "Curcumin downregulates homeobox gene NKX3. 1 in prostate cancer cell LNCaP." *Acta pharmacologica sinica* 28, no. 3 (2007): 423–430.

Zhang S, Liu Y, Wang X, Yang L, Li H, Wang Y, Liu M, et al. SARS-CoV-2 binds platelet ACE2 to enhance thrombosis in COVID-19. *Journal of Hematology & Oncology* 13, (2020):120.

Zhou, Fei, Ting Yu, Ronghui Du, Guohui Fan, Ying Liu, Zhibo Liu, Jie Xiang et al. "Clinical course and risk factors for mortality of adult inpatients with COVID-19 in Wuhan, China: A retrospective cohort study." *The lancet* 395, no. 10229 (2020): 1054–1062.

Zupin, Luisa, Francesco Fontana, Libera Clemente, Violetta Borelli, Giuseppe Ricci, Maurizio Ruscio, and Sergio Crovella. "Optimization of anti-SARS-CoV-2 treatments based on curcumin, used alone or employed as a photosensitizer." *Viruses* 14, no. 10 (2022): 2132.

18 The Impact of Traditional Homeopathic Medicines in COVID-19 Management

*Saurabh Sarkar, Prem Rajak, Abhijit Ghosh, Moutushi Mandi,
Abhratanu Ganguly, Sayantani Nanda, and Kanchana Das*

18.1 INTRODUCTION

The Coronavirus disease-19 (COVID-19) pandemic, caused by Severe Acute Respiratory Syndrome Coronavirus-2 (SARS-CoV-2), was initially identified in China in the last month of 2019 and swiftly spread to 213 nations worldwide. Vaccines with high efficacy have been formulated by several research institutions, and immunization has been continuing in multiple nations. Nevertheless, it should be noted that immunizations may not offer complete immunity against recitative infections. Hence, even post-vaccination, individuals must adhere to the prescribed protocols set forth by the World Health Organization (WHO) and regional authorities to avoid the possibility of repetitive infection. Numerous African, European, as well as Asian nations are witnessing successive surges of COVID-19 as a result of the introduction of novel viral strains like Alpha, Beta, Gamma, Epsilon, Epsilon, and Delta, etc (Rajak et al., 2021a). These variants exhibit significant divergence, with each displaying a distinct combination of mutations that may have biological significance. As a result, they can evade the immune defenses acquired from a prior infection. The disease is a persistent health crisis that impacts people of all age groups worldwide. At first, it was reported that the illness primarily impacted the internal morphology of the lungs. Recent research has elucidated the effect of COVID-19 on the cardiovascular and reproductive systems of both males and females (Rajak et al., 2021b, Rajak et al., 2022). Furthermore, COVID-19 impacts akin to altered renin-angiotensin homeostasis, oxidative stress, elevated body temperature, cytokine storm, as well as mental stress, could propagate several other physiological and medical complications (Rajak et al., 2021c).

There are several medical systems like the Allopathic System of Medicine, Homeopathy, Ayurveda, Unani Medicine, Chinese Medicine, Alternative Medicine, etc., to facilitate different remedies for patients worldwide (WHO 2001). All the diverse systems of medicine might follow diverse diagnosis procedures, philosophies, choices of medicine (drug), methodology of treatment, application and administration of medicine (drug) but the primary objective of all the systems of medicine is to cure the body and mind as well as restore to health. The world has witnessed huge co-morbid casualties and severely infected COVID-19 patients during this COVID-19 pandemic resulting in the breakdown of medical systems and facilities. Existing different medical and healing systems (Allopathic Medicine, Homeopathy, Ayurveda, Unani Medicine, Chinese Medicine, and Alternative Medicine) served their best in fighting against COVID-19 with their limitations. Though Allopathic medicines have recently accelerated their pace, the importance and impact of various traditional and herbal medicines (e.g., Homeopathic medicines) have not faded. Among all the systems of medicine, the Homeopathy system of medicine is the rapidly spreading and second-most extensively utilized traditional and complementary system of medicine facilitating more than 200 million people around the globe (Black, 1994). Traditional Homeopathic systems of medicine have contributed to effective protection, prevention as well as remedial solutions for COVID-19 management during this pandemic period.

DOI: 10.1201/9781003452621-18

18.2 STRUCTURE OF SARS-CoV-2

SARS-CoV-2 is a retrovirus classified under the family *Coronaviridae*. The virus is spherical and possesses a diameter ranging from 50 to 200 nanometers. The genome is composed of polycistronic positive-sense single-stranded RNA, measuring approximately 29,881 base pairs in length and encoding about 9860 amino acids. The RNA is enveloped by both 5′ and 3′ untranslated regions (UTR). Chan et al. (2020) reported that the viral genome has 11 open reading frames (ORFs) responsible for encoding different types of proteins, including non-structural (NSP) proteins, structural proteins like spike (S), envelop (E), membrane (M), nucleocapsid (N), as well as auxiliary proteins. The N protein enhances the robustness and stability of the RNA within the genome. Moreover, it impedes the interferon (IFN) activity and RNA-silencing in the host cells (Cui et al., 2015). The M glycoprotein has been recognized as the primary envelope protein that maintains the spherical structure of the virion (Neuman et al., 2011). The E protein plays a crucial role in the virus's assembly, release, and ability to cause disease (Nieto-Torres et al., 2014). Wrapp et al. (2020) discovered that the S (spike) protein arranges itself in a group of three on the outer layer of the virus particle, creating a distinct appearance like a crown, also known as a "corona". The S protein is made up of a trimeric S1 region located at the top and binds to the host angiotensin-converting enzyme-2 (ACE-2) receptor. The receptor-binding-domain (RBD) of the virus undergoes regular transitions between an upright conformation, which allows it to adhere to receptors, and a prone conformation, which helps it elude the immune system (Shang et al., 2020). The fusion peptide (FP) in the S2 protein consists of 5–25 amino acids and is referred to as the heptad-repeat (HR1 and HR2) regions. These regions are implicated in the fusion process between the viral envelope and the host cell membrane (Henning and Pillat, 2020). According to Romano et al. (2020), the ORF1a and ORF1ab are the main units involved in transcription and are responsible for producing two important replicase polyproteins, namely 1a (PP1a) and 1ab (PP1ab). The replicase transcriptase complex (RCT) consists of Nsp1–16 that is enclosed within the largest polyprotein, pp1ab (Rajak et al., 2021a).

18.3 CELL-INVASION AND INFECTION MECHANISM OF SARS-CoV-2

SARS-CoV-2 enters specific cells by participating in a molecular interaction between the viral spike glycoprotein and the ACE-2. SARS-CoV-2 interacts with CD147 receptors to enter a particular cell (Henning and Pillat, 2020). Other elements, such as lysosomal cathepsin L (CTSL), neuropilin-1 (NRP-1), and cathepsin B (CTSB), can further facilitate the virus's penetration into the host cell. The binding of S1 to the ACE-2 receptor increases the proteolytic activation of the PPC motif at the S1/S2 region. Proteolytic activation can alter the structure of S2, leading to the exposure of FP. This structural alteration allows for the penetration of the plasma membrane by adopting hairpin-loop conformation. Subsequently, HR1 and HR2 undergo folding, resulting in the formation of a stable fusion core consisting of a 6-helical antiparallel bundle (Xia et al., 2020). These essential steps facilitate the proximity of the viral membrane to the host cell membrane, enabling the fusing of the two membranes and the transmission of the viral genetic material, referred to as genomic RNA, into the host cell. The transfer can take place either directly into the host cell or by a process known as endocytosis, as depicted in Figure 18.1. Upon entry into the cell, the ribosome of the host decodes the genetic information of the virus, resulting in the production of enzymes such as proteases, helicase, and RNA-dependent RNA polymerage (RdRP). Viral enzymes synthesize fresh viral genomes, messenger RNAs, and proteins to produce further viral particles. Subsequently, these particles are expelled from infected cells via exocytosis to contaminate and invade fresh cells. Similar to other RNA viruses, like those examined by Cheng et al. (2018) in 2018, SARS-CoV-2 has the ability to utilize host proteins, such as DDX5 RNA helicases, to augment its replication process. SARS-CoV-2 virions replicate, package, and then exit the cytosolic compartment of the cell, enabling them to infect neighboring cells (Rajak et al., 2021b).

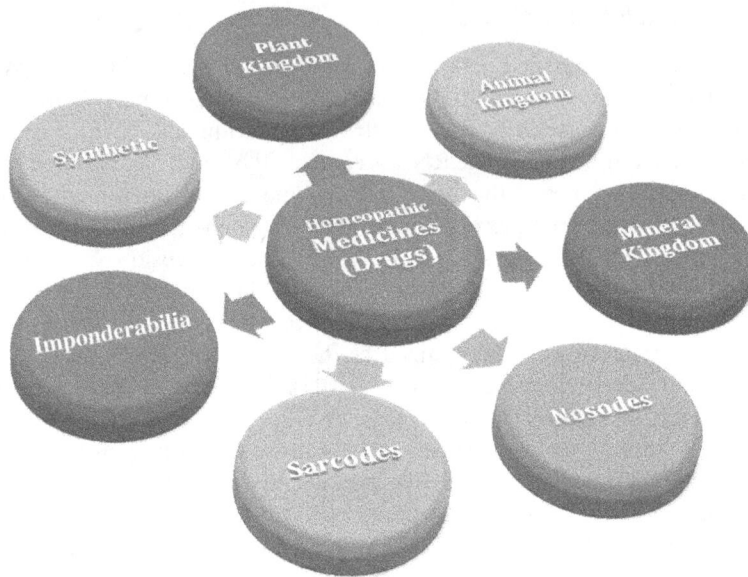

FIGURE 18.1   Types of medicines (drugs) used in Homeopathy for COVID-19 management.

18.4   IMMUNE RESPONSE OF OUR BODY AGAINST SARS-CoV-2 INFECTION

The innate immune system of our body offers the most effective protection against coronavirus infection. The causative pathogen, SARS-CoV-2 triggers several elements of the innate as well as adaptive immune system that are concerned with the removal of virus particles from the body.

18.4.1   INNATE IMMUNE RESPONSE (IIR)

The IIR acts as the primary defense against viral infection. COVID-19 triggers the population of myeloid lineage immunocytes. Various investigations have reported that individuals with severe COVID-19 exhibited elevated levels of activated mast cells, neutrophil-to-lymphocyte ratio, and blood neutrophil count. These findings indicate that these immune cells are involved in combating the SARS-CoV-2 virus (Liu et al., 2020a; Zhou et al., 2020c). The complement system is essential for IIR. In acute patients of COVID-19, there is an increase in the levels of complement proteins in the blood compared to moderate cases and healthy individuals (Gao et al., 2020; Xiong et al., 2020). IFNs exert a potent antiviral immune response. Notably, infection with SARS-CoV-2 hinders the generation of IFN, which is essential for sustaining the innate defense system (Acharya et al., 2020). Concomitantly, infection with SARS-CoV-2 stimulates a substantial release of proinflammatory cytokines, leading to the development of cytokine storms and pyroptosis. The increased synthesis of proinflammatory chemokines, explicitly interleukin (IL) like IL-6, IL-2, IL-1β, IL-7, IP-10, granulocyte-colony-stimulating factor (G-CSF), macrophage inflammatory protein-1 (MIP-1), monocyte chemoattractant protein-1 (MCP-1), IL1RA, and tumor necrosis factor-α (TNF-α) leads to the over-activation of the innate immune system, hence significantly exacerbating immunological disorders in individuals with COVID-19 as suggested by Blanco-Melo et al., 2020; Chu et al., 2020; and Huang et al., 2020. It is imperative to preserve a strong innate immune response to restrict the progression of illness caused by SARS-CoV-2 (Rajak et al., 2021c).

18.4.2 Adaptive Immune Responses (AIR)

The change from the innate (IIR) to adaptive immune response (AIR) is of utmost importance in the clinical development of coronavirus illness. T-lymphocytes coordinate the AIR (adaptive immune responses) necessary for eradicating cells that have been infiltrated by viruses. The immune response as a result of the SARS-CoV-2 virus is controlled by T-cells that secrete INF-γ, natural killer (NK) cells, and T-cells that specifically target the spike RBD (Ni et al., 2020). An optimal aggregation of cytotoxic CD8 cells is essential for effectively eliminating virally infected cells from our bodies (Li et al., 2020). The progression of illness and negative outcomes in infected patients is linked to decline stimulation of B-cells and dendritic cells (DCs) (Wang et al., 2020a). The resurrection of COVID-19 infection relies heavily on the immunological response mediated by antibodies. Zhou et al. (2020a) discovered that infected individuals exhibited a specific antibody response to the nucleocapsid protein (NP), indicating their direct participation in the humoral immune response. Ni et al. (2020) discovered the existence of NP- and spike RBD-specific IgM and IgG1 antibodies in the blood samples of recently discharged patients. Prior to clinical trials, laboratory tests detected the existence of neutralizing IgA antibodies in bronchoalveolar lavages during the testing of anti-SARS vaccinations (Lu et al., 2010). The presence of IgG antibodies can be identified after the manifestation of warning signs (Long et al., 2020). Hence, the aforementioned data suggest that a strong population of varied immune cells as well as antibodies is crucial for eliminating the SARS-CoV-2 infection. Individuals with COVID-19 who have weakened humoral and cell-mediated immunity are at risk of experiencing more severe diseases, pulmonary immunopathology, as well as potentially lethal results (Rajak et al., 2022).

18.5 COVID-19-MEDIATED RESPIRATORY ISSUES

Since December 2019, the globe has endured a global health crisis known as COVID-19 and its subsequent consequences. COVID-19 infections elicit robust inflammatory reactions that lead to severe pneumonitis, dyspnea, bronchitis, and respiratory failure in patients (Greenland et al., 2020). Infected individuals commonly exhibit a high incidence of severe lung injury, which is defined by the presence of "ground glass opacities". In addition to respiratory difficulties, the majority of COVID-19 individuals also exhibit additional illnesses (Mokhtari et al., 2020). Typically, when virions enter by the nose or mouth, they initially settle in the broncho-pulmonary epithelium and contribute to lung diseases caused by necroinflammation (Rajak et al., 2021a). The involvement of myeloid cells and the complement system is significant in limiting the spread of SARS-CoV-2 infection. Nevertheless, a repressed or delayed IFN response might weaken the initial control of the virus, resulting in a significant release of proinflammatory cytokines. These harmful effects contribute to respiratory diseases such as bronchitis, pneumonia as well as acute respiratory distress syndrome (ARDS). SARS-CoV-2 can disseminate to different organs through the bloodstream by infiltrating the endothelium layer of blood vessels, originating from the lungs (Rajak et al., 2021b; Varga et al., 2020). Different organs, such as the heart, liver, kidney, testis, and ovary, express various SARS-CoV-2 receptors for entrance such as transmembrane serine protease 2 (TMPRSS2), cathepsin L, furin, CD26, ezrin, and cyclophilins as well as ACE-2. These receptors are associated with different warning signs of infection in COVID-19 patients (Rajak et al., 2022).

18.6 COVID-19-MEDIATED CARDIOVASCULAR ISSUES

There is a strong link between viral infection and myocarditis. COVID-19-infected patients who were hospitalized in China had cardiac damage. This was demonstrated by elevated levels of high-sensitivity cardiac troponin I (hs-cTnI) as well as abnormal electrocardiogram (ECG) findings as suggested by Wang et al., (2020b). Furthermore, the National Health Commission of China documented that about 12% of COVID-19-infected patients without a preexisting cardiovascular condition exhibited high troponin levels or suffered cardiac arrest during their hospitalization (Zheng et al., 2020). Moreover, an increased concentration of troponin-I is linked to cardiac injury and has been demonstrated to heighten the mortality risk in patients who have tested positive for SARS-CoV-2 infection (Clerkin et al., 2020; Shi et al., 2020). Infection can cause a modest to moderate decline in heart function, leading to a reduction in the ejection fraction (Wolfler et al., 2020). Furthermore, SARS-CoV-2 infiltrates the endothelium of blood vessels and stimulates the generation of endothelins, which results in the development of blood clots that might potentially cause detrimental impacts on the circulatory system (Varga et al., 2020). Hence, the aforementioned works unequivocally demonstrate that SARS-CoV-2 could play a role in the onset of cardiovascular conditions in patients. Cytokines are essential for maintaining the structural as well as functional integrity of various organs. However, when proinflammatory cytokines are generated at excessive levels, they can modify the cellular microenvironment and severely affect the functioning of various organs, such as the heart and the cardiovascular system as a whole. Excessive and prolonged production of cytokines is linked to the development of several cardiac diseases, including heart failure and adverse alterations in the anatomy of the left ventricle (Hohensinner et al., 2011). Elevated concentrations of TNF-α and IL-6 in the bloodstream may serve as possible indicators of increased mortality risk associated with heart failure (Deswal et al., 2001). C-reactive proteins (CRP) are the main markers for both inflammation and myocardial infarction. According to Lagrand et al. (1999), the activation of the complement system allows them to directly engage with atherosclerotic arteries or ischemic myocardium, increasing inflammation and thrombosis.

18.7 COVID-19-MEDIATED REPRODUCTIVE ISSUES AND ART

The SARS-CoV-2 virus specifically affects the heart, digestive system, liver, skin, and kidneys, causing damage to these organs. It accomplishes this by utilizing ACE-2, TMPRSS2, as well as other components to invade and inhabit the host cells. Therefore, organs that express these entry factors are more susceptible to COVID-19-induced damage and dysfunction. The presence of these entry factors in gonads and other accessory reproductive organs has been reported in studies conducted by Liu et al. (2020b) and Larasati et al. (2020). Hence, this virus has the potential to infiltrate reproductive organs and disturb the reproductive well-being of those who are affected. Recent research has found that adults who contract COVID-19 may experience inflammation in both testicles (bilateral orchitis), as well as reduced sperm quality and delayed sperm production (spermatogenesis). In children, the infection can lead to inflammation in the testicles and epididymis (orchiepididymitis). These findings were reported by Gagliardi et al. (2020), Holtmann et al. (2020), Li et al. (2020) as well as Bridwell et al. (2021). In addition, women of reproductive age who contract COVID-19 experience a reduction in menstrual flow and an elongation of their menstrual cycle, which could potentially result in future infertility (Li et al., 2021). Past studies have documented occurrences of spontaneous miscarriage and premature delivery in pregnant women who have been infected with the coronavirus (Wong et al., 2004). Multiple investigations have demonstrated an increased susceptibility to maternal mortality in pregnant individuals with COVID-19 (Matar et al., 2021). Additionally, the transmission of viral particles from infected mothers to their babies has been seen (Dong et al., 2020). The COVID-19 pandemic has caused a postponement of medical operations related to Assisted Reproductive Technology (ART), a commonly utilized method for treating infertility globally. This delay has resulted in heightened levels of anxiety and sadness among individuals receiving ART (Tokgoz et al., 2020). Both mental health as well as depression are recognized as potential risk factors for infertility. Consequently, populations with elevated levels of IL-6 and TNF have reduced fertility rates, as evidenced by greater levels of despair and anxiety, compared to those with normal levels of these inflammatory markers (Tanacan et al., 2021). Overproduction of IL6 and TNF has been linked to negative pregnancy outcomes, including premature delivery, early rupture of the membrane, as well as pregnancy loss (Qiu et al., 2018). COVID-19 mediators, including disruptions in the renin-angiotensin system, oxidative stress, cytokine storm, high body temperature (fever), and mental stress significantly impact the reproductive organs of both males and females (Rajak et al., 2021b, 2021c).

18.8   HOMEOPATHY SYSTEM OF MEDICINE (TRADITIONAL SYSTEM OF MEDICINE)

A German physician, Dr. Christian Fredrick Samuel Hahnemann (1755–1843), discovered Homeopathy while completing his medical degree in Allopath (1779) from Erlangen (Loudon, 2006). During his first 15 years, he was a practitioner of an allopathic system of medicine and witnessed the brutality of several allopathic treatment procedures as well as surgery. He searched for alternate ways to cure his patients through gentle therapeutic procedures. Initially, he consumed normal doses of cinchona or "the bark" (i.e. quinine) regularly as a part of his experiment, and found all the symptoms (in mild degree) of intermittent fever (malaria) without that disease (Loudon, 2006). This experimental result was first published by Hahnemann in 1796 as an *Essay on a New Principle for Ascertaining the Curative Power of Drugs*, and later on, that hypothesis was summarized in his remarkable work, *The Organon of the Healing Art* (1810) (Gevitz, 1993).

The American Institute of Homeopathy, founded in the late 1800s, was among the pioneering institutions to acknowledge the value of Homeopathy and its role alongside conventional medicine in the effective treatment of patients. Subsequently, Homeopathy was disseminated throughout the United Kingdom and the majority of European nations. Recent evidence has shown that Homeopathy is an integral part of the healthcare system in the Indian subcontinent and serves as an alternative medicine, providing healthcare services to a significant population (Harrison, 1806). The Ministry of AYUSH in India experienced a consistent annual growth rate of 26.3% in 2022. Homeopathy users, comprising approximately 100 million patients in India, hold the second place in the medical industry (Loudon, 1987). Multiple types of research indicate that the current regular usage of homeopathy exceeds 200 million individuals. Currently, homeopathy is being practiced in more than 100 nations where it has legal recognition as an independent system of medicine. Additionally, it is considered a part of complementary and alternative medicine in more than 50 countries. The WHO currently recognizes Homeopathy as the rapidly expanding and second most prevalent traditional and complementary medical system globally (Black, 1994; Loudon, 1987).

Homeopathy has been recognized as an alternative system of medicine, depending on the similia principle "*similia similibus curentur*" which demonstrates diseases can be cured by symptom similarity (Hahnemann, 1901, 1991). The word "Homeopathy" is been derived from two Greek words: "*homeos*", meaning "similar" and "*pathos*", meaning "suffering". Thus, it can be explained as the therapeutic method of treating diseases based on symptom similarity (Mukherjee and Wahile, 2006). Hahnemann believed that the illness of a patient could be cured by providing a medicine that is capable of producing similar symptoms of that same illness in a healthy person but to a lesser degree. Thus, if a patient was suffering from acute nausea, he should be prescribed a medicine that would provoke mild nausea in a healthy person. Hahnemann stated this process of selection of appropriate remedies (choice of drug) by the method of "proving". This ascertained his renowned aphorism, "like cures like", which is commonly called the "principle of similars". He also tried to explain his idea through the concept of cowpox vaccination used by Dr. Jenner to prevent smallpox as an example (Boericke, 2004, 2010). The therapeutic approach of homeopaths is concerned with "holistic medicine" as well as the "holistic approach of treatment" for the overall healing of the body, mind, and soul (Mondal 2006). Thus, the classical homeopathy system of medicine (Traditional medicine) has been suggested as an attractively safe, simple, easy to understand, and patient-centric system of medicine for better management of COVID-19 patients worldwide.

18.9   THE ART AND SCIENCE OF HOMEOPATHIC MEDICINES (DRUGS)

Medicines in Homeopathy include different types of drugs sourced from plants, animals, microbes, and different secretions from animals including poison glands, milk, etc. The particular "choice of drug" should be administered after carefully observing the symptoms of the patient to avoid any unusual toxicological effects (Boericke and Hahnemann, 2020). The art and science of Homeopathic drugs exist on three key pillars:

i. **Homeopathic pharmacy:** Homeopathic pharmacy is the art and science of collecting, compounding, combining, preparing, preserving, and standardizing drugs and medicines from vegetables, animals, the mineral kingdom, certain physiological substances, and morbid substances following the Homeopathy principle (as given in the *Organon of Medicine, Materia Medica Pura,* and *Chronic Diseases*) and also dispensing medicines according to the prescriptions of the Homeopathic physicians. Monopharmacy, pharmacodynamics, and potentization are the basis of Homeopathic pharmacy (Banerjee, 2006; Kent, 1993).

ii. **Homeopathic pharmacopoeia:** Pharmacopoeia originated from two Greek words "*pharmakon*" meaning "a drug" and "*poies*" meaning "to make". It is the standard authoritative book, containing a list of drugs, medicines, habits, descriptions, collections, and identification of drugs. It also provides direction for the preparation, combination, compounding, and standardization. It also contains information about external applications for posology and a monograph of drugs. It is the theoretical portion of pharmacy that is officially published by the authority, i.e. the government of a country or any medical or pharmaceutical society, either constituted or authorized by the government; and revised at regular interference intervals. There are different pharmacopoeia present in this world, i.e. German Homeopathic Pharmacopoeia (G.H.P., 1825), British Homeopathic Pharmacopoeia (B.H.P., 1870), Homeopathic Pharmacopoeia of the United States (1897), Homeopathic Pharmacopoeia of India (H.P.I., 1971), French Homeopathic Pharmacopoeia (P.H.F.), etc (Banerjee, 2006, Department of AYUSH, 2013).

iii. **Homeopathic pharmacology:** Homeopathic pharmacology is the science of drugs which includes pharmacy, therapeutics, and *Materia Medica,* which is derived from two Greek words, "*pharmakon*" ("drug") and "*logos*", which is "knowledge". It is the branch of medical science dealing with the preparation, uses, and effects of drugs and medicines, i.e. the science and theory of pharmacy (Banerjee, 2006). In Homeopathy, human proving is the most preferred source of *Materia Medica*, as both subjective and objective symptoms can be elicited (Clarke, 2014).

18.10   MANAGEMENT OF COVID-19 WITH HOMEOPATHY MEDICINES (DRUGS)

The diversified drugs used in the Homeopathic system of medicines are prepared from a wide range of natural sources such as plants, microbes, minerals, chemicals, animals as well as animals' venoms, secretions fluids, etc. The safety, quality, and therapeutic efficacy of drugs against COVID-19 chiefly depend on their purity and authenticity. Homeopathic drugs used for COVID-19 patients could be divided based on their sources as nosodes, Plant Kingdom, sarcodes, Animal Kingdom, imponderabilia, synthetic source, and Mineral Kingdom (Figure 18.1) (Banerjee, 2006; Biswas et al., 2018, 2019).

18.10.1   Homeopathy Medicines from the Plant Kingdom Used in COVID-19 Management

The plant kingdom is a major source of Homeopathic medicines (drugs). More than 60% of all Homeopathic drugs could be derived from a plant (herbal) source. It is assumed that homeopathy uses different plant/plant parts (like leaves, flowers, stems, bark, roots, etc., including whole plants) as medicines for various diseases and illnesses including COVID-19 cases (Table 18.1). About 300 plant species are thought to be valuable for medicinal preparation in homeopathy and more than

TABLE 18.1

Different Sources and Types of Plants Used as Homeopathic Medicines for COVID-19 Management

Types	Name of Plant and Medicine (Drug)	Common Name	Season for Collection	Medicinal uses in Homeopathy for COVID-19 Management
Whole plant including root	Aconitum napellus	Monkshood	Beginning to flowering	Primary medicine for inflammation and infection, nasal coryza, malaise exposure to cold and fever
	Belladonna	Deadly nightshade	Beginning to flowering	Primary medicine for inflammation and infection, asthma, hemorrhoids, antibiotic property, any exposure to cold and fever with body pain
	Dulcamara	Woody nightshade	Before flowering, plants growing where rootlets run into water	Primary drug use in sudden change of weather, cold and fever, pneumonia, psoriasis, asthma, cold, congested headache, stuffy nose
	Pulsatilla nigricans	Wind flower	When in flower	Yellow sordid, sinus, second stage in cold, loss of smell, back pain, male and female reproductive disorders of COVID-19 infection
Whole plant minus the root	Alfalfa	Medicago latura	Freshly dried	Reduces cholesterol, inflammation, and weakness; detoxifies body, controls diabetes due to COVID-19. Increases appetite
	Lobelia inflata	Indian tobacco	Collected in dry seasons without wash	Cold, asthma, bronchitis, whooping cough, dyspnea, chest infection, gastric disorders, and dyspepsia
	Ocimum sanctum	Brazilian alfavaca	Freshly dried	Coryza, nasal cold, elevated body temperature, mouth ulcer, loose motion, antioxidant, and anti-inflammatory properties
Roots	Aralia racemosa	Aralia	Freshly dried root	Asthma, sneezing, nasal congestion cough aggravated infection, potent detoxifier
	Bryonia alba	Wild hops	Freshly dried	Cold or flu symptoms, dry pneumonitis, cough, headache, back pain, constipation, during any stage of catarrhal fever, pneumonia
	Ipecacuanha	Ipecac root	Collected dry, in spring	Dry cough and vomit, bronchitis, mucous infection, asthma condition
	Rauwolfia serpentina	Rauwolfia	Dried root	Regulates and controls blood pressure in COVID-19 patients
	Senega	Snakewort	Dried root	With Bryonia, good for pneumonia, respiratory disorder, and chronic bronchitis, throat, nose, and chest inflammation
Bark	Azadirachta indica	Margosa bark	At least 2 years old	Herbal and Homeopathic medicine having antioxidant and anti-inflammatory properties
	Cinchona officinalis	Peruvian bark	Dried	Post-Covid recovery with alfalfa, gastric disorders, belly-ache, anti-inflammatory
Leaves	Digitalis purpurea	Foxglove	Second year's growth	Cardiovascular use, cyanosis, depletion of oxygen in blood, edema, atrial arrhythmias
	Rhus toxicodendron	Poison oak	Recently dried, Havana quality preferred	Fever, body ache (during flu), inflammation of glands, immunity booster, mucous membrane infection
	Eucalyptus globulus	Blue gum	Fresh laves	Respiratory tract infection, influenza, high fever, chronic bronchitis and asthma, antiviral, antiseptic, and analgesic

(Continued)

TABLE 18.1   *(Continued)*

Different Sources and Types of Plants Used as Homeopathic Medicines for COVID-19 Management

Types	Name of Plant and Medicine (Drug)	Common Name	Season for Collection	Medicinal uses in Homeopathy for COVID-19 Management
Flowers & Leaves	*Sambucus nigra*	Elder	Freshly dried	Cough and cold, second and third stage of pneumonia, asthma, colic, dry coryza
Flowering Tops and Leaves	*Grindelia robusta*	Rosin wood	Freshly dried	Second and third stage of pneumonia, obstruction in nasal passage, adhere cold in throat repeated hawking, chronic bronchial asthma, chronic obstruction pulmonary disease
Spores	*Lycopodium clavatum*	Club moss	Freshly dried	Immune booster in any stage, chronic lung and bronchial disorders, liver complications
Seeds	*Avena sativa*	Oat straw	Dried	Neuro-tonic, insomnia, fatigue, anxiety, headache, mental illness due to COVID-19
	Nux vomica	Poison nut	Dried	Any form of gastric problem, immune booster, antioxidant, and anti-inflammatory in COVID-19
	Sabadilla	Cevadilla seeds	Dried	Rhinitis, asthma, cough and cold, runny nose, sneezing, throat infection, infection in mucous membrane, coryza

50 plants are traditionally recognized as drug sources in homeopathy including fungi, mushrooms, weeds, herbs, etc., that can be very useful for fighting SARS-CoV-2 infection, inflammation, and other critical symptoms (Banerjee, 2006).

Aconitum napellus, Bryonia alba, and Dulcamara are primary medicines (drugs) for inflammation and infection, asthma, bronchitis, pneumonia, exposure to cold, and fever in response to COVID-19. *Rhus toxicodendron*, and Belladonna show some antiviral and antibiotic properties and help in minimizing flu-like symptoms, fever with body pain, etc. *Eucalyptus globulus*, Sabadilla as an antiviral, antiseptic, analgesic medicine could cure any type of respiratory tract infection (RTI), influenza, high fever, chronic bronchitis, asthma, rhinitis, and coryza. *Sambucus nigra* and *Grindelia robusta* are used in advanced stages of pneumonia (second and third stage), obstruction in nasal passage, adhere cold in the throat, repeated hawking, chronic bronchial asthma, chronic obstruction pulmonary disease (COPD), etc. *Lycopodium clavatum* and alfalfa as immune boosters prevent infection, inflammation, and weakness, detoxify the body, and control liver complications and diabetes in COVID-19 patients.

18.10.2   HOMEOPATHY MEDICINES FROM THE ANIMAL KINGDOM USED IN COVID-19 MANAGEMENT

The Animal Kingdom has served as another source of medicines (drugs) used in COVID-19 management (Table 18.2). The drugs are made up of living or dried entire animals, animal components, various secretions like milk and milk products (lacs) as well as venoms of poisonous animals (ophiotoxins) (Biswas et al., 2018, 2019). *Apis mellifica* and *Latrodectus hasselti* are used in anti-inflammatory, anti-allergic medicines that could heal analgesic pain, angina, edema, sore throat, tonsillitis, and pneumonia. Some anti-anaphylactic and anti-asthmatic drugs from *Blatta orientalis* and *Naja tripudians* are used in asthma, chronic bronchitis and bronchiolitis, acute and chronic endocarditis, and analgesic pain. *Spongia tosta* and *Pulmo vulpis* are used for any type of RTI, spasmodic dry cough and sore throat, wheezing, asthma, bronchitis, catarrh, lung infection, edema.

TABLE 18.2

Different Sources and Types of Animals Used as Homeopathic Medicines for COVID-19 Management

Types	Name of Animal and Drug	Common Name	Class	Phylum	Medicinal uses in Homeopathy for COVID-19 Management
			Whole Living Animals		
Whole body of living animals are crushed with rectified spirit	*Apis mellifica*	Hive bee	Insecta	Arthropoda	Anti-inflammatory, anti-allergic, sore throat, tonsillitis, pneumonia, and meningitis
	Latrodectus hasselti	Black spider (South Wales)	Arachnida	Arthropoda	Analgesic pain, angina, advanced cardiac problems, regulates blood pressure and edema
			A. Whole Dried Animals		
Whole Dried Animals	*Blatta orientalis*	Indian cockroach (India)	Insecta	Arthropoda	In any form of asthma, chronic bronchitis and bronchiolitis. Acts as anti-anaphylactic and anti-asthmatic drug
			B. Different Parts, Secretions of Animals		
Skeletons	*Spongia tosta*	Roasted sponge (Syria; Greece)	Calcarea or Calcispongiae	Porifera	Respiratory tract infection, spasmodic dry cough and sore throat, wheezing, anti-inflammatory
Juices	*Sepia succus*	Cuttlefish (India, Europe)	Cephalopoda	Mollusca	Mood swing/Mental health, antidepression, vaginitis, cure epithelioma of lips, nose, and skin in COVID-19 cases
Gall-bladder	Fel tauri	Fresh gall of horse, calves	Mammalia	Chordata	Inflamed and enlarged liver in COVID-19 patients, control cholesterol, asthma, diarrhea
Lung	*Vulpis pulmo/ Pulmo vulpis*	Fresh lung of wolf or fox	Mammalia	Chordata	Asthma, bronchitis, catarrh, lung infection, edema, respiratory tract infection
Oil	*Oleum jecoris* aselli	Fresh liver of the Cod, *Gadus morhua*	Osteichthyes	Chordata	Immune booster, anti-inflammatory, COVID-19 infected hepatic and liver disorders, sore throat, atrophy
			C. Venoms		
Snake Poison: obtained by pressing the poison gland	*Naja tripudians*	Gokshura (India)	Reptilia	Chordata	Acute and chronic endocarditis, analgesic pain, asthma, improves cardiovascular and respiratory health, prevents heart attack, and anti-inflammatory

Moreover, *Oleum jecoris aselli* acts as an immune booster, and anti-inflammatory in COVID-19-infected patients for hepatic and liver disorders, sore throat, atrophy, etc.

18.10.3 HOMEOPATHY MEDICINES FROM THE MINERAL KINGDOM USED IN COVID-19 MANAGEMENT

The Mineral Kingdom includes metals, non-metals, metalloids, inorganic and organic salts, minerals, inorganic and organic acids, mineral spring water, organic mixtures, etc., which serve as prominent remedies for various symptoms of COVID-19 (Table 18.3) (Banerjee, 2006). Acids are compounds with a hydrogen atom that could be substituted by metals. Metalloids could be defined as elements that have both metal as well as non-metal properties. Example: arsenic antimony. Salts include amalgamation of various previously mentioned components, apart from Paladium and Tellurium (Banerjee, 2006). Iridium mettalicum and ferrum metallicum are known for increasing oxygen saturation in blood and prevent anemia and weakness in COVID-19 patients. Arsenic and phosphorus (calcarea phosphorica) are some remarkable anti-allergic immune boosters, which are effective in pneumonia, bronchitis, COPD, cough and cold, sore throat, etc. Arsenicum iodatum, antimonium arsenicosum, antimonium tartaricum, and potassium/kalium bichromate are chief drugs used as medicines in coryza, common cough, cold and flu, loss of smell, breathlessness, catarrhal pneumonia with influenza, rattling cough, chronic RTI, and sinusitis conditions.

18.10.4 HOMEOPATHY MEDICINES FROM SARCODES USED IN COVID-19 MANAGEMENT

The term *"sarcode"* means "fleshy" in Greek. Sarcodes include the protoplasm of fauna and are collected from the secretions of the vigorous ductless glands or endocrine system of existing human organs or as well as minor animals and could be used in COVID-19 management (Table 18.4). The secretions are mostly hormones. Hormones are proteins (e.g. insulin), "steroids" (e.g. cortisone), or comparatively basic organic entities (e.g. adrenaline) like cholesterinum, adrenalinum, insulinum, Fel Tauri, pancreatinum, pituitarinum, pepsinum, thyreoidinum, etc (Banerjee, 2006; Seca and Pinto, 2019). Adrenalium, insulinum, pepsinum, and Fel Tauri are some examples of sarcodes used in various health conditions in COVID-19-infected patients (Table 18.4).

18.10.5 HOMEOPATHY MEDICINES FROM NOSODES USED IN COVID-19 MANAGEMENT

The name "nosode" comes from two Greek words, *"noses"* meaning disease and *"cidos"* meaning appearance. Nosodes refer to the utilization of a disease's pathogen or a byproduct of a similar

---

TABLE 18.3

Different Sources and Types of Minerals Used as Homeopathic Medicines for COVID-19 Management

Medicine (Mineral) Name	Medicinal uses in Homeopathy for COVID-19 Management
	I. Metals
Iridium mettalicum	Prevents oxygen saturation, anemia and weakness in COVID-19 patients, cough
Ferum metallicum	Prevents oxygen saturation, anemia and weakness, hypertrophy of the heart muscle
	II. Non-metals
Phosphorus	Immune booter, prevents pneumonia, bronchitis, gastric problem, hemorrhage, chronic obstruction pulmonary disease, anxiety
	III. Metalloids
Arsenic	Immune booster in any form stage of COVID-19, cough and cold, anti-allergic
	IV. Salts, Compounds
Arsenicum iodatum	Coryza, fever, allergy, pneumonia, diarrhea and heart attack, anti-inflammatory, loss of smell, common cold, chronic respiratory tract infection
Antimonium arsenicosum	Breathlessness, cough, catarrhal pneumonia with influenza, cardiac weakness
Antimonium tartaricum	Rattling cough, vomiting, and bronchitis relief by left-side lying, reproductive health
Calcarea phosphorica	Immunity booster works similarly to vitamin D in COVID-19, sore throat
Potassium/Kalium bichromate	Sinusitis, loss of smell, respiratory tract infection, sore throat, common flu, cough and cold, nasal discharge, sinus congestion

TABLE 18.4

Different Sarcodes Used as Homeopathic Medicines for COVID-19 Management

Name of Medicine	Name of Sarcode	Source	Medicinal uses in Homeopathy for COVID-19 Management
		Sarcodes From Healthy Secretions, i.e. Hormones	
Adrenalium	Adrenalin/ Epinephrine	Hormones are produced by the suprarenal renal gland. Could be prepared synthetically from Epinephrine hydrochloride salt	Regulates pulse rate with chest pain, anxiety, contraction of heart muscle, acute congestion in lung, asthma, arteriosclerosis
Insulinum	Insulin	A pancreatic hormone produced in the ß-cells of the Islets of Langerhans. Controls sugar metabolism in the body	COVID-19 infected diabetic patients with skin symptoms, antihyperglycemic, ulceration
Pepsin	Pepsinum	A digestive enzyme produced in the stomach, which converts proteins into peptones. It is procured from the stomach of sheep or calves	Indigestion, loss of appetite, any type of stomach disorder, and stomach pain due to COVID-19 infection
Fel tauri	Fel tauri	Prepared from fresh ox gall	Inflamed and enlarged liver in COVID-19 patients, controls cholesterol, asthma, diarrhea

infection for healing. Nosodes are pharmaceutical substances obtained from infected tissues, organs, excretions of live beings, and microbes (such as bacteria and viruses). Nosodes are drugs derived from disease-causing agents, such as disease-producing entities, disease commodities, or diseased sections of plants, minor animals, or human beings (Table 18.5) (Allen, 2007, 2008). Nosodes, as defined by Dr. Dewey, are the pathological substances derived from diseases that are used as cures. Pneumococcinum, pyrogenium, bacillinum, and influenzinum are Homeopathic medicines with antiviral, antibacterial, analgesic, and antipyretic properties that are frequently prescribed for any type and stage of pneumonia, RTI, chronic catarrhal condition. They also prevent all types of cold and flu symptoms. Thus, these nosodes were suggested to COVID-19 patients throughout the pandemic.

TABLE 18.5

Different Sources and Types of Nosodes Used as Homeopathic Medicines for COVID-19 Management

Source	Disease Product	Medicine (Drug)	Medicinal uses in Homeopathy for COVID-19 Management
Bacteria	From a gram (+) bacteria, *Diplococcus pneumoniae*	Pneumococcinum	Controls any type and stage of pneumonia
Plant Kingdom	From a fungus growing on the stem or grains of Indian corn	*Ustilago maydis*	COVID-19-infected sexual organ disorders and infertility, uterus with dark hemorrhages, clotted oozing, ovaritis, menstrual problems
Animal Kingdom	From decomposed lean beef	Pyrogenium	Any form of infection with fever
Human	A maceration of a typical tuberculous lung	Bacillinum	Controls respiratory tract infection, chronic catarrhal condition, antibacterial, antiviral
	The nosode of influenza	Influenzinum	Prevents all types of cold and flu symptoms, analgesic, antipyretic

18.10.6 HOMEOPATHY MEDICINES FROM IMPONDERABILIA USED IN COVID-19 MANAGEMENT

Imponderabilia medicines are prepared from the immaterial power of energy available from natural and physical reactions (Wadhwani, 2011). Hahnemann demonstrated in the footnote of (aphorism) 280 in the *Organon* that, "even imponderable agencies, can produce most violent medicinal effects upon man". Dr. H. W. Allen depicts their method of preparation and symptoms in his *Materia Medica of Nosodes* (Allen, 2002, 2008). Elizabeth Wright demonstrates several sources of medicines in her book *A Brief Study Course in Homeopathy*, i.e. the *Magnetis poli ambo* (normal magnet), the *Magnetis polus australis* (south pole of magnet) and the *Magnetis polus arcticus* (north pole of magnet) as *Natural* sources whereas the sun's ray (sol), electricity (electricians), radium bromide (radium) X-rays (X-ray), and artificial magnets (*Magnetis artificialis*) as *Artificial* sources (Clarke, 2014).

18.10.7 HOMEOPATHY MEDICINES FROM SYNTHETIC SOURCES USED IN COVID-19 MANAGEMENT

Synthetic or tautopathic medicines (drugs) are medicines (drugs) in homeopathy sourced via a variety of synthetic procedures. The number and variety of those drugs are very low but the importance, application, impact as well and action of those drugs on patients suffering from various critical diseases including COVID-19 are immense during their therapeutic remedies. The examples of the various synthetic medicines (drugs) with their chemical formulas used for COVID-19 patients are illustrated in Table 18.6 (Banerjee, 2006). Penicillin (benzylpenicillin sodium), streptomycin sulphate, and streptomycin (dihydrosulphate) are some important examples of synthetic medicines (drugs) and are thought to cure various types of throat infections in COVID-19 patients.

18.11 CRITICISM AND LIMITATIONS OF HOMEOPATHIC MEDICINES IN COVID-19 MANAGEMENT

Homeopathy has been extensively criticized all over the world by the hardcore scientific community due to lack of scientific establishment, explanation, and exploration. This system of medicine mainly builds on the "Homeopathic Philosophy" (Organon of Medicine) with a fundamental law of homeopathy i.e. "Law of similar" (Subramanian, 2004). The application of Homeopathy (Materia Medica) lacks proper scientific experiments and explanations (Kent, 2005). Thus, critics describe the Homeopathic system of medicine as a "pseudoscience" having a "placebo effect". The method of preparing various drugs follows a process called "potentization" (99 parts of rectified spirit [90% ethyl alcohol] diluted with 1 part of the mother tincture and given 10 "succussion" or vigorous jerks to formulate IC potency) has received the greatest criticism. Thus, at high dilutions, few molecules of the mother tincture or stock solution could be found beyond 12 C potency (ahead of Avogadro's limit or 10^{-23}) (Anagnostatos, 1994; Barnard, 1965; Davenas et al., 1988; Smith and Boericke, 1968).

TABLE 18.6

Some Synthetic Drugs (Tautopathic) Used as Homeopathic Medicines for COVID-19 Management

Name of the Synthetic Medicine/Drugs (Tautopathic)	Medicinal uses in Homeopathy for COVID-19 Management
Penicillin (benzylpenicillin sodium) $C_{16}H_{17}N_2O_4SNa$ Streptomycin sulphate $(C_{16}H_{39}O_{12}N_7)_2, 3H_2SO_4$ Streptomycin (dihydrosulphate)	Throat infection

---

Critics point out the confirmed placebo effect of Homoeopathic drugs when they are used in a higher potency (>30 C) and homeopaths claim to heal the patients with their therapeutic remedies. The process of the purification of medicine (drug) has its limitations (WHO, 2009). Any mother tincture (drug) recovered from the environment in "crude form" having several other unknown materials could interrupt or lower the effectiveness of the "choice of drug". Unfortunately, the uniformity of medicine (drug) collected from different sources has been lost due to different levels of adulteration within it, making the medicine (drug) irregularly active against its use (Rehman and Ahmad, 2017). Finally, classical Homeopathy is thought to be considered a holistic approach to any symptomatic treatment. This system of medicine follows a "single drug" protocol, which demonstrates Homeopathic remedy could be achieved with the application of a single drug, i.e. "choice of drug" which could be enough to heal the patient for single or multiple symptoms (Boericke and Hahnemann, 2020; Hahnemann, 1991; Mondal, 2006; Subramanian, 2004). Every drug (medicine) has some specific characteristics as well as functions. Choice of drug could heal similar related symptoms but it is hardly believed that a single drug could cure all the unrelated symptoms within a human body. The "drug proving" system of Homeopathy is not an outcome of scientific and experimental procedure, but rather of philosophy. Proper scientific research is somewhat lacking in the homeopathy system of medicine and, thus, has to face major criticism.

18.12 CONCLUSION

Homeopathy is an important integrative system of medicine that may treat numerous crucial diseases with its philosophy. It is a comparatively safe, gentle, and natural system of therapy in the human healthcare system (Table 18.7). Homeopathic medicines (drugs) are extremely affordable as well as easily available in the market and they help in healing acute, chronic, as well as complicated non-surgical diseased conditions including COVID-19. Several Homeopathic medicines are effective against various symptoms of inflammatory manifestations, pulmonary disorders, cardiovascular injuries, reproductive health hazards and infertility, mental stress, etc., in COVID-19 patients. The appropriate drug administration endorses Homeopathy as one of the best choices of alternative therapy (Traditional medicine) due to its reasonable, cost-effective, and secure therapeutic remedies for better management of COVID-19. The Homeopathy system of medicine has many criticisms and limitations, but authentic scientific research and technological approaches in Traditional Homeopathic medicines might contribute a great curative and preventive aid for physiological disorders as well as rehabilitative aspects for better management of COVID-19.

TABLE 18.7

Key Homeopathic Medicines Used in COVID-19 Management

Symptoms/Diseases	Key Homeopathic Medicines Used for Symptomatic Treatment of COVID-19 Patients
Asthma	Hepar Sulphur, kali nitricum, *Lachesis mutus, Spongia tosta, Blata orientalis, Aspidosperma quebracho, Eucalyptus globulus, Arsenicum album*
Anemia	Iridium mettalicum, manganum aceticum, *Rubia tinctorum, Calcarea phosphorica,* ferrum metallicum, ferrum muriaticum, *China (Chinchona) officinalis*
Back-Ache	*Aesculus hippocastanum,* cobaltum mettalicum, gambogia, *Rhus toxicodendron*
Bronchitis	Tuberculinum aviaire, *Copaiva officinalis, Eucalyptus globulus, Narcissus pseudonarcissus, Bryonia alba,* Senega
Inflammation/Burning senses	Phosphorus, *Bryonia alba, Iris versicolor, Urtica urens,* Belladonna
Catarrhs	*Allium cepa, Euphrasia officinalis,* Sumbul, *Sabadilla officinalis, Ocimum sanctum*
Colic	*Bovista lycoperdon,* stannum mettalicum, nux moschata, colocynthis, magesium phosphoricum

(Continued)

TABLE 18.7 *(Continued)*
Key Homeopathic Medicines Used in COVID-19 Management

Symptoms/Diseases	Key Homeopathic Medicines Used for Symptomatic Treatment of COVID-19 Patients
Constipation	Ammonium muriaticum, *Bryonia alba*, mezerium, nux vomica, opium
Cough	*Rumex crispus, Capsicum annum, Justicia adhatoda* (Alba), *Spongia tosta, Drosera rotundifolia, Mentha piperita*
Debility	*Abies canadensis, Alstonia constricta, Galega officinalis, China (Chinchona) officinalis,* Alfalfa, *Avena sativa*
Diarrhea	Acid phos (acidum phosphoricum), *Elaterium officinarum, Podophyllum peltatum, Pulsatilla nigricans, Aloe socotrina, China (Chinchona) officinalis*
Dysentery	*Arnica montana, Chaparro amargoso, Colchicum autumnale,* mercurius solubilis, *Aloe socotrina,* mercurius corrosivus
Dyspepsia	Natrum carbonicum, abies nigra, *Lobelia inflata,* nux vomica, *Colchicum autumnale, China (Chinchona) officinalis, Pulsatilla nigricans,* carbo vegetabilis, *Lycopodium clavatum*
Dyspnea	Coca, natrum salicylicum, *Eucalyptus globulus*
Dropsy/Edema	Terebinthinae oleum, *Apocynum cannabium, Liatris spicata, Mercurius sulphuricus, Apis mellifica,* secale cornutum, *Kalmia latifolia,* baryta carbonica, aceticum acidum, *Apocyanum cannabinum, Digitalis purpurea*
Headache	*Iris versicolor,* niccolum mettalicum, *Valeriana officinalis,* Belladonna, *Gelsimium sempervirens,* glonoin (glycerinus nitrate)
Cardiovascular disorders	Phaseolus nanus, ammonium carbonicum, *Crataegus oxyacantha, Naja tripudians, Strophanthus hispidus, Digitalis purpurea, Cactus grandiflorus, Terminalia arjuna*
Reproductive problems/ Infertility	Magnesia phosphorica, chamomilla, *Sabina officinalis, Drymis winteri, Pulsatilla nigricans, Ignatia amara, Actea racemosa,* lachesis mutus, sulfur, phosphorus, sepia, sulphoricum acidum, nux vomica
Mental Health issues	Natrum muriaticum, *Ignatia amara,* kali phos (potassium/kalium phosphoricum), *Coffea,* natrum sulphuricum, aurum metallicum, *Sepia officinalis*
Anxiety/Stress management/Nerve relaxation	*Aconitum napellus,* argentum album, arsenicum album, calcarea carbonica, kali phos (potassium/kalium phosphoricum), *Lycopodium clavatum,* phosphorus, *Pulsatilla nigricans,* nux vomica, *Ignatia amara*

REFERENCES

Acharya, D., Liu, G., Gack, M.U., 2020. Dysregulation of type I interferon responses in COVID-19. Nat. Rev. Immunol. 20, 397–398.

Allen, H.C., 2002. Keynotes and Characteristics with Comparisons of Some of the Leading Remedies of the Materia Medica with Bowel Nosodes. B. Jain Publishers, India.

Allen, H.C., 2007. Materia Medica of Nosodes with Provings of the X-Ray. B. Jain Publishers, India.

Allen, H.C., 2008. Nosodes -The Materia Medica of Some More Important Remedies. B. Jain Publishers, India.

Anagnostatos, G., 1994. Small Water Clusters (Clathrates) in the Preparation Process of Homoeopathy, In: Endler, P.C., Schulte, J. (eds), Ultra High Dilution (pp. 121–128). Springer.

Banerjee, D., 2006. Augmented Textbook of Homoeopathic Pharmacy, 2nd edition. B. Jain Publishers, India.

Barnard, G., 1965. Microdose paradox-a new concept. J. Am. Inst. Homeopath. 58, 205–212.

Biswas, B., Jhansi, S., Potu, R., Patel, S., Nagaraju, M., Arya, R., et al. 2018. Physicochemical study of the homoeopathic drug, Blatta orientalis, Indian J. Res. Homoeopath. 12, 125–131.

Biswas, B., Sundaram, E.N., Jhansi, S., Patel, S., Khurana, A., Manchanda, R.K., 2019. A review on animal-based homoeopathic drugs and their applications in biomedicine. Indian J. Res. Homoeopathy 13, 159–176.

Black, D., 1994. Complementary Medicine. In: Walter, J., Walton, L., Jeremiah, A., Barondess, J.A., Lock, S. (eds), The Oxford Medical Companion. Oxford: Oxford University Press.

Blanco-Melo, D., Nilsson-Payant, B.E., Liu, W.C., Uhl, S., Hoagland, D., Møller, R., Jordan, T.X., Oishi, K., Panis, M., Sachs, D., Wang, T.T., Schwartz, R.E., Lim, L.M., Albrecht, R.A., tenOever, B.R., 2020. Imbalanced host response to SARS-CoV-2 drives development of COVID-19. Cell 181, 1036–1045.e9.

Boericke, W., 2004. Homeopathic Materia Medica (Vol-II). Kessinger Publishing.

Boericke, W., 2010. Homeopathic Materia Medica (Vol-I). Kessinger Publishing.

Boericke, W., Hahnemann, S., 2020. Organon of Medicine: The Cornerstone of Homeopathy. E-Artnow Publishing.

Bridwell, R.E., Merrill, D.R., Griffith, S.A., Wray, J., Oliver, J.J., 2021. A coronavirus disease 2019 (COVID-19) patient with bilateral orchitis. Am. J. Emerg. Med. 42 (260), 260.e3–260.e5.

Chan, J.F., Kok, K.H., Zhu, Z., Chu, H., To, K.K., Yuan, S., Yuen, K.Y., 2020. Genomic characterization of the 2019 novel human-pathogenic coronavirus isolated from a patient with atypical pneumonia after visiting Wuhan. Emerg. Microb. Infect. 9, 221–236.

Cheng, W., Chen, G., Jia, H., He, X., Jing, Z., 2018. DDX5 RNA helicases: emerging roles in viral infection. Int. J. Mol. Sci. 19, 1122.

Chu, H., Chan, J.F., Wang, Y., Yuen, T.T., Chai, Y., Hou, Y., Shuai, H., Yang, D., Hu, B., Huang, X., Zhang, X., Cai, J., Zhou, J., Yuan, S., Kok, K., To, K.K., Chan, I.H., Zhang, A.J., Sit, K., Au, W., Yuen, K.Y., 2020. Comparative replication and immune activation profiles of SARS-CoV-2 and SARS-CoV in human lungs: an ex vivo study with implications for the pathogenesis of COVID-19. Clin. Infect. Dis. 71, 1400–1409.

Clarke, J.H., 2014. Dictionary of Practical Materia Medica [Large Print]. B. Jain Publishers, India.

Clerkin, K.J., Fried, J.A., Raikhelkar, J., Sayer, G., Griffin, J.M., Masoumi, A., Jain, S.S., Burkhoff, D., Kumaraiah, D., Rabbani, L., Schwartz, A., Uriel, N., 2020. COVID-19 and cardiovascular disease. Circulation 141, 1648–1655.

Cui, L., Wang, H., Ji, Y., Yang, J., Xu, S., Huang, X., Wang, Z., Qin, L., Tien, P., Zhou, X., Guo, D., Chen, Y., 2015. The nucleocapsid protein of coronaviruses acts as a viral suppressor of RNA silencing in mammalian cells. J. Virol. 89, 9029–9043.

Davenas, E., Beauvais, F., Amara, J., Oberbaum, M., Robinzon, B., Miadonnai, A. et al., 1988. Human basophil degranulation triggered by very dilute antiserum against IgE, Nature 333, 30.

Deswal, A., Petersen, N.J., Feldman, A.M., Young, J.B., White, B.G., Mann, D.L., 2001. Cytokines and cytokine receptors in advanced heart failure: an analysis of the cytokine database from the Vesnarinone trial (VEST). Circulation 103, 2055–2059.

Dong, L., Tian, J., He, S., Zhu, C., Wang, J., Liu, C., et al. 2020. Possible vertical transmission of SARS-CoV-2 from an infected mother to her newborn. JAMA 323, 1846–1848.

Department of AYUSH (Drug Control Cell), 2013. Essential Drugs List (EDL) Homeopathy, Ministry of Family and Welfare, Government of India.

Gagliardi, L., Bertacca, C., Centenari, C., Merusi, I., Parolo, E., Ragazzo, V., et al. 2020. Orchiepididymitis in a boy with COVID-19. Pediatr. Infect. Dis. J. 39, e200–e202.

Gao, T., Hu, M., Zhang, X., Li, H., Zhu, L., Liu, H., Dong, Q., Zhang, Z., Wang, Z., Hu, Y., Fu, Y., Jin, Y., Li, K., Zhao, S., Xiao, Y., Luo, S., Li, L., Zhao, L., Liu, J., Zhao, H., Liu, Y., Yang, W., Peng, J., Chen, X., Li, P., Liu, Y., Xie, Y., Song, J., Zhang, L., Ma, Q., Bian, X., Chen, W., Liu, X., Mao, Q., Cao, C., 2020. Highly pathogenic coronavirus N protein aggravates lung injury by MASP-2-mediated complement over-activation. medRxiv. https://doi.org/10.1101/2020.03.29.20041962.

Gevitz, N., 1993. Unorthodox medical theories, In: Bynum W.F., Porter R (eds), Companion Encyclopaedia of the History of Medicine (pp. 603–633) London: Routledge.

Greenland, J.R., Michelow, M.D., Wang, L., London, M.J., 2020. COVID-19 infection: implications for perioperative and critical care physicians. Anesthesiology 132, 1346–1361.

Hahnemann, S., 1901. Organon of Medicine. Boericke & Tafel, America: Philadelphia.

Hahnemann, S., 1991. Organon of Medicine, 6th edition. (Translated by William Boericke), pp. 243–253, B Jain Publishers, India.

Harrison, E., 1806. The Ineffective State of the Practice of Physic. London, UK.

Henning, U., Pillat, M.M., 2020. CD147 as a target for COVID-19 treatment: suggested effects of azithromycin and stem cell engagement. Stem. Cell. Rev. Rep. 16, 434–440.

Hohensinner, P.J., Niessner, A., Huber, K., Weyand, C.M., Wojta, J., 2011. Inflammation and cardiac outcome. Curr. Opin. Infect. Dis. 24, 259–264.

Holtmann, N., Edimiris, P., Andree, M., Doehmen, C., Baston-Buest, D., Adams, O., et al. 2020. Assessment of SARS-CoV-2 in human semen-a cohort study. Fertil. Steril. 114, 233–238.

Huang, C., Wang, Y., Li, X., Ren, L., Zhao, J., Hu, Y., Zhang, L., Fan, G., Xu, J., Gu, X., Cheng, Z., Yu, T., Xia, J., Wei, Y., Wu, W., Xie, X., Yin, W., Li, H., Liu, M., Xiao, Y., Gao, H., Guo, L., Xie, J., Wang, G., Jiang, R., Gao, Z., Jin, Q., Wang, J., Cao, B., 2020. Clinical features of patients infected with 2019 novel coronavirus in Wuhan, China. Lancet 395, 497–506.

Kent, J.T., 1993. Lecture on Homeopathic Philosophy. North Atlantic Books, U.S.

Kent, J.T., 2005. Lectures on Homeopathic Materia Medica. B. Jain Publishers, India.

Lagrand, W.K., Visser, C.A., Hermens, W.T., Niessen, H.W., Verheugt, F.W., Wolbink, G.J., Hack, C.E., 1999. C-reactive protein as a cardiovascular risk factor: More than an epiphenomenon? Circulation 100, 96–102.

Larasati, T., Noda, T., Fujihara, Y., Shimada, K., Tobita, T., Yu, Z., et al. 2020. Tmprss12 is required for sperm motility and uterotubal junction migration in mice†. Biol. Reprod. 103, 254–263.

Li, K., Chen, G., Hou, H., Liao, Q., Chen, J., Bai, H., et al. 2021. Analysis of sex hormones and menstruation in COVID-19 women of child-bearing age. Reprod. Biomed. Online 42, 260–267.

Li, G., Fan, Y., Lai, Y., Han, T., Li, Z., Zhou, P., Pan, P., Wang, W., Hu, X., Zhang, O., Wu, J., 2020. Coronavirus infections and immune responses. J. Med. Virol 92 (4), 424–432.

Liu, X., Chen, Y., Tang, W., Zhang, L., Chen, W., Yan, Z., et al. 2020b. Single-cell transcriptome analysis of the novel coronavirus (SARS-CoV-2) associated gene ACE2 expression in normal and non-obstructive azoospermia (NOA) human male testes. Sci. China Life. Sci. 63, 1006–1015.

Liu, X., Zhang, R., He, G., 2020a. Hematological findings in coronavirus disease 2019: indications of progression of disease. Ann. Hematol 99, 1421–1428.

Li, H., Xiao, X., Zhang, J., Zafar, M.I., Wu, C., Long, Y., et al. 2020. Impaired spermatogenesis in COVID-19 patients. EClinicalMedicine 28, 100604.

Long, Q.X., Tang, X.J., Shi, Q.L., Li, Q., Deng, H.J., Yuan, J., Hu, J.L., Xu, W., Zhang, Y., Lv, F.J., Su, K., Zhang, F., Gong, J., Wu, B., Liu, X.M., Li, J.J., Qiu, J.F., Chen, J., Huang, A.L., 2020. Clinical and immunological assessment of asymptomatic SARS-CoV-2 infections. Nat. Med. 26, 1200–1204.

Loudon, I., 1987. The Vile Race of Quacks with which this Country is Infested. In: Bynum WF, Porter R (eds), Medical Fringe and Medical Orthodoxy 1750–1850. London: Croom Helm, 42.

Loudon, I., 2006, A brief history of homeopathy, J. R. Soc. Med., 99(12), 607–610. doi: 10.1258/jrsm.99.12.607

Lu, B., Huang, Y., Huang, L., Li, B., Zheng, Z., Chen, Z., Chen, J., Hu, Q., Wang, H., 2010. Effect of mucosal and systemic immunization with virus-like particles of severe acute respiratory syndrome coronavirus in mice. Immunology 130, 254–261.

Matar, R., Alrahmani, L., Monzer, N., Debiane, L.G., Berbari, E., Fares, J., et al. 2021. Clinical presentation and outcomes of pregnant women with Coronavirus Disease 2019: a systematic review and meta-analysis. Clin. Infect. Dis. 72, 521–533.

Mokhtari, T., Hassani, F., Ghaffari, N., Ebrahimi, B., Yarahmadi, A., Hassanzadeh, G., 2020. COVID-19 and multiorgan failure: a narrative review on potential mechanisms. J. Mol. Histol. 51, 613–28.

Mondal, D.T.C., 2006. Potency, Spirit of the Organon, 2nd edition, pp, 224–233, B. Jain Publishers, India.

Mukherjee, P.K., Wahile, A., 2006. Integrated approaches towards drug development from Ayurveda and other Indian system of medicines. J. Ethnopharmacol. 103, 25–35.

Neuman, B.W., Kiss, G., Kunding, A.H., Bhella, D., Baksh, M.F., Connelly, S., Droese, B., Klaus, J.P., Makino, S., Sawicki, S.G., Siddell, S.G., Stamou, D.G., Wilson, I.A., Kuhn, P., Buchmeier, M.J., 2011. A structural analysis of M protein in coronavirus assembly and morphology. J. Struct. Biol. 174, 11–22.

Nieto-Torres, J.L., DeDiego, M.L., Verdiá-Báguena, C., Jimenez-Guardeño, J.M., Regla- Nava, J.A., Fernandez-Delgado, R., Castaño-Rodriguez, C., Alcaraz, A., Torres, J., Aguilella, V.M., Enjuanes, L., 2014. Severe acute respiratory syndrome coronavirus envelope protein ion channel activity promotes virus fitness and pathogenesis. PLoS Pathog. 10, e1004077 https://doi.org/10.1371/journal.ppat.1004077.

Ni, L., Ye, F., Cheng, M.L., Feng, Y., Deng, Y.Q., Zhao, H., Wei, P., Ge, J., Gou, M., Li, X., Sun, L., Cao, T., Wang, P., Zhou, C., Zhang, R., Liang, P., Guo, H., Wang, X., Qin, C.F., Chen, F., Dong, C., 2020. Detection of SARS-CoV-2-specific humoral and cellular immunity in COVID-19 convalescent individuals. Immunity 52, 971–977.e3.

Qiu, X., Zhang, L., Tong, Y., Qu, Y., Wang, H., Mu, D., 2018. Interleukin-6 for early diagnosis of neonatal sepsis with premature rupture of the membranes: a meta-analysis. Medicine (Baltimore) 97, e13146.

Rajak, P., Ganguly, A., Sarkar, S., Mandi, M., Dutta, M., Podder, S., Khatun, S., Roy, S., 2021a. Immunotoxic role of organophosphates: an unseen risk escalating SARS-CoV- 2 pathogenicity. Food Chem. Toxicol. 149, 112007. https://doi.org/10.1016/j.fct.2021.112007.

Rajak, P., Roy, S., Dutta, M., Podder, S., Sarkar, S., Ganguly, A., Mandi, M., Khatun, S., 2021b. Understanding the cross-talk between mediators of infertility and COVID-19. Reprod. Biol. 21, 100559. https://doi.org/10.1016/j.repbio.2021.100559.

Rajak, P., Roy, S., Pal, A.K., Paramanik, M., Dutta, M., Podder, S., Sarkar, S., Ganguly, A., Mandi, M., Dutta, A., Das, K., Ghanty, S., Khatun, S., 2021c. In silico study reveals binding potential of rotenone at multiple sites of pulmonary surfactant proteins: a matter of concern. Curr Res Toxicol. 4, 411–423. https://doi.org/10.1016/j.crtox.2021.11.003.

Rajak, P., Roy, S., Podder, S., Dutta, M., Sarkar, S., Ganguly, A., Mandi, M., Dutta, A., Nanda, S., Khatun, S., 2022. Synergistic action of organophosphates and COVID-19 on inflammation, oxidative stress, and renin-angiotensin system can amplify the risk of cardiovascular maladies. Toxicol. Appl. Pharmacol. 456, 116267. https://doi.org/10.1016/j.taap.2022.116267

Rehman, T., Ahmad, S., 2017. Introduction of homeopathy and homeopathic medicines: A review. Int. J. Hom. Sci. 1(1), 21–25.

Romano, M., Ruggiero, A., Squeglia, F., Maga, G., Berisio, R., 2020. A structural view of SARS-CoV-2 RNA replication machinery: RNA synthesis, proofreading and final capping. Cells 9, 1267.

Seca, A.M.L., Pinto, D.C.G.A., 2019. Biological potential and medical use of secondary metabolites. Medicines (Basel) 6(2), 66. doi: 10.3390/medicines6020066

Shang, J., Wan, Y., Luo, C., Ye, G., Geng, Q., Auerbach, A., Li, F., 2020. Cell entry mechanisms of SARS-CoV-2. Proc. Natl. Acad. Sci. U.S.A. 117, 11727–11734.

Shi, S., Qin, M., Shen, B., Cai, Y., Liu, T., Yang, F., Gong, W., Liu, X., Liang, J., Zhao, Q., Huang, H., Yang, B., Huang, C., 2020. Association of cardiac injury with mortality in hospitalized patients with COVID-19 in Wuhan, China. JAMA Cardiol. 5, 802–810.

Smith, R., Boericke, G., 1968. Changes caused by succussion on NMR patterns and bioassay of bradykinin triacetate succussions and dilutions. Journal of the American Institute of Homoeopathy 61, 197–212.

Subramanian, R., 2004. Textbook of Homeopathic Philosophy. B. Jain Publishers, India.

Tanacan, A., Yazihan, N., Erol, S.A., Anuk, A.T., Yucel Yetiskin, F.D., Biriken, D., et al. 2021. The impact of COVID-19 infection on the cytokine profile of pregnant women: a prospective case-control study. Cytokine 140, 155431.

Tokgoz, V.Y., Kaya, Y., Tekin, A.B., 2020. The level of anxiety in infertile women whose ART cycles are postponed due to the COVID-19 outbreak. J. Psychosom. Obstet. Gynaecol. 43(2), 1–8.

Tournier, A., Klein, S.D., Wu"rtenberger, S., Wolf, U., Baumgartner, S., 2019. Physicochemical investigations of homeopathic preparations: A systematic review and bibliometric analysis—part 2. J. Altern. Complement. Med. 25(9), 890–901. doi: 10.1089/acm.2019.0064

Varga, Z., Flammer, A.J., Steiger, P., Haberecker, M., Andermatt, R., Zinkernagel, A.S., Mehra, M.R., Schuepbach, R.A., Ruschitzka, F., Moch, H., 2020. Endothelial cell infection and endotheliitis in COVID-19. Lancet 395(10234), 1417–1418.

Wadhwani, G.G., 2011. Imponderabilia in homeopathic practice. AJHM. 104, 131.

Wang, F., Hou, H., Luo, Y., Tang, G., Wu, S., Huang, M., Liu, W., Zhu, Y., Lin, Q., Mao, L., Fang, M., Zhang, H., Sun, Z., 2020a. The laboratory tests and host immunity of COVID-19 patients with different severity of illness. JCI Insight 5, e137799. https://doi.org/10.1172/jci.insight.137799.

Wang, D., Hu, B., Hu, C., Zhu, F., Liu, X., Zhang, J., Wang, B., Xiang, H., Cheng, Z., Xiong, Y., Zhao, Y., Li, Y., Wang, X., Peng, Z., 2020b. Clinical characteristics of 138 hospitalized patients with 2019 novel coronavirus-infected pneumonia in Wuhan, China. JAMA. 323, 1061–1069. https://doi.org/10.1001/jama.2020.1585.

WHO, 2001. Programme on Traditional Medicine; Legal status of traditional medicine and complementary/alternative medicine: a worldwide review. World Health Organization. https://apps.who.int/iris/handle/10665/42452

WHO, 2009. Safety issues in the preparation of homeopathic medicines. https://iris.who.int/bitstream/handle/10665/44238/9789241598842_eng.pdf?sequence=1

Wolfler, A., Mannarino, S., Giacomet, V., Camporesi, A., Zuccotti, G., 2020. Acute myocardial injury: a novel clinical pattern in children with COVID-19. Lancet Child Adolesc. Health 4, e26–e27.

Wong, S.F., Chow, K.M., Leung, T.N., Ng, W.F., Ng, T.K., Shek, C.C., et al. 2004. Pregnancy and perinatal outcomes of women with severe acute respiratory syndrome. Am. J. Obstet. Gynecol. 191, 292–297.

Wrapp, D., Wang, N., Corbett, K.S., Goldsmith, J.A., Hsieh, C.L., Abiona, O., Graham, B.S., McLellan, J.S., 2020. Cryo-EM structure of the 2019-nCoV spike in the prefusion conformation. Science 367, 1260–1263. https://doi.org/10.1126/science.abb2507.

Xia, S., Zhu, Y., Liu, M., Lan, Q., Xu, W., Wu, Y., Ying, T., Liu, S., Shi, Z., Jiang, S., Lu, L., 2020. Fusion mechanism of 2019-nCoV and fusion inhibitors targeting HR1 domain in spike protein. Cell. Mol. Immunol. 17, 765–767. https://doi.org/10.1038/s41423-020-0374-2.

Xiong, Y., Liu, Y., Cao, L., Wang, D., Guo, M., Jiang, A., Guo, D., Hu, W., Yang, J., Tang, Z., Wu, H., Lin, Y., Zhang, M., Zhang, Q., Shi, M., Liu, Y., Zhou, Y., Lan, K., Chen, Y., 2020. Transcriptomic characteristics of bronchoalveolar lavage fluid and peripheral blood mononuclear cells in COVID-19 patients. Emerg. Microbes Infect. 9, 761–770. https://doi.org/10.1080/22221751.2020.1747363.

Zheng, Y.Y., Ma, Y.T., Zhang, J.Y., Xie, X., 2020. COVID-19 and the cardiovascular system. Nat. Rev. Cardiol. 17, 259–260.

Zhou, Y., Fu, B., Zheng, X., Wang, D., Zhao, C., qi, Y., Sun, R., Tian, Z., Xu, X., Wei, H., 2020c. Pathogenic T cells and inflammatory monocytes incite inflammatory storm in severe COVID-19 patients. Natl. Sci. Rev. 7, 998–1002. https://doi.org/10.1093/nsr/nwaa041.

Zhou, P., Yang, X.L., Wang, X.G., Hu, B., Zhang, L., Zhang, W., Si, H.R., Zhu, Y., Li, B., Huang, C.L., Chen, H.D., Chen, J., Luo, Y., Guo, H., Jiang, R.D., Liu, M.Q., Chen, Y., Shen, X.R., Wang, X., Zheng, X.S., Zhao, K., Chen, Q.J., Deng, F., Liu, L.L., Yan, B., Zhan, F.X., Wang, Y.Y., Xiao, G.F., Shi, Z.L., 2020a. A pneumonia outbreak associated with a new coronavirus of probable bat origin. Nature 579, 270–273. https://doi.org/10.1038/s41586-020-2012-7.

19 The Impact of Traditional Herbal Medicine on COVID-19 Management in the Post-Pandemic Era

Rajesh Kumari

19.1 INTRODUCTION

Through the ages, herbal and traditional medicines have been recommended to cure different human and animal diseases around the world. Most of the herbal and traditional medicines are prepared either by plant extracts or plant ashes and have phytochemicals as effective ingredients to treat an ailment [1, 2]. The COVID-19 pandemic was a global emergency and the severe acute respiratory syndrome coronavirus 2 (SARS-CoV-2) pathogen (RNA virus) was the infectious agent. The term "corona" for coronaviruses indicates a crown-like structure that is seen under electron microscopy. SARS-CoV-2 has multiple variants with different severity of infections, which pose a tremendous challenge for virus diagnosis and utilizing effective treatment strategies against this deadly virus [3, 4]. The α, β, γ, δ, and Omicron variants were identified with different subvariants. The Omicron variant was reported to have more than nine subvariants including BA.2. BA.4, BA.5, BQ.1, XBB.1.16, and FE.1, etc [3–5]. Developing mRNA-based vaccines was of the utmost importance to control the spread and reduce the loss of human lives during the COVID-19 pandemic, and vaccine hesitancy was predicted to pose a bigger risk for new variants [6].

19.2 SARS-CoV-2 EPIDEMIOLOGY

In December 2019, the local hospitals reported a different type of pneumonia with high fever, chest tightness, and respiratory distress the Wuhan, China. The patients were either sellers or frequent visitors of the seafood wholesale market in Wuhan. For reasons unknown, bronchoalveolar-lavage samples were collected and the genome was isolated and sequenced. The genome sequence showed 85% identity with bat SARS-like CoV. Further, the genome sequence from three patients was submitted and the phylogenetic analysis established that this new virus belonged to the genus *Betacoronavirus*, which mainly infected humans and bats. Later, based on taxonomy nomenclature, the *Coronaviridae* study group named it "SARS-CoV-2" [7, 8]. The World Health Organization (WHO) announced the SARS-CoV-2 infection as a public health emergency worldwide on 30 January 2020, and later in the same year, the WHO announced COVID-19 as a pandemic disease [9].

By late February 2020, the rapid rise in SARS-CoV-2 cases by human-to-human transmission was evident as the virus was already transmitted from China to every continent on the Earth. Italy, London, and New York were the earliest affected cities in the world [10].

The risk, spread and severity of COVID-19 was highly heterogeneous. The first responders in the health care systems were at highest risk of infection followed by people in public-facing jobs as well as people living in poorly ventilated and crowded spaces. Social inequities, socioeconomic background, ethnicity, and cultural traditions were some of the major determinations of SARS-CoV-2 spread and related outcomes as the death toll and spread of infection were accelerated in poor sanitation and improper healthcare situations [11]. As it is known that viruses can live on any surface and also in the air, studies on the transmission of SARS-CoV-2 also reported that the virus can remain alive in aerosols and surfaces after many hours of inoculation, however, the research found that the spread of SARS-CoV-2 was highly heterogeneous and infected patients were the major source of virus transmission [12]. The rate of spread was variable around the globe, but the convergence of the data suggested that during the first wave of infection, 10% of population infected 80% as a result of secondary infection [13]. Soon after the emergence, the adaptive evolution was reported in the viral genome and the subsequential emergence, spread, and reduction in SARS-CoV-2 from 614D, 614G, α, β, γ, δ, Omicron, Lambda, and Mu was seen from March 2020 to November 2021. This evolution and emergence of new strains and substrains were greatly correlated with better transmissibility and higher pathogenesis [10, 14]. The variants were divided into four classes, namely variants of concern, variants being monitored, variants of interest, and variants of high consequence [15]. The classification of a particular variant changes as time passes and the variant becomes less harmful [15]. As of 18 December 2023, XBB.1.5 Omicron subvariant is the variants of interest class. According to WHO statistics, COVID-19 spread into 220 countries and territories globally. As of 17 December 2023, 772,838,745 COVID-19 cases were reported with 6,988,679 deaths worldwide with the highest casualties in the United States, Brazil, and India, respectively (https://data.who.int/dashboards/covid19/deaths?n=c).

19.3   VIRAL ASSEMBLY OF SARS-CoV-2

SARS-CoV-2 has a varied length of the genome from 29.8 to 29.9 kilobase (kb), having the largest two-thirds portion encoding for open reading frame 1ab (ORF1ab) polyproteins and one-third encoding for structural proteins. The remaining encodes for six accessory proteins [16]. Overall, the coronavirus genome encodes a total of 29 proteins, involved in various functions such as surface attachment, invasion, and unloading along with viral genome replication, packaging, and release from the host cells [17].

As the name of coronaviruses suggests, they have a crown-like structure protruding from the capsule membrane. These structures are made up of spike (S) proteins, which interact with the angiotensin-converting enzyme 2 (ACE2) receptor and start virus-host communication for viral entry [18].

The human proteases cleave the S protein into S1 and S2 fragments and this cleavage occurs before viral and host membrane fusion. The S1 fragment contains N-terminal binding, receptor binding, and C-terminal domains, which further help in attachment and conformational changes to facilitate viral invasion [19]. A range of antiviral targets against S proteins was designed [20]. Alpha, Beta, Gamma, and Delta viruses have mutations in the N-terminal domain of the S1 fragment of the S protein [17].

The nucleocapsid (N) protein is a multifunctional protein that shares 90% similarity with the N protein of SARS-CoV. This protein helps in viral assembly and replication as well as participates in modulating the host immune response [21]. N protein interacts with the viral genome and forms RNA ribonucleoprotein complex and the N terminal domain interacts with nCoV396 monoclonal antibody in COVID-19 infected patients [22, 23].

The envelope (E) protein is 75 amino acids long and helps in viral assembly and release by forming the selective cation channels in the membrane. The conductivity of these channels is regulated by membrane lipid composition [24].

The membrane glycoprotein (M) of SARS-CoV-2 is 222 amino acids long and shares 90.5% sequence similarity with the M protein of SARS-CoV. M protein is the most abundant protein, forms the virus, and has transmembrane and cytoplasmic domains. This protein also helps in trafficking and packaging of other proteins thus it is required for SARS-CoV-2 assembly and release [25, 26].

Other than the previously mentioned structural proteins, nonstructural proteins (Nsps) are synthesized by the SARS-CoV-2 genome. Most of the antiviral drugs target the spike protein, which is highly prone to mutations, so researchers are considering Nsps as new targets for antiviral therapies.

The integral membrane proteins Nsp3, Nsp4, and Nsp6 participate in rearranging the membrane of the endoplasmic reticulum to exploit it for double-membrane vesicle formation for assisting in viral replication and host immune evasion [27–30].

Nsp1 is a 180-amino-acid-long virulent protein having N and C-terminal domains. The C-terminal domain interacts with the 40s ribosome subunits and hijacks the host translation machinery for viral mRNA cleavage and translation. Some inhibitors that break this interaction between the 40s ribosome and C-terminal domain of Nsp1 are under consideration as new antiviral therapies [31–33].

Nsp2 is a long protein with 638 amino acids with unknown functions, but the structural analysis suggests that this protein might bind to nucleic acids. Later studies found that Nsp2 regulates mitochondrial biogenesis and intracellular signaling pathways by binding with actin-nucleation-promoting WASH protein, prohibitin 1 (PHB1), and PHB2 of the host [34, 35].

Nsps are produced after the splicing of polyproteins pp1a and pp1ab by the two viral proteases, namely, papain-like protease and main-protease. The overall function of Nsps is to support viral genome replication and translation of viral constituent proteins [36].

RNA-dependent RNA polymerase (RdRp) is required for viral replication and is highly essential for SARS-CoV-2 pathogenesis. Remdesivir, an antiviral drug, targets RdRp and thus inhibits virus proliferation [37]. The ORF3a, ORF3b, ORF6, ORF7a, ORF7b, ORF8, ORF9b, etc. other nine accessory proteins are also present in this virus [38].

19.4 MECHANISMS OF SARS-CoV-2 PATHOPHYSIOLOGY

The SARS-CoV-2 enters the human body by coughing, sneezing, or aerosol or by touching a contaminated surface [39]. This infection causes acute respiratory distress syndrome (ARDS). Viruses first attack the lung's alveolar epithelial cells by binding with the ACE2 receptor; this pathological condition is also called as diffuse alveolar damage and is characterized by excess surfactant, dying cells, and inflammation [40, 41]. In further damage, a hyaline membrane with fibrin-rich spots is formed, which results in fluid accumulation and diminished oxygen supply in the alveoli [42]. Fibrin thrombi formation followed by fibrin degradation and fibrinolysis is one more condition reported from COVID-19 patients. The patients also have very low platelets, as the platelets were used in coagulation and clotting in severe infection [43]. The anticoagulation therapy was effective in hospitalized patients in preventing deaths, thus coagulation is an essential mechanism in SARS-CoV-2 pathophysiology [44]. The pyroptotic macrophages secrete an excess of cytokines and chemokines, which attract other immune cells,–a condition called the cytokine storm [45]. As kidney and heart cells also express the ACE2 receptor, SARS-CoV-2 attacks these vital organs, which leads to multiorgan infection and complete systemic failure [46, 47].

19.5 LONG COVID OR POST-COVID SYNDROME

Defining the post-COVID syndrome (long COVID) or post-COVID conditions is a challenge due to the heterogeneity present in symptoms and health complaints after the treatment of initial infection of SARS-CoV-2. The WHO and the government of Canada announced that the symptoms occurring after 3 months of COVID-19 treatment should be classified as "post-COVID conditions" whereas the National Institute for Clinical Excellence (NICE) defined a 12-week or longer illness after initial recovery as "long COVID" and a condition that develops during COVID-19 and persists more than 12 weeks as a post-COVID syndrome. The US Centers for Disease Control and Prevention on the other hand took the term "post-COVID conditions" for all-inclusive conditions faced by COVID-19 patients after 4 weeks of infection. The Robert Koch Institute (RKI) differentiates the term long COVID from post-COVID conditions or post-COVID syndrome by the duration. The symptoms that were seen after 4 weeks of the acute phase of sickness were considered "long COVID" but if the condition exceeded 12 weeks or longer, it was defined as a post-COVID syndrome or post-COVID conditions [48].

In this chapter, we will be using long COVID for all the above-mentioned conditions. The global epidemiology of long COVID was around 43% with around 51% in Asia, 44% in Europe, and 31% in the United States. The statistics of long COVID was higher in hospitalized patients and was estimated at around 54% as compared to 34% in nonhospitalized patients [48, 49].

The heterogeneity in the clinical symptoms of COVID-19 patients was the greatest challenge to understand and address for the whole scientific community worldwide. More than 100 symptoms were reported and COVID-19 was accepted as a multiorgan contagious disease [45, 48, 49]. The major reported clinical symptoms in the respiratory system were hypoxia, dry cough, and decreased lung filling capacity. In the cardiovascular system, myocardial fibrosis, cardiomyopathy, tachycardia and chest pain were frequently reported. Similarly, in the blood circulation, they included venous thromboembolism and persistent hypercoagulability. In the gastrointestinal and liver system, gut microbiome alteration, diarrhea, vomiting, and abdominal pain were present. Kidney dysfunction and altered eGFR were present in the renal systems. In the nervous system, sleep disturbance, anxiety, depression, fatigue, and myalgia symptoms were reported. Type 2 diabetes, hyperthyroidism and anorexia along with muscle weakness were reported in the endocrine dysfunction [48, 49].

The long COVID symptoms were also heterogeneous among patients who recovered from COVID-19 after different durations. The most common symptoms observed in the US population and other parts of the world were fatigue, muscle pain, respiratory issues, cognitive impairment, cardiovascular and other metabolic diseases, such as digestive issues and diabetes [50, 51].

During the pandemic, when SARS-CoV-2 infection was at its peak with very limited treatment options, the WHO initiated efforts to repurpose the existing drugs and use herbal and traditional medicines. Among these, the collaboration with African, Indian and Chinese health systems had a great impact on controlling COVID-19. The subsequent sections describe the role of traditional and herbal medicines in detail (https://www.who.int/news-room/feature-stories/detail/who-africa-cdc-push-for-covid-19-traditional-medicine-research-in-africa).

19.6 APPLICATIONS OF THE TRADITIONAL AND HERBAL MEDICINE IN CHINA

Chinese herbal and traditional medicine (CHM) has a history of 2000 years. CHM played a crucial role in fighting the SARS-CoV-2 pandemic. Most of the molecular mechanisms underlying these therapies are unknown, though significant efforts are made to reveal the name and mechanism of drug compounds [52–55]. A few detailed studies from CHM are mentioned below:

Xuanfei Baidu Decoction (XFBD) is a "three medicine three prescription" formula with 13 Chinese medicinal herbs. Originally, this decoction mixes four traditional therapies, namely, Maxing Shigan, Qianjin Weijing, Tingli Dazao Xiefei, and Maxing Yigan decoctions. The Chinese herbs *Ephedra sinica* Stapf (Ma Huang; stem), *Prunus armeniaca* L (Xing Ren; seed) and *Gypsum fibrosum* (Sheng Shi Gao), etc. were used in these preparations. A treatment with XFBD in different clinical trials with COVID-19 patients having mild, moderate, or severe infection showed a significant improvement in hematological markers, reduced fever, cough, fatigue, and inflammation with shorter hospital stays due to the antiviral and immunomodulatory properties of XFBD decoction [56, 57].

The China Food and Drug Administration has permitted the use of Lianhua Qingwen capsules/granules having rhein, forsythoside A, etc., for mild COVID-19 patients. The chemical compounds of these granules reduced SARS-CoV-2 infection by hindering the spike protein interaction with the ACE2 receptor [58, 59].

Cardiac dysfunction was one of the common co-morbidities in the COVID-19 pandemic. The only approved Chinese medicine to treat cardiac abnormalities is Xuebijing injection, which has antistressor, anti-inflammation, and antiplaque formation properties. The molecular docking studies showed that the ACE2, ACE7, CCL2, CXCL8, FOS, IFNB1, etc., are molecular targets of the Xuebijing injection. The salvianolic acid B, rutin, and anhydrosafflor yellow B are three compounds in the Xuebijing formulation that bind the ACE2 receptor on the human lung epithelium and the 3C-like proteinase and S protein of SARS-CoV-2 [60, 61].

The National Health Commission of China recommended Qingfei Paidu decoction, made from *Cinnamomum cassia*, *Pogostemon cablin*, *Ephedra sinica* and *Bupleurum chinense*, etc., as major herbs for mild, moderate, and severe infection. The three active ingredients scutellarin, hesperetin, and oroxylin A of Qingfei Paidu decoction inhibit ACE2 receptor binding and further stop the cytokine storm by reducing the secretion of CXCL-10, IL-1β, and IL-6, thus, possessing significant antiviral, immunomodulatory, and anti-inflammatory properties [62, 63].

Yiqi Yangyin granules were reported to alleviate SARS-CoV-2-mediated fatigue, fever, and dry mouth symptoms. This mix contains Baizhu (*Atractylodis macrocephalae*), Dangshen (*Codonopsis radix*), Baihe (*Lilii bulbus*), and Maidong (*Ophiopogonis radix*) as herbal components [64].

Huashi Baidu formula, a part of three Chinese patent medicines and three TCM prescriptions, inhibits SARS-CoV-2 binding at the ACE2 receptor, resulting in reduced infection. Pharmacology network analysis and molecular docking showed that the chemical compounds quercetin, isotrifoliol, ellagic acid, and baicalein have an excellent binding affinity with ACE2, ADAM17, and the 3C-protease of SARS-CoV-2, which directly results in reduced binding with the receptor to inhibit viral entry [57, 65, 66].

Another similar herbal medicine, including kaempferol, wogonin, baicalein, quercetin, and stigmasterol chemical compounds, known as the Shufeng Jiedu capsule, is considered as a Chinese patented treatment for COVID-19 infection. The active ingredients of the Shufeng Jiedu capsule interact with different molecular pathways involved in PI3K-Akt, toll-receptor signaling and in regulating MAPK and cAMP signaling pathways, and show high binding with 3C-protease and ACE2 receptor. The simulation studies indicated that ACE2, 2OFZ, and 1SSK are the potential targets of the above-mentioned ingredients to combat SARS-CoV-2 infection [67, 68].

Maxing Shigan decoction which is also a part of Chinese patented medicine has different active chemicals including Gancao, Kuxingren, Amygdalin, euchrenone, glycyrrhizin, and glycyrol. These chemicals synergistically target JAK-STAT and PI3K/AKT signaling pathways and give antiviral and anti-inflammatory properties to Maxing Shigan decoction via probable molecular binding with ACE2, main protease, and RdRp proteins of SARS-CoV-2 [69].

YuPing Feng San decoction containing Guanghuoxiang (*Pogostemonis herba*), Banxia (*Pinelliae rhizoma*), Fangfeng (*Saposhnikoviae radix*), Fuling (Poria), Shengjiang (*Zingiberis rhizoma* Recens), Cangzhu (*Atractylodis rhizoma*), Zisugeng (*Perillae caulis*), Huangqi (*Astragali radix*), Sharen (*Amomi fructus*) and Baizhu (*Atractylodis macrocephalae* Rhizoma) components were tested in mild and moderate COVID-19 patients for 14 days and results showed that YuPing Feng San decoction reduced the severity of disease by increasing lymphocytes and shortening hospitalization [70].

Danggui Shaoyao powder was used to treat severe patients. This has Zexie (*Alismatis rhizoma*), Danggui (*Angelicae sinensis* Radix), Baizhu (*Atractylodis macrocephalae* Rhizoma), Chuanxiong (*Chuanxiong rhizoma*), and Baishao (*Paeoniae radix* Alba) as herbal ingredients and helped in reducing fever, cough, and fatigue and also increase lymphocyte counts in COVID-19 patients [64].

Ganlu Xiaodu decoction, Gegen Qinlian pill, and HuoXiang Zhengqi powder were used to reduce moderate and severe problems of COVID-19 and these traditional medicines helped in improving SARS-CoV-2 clinical symptoms [71–73].

Yidu-toxicity blocking lung decoction reduced severe pulmonary inflammation by decreasing IL-6 and tumor necrosis factor (TNF)-α helped in curing severe SARS-CoV-2 infection in pneumonia patients. The formula has 11 chemical and herbal ingredients including Kuxungren, Shenghigao, Gualou, etc [74].

Shuanghuanglian oral liquid is a mixture of three plants including *Lonicera japonica*, *Forsythia suspense,* and *Scutellaria baicalensis*. In a randomized open-label trial along with standard care in SARS-CoV-2 infected patients, the treatment reduced inflammation and didn't have any adverse effects [75].

Another CHM, Dang Gui Long Hui Pill, contains indirubin, which is effective in boosting immunity and reducing COVID-19 infection. *In vitro* studies with human macrophages showed that indirubin halts viral replication when macrophages are infected with influenza H5N1 virus [76, 77].

Shengmai injections are made from Wuweizi (*Schisandrea chinensis* Fructus), Hongshen, Maidong (*Ophiopogonis radix*), and (*Ginseng radix* Et Rhizoma Rubra) and tested for their efficacy in reducing SARS-CoV-2 infection. The clinical data suggested that the Shengmai injection in severe and moderate infections of COVID-19 cured dry throat and fatigue, reduced inflammation, and increased oxygen saturation [64, 78].

Reduning injection was tested in mild SARS-CoV-2 infection for 14 days and it was shown that the injection relieved COVID-19 patients from cold, fever, and cough along with reducing the hospitalization duration. This injection was a combination of three herbs namely Zhizi (*Gardeniae fructus*), Jinyinhua (*Lonicerae japonicae* Flos) and Qinghao (*Artemisiae annuae* Herba) [79].

Thus far, we have discussed studies on herbal connections or multiple mixtures. Next, we will describe a few single-herbal treatments reported to be effective in treating SARS-CoV-2 infection.

Ephedra sinica Stapf also known as Mahuang has ephedrine, pseudoephedrine, and methylephedrine as active ingredients. These compounds synergistically inhibit ACE2 and spike protein interaction, which leads to the restricted entry of SARS-CoV-2. Mahuang-Kuxingren mixture was also studied to treat SARS-CoV-2 infection by targeting 3C-protease and ACE2 receptor binding [80]. *Scutellariae radix* (Huangqin) is known as an herbal antibiotic that has been suggested to combat COVID-19 infection. The active ingredients baicalein and scutellarin inhibit S protein binding with ACE2 receptor by modulating MAPK and NF-κB molecular pathways [81–83]. The mixture of *Acacia catechu* and *S.cutellariae radix* is also used to reduce pulmonary infections in China. By molecular docking, the active components like quercetin, Fisetin(1-), kaempferol, Wogonin, Beta-sitosterol, etc., were shown to interact with TNF, PI3K-Akt and IL-17 signaling pathways. FRET-based analysis showed that baicalein, fisetin(1-), etc., have a high binding affinity with 3C-protease of SARS-CoV-2, indicating the potential antiviral mechanisms of the combinations [84]. The antiviral properties of *L. japonica* (Jinyinhua), and *Houttuynia cordata* (Yuxingcao) are also elucidated. Jinyinhua is used as a part of tea and beverages, whereas Yuxingcao is known to treat pneumonia. Both these herbs block ACE2 receptor interaction with the S protein of SARS-CoV-2 [85]. Other Chinese herbs such as *Salvia miltiorrhiza* (Danshen), *Forsythia suspensa* (Lianqiao), and *Andrographis paniculata* (Chuanxinlian) were also studied for their antiviral potential against COVID-19 infection [86, 87].

In post-hospitalization in COVID-19 patients with severe lung inflammation, the decoction of CHM reduced the symptoms and fostered a fast recovery [88]. Traditional Chinese medicine (TCM) solely improved pulmonary fibrosis and lung inflammation in a 61-year-old female, whereas a similar study in combination with modern medicine was thought to improve a patient's life in a more efficient way [89, 90].

One of the important studies on the Shenhuang granule with different plant extracts including *Panax ginseng* (Renshen) root, *Rheum palmatum L.* stem (Dahuang), etc., was found a beneficial integrative therapy which increased lymphocytes and reduced mortality rates [91].

In Hong Kong, a study during the Omicron wave with CHM was conducted with patient's diagnosed positive within 10 days of SARS-CoV-2 infection. The patients who were selected for CHM not only showed faster recovery but had reduced COVID-19 clinical symptoms such as fatigue, diarrhea, headache, muscle pain, and other cold-related problems [92].

The traditional Taiwan Chingguan Yihau (NRICM101) was formulated with 10 herbs including Fineleaf Nepeta *(Nepeta tenuifolia)*, Mongolian Snakegourd fruit *(Trichosanthes kirilowii)*, Scutellaria root *(S. baicalensis)*, Saposhnikovia root *(Saposhnikovia divaricata)*, Heartleaf Houttuynia *(H. cordata)*, Peppermint herb *(Mentha haplocalyx)*, Magnolia Bark *(Magnolia officinalis)*, Indigowoad root *(Isatis indigotica)*, Mulberry leaf *(Morus alba)* and baked Licorice root *(Glycyrrhiza glabra)* has strong anti-inflammatory potential with antiviral activity. The different chemical ingredients of NRICM101 blocked spike protein interaction with the ACE2 receptor, inhibited 3CL protease activity and reduced viral invasion and proliferation [93].

19.7   USE OF HERBAL AND TRADITIONAL MEDICINE IN INDIA

During the early outbreak of COVID-19, the Indian government directed the AYUSH ministry to start a practical trial to combat the pandemic. As the name suggests, AYUSH is the traditional medicine science in India and neighboring countries, combining Ayurveda, Yoga, Unani, Siddha, and Homeopathy [94]. The Sidhha medicine (containing Thalisathi churnam, Nilavembu kudineer, Amukkura churnam, and Kabasura Kudineer, etc.) was tried as an adjunct therapy along with standard care management, and the study showed that the addition of Siddha significantly improved oxygen level in the blood, improved recovery speed, and reduced mortality compared to standard care alone [95]. Molecular docking studies found that the Kabasura Kudineer, a type of Siddha herbal medicine and a formulation of *Justicia adhatoda* L., *A. paniculata*, Carica Papaya, *Ocimum tenuiflorum*, and *Melia azedarach* (JACOM) have nine chemical compounds that showed high affinity binding with the SARS-CoV-2 S protein. Out of 37 screened compounds, quercetin from JACOM and chrysoeriol and luteolin from Kabasura Kudineer showed a high-affinity strong interaction with the SARS-CoV-2 S protein, having the highest dock score of $\geq$ -11.00 [96].

During the COVID-19 crisis, the AYUSH ministry recommended the use of herbal drinks (kadha) from different medicinal plants (Munakka [*Vitis vinifera*], Tulsi [*O. tenuiflorum* L.], Kalimirch [*Piper nigrum*], Dalchini [*Cinnamomum verum*] and Shunthi [*Zingiber officinale*]) as an immunity booster. Further, the bioinformatics analysis of the constituents of this herbal drink showed that it regulates inflammatory and immune signaling pathways including HIF-1, p53, Wnt, NF-kappa B, IL-17, and TNF, etc. [97]. In other recommendations as immunity boosters, the use of certain spices such as *Coriandrum sativum* and *Curcuma longa* L. etc., was suggested along with *Tinospora cordifolia*, *Emblica officinalis* Gaertn., with honey, and Ashwagandha Churna [98]. Long COVID not only harmed physical wellbeing but the disease took a great toll on mental wellbeing. More than 40% of SARS-CoV-2 infected patients suffered from different mental stresses including sleep disorders and cognition impairment [99]. One of the Indian herbal formulations containing Brahmi (*Bacopa monnieri*), Ashwagandha (*Withania somnifera*), Giloy (*T. cordifolia*), Shankhpushpi (*Convolvulus prostratus* Forssk.), Tulsi (*O. tenuiflorum*) and Malkangni (*Celastrus paniculatus* Willd.), etc. was shown to be effective for long COVID associated mental health issues. These herbs cumulatively showed antioxidant, neuroprotective, tranquilizer, sedative, and antidepressant properties that help in reducing many sleep and depression-related problems [94, 100].

The warm extracts of *T. cordifolia*, *A. paniculata*, *Cydonia oblonga*, *Zizyphus jujube*, *Cordia myxa* and arsenicum album were recommended in treating cold, high fever and modulating immune response along with muscle relaxation during SARS-CoV-2 infection [101].

In other patient trials in Ayurveda medicine, the combination of Tulsi Ghanvati (*Ocimum sanctum*), Giloy Ghanvati (*T. cordifolia*), Ashwagandha (*W. somnifera*) and Swasari Ras (traditional herbo-mineral formulation) was given and 100% recovery was observed compared to 60% recovery in patients with the placebo group. TNF-α and IL-6-like inflammatory cytokines were reduced within the formulation-treated patients [102].

Ayurvedic Bhasma preparations containing Swarna Bhasma, Rajata Bhasma, and Tamra Bhasma in the form of nanoparticles could act as novel antivirals, as these were shown to pose anti-inflammatory and immunomodulatory activities and, thus, could be an effective treatment against COVID-19 [103].

In the search for inhibitors against SARS-CoV-2 main protease, two phytochemicals from *W. somnifera* (Ashwagandha), one from *T. cordifolia* (Giloy), and three from *O. sanctum* (Tulsi) were shown to possess drug-like properties and could be useful stable compounds for further studies [104].

One of the biggest clinical trials for Ayurvedic medicines as an immunity booster was performed with almost 80,000 Delhi (Capital City) police personnel. The immunity enhancer "AYURAKSHA" kits (containing Ayush Kwath, Sanshamani Vati, and Anu Taila) were distributed and suggested to go through the formulation for 60 days (first phase) and 20 days (second phase). The data suggested reduced severity, less risk of liver disease, better glycemic control, and balanced lymphocytes with proper cytokines secretion. The trial group had significantly lower immunoglobulin G (IgG) (17.5%) compared to the control (39.4%) with a lower risk and less severity of COVID-19 infection (55.6%) [105]. Another similar study was conducted for AYUSH-64, an herbal formulation designed by the Central Council for Research in Ayurvedic Sciences, Ministry of AYUSH, India. AYUSH-64 contains four herbal plants including Kuberaksha (*Caesalpinia crista*), Saptaparna (*Alstonia scholaris*), Kiratatikta (*Swertia chirata*), and Katuki (*Picrorhiza kurroa* Royle) having anti-inflammatory and antiviral properties [106]. The smaller clinical trials in different parts of India with 41, 60, and 80 patients and 14 days of treatments revealed that the AYUSH-64 treatment lowered the progression and severity of SARS-CoV-2 infection, but the results were not conclusive regarding the mortality and invasive ventilators due to enrollment of a small number of patients [107–109]. Later, a bigger clinical trial on AYUSH-64 was conducted with 49,770 participants. This study showed better and stronger outcomes regarding COVID-19 management and can be recommended as a standalone or adjunct therapy with standard allopathic care [110].

Another formulation of Ayurvedic medicine known as AYUSH-Kwath is made up of four medicinal plants, namely, Marich (*P. nigrum*), Tulsi (*O. sanctum*), Sunthi (*Z. officinale*), and Dalchini (*C. zeylanicum*) as its main constituents. AYUSH-Kwath also has antiviral and anti-inflammatory activities and can be a potential treatment for the initial stage of COVID-19 [111].

Overall, the meta-analysis suggested that AYUSH interventions have therapeutic benefits against SARS-CoV-2 infection. AYUSH interventions reduce the severity, symptoms, and inflammation and thus improve the overall clinical conditions of COVID-19 patients requiring mild-to-moderate care [112].

19.8 USE OF HERBAL AND TRADITIONAL MEDICINE IN OTHER PARTS OF THE WORLD

The uncontrollable spread of SARS-CoV-2 infection brought the world together to utilize all possible ways to reduce the losses of human life and the world economy. In this section, we summarize the use of herbal medicines in other parts of the world.

Zofa herbal syrup, containing morning glory, caper, Zante currant, *Echium vulgare* (viper's bugloss), *G. glabra* (licorice), *Z. jujuba* (jujube), cucumber seed, fig, cordia fruit, nut mace, quince, maidenhair, tragacanth gum, marshmallow, melon seed, acacia gum, sweet violet, mallow, *Hyssopus officinalis* (hyssop), stevia, and honey, was tested in Iranian COVID-19 patients. The patients who used Zofa syrup showed better recovery and fewer deaths. Though the sample size was small, the study warned of more and deeper experimental trials [113].

Black seed (*Nigella sativa*) have antiviral properties widely tested against HIV infection, function as antioxidants in obese people and post-menopausal women, and reduce oxidative stress in diabetic patients [114, 115]. A randomization and blind study with 262 mild patients reported no significant effect of *N. sativa* in improving disease outcomes of COVID-19, whereas two other trials (one open-label randomized and another double-blind, placebo-controlled randomized clinical trial) showed that the use of *N. sativa* significantly improved symptoms of SARS-CoV-2 infection and has a beneficial effect on mortality rate and hospitalization duration. This study warned that more clinical trials with a high number of patients to use *N. sativa* as an antiviral and anti-inflammatory herbal medicine to combat COVID-19 [116–119] were needed. *N. sativa* has many chemical ingredients such as thymoquinone, nigellide, and α *hederin,* etc., and the molecular docking and network pharmacology work identified the 3C-protease of SARS-CoV-2 as a binding partner for nigellide and α hederin, suggesting the potent antiviral activities of *N. sativa* [120–122].

In a community-based study in the Jimma zone of Ethiopia, it was found that 46% of COVID patients used herbal medicines including garden cress, garlic, lemon, ginger, and Damakase (*Ocimum lamiifolium*) as the most commonly used components [123].

In Morocco, almost 20 herbal plants were exploited for the treatment of COVID-19 including *Azadirachta indica*, *Eucalyptus globulus* Labill., and *Ziziphus lotus*. Molecular docking studies with Moroccan herbal plants identified three chemical molecules, namely, crocin, digitoxigenin, and β-eudesmol, which were potent inhibitors of the 3C main protease of SARS-CoV-2 [124, 125].

After China, Mexico has the highest use of traditional and herbal medicines. A computational study with 100 chemical ingredients from Mexican herbal products identified 10 compounds including aucubin, emodin anthrone, quercetin, riolozatrione, aesculin, cichoriin, monocaffeoyl tartaric acid, luteolin, matricin, and kaempferol with high-affinity binding against SARS-CoV-2. Among these 10, three compounds–quercetin, riolozatrione and cichoriin–target RdRp, Nsp14, and Nsp3 proteins of SARS-CoV-2, but only cichoriin was found to be safe for human use [126, 127].

In a Brazilian study with 497 secondary plant metabolites mainly belonging to alkaloids, triterpenes, and phenolic compounds, it was found that triterpenes and phenolic compounds had higher interaction affinity with SARS-CoV-2 components. These results were based on *in silico* analysis [128].

A similar computational analysis of chemical compounds in Ecuadorian plants identified isorhamnetin, isorhamnetin-3-O-(6-Orhamnosyl-galactoside), benzoic acid, 2-(ethylthio)-ethyl ester, etc., as potent inhibitors of SARS-CoV-2 main protease and papain-like proteases. Anthocyanin derivative also showed potential inhibition on SARS-CoV-2 main protease [129, 130].

In a Taiwanese study on herbal medicine, *Sambucus FormosanaNakai* was tested for antiviral potential. *S. FormosanaNakai* is a plant from *Adoxaceae* family, containing oleanolic acid, phenolic acid, sambuculin A, α-amyrin, β-sitosterol, β-amyrin, ursolic acid, glycyrrhetic acid, and pomolic acid, etc., as chemical ingredients. The antiviral potential of *S. FormosanaNakai* was tested against human coronavirus NL63 and the extract was effective at low IC_{50} values. Previously, the antiviral activity was also reported against the influenza virus [131, 132].

Japanese herbal medicines are also known as Kampo medicines, which originated in China but are mostly used in Japan. Hochuekkito, goreisan, tokishakuyakusan, keishibukuryogan, rikkunshito, ryokeijutsukanto, kakkonto, juzentaihoto, ninjin'yoeito, and hangekobokuto were the most prescribed Kampo medicines to treat fatigue after long COVID [133]. Kakkonto with shosaikotokakikyosekko is one of the Japanese traditional medicines. A clinical study reported that Kakkonto was highly effective in fever reduction and in the suppression of COVID-19 progression. No side effects were reported in the treated group [134]. The oral administration of Maoto, a Japanese herbal medicine, was found to be effective against post-exposure prophylaxis after nosocomial SARS-CoV-2 infection [135].

Kakkontokasenkyushin'I, another Kampo medicine, was used to treat mild COVID-19. This medicine helped in relieving inflammation and congestion in infected patients [136]. Bangladeshi herbal medicines are proposed as immunomodulators and adjunct therapies to reduce the severity of the SARS-CoV-2 infection [137].

19.9   CONCLUSION WITH PERSPECTIVES

The world is still living with SARS-CoV-2, though due to the vaccine development, the burden and casualties are significantly reduced. The emergence of new viral variants and subvariants is the current as well as future global challenge. The heterogeneous response against different SARS-CoV-2 vaccines and available treatments demands the search for new treatments. The use of herbal and traditional medicines is a potent alternative for exploring effective and affordable antiviral therapies worldwide. As herbal and traditional medicines are used widely for different metabolic and infectious diseases, modern use will not only provide more treatment options but also help in overcoming the burden of drug resistance. Though significant efforts are made to identify the underlying molecular mechanisms, the pharmacodynamics and pharmacokinetics are still under investigation with respect to herbal and traditional medicines. Similarly, the lack of concrete statistical data on the use and effectiveness of herbal and traditional medicines against COVID-19 makes it difficult to design worldwide health strategies. A comprehensive approach is needed to translate the *in silico* research into human clinical trials to develop herbal and traditional medicines as equally effective treatments against SARS-CoV-2 infections. New therapeutic options such as herbal and traditional medicines, as an added therapy or single-use, have to be discovered to enhance immunity, improve health conditions, and provide affordable cures against emerging variants of SARS-CoV-2.

REFERENCES

1. Komolafe, K., et al., *Coronavirus disease 2019 and herbal therapy: Pertinent issues relating to toxicity and standardization of phytopharmaceuticals.* Rev Bras Farmacogn, 2021. **31**(2): p. 142–161.
2. Kothandan, R., et al., *Virtual screening of phytochemical compounds as potential inhibitors against SARS-CoV-2 infection.* Beni Suef Univ J Basic Appl Sci, 2021. **10**(1): p. 9.
3. Aleem, A., A.B. Akbar Samad, and S. Vaqar, *Emerging Variants of SARS-CoV-2 and Novel Therapeutics Against Coronavirus (COVID-19),* in *StatPearls.* 2023: Treasure Island (FL) ineligible companies. Disclosure: Abdul Bari Akbar Samad declares no relevant financial relationships with ineligible companies. Disclosure: Sarosh Vaqar declares no relevant financial relationships with ineligible companies.
4. Kumari, R., et al., *A novel rolling circle amplification-based detection of SARS-CoV-2 with multi-region padlock hybridization.* Diagnostics (Basel), 2022. **12**(9): p. 2252.
5. Vitiello, A., A. Zovi, and G. Rezza, *New emerging SARS-CoV-2 variants and antiviral agents.* Drug Resist Updat, 2023. **70**: p. 100986.
6. Jing, S., et al, *Vaccine hesitancy promotes emergence of new SARS-CoV-2 variants.* J Theor Biol, 2023. **570**: p. 111522.
7. Zhu, N., et al., *A novel coronavirus from patients with pneumonia in China, 2019.* N Engl J Med, 2020. **382**(8): p. 727–733.
8. Coronaviridae Study Group of the International Committee on Taxonomy of Viruses., *The species severe acute respiratory syndrome-related coronavirus: Classifying 2019-nCoV and naming it SARS-CoV-2.* Nat Microbiol, 2020. **5**(4): p. 536–544.
9. Li, X., et al., *Transmission dynamics and evolutionary history of 2019-nCoV.* J Med Virol, 2020. **92**(5): p. 501–511.
10. Koelle, K., et al., *The changing epidemiology of SARS-CoV-2.* Science, 2022. **375**(6585): p. 1116–1121.
11. Chen, J.T., and N. Krieger, *Revealing the unequal burden of COVID-19 by income, Race/Ethnicity, and household crowding: US county versus zip code analyses.* J Public Health Manag Pract, 2021. **27**(**Suppl 1, COVID-19 and Public Health: Looking Back, Moving Forward**): p. S43–S56.
12. Meyerowitz, E.A., et al., *Transmission of SARS-CoV-2: A review of viral, host, and environmental factors.* Ann Intern Med, 2021. **174**(1): p. 69–79.
13. Endo, A., et al., *Estimating the overdispersion in COVID-19 transmission using outbreak sizes outside China.* Wellcome Open Res, 2020. **5**: p. 67.
14. Volz, E., et al., *Evaluating the effects of SARS-CoV-2 spike mutation D614G on transmissibility and pathogenicity.* Cell, 2021. **184**(1): p. 64–75.e11.
15. Dhama, K., et al., *Global emerging Omicron variant of SARS-CoV-2: Impacts, challenges and strategies.* J Infect Public Health, 2023. **16**(1): p. 4–14.
16. Khailany, R.A., M. Safdar, and M. Ozaslan, *Genomic characterization of a novel SARS-CoV-2.* Gene Rep, 2020. **19**: p. 100682.
17. Yan, W., et al., *Structural biology of SARS-CoV-2: Open the door for novel therapies.* Signal Transduct Target Ther, 2022. **7**(1): p. 26.
18. Walls, A.C., et al., *Structure, function, and antigenicity of the SARS-CoV-2 spike glycoprotein.* Cell, 2020. **181**(2): p. 281–292.e6.
19. Lan, J., et al., *Structure of the SARS-CoV-2 spike receptor-binding domain bound to the ACE2 receptor.* Nature, 2020. **581**(7807): p. 215–220.
20. Vankadari, N., *Arbidol: A potential antiviral drug for the treatment of SARS-CoV-2 by blocking trimerization of the spike glycoprotein.* Int J Antimicrob Agents, 2020. **56**(2): p. 105998.
21. Lu, X., et al., *SARS-CoV nucleocapsid protein antagonizes IFN-beta response by targeting initial step of IFN-beta induction pathway, and its C-terminal region is critical for the antagonism.* Virus Genes, 2011. **42**(1): p. 37–45.
22. Guo, Y., et al., *Crystal structure of the core region of hantavirus nucleocapsid protein reveals the mechanism for ribonucleoprotein complex formation.* J Virol, 2016. **90**(2): p. 1048–61.

23. Kang, S., et al., *A SARS-CoV-2 antibody curbs viral nucleocapsid protein-induced complement hyper-activation*. Nat Commun, 2021. **12**(1): p. 2697.

24. Verdia-Baguena, C., et al., *Coronavirus E protein forms ion channels with functionally and structurally-involved membrane lipids*. Virology, 2012. **432**(2): p. 485–494.

25. de Haan, C.A., et al., *Mapping of the coronavirus membrane protein domains involved in interaction with the spike protein*. J Virol, 1999. **73**(9): p. 7441–7452.

26. Li, J., et al., *Virus-host interactome and proteomic survey reveal potential virulence factors influencing SARS-CoV-2 pathogenesis*. Med, 2021. **2**(1): p. 99–112.e7.

27. Hagemeijer, M.C., et al., *Membrane rearrangements mediated by coronavirus nonstructural proteins 3 and 4*. Virology, 2014. **458-459**: p. 125–135.

28. Angelini, M.M., et al., *Severe acute respiratory syndrome coronavirus nonstructural proteins 3, 4, and 6 induce double-membrane vesicles*. mBio, 2013. **4**(4): p. e00524-13.

29. Thomas, S., *Mapping the nonstructural transmembrane proteins of severe acute respiratory syndrome coronavirus 2*. J Comput Biol, 2021. **28**(9): p. 909–921.

30. Rogstam, A., et al., *Crystal structure of non-structural protein 10 from severe acute respiratory syndrome coronavirus-2*. Int J Mol Sci, 2020. **21**(19): p. 7375.

31. Yuan, S., et al., *Nonstructural protein 1 of SARS-CoV-2 is a potent pathogenicity factor redirecting host protein synthesis machinery toward viral RNA*. Mol Cell, 2020. **80**(6): p. 1055–1066.e6.

32. Vankadari, N., N.N. Jeyasankar, and W.J. Lopes, *Structure of the SARS-CoV-2 Nsp1/5'-untranslated region complex and implications for potential therapeutic targets, a vaccine, and virulence*. J Phys Chem Lett, 2020. **11**(22): p. 9659–9668.

33. Clark, L.K., T.J. Green, and C.M. Petit, *Structure of nonstructural protein 1 from SARS-CoV-2*. J Virol, 2021. **95**(4): p. e02019-20.

34. Ma, J., et al., *Structure and function of N-terminal zinc finger domain of SARS-CoV-2 NSP2*. Virol Sin, 2021. **36**(5): p. 1104–1112.

35. Cornillez-Ty, C.T., et al., *Severe acute respiratory syndrome coronavirus nonstructural protein 2 interacts with a host protein complex involved in mitochondrial biogenesis and intracellular signaling*. J Virol, 2009. **83**(19): p. 10314–10318.

36. Helmy, Y.A., et al., *The COVID-19 pandemic: A comprehensive review of taxonomy, genetics, epidemiology, diagnosis, treatment, and control*. J Clin Med, 2020. **9**(4): p. 1225.

37. Jiang, Y., W. Yin, and H.E. Xu, *RNA-dependent RNA polymerase: Structure, mechanism, and drug discovery for COVID-19*. Biochem Biophys Res Commun, 2021. **538**: p. 47–53.

38. Michel, C.J., et al., *Characterization of accessory genes in coronavirus genomes*. Virol J, 2020. **17**(1): p. 131.

39. Vallamkondu, J., et al., *SARS-CoV-2 pathophysiology and assessment of coronaviruses in CNS diseases with a focus on therapeutic targets*. Biochim Biophys Acta Mol Basis Dis, 2020. **1866**(10): p. 165889.

40. Cardinal-Fernandez, P., et al., *Acute respiratory distress syndrome and diffuse alveolar damage. New insights on a complex relationship*. Ann Am Thorac Soc, 2017. **14**(6): p. 844–850.

41. Carsana, L., et al., *Pulmonary post-mortem findings in a series of COVID-19 cases from northern Italy: A two-centre descriptive study*. Lancet Infect Dis, 2020. **20**(10): p. 1135–1140.

42. Gomez-Mesa, J.E., et al., *Thrombosis and coagulopathy in COVID-19*. Curr Probl Cardiol, 2021. **46**(3): p. 100742.

43. Lippi, G., M. Plebani, and B.M. Henry, *Thrombocytopenia is associated with severe coronavirus disease 2019 (COVID-19) infections: A meta-analysis*. Clin Chim Acta, 2020. **506**: p. 145–148.

44. Rentsch, C.T., et al., *Early initiation of prophylactic anticoagulation for prevention of coronavirus disease 2019 mortality in patients admitted to hospital in the United States: Cohort study*. BMJ, 2021. **372**: p. n311.

45. Zhang, J., et al., *Pyroptotic macrophages stimulate the SARS-CoV-2-associated cytokine storm*. Cell Mol Immunol, 2021. **18**(5): p. 1305–1307.

46. Puelles, V.G., et al., *Multiorgan and renal tropism of SARS-CoV-2*. N Engl J Med, 2020. **383**(6): p. 590–592.

47. Channappanavar, R., et al., *Dysregulated type I interferon and inflammatory monocyte-macrophage responses cause lethal pneumonia in SARS-CoV-infected mice*. Cell Host Microbe, 2016. **19**(2): p. 181–93.

48. Scharf, R.E., and J.M. Anaya, *Post-COVID syndrome in adults - An overview*. Viruses, 2023. **15**(3): p. 675.

49. Chen, C., et al., *Global prevalence of post-coronavirus disease 2019 (COVID-19) condition or long COVID: A meta-analysis and systematic review*. J Infect Dis, 2022. **226**(9): p. 1593–1607.

50. Jegal, K.H., et al., *Herbal medicines for post-acute sequelae (fatigue or cognitive dysfunction) of SARS-CoV-2 infection: A phase 2 pilot clinical study protocol.* Healthcare (Basel), 2022. **10**(10): p. 1839.

51. Crook, H., et al., *Long covid-mechanisms, risk factors, and management.* BMJ, 2021. **374**: p. n1648.

52. Li, L., et al., *Potential treatment of COVID-19 with traditional Chinese medicine: What herbs can help win the Battle with SARS-CoV-2?* Engineering (Beijing), 2022. **19**: p. 139–152.

53. Jin, D., et al., *Contribution of Chinese herbal medicine in the treatment of coronavirus disease 2019 (COVID-19): A systematic review and meta-analysis of randomized controlled trials.* Phytother Res, 2023. **37**(3): p. 1015–1035.

54. Xu, J., and Y. Zhang, *Traditional Chinese medicine treatment of COVID-19.* Complement Ther Clin Pract, 2020. **39**: p. 101165.

55. Wang, W.Y., et al., *Contribution of traditional Chinese medicine to the treatment of COVID-19.* Phytomedicine, 2021. **85**: p. 153279.

56. Meng, T., et al., *Xuanfei Baidu decoction in the treatment of coronavirus disease 2019 (COVID-19): Efficacy and potential mechanisms.* Heliyon, 2023. **9**(9): p. e19163.

57. He, Q., et al., *Herbal medicine in the treatment of COVID-19 based on the gut-lung axis.* Acupunct Herb Med, 2022. **2**(3): p. 172–183.

58. Zhang, B., and F. Qi, *Herbal medicines exhibit a high affinity for ACE2 in treating COVID-19.* Biosci Trends, 2023. **17**(1): p. 14–20.

59. Chen, X., et al., *Identifying potential anti-COVID-19 pharmacological components of traditional Chinese medicine Lianhuaqingwen capsule based on human exposure and ACE2 biochromatography screening.* Acta Pharm Sin B, 2021. **11**(1): p. 222–236.

60. He, D.D., et al., *Network pharmacology and RNA-sequencing reveal the molecular mechanism of Xuebijing injection on COVID-19-induced cardiac dysfunction.* Comput Biol Med, 2021. **131**: p. 104293.

61. Xing, Y., et al., *Traditional Chinese medicine network pharmacology study on exploring the mechanism of Xuebijing injection in the treatment of coronavirus disease 2019.* Chin J Nat Med, 2020. **18**(12): p. 941–951.

62. Zhong, L.L.D., et al., *Potential targets for treatment of coronavirus disease 2019 (COVID-19): A review of Qing-Fei-Pai-Du-Tang and its major herbs.* Am J Chin Med, 2020. **48**(5): p. 1051–1071.

63. Li, Y., et al., *Identification of phytochemicals in Qingfei Paidu decoction for the treatment of coronavirus disease 2019 by targeting the virus-host interactome.* Biomed Pharmacother, 2022. **156**: p. 113946.

64. Yang, Z., et al., *Traditional Chinese medicine against COVID-19: Role of the gut microbiota.* Biomed Pharmacother, 2022. **149**: p. 112787.

65. Tao, Q., et al., *Network pharmacology and molecular docking analysis on molecular targets and mechanisms of Huashi Baidu formula in the treatment of COVID-19.* Drug Dev Ind Pharm, 2020. **46**(8): p. 1345–1353.

66. Cai, Y., M. Zeng, and Y.Z. Chen, *The pharmacological mechanism of Huashi Baidu Formula for the treatment of COVID-19 by combined network pharmacology and molecular docking.* Ann Palliat Med, 2021. **10**(4): p. 3864–3895.

67. Simayi, J., et al., *Analysis of the active components and mechanism of Shufeng Jiedu capsule against COVID-19 based on network pharmacology and molecular docking.* Medicine (Baltimore), 2022. **101**(1): p. e28286.

68. Zhuang, Z., et al., *Exploring the potential mechanism of Shufeng Jiedu capsule for treating COVID-19 by comprehensive network pharmacological approaches and molecular docking validation.* Comb Chem High Throughput Screen, 2021. **24**(9): p. 1377–1394.

69. Li, Y., et al., *Potential effect of Maxing Shigan decoction against coronavirus disease 2019 (COVID-19) revealed by network pharmacology and experimental verification.* J Ethnopharmacol, 2021. **271**: p. 113854.

70. Chien, T.J., et al., *Therapeutic effects of herbal-medicine combined therapy for COVID-19: A systematic review and meta-analysis of randomized controlled trials.* Front Pharmacol, 2022. **13**: p. 950012.

71. Xiao, M., et al., *Efficacy of Huoxiang Zhengqi dropping pills and Lianhua Qingwen granules in treatment of COVID-19: A randomized controlled trial.* Pharmacol Res, 2020. **161**: p. 105126.

72. Wei, X., et al., *Gegen Qinlian pills alleviate carrageenan-induced thrombosis in mice model by regulating the HMGB1/NF-kappaB/NLRP3 signaling.* Phytomedicine, 2022. **100**: p. 154083.

73. Zhuang, J., et al., *Efficacy and safety of integrated traditional Chinese and Western medicine against COVID-19: A systematic review and meta-analysis.* Phytother Res, 2022. **36**(12): p. 4371–4397.

74. Zhao, J., et al., *Yidu-toxicity blocking lung decoction ameliorates inflammation in severe pneumonia of SARS-COV-2 patients with Yidu-toxicity blocking lung syndrome by eliminating IL-6 and TNF-a.* Biomed Pharmacother, 2020. **129**: p. 110436.

75. Ni, L., et al., *Effects of Shuanghuanglian oral liquids on patients with COVID-19: A randomized, open-label, parallel-controlled, multicenter clinical trial.* Front Med, 2021. **15**(5): p. 704–717.

76. Chan, M.C., et al., *Indirubin-3'-oxime as an antiviral and immunomodulatory agent in treatment of severe human influenza virus infection.* Hong Kong Med J, 2018. **24 Suppl 6**(5): p. 45–47.

77. Buabeid, M., et al., *Therapeutic uses of traditional Chinese medicines against COVID-19.* Infect Drug Resist, 2021. **14**: p. 5017–5026.

78. Zhu, X.B., et al., *Chinese herbal injections for coronavirus disease 2019 (COVID-19): A narrative review.* Integr Med Res, 2021. **10**(4): p. 100778.

79. Ma, Q., et al., *Efficacy and safety of ReDuNing injection as a treatment for COVID-19 and its inhibitory effect against SARS-CoV-2.* J Ethnopharmacol, 2021. **279**: p. 114367.

80. Li, X., et al., *Chemical composition and pharmacological mechanism of ephedra-glycyrrhiza drug pair against coronavirus disease 2019 (COVID-19).* Aging (Albany NY), 2021. **13**(4): p. 4811–4830.

81. Lv, Y., et al., *Screening and evaluation of anti-SARS-CoV-2 components from Ephedra sinica by ACE2/CMC-HPLC-IT-TOF-MS approach.* Anal Bioanal Chem, 2021. **413**(11): p. 2995–3004.

82. Mei, J., et al., *Active components in Ephedra sinica stapf disrupt the interaction between ACE2 and SARS-CoV-2 RBD: Potent COVID-19 therapeutic agents.* J Ethnopharmacol, 2021. **278**: p. 114303.

83. Liu, J., et al., *Integrated network pharmacology analysis, molecular docking, LC-MS analysis and bioassays revealed the potential active ingredients and underlying mechanism of Scutellariae radix for COVID-19.* Front Plant Sci, 2022. **13**: p. 988655.

84. Feng, T., et al., *Exploration of molecular targets and mechanisms of Chinese medicinal formula Acacia Catechu -Scutellariae Radix in the treatment of COVID-19 by a systems pharmacology strategy.* Phytother Res, 2022. **36**(11): p. 4210–4229.

85. Liu, J., et al., *Network pharmacology and molecular docking elucidate the underlying pharmacological mechanisms of the herb Houttuynia cordata in treating pneumonia caused by SARS-CoV-2.* Viruses, 2022. **14**(7): p. 1588.

86. Hu, S., et al., *Three salvianolic acids inhibit 2019-nCoV spike pseudovirus viropexis by binding to both its RBD and receptor ACE2.* J Med Virol, 2021. **93**(5): p. 3143–3151.

87. Xie, R., et al., *Deciphering the potential anti-COVID-19 active ingredients in Andrographis paniculata (Burm. F.) Nees by combination of network pharmacology, molecular docking, and molecular dynamics.* RSC Adv, 2021. **11**(58): p. 36511–36517.

88. Li, L., et al., *Effects of Chinese medicine on symptoms, syndrome evolution, and lung inflammation absorption in COVID-19 convalescent patients during 84-day follow-up after Hospital discharge: A prospective cohort and nested case-control study.* Chin J Integr Med, 2021. **27**(4): p. 245–251.

89. Zhi, N., et al., *Treatment of pulmonary fibrosis in one convalescent patient with corona virus disease 2019 by oral traditional Chinese medicine decoction: A case report.* J Integr Med, 2021. **19**(2): p. 185–190.

90. Lu, Z.H., et al., *Efficacy of the combination of modern medicine and traditional Chinese medicine in pulmonary fibrosis arising as a sequelae in convalescent COVID-19 patients: A randomized multicenter trial.* Infect Dis Poverty, 2021. **10**(1): p. 31.

91. Zhou, S., et al., *Traditional Chinese medicine shenhuang granule in patients with severe/critical COVID-19: A randomized controlled multicenter trial.* Phytomedicine, 2021. **89**: p. 153612.

92. Zhang, J., et al., *Effectiveness of Chinese herbal medicine in patients with COVID-19 during the Omicron wave in Hong Kong: A retrospective case-controlled study.* Am J Chin Med, 2023. **51**(7): p. 1615–1626.

93. Tsai, K.C., et al., *A traditional Chinese medicine formula NRICM101 to target COVID-19 through multiple pathways: A bedside-to-bench study.* Biomed Pharmacother, 2021. **133**: p. 111037.

94. Mukherjee, P.K., et al., *Role of medicinal plants in inhibiting SARS-CoV-2 and in the management of post-COVID-19 complications.* Phytomedicine, 2022. **98**: p. 153930.

95. Christian, G.J., et al., *Safety and efficacy of Siddha management as adjuvant care for COVID-19 patients admitted in a tertiary care hospital - An open-label, proof-of-concept randomized controlled trial.* J Ayurveda Integr Med, 2023. **14**(2): p. 100706.

96. Kiran, G., et al., *In silico computational screening of Kabasura Kudineer - Official Siddha Formulation and JACOM against SARS-CoV-2 spike protein.* J Ayurveda Integr Med, 2022. **13**(1): p. 100324.

97. Khanal, P., et al., *Network pharmacology of AYUSH recommended immune-boosting medicinal plants against COVID-19.* J Ayurveda Integr Med, 2022. **13**(1): p. 100374.

98. Gupta, P.K., Sonewane, K., Rajan, M., Patil, N.J., Agrawal, T., Banerjee, E.R., Chauhan, N.S., and Kumar, A., Scientific rationale of Indian AYUSH Ministry advisory for COVID-19 prevention, prophylaxis, and immunomodulation. Adv Tradit Med (ADTM). 2023. **23**(2): p. 321–345.

99. Kaseda, E.T., and A.J. Levine, *Post-traumatic stress disorder: A differential diagnostic consideration for COVID-19 survivors.* Clin Neuropsychol, 2020. **34**(7–8): p. 1498–1514.

100. Kumar, A., et al., *Role of herbal medicines in the management of patients with COVID-19: A systematic review and meta-analysis of randomized controlled trials.* J Tradit Complement Med, 2022. **12**(1): p. 100–113.

101. Vellingiri, B., et al., *COVID-19: A promising cure for the global panic.* Sci Total Environ, 2020. **725**: p. 138277.

102. Devpura, G., et al., *Randomized placebo-controlled pilot clinical trial on the efficacy of ayurvedic treatment regime on COVID-19 positive patients.* Phytomedicine, 2021. **84**: p. 153494.

103. Sarkar, P.K., and C. Das Mukhopadhyay, *Ayurvedic metal nanoparticles could be novel antiviral agents against SARS-CoV-2.* Int Nano Lett, 2021. **11**(3): p. 197–203.

104. Shree, P., et al., *Targeting COVID-19 (SARS-CoV-2) main protease through active phytochemicals of ayurvedic medicinal plants - Withania somnifera (Ashwagandha), Tinospora cordifolia (Giloy) and Ocimum sanctum (Tulsi) - a molecular docking study.* J Biomol Struct Dyn, 2022. **40**(1): p. 190–203.

105. Nesari, T., et al., *AYURAKSHA, a prophylactic Ayurvedic immunity boosting kit reducing positivity percentage of IgG COVID-19 among frontline Indian Delhi police personnel: A non-randomized controlled intervention trial.* Front Public Health, 2022. **10**: p. 920126.

106. Panda, A.K., et al., *AYUSH- 64: A potential therapeutic agent in COVID-19.* J Ayurveda Integr Med, 2022. **13**(2): p. 100538.

107. Thakar, A., et al., *Impact of AYUSH 64 as an adjunctive to standard of care in mild COVID 19 - An open-label randomized controlled pilot study.* J Ayurveda Integr Med, 2022. **13**(3): p. 100587.

108. Singh, H., et al., *AYUSH-64 as an adjunct to standard care in mild to moderate COVID-19: An open-label randomized controlled trial in Chandigarh, India.* Complement Ther Med, 2022. **66**: p. 102814.

109. Reddy, R.G., et al., *AYUSH-64 as an add-on to standard care in asymptomatic and mild cases of COVID-19: A randomized controlled trial.* Ayu, 2020. **41**(2): p. 107–116.

110. Srikanth, N., et al., *Disease characteristics, care-seeking behavior, and outcomes associated with the use of AYUSH-64 in COVID-19 patients in home isolation in India: A community-based cross-sectional analysis.* Front Public Health, 2022. **10**: p. 904279.

111. Gautam, S., et al., *Immunity against COVID-19: Potential role of Ayush Kwath.* J Ayurveda Integr Med, 2022. **13**(1): p. 100350.

112. Thakar, A., et al., *AYUSH (Indian system of medicines) therapeutics for COVID-19: A living systematic review and meta-analysis (First update).* J Integr Complement Med, 2023. **29**(3): p. 139–155.

113. Ghazvini, A., et al., *Effects of Iranian herbal Zofa((R)) syrup for the management of clinical symptoms in patients with COVID-19: A randomized clinical trial.* Avicenna J Phytomed, 2023. **13**(5): p. 500–512.

114. Namazi, N., et al., *Oxidative stress responses to Nigella sativa oil concurrent with a low-calorie diet in obese women: A randomized, double-blind controlled clinical trial.* Phytother Res, 2015. **29**(11): p. 1722–8.

115. Kaatabi, H., et al., *Nigella sativa improves glycemic control and ameliorates oxidative stress in patients with type 2 diabetes mellitus: Placebo controlled participant blinded clinical trial.* PLoS One, 2015. **10**(2): p. e0113486.

116. Bin Abdulrahman, K.A., et al., *The effect of short treatment with Nigella sativa on symptoms, the cluster of differentiation (CD) profile, and inflammatory markers in mild COVID-19 patients: A randomized, double-blind controlled trial.* Int J Environ Res Public Health, 2022. **19**(18): p. 11798.

117. Koshak, A.E., et al., *Nigella sativa for the treatment of COVID-19: An open-label randomized controlled clinical trial.* Complement Ther Med, 2021. **61**: p. 102769.

118. Ashraf, S., et al., *Honey and Nigella sativa against COVID-19 in Pakistan (HNS-COVID-PK): A multicenter placebo-controlled randomized clinical trial.* Phytother Res, 2023. **37**(2): p. 627–644.

119. Pakkir Maideen, N.M., et al., *Potential of black seeds (Nigella sativa) in the management of long COVID or post-acute sequelae of COVID-19 (PASC) and persistent COVID-19 symptoms - An insight.* Infect Disord Drug Targets, 2023. **23**(4): p. e230223213955.

120. Liu, Y.X., et al., *Prevention, treatment and potential mechanism of herbal medicine for Corona viruses: A review.* Bioengineered, 2022. **13**(3): p. 5480–5508.

121. Rahman, M.T., *Potential benefits of combination of Nigella sativa and Zn supplements to treat COVID-19.* J Herb Med, 2020. **23**: p. 100382.

122. Xu, H., et al., *Computational and experimental studies reveal that thymoquinone blocks the entry of coronaviruses into in vitro cells.* Infect Dis Ther, 2021. **10**(1): p. 483–494.

123. Umeta Chali, B., et al., *Traditional medicine practice in the context of COVID-19 pandemic: Community claim in Jimma zone, Oromia, Ethiopia.* Infect Drug Resist, 2021. **14**: p. 3773–3783.

124. Chaachouay, N., A. Douira, and L. Zidane, *COVID-19, prevention and treatment with herbal medicine in the herbal markets of sale prefecture, North-Western Morocco.* Eur J Integr Med, 2021. **42**: p. 101285.

125. Aanouz, I., et al., *Moroccan Medicinal plants as inhibitors against SARS-CoV-2 main protease: Computational investigations.* J Biomol Struct Dyn, 2021. **39**(8): p. 2971–2979.

126. Rivero-Segura, N.A., and J.C. Gomez-Verjan, *In silico screening of natural products isolated from Mexican herbal medicines against COVID-19.* Biomolecules, 2021. **11**(2): p. 216.

127. Mata, R., et al., *Chemistry and biology of selected Mexican medicinal plants.* Prog Chem Org Nat Prod, 2019. **108**: p. 1–142.

128. Amparo, T.R., et al., *In silico approach of secondary metabolites from Brazilian herbal medicines to search for potential drugs against SARS-CoV-2.* Phytother Res, 2021. **35**(8): p. 4297–4308.

129. Tejera, E., et al., *Computational modeling predicts potential effects of the herbal infusion "horchata" against COVID-19.* Food Chem, 2022. **366**: p. 130589.

130. Fakhar, Z., et al., *Anthocyanin derivatives as potent inhibitors of SARS-CoV-2 main protease: An in-silico perspective of therapeutic targets against COVID-19 pandemic.* J Biomol Struct Dyn, 2021. **39**(16): p. 6171–6183.

131. Weng, J.R., et al., *Antiviral activity of Sambucus FormosanaNakai ethanol extract and related phenolic acid constituents against human coronavirus NL63.* Virus Res, 2019. **273**: p. 197767.

132. Krawitz, C., et al., *Inhibitory activity of a standardized elderberry liquid extract against clinically-relevant human respiratory bacterial pathogens and influenza A and B viruses.* BMC Complement Altern Med, 2011. **11**: p. 16.

133. Tokumasu, K., et al., *Application of Kampo medicines for treatment of general fatigue due to long COVID.* Medicina (Kaunas), 2022. **58**(6): p. 730.

134. Takayama, S., et al., *Multicenter, randomized controlled trial of traditional Japanese medicine, kakkonto with shosaikotokakikyosekko, for mild and moderate coronavirus disease patients.* Front Pharmacol, 2022. **13**: p. 1008946.

135. Nabeshima, A., et al., *Maoto, a traditional herbal medicine, for post-exposure prophylaxis for Japanese healthcare workers exposed to COVID-19: A single center study.* J Infect Chemother, 2022. **28**(7): p. 907–911.

136. Takayama, S., et al., *Treatment of COVID-19-related olfactory disorder promoted by Kakkontokasenkyushin'i: A case series.* Tohoku J Exp Med, 2021. **254**(2): p. 71–80.

137. Rahman, M.M., et al., *An insight into COVID-19 and traditional herbs: Bangladesh Perspective.* Med Chem, 2023. **19**(4): p. 361–383.

20 Molecular and Bioinformatics Tools for the Analysis of Anti-SARS-CoV-2 Traditional Herbal Medicines

Sora Yasri and Viroj Wiwanitkit

20.1 INTRODUCTION

The severe acute respiratory syndrome coronavirus, commonly known as COVID-19 or SARS-CoV-2, is a devastating viral illness. The coronavirus, SARS-CoV-2, which was originally discovered in 2019, has grown to be a serious global threat. Usually, this virus spreads by passing on contaminated respiratory droplets or secretions from one person's respiratory system to another's. This uncommon respiratory virus can cause a range of clinical problems, including mortality. Although efforts are being made in every nation, coronavirus disease 2019 (COVID-19) has not yet been contained [1, 2]. Immediate action is needed to fight this respiratory virus. The two main types of treatment are symptomatic and supportive therapies. The genuine effectiveness and safety of current therapeutic modalities must be investigated because there is currently no evidence-based therapy for COVID-19. It will be critical to assess the genuine efficacy and safety of various therapeutic options until sufficient multi-site clinical data are available. There aren't many treatment options due to the recent nature of the condition. According to reports as of July 2020, the Americas and Europe appear to have the highest prevalence of the disease. Most SARS-CoV-2 patients will have mild to moderate respiratory symptoms or none at all, and they will recover without treatment. Nevertheless, to address the severe cases that necessitate hospitalization, an efficient antiviral medication is still required. COVID-19, which is caused by the SARS-CoV-2 virus, was first found in China. Early COVID-19 symptoms include fever, a dry cough, and dyspnea and are similar to those of other viral respiratory infections including the flu. Because of this, COVID-19 diagnosis based on anamnesis is still difficult. Although it has been noticed that the incubation can range from 0 to 24 days, the typical incubation period is 15 days [1, 2]. The viral infection may cause respiratory problems and a fever. The COVID-19 virus has claimed millions of lives. Despite the widespread immunization against the virus, it is still a significant threat. It is critical to continue conducting fresh research and developing new drugs to address this persistent public health concern.

"Alternative COVID cures" are widely used. As of 2019, the database for such treatments was small, if not nonexistent. Strong, unsupported claims regarding the benefits and dangers of herbal remedies, however, are frequently made, giving COVID-19 patients genuine and misplaced hopes or considerable worries. Other investigations are being conducted to identify plants and other natural compounds that may be effective in the treatment of coronavirus disease [3]. Through a range of components that work on several targets and pathways, traditional medicine plays antiviral, anti-inflammatory, and immunoregulatory functions in the treatment of COVID-19. Traditional medicine has also been shown in clinical experiments to be effective in treating patients. Recently published results provide significant and useful information for the development of COVID-19 and other antiviral infectious disease medications [3]. Traditional medicine can assist patients in managing COVID-19 by providing them with complementary care. Numerous conventional medications and therapies have shown promise in improving COVID-19 symptoms and increasing the immune

DOI: 10.1201/9781003452621-20

system. For instance, vitamin C, ginger, and turmeric are dietary supplements and herbal remedies that can boost immunity and stave off disease. Additionally, traditional methods like acupuncture and meditation can aid in the reduction of stress and anxiety, which can enhance overall health and recovery. Traditional medicine must only be used under the supervision of a healthcare professional and in conjunction with therapies that are backed by scientific research. Following the discovery of COVID-19, the first publication by Sriwijitalai and Wiwanitkit showed that using specific natural product regimens against COVID-19 may have some benefits. Several classical herbal plants are confirmed for advantages in counteracting pathogenic viruses [4].

Many classical herbs have already been demonstrated to be effective in the treatment of viral respiratory infections. These findings are helpful for future research into various herbal treatments and their potential value in the management of COVID-19. Many of these herbs have already been demonstrated to be effective in the treatment of viral respiratory infections. In the modern world, complementary and alternative medicine does have a place. Today, a number of conventional medical systems are accessible. Historical herbs are widely used in traditional medicine, and it's fascinating to observe how these venerable plants are used to treat a variety of ailments. There is current research that supports the notion that many plants are helpful in the treatment of a number of diseases [5, 6].

20.2   TRADITIONAL HERBAL MEDICINES FOR COVID-19 MANAGEMENT

Traditional medicine refers to a variety of medicinal procedures that have been practiced for millennia in many cultures around the world. Herbal medicine, acupuncture, massage, and other natural therapies are examples of such methods. The primary distinction between traditional medicine and traditional herbal medicine is that the former may incorporate activities other than the use of plants, whilst the latter concentrates solely on the use of herbs and natural cures. Both types of medicine can be successful, and many patients prefer to combine conventional therapies. Traditional medicine refers to the knowledge, skills, and practices based on indigenous theories, beliefs, and experiences utilized in the maintenance of health and the prevention, diagnosis, improvement, or treatment of physical and mental illness. Herbalism, acupuncture, manual treatments, and energy healing are examples of complementary therapies. The use of plants or plant extracts to treat medical issues is referred to as traditional herbal medicine. This method of medicine has been utilized for centuries and is still widely used today.

Global public health continues to be significantly impacted by traditional herbal treatment. Many people still use traditional herbal treatments as their main source of healthcare all around the world. For those residing in low- and middle-income nations, traditional herbal medicine is frequently more accessible and affordable than modern medicine. Traditional medicine can also offer insightful analysis and helpful solutions for a range of health issues that modern medicine might not be able to address. It's crucial to remember that conventional herbal medicine should be used in conjunction with modern medicine, and that any usage of herbal medicines needs to be reviewed with a healthcare professional. For ages, traditional herbal therapy has been utilized to treat a wide range of illnesses and conditions. As the COVID-19 pandemic continues, many people are turning to conventional herbal treatment to strengthen their immune systems and perhaps lessen virus-related symptoms. Some herbs and natural therapies have been shown to have antiviral and immune-boosting effects, while there is little scientific evidence to support the effectiveness of conventional herbal medication in treating COVID-19 in particular. For instance, it has been discovered that ginger contains immune-boosting and anti-inflammatory characteristics whereas garlic has antiviral capabilities. Overall, traditional herbal medicine can help to maintain the immune system and overall health during the COVID-19 pandemic, but it should be taken in concert with other preventative measures such as social isolation, mask use, and basic hygiene.

According to the evidence from the included trials, incorporating traditional medicine with COVID-19 treatment may improve clinical outcomes when compared to conventional Western medical therapy alone [7]. The traditional medicine evidence for COVID-19 was generally of moderate to low quality [7]. Clinical decision-making can benefit from using meta-analyses of the use of conventional medicine in the treatment of COVID-19 by taking into consideration the experiences of clinical experts, medical policies, and other factors [7]. According to Ding et al. [8], studies have demonstrated that Traditional Chinese Medicine (TCM) treatments can dramatically reduce clinical symptoms. To find new targeted treatments for COVID-19, Ding et al. suggested that future studies may have a high translational value [8].

Numerous drugs have already been recommended for the treatment of illness for ages. At present, traditional herbal therapy has been utilized to treat a wide range of illnesses and conditions. They may be used to treat infectious illnesses and a small number of them are now being investigated in clinical settings. Both Indian Ayurveda and TCM have made extensive use of plant extracts or herbal remedies. Additionally, it has been applied to suppress the actions of particular genes and proteins to facilitate the regulation of signaling pathways [9]. Because natural treatments have been shown to have amazing bioactivities, they are regarded as an accessible and secure treatment for lung disease. For their potential to treat various lung illnesses, natural plant metabolites (such as flavonoids, alkaloids, and terpenoids) have been investigated *in vitro*, *in vivo*, and computationally [8]. Natural metabolites that have been repurposed for use in different lung diseases should be evaluated more by advanced computational applications, and experimental models in the biological system, and need to be validated by clinical trials to find potential treatments for the most difficult lung diseases, particularly SARS-CoV-2 [9]. Traditional medicine has been used for many years to treat a wide range of medical conditions, including infectious diseases, however, at this time, there is no scientific evidence to support its use in the management of COVID-19. Utilizing COVID-19 treatments and therapies that are supported by evidence is strongly advised by the World Health Organization (WHO) and medical professionals worldwide. Traditional medical techniques that may be effective include herbal remedies, acupuncture, and breathing exercises.

20.3   BIOINFORMATICS APPLICATIONS FOR MANAGEMENT OF COVID-19

Drugs are critical in combating infectious disease epidemics. In some cases, they can even heal the infection. They have the ability to treat and relieve symptoms. In order to develop novel treatments against new diseases, researchers typically begin by studying the biology of the virus that causes the illness, seeking out druggable flaws in the pathogen's structure, metabolism, or reproduction mechanism. Once a potential target has been identified, thousands of molecules can be examined to discover if any of them have the desired effect. Although it takes time, this method is required to ensure the drug's efficacy and safety. Drug discovery has become easier in recent years as a result of technological advancements such as genomics and machine intelligence. These strategies allow researchers to screen large quantities of compounds for new medicinal targets more quickly and effectively. Drug development, in general, is a key component of the response to infectious disease outbreaks. By identifying novel treatments that target emerging pathogens, it can help to safeguard public health and prevent disease transmission.

An important technology that makes use of a computational approach for the management of problems in medicine and public health is bioinformatics. Bioinformatics is an interdisciplinary field that analyzes and interprets biological data by combining biology, computer science, and statistics. It entails the creation of algorithms and software tools for the analysis and storage of huge amounts of biological data, such as DNA sequences, protein structures, and gene expression data. The study of large-scale biological data, such as genomics, transcriptomics, proteomics, and metabolomics, is referred to as omics. The study of an organism's entire genome is known as genomics, whereas transcriptomics is concerned with the study of all RNA transcripts produced by the genome. The study of all the proteins created by the genome is known as proteomics, while the study of an organism's metabolites is known as metabolomics. Bioinformatics is essential for processing the massive volumes of data generated by omics studies. The data can be interpreted by bioinformaticians to learn more about biological systems and processes by applying computer tools and algorithms. In a nutshell, omics refers to the several kinds of data being evaluated, whereas bioinformatics is the study of biological data.

Bioinformatics tools can be quite beneficial in the treatment of COVID-19. They can be used to examine the genome of the SARS-CoV-2 virus, identify potential therapeutic targets, and produce vaccines. Bioinformatics tools can also be used to evaluate clinical data to improve patient care and treatment. Furthermore, these techniques can aid in the tracking and monitoring of the virus's spread, which is critical for public health activities. COVID-19 can be regulated in a variety of ways using omics disciplines such as genomics, proteomics, and metabolomics. Here are two such cases. Using genomics, the genetic sequence of the SARS-CoV-2 virus, which produces COVID-19, may be discovered. This information can be utilized to develop vaccines and antiviral drugs. When it comes to proteomics, it can be utilized to look at the proteins produced by the SARS-CoV-2 virus and their interactions with human cells. This knowledge can be used to develop personalized medications that prevent the virus from infecting human cells. It can be employed in metabolomics to evaluate the metabolic alterations produced by COVID-19 infection. This information can be used to develop biomarkers that can be used to diagnose and treat disease (Table 20.2) [10–12]. Overall, omics sciences can aid in the development of more potent COVID-19 medications and interventions as well as a better understanding of the disease's biology. It is reasonable to conclude that omics technology, which has been significantly enhanced and expanded in recent decades, is capable of high-throughput detection and analysis of target samples. COVID-19 pathogenesis, potential treatment targets, and diagnostic approaches have all been thoroughly explored using multi-omics-based technologies [10].

Although omics methods such as genomics, transcriptomics, proteomics, and metabolomics have made substantial advances in our understanding of COVID-19, their use is currently limited. One of the major challenges is the cost and time involved in creating the huge datasets required for omics research. As a result, smaller labs or institutions with fewer resources may find it difficult to do this type of study. Furthermore, understanding omics data may be challenging and complicated, necessitating specialized knowledge. Another disadvantage is a lack of consistency in sample collection and processing, which may result in variability and inconsistent data output. As a result, comparing study data and reaching strong conclusions may be difficult. Omics approaches remain a crucial tool for COVID-19 research and management despite these limitations, and work is still being done to resolve these problems and improve the utilization of these technologies.

20.4 MOLECULAR TOOL APPLICATIONS FOR MANAGEMENT OF COVID-19

Techniques known as molecular tools are used to examine the molecular makeup and operation of biological systems. The detection, diagnosis, and monitoring of diseases at the genetic and molecular level are done in clinical medicine and public health using molecular methods. Important molecular tools are presented in Table 20.1. Because they enable the rapid and precise diagnosis of infectious disorders, molecular techniques are particularly significant in infectious medicine. By detecting its genetic material, such as DNA or RNA, molecular instruments can identify the exact bacteria causing the infection. This is especially effective for identifying infectious pathogens that are difficult to culture or are present in small numbers in a patient's sample. Molecular technologies also enable the quick discovery of antibiotic resistance markers, which can aid in treatment decisions and minimize antibiotic misuse [13, 14]. Overall, molecular tools are critical in the diagnosis, treatment, and prevention of infectious diseases [13, 14].

Molecular tools are critical in COVID-19 management because they enable the virus's rapid and accurate detection. Polymerase chain reaction (PCR) tests, for example, use molecular technologies to detect virus genetic material in a patient's sample. Traditional viral culture methods, which might take days to generate findings, are far faster and more precise. Furthermore, molecular technologies can be utilized to investigate the virus's genetic sequence, which will aid scientists in developing

TABLE 20.1

Examples of Important Molecular Tools

Examples	Details
Polymerase chain reaction (PCR)	PCR is a technique for amplifying DNA that involves producing numerous copies of a given DNA sequence. This method detects and diagnoses viral and bacterial illnesses such as HIV, hepatitis B and C, and TB.
Next-generation sequencing (NGS)	NGS is a technique for high-throughput sequencing that can sequence vast amounts of DNA or RNA in a single run. This method is used to detect and diagnose genetic problems, as well as to determine the genetic etiology of diseases.
Microarrays	Microarrays are used to quantify gene expression levels in cells by detecting mRNA levels. This method is used to uncover genetic markers linked to diseases and to design targeted medicines.
CRISPR/Cas9	CRISPR/Cas9 is a gene-editing technique that targets and cuts specific DNA sequences using a bacterial defensive mechanism. This approach has transformed genetic research and holds the promise of curing hereditary illnesses.

targeted therapies and vaccines [15, 16]. Overall, molecular techniques are an important aspect of our attempts to manage and control COVID-19 dissemination [15, 16].

20.5 ADVANTAGES OF MOLECULAR AND BIOINFORMATICS TOOLS IN TRADITIONAL HERBAL MEDICINES

Medical research can make use of both molecular and bioinformatics techniques. The molecular tools are used in either *in vitro* or *in vivo* studies, whilst the bioinformatics tools are used in *in silico* studies. Both types of tools can be employed to accomplish success in research and development (Figure 20.1). The active compounds present in herbs can be identified and characterized using molecular techniques in traditional herbal medicine. By analyzing the molecular structure of these compounds, researchers can gain a better understanding of their pharmacological properties and modes of action. The efficacy and safety of currently available herbal pharmaceuticals can then be enhanced using this knowledge, and innovative drugs based on natural constituents

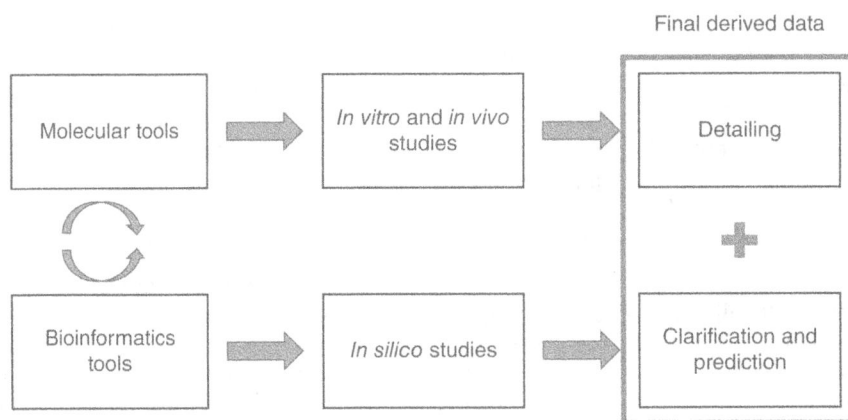

FIGURE 20.1 Conceptual framework for applying molecular and bioinformatics tools for the analysis of anti-SARS-CoV-2 traditional herbal medicines.

can also be developed. To further ensure the quality and safety of herbal products, molecular techniques can be employed to authenticate the identification of the herbs and find any contamination or adulteration.

Likewise, bioinformatic methods can be quite beneficial in traditional herbal medicine. They can aid in our understanding of plant active chemicals and their modes of action, as well as in identifying prospective therapeutic targets and predicting the success of herbal medicines. Bioinformatic methods, for example, can be used to examine plant DNA sequences to identify genes involved in the creation of bioactive chemicals. Researchers can also produce new and more effective herbal treatments by combining knowledge from traditional plant applications with contemporary bioinformatics approaches. Overall, bioinformatic methods can assist us in realizing the full potential of traditional herbal therapy and developing innovative treatments for a variety of health problems.

When it comes to evaluating traditional herbal medications, molecular and bioinformatics technologies have various advantages [10–16]. For starters, it aids in the discovery of active substances. Using molecular methods, researchers may identify active components in traditional herbal treatments. This can aid in a better understanding of how certain medications work. Second, it aids in quality control. The quality of herbal medications can be assessed using bioinformatics technologies. This can aid in the detection of adulteration or contamination, ensuring that the medication is of high quality. Third, it aids in medication interaction prediction. Researchers can identify potential medication interactions between traditional herbal treatments and modern drugs using molecular and bioinformatics technologies. This can aid in avoiding negative consequences. Fourth, it is in charge of new drug development. At the moment, molecular and bioinformatics methods can be utilized to generate new medicines based on traditional herbal remedies. Researchers can design new medications that are more effective and have fewer adverse effects by finding active chemicals and studying their mechanism of action.

20.6 MOLECULAR TOOLS FOR THE ANALYSIS OF ANTI-SARS-CoV-2 TRADITIONAL HERBAL MEDICINES

There are various benefits to using molecular technologies for the investigation of traditional herbal remedies for SARS-CoV-2. First, it enables researchers to pinpoint the precise compounds in herbs that are antiviral against SARS-CoV-2. For those who have the virus, this knowledge can be used to create specialized medications and treatments. Second, molecular analysis helps us comprehend the chemical makeup of conventional herbal medicines more precisely and accurately. This is crucial since these medications frequently contain a number of different chemicals, and determining which ones are in charge of the antiviral effects can assist in increasing their effectiveness and safety. Last but not least, molecular technologies can support efforts to guarantee the purity and quality of herbal medications. Scientists can confirm the identity and provenance of the plants used by employing DNA-based technologies, preventing fraud and adulteration.

The details of how important molecular tools play a role in the analysis of anti-SARS-CoV-2 traditional herbal medicines are provided next.

1. PCR

   PCR is an extremely effective method for detecting and analyzing DNA and RNA in biological materials. When it comes to evaluating traditional herbal medicines for anti-SARS-CoV-2 capabilities, PCR can be extremely useful in identifying the precise chemicals or genetic markers that may be responsible for any reported antiviral effects [17, 18]. PCR can be used to amplify specific DNA or RNA sequences in a sample, which can then be evaluated to see if there are any genetic markers or other molecules of interest present.

This can assist researchers in determining which components in a specific herbal medication may be responsible for its possible antiviral activities [17, 18].

The fact that PCR is extremely sensitive and specific, allowing for the detection and analysis of even very minute amounts of genetic material, is one of the key benefits of utilizing it for this type of analysis. This can be especially helpful when working with herbal remedies that are complicated or poorly defined and may contain a vast variety of various chemicals. Overall, PCR can be a useful tool for scientists trying to better understand the possible antiviral effects of conventional herbal remedies and can aid in the identification of particular chemicals or genetic markers that may call for additional research [19].

2. Next-generation sequencing (NGS)

For the investigation of traditional herbal remedies with potential anti-SARS-CoV-2 efficacy, NGS is a potent technique. Large volumes of genetic data may be quickly and accurately analyzed using NGS, making it possible to pinpoint these drugs' active ingredients and methods of action. We'll go through a few benefits of utilizing NGS to analyze traditional herbal remedies in more detail. It offers high-throughput, to start. In order to identify potential active chemicals more quickly, NGS can process enormous amounts of data in a brief amount of time. Second, NGS is very sensitive and capable of detecting very small amounts of genomic variation, which is crucial for locating the active ingredients in conventional herbal remedies. Third, NGS can produce extremely accurate data, which is crucial for identifying the active ingredients in conventional herbal remedies. Fourth, in terms of cost-effectiveness, NGS is becoming more affordable, making it more available to researchers who want to examine conventional herbal remedies. NGS can also be used to identify the active ingredients and modes of action of conventional herbal remedies with potential anti-SARS-CoV-2 efficacy. This may aid in the creation of novel antiviral therapies for COVID-19 and other viral diseases. The combination of NGS and bioinformatics tools is extremely beneficial in better understanding traditional herbal medicine therapy on clinical difficulties caused by COVID-19. A recent study on the molecular mechanism of Xuebijing injection on COVID-19-induced heart dysfunction is a good example [20].

3. Microarray

Microarray technology is an effective method for studying gene expression patterns in cells or tissues. When it comes to analyzing anti-SARS-CoV-2 traditional herbal treatments, microarrays can help discover active chemicals and their mechanisms of action. It is capable of high-throughput analysis. Microarrays can simultaneously examine hundreds of genes, offering a comprehensive picture of the molecular pathways involved in the response to traditional herbal remedies. Microarrays can also detect small variations in gene expression, which can help identify the precise target genes of anti-SARS-CoV-2 drugs. Furthermore, microarrays can be programmed to target specific groups of genes or pathways, allowing researchers to adapt their analysis to a specific research question. Finally, microarrays can be utilized to quickly and affordably evaluate a variety of substances for possible anti-SARS-CoV-2 action. Overall, microarrays can offer insightful information on the modes of action of conventional herbal remedies against SARS-CoV-2, assisting in the identification of prospective candidates for more research and development. The article by Chen et al. is a nice illustration of research that makes use of microarrays [21]. The primary drug targets for the SARS-CoV-2 virus and the host's anti-viral, anti-inflammatory response targets were assembled into a COVID-19 protein microarray by Chen et al. [21]. A number of quality control tests were conducted, and the results indicated that it might be used to screen bioactive natural materials for therapeutic targets. By creating this microarray, a useful tool for studying the molecular pharmacology of natural compounds will be made available [21].

TABLE 20.2

Potential Roles of CRISPR/Cas9 for the Analysis of anti-SARS-CoV-2 Traditional Herbal Medicines

Examples	Details
Identifying key molecular targets	To investigate the molecular targets of herbal remedies, CRISPR/Cas9 can be utilized to knock out specific genes in SARS-CoV-2 or human cells. This information can assist us in better understanding how these medications work and identifying prospective pharmacological targets.
Creating cellular models	CRISPR/Cas9 technology can be utilized to generate cell lines that produce specific viral proteins or cellular receptors required for SARS-CoV-2 infection. These cell lines can be used to assess herbal medicines' impact on viral propagation or host response.
Optimizing active compounds	We can create plant cells to produce larger levels of certain active chemicals known to have anti-SARS-CoV-2 characteristics using CRISPR/Cas9 genome editing. This can aid in the development of more effective herbal medications.

4. CRISPR/Cas9

A cutting-edge genome editing technique called CRISPR/Cas9 can be used to examine conventional herbal remedies that might be anti-SARS-CoV-2 in nature. Table 20.2 lists several potential applications for CRISPR/Cas9 in this situation. All things considered, CRISPR/Cas9 is an effective technology that can aid in our understanding of the mechanisms of action of conventional herbal remedies and help us maximize their therapeutic potential.

20.7 BIOINFORMATICS TOOLS FOR THE ANALYSIS OF ANTI-SARS-CoV-2 TRADITIONAL HERBAL MEDICINES

When it comes to studying traditional herbal remedies for SARS-CoV-2, bioinformatics has several benefits. In the beginning, it can offer quicker and more precise analyses. Large datasets may be processed with the use of bioinformatics tools, target proteins can be identified, and the effectiveness of various herbal remedies against SARS-CoV-2 can be predicted. When compared to conventional procedures, this can result in significant time and resource savings. Second, it can help us better understand how things work. Researchers can use bioinformatics methods to identify the molecular mechanisms by which various herbal medicines act. This can shed light on the underlying mechanisms of action and assist researchers in making the best use of them. Third, it improves medication development precision. By identifying active chemicals in herbal remedies, bioinformatics techniques can assist researchers in developing more precise and effective treatments. This could lead to more focused medicines with fewer side effects. Overall, bioinformatics technologies provide a formidable toolkit for studying traditional herbal remedies for SARS-CoV-2 treatment, and they can provide useful insights into their mechanisms of action and possible therapeutic advantages.

The details of how important bioinformatics tools play roles in the analysis of anti-SARS-CoV-2 traditional herbal medicines are provided next.

1. Genomics

Comparative genomics is the study and comparison of how different genomes might relate to each other using computer techniques. In essence, comparing genomes can reveal similarities, and microbiology routinely uses this concept [12, 22]. Genomic research is the study of genomes, or the complete collection of DNA that makes up an organism's genetic makeup. The genetic makeup of the SARS-CoV-2 virus, which causes COVID-19, can be analyzed by genomics. Understanding the virus's genetic structure can aid researchers in identifying potential treatment targets. One method that genomics can be employed in the search for anti-COVID-19 drugs is by developing treatments that specifically target specific proteins produced by the virus. Then, they can utilize drugs that precisely target these proteins to inhibit the virus from operating as it ought to. Another way that genomics can be used in the search for anti-COVID-19 drugs is by looking at the DNA of those who have the disease. By analyzing the genetic profiles of these patients, researchers can identify hereditary components that can make a person more susceptible to the disease or result in severe symptoms. This knowledge can aid in the development of medicines that are individually tailored to each patient's genetic makeup.

Genomics can help with the analysis of anti-SARS-CoV-2 traditional herbal treatments. Researchers can pinpoint specific genetic targets important for the virus's ability to infect and reproduce by sequencing the genome of the SARS-CoV-2 virus. This data can then be used to screen traditional herbal remedies for antiviral efficacy against SARS-CoV-2. Furthermore, genomics can be utilized to identify active ingredients in herbal treatments and establish how they work against viruses. This understanding can assist researchers in developing more effective COVID-19 treatments using traditional herbal remedies. Xia et al. published a good example of genomics research. The Traditional Chinese Medicine Systems Pharmacology database and seven disease-gene databases were used in this work to identify a Qingwen capsule (LQC) target and COVID-19-related gene set. The putative mechanism was discovered using gene ontology (GO), the Kyoto Encyclopedia of Genes and Genomes (KEGG) enrichment analysis, and a protein-protein interaction (PPI) network [23]. According to Xia et al., the network pharmacological technique incorporates molecular docking to uncover the molecular mechanism of LQC. AKT1 is a possible therapeutic target for reducing tissue damage and aiding in the elimination of viral infection [23].

2. Pharmacogenomics

A field of study called pharmacogenomics looks at how a person's genetic makeup affects how they respond to drugs. It entails using genetic data to develop personalized treatment plans. Pharmacogenomics can be very helpful in the search for innovative anti-COVID drugs by identifying which drugs work best for particular people based on their genetic makeup [24, 25]. By analyzing genetic variants in patients, researchers can anticipate how different people would respond to specific drugs [24, 25]. This knowledge can be put to use to produce more customized and effective treatments [24, 25]. To ensure patient safety, this approach can also help identify potential side effects or harmful drug combinations. Pharmacogenomics has the potential to help in the fight against COVID-19 and significantly enhance drug development and treatment outcomes [24, 25].

Pharmacogenomics can also be used to analyze anti-SARS-CoV-2 traditional herbal remedies. Researchers can identify specific genes responsible for variances in the efficiency or side effects of traditional herbal treatments by examining genetic variations in people. This information can be used to determine which people are most likely to benefit from the usage of traditional herbal medicines for COVID-19 treatment. Researchers can also optimize the administration and dosage of traditional herbal medicines by identifying the genetic differences that influence responsiveness to them. Overall, pharmacogenomics can offer useful insights into the individualized and targeted use of traditional herbal remedies to treat SARS-CoV-2 illness. The molecular mechanism of Taiwan Chingguan Yihau (NRICM101) was comprehensively studied in Singh and Yang's paper using exploratory bioinformatics and pharmacodynamics techniques [26]. Singh and Yang discovered 434 shared interactions between NRICM101 and COVID-19-related genes/proteins. In the KEGG analysis, the 434 frequent interacting genes/proteins exhibited the strongest correlations with the interleukin -17 signaling pathway for the NRICM101 network pharmacology [26].

3. Proteomics

Proteomics approaches are quite useful in the treatment of COVID-19. By studying the proteins connected to the virus and the host response, we can learn more about disease mechanisms and identify potential targets for treatments and vaccinations. Proteomics, for example, can aid in the identification of viral proteins that cause human cells to become infected, which can benefit the development of antiviral drugs. The identification of proteins involved in the host immune system's response to the virus can contribute to the development of immunotherapies. Furthermore, by monitoring changes in protein expression levels over time, proteomics can be utilized to evaluate disease progression and therapy efficacy. Furthermore, proteomics can be utilized to detect changes in protein expression levels across time. The investigation of traditional herbal remedies that are anti-SARS-CoV-2 can benefit greatly from proteomics. Researchers can pinpoint the proteins in these drugs that might have antiviral efficacy against SARS-CoV-2 by using proteomic approaches. The mechanisms through which these herbal remedies may be helpful against SARS-CoV-2 can potentially be studied using proteomics. Researchers may be able to improve these medications' therapeutic efficacy by better understanding how they function. Proteomics can also be employed to pinpoint prospective pharmacological targets for the creation of fresh antiviral treatments. Researchers can create medicines that target the proteins that interact with SARS-CoV-2 by identifying the proteins in question, which may help them create more potent COVID-19 therapies. Overall, proteomics can provide vital insights into the processes behind traditional herbal medicines' antiviral activity and can be a useful weapon in the fight against SARS-CoV-2. Guo et al. report an intriguing new development using proteomics. Guo et al [27] created the TCMATCOV prediction platform to forecast the efficacy of TCM's anti-coronavirus pneumonia effect, which is based on an interaction network that mimics the illness network of COVID-19 [27]. PPIs of differentially expressed genes in mouse pneumonia caused by SARS-CoV and cytokines specifically up-regulated by COVID-19 were employed to build this COVID-19 network model [27]. To forecast probable medication effects, TCMATCOV used a quantitative evaluation technique of illness network disruption after a multi-target pharmacological attack [27].

4. Metabolomics

Metabolomics is the study of metabolites, which are tiny compounds produced by cells in the body as they execute various metabolic functions. Metabolomics provides insight into the metabolic processes that occur in the body and can be used to identify new pharmaceutical targets and illness biomarkers [12, 15, 16, 22]. In the instance of COVID-19, metabolomics can be used to determine the metabolic signature of the virus and the host cells it infects. This information can be utilized to identify potential treatment targets that could either inhibit the virus or boost the immune system's response to the virus [12, 15, 16, 22]. Metabolomics can potentially be used in drug repurposing activities to examine the efficacy of existing medications against COVID-19.

Metabolomics can be used to analyze traditional herbal treatments that are being studied for their potential usefulness against SARS-CoV-2. Researchers can use metabolomics to discover specific metabolites that are altered in response to a given medicine or treatment. In the case of traditional herbal medicines, metabolomics can aid in identifying the specific active chemicals found in the plants as well as determining how these compounds interact with the body. This can provide vital insights into the potential mechanisms of action of these herbal medications while also assisting in the identification of any potential hazards or side effects.

20.8 CONCLUSIONS

Traditional herbal therapies used to combat SARS-CoV-2 require the use of molecular and bioinformatics methods. Finding active chemicals is the key benefit and role of these instruments.

Bioinformatics technologies can help identify bioactive components in traditional herbal medicines. This allows researchers to study the effects of the chemicals on the virus and determine which compounds may be useful in healing the infection. The safety and toxicity of potential drugs can also be predicted via molecular modeling. This allows researchers to identify potential side effects and devise treatments that are as safe as possible for patients. In addition, molecular modeling can be used to predict which compounds will attach to specific proteins in the virus. This knowledge can be used to develop drugs that target these proteins and impede viral replication. Bioinformatics tools can be used to compare the genetic makeup of different SARS-CoV-2 strains.

REFERENCES

1. Hsia W. Emerging new coronavirus infection in Wuhan, China: Situation in early 2020. Case Study Case Rep. 2020;10:8–9.
2. Yasri S, Wiwanitkit V. Editorial: Wuhan coronavirus outbreak and imported case. Adv Trop Med Pub Health Int. 2019;9:1–2.
3. Ling CQ. Traditional Chinese medicine is a resource for drug discovery against 2019 novel coronavirus (SARS-CoV-2). J Integr Med. 2020;18:87–88.
4. Sriwijitalai W, Wiwanitkit V. Herbs that might be effective for the management of COVID-19: A bioinformatics analysis on anti-tyrosine kinase property. J Res Med Sci. 2020;25:44.
5. Singh PK, Rawat P. Evolving herbal formulations in management of COVID-19. J Ayurveda Integr Med. 2017;8:207–210.
6. Chawla P, Yadav A, Chawla V. Clinical implications and treatment of COVID-19. Asian Pac J Trop Med. 2014;7:169–178.
7. Wu HT, Ji CH, Dai RC, Hei PJ, Liang J, Wu XQ, Li QS, Yang JC, Mao W, Guo Q. Traditional Chinese medicine treatment for COVID-19: An overview of systematic reviews and meta-analyses. J Integr Med. 2022 Sep;20(5):416–426.
8. Ding X, Fan LL, Zhang SX, Ma XX, Meng PF, Li LP, Huang MY, Guo JL, Zhong PZ, Xu LR. Traditional Chinese medicine in treatment of COVID-19 and viral disease: Efficacies and clinical evidence. Int J Gen Med. 2022 Nov 28;15:8353–8363.
9. Rahman MM, Bibi S, Rahaman MS, Rahman F, Islam F, Khan MS, Hasan MM, Parvez A, Hossain MA, Maeesa SK, Islam MR, Najda A, Al-Malky HS, Mohamed HRH, AlGwaiz HIM, Awaji AA, Germoush MO, Kensara OA, Abdel-Daim MM, Saeed M, Kamal MA. Natural therapeutics and nutraceuticals for lung diseases: Traditional significance, phytochemistry, and pharmacology. Biomed Pharmacother. 2022 Jun;150:113041.
10. Lewin R. National academy looks at human genome project, sees progress. Science. 1987;235:747–748.
11. Yu U, Lee SH, Kim YJ, Kim S. Bioinformatics in the post-genome era. J Biochem Mol Biol. 2004;37:75–82.
12. Wiwanitkit V. Utilization of multiple "omics" studies in microbial pathogeny for microbiology insights. Asian Pac J Trop Biomed. 2013 Apr;3(4):330–333.
13. Foxman B. Infectious diseases. IARC Sci Publ. 2011;163:421–440.
14. Sellon RK. Update on molecular techniques for diagnostic testing of infectious disease. Vet Clin North Am Small Anim Pract. 2003 Jul;33(4):677–693.
15. Sreepadmanabh M, Sahu AK, Chande A. COVID-19: Advances in diagnostic tools, treatment strategies, and vaccine development. J Biosci. 2020;45(1):148.
16. Tsai SC, Lu CC, Bau DT, Chiu YJ, Yen YT, Hsu YM, Fu CW, Kuo SC, Lo YS, Chiu HY, Juan YN, Tsai FJ, Yang JS. Approaches towards fighting the COVID-19 pandemic (review). Int J Mol Med. 2021 Jan;47(1):3–22.
17. Ji Z, Hu H, Qiang X, Lin S, Pang B, Cao L, Zhang L, Liu S, Chen Z, Zheng W, Liu C, Wang H, Zhang J. Traditional Chinese medicine for COVID-19: A network meta-analysis and systematic review. Am J Chin Med. 2022;50(4):883–925.
18. Luo X, Ni X, Lin J, Zhang Y, Wu L, Huang D, Liu Y, Guo J, Wen W, Cai Y, Chen Y, Lin L. The add-on effect of Chinese herbal medicine on COVID-19: A systematic review and meta-analysis. Phytomedicine. 2021 May;85:153282.
19. Nagendran M, John J, Annamalai K, Gandhi Sethuraman MI, Balamurugan N, Rajendran HK, Deen Fakrudeen MA, Chandrasekar R, Ranjan S, Padmanaban VC. Can human overcome viral hijack-? Comprehensive review on COVID-19 in the view of diagnosis and mitigation across countries. J Drug Deliv Sci Technol. 2021 Feb;61:102120.

20. He DD, Zhang XK, Zhu XY, Huang FF, Wang Z, Tu JC. Network pharmacology and RNA-sequencing reveal the molecular mechanism of Xuebijing injection on COVID-19-induced cardiac dysfunction. Comput Biol Med. 2021 Apr;131:104293.

21. Chen P, Zeng Z, Du H. Establishment and validation of a drug-target microarray for SARS-CoV-2. Biochem Biophys Res Commun. 2020 Sep 10;530(1):4–9.

22. Wiwanitkit V. Application of bioinformatics in tropical medicine. Asian Pac J Trop Med. 2008 Dec;1(4):72–75.

23. Xia QD, Xun Y, Lu JL, Lu YC, Yang YY, Zhou P, Hu J, Li C, Wang SG. Network pharmacology and molecular docking analyses on Lianhua Qingwen capsule indicate Akt1 is a potential target to treat and prevent COVID-19. Cell Prolif. 2020;53(12):e12949.

24. Takahashi T, Luzum JA, Nicol MR, Jacobson PA. Pharmacogenomics of COVID-19 therapies. NPJ Genom Med. 2020 Aug 18;5:35.

25. Fricke-Galindo I, Falfán-Valencia R. Pharmacogenetics approach for the improvement of COVID-19 treatment. Viruses. 2021 Mar 5;13(3):413.

26. Singh S, Yang YF. Pharmacological mechanism of NRICM101 for COVID-19 treatments by combined network pharmacology and pharmacodynamics. Int J Mol Sci. 2022 Dec 6;23(23):15385. doi: 10.3390/ijms232315385

27. Guo FF, Zhang YQ, Tang SH, Tang X, Xu H, Liu ZY, Huo RL, Li D, Yang HJ. TCMATCOV–a bioinformatics platform to predict efficacy of TCM against COVID-19. Zhongguo Zhong Yao Za Zhi. 2020 May;45(10):2257–2264.

21 Risk Assessment of COVID-19-Infected Patients and Their Dietary Modulation

Ravindra Verma, Ashwin Kotnis,
Prakash S. Bisen, and Mònica Bulló

21.1 INTRODUCTION

Worldwide, the Severe Acute Respiratory Syndrome Coronavirus-2 (SARS-CoV-2) pandemic spread rapidly. Public health concerns were raised during the pandemic due to the emergence of new lineages carrying mutations in the spike (S) protein. It was found that the Alpha variant originated in the United Kingdom, Beta in South Africa, Gamma in Brazil, Delta in India, Omicron in South Africa, Mu in Colombia, Epsilon in the United States, and Lambda in Peru (Wolf et al., 2023). Like many infectious diseases, COVID-19 affects the inflammatory response pathway. Although diet could play a role in the etiology of COVID-19, most of the research on diet and nutrition is involved in managing COVID-19 symptoms (Khan, 2021; Liu et al., 2021). A healthy immune response can be triggered by food, thereby preventing and reducing the incidence of pathogenic diseases. Our daily food supplements contain a variety of functional foods with antiviral properties that can improve human immunity during COVID-19 (Verma & Bagel, 2022). The human body has two types of immune systems: innate and adaptive. An innate immune system is thought to have preceded the adaptive immune response in multicellular organisms. All multicellular organisms have an innate immune system, whereas vertebrates have adaptive immunity (Medzhitov & Janeway, 1997). Infection activates immune cells, leading to their proliferation, differentiation, and ultimately death. There are multiple organ systems and cells that are involved in the immune response. When immune cells respond to infection, they also need to decide whether to function as effectors triggering inflammation or suppressors inhibiting it. A type of infection, such as a virus, a bacterial infection, or a parasite, requires specific effector cells with specific abilities. As a result, immunity depends on rapid changes in cell fate. Although signaling pathways take detours through metabolic machinery to ensure metabolism cooperates in the cell-fate decision, recent research has discovered that cell-fate decisions play a crucial role in directing signaling from the surface to the nucleus (Pearce et al., 2013; Weinberg et al., 2015). Immunity is significantly influenced by nutrition and diet. Diet not only affects immunity development but also maintains immunity(Pahwa & Sharan, 2022). In COVID-19, the lungs are severely affected due to alveoli collapse, diffuse alveolar damage, and the formation of hyaline membranes. Human angiotensin-converting enzyme II (ACE2) receptors are used by both SARS-CoV and SARS-CoV-2 for virus entry and infectivity. In addition to lung epithelia, SARS-CoV-2 can also infect intestinal and other epithelial cells, reproducing actively and producing virus *de novo* (Osuchowski et al., 2021). Activating the immune system and producing antibodies are the only ways for a patient to recover. In several studies, food additives have been evaluated for their potential to prevent viral infections like SARS-CoV-2 (Rahman et al., 2021; Singh et al., 2020). A two-phase treatment approach is recommended for COVID-19 patients, based on Shi et al.(2020): immune protection followed by inflammation-driven damage. As a result of COVID-19's genetic profile, two types of therapy will be offered: immune boosting and inflammatory suppression. The immune storm (cytokine storm) that results in organ damage is caused by COVID-19. The immune system overreacts during a cytokine storm to prevent the replication of

DOI: 10.1201/9781003452621-21

viruses. In one study (https://clinicaltrials.gov/ct2/show/NCT04790240), herbs were found to inhibit inflammation and direct T cells to kill the COVID-19 virus.

21.2   RISK ASSESSMENT OF COVID-19 INFECTED PATIENTS

Coronavirus infection is not fully understood; some people are critically infected, while others are not (Verma et al., 2021). Perhaps the patient's overall immunity is the dominant factor. The risks associated with a pathogen are assessed in a formal risk assessment, which looks at the pathogen's interactions with humans and the environment as well. For biomedical research, risk assessment is based on a number of well-defined criteria, including pathogens, humans, and contexts or environments that influence public-health policy decisions, including healthcare decisions (Schröder, 2020). The National Institutes of Health (NIH) and the World Health Organization (WHO) categorize pathogens and toxins produced by certain pathogens into four risk groups based on the severity of diseases caused by specific pathogens, the adverse effects of certain pathogens on communities, and the availability of preventative measures and treatments (https://osp.od.nih.gov, https://www.phe.gov). Pathogens classified as Risk Group 3 include Middle East respiratory syndrome corona virus (MERS)-CoV and SARS-CoV-1. It is apparent that SARS-CoV-2 poses a very high risk to the community due to its rapid spread around the world, as well as its impact on human health and the global economy. The high risk SARS-CoV-2 poses to the community make it comparable to pathogens of Risk Group 4. Since no Risk Group has yet been assigned to SARS-CoV-2, the NIH has developed interim guidelines for handling specimens that may contain the virus (https://osp.od.nih.gov/policies/biosafety-and-biosecurity-policy/interim-laboratory-biosafety-guidance-for-research-with-sars-cov-2-and-ibc-requirements-under-the-nih-guidelines/). It has been found that influenza and Coronavirus outbreaks are associated with two factors: proximity to livestock and wild animals, which facilitates interspecies transmission, and travel. COVID-19 outbreaks differ, however, from those of SARS and MERS due to a number of factors. Infection rates were recently estimated to be 5.7 for SARS-CoV-2, viral shed peaked before symptoms appeared, and a large number of asymptomatic carriers were still infectious (Schröder, 2020). COVID-19 can easily spread undetected in societies with high population density and mobility.

21.3   MOLECULAR MECHANISM OF SARS-CoV-2 INTERACTION

The surface spike (S) glycoprotein on Coronavirus envelopes is responsible for attaching to host cells and mediating membrane fusion between the host and virus. As mentioned previously, the spike protein contains two regions—S1 and S2. S1 contains a host cell receptor binding domain (RBD), while S2 contains a membrane fusion domain. The S1 region also has an N-terminal domain (NTD) and three C-terminal domains (CTD1, CTD2, and CTD3) (He et al., 2020). SARS-CoV-2 attaches to the human host cells by binding to the ACE2 by the RBD in the CTD1 of the S1 region (Jackson et al., 2022; Yan et al., 2021). In conclusion, RBD-ACE2 interactions are needed for human infection by SARS-CoV-2. Xu et al. have used Molecular Operating Environment (MOE-2019) to calculate the binding free energies between the Spike-RBD protein and human ACE2, showing that the binding free energy between SARS-CoV-2 RBD and human ACE2 was -50.6 kcal/mol, whereas that between SARS-CoV RBD and human ACE2 was -78.6 kcal/mol (Xu et al., 2020) (Figure 21.1).

21.4   VACCINATION AND SARS-CoV-2

Many RNA viruses, including SARS-CoV-2, are enveloped with a major surface glycoprotein (in this case the Spike protein) that binds to angiotensin-converting enzyme 2 (ACE2), the virus' cell receptor, and is the main target of antibodies that neutralize it. The most potent neutralizing antibodies target the RBD of the S-protein (Kleanthous et al., 2021). ACE2 and antibodies can interact differently with the S-protein when it is altered by mutations, as mutations in the S-protein can change their interaction (Barrett et al., 2022) (Figure 21.2).

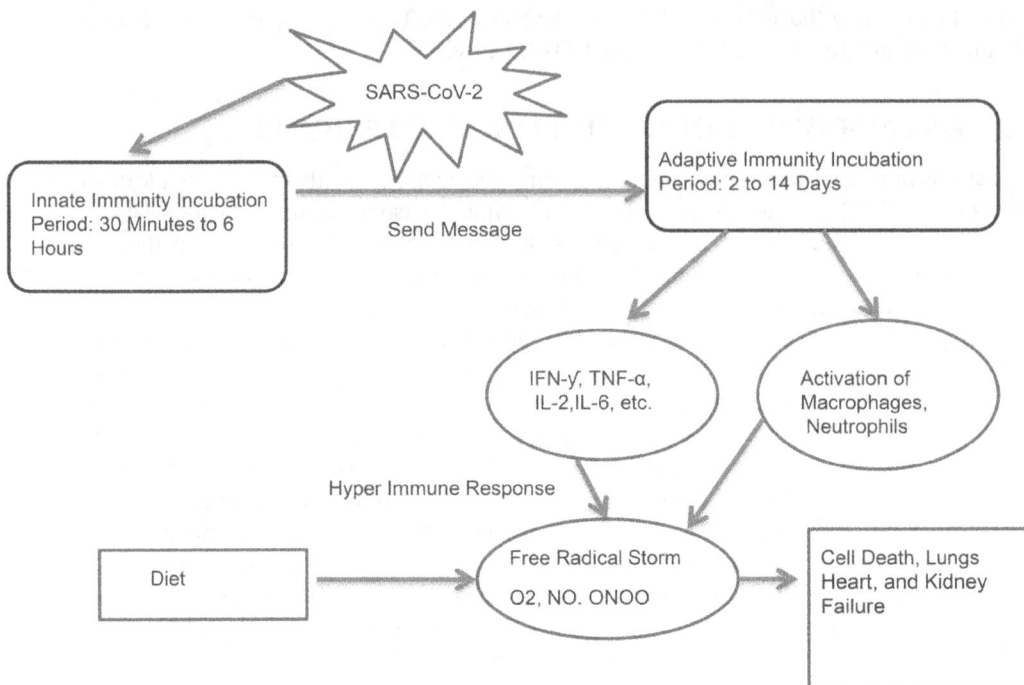

FIGURE 21.1   Immune response activation against SARS-CoV-2.

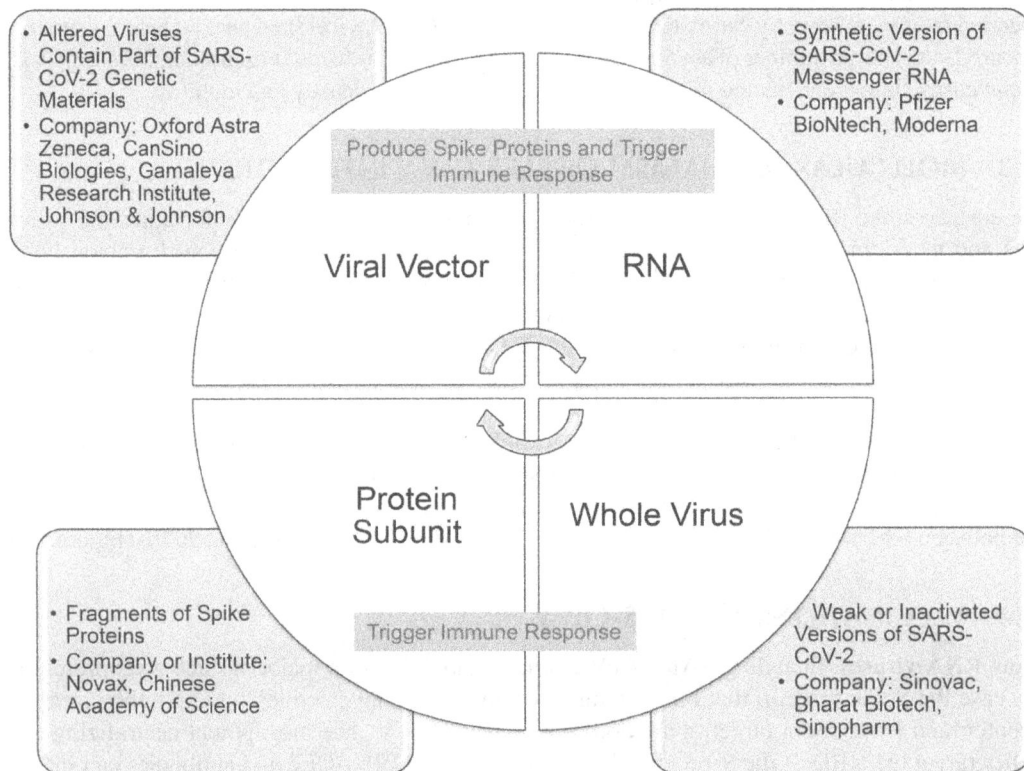

FIGURE 21.2   Vaccine technology for COVID-19.

SARS-CoV-2 vaccines have been developed by some companies using mRNA, while others use inactive cells from the virus. Linares-Fernández et al.(2020) demonstrate that mRNA vaccines activate antigen-presenting cells (APCs) or inhibit mRNA translation. SARS-CoV-2 infection can only be reversed by activating an immune response. To overcome the infection ofSARS-CoV-2, intracellular signaling pathways that modulate the host immune response are most relevant for drug investigations (Catanzaro et al., 2020). Inflammatory symptoms and viral load influence the host immune response (Bartsch et al., 2021). There is no current evidence that immunity from symptomatic disease can last beyond the current follow-up period. However, vaccination can provide protection against symptomatic disease at least for the time being (Krammer, 2021). The prevalence of reinfection is lower in people who have been infected than in those who have not (Hall et al., 2021; Hansen et al., 2021; Harvey et al., 2021; Lumley et al., 2021).

21.5 DIETARY MODULATION OF COVID-19-INFECTED PATIENTS

A significant relationship exists between nutritional status, immune response, and severity of COVID-19 clinical manifestations, confirming that nutrition status, immune response, and severity are all related. Across all stages of the disease phase, COVID-19 has shown that this relationship is critical, especially in obese, undernourished, or older patients with poor prognosis. The immune system is largely responsible for malnutrition as well as its consequences, such as muscle loss and impaired physical function. Even long-term intensive care can lead to malnutrition. It has been demonstrated that SARS-CoV-2 infections can aggravate catabolic processes and cachexia as a result of altered inflammatory responses. As a result, malnutrition may worsen and patients may lose their independence in daily life, suffer from depression or disability, and generally have a lower quality of life after leaving the ICU (de AraújoMorais et al., 2021; Mentella et al., 2021). Dysfunction of the immune system and metabolic diseases have been linked to alterations in nutritional status. The immune system is altered by these conditions, which makes viral infections like COVID-19 more likely to spread. COVID-19 can be caused by a variety of risk factors, including unbalanced diets. In addition to facilitating adequate nutritional status and providing optimal amounts of micronutrients, eating a balanced diet could provide beneficial outcomes both in prevention and treatment (Renata et al., 2023). It may be possible to develop an impaired nutritional status as a result of changes in dietary habits and lifestyle parameters as a result of quarantine and social isolation. Infections and transmission of COVID-19 are increased by obesity and related comorbidities, which result in physiological alterations. As the COVID-19 pandemic is not expected to end soon, people should be encouraged to improve their lifestyles to lessen their risk in both the current and likely subsequent waves. A healthy lifestyle can not only help patients with COVID-19 ensure optimal immune response but also prevent and/or treat undernutrition, obesity, and obesity-related comorbidities (Silverio et al., 2021).

21.6 SELECTED FUNCTIONAL FOODS AND MEDICINAL PLANTS FOR COVID-19 INFECTED PATIENTS

The immune response to pathogenic viruses can be modulated by functional food additives (Verma & Bagel, 2022). Adaptive as well as inherent immune responses to SARS-CoV-2 are affected by food additives (Wheatley et al., 2021).

21.6.1 Turmeric

Researchers have shown that curcumin enhances the production of antibody (IgG), promotes T-cell proliferation induced by Polyhydroxyalkanoate (PHA), produces interleukin (IL)-2, promotes delayed-type hypersensitivity, and activates natural killer cells (South et al., 1997; Yadav et al., 2005;

Zahedipour et al., 2020). According to a study, curcumin modulates levels of IL-1, IL-6, and tumor necrosis factor (TNF)-α (Brendler et al., 2021). Non-structural protein 9 (Nsp9), the replication protein of SARS-CoV-2, and curcumin have multiple reciprocals, which confirms the protective effects of curcumin against inflammation caused by COVID-19 (Kumar et al., 2021). Inflammatory immune responses are modulated by curcumin, which inhibits Toll-like receptors like TLR2, TLR4, and TLR9 (Boozari et al., 2019; Dhar & Bhattacharjee, 2021). ACE gene expression can be increased or decreased by curcumin (Fazal et al., 2015). The antiviral and anti-inflammatory properties of curcumin have been reported to inhibit the spread of Coronavirus, an emerging virus. Curcumin modulates intracellular signaling pathways (Zahedipour et al., 2020).

21.6.2 CINNAMON

In bronchoalveolar lavage fluid, cinnamon's main constituent, trans-cinnamaldehyde (TCA), inhibits influenza A/PR/8 virus maturation. An herbal formulation containing *Cinnamomi cortex* was recommended for acute respiratory viral infections (Hayashi et al., 2007). According to Badger-Emeka et al. (2020), cinnamaldehyde is a standard allergen, which causes the release of histamine and stimulates the immune system.

21.6.3 GINGER

The therapeutic and pharmaceutical properties of ginger have long been known. Numerous health problems can be treated with them, including viral infections and gastric ulcers. The compounds shogaol and gingerol may provide relief from cold symptoms including inflammation, fever, pain, and coughing. The clinical and pharmaceutical applications and properties of ginger compounds such as gingerol, shogaols, zingiberene, zingerone, and paradols are receiving attention. In addition to (NF-κB) and immune activation, ginger inhibits many other cellular processes during inflammatory responses. The interaction of ginger compounds with Coronavirus spike proteins has become increasingly clear. The SARS-CoV-2 virus needs both papain-like protease(PLpro) and chymotrypsin-like protease(3CLPro) to survive and reproduce(Li & De Clercq, 2020). In a docking study, 8-Gingerol, 10-Gingerol, and 6-Gingerol have high binding affinity for PLpro, suggesting they may represent potent inhibitors of this enzyme (Goswami et al., 2020). A study showed that ginger components are highly compatible with CoV-2's main protease (Mpro), the virus involved in the transmission of SARS (Jahan et al., 2021).

21.6.4 *GLYCYRRHIZA GLABRA* (LICORICE)

A traditional use of *licorice infusion* is to treat respiratory conditions, digestive disorders, and viral infections (Fiore et al., 2005). It contains immune stimulants, antioxidants, and free radical scavengers (Cheel et al., 2010). Pompei et al. (1979) reported that DNA and RNA viruses could be replicated more efficiently with extracts of glycyrrhizic acid. Glycyrrhizin is a potential inhibitor of SARS-CoV-2, inhibiting the virus' main protease Mpro (van de Sand et al., 2021). It has a noticed effect on TNF-α, IL-6, iNOS, and COX-2 levels (Wang et al., 2015). According to Zhang and Liu (2020), the Chinese used glycyrrhizin to treat COVID-19. The glycyrrhizin binds to the receptor ACE2, which prevents the entry of SARS-CoV-2 into the cell (Gomaa & Abdel-Wadood, 2021;Gowda et al., 2021; Luo et al., 2020).

21.6.5 *PHYLLANTHUS EMBLICA*

Phyllanthus emblica was reported to stimulate lymphocyte proliferation and scavenge nitric acid (NO) (Kumaran & Karunakaran, 2006; Ram et al., 2002). *P. emblica* may possess evidence-based drug glycoproteins (EBDgp) that may be effective for treating SARS-CoV-2 RNA-dependent RNA polymerase

(RdRp) (Pandey & Lokhande, 2021). It has been found that *P. emblica* contains phytocompounds that act as antimicrobials, antioxidants, anti-inflammatory, analgesics and antipyretics, adaptogenic, hepato-protectives, antitumor, and antiulcerogenic (Estari et al., 2012; Gaire & Subedi, 2014; Yadav et al., 2017).

21.6.6 GARLIC

In addition to neutralizing viruses, garlic is a natural antibiotic. Organosulfurs (such as allicin) and flavonoids (such as quercetin) modulate immunity (Khubber et al., 2020). In fresh garlic, S-allyl-L-cysteine sulfoxide (alliin) and mutant cysteine derivatives are major components (Arreola et al., 2015). In a study conducted by Feng et al. (2012), allicin from garlic augments interferon alpha (IFN-α) - and stimulates CD4+ T-cell and macrophage expansion in malaria infections. Aside from stimulating lymphocyte proliferation and macrophage phagocytosis, it is also believed that garlic promotes the growth of lymphokine-activated killer cell functions and the activity of natural killer cells by high levels of lymphokines (Lamm & Riggs, 2001).

21.7 CONCLUSION

A variety of phytocompounds are found in the previously mentioned plants, and they inhibit the growth and spread of pathogenic bacteria. Food additives not only enhance immunity but also promote health. A number of studies have found that functional food additives stimulate the immune system, but the mechanism by which this occurs is unclear. As well as understanding how additives in food affect immune responses and the behavior of Coronavirus, it is crucial to understand how food can be used to prevent COVID-19 disease. A number of studies have concluded that food additives influence the adaptive immune system as well as the inherent immune system. An adaptive immune response enhances the body's ability to fight SARS-CoV-2-induced inflammation. In addition to CD8+ cytotoxic T cells and antibodies, alpha interferon, TNF-α, IL-12, and IL-6 are secreted as a result of the virus entering the cell through the ACE2 receptor and being recognized by IL-6 in the endosome. Detecting and countering pathogenic attacks, B cell receptors provide long-term protection by producing antibodies against pathogens. To develop a drug suitable for Coronavirus, experimental studies and simulations of docked proteins would need to be carried out.

REFERENCES

Arreola, R., Quintero-Fabián, S., López-Roa, R. I., Flores-Gutiérrez, E. O., Reyes-Grajeda, J. P., Carrera-Quintanar, L., & Ortuño-Sahagún, D. (2015). Immunomodulation and anti-inflammatory effects of garlic compounds. *Journal of Immunology Research*, *2015*(1), p. 401630.

Badger-Emeka, L.I., Emeka, P.M., Thirugnanasambantham, K., & Ibrahim, H. I. M. (2020). Anti-allergic potential of cinnamaldehyde via the inhibitory effect of histidine decarboxylase (HDC) producing klebsiella pneumonia. *Molecules*, *25*(23), 5580.

Barrett, A. D., Titball, R. W., MacAry, P. A., Rupp, R. E., von Messling, V., Walker, D. H., & Fanget, N. V. (2022). The rapid progress in COVID vaccine development and implementation. *NPJ Vaccines*, *7*(1), 20.

Bartsch, Y. C., Fischinger, S., Siddiqui, S. M., Chen, Z., Yu, J., Gebre, M., ... Alter, G. (2021). Discrete SARS-CoV-2 antibody titers track with functional humoral stability. *Nature Communications*, *12*(1), 1018.

Boozari, M., Butler, A. E., & Sahebkar, A. (2019). Impact of curcumin on toll-like receptors. *Journal of Cellular Physiology*, *234*(8), 12471–12482.

Brendler, T., Al-Harrasi, A., Bauer, R., Gafner, S., Hardy, M. L., Heinrich, M., ... Williamson, E. M. (2021). Botanical drugs and supplements affecting the immune response in the time of COVID-19: Implications for research and clinical practice. *Phytotherapy Research*, *35*(6), 3013–3031.

Catanzaro, M., Fagiani, F., Racchi, M., Corsini, E., Govoni, S., & Lanni, C. (2020). Immune response in COVID-19: Addressing a pharmacological challenge by targeting pathways triggered by SARS-CoV-2. *Signal Transduction and Targeted Therapy*, *5*(1), 84.

Cheel, J., Van Antwerpen, P., Tůmová, L., Onofre, G., Vokurková, D., Zouaoui-Boudjeltia, K.... Nève, J. (2010). Free radical-scavenging, antioxidant and immunostimulating effects of a licorice infusion (*Glycyrrhiza glabra L.*). *Food Chemistry*, *122*(3), 508–517.

de Araújo Morais, A. H., de Souza Aquino, J., da Silva-Maia, J. K., de Lima Vale, S. H., Maciel, B. L. L., & Passos, T. S. (2021). Nutritional status, diet and viral respiratory infections: Perspectives for severe acute respiratory syndrome coronavirus 2. *British Journal of Nutrition, 125*(8), 851–862.

Dhar, S., & Bhattacharjee, P. (2021). Promising role of curcumin against viral diseases emphasizing COVID-19 management: A review on the mechanistic insights with reference to host-pathogen interaction and immunomodulation. *Journal of Functional Foods, 82*, 104503.

Estari, M., Venkanna, L., Sripriya, D., & Lalitha, R. (2012). Human immunodeficiency virus (HIV-1) reverse transcriptase inhibitory activity of *Phyllanthus emblica* plant extract. *Biology and Medicine, 4*(4), 178.

Fazal, Y., Fatima, S. N., Shahid, S. M., & Mahboob, T. (2015). Effects of curcumin on angiotensin-converting enzyme gene expression, oxidative stress and anti-oxidant status in thioacetamide-induced hepatotoxicity. *Journal of the Renin-Angiotensin-Aldosterone System, 16*(4), 1046–1051.

Feng, Y., Zhu, X., Wang, Q., Jiang, Y., Shang, H., Cui, L., & Cao, Y. (2012). Allicin enhances host pro-inflammatory immune responses and protects against acute murine malaria infection. *Malaria Journal, 11*, 1–9.

Fiore, C., Eisenhut, M., Ragazzi, E., Zanchin, G., &Armanini, D. (2005). A history of the therapeutic use of liquorice in Europe. *Journal of Ethnopharmacology, 99*(3), 317–324.

Gaire, B. P., & Subedi, L. (2014). Phytochemistry, pharmacology and medicinal properties of *Phyllanthus emblica Linn. Chinese Journal of Integrative Medicine*, 1–8. https://doi.org/10.1007/s11655-014-1984-2

Gomaa, A. A., & Abdel-Wadood, Y. A (2021). The potential of glycyrrhizin and licorice extract in combating COVID-19 and associated conditions. *Phytomedicine Plus, 1*(3), 100043.

Goswami, D., Kumar, M., Ghosh, S.K., & Das, A. (2020). Natural product compounds in *Alpinia officinarum* and ginger are potent SARS-CoV-2 papain-like protease inhibitors. *Theoretical and Computational Chemistry (Preprint).* doi.10.26434/chemrxiv.12071997.v1

Gowda, P., Patrick, S., Joshi, S. D., Kumawat, R. K., & Sen, E. (2021). Glycyrrhizin prevents SARS-CoV-2 S1 and Orf3a induced high mobility group box 1 (HMGB1) release and inhibits viral replication. *Cytokine, 142*, 155496.

Hall, V. J., Foulkes, S., Charlett, A., Atti, A., Monk, E. J., Simmons, R., … Cowley, A. (2021). SARS-CoV-2 infection rates of antibody-positive compared with antibody-negative health-care workers in England: A large, multicentre, prospective cohort study (SIREN). *The Lancet, 397*(10283), 1459–1469.

Hansen, C. H., Michlmayr, D., Gubbels, S. M., Mølbak, K., & Ethelberg, S. (2021). Assessment of protection against reinfection with SARS-CoV-2 among 4 million PCR-tested individuals in Denmark in 2020: A population-level observational study. *The Lancet, 397*(10280), 1204–1212.

Harvey, R. A., Rassen, J. A., Kabelac, C. A., Turenne, W., Leonard, S., Klesh, R., … Penberthy, L. T. (2021). Association of SARS-CoV-2 seropositive antibody test with risk of future infection. *JAMA Internal Medicine, 181*(5), 672–679.

Hayashi, K., Imanishi, N., Kashiwayama, Y., Kawano, A., Terasawa, K., Shimada, Y., &Ochiai, H. (2007). Inhibitory effect of cinnamaldehyde, derived from *Cinnamomicortex*, on the growth of influenza A/PR/8 virus in vitro and in vivo. *Antiviral Res, 74*(1), 1–8.

He, J., Tao, H., Yan, Y., Huang, S. Y., &Xiao, Y. (2020). Molecular mechanism of evolution and human infection with SARS-CoV-2. *Viruses, 12*(4), 428.

Jackson, C. B., Farzan, M., Chen, B., & Choe, H. (2022). Mechanisms of SARS-CoV-2 entry into cells. *Nature Reviews Molecular Cell Biology, 23*(1), 3–20.

Jahan, R., Paul, A.K., Bondhon, T.A., Hasan, A., Jannat, K., Mahboob, T., Nissapatorn, V., Pereira, M.L., Wiart, C., Wilairatana, P., & Rahmatullah, M. (2021). *Zingiber officinale*: Ayurvedic uses of the plant and in silico binding studies of selected phytochemicals with Mpro of SARS-CoV-2. *Natural Product Communications, 16*(10), 1934578X211031766.

Khan, E. (2021). Heart failure and COVID-19: Synergism of two inflammatory conditions?*British Journal of Community Nursing, 26*(1), 18–25.

Khubber, S., Hashemifesharaki, R., Mohammadi, M., & Gharibzahedi, S. M. T. (2020). Garlic (*Allium sativum L.*): A potential unique therapeutic food rich in organosulfur and flavonoid compounds to fight with COVID-19. *Nutrition Journal, 19*, 1–3.

Kleanthous, H., Silverman, J. M., Makar, K. W., Yoon, I. K., Jackson, N., & Vaughn, D. W. (2021). Scientific rationale for developing potent RBD-based vaccines targeting COVID-19. *NPJ Vaccines, 6*(1), 128.

Krammer, F. (2021). Correlates of protection from SARS-CoV-2 infection. *The Lancet, 397*(10283), 1421–1423.

Kumaran, A., & Karunakaran, R. J. (2006). Nitric oxide radical scavenging active components from *Phyllanthus emblica L. Plant Foods for Human Nutrition, 61,* 1–5.

Kumar, M., Sodhi, K. K., & Singh, D. K. (2021). Addressing the potential role of curcumin in the prevention of COVID-19 by targeting the Nsp9 replicase protein through molecular docking. *Archives of Microbiology, 203,* 1691–1696.

Lamm, D. L., & Riggs, D. R. (2001). Enhanced immunocompetence by garlic: Role in bladder cancer and other malignancies. *The Journal of Nutrition, 131*(3), 1067S–1070S.

Li, G., & De Clercq, E. (2020). Therapeutic options for the 2019 novel coronavirus (2019-nCoV). *Nature Reviews Drug Discovery, 19*(3), 149–150.

Linares-Fernández, S., Lacroix, C., Exposito, J. Y., & Verrier, B. (2020). Tailoring mRNA vaccine to balance innate/adaptive immune response. *Trends in Molecular Medicine, 26*(3), 311–323.

Liu, H., Zhou, L., Wang, H., Wang, X., Qu, G., Cai, J., & Zhang, H. (2021). Malnutrition is associated with hyperinflammation and immunosuppression in COVID-19 patients: A prospective observational study. *Nutrition in Clinical Practice, 36*(4), 863–871.

Lumley, S. F., O'Donnell, D., Stoesser, N. E., Matthews, P. C., Howarth, A., Hatch, S. B., … Eyre, D. W. (2021). Antibody status and incidence of SARS-CoV-2 infection in health care workers. *New England Journal of Medicine, 384*(6), 533–540.

Luo, P., Liu, D., & Li, J. (2020). Pharmacological perspective: Glycyrrhizin may be an efficacious therapeutic agent for COVID-19. *International Journal of Antimicrobial Agents, 55*(6), 105995.

Medzhitov, R., & Janeway, C. A., Jr. (1997). Innate immunity: Impact on the adaptive immune response. *Current Opinion in Immunology, 9*(1), 4–9.

Mentella, M. C., Scaldaferri, F., Gasbarrini, A., & Miggiano, G. A. D. (2021). The role of nutrition in the COVID-19 pandemic. *Nutrients, 13*(4), 1093.

Osuchowski, M. F., Winkler, M. S., Skirecki, T., Cajander, S., Shankar-Hari, M., Lachmann, G.,… Rubio, I. (2021). The COVID-19 puzzle: Deciphering pathophysiology and phenotypes of a new disease entity. *The Lancet Respiratory Medicine, 9*(6), 622–642.

Pahwa, H., & Sharan, K. (2022). Food and nutrition as modifiers of the immune system: A mechanistic overview. *Trends in Food Science & Technology, 123,* 393–403.

Pandey, K., Lokhande, K. B., Swamy, K. V., Nagar, S., & Dake, M. (2021). In silico exploration of phytoconstituents from *Phyllanthus emblica* and *Aegle marmelos* as potential therapeutics against SARS-CoV-2 RdRp. *Bioinformatics and Biology Insights, 15,* 11779322211027403.

Pearce, E. L., Poffenberger, M. C., Chang, C. H., & Jones, R. G. (2013). Fueling immunity: Insights into metabolism and lymphocyte function. *Science, 342*(6155), 1242454.

Pompei, R., Flore, O., Marccialis, M. A., Pani, A., & Loddo, B. (1979). Glycyrrhizic acid inhibits virus growth and inactivates virus particles. *Nature, 281*(5733), 689–690.

Rahman, M. M., Mosaddik, A., & Alam, A. K. (2021). Traditional foods with their constituent's antiviral and immune system modulating properties. *Heliyon, 7*(1), e05957.

Ram, M. S., Neetu, D., Yogesh, B., Anju, B., Dipti, P., Pauline, T., … Selvamurthy, W. (2002). Cyto-protective and immunomodulating properties of Amla (*Emblica officinalis*) on lymphocytes: An in-vitro study. *Journal of Ethnopharmacology, 81*(1), 5–10.

Renata, R. B. N., Arely, G. R. A., Gabriela, L. M. A., & Esther, M. L. M. (2023). Immunomodulatory role of microelements in COVID-19 outcome: A relationship with nutritional status. *Biological Trace Element Research, 201*(4), 1596–1614.

Schröder, I. (2020). COVID-19: A risk assessment perspective. *ACS Chemical Health & Safety, 27*(3), 160–169.

Science Safety Security. https://www.phe.gov/s3/BioriskManagement/biosafety/Pages/Risk-Groups.aspx (accessed June 14, 2023).

Shi, Y., Wang, Y., Shao, C., Huang, J., Gan, J., Huang, X., … Melino, G. (2020). COVID-19 infection: The perspectives on immune responses. *Cell Death & Differentiation, 27*(5), 1451–1454.

Silverio, R., Gonçalves, D. C., Andrade, M. F., & Seelaender, M. (2021). Coronavirus disease 2019 (COVID-19) and nutritional status: The missing link?*Advances in Nutrition, 12*(3), 682–692.

Singh, P., Tripathi, M. K., Yasir, M., Khare, R., Tripathi, M. K., & Shrivastava, R. (2020). Potential inhibitors for SARS-CoV-2 and functional food components as nutritional supplement for COVID-19: A review. *Plant Foods for Human Nutrition, 75,* 458–466.

South, E. H., Exon, J. H., & Hendrix, K. (1997). Dietary curcumin enhances antibody response in rats. *Immunopharmacology and Immunotoxicology, 19*(1), 105–119.

van de Sand, L., Bormann, M., Alt, M., Schipper, L., Heilingloh, C. S., Steinmann, E., ... Krawczyk, A. (2021). Glycyrrhizin effectively inhibits SARS-CoV-2 replication by inhibiting the viral main protease. *Viruses*, *13*(4), 609.

Verma, R., & Bagel, M. P. (2022). Role of functional food additives in regulating the immune response to COVID-19. *Current Biotechnology*, *11*(3), 230–239.

Verma, R., Misra, V., Tiwari, D., & Bisen, P. S. (2021). Potential of selected Indian herbs for COVID-19. *Current Traditional Medicine*, *7*(4), 470–477.

Wang, L., Yang, R., Yuan, B., Liu, Y., & Liu, C. (2015). The antiviral and antimicrobial activities of licorice, a widely-used Chinese herb. *Acta PharmaceuticaSinica B*, *5*(4), 310–315.

Weinberg, S. E., Sena, L. A., & Chandel, N. S. (2015). Mitochondria in the regulation of innate and adaptive immunity. *Immunity*, *42*(3), 406–417.

Wheatley, A. K., Juno, J. A., Wang, J. J., Selva, K. J., Reynaldi, A., Tan, H. X., ... Kent, S. J. (2021). Evolution of immune responses to SARS-CoV-2 in mild-moderate COVID-19. *Nature Communications*, *12*(1), 1162.

Wolf, J. M., Wolf, L. M., Bello, G. L., Maccari, J. G., & Nasi, L. A. (2023). Molecular evolution of SARS-CoV-2 from December 2019 to August 2022. *Journal of Medical Virology*, *95*(1), e28366.

Xu, X., Chen, P., Wang, J., Feng, J., Zhou, H., Li, X., ... Hao, P. (2020). Evolution of the novel coronavirus from the ongoing Wuhan outbreak and modeling of its spike protein for risk of human transmission. *Science China Life Sciences*, *63*, 457–460.

Yadav, V. S., Mishra, K. P., Singh, D. P., Mehrotra, S., & Singh, V. K. (2005). Immunomodulatory effects of curcumin. *Immunopharmacology and Immunotoxicology*, *27*(3), 485–497.

Yadav, S. S., Singh, M. K., Singh, P. K., & Kumar, V. (2017). Traditional knowledge to clinical trials: A review on therapeutic actions of *Emblica officinalis*. *Biomedicine & Pharmacotherapy*, *93*, 1292–1302.

Yan, R., Zhang, Y., Li, Y., Ye, F., Guo, Y., Xia, L., ... Zhou, Q. (2021). Structural basis for the different states of the spike protein of SARS-CoV-2 in complex with ACE2. *Cell Research*, *31*(6), 717–719.

Zahedipour, F., Hosseini, S. A., Sathyapalan, T., Majeed, M., Jamialahmadi, T., Al-Rasadi, K., ... Sahebkar, A. (2020). Potential effects of curcumin in the treatment of COVID-19 infection. *Phytotherapy Research*, *34*(11), 2911–2920.

Zhang, L., & Liu, Y. (2020). Potential interventions for novel coronavirus in China: A systematic review. *Journal of Medical Virology*, *92*(5), 479–490.

22 Clinical Trials on Plant-Derived Medicines and Drug Supplements for COVID-19

Current Achievements and Future Directions

Pankaj Jinta, Priyanka Nagu, Chetna Jhagta, Shivani Rana, Minaxi Sharma, Vineet Mehta, and Kandi Sridhar

22.1 INTRODUCTION

The COVID-19 pandemic posed significant challenges to public health systems worldwide and has also brought about unprecedented disruptions to the daily lives of individuals and communities. The rapid spread of the virus has overwhelmed healthcare infrastructure in many countries, leading to a surge in hospitalizations and fatalities (Omotayo et al., 2024). The cause of COVID-19 is attributed to the Severe Acute Respiratory Syndrome Coronavirus-2 (SARS-CoV-2). It is believed to have originated from bats and possibly transmitted to humans through an intermediate animal host. COVID-19 symptoms vary widely and can range from mild to severe. Common symptoms include fever, cough, and shortness of breath, while some individuals may experience loss of taste or smell, fatigue, muscle aches, and gastrointestinal issues (Al-Taee et al., 2024). The exact origin and initial transmission events are still under investigation, but it is thought to have started in a seafood market in Wuhan, China. Certain groups, such as older adults and those with underlying health conditions like heart disease, diabetes, or compromised immune systems, are at a higher risk of severe illness or complications (Nagu et al., 2021). The high transmissibility of the virus has contributed to its rapid global spread and the subsequent impact on public health, economies, and societies worldwide. Understanding the origins and transmission of the COVID-19 virus is crucial for implementing effective public health measures, developing vaccines, and addressing future pandemic threats (WHO, 2020).

Ongoing research and collaboration among scientists, healthcare experts, and government agencies are essential for advancing our knowledge of the virus and its underlying causes. Several variants of the SARS-CoV-2 virus, responsible for COVID-19, have emerged, each with unique characteristics. The Alpha variant (B.1.1.7), initially detected in the United Kingdom, was notable for its increased transmissibility and has since spread to multiple countries (Braybrook et al., 2021). The Beta variant (B.1.351), originating in South Africa, raised concerns due to its potential impact on vaccine effectiveness and has been identified in various regions globally. The Gamma variant (P.1), first identified in Brazil, shares mutations with the Beta variant and is associated with heightened transmissibility, having been reported in multiple countries (Hirahata et al., 2024). The Delta variant (B.1.617.2), initially found in India, gained global prominence for its heightened transmissibility and potential immune evasion, leading to a surge in cases worldwide (Vieira et al., 2024). The Omicron variant (B.1.1.529), first identified in Botswana and South Africa, caused international concern due to its numerous spike protein mutations, increased transmissibility, potential for reinfections, and uncertainties about vaccine efficacy. These variants underscore the importance of ongoing surveillance and adaptation of public health strategies to mitigate the impact of evolving viral strains (Chenchula et al., 2024).

Globally, rigorous efforts have been dedicated to managing, treating, and mitigating the impact and transmission of SARS-CoV-2. As of 2023, more than 30 vaccines have received approval and are actively being administered in widespread immunization campaigns (Machado et al., 2022). Additionally, the US Food and Drug Administration (FDA) has reported the approval of both individual agents, including antivirals, immunomodulators, cell and gene therapies, and neutralizing antibodies, as well as combination agents for the treatment of COVID-19 (Mishra et al., 2024). Notably, antiviral investigational drugs like remdesivir and other repurposed drugs have obtained emergency use authorization specifically for treating COVID-19. Immunomodulators such as dexamethasone have also been recommended, demonstrating a significant reduction in mortality rates among hospitalized patients, particularly those requiring oxygen or ventilation (Chaudhury et al., 2024). These advancements reflect a concerted global commitment to combating the pandemic through a multifaceted approach encompassing vaccination, therapeutics, and innovative treatments. This pandemic has shifted the focus of the scientific community to the use of herbal medicines as these were suggested to improve the immune system, have antimicrobial properties, and have been used globally as a part of traditional medicines.

22.2   HISTORICAL ASPECT OF PLANT-DERIVED MEDICINES

The historical use of plant-derived medicines dates back thousands of years, with various cultures around the world relying on botanical remedies for healing. *Ginkgo biloba* boasts a rich history in Traditional Chinese Medicine (TCM) where it has been employed to enhance cognitive function and address various health issues including significant antiviral properties (Shahrajabian et al., 2023). In Ayurveda medicine, turmeric has been celebrated for its anti-inflammatory properties and has been utilized for various conditions. Its active compound, curcumin, has been a cornerstone of traditional medicine in India for centuries for various microbial infections (Rizvi et al., 2022). Historically, willow bark has been utilized by ancient civilizations such as the Greeks and Egyptians to address pain and reduce fever. The active compound in willow bark is salicin, a natural precursor to aspirin, making it a traditional remedy with analgesic and anti-inflammatory properties (Mahdi et al., 2006). By the end of the 19th century, extracts and derivatives like morphine and codeine gained widespread use. The active compounds in opium poppy, namely morphine and codeine, have played pivotal roles in the development of modern analgesics and pain management (Biharee et al., 2024).

The COVID-19 pandemic has sparked interest in exploring alternative treatments and preventive strategies, including plant-derived medicines and supplements. These natural products have been traditionally used for their therapeutic properties and are believed to possess antiviral, immunomodulatory, and anti-inflammatory effects (Alhazmi et al., 2021; Hu et al., 2024; Solanki et al., 2024). Moreover, plant-derived medicines and supplements are often perceived as safe and more accessible than pharmaceutical drugs, which led to an increased interest in their use during the pandemic (Antonio da Silva et al., 2020). Several drugs that lay the foundation of contemporary pharmacology have their roots in plant sources. In the past, a significant proportion of the few effective drugs available were derived from plants. Notable examples include aspirin, whose origins trace back to willow bark, quinine extracted from cinchona bark, digoxin sourced from the foxglove plant, morphine derived from the opium poppy, etc. (Badshah et al., 2023; Marafeli et al., 2023; Miller et al., 2023; Tapia et al., 2023). These plant-derived compounds served as the initial building blocks for pharmaceutical advancements, laying the groundwork for the development of a wide array of medicines that continue to play pivotal roles in healthcare today. The interest in plant-derived medicines and supplements as alternative treatments and preventive strategies in response to the COVID-19 pandemic reflects a growing awareness of the potential benefits of natural products. Traditional medicine systems across various cultures have long relied on plant-based remedies for managing a wide range of health conditions, including viral infections (Mariappan et al., 2023). Research into the antiviral properties of specific plant compounds and their immunomodulatory effects has surged, prompting a deeper exploration of their potential role in combating viral infections, including COVID-19 (Mohammed et al., 2023). The accessibility, perceived safety, and historical use of plant-derived medicines present a compelling rationale for their investigation as potential adjuncts to conventional treatments and preventive measures. By delving into the scientific basis of these traditional remedies and evaluating their efficacy in the context of COVID-19, insights can be gained into the feasibility of integrating plant-based interventions into public health strategies (England et al., 2023). Further, examining the cultural and socioeconomic implications of utilizing plant-derived medicines and supplements during the pandemic can provide valuable perspectives on healthcare behaviors, accessibility of resources, and the diversification of therapeutic options. Understanding the rationalization behind the exploration of these natural products offers a comprehensive view of the dynamic landscape of global health responses to the COVID-19 crisis (Viroli et al., 2023).

22.3 IMPORTANCE OF CLINICAL TRIALS IN EVALUATING THEIR EFFICACY AND SAFETY

Plant-based medicines have been employed for centuries, harnessing the pharmacological effects of various chemical constituents known as phytoconstituents. The global receipt and application of plant-based medicines stress the historical significance and perceived efficacy. Nevertheless, concerns regarding the safety and efficacy of these remedies persist due to the limited availability of comprehensive pharmacokinetic, pharmacological, and clinical data for a majority of plant-based medicinal products (Pires et al., 2023). The existing gap in meeting statutory necessities for research on plant drugs amplifies the challenge of effectively regulating these products. Numerous clinical studies are in progress, however, there is yet to be an approved therapy or established drug specifically proven for the effective treatment of COVID-19 (Nile and Kai, 2021). To ensure public safety and foster greater confidence in the therapeutic use of plant-based medicines, it becomes imperative to bridge this gap through robust scientific research, standardized testing methodologies, and adherence to regulatory standards. Only through a more rigorous and comprehensive understanding of the pharmacological properties and safety profiles of plant-based medicines can regulatory frameworks be strengthened, providing healthcare practitioners and consumers with the necessary information for informed decision-making and safe usage (Alam et al., 2021).

Clinical trials involving herbal products pose unique challenges and concerns that warrant careful consideration. One prominent issue is the potential for product adulteration, where the authenticity and purity of herbal remedies may be compromised (Alam et al., 2021). Instances of documented adulteration have raised concerns about the reliability of trial results, emphasizing the need for rigorous quality control measures to ensure the integrity of the studied products. Another significant concern involves the complex interactions between herbal remedies and other medications or substances. Unlike pharmaceutical drugs, herbal products contain myriad compounds, and their interactions with conventional treatments are often poorly understood. This lack of clarity poses challenges in predicting and managing potential adverse effects or altered efficacy when herbal remedies are combined with other entities (Upton et al., 2020). Reproductive and organ toxicity data for herbal products may be limited, and comprehensive information on potential risks is often lacking. Understanding the impact on reproductive health and organ systems is crucial for ensuring patient safety, making it imperative to conduct thorough toxicity assessments during clinical trials (Parveen et al., 2015). Complicating matters further, different plant species may share the same name in local languages, exemplified by "Brahmi," which can refer to both *Centella asiatica* and *Bacopa monnieri* (Singh et al., 2014). This diversity and potential lack of standardization pose significant challenges when evaluating the effectiveness and safety of herbal medicines in clinical settings. Prior dose finding is another concern specific to herbal product trials (Parveen et al., 2015). The traditional use of many herbal remedies might not be backed by systematic dose optimization studies. As a result, incomplete or inadequate information about optimal dosage levels can affect the efficacy and safety assessments in clinical trials. The absence of well-established dose-response relationships complicates the interpretation of trial outcomes and hinders the development of evidence-based guidelines for herbal product use (Fong et al., 2006). In addressing these concerns, it becomes essential for clinical trials with herbal products to implement stringent quality control measures, conduct in-depth investigations into potential interactions, prioritize comprehensive toxicity assessments, and systematically explore optimal dosage ranges. By addressing these challenges, researchers can enhance the reliability and applicability of clinical trial findings, ultimately contributing to a more robust understanding of the efficacy and safety of herbal remedies in therapeutics. Overcoming challenges in clinical trials can be achieved by implementing the latest methodologies and adhering to updated guidelines (Singh et al., 2014). Leveraging modern manufacturing techniques and utilizing approved guidelines permits the manufacture of drugs that can effectively solve the problems associated with herbs, this in turn, facilitates the implementation of striking methods in clinical trials (Vaidya and Devasagayam, 2007). The careful planning of a scientifically sound clinical study design is crucial for the success of any clinical trial, especially in the context of herbal medicines. To ensure reliable results, double-blind experiments should be conducted with an adequate number of selected patients, ideally following the established standards for clinical trial methodologies applied in the development of new drugs (Itokawa et al., 2008).

22.4 EMERGING INTEREST IN NATURAL PRODUCTS FOR ANTIVIRAL PROPERTIES AND COVID-19

The field of antiviral drug development witnessed a significant milestone in 1960 when the inaugural antiviral agent, acyclovir, made its debut in clinical settings. The landscape of antiviral research underwent a transformative shift with the onset of the AIDS epidemic in 1981 and subsequent pandemics in subsequent years (Upadhyay, 2023). Non-antiviral agents with antimicrobial properties, deemed safe for human use, are readily accessible in the market to combat a diverse array of bacterial pathogens. In stark contrast, the availability of antiviral drugs remains limited. This discrepancy arises from the comprehensive knowledge surrounding bacterial systems, their molecular infection mechanisms, and pathogenesis (Tulp et al., 2022). In contrast, the process through which viruses infiltrate host cells remains shrouded in mystery. Plants serve as a rich reservoir of novel compounds with potential therapeutic applications against diseases. Despite this abundance, only a limited number of plants have progressed to clinical trials, demonstrating positive outcomes that warrant commercial exploitation (Chatterjee, 2023).

Salvia miltiorrhiza, rich in flavonoids like tanshinone, exhibits multiple health benefits such as inhibiting protease activity like papain-like protease (PL^{pro}) and 3-chymotrypsin-like cysteine protease (3CL^{pro}), anti-inflammatory effects, and increased serum antioxidant activities (Bafandeh et al., 2023). Thymoquinone, found in *Nigella sativa*, exhibits myriad health benefits, serving as an antioxidant, immune regulator, and anti-inflammatory agent. Beyond its general health-promoting properties, thymoquinone also plays a crucial role in preventing SARS-CoV-2 entry and inhibiting viral replication (Rajamanickam et al., 2023). *Zingiber* and *Curcuma*, both containing polyphenols like diarylheptanoids, exhibit immune-enhancing properties and ACE-2-Ang-(1–7)-Mas pathway activation for *Zingiber*, and PL^{pro} inhibition, anti-inflammatory, and antimicrobial effects for *Curcuma* (Badrunanto et al., 2023). Quercetin, present in various fruits and vegetables such as apples, citrus fruits, and green leafy vegetables, showcases a broad spectrum of health benefits. Its antioxidant, anti-inflammatory, anticancer, antiviral, and immunomodulatory properties make it a valuable component in the fight against diseases. Quercetin's ability to inhibit M^{pro} and viral entry into host cells positions it as a potential defense against viral infections, including SARS-CoV-2 (Umar et al., 2023). Ellagic acid, found in raspberries, pomegranate, mango, walnuts, almonds, green tea, and *Momordica charantia*, is an antioxidant and antiproliferative compound. It inhibits the M^{pro}, preventing viral attachment and internalization into the host cells (Rajak and Ganguly, 2023). Ursolic acid, derived from plants like *Mimusops caffra, Ilex paraguariensis*, and *Glechoma hederacea*, contributes to health with its anti-inflammatory, antibacterial, antioxidant, and anti-diabetic properties. Its potency in blocking the M^{pro} enzyme positions it as a potential therapeutic agent against various ailments (Jaber, 2023). Caffeic acid acts by inhibiting virus attachment to host cells and binding to $3CL^{pro}$, thereby inhibiting viral replication (Pojtanadithee et al., 2024). *Ocimum sanctum* (Tulsi) showcases a plethora of medicinal properties attributed to compounds like dihydrodieuginol B and Tulsinol A–G, encompassing antiviral activity (Bhattacharya et al., 2024). Vanillin, sourced from vanilla bean, demonstrates antimicrobial and antioxidant properties. Its inhibition of M^{pro} showcases its potential role in countering coronaviral infections (Hyun et al., 2023). Thymol, present in *Thymus vulgaris, Ocimum, Origanum,* and *Monarda* genera and members of the *Verbenaceae, Ranunculaceae, Scrophulariaceae,* and *Apiaceae* families, offers antioxidant, local anesthetic, anticarcinogenic, antinociceptive, cicatrizing, antiseptic, and immunomodulatory effects. Thymol inhibits the viral spike protein, preventing SARS-CoV-2 entry, and serves as a potent disinfectant (Nadi et al., 2023). Rosmarinic acid, found in rosemary, perilla, sage, mint, and basil, exhibits antispasmodic, analgesic, antirheumatic, diuretic, and antiepileptic properties. Its inhibition of viral entry and replication further underscores its potential in combating viral infections (Yao et al., 2023). Figure 22.1 depicts the various targets of herbal medicines for their beneficial effects against COVID-19.

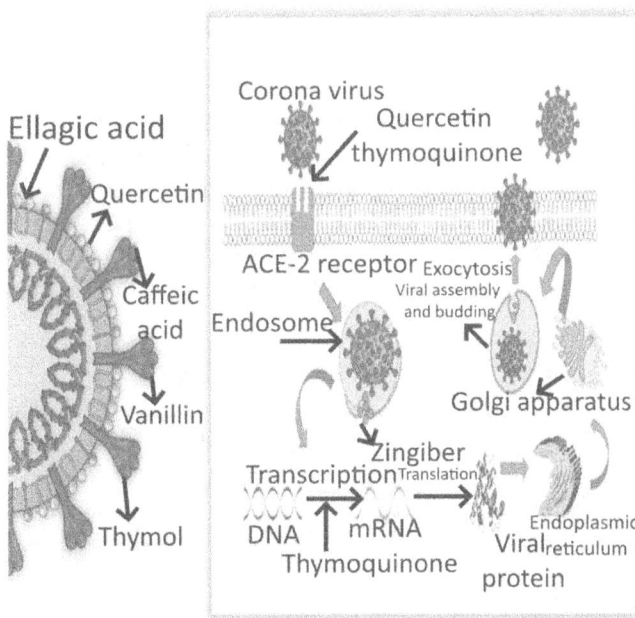

FIGURE 22.1 Life cycle of coronavirus in humans and various targets of herbal medicines. (Illustration constructed by the authors using BioRender, Toronto, Canada.)

22.5   CURRENT STATUS OF COVID-19 TREATMENT

Currently, several options are being explored as a therapeutic intervention for COVID-19. This mainly includes the use of antiviral agents and host-targeting therapies. Antiviral agents against COVID-19 mainly include polymerase inhibitors, protease inhibitors, inhibitors of nucleoside and nucleotide reverse transcriptase, entry and uncoating inhibitors, and other antivirals. The treatments targeting the host include neutralizing antibody therapy, Janus kinase (JAK) inhibitors, and steroids (Yuan et al., 2023).

22.5.1   ANTIVIRAL AGENTS

22.5.1.1   RNA Dependent RNA Polymerases (RdRp) Inhibitors

RNA polymerase is an important enzyme that plays a crucial role in viral replication. Various drug treatments aimed at inhibiting this enzyme include remdesivir, a well-known FDA-approved antiviral drug for COVID-19 treatment (Wang et al., 2020; Williamson et al., 2020). However, its efficacy against SARS-CoV-2 remains a subject of debate due to mixed results observed in hospitalized patients during the COVID-19 pandemic. Despite this uncertainty, remdesivir has demonstrated synergistic effects when used in combination with baricitinib, a drug commonly prescribed for rheumatoid arthritis. Baricitinib (Kalil et al., 2021) acts by suppressing the immune system, and the combined treatment has shown potential in increasing the recovery time for hospitalized patients. Another important drug in this context is favipiravir, a selective antiviral drug that targets polymerases. Several studies suggest its beneficial effects against SARS-CoV-2, however, it is associated with the same limitations as remdesivir (Delang et al., 2018).

22.5.1.2   Protease Inhibitors

The protease enzyme is a key enzyme that is involved in viral protein replication. Lopinavir antiviral drugs used in the treatment of HIV (Yao et al., 2020), in combination with ritonavir and arbidol, show effective results in hospitalized patients with pneumonia-related symptoms (Cao et al., 2020; RECOVERY Collaborative Group, 2020; Wang et al., 2020). Nirmatrelvir, an inhibitor of the SARS-CoV-2 main protease, has shown promise in a Phase 2–3 trial (Owen et al., 2021).

22.5.1.3   Inhibitors of Nucleoside and Nucleotide Reverse Transcriptase

In July 2021, China's National Medical Products Administration (NMPA) approved azvudine (FNC), a nucleoside reverse transcriptase inhibitor with broad-spectrum antiviral activity, for the management of AIDS (Ren et al., 2020; Yu and Chang, 2020; Zhang et al., 2021). On 25 July 2022, the NMPA conditionally approved azvudine to treat common COVID-19 in adults. Another small molecule active against coronavirus including SARS-CoV-2 is molnupiravir. It is a prodrug of N-hydroxycytidine. It was reported that molnupiravir was active against the variants Delta, Gamma, and Mu of SARS-CoV-2 (Jayk Bernal et al., 2022).

22.5.1.4   Fusion Inhibitors (Entry and Un-Coating Inhibitors)

Amantadine is known for its role in treating influenza A by inhibiting viral replication (Araujo et al., 2020). It has been reported in several studies that it also shows its protective action against SARS-CoV-2 by preventing the release of viral RNA into cells (fusion) (Smieszek et al., 2020). Enfuvirtide, an HIV-1 fusion inhibitor peptide, also reported the protective action against SARS-CoV-2. Some clinical trial studies are still pending to validate the effect of enfuvirtide (Ahmadi et al., 2022).

22.5.1.5   Other Antivirals

Azithromycin, a macrolide, has antiviral, and anti-inflammatory properties but superiority trials conducted by the UK showed that it did not reduce the risk of hospital admission and death among mild to moderate COVID-19 (Oliver and Hinks, 2021). Antimalarial drugs hydroxychloroquine and chloroquine have been suggested to treat COVID-19 and underwent numerous trials (Ferner and Aronson, 2020). Some studies indicated that these antimalarial drugs did not demonstrate significant improvement in patients with mild to moderate symptoms (Boulware et al., 2020). Interferons (IFNs) including type I (IFN-I) and type III (IFN-III) are inflammatory mediators induced by SARS-CoV-2. IFNs remain controversial, with evidence suggesting their effectiveness in improving prognosis, especially when administered early (Sposito et al., 2021).

22.5.2 Targeting Host

22.5.2.1 Convalescent Plasma Therapy

Convalescent plasma therapy is the use of "super blood" from a recovered patient to a person suffering from severe SARS-CoV-2 infection. Several trials were conducted to determine the effectiveness of plasma therapy (Li et al., 2020). However, the results are still not clear or still controversial. In 2019, an early Chinese study suggested that transfusion shows no statistical difference compared to the standardized treatment. This trial was performed on only 103 patients. In 2020, a Swiss Phase 1 clinical trial was performed on 30 patients demonstrating that no adverse event was reported, however, faster virus clearance was observed (Marconato et al., 2022). Spanish trials in 2020 suggested that there was no significant decrease in viral load and illness progression. This discrepancy may be due to differences in the sample size (Alemany et al., 2022). Therefore, there is still no clarity on the therapeutic benefits of this therapy, however, some ongoing trials like the COVID-19 trial and REMAP-CAP trial may add some clarity on the effectiveness of plasma therapy.

22.5.2.2 Neutralizing Monoclonal and Polyclonal Antibody Therapies

LY-CoV555 (bamlanivimab) was the first effective monoclonal antibody found to be effective against COVID-19 (Jones et al., 2021; Matz et al., 2022). It shows a strong affinity for Angiotensin-converting enzyme 2 (ACE-2) receptors and reduces the viral load in hospitalized patients. Studies also reported that it prevents illness progression in combination with etesevimab (Gottlieb et al., 2021). Trials showed that due to a reduction in viral load, the combination was authorized for emergency use but later retracted for the omicron variant (Iketani et al., 2022). It has been reported that bebtelovimab is the only neutralizing monoclonal antibody that did not show resistance against the omicron variant. It is the only neutralizing monoclonal antibody approved by the FDA for emergency use in the case of the Omicron variant (Iketani et al., 2022).

22.5.2.3 Janus Kinase (JAK) Inhibitors

Baricitinib is a JAK1/JAK2 inhibitor used for the treatment of rheumatoid arthritis shows anti-COVID properties by disrupting cytokine signaling pathways and viral endocytosis (Hoang et al., 2021). The FDA has authorized baricitinib for emergency use in combination with remdesivir for severe patients who require supplemental oxygen, ventilation, etc., and can also be used as a stand-alone therapy (Smolen et al., 2020; Stebbing et al., 2020). Ruxolitinib is another JAK1 and JAK2 protein kinase inhibitor with antiviral potency against HIV and Epstein–Barr virus infections reported to suppress cytokine production associated with SARS-CoV-2 infection (Marconi et al., 2022; Zhang et al., 2022). Due to its immunomodulatory effect, it is successfully employed in patients with COVID-19 (Yan et al., 2021). Clinical studies showed that it shows a beneficial effect on COVID-19-associated pneumonia but did not show any beneficial effect against COVID-19-associated Acute Respiratory Distress Syndrome (ARDS). Likewise, tofacitinib is an orally available JAK inhibitor used to treat autoimmune disease and rheumatoid arthritis (Mease et al., 2017; Pillaiyar and Laufer, 2022). Clinical trial studies indicated that it shows promising results in COVID-19-associated pneumonia by inhibiting inflammatory cytokines (Guimaraes et al., 2021). Similarly, nezulcitinib is an inhaled, lung-selective, pan-JAK inhibitor that could be a promising therapy for cytokine-driven lung inflammation in COVID-19 patients (Singh et al., 2021).

22.5.2.4 Steroids

Some steroidal therapy has been proved successful in the prevention of disease progression in severe COVID-19 cases. These drugs include dexamethasone, budesonide, ciclesonide, and glucocorticoids (hydrocortisone and methylprednisolone). These drugs have demonstrated some promising effects on severe COVID-19 patients by their inherited anti-inflammatory actions. However, none of them can be used as a first-line therapy for COVID-19 as they are focused on providing symptomatic relief rather than being a potential antiviral agent (Batiha et al., 2022).

22.5.2.5 COVID Vaccination

During the COVID-19 pandemic, major competition between the pharmaceutical giants had been raised on the discovery of producible vaccines. Many vaccines have been identified for COVID-19 treatments. These vaccines include Pfizer (BNT162b2) and Moderna (mRNA-1273) (Thompson et al., 2021), Janssen/Johnson & Johnson (Ad26.COV2.S) (Sadoff et al., 2021), AstraZeneca (ChAdOx1 nCoV-19/AZD1222) (Voysey et al., 2021), Covaxin (BBV152) (Ella et al., 2021), Novavax (NVX-cov2373) (Dunkle et al., 2022), Sinovac (Tanriover et al., 2021), Sinopharm (WIV04 & HB02) (Xia et al., 2020), Sputnik V (Logunov et al., 2020), and CanSino Biologics (Ad5-based COVID-19 vaccine) (Zhu et al., 2020). Although these vaccines are effective against SARS-CoV-2, limited data is available on vaccine efficacy in high-risk groups, such as cancer patients, leading to a lack of specific vaccination guidance for these individuals. High-risk groups, including those with obesity, chronic obstructive pulmonary disease, chronic kidney disease, chronic liver disease, diabetes mellitus, and cardiovascular diseases, may have attenuated vaccine protection, posing a significant risk to this population.

22.6 PLANT-DERIVED MEDICINES WITH ANTIVIRAL PROPERTIES

Medicines of plant origin offer a rich source of diverse bioactive compounds that exhibit potential pharmacological actions. The use of medicinal plants as therapeutic agents has been practiced since ancient times and is well described in the oldest system of medicines, i.e. Ayurveda, Siddha, Unani, and TCM. The prominence of plants as medicinal agents in the search for lead therapeutic agents for drug discovery is attributed to the structural and chemical variability of plants (Thomas et al., 2021). In today's scenario, the majority of synthetic drugs are discovered by using natural products as precursors, intermediate, and scaffold fingerprinting. This is illustrated by an example of shikimic acid (a natural compound) obtained from the Chinese star anise plant (*Illicium verum*, family *Schisandraceae*) is used as an intermediate in the synthesis of a synthetic antiviral drug oseltamivir (Singh et al., 2020). In recent years, the number of extensive studies on plant extracts and isolated phytoconstituents have been reported for antiviral properties against varied viral strains. Some of the major phytochemical classes with preclinical antiviral properties are discussed next.

22.6.1 ESSENTIAL OILS

The essential or volatile oils derived from plants are composed of terpenes, which include monoterpenes, sesquiterpenes, diterpenes, triterpenes, etc. The antiviral activity of aromatic compounds of herbal extracts has been demonstrated in several studies against strains of influenza virus, herpes virus, yellow fever virus, HIV, and coronavirus (Ma and Yao, 2020). An *in silico* study reported on eucalyptol (1,8-cineole) a monoterpene from the essential oil of *Eucalyptus globulus,* family *Myrtaceae,* for its potent effectiveness against coronavirus that is mediated through the inhibition of M^{pro} (Sharma et al., 2020). Ravintsara oil obtained from the *Cinnamomum camphora,* which is rich in camphor and contains a higher concentration of eucalyptol, is suggested to be effective against coronavirus (Tine et al., 2023). Computational studies on 12 phytoconstituents of lemon balm essential oil obtained from *Melissa officinalis* stated that the highest proportion of luteolin-7-glucoside-3′-glucuronide, melitric acid-A, and quadranoside-III showed excellent affinity against the

M^{pro} and spike protein targets of coronavirus, which suggests potential antiviral activity (Prasanth et al., 2021). In another *in silico* study, the constituents melitric acid-A and salvanolic acid A were found to be potential inhibitors of M^{pro} target of COVID-19 (Elekofehinti et al., 2021). The constituents of garlic (*Allium sativum*) *Amaryllidaceae*, and onion (*Allium cepa*) *Amaryllidaceae* were tested in a computational study using the approach of molecular docking. Among all the volatile constituents the sulphur-containing compounds, i.e. allin (S-allylcysteine sulfoxide), S-propyl cysteine, S-allylcysteine, S-allylmercaptocysteine, S-ethylcysteine, S-methylcysteine, and S-propyl L-cysteine, were the suggested potential candidates for inhibition of M^{pro} (Pandey et al., 2021). In another study, garlic aqueous extract was studied on embryonic eggs against coronavirus and the extract showed inhibitory effects on viral replication (Rouf et al., 2020). The essential oil of *Laurus nobilis* contains artecanin, which is revealed as a potent M^{pro} inhibitor along with an excellent pharmacokinetic profile (Al-Shuhaib et al., 2023). The rosemary water extract (*Rosmarinus officinalis*) contains rosmarinic acid, caffeic acid, tuberonic acid hexoxide, eucalyptol, etc. The extract and its two main constituents, rosmarinic acid and caffeic acid, exhibit a potential inhibitory effect on the interaction of the spike protein of COVID-19 and ACE-2 (Yao et al., 2023). In another *in vitro* study 1,8-cineole constituent of rosemary essential oil was explored for its ACE-2 and LOX-5 inhibitory activity and showed positive results. Inhibitory action on ACE-2 is higher as compared to the LOX-5, suggesting the potential role of 1,8-cineole against coronavirus infection (Demirci et al., 2022). Thyme (*T. vulgaris*) consists of thymol and carvacrol as its main constituents and is traditionally known for its antiviral, antibacterial, and antifungal activities. The molecular mechanism responsible for the antiviral activity is mediated through the suppression of pro-inflammatory cytokines, particularly TNF-α, IL-6, IL-7, and IL-8. Further, carvacrol is also known to have ACE-2 inhibitory effects (Nadi et al., 2023). A furanocoumarin, bergamottin from bergamot oil (*Citrus bergamia*) acts against coronavirus by blocking the membrane fusion through spike protein interaction, which ultimately affects the synthesis of viral RNA (Zhou et al., 2022). Geranium, citronellal and limonene, monoterpenes obtained from lemon oil (*Citrus limon*) possess significant blocking action on ACE-2, which is the key enzyme that plays a major role in the invasion of viral particle in the host cell (Kumar et al., 2020). The essential oils from different species of *Salvia* species are known to contain 1,8-cineole, camphor, α-thujone, β-caryophyllene, etc. The oil was examined in an *in vitro* assay to assess the ACE-2 inhibitory action, which indicated a positive response and suggested the antiviral activity against coronavirus (Karadağ et al., 2022). Cannabinoids (cannabidiol) from *Cannabis sativa* were screened for antiviral potential against COVID-19 in an *in silico* study that showed valuable results against viral replication that mediated through SARS-COV-2 M^{pro} (Altyar et al., 2022).

22.6.2 Saponins

Glycyrrhizin is a triterpenoid saponin obtained from the roots of licorice (*Glycyrrhiza glabra*). This phytoconstituent proved to exhibit antiviral activity against COVID-19, which was reported to act via diverse mechanisms such as increased production of macrophages, altered cellular signaling pathways, transcription factors, and inhibition of ACE-2 (Chrzanowski et al., 2021). Ginsenosides, a triterpene saponin from the roots of ginseng (*Panax ginseng*) showed promising antiviral potential to combat coronavirus. The plant has been known as an immunomodulator since ancient times and, therefore, acts in viral infection by modulating the acquired and natural immunity (Ratan et al., 2022). Another *in vitro* study demonstrated the antiviral potential of Korean ginseng against human coronavirus OC43 by decreased expression of viral protein and increased expression of interferon genes (Jeong et al., 2023).

22.6.3 Flavonoids

Baicalin (a glycosyloxyflavone) is obtained from *Scutellaria baicalensis*. In an *in silico* study, the compound was reported to modulate 3CL^{pro} and PL^{pro} targets, thereby producing its beneficial effects against coronavirus (Lin et al., 2021). The targeted action of baicalin on 3CL^{pro} is also confirmed in another *in vitro* study (Liu et al., 2021). Some studies suggest the antiviral action is mediated through the interference of mitochondrial oxidative phosphorylation that ultimately impairs viral replication (Huang et al., 2020). Other flavonoids such as astragalin, luteolin, quercetin, kaempferol, apigenin, naringenin, catechin, and epigallocatechin have demonstrated promising antiviral properties against COVID-19 by blocking the M^{pro} target in certain docking studies (Alzaabi et al., 2021).

22.6.4 Lignans

Two lignans, paronychiarabicine A and B, constituents of *Paronychia arabica*, were screened in a computational approach of molecular docking, which concluded that both the lignan compounds are effective in the management of SARS-COV-2 infection. Another novel megastigmane paronychiarabicastigmane A from the same plant was proved to be a potent candidate for the COVID-19 treatment (Elshamy et al., 2021).

22.6.5 Sesquiterpene Lactones

Artemisia annua contains artemisinin as a main constituent that has shown antiviral activity against SARS-COV-2 and its variants (UK and South African variants) in an *in vitro* study. The mechanism of action is thought to be mediating through the inhibition of viral replication by interfering with the viral entry in the host cell (Fielding et al., 2020; Nair et al., 2021).

22.7 CLINICAL TRIALS OVERVIEW

Before being released to the market, novel medications such as drugs, therapies, and medical devices undergo a series of clinical trial studies. These studies are conducted to investigate their safety, effectiveness, and overall suitability for the prevention, treatment, and diagnosis of various medical conditions (Kandi and Vadakedath, 2021). A clinical trial is a type of research study that involves the participation of human subjects to evaluate the safety and effectiveness of the medical intervention. These clinical research studies are conducted in multiple phases, which begin after obtaining pertinent preclinical data and receiving approval from the health authority and ethics committee. The phases of clinical trials include Phase 0, Phase 1, Phase 2, Phase 3, and Phase 4 (Le-Rademacher et al., 2023).

Phase 0, also known as micro-dosing studies, is an exploratory investigational study that examines how the drug functions in humans and determines the appropriate dose for Phase 1 studies. Phase 1 is a non-therapeutic trial involving the participation of fewer than 50 healthy subjects. This phase is conducted to find out the maximum tolerated dose, safe dose range, and pharmacokinetics of a new treatment. Phase 2 is an exploratory study that recruits approximately 5–100 patients, determines the appropriate dosage, and assesses the therapeutic effects of the drug (Van Norman, 2019). Phase 3, also referred to as the pivotal trial, is conducted with 300–3000 patients to validate the preliminary data collected in previous trials. This phase is aimed at providing statistically significant and scientifically reliable proof regarding the treatment (Umscheid et al., 2011).

22.8 CHALLENGES SPECIFIC TO CLINICAL TRIALS FOR NATURAL PRODUCTS

Clinical trials focusing on natural products encounter more challenges when compared to synthetic compounds. Natural products usually contain more than one phytoconstituent that works as an active constituent. This complex natural drug poses a great challenge in formulation and clinical studies (Zhu et al., 2022). Clinical research studies are typically more expensive than preclinical studies (Truebel and Thurston, 2020). Most companies involved in the production of natural products usually do not have patent protection, with a few exceptions. As a result, they generally lack the pricing power required to pay for clinical trials (Atanasov et al., 2021). In preclinical studies, a high amount of natural product is typically administered. However, replicating such settings in humans is sometimes not feasible, especially for the drugs that are administered orally because of the minimal bioavailability (Paller et al., 2016). While conducting clinical trials on herbal medicine, creating a homogenous group of subjects poses a significant challenge due to the different perspectives of herbal and modern medicine. Consequently, it becomes difficult to establish the eligibility criteria for clinical studies involving herbal intervention (Parveen et al., 2015). Additionally, selecting a control group for an herbal medicine trial is a big challenge. The control group should be like the intervention group in all aspects, e.g. color, odor, and taste; however, it is very difficult in the case of an herbal medicine trial to have a similar intervention and control group. These difficulties are also experienced in making a placebo because herbal drugs have a specific aroma that cannot be matched with a placebo (Goyal, 2017).

Randomization and blinding are the most effective ways to minimize bias and remove placebo effects. However, maintaining a double-blind setup for natural medicine is difficult since natural remedies include lifestyle recommendations, detailed explanations, and counseling (Misra, 2012). Other challenges related to clinical trials for natural drugs include the non-availability of suitable subjects required, quality control issues, and amount of active phytoconstituents depending upon the cultivation, storage, and geographical condition (Jachak, 2007).

22.9 REGULATORY CONSIDERATIONS FOR PLANT-DERIVED MEDICINES

The regulatory frameworks for plant-based medicine and dietary supplements are less comprehensive when compared to pharmaceutical products. These regulatory guidelines vary from country to country since national regulatory authorities define and classify plant-based products and dietary supplements differently (Thakkar et al., 2020). Dietary supplement products are regulated in the United States by the FDA through the Dietary Supplement Health and Education Act (DSHEA) of 1994 (Denham, 2011). The FDA has a distinct regulatory approach for plant-based medicines and these medicines must adhere to the premarket approval procedure. In 2003, the FDA established the Botanical Review Team (BRT) to review the pharmacognosy of all submitted plant-based medicines. BRT aims to examine all aspects, including raw material used, phytoconstituent profile, and previous human exposure for all original Investigational New Drug (IND) and New Drug Application (NDA) submissions (Wu et al., 2020). The Department of AYUSH, Central Council of Indian Medicine, and Drug and Cosmetics Act (D&C Act) of 1940 regulate herbal medicine in India. Herbal drugs fall under the D&C Act of 1940 and Drugs and Cosmetics Rules, 1945 (Sahoo and Manchikanti, 2013).

The regulatory framework encounters multiple challenges due to the diverse nature and traditional usage associated with herbal medicine. Additionally, lack of trained personnel, latest techniques, patent protection, comprehensive databases, and irrational use are other factors that pose challenges (Sharma, 2015). There is a need to update the guidelines for the assessment of the safety and potential risks of the drug supplement and herbal drugs. Manufacturers of plant-derived products and supplements are required to comply with Good Agricultural and Collection Practices (GACP), Good Manufacturing Practices (GMP), and good laboratory practices to produce good quality products (Gavali, 2017). Furthermore, they should maintain documentation on safety assessment and product development. The national regulatory authority should establish clear guidelines for obtaining a license for the distribution, import, and manufacturing of plant-based products to guarantee efficacy, safety, and quality (Sahoo et al., 2010). Also, dedicated pharmacovigilance centers should be established to monitor the adverse drug reactions for plant-based medicine (Skalli and Bencheikh, 2012).

22.10 ONGOING CLINICAL TRIALS OF PLANT-DERIVED MEDICINE AND DIETARY SUPPLEMENTS

The summary of ongoing trials of dietary supplements and plant-derived medicine for COVID-19 is given in Tables 22.1 and 22.2, respectively. The majority of the clinical trials are randomized controlled trials, interventions composed of multiple herbs and funded by a government body.

TABLE 22.1

Ongoing Clinical Trials of Dietary Supplements for COVID-19

NCTID	Phase	Intervention	Study Description	Route	Primary Outcome
NCT04385940	Phase 3	Vitamin D3	Randomized, double-blind, controlled 3-week study of vitamin D in 12 COVID-19 patients	Oral	Recovery of COVID-19 symptoms
NCT04536298	Phase 3	Vitamin D	Randomized, placebo-controlled clinical trial in 2024 COVID-19 patients	Oral	Rate of seeking healthcare visits for symptoms and deaths in participants newly diagnosed with COVID-19
NCT05618574	Phase 3	Beet-It nitrate beverage	Randomized, parallel controlled trial in 40 COVID-19 patients	Oral	Fatigability and walking efficiency
NCT05157243	Phase 3	Vitamin Super B-Complex	A randomized, parallel study of 2000 mild or moderate COVID-19 patients	Oral	Degree of COVID-19 progression or death because of any reason
NCT04359680	Phase 3	Vitamin Super B-Complex	A double-blind randomized, placebo-controlled trial in 1407 participants	Oral	Percentage of individuals exhibiting laboratory-confirmed COVID-19 symptoms and symptomatic laboratory-confirmed Viral Respiratory Illness (VRI) determination
NCT05157269	Phase 3	Vitamin Super B-Complex	A double-blind randomized, placebo-controlled trial in 600 mild COVID-19 patients	Oral	Duration of sustained recovery from COVID-19
NCT05277155	Phase 2/ Phase 3	BEJO Red Ginger extract	A randomized, double-blind, placebo-controlled study in 169 COVID-19 patients with mild symptoms	Oral	Time required for the conversion of nucleic acid
NCT04335084	Phase 2	Vitamin C, Vitamin D	A double-blind randomized, placebo-controlled study in 600 subjects for the prevention of COVID-19 symptoms	–	COVID-19 symptoms prevention recorded and safety
NCT05623007	Phase 2	*Lactobacillus rhamnosus*, *Bifidobacterium animalis* subsp. lactis, and *Lactobacillus acidophilus*	A randomized, parallel controlled study in 440 adolescents	–	Change in BMI for age z-scores and concentration of specific Immunoglobulin G (IgG) targeting SARS-COV-2 titer
NCT04404218	Phase 2	Açaí palm berry extract	A randomized, parallel, placebo-controlled trial in 480 patients with COVID-19	Oral	7-point ordinal symptom scale
NCT05703074	Phase 2	Nicotinamide riboside (NR)	Randomized controlled, 2 x 2 factorial trial in 310 long COVID patients	Oral	Health-related quality of life

(Continued)

TABLE 22.1 *(Continued)*

Ongoing Clinical Trials of Dietary Supplements for COVID-19

NCTID	Phase	Intervention	Study Description	Route	Primary Outcome
NCT06054438	Phase 2	Cordyceps	Randomized, crossover study in 110 post-COVID-19 disorders	Oral	Yorkshire Rehabilitation Scale (C19-YRSm) for COVID-19 as declared by individuals
NCT04849637	Phase 2	Virgin coconut oil	Randomized controlled trial in 74 hospitalized COVID-19 Patients	Oral	Duration of hospitalization and symptoms recovery time
NCT05465798	Phase 2	MC 3 x 3 (beta-Glucans)	A randomized, parallel study in 60 hospitalized COVID-19 patients	Oral	Duration of symptoms, severity, and clinical recovery
NCT06108297	Phase 1	Lithium aspartate	Open-label, dose-finding study in 40 long COVID patients	Oral	Fatigue severity, brain fog severity
NCT05371288	Early Phase 1	NAC (N-acetyl cysteine), alpha lipoic acid (ALA), liposomal glutathione (GSH)	A randomized double-blind study in 50 long COVID-19 patients	Oral	Symptoms changes assessment, quality of life change, severity of symptoms questionnaire, and change in time to clinical recovery
NCT04703036	Early Phase 1	Glycine, N-acetylcysteine, alanine	Randomized, placebo-controlled, double-blind design study of 64 COVID-19 patients	–	Change in glutathione, Interleukein-6 concentrations, and ordinal scale

TABLE 22.2

Ongoing Clinical Trials of Plant-Derived Medicine for COVID-19

NCTID	Phase	Intervention	Study Description	Route	Primary Outcome
NCT04802382	Phase 3	CimetrA (artemisinin, curcumin, boswellia, and vitamin C)	Double-blinded, randomized clinical study to evaluate the CimetrA effects in 252 patients	Oral	Clinically significant improvement within intervention groups
NCT06175468	Phase 3	Formosa 1-Breath Free (NRICM101)	Double-blinded randomized, parallel, placebo-controlled trial in 170 subjects with COVID-19 symptoms or influenza-like symptoms	Oral	Evaluating the reduction in sore throat symptoms in subjects experiencing COVID-19 or influenza-like symptoms
NCT05890534	Phase 3	Pycnogenol (*Pinus atlantica*)	Randomized, quadruple-blinded, placebo-controlled trial in 150 post-COVID-19 subjects	Oral	Change in health status using EQ-Visual Analogue Scale
NCT05684952	Phase 2	Shenlingcao oral liquid (American ginseng, *Ganoderma lucidum*, rose, and fermented cordyceps powder)	Double-blinded, randomized, parallel, placebo-controlled study in 152 long COVID associated fatigue	Oral	Evaluating the changes in scores of the Chalder fatigue scale

(Continued)

TABLE 22.2 *(Continued)*
Ongoing Clinical Trials of Plant-Derived Medicine for COVID-19

NCTID	Phase	Intervention	Study Description	Route	Primary Outcome
NCT05787327	Phase 2	Yinqiaosan-Maxingganshitang granules	Double-blinded randomized, parallel, placebo-control study in 96 mild and moderate COVID-19 patients	Oral	The time required for body temperature to return to normal
NCT05360004	Phase 2	*Astragali radix, Codonopsis radix, Atractylodis rhizoma, Atractylodis macrocephala* Rhizoma, *Glycyrrhizae radix* et Rhizoma, *Lonicerae japonicae* Flos, *Mori folium*, Poria, *Saposhnikovia radix, Schizonepetae herba*	A randomized, open-labeled, controlled study in 652 subjects to evaluate Chinese Herbal Medicine for the prevention of COVID-19 transmission	Oral	Percentage of subjects who test COVID-19 positive
NCT04622865	Phase 2	Isoquercetin	Triple-blind randomized, parallel study in 200 moderate and severe COVID-19 patients	Oral	Evaluating the clinical status on 7-point ordinal scale on day 15
NCT06095258	Phase 2	Traditional Chinese Medicine	A randomized, parallel controlled study in 162 long COVID patients	–	Quality of Life using a 5-level EQ-5D version
NCT03944447	Phase 2	Cannabis	Open-label, non-randomized, parallel study in 200,000 patients with various chronic medical conditions	–	The rate of Covid-19 infection and severity of symptoms in cannabis users and the general population will be compared
NCT05897203	Phase 2	TazCoV (herbal syrup), Vidicine (herbal syrup)	Open-label, randomized, control study in 510 subjects with COVID-19 and other respiratory viral infections	Oral	Severe, moderate, and mild serious adverse events, time in symptoms recovery, duration of viral clearance, duration of TazCoV and Vidicine to reach maximum concentration and transition from severe to critical acute respiratory infections
NCT04667247	Phase1/2	FoTv (*Fomitopsis officinalis* and *Trametes versicolor*)	Double-blinded, randomized, parallel, placebo control trial in 54 COVID-19 patients	Oral	Adverse event, level of creatinine, aspartate transaminase, alanine transaminase, and prothrombin and partial thromboplastin, and time in hospital for patients

(Continued)

TABLE 22.2   *(Continued)*
Ongoing Clinical Trials of Plant-Derived Medicine for COVID-19

NCTID	Phase	Intervention	Study Description	Route	Primary Outcome
NCT04939415	Phase 1	Qing Fei Pai Du Tang (mQFPD) and organic brown rice	A randomized, double-blind, placebo-controlled study in 60 COVID-19 outpatients with mild to moderate symptoms	Oral	The frequency of treatment-related adverse events and the intervention's viability
NCT05839236	Phase 1	*Isatis indigotica* L. roots, *Forsythia suspensa*, Gypsum, Common Anemarrhena, reed rhizome, patchouli, tatarinow sweetflag rhizome, and *Curcuma aromatica* are combined with dextrin, sodium cyclamate, patchouli oil, peppermint oil, and angelica dahurica tincture	Open-label study in long COVID-19	–	Change in total cholesterol, triglycerides, HDL-C, LDL-C, apolipoprotein A-I, apolipoprotein B, lipopoliprotein A, eosinophil absolute number, eosinophil percentage, basophil absolute number, basophil percentage, mean corpuscular volume, mean corpuscular hemoglobin, concentration of mean corpuscular hemoglobin, red cell distribution, plateletcrit, platelet distribution and mean platelet volume

The clinical trials are undergoing Chinese traditional medicine-based therapies and a combination of herbs as immune system modulators for the prophylaxis and treatment of COVID-19. The outcomes for the trials encompass a range of measures, including recovery, severity, and duration of the symptoms, laboratory-confirmed symptoms, and disease progression.

For future clinical trial studies focusing on dietary supplements and plant-based medicine, it is essential to improve the methodological quality to thoroughly evaluate the clinical benefits and potential harms. To achieve this, it is important to ensure that the study design is rigorous, incorporating randomization and double-blinding. It has been seen that pharmaceutical companies are not proactively engaged in clinical trials for plant-based medicine due to various constraints, like patent protection and irrational use. Therefore, to increase the status of clinical studies for plant-based medicine, it is important to encourage and support pharmaceutical companies by providing funds from the government and various schemes.

22.11   CONCLUSION AND FUTURE DIRECTIVES

Several challenges and limitations have been faced during the treatment of COVID-19. The main challenges are balancing benefits and risks together, as in the case of corticosteroids. They are mainly used to cure inflammation but may lead to several complications including delayed viral clearance and increased infection risk. Despite this, ensuring effective respiratory support, the risk of local outbreaks post-herd immunity (HDI) threshold, uncertain infection and case fatality rates, and antibiotic use for co-infections are some significant challenges (Sarkar et al., 2020).

In spite of the current or modern treatment approaches like antiviral, plasma therapy, vaccines, etc., some countries such as India and China are still searching for alternative and complementary approaches for COVID-19 treatment by performing research on traditional medicines, herbal medicines, Chinese medicines, etc. (Shankar et al., 2020). However, there is a lack of strong evidence behind the effectiveness of these medicines. Governments in India and other countries are implementing these approaches for prevention and symptom relief, often alongside conventional medicine. Madagascar promoted COVID-ORGANICS for immunity boosting (Dew and Liyanagunawardena, 2023), though its effectiveness remains contested. Traditional African herbs and Unani medicine (fumigants) are also explored for managing the virus. Yoga and Ayurveda are seen as effective in managing stress and anxiety related to the pandemic. Some studies reported that homeopathic medicines like arsenicum album and Bryonia in treating flu-like symptoms like fever, cough, and headache, including during the H1N1 flu pandemic (Jain et al., 2023). While some reports suggest positive outcomes with combinations of homeopathy and modern medicine (hydroxychloroquine) or Ayurvedic preparations, these lack specific details or rigorous validation (Nayak et al., 2023; Sinha et al., 2021).

Despite several promising outcomes in preclinical trials to ensure the effectiveness of these medicines, there remain challenges in standardizing and controlling the quality of herbal medicines due to the lack of standard protocols and strict quality check norms. Lack of a strong scientific background, significant number of participants, no proper follow-up, absence of a control group, and limited mechanistic understanding while conducting clinical trials make it challenging to meet the standards. Unless research adheres to strict standard quality protocols, no effective plant-derived medicine will be identified. There is a need to perform broad, high-quality, and rigorous human trials to obtain a better outcome and effective treatment.

The one disease, one target, one drug approach is a conventional approach. There is an urgent need to change this framework because one drug can aim at only one specific target, however, multi-targeted drugs can aim at multiple targets simultaneously and can provide effective outcomes. Traditional medicines have limited side effects and they can produce synergetic effects with other drugs. Exploring their synergistic mechanism with modern medicine can pave the way for the discovery of new drugs and improved treatments. Integrating traditional medicine with modern medicine offers opportunities for collaboration, providing insights and inspiration for innovative healthcare solutions.

ACKNOWLEDGEMENT

The authors acknowledge Government College of Pharmacy, Rohru, District Shimla, Himachal Pradesh, India, for providing the resources for this manuscript.

Conflict of Interest
The authors declare no conflict of interests with respect to any aspect of this work.

REFERENCES

Ahmadi K, Farasat A, Rostamian M, Johari B, Madanchi H. Enfuvirtide, an HIV-1 fusion inhibitor peptide, can act as a potent SARS-CoV-2 fusion inhibitor: An in silico drug repurposing study. Journal of Biomolecular Structure and Dynamics. 2022;40(12):5566–76.

Alam S, Sarker MM, Afrin S, Richi FT, Zhao C, Zhou JR, Mohamed IN. Traditional herbal medicines, bioactive metabolites, and plant products against COVID-19: Update on clinical trials and mechanism of actions. Frontiers in Pharmacology. 2021;12:671498.

Alemany A, Millat-Martinez P, Corbacho-Monne M, Malchair P, Ouchi D, Ruiz Comellas A, et al. High-titre methylene blue-treated convalescent plasma as an early treatment for outpatients with COVID-19: A randomised, placebo-controlled trial. The Lancet Respiratory Medicine. 2022;10(3):278–88.

Alhazmi HA, Najmi A, Javed SA, Sultana S, Al Bratty M, Makeen HA, Meraya AM, Ahsan W, Mohan S, Taha MM, Khalid A. Medicinal plants and isolated molecules demonstrating immunomodulation activity as potential alternative therapies for viral diseases including COVID-19. Frontiers in Immunology. 2021;12:637553.

Al-Shuhaib MB, Hashim HO, Al-Shuhaib JM, Obayes DH. Artecanin of Laurus nobilis is a novel inhibitor of SARS-CoV-2 main protease with highly desirable druglikeness. Journal of Biomolecular Structure and Dynamics. 2023Apr 13;41(6):2355–67.

Al-Taee SK, Mustafa ES, Al-Saidya AM. Transmission, pathological mechanism and pathogenesis of SARS-1 and SARS-2. Egyptian Journal of Veterinary Sciences. 2024;55(3):747–58.

Altyar AE, Youssef FS, Kurdi MM, Bifari RJ, Ashour ML. The role of *Cannabis sativa L.* as a source of cannabinoids against coronavirus 2 (SARS-CoV-2): An in silico study to evaluate their activities and ADMET properties. Molecules. 2022;27(9):2797.

Alzaabi MM, Hamdy R, Ashmawy NS, Hamoda AM, Alkhayat F, Khademi NN, Al Joud SM, El-Keblawy AA, Soliman SS. Flavonoids are promising safe therapy against COVID-19. Phytochemistry Reviews. 2021;21:291–312.

Antonio da Silva A, Wiedemann LS, Veiga-Junior VF. Natural products' role against COVID-19. RSC Advances. 2020;10(39):23379–93.

Araujo R, Aranda-Martinez JD, Aranda-Abreu GE. Amantadine treatment for people with COVID-19. Archives of Medical Research. 2020;51(7):739–40.

Atanasov AG, Zotchev SB, Dirsch VM, Supuran CT. Natural products in drug discovery: Advances and opportunities. Nature Reviews Drug Discovery. 2021;20(3):200–16.

Badrunanto B, Awaluddin F, Wahyudi ST, Wahyuni WT, Batubara I. In-silico approaches potential compounds in ginger (Zingiberofficinale) as inhibitors of membrane, envelope, nucleocapsid, Plpro, and Helicase proteins of the SARS-CoV-2. In AIP Conference Proceedings 2023; Vol. 2614, No. 1. AIP Publishing.

Badshah I, Anwar M, Murtaza B, Khan MI. Molecular mechanisms of morphine tolerance and dependence; novel insights and future perspectives. Molecular and Cellular Biochemistry. 2023:1–29.

Bafandeh S, Khodadadi E, Ganbarov K, Asgharzadeh M, KöseSamadiKafil H. Natural products as a potential source of promising therapeutics for COVID-19 and viral diseases. Evidence-Based Complementary and Alternative Medicine. 2023;2023(1):5525165.

Bhattacharya R, Bose D, Maqsood Q, Gulia K, Khan A. Recent advances on the therapeutic potential with *Ocimum* species against COVID-19: A review. South African Journal of Botany. 2024;164:188–99.

Biharee A, Chaudhari L, Bhartiya S, Kori SK, Chaudhary A, Dubey D, Yadav A. A comprehensive study on natural products and their bioactive constituents to cure respiratory diseases. The Natural Products Journal. 2024;14(2):32–70.

Boulware DR, Pullen MF, Bangdiwala AS, Pastick KA, Lofgren SM, Okafor EC, et al. A randomized trial of hydroxychloroquine as postexposure prophylaxis for covid-19. The New England Journal of Medicine. 2020;383(6):517–25.

Braybrook E, Pandey S, Vryonis E, Anderson NR, Young L, Grammatopoulos DK. Screening for the alpha variant of SARS-CoV-2 (B. 1.1. 7) the impact of this variant on circulating biomarkers in hospitalised patients. medRxiv. 2021:2021–06. https://doi.org/10.1101/2021.06.18.21258699

Cao B, Wang Y, Wen D, Liu W, Wang J, Fan G, et al. A trial of lopinavir-ritonavir in adults hospitalized with severe covid-19. The New England Journal of Medicine. 2020 382(19):1787–99.

Chatterjee S. "The virus world in deep time." In From Stardust to First Cells: The Origin and Evolution of Early Life 2023; pp. 171–91. Cham: Springer International Publishing.

Chaudhury S, Kaur P, Gupta D, Anand P, Chaudhary M, Tiwari S, Mittal A, Gupta J, Kaur S, Singh VD, Dhawan D. Therapeutic management with repurposing approaches: A mystery during COVID-19 outbreak. Current Molecular Medicine. 2024, 24(6):712–33.

Chenchula S, Chandra MB, Adusumilli MB, Ghanta SN, Bommasani A, Kuttiappan A, Padmavathi R, Amerneni KC, Chikatipalli R, Ghanta MK, Reddy SS. Immunogenicity, clinical efficacy and safety of additional second COVID-19 booster vaccines against omicron and its subvariants: A systematic review. Reviews in Medical Virology. 2024;34(1):e2515.

Chrzanowski J, Chrzanowska A, Graboń W. Glycyrrhizin: An old weapon against a novel coronavirus. Phytotherapy Research. 2021;35(2):629–36.

Delang L, Abdelnabi R, Neyts J. Favipiravir as a potential countermeasure against neglected and emerging RNA viruses. Antiviral Research. 2018;153:85–94.

Demirci F, Karadağ AE, Biltekin SN, Demirci B. In vitro ACE2 and 5-LOX inhibition of *Rosmarinus officinalis L.* essential oil and its major component 1, 8-cineole. Records of Natural Products. 2022;16(2):194–9

Denham BE. Dietary supplements—Regulatory issues and implications for public health. JAMA. 2011; 306(4):428–9.

Dew K, Liyanagunawardena S. Traditional medicine and global public health. In Handbook of Social Sciences and Global Public Health 2023; pp. 221–37. Cham: Springer International Publishing.

Dunkle LM, Kotloff KL, Gay CL, et al. Efficacy and safety of NVX-CoV2373 in adults in the United States and Mexico. The New England Journal of Medicine. 2022;386(6):531–43.

Elekofehinti OO, Iwaloye O, Famusiwa CD, Akinseye O, Rocha JB. Identification of main protease of coronavirus SARS-CoV-2 (Mpro) inhibitors from Melissa officinalis. Current Drug Discovery Technologies. 2021;18(5):38–52.

Ella R, Vadrevu KM, Jogdand H, et al. Safety and immunogenicity of an inactivated SARS-CoV-2 vaccine, BBV152: A double-blind, randomised, phase 1 trial. The Lancet Infectious Diseases. 2021;21(5):637–46.

El-Saber Batiha, G., Al-Gareeb, A.I., Saad, H.M. et al. COVID-19 and corticosteroids: a narrative review. Inflammopharmacol 30, 1189–1205 (2022). https://doi.org/10.1007/s10787-022-00987-z

Elshamy AI, Mohamed TA, Ibrahim MA, Atia MA, Yoneyama T, Umeyama A, Hegazy ME. Two novel oxetane containing lignans and a new megastigmane from Paronychia arabica and in silico analysis of them as prospective SARS-CoV-2 inhibitors. RSC Advances. 2021;11(33):20151–63.

England C, TrejoMartinez J, PerezSanchez P, Karki U, Xu J. Plants as biofactories for therapeutic proteins and antiviral compounds to combat COVID-19. Life. 2023;13(3):617.

Ferner RE, Aronson JK. Chloroquine and hydroxychloroquine in covid-19. BMJ. 2020;369:m1432.

Fielding BC, da Silva Maia Bezerra Filho C, Ismail NS, Sousa DP. Alkaloids: Therapeutic potential against human coronaviruses. Molecules. 2020;25(23):5496.

Fong HH, Pauli GF, Bolton JL, Van Breemen RB, Banuvar S, Shulman L, Geller SE, Farnsworth NR. Evidence-based herbal medicine: Challenges in efficacy and safety assessments. In Current review of Chinese medicine: Quality control of herbs and herbal material. 2006;11–26. (https://doi.org/10.1142/9789812774019_0002)

Gavali J. WHO (world health organization) guidelines for standardization of herbal drugs. Education. 2017;2020.

Gottlieb RL, Nirula A, Chen P, Boscia J, Heller B, Morris J, et al. Effect of bamlanivimab as monotherapy or in combination with etesevimab on viral load in patients with mild to moderate COVID-19: A randomized clinical trial. JAMA. 2021;325(7):632–44.

Goyal M. Clinical trials in ayurveda: Issues, challenges and approaches. Ayu. 2017;38(1–2):1.

Guimaraes PO, Quirk D, Furtado RH, Maia LN, Saraiva JF, Antunes MO, et al. Tofacitinib in patients hospitalized with covid-19 pneumonia. The New England Journal of Medicine. 2021;385(5):406–15.

Hirahata T, ulQuraish R, ulQuraish A, ulQuraish S. A review of SARS-CoV-2 virology, vaccines, variants and their impact on the COVID-19 pandemic. Reviews and Research in Medical Microbiology. 2024:10–97. https://journals.lww.com/revmedmicrobiol/fulltext/9900/a_review_of_sars_cov_2_virology,_vaccines,.83.aspx

Hoang TN, Pino M, Boddapati AK, Viox EG, Starke CE, Upadhyay AA, Gumber S, Nekorchuk M, Busman-Sahay K, Strongin Z, Harper JL. Baricitinib treatment resolves lower-airway macrophage inflammation and neutrophil recruitment in SARS-CoV-2-infected rhesus macaques. Cell. 2021;184(2):460–475.

Hu Y, Zhang X, Shan LP, Liu L, Chen J. The antiviral activity of currently used medicinal plants in aquaculture and structure–activity relationship of their active ingredients. Reviews in Aquaculture. 2024;16(1):154–73.

Huang S, Liu YE, Zhang Y, Zhang R, Zhu C, Fan L, Pei G, Zhang B, Shi Y. Baicalein inhibits SARS-CoV-2/VSV replication with interfering mitochondrial oxidative phosphorylation in a mPTP dependent manner. Signal Transduction and Targeted Therapy. 2020;5(1):266.

Hyun SW, Han S, Son JW, Song MS, Kim DA, Ha SD. Development and efficacy assessment of hand sanitizers and polylactic acid films incorporating caffeic acid and vanillin for enhanced antiviral properties against HCoV-229E. Virology Journal. 2023;20(1):194.

Iketani S, Liu L, Guo Y, Liu L, Chan JF, Huang Y, et al. Antibody evasion properties of SARS-CoV-2 omicron sublineages. Nature. 2022;604(7906):553–6.

Itokawa H, Morris-Natschke SL, Akiyama T, Lee KH. Plant-derived natural product research aimed at new drug discovery. Journal of Natural Medicines. 2008;62:263–80.

Jaber SA. The effect of a small polyphenolic and terpenoids phytochemical constituent on curing and preventing of Covid-19 infections. Pharmacia. 2023;70(3):665–72.

Jachak SM, Saklani A. Challenges and opportunities in drug discovery from plants. Current Science. 2007;92(9):1251–7.

Jain A, Sharma A, Bais S, Kaur S, Verma A, Sinha P, Gupta B, Joshi V, Sharma R, Dubey A, Vyas A. Traditional herbal medicines: A prospective panacea for SARS CoV 2. Practice and Re-Emergence of Herbal Medicine. 2023;1:25.

Jayk Bernal A, Gomes da Silva MM, Musungaie DB, Kovalchuk E, Gonzalez A, Delos Reyes V, et al. Molnupiravir for oral treatment of covid-19 in nonhospitalized patients. The New England Journal of Medicine. 2022;386(6):509–20.

Jeong CH, Kim J, Kim BK, Dan KB, Min H. Antiviral effects of Korean Red Ginseng on human coronavirus OC43. Journal of Ginseng Research. 2023;47(2):329–36.

Jones BE, Brown-Augsburger PL, Corbett KS, Westendorf K, Davies J, Cujec TP, et al. The neutralizing antibody, LY-CoV555, protects against SARS-CoV-2 infection in nonhuman primates. Science Translational Medicine. 2021;13(593):eabf1906.

Kalil AC, Patterson TF, Mehta AK, Tomashek KM, Wolfe CR, Ghazaryan V, et al. Baricitinib plus remdesivir for hospitalized adults with covid-19. The New England Journal of Medicine. 2021;384(9):795–807.

Kandi V, Vadakedath S. Clinical research: An overview of study types, designs, and their implications in the public health perspective. American Journal of Clinical Medicine Research. 2021;9(2):36–42.

Karadağ A, Biltekin S, Demirci B, Demirci F. In vitro ACE2 enzyme inhibitory activity evaluation of different Salvia essential oils. ACTA Pharmaceutica Sciencia. 2022;60(2):117–124.

Kumar KS, Vani MG, Wang CS, Chen CC, Chen YC, Lu LP, Huang CH, Lai CS, Wang SY. Geranium and lemon essential oils and their active compounds downregulate angiotensin-converting enzyme 2 (ACE2), a SARS-CoV-2 spike receptor-binding domain, in epithelial cells. Plants. 2020;9(6):770.

Le-Rademacher J, Gunn H, Yao X, Schaid DJ. Clinical trials overview: From explanatory to pragmatic clinical trials. Mayo Clinic Proceedings. 2023;98(8):1241–1253. Elsevier.

Li L, Zhang W, Hu Y, Tong X, Zheng S, Yang J, et al. Effect of convalescent plasma therapy on time to clinical improvement in patients with severe and life-threatening COVID-19: A randomized clinical trial. JAMA. 2020;324(5):460–70.

Liu H, Ye F, Sun Q, Liang H, Li C, Li S, Lu R, Huang B, Tan W, Lai L. *Scutellaria baicalensis* extract and baicalein inhibit replication of SARS-CoV-2 and its 3C-like protease in vitro. Journal of Enzyme Inhibition and Medicinal Chemistry. 2021;36(1):497–503.

Logunov DY, Dolzhikova IV, Zubkova OV, et al. Safety and immunogenicity of an rAd26 and rAd5 vector-based heterologous prime-boost COVID-19 vaccine in two formulations: Two open, non-randomised phase 1/2 studies from Russia. Lancet. 2020;396(10255):887–97.

Ma L, Yao L. Antiviral effects of plant-derived essential oils and their components: An updated review. Molecules. 2020;25(11):2627.

Machado BA, Hodel KV, Fonseca LM, Pires VC, Mascarenhas LA, da Silva Andrade LP, Moret MA, Badaró R. The importance of vaccination in the context of the COVID-19 pandemic: A brief update regarding the use of vaccines. Vaccines. 2022;10(4):591.

Mahdi JG, Mahdi AJ, Mahdi AJ, Bowen ID. The historical analysis of aspirin discovery, its relation to the willow tree and antiproliferative and anticancer potential. Cell Proliferation. 2006;39(2):147–55.

Marafeli ÉA, Chibli LA, Rocha JP, de Assis RM, Pinto JE, de Pádua RM, Kreis W, Munkert J, Braga FC, Bertolucci SK. Photoconverting nets affect plant growth and levels of antiviral glucoevatromonoside and total cardenolides in *Digitalis mariana ssp. heywoodii* (P. Silva and M. Silva) Hinz. Industrial Crops and Products. 2023;204:117348.

Marconato M, Abela IA, Hauser A, Schwarzmüller M, Katzensteiner R, Braun DL, et al. Antibodies from convalescent plasma promote SARS-CoV-2 clearance in individuals with and without endogenous antibody response. Journal of Clinical Investigation. 2022;132(12):e158190.

Marconi VC, Moser C, Gavegnano C, Deeks SG, Lederman MM, Overton ET, et al. Randomized trial of ruxolitinib in antiretroviral-treated adults with human immunodeficiency virus. Clinical Infectious Diseases. 2022;74(1):95–104.

Mariappan B, Kaliyamurthi V, Binesh A. Medicinal plants or plant derived compounds used in aquaculture. In Recent advances in aquaculture microbial technology 2023; pp. 153–207. Academic Press.

Matz AJ, Qu L, Karlinsey K, Zhou B. MicroRNA-regulated b cells in obesity. Immunometabolism (Cobham). 2022;4(3):e00005.

Mease P, Hall S, FitzGerald O, van der Heijde D, Merola JF, Avila-Zapata F, et al. Tofacitinib or adalimumab versus placebo for psoriatic arthritis. The New England Journal of Medicine. 2017;377(16):1537–50.

Miller LH, Rojas-Jaimes J, Low LM, Corbellini G. What historical records teach us about the discovery of quinine. The American Journal of Tropical Medicine and Hygiene. 2023;108(1):7.

Mishra R, Chaudhary K, Mishra I. Weapons and strategies against COVID-19: A perspective. Current Pharmaceutical Biotechnology. 2024;25(2):144–58.

Misra S. Randomized double blind placebo control studies, the "Gold standard" in intervention based studies. Indian Journal of Sexually Transmitted Diseases and AIDS. 2012;33(2):131.

Mohammed FS, Uysal I, Sevindik M. A review on antiviral plants effective against different virus types. Prospects in Pharmaceutical Sciences. 2023;21(2):1–21.

Nadi A, Shiravi AA, Mohammadi Z, Aslani A, Zeinalian M. Thymus vulgaris, a natural pharmacy against COVID-19: A molecular review. Journal of Herbal Medicine. 2023;38:100635.

Nagu P, Parashar A, Behl T, Mehta V. CNS implications of COVID-19: A comprehensive review. Reviews in the Neurosciences. 2021;32(2):219–34.

Nair MS, Huang Y, Fidock DA, Polyak SJ, Wagoner J, Towler MJ, Weathers PJ. *Artemisia annua L.* extracts inhibit the in vitro replication of SARS-CoV-2 and two of its variants. Journal of Ethnopharmacology. 2021;274:114016.

Nayak D, Devarajan K, Pal PP, Ponnam HB, Jain N, Shastri V, Bawaskar R, Chinta R, Khurana A, COVID-19 Study Group. Efficacy of Arsenicum album 30C in the prevention of COVID-19 in individuals residing in containment areas: A prospective, multicenter, cluster-randomized, parallel-arm, community-based, open-label study. Complementary Medicine Research. 2023;30(5):375–85.

Nile SH, Kai G. Recent clinical trials on natural products and traditional Chinese medicine combating the COVID-19. Indian Journal of Microbiology. 2021;61:10–5.

Oliver ME, Hinks TSC. Azithromycin in viral infections. Reviews in Medical Virology. 2021;31(2):e2163.

Omotayo O, Muonde M, Olorunsogo TO, Ogugua JO, Maduka CP. Pandemic epidemiology: A comprehensive review of covid-19 lessons and future healthcare preparedness. International Medical Science Research Journal. 2024;4(1):89–107.

Owen DR, Allerton CMN, Anderson AS, Aschenbrenner L, Avery M, Berritt S, et al. An oral SARS-CoV-2 m(pro) inhibitor clinical candidate for the treatment of COVID-19. Science. 2021;374(6575):1586–93.

Paller CJ, Denmeade SR, Carducci MA. Challenges of conducting clinical trials of natural products to combat cancer. Clinical Advances in Hematology & Oncology: H&O. 2016;14(6):447.

Pandey P, Khan F, Kumar A, Srivastava A, Jha NK. Screening of potent inhibitors against 2019 novel coronavirus (Covid-19) from Allium sativum and Allium cepa: An in silico approach. Biointerface Research in Applied Chemistry. 2021;11(1):7981–93.

Parveen A, Parveen B, Parveen R, Ahmad S. Challenges and guidelines for clinical trial of herbal drugs. Journal of Pharmacy & Bioallied Sciences. 2015;7(4):329.

Pillaiyar T, Laufer S. Kinases as potential therapeutic targets for anti-coronaviral therapy. Journal of Medicinal Chemistry. 2022;65(2):955–82.

Pires SM, Reis RS, Cardoso SM, Pezzani R, Paredes-Osses E, Seilkhan A, Ydyrys A, Martorell M, SönmezGürer E, Setzer WN, AbdullRazis AF. Phytates as a natural source for health promotion: A critical evaluation of clinical trials. Frontiers in Chemistry. 2023;11:1174109.

Pojtanadithee P, Isswanich K, Buaban K, Chamni S, Wilasluck P, Deetanya P, Wangkanont K, Langer T, Wolschann P, Sanachai K, Rungrotmongkol T. A combination of structure-based virtual screening and experimental strategies to identify the potency of caffeic acid ester derivatives as SARS-CoV-2 3CLpro inhibitor from an in-house database. Biophysical Chemistry. 2024;304:107125.

Prasanth DS, Murahari M, Chandramohan V, Bhavya G, Lakshmana Rao A, Panda SP, Rao GK, Chakravarthi G, Teja N, Suguna Rani P, Ashu G. In-silico strategies of some selected phytoconstituents from Melissa officinalis as SARS CoV-2 main protease and spike protein (COVID-19) inhibitors. Molecular Simulation. 2021;47(6):457–70.

Rajak P, Ganguly A. In silico study unfolds inhibitory potential of epicatechingallate against SARS-CoV-2 entry and replication within the host cell. Mechanobiology in Medicine. 2023;1(2):100015.

Rajamanickam K, Rathinavel T, Periyannan V, Ammashi S, Marimuthu S, Nasir Iqbal M. Molecular insight of phytocompounds from Indian spices and its hyaluronic acid conjugates to block SARS-CoV-2 viral entry. Journal of Biomolecular Structure and Dynamics. 2023;41(15):7386–405.

Ratan ZA, Mashrur FR, Runa NJ, Kwon KW, Hosseinzadeh H, Cho JY. Ginseng, a promising choice for SARS-COV-2: A mini review. Journal of Ginseng Research. 2022;46(2):183–7.

RECOVERY Collaborative Group. Lopinavir-ritonavir in patients admitted to hospital with COVID-19 (RECOVERY): A randomised, controlled, open-label, platform trial. Lancet. 2020;396(10259): 1345–52.

Ren Z, Luo H, Yu Z, Song J, Liang L, Wang L, et al. A randomized, open-label, controlled clinical trial of azvudine tablets in the treatment of mild and common COVID19, a pilot study. Advanced Science (Weinh). 2020;7(19):e2001435.

Rizvi SA, Einstein GP, Tulp OL, Sainvil F, Branly R. Introduction to traditional medicine and their role in prevention and treatment of emerging and Re-emerging diseases. Biomolecules. 2022;12(10):1442.

Rouf R, Uddin SJ, Sarker DK, Islam MT, Ali ES, Shilpi JA, Nahar L, Tiralongo E, Sarker SD. Antiviral potential of garlic (*Allium sativum*) and its organosulfur compounds: A systematic update of pre-clinical and clinical data. Trends in Food Science & Technology. 2020;104:219–34.

Sadoff J, Gray G, Vandebosch A, et al. Safety and efficacy of single-dose Ad26.COV2.S vaccine against covid-19. The New England Journal of Medicine. 2021;384(23):2187–201.

Sahoo N, Manchikanti P, Dey S. Herbal drugs: Standards and regulation. Fitoterapia. 2010;81(6):462–71.

Sahoo N, Manchikanti P. Herbal drug regulation and commercialization: An Indian industry perspective. The Journal of Alternative and Complementary Medicine. 2013;19(12):957–63.

Sarkar C, Mondal M, Torequl Islam M, Martorell M, Docea AO, Maroyi A, Sharifi-Rad J, Calina D. Potential therapeutic options for COVID-19: Current status, challenges, and future perspectives. Frontiers in Pharmacology. 2020;11:572870.

Shahrajabian MH, Kuang Y, Cui H, Fu L, Sun W. Metabolic changes of active components of important medicinal plants on the basis of traditional Chinese medicine under different environmental stresses. Current Organic Chemistry. 2023;27(9):782–806.

Shankar A, Dubey A, Saini D, Prasad CP. Role of complementary and alternative medicine in prevention and treatment of COVID-19: An overhyped hope. Chinese Journal of Integrative Medicine. 2020;26(8):565.

Sharma AD, Inderjeet KA. Molecular docking and pharmacokinetic screening of eucalyptol (1, 8 cineole) from eucalyptus essential oil against SARS-CoV-2. Notulae Scientia Biologicae. 2020 Sep 29;12(3):536–45.

Sharma S. Current status of herbal product: Regulatory overview. Journal of Pharmacy & Bioallied Sciences. 2015;7(4):293.

Singh D, Bogus M, Moskalenko V, Lord R, Moran EJ, Crater GD, et al. A phase 2 multiple ascending dose study of the inhaled pan-JAK inhibitor nezulcitinib (TD-0903) in severe COVID-19. European Respiratory Journal. 2021;58(4):2100673.

Singh HP, Sharma S, Chauhan SB, Kaur I. Clinical trials of traditional herbal medicines in India: Current status and challenges. International Journal of Pharmacognosy. 2014;1:415–21.

Singh P, Gupta E, Mishra N, Mishra P. Shikimic acid as intermediary model for the production of drugs effective against influenza virus. In Phytochemicals as Lead compounds for new drug discovery 2020; pp. 245–56. Elsevier.

Sinha K, Chaudhury SS, Sharma P, Ruidas B. COVID-19 rhapsody: Rage towards advanced diagnostics and therapeutic strategy. Journal of Pharmaceutical Analysis. 2021;11(5):529–40.

Skalli S, Bencheikh RS. Safety monitoring of herb-drug interactions: A component of pharmacovigilance. Drug Safety. 2012;35:785–91.

Smieszek SP, Przychodzen BP, Polymeropoulos MH. Amantadine disrupts lysosomal gene expression: A hypothesis for COVID19 treatment. The International Journal of Antimicrobial Agents. 2020;55(6):106004.

Smolen JS, Landewe RBM, Bijlsma JWJ, Burmester GR, Dougados M, Kerschbaumer A, et al. EULAR recommendations for the management of rheumatoid arthritis with synthetic and biological disease-modifying antirheumatic drugs: 2019 update. Annals of the Rheumatic Diseases. 2020;79(6):685–99.

Solanki N, Gupta G, Chellappan DK, Singh SK, Gulati M, Paudel KR, Hansbro PM, Dua K, Bhan S, Saini M, Dureja H. Boswellic acids: A critical appraisal of their therapeutic and nutritional benefits in chronic inflammatory diseases. Endocrine, Metabolic & Immune Disorders-Drug Targets (Formerly Current Drug Targets-Immune, Endocrine & Metabolic Disorders). 2024;24(1):116–29.

Sposito B, Broggi A, Pandolfi L, Crotta S, Clementi N, Ferrarese R, et al. The interferon landscape along the respiratory tract impacts the severity of COVID-19. Cell. 2021;184(19):4953–68 e16.

Stebbing J, Phelan A, Griffin I, Tucker C, Oechsle O, Smith D, et al. COVID-19: Combining antiviral and anti-inflammatory treatments. The Lancet Infectious Diseases. 2020;20(4):400–2.

Tanriover MD, Doğanay HL, Akova M, et al. Efficacy and safety of an inactivated whole-virion SARS-CoV-2 vaccine (CoronaVac): Interim results of a double-blind, randomised, placebo-controlled, phase 3 trial in Turkey. Lancet. 2021;398(10296):213–22.

Tapia N, Hanau J, Shliozberg J, Poretsky L. Inflammation: Pathogenesis and biological markers. In Obesity, Diabetes and Inflammation: Molecular Mechanisms and Clinical Management 2023 Sep 27; pp. 1–13. Cham: Springer International Publishing.

Thakkar S, Anklam E, Xu A, Ulberth F, Li J, Li B, Hugas M, Sarma N, Crerar S, Swift S, Hakamatsuka T. Regulatory landscape of dietary supplements and herbal medicines from a global perspective. Regulatory Toxicology and Pharmacology. 2020;114:104647.

Thomas E, Stewart LE, Darley BA, Pham AM, Esteban I, Panda SS. Plant-based natural products and extracts: Potential source to develop new antiviral drug candidates. Molecules. 2021;26(20):6197.

Thompson MG, Stenehjem E, Grannis S, et al. Effectiveness of Covid-19 vaccines in ambulatory and inpatient care settings. The New England Journal of Medicine. 2021;385(15):1355–71.

Tine Y, Diédhiou A, Diallo A, Ndoye I, Baldé M, Gaye C, Ndiaye B, Wélé A, Fall D. Chemotaxonomic study of the covid-organics of Madagascar based on the chemical composition of their essential oils. International Journal of Organic Chemistry. 2023;13(2):50–6.

Truebel H, Thurston T. Danger in the valley of death: How the transition from preclinical research to clinical trials can impact valuations. Drug Discovery Today. 2020;25(12):2089–94.

Tulp OL, Sainvil F, Awan AR, Sciranka AA. An editorial overview and perspective: Can the attenuated omicron variant of covid-19 virus resolve the pandemic in 2022. SOJ Complementary and Emergency Medicine. 2022;2(1):1–0.

Umar AK, Zothantluanga JH, Luckanagul JA, Limpikirati P, Sriwidodo S. Structure-based computational screening of 470 natural quercetin derivatives for identification of SARS-CoV-2 Mpro inhibitor. PeerJ. 2023;11:e14915.

Umscheid CA, Margolis DJ, Grossman CE. Key concepts of clinical trials: A narrative review. Postgraduate Medicine. 2011;123(5):194–204.

Upadhyay RK, Upadhyay Y. Anti-HIV natural products from medicinal plants: A review. International Journal of Green Pharmacy (IJGP). 2023;17(1):27–39. https://www.researchgate.net/publication/370231629_Anti-HIV_natural_products_from_medicinal_plants_A_review

Upton R, David B, Gafner S, Glasl S. Botanical ingredient identification and quality assessment: Strengths and limitations of analytical techniques. Phytochemistry Reviews. 2020;19(5):1157–77.

Vaidya AD, Devasagayam TP. Current status of herbal drugs in India: An overview. Journal of Clinical Biochemistry and Nutrition. 2007;41(1):1–1.

Van Norman GA. Phase II trials in drug development and adaptive trial design. JACC: Basic to Translational Science. 2019;4(3):428–37.

Vieira DF, Bandeira DM, da Silva MA, de Almeida AL, Araújo M, Machado AB, Tort LF, Nacife VP, Siqueira MM, Motta FC, Pauvolid-Corrêa A. Comparative analysis of SARS-CoV-2 variants Alpha (B. 1.1. 7), Gamma (P. 1), Zeta (P. 2) and Delta (B. 1.617. 2) in Vero-E6 cells: Ultrastructural characterization of cytopathology and replication kinetics. The Brazilian Journal of Infectious Diseases. 2024;28(1):103706.

Viroli G, Kalmpourtzidou A, Cena H. Exploring benefits and barriers of plant-based diets: Health, environmental impact, food accessibility and acceptability. Nutrients. 2023;15(22):4723.

Voysey M, Costa Clemens SA, Madhi SA, et al. Single-dose administration and the influence of the timing of the booster dose on immunogenicity and efficacy of ChAdOx1 nCoV-19 (AZD1222) vaccine: A pooled analysis of four randomised trials. Lancet. 2021;397(10277):881–91.

Wang M, Cao R, Zhang L, Yang X, Liu J, Xu M, et al. Remdesivir and chloroquine effectively inhibit the recently emerged novel coronavirus (2019-nCoV) in vitro. Cell Research. 2020;30(3):269–71.

Wang Z, Chen X, Lu Y, Chen F, Zhang W. Clinical characteristics and therapeutic procedure for four cases with 2019 novel coronavirus pneumonia receiving combined Chinese and Western medicine treatment. BioScience Trends. 2020;14(1):64–8.

WHO Rapid Evidence Appraisal for COVID-19 Therapies (REACT) Working Group, Sterne JAC, Murthy S, Diaz JV, Slutsky AS, Villar J, et al. Association between administration of systemic corticosteroids and mortality among critically ill patients with COVID-19: A meta-analysis. JAMA. 2020;324(13):1330–41.

Williamson BN, Feldmann F, Schwarz B, Meade-White K, Porter DP, Schulz J, et al. Clinical benefit of remdesivir in rhesus macaques infected with SARS-CoV-2. Nature. 2020;585(7824):273–6.

Wu C, Lee SL, Taylor C, Li J, Chan YM, Agarwal R, Temple R, Throckmorton D, Tyner K. Scientific and regulatory approach to botanical drug development: A US FDA perspective. Journal of Natural Products. 2020;83(2):552–62.

Xia S, Duan K, Zhang Y, et al. Effect of an inactivated vaccine against SARS-CoV-2 on safety and immunogenicity outcomes: Interim analysis of 2 randomized clinical trials. JAMA. 2020;324(10):951–60.

Yan B, Freiwald T, Chauss D, Wang L, West E, Mirabelli C, et al. SARS-CoV-2 drives JAK1/2-dependent local complement hyperactivation. Science Immunology. 2021;6(58):eabg0833.

Yao TT, Qian JD, Zhu WY, Wang Y, Wang GQ. A systematic review of lopinavir therapy for SARS coronavirus and MERS coronavirus-a possible reference for coronavirus disease-19 treatment option. Journal of Medical Virology. 2020;92(6):556–63.

Yao Y, Choe U, Li Y, Liu Z, Zeng M, Wang TT, Sun J, Wu X, Pehrsson P, He X, Zhang Y. Chemical composition of rosemary (*Rosmarinus officinalis L.*) Extract and its inhibitory effects on SARS-CoV-2 spike Protein–ACE2 interaction and ACE2 activity and free radical scavenging capacities. Journal of Agricultural and Food Chemistry. 2023;71(48):18735–45.

Yu B, Chang J. Azvudine (FNC): A promising clinical candidate for COVID-19 treatment. Signal Transduction and Targeted Therapy. 2020;5(1):236.

Yuan Y, Jiao B, Qu L, Yang D, Liu R. The development of COVID-19 treatment. Frontiers in Immunology. 2023 Jan 26;14:1125246.

Zhang JL, Li YH, Wang LL, Liu HQ, Lu SY, Liu Y, et al. Azvudine is a thymushoming anti-SARS-CoV-2 drug effective in treating COVID-19 patients. Signal Transduction and Targeted Therapy. 2021;6(1):414.

Zhang Q, Zhao YZ, Ma HH, Wang D, Cui L, Li WJ, et al. A study of ruxolitinib response-based stratified treatment for pediatric hemophagocytic lymphohistiocytosis. Blood. 2022;139(24):3493–504.

Zhou M, Liu Y, Cao J, Dong S, Hou Y, Yu Y, Zhang Q, Zhang Y, Jia X, Zhang B, Xiao G. Bergamottin, a bioactive component of bergamot, inhibits SARS-CoV-2 infection in golden Syrian hamsters. Antiviral Research. 2022;204:105365.

Zhu FC, Guan XH, Li YH, et al. Immunogenicity and safety of a recombinant adenovirus type-5-vectored COVID-19 vaccine in healthy adults aged 18 years or older: A randomised, double-blind, placebo-controlled, phase 2 trial. Lancet. 2020;396(10249):479–88.

Zhu Y, Ouyang Z, Du H, Wang M, Wang J, Sun H, Kong L, Xu Q, Ma H, Sun Y. New opportunities and challenges of natural products research: When target identification meets single-cell multiomics. Acta Pharmaceutica Sinica B. 2022;12(11):4011–39.

23 The Potential of Plant Secondary Metabolites as a Defense against COVID-19 Infections

*Ankur Das, Raja Ahmed, Khaleda Begum,
Suraiya Akhtar, Amartya Chakraborty,
Tanisha Neog, Abhijit Sarma, and Sofia Banu*

23.1 INTRODUCTION

The global effect of COVID-19, triggered by the novel beta coronavirus SARS-CoV-2 has been profound and unparalleled (Atzrodt et al. 2020). Originating in Wuhan, China, in December 2019, the virus swiftly traversed continents, prompting the World Health Organization (WHO) to declare it a pandemic on March 11, 2020 (Cucinotta & Vanelli 2020, Zhu et al. 2020). Mainly transmitted through respiratory droplets, the virus manifests with symptoms ranging from mild respiratory issues to severe pneumonia, posing heightened risks for older adults and individuals with compromised health conditions (Jayaweera et al. 2020, Patgiri et al. 2022). Governments responded by implementing diverse measures, including lockdowns and travel restrictions, while the scientific community collaborated to expedite the development and distribution of vaccines (Bok et al. 2021, Zhou et al. 2022). The pandemic precipitated shifts in societal norms, such as increased telecommuting and digital solutions for education and healthcare (De et al. 2020). However, economic repercussions were substantial, with widespread job losses and disruptions to global supply chains. Healthcare systems faced unprecedented challenges, spurring innovations in healthcare delivery. The global community grappled with vaccine distribution challenges, equity issues, and vaccine hesitancy (Khairi et al. 2022). Evolving situations, including the emergence of virus variants, added complexity to ongoing response efforts (Markov et al. 2023). A number of SARS-CoV-2 variations have been documented, with only a few identified as variants of concern (VOCs) by the WHO. These VOCs display traits such as heightened transmissibility or virulence, decreased susceptibility to neutralization by antibodies acquired through either natural infection or vaccination, evasion of detection, or a reduction in the efficacy of treatments or vaccines (Cascella et al. 2023, WHO 2023). Variants with genetic markers linked to notable alterations are termed variants of interest (VOIs) (Aleem et al. 2023). Ultimately, the COVID-19 crisis highlighted the imperative of global solidarity, preparedness, and collaborative strategies in addressing unforeseen public health challenges, prompting a comprehensive re-evaluation of healthcare infrastructures and strategies on a global scale. The trajectory of COVID-19 confirmed cases has exhibited fluctuations since its emergence. Within a few months of the initial outbreak, there were elevated cases (Sahu et al. 2020), and subsequent spikes were observed in April and October 2021 (Agarwala et al. 2022). Notably, in January 2022, confirmed cases experienced a substantial surge (Assefa et al. 2022). According to the WHO report as of January 25, 2022, the global tally of confirmed cases reached 364 million with a tragic toll of 5 million deaths (WHO 2022). These figures indicated the dynamic nature of the pandemic, with periods of mitigation and resurgence, spotlighting the ongoing challenges in managing and controlling the impact on a global scale.

DOI: 10.1201/9781003452621-23

Various approaches and strategies have been explored for the treatment of COVID-19, and drug repurposing is among them. In addition, the pandemic has spurred an extraordinary effort in vaccine development, resulting in the authorization of multiple vaccines for emergency use and subsequent approval for general distribution (Koritala et al. 2021). Understanding the structural, and functional features, post-infection effects, and transmission mechanism of SARS-CoV-2 has opened up many aspects, which is crucial for developing targeted therapeutic interventions against COVID-19 (Das et al. 2021, Eslami et al. 2022). The SARS-CoV-2 pandemic has prompted research teams worldwide to expedite the development of vaccines to mitigate this urgent global challenge. Currently, research organizations across the globe are immersed in the development of more than 90 vaccines (Sarwar et al. 2020). These encompass various approaches, viz., inactivated and weakened virus vaccines, viral vector-based vaccines (with replicating or non-replicating viral vectors), nucleic acid vaccines (in either DNA or RNA form), protein-based vaccines (such as protein subunit vaccines), and virus-like particle approaches (Callaway 2020). Various methodologies are being employed, with a predominant focus on the spike (S) glycoprotein the primary stimulator for neutralizing antibodies, exposed on the virus's surface (Dhama et al. 2020). Also, S protein centered vaccines are responsible for the production of antibodies that inhibit both viral receptor attachments and the virus genome's uncoating event (Dhama et al. 2020). Additionally, immuno-informatics has recently played a role in identifying major B-cell epitopes and cytotoxic T lymphocytes (CTLs) (Sarwar et al. 2020). Despite these concerted efforts, no authorized vaccine or drug is available that could completely eradicate this deadly disease. Numerous antiviral drugs have undergone screening to assess their effectiveness against COVID-19 since the final recognition of a drug entails a multi-step process (Sarwar et al. 2020). In addition, antiviral medications have been repurposed, including remdesivir, ivermectin, ebselen, chloroquine, molnupiravir, bebtelovimab, sotrovimab favipiravir, etc. (Govender & Chuturgoon 2022). Notably, such antiviral drugs including remdesivir have been shown to cause serious side effects and post-treatment complications (Izcovich et al. 2022).

Nevertheless, natural therapeutics, with a long history of treating respiratory infections, include various herbal drugs and nutritional additives that have gained approval in treating COVID-19 (Chavda et al. 2022). Leveraging plant-based biopharmaceuticals presents efficient and cost-effective approaches to fortify defense against emerging infectious diseases like COVID-19 (Dhama et al. 2020, Uvarova et al. 2022). Natural medicines derived from plants offer a safe alternative, given their extensive ethno-pharmaceutical history, reducing the likelihood of cross-reactivity (Bhar et al. 2022). Plant secondary metabolites (PSMs), a diverse array of organic compounds within plants, have emerged as promising alternatives owing to their potential antiviral, anti-inflammatory, and immunomodulatory properties (Vaou et al. 2021). These compounds are renowned for their adaptability and evolutionary significance in defense mechanisms and represent valuable resources for drug development (Vivekanandhan et al. 2021). Their demonstrated antiviral activities and capacity to modulate immune responses suggest a potential role in managing severe COVID-19 cases by preventing excessive inflammation and cytokine storms (Soy et al. 2021). The varied origins and intricate chemical structures of secondary metabolites present a wide spectrum of potential drug candidates, and the incorporation of traditional medicine and ethnobotanical knowledge enhances the exploration of these natural resources (Dias et al. 2012, Nasim et al. 2022). Moreover, the complexity inherent in secondary metabolites may reduce the risk of drug resistance, adding to their allure as subjects of ongoing research in the global endeavor to combat the pandemic (Alipour et al. 2024). Natural plant-derived metabolites have gained significant attention as prophylactic agents against COVID-19, with much of the research conducted virtually due to extensive health risks (Bhar et al. 2023). Notably, epigallocatechin, predominantly found in green tea, has been identified as a successful inhibitor of SARS-CoV-2 infection (Henss et al. 2021). Another remarkable compound, curcumin, has demonstrated an attenuation effect through pro-inflammatory angiotensin II-AT1 signaling pathways (Manoharan et al. 2020). Certain PSMs have shown promise as anti-COVID-19 agents, capable of blocking essential proteins such as papain-like protease (PL^{pro}), main protease (M^{pro}), angiotensin-converting enzyme 2 (ACE2), and RNA-dependent RNA polymerase

(RdRp), which are required for the survival of the virus. Green tea, comprising epigallocatechin gallate (EGCG), along with bis-benzylisoquinoline alkaloids and neferine, are cited as effective coronavirus entry inhibitors, displaying high efficacy in inhibiting infection from new variants (Mohammadi et al. 2022). These findings highlight the potential of natural compounds in the ongoing efforts to combat COVID-19.

Therefore, drawing on historical evidence of natural products demonstrating antiviral properties, this chapter offers a comprehensive overview of the recent studies related to the utilization of PSMs as antivirals against SARS-CoV-2. Additionally, the chapter summarizes their mechanisms of action and explores the scientific methodologies employed in conducting these investigations.

23.2 HISTORICAL OVERVIEW OF PSMs AND THEIR TRADITIONAL USES

The plant kingdom showcases an abundant variety of PSMs, encompassing terpenoids, flavonoids, alkaloids, phenolic compounds, glycosides, lignans, steroids, chromones, and numerous others (Elshafie et al. 2023). Dating back to ancient thinkers such as Hippocrates and civilizations such as the Sumerians, PSMs have remained a cornerstone in traditional medicine for centuries, with historical evidence tracing their use as far back as 2600 BC (Elshafie et al. 2023, Twaij & Hasan 2022). These compounds exhibit a diverse range of properties, including, antiviral antidiabetic, antioxidant, and anticancer effects (Jain et al. 2019), while also fulfilling roles in poisoning (Twaij & Hasan 2022). Herbal medicines, which are derived from these metabolites have been found to be beneficial in healthcare (Teoh 2016). According to the WHO, more than 80% of the global population continues to rely on herbal medicine for fundamental healthcare requirements (Jain et al. 2019). The utilization of plants with medicinal properties against viruses has a long history, with numerous herbal preparations such as decoctions, leaf powders, pastes, infusions, and pills documented for their efficacy for viral infections (Akram et al. 2018, Khan et al. 2021). There exists a rich repository of plants with medicinal properties effective against human viruses (Mukhtar et al. 2008). Several natural immune boosters have proven beneficial for preventing and alleviating symptoms related to COVID-19 (Khanna et al. 2021). Notably, extracts and the metabolites of quite a number of plants have been demonstrated to be beneficial in the treatment of various associated diseases (Gasmi et al. 2023, Khan et al. 2021, Malekmohammad & Kopaei 2021). Furthermore, specific extracted compounds, i.e., quercetin, kaempferol, scutellarin, ursolic acid, baicalin, glycyrrhizin, chloroquine, etc., are identified as promising candidates for treating symptoms associated with SARS-CoV-2 infection (Jamshidnia et al. 2022, Kashyap et al. 2022). In addition, crude extracts or purified compounds derived from medicinal plants and herbs have demonstrated significant inhibitory effects against coronaviruses (Adhikari et al. 2021, Jang et al. 2023, Khan et al. 2021, Puttaswamy et al. 2020, Wink 2020). A group of natural spice and herbal secondary metabolites has been investigated for their inhibition efficacy against multiple target proteins of SARS-CoV-2 along with human ACE2 (hACE2). Computational analysis allowed the ranking of 69 natural spice and herbal secondary metabolites, along with 10 known drugs, for their inhibition efficacy against these proteins (Gupta et al. 2020, Singh et al. 2021). Several phytochemicals, including EGCG, aloe-emodin, quercetin, rhoifolin, hesperetin, sinigrin, and 3β-friedelanol from various plant sources, demonstrate antiviral activity against SAR-CoV-2 (Singh et al. 2021). Studies on Chinese medicinal herbs revealed three herbs out of a total of 312 that block the interaction between SARS-CoV-2 S protein and ACE2 (Fuzimoto & Isidoro 2020). Glycyrrhizin extracted from *Glycyrrhiza glabra* has been reported for its anti-coronavirus role, highlighting the importance of PSMs against coronaviruses (Khan et al. 2021). Specific lignoids and abietane-type diterpenoids, found in *Cryptomeria japonica*, exhibit potent SARS-CoV-2 inhibiting activities (Jamal 2022). Extracts of *Forsythiae fructus* based on traditional medicine are effective treatments for COVID-19 (Khan et al. 2021).

Owing to the user-friendly and less toxic nature of PSMs, investigating plant-based herbal medicines or edible plant parts rich in antiviral phytochemicals have emerged as promising strategies for addressing the challenges posed by the COVID-19 pandemic (Puttaswamy et al. 2020).

These compounds have gathered attention because of their antiviral activity and have found extensive applications in pharmaceutical industries (Giordano et al. 2023, Ozyigit et al. 2023). Plant cells secrete these metabolites, which have been harnessed in the development of antiviral drugs through various biotechnological approaches and tools (Jamal 2022, Panigrahi et al. 2023), aiming to address a range of viral diseases, including COVID-19, AIDS, hepatitis, and herpes (Tomas et al. 2022). In summary, medicinally important plants and secondary metabolites have significant potential in addressing viral infections, particularly to the urgent challenge posed by SARS-CoV-2.

23.3 PSMs AND THEIR ANTIVIRAL PROPERTIES

The plant kingdom showcases an abundant variety of PSMs, including terpenoids, flavonoids, alkaloids, phenolic compounds, glycosides, lignans, steroids, chromones, and numerous others (Elshafie et al. 2023). PSMs are activated in response to specific environmental stress conditions, such as attacks by pathogens or herbivores (Teoh 2016). These compounds act as a defensive mechanism against various microorganisms, it stimulates fruiting and flowering, serve as signaling molecules, and facilitate communication between plants and their surrounding environment (Xu et al. 2023). All forms of metabolites loaded in the extracts of plants in some way or another exhibit excellent forms of pharmacological activities. In the recent scenario, the COVID-19 pandemic is regarded as the most widespread disease to use PSMs as a source of several antiviral drugs, which were generally regarded as safe and were less toxic in nature (Dhamija & Mangla 2022). Most of the PSMs that showed a great deal of antiviral activities and properties include terpenoids, flavonoids, alkaloids, phytoalexins, phenanthrenes, coumarins, tannins, etc., to name a few (Giordano et al. 2023). Among the widespread PSMs, the largest class among them is the phenolics, which includes the subclasses of phenols, flavonoids, coumarins, quinones, and polyphenols, which show a wide range of pharmacological activities, especially their antiviral activities (Teoh, 2016, Guerriero et al. 2018). PSMs are known for their antiviral properties, which have proven effective in inhibiting viruses by interacting with crucial proteins and key molecules involved in essential viral processes, including replication, cleavage and processing, release, and pathogenesis (Bhuiyan et al. 2020, Gupta et al. 2020). Researchers were actively working to identify and develop plant-derived drug molecules targeting key viral pathogenicity factors, such as the S protein-hACE2-mediated viral entry, main protease (Mpro), papain-like protease 2 (PLP2), RdRp, and SARS-CoV-2 helicase (Raj et al. 2021). Also, these PSMs were seen to hinder the cell signaling pathways of the affected cell and inhibit the overall viral replication, thus preventing the spread of the virus to other cells (Bhuiyan et al. 2020, Giordano et al. 2023).

Serious diseases like COVID-19, which is caused by the SARS-CoV-2 virus, made researchers focus on prioritizing the importance of PSMs (Pal & Lal 2023). PSMs were reported to contain certain compounds under specific classes that were more useful than synthetic drugs against viral proteins (Khan et al. 2021). Previously, with the aim of repurposing, *in vitro* studies and clinical trials explored the efficacy of hydroxychloroquine, an antimalarial drug, when combined with azithromycin, a broad-spectrum antibacterial drug. This combination could reduce the severity of COVID-19, but negative effects were observed after treatment (Mégarbane & Scherrmann 2020). In addition, various off-label medicines, such as lopinavir-ritonavir, remdesivir, ribavirin, favipiravir, etc., were suggested as potential investigational drugs for treating COVID-19 (Teoh et al. 2020) but with side-effects (Izcovich et al. 2022).

23.4 EVIDENCE OF PSMs AS anti-SARS-CoV-2

Amidst the global COVID-19 pandemic, there has been an increasing focus on investigating the antiviral capabilities of natural compounds against SARS-CoV-2 (Giordano et al. 2023). Notably, numerous studies have indicated that the metabolites produced by plants exhibit significant effects on SARS-CoV-2 (Khalifa et al. 2023). This exploration demonstrates the valuable contributions that natural compounds, derived from plant sources, may offer in the ongoing efforts to develop effective therapeutics against the virus (Giordano et al. 2023, Yang et al. 2018). The PSMs and their corresponding target proteins of SARS-CoV-2, sourced from the literature, have been summarized in Table 23.1.

TABLE 23.1

Plant Secondary Metabolites and Their Target Proteins in SARS-CoV-2

Sl No.	Compound	Class	Activity Against SARS-CoV-2	References
1	Lycorine	Alkaloids	Inhibition with M^{pro}, PL^{pro}, and RdRp	Khalifa et al. 2023, Gurung et al. 2020
2	Narciprimine		Inhibition with M^{pro}	Alrasheid et al. 2021
3	Capsaicin			
4	Anisotine		Inhibition of $3CL^{pro}$, S protein, and RdRp	Ghosh et al. 2021
5	Berbamine		Inhibition of envelope protein	Xia et al. 2021
6	Ajmalicine		Inhibition of Nsp15	Kumar et al. 2021
7	Reserpine			
8	Taspine			
9	Fumigatoside E		Inhibition of ACE2	Youssef et al. 2021
10	Norquinadoline A		Inhibition of ACE2 and PL^{pro}	Ismail et al. 2021
11	Epiisopiloturine		Interaction of $3CL^{pro}$	De Sá et al. 2021
12	Chelerythrine		Interaction with $3CL^{pro}$ and RdRp	Wink 2020
13	Dicentrine			
14	Palmatine			
15	Berberrubine			
16	Berberine			
17	Chelidonine			
17	Sanguinarine			
19	Berbamine			
20	Coptisine			
21	Jatrorrhizine			
22	Berberine		Interaction of $3CL^{pro}$	Chowdhury 2020
23	Tetrahydropalmatine			
24	Tryptanthrine			
25	Indirubin			
26	Indigo			
27	Indican			Narkhede et al. 2020
28	Thalimonine			Garg and Roy 2020
29	Sophaline D			
30	Noscapine			Kumar et al. 2020
31	Cepharanthine		Inhibition of RdRp	Ruan et al. 2020
32	Piperine		Inhibition of S glycoprotein	
33	Thebaine		Interaction of ACE2	Maurya et al. 2020
34	18-hydroxy-3-epi-alphayohimbine		Inhibition of $3CL^{pro}$	Gurung et al. 2020
35	Alloyohimbine			
36	Asparagamine A			
37	Vincapusine			
38	Sophoridine			

(Continued)

TABLE 23.1 *(Continued)*

Plant Secondary Metabolites and Their Target Proteins in SARS-CoV-2

Sl No.	Compound	Class	Activity Against SARS-CoV-2	References
39	Homoharringtonine			Choy et al. 2020
40	10-hydroxyusambarensine			Gyebi et al. 2020
41	Cryptoquindoline			
42	Cryptospirolepine			
43	Chrysopentamine			
44	Speciophylline			Yepes-Pérez et al. 2020
45	Cadambine			
46	Schizanthine Z			Alfaro et al. 2020
47	Schizanthine Y			
48	Quinine			Gendrot et al. 2020
49	Cryptomisrine		Inhibition of 3CL^{pro} and RdRp	Borquaye et al. 2020
50	Cryptospirolepine			
51	Cryptoquindoline			
52	Emetine			Bleasel & Peterson
53	Ipecac			2020
54	Ergotamine			Gül et al. 2020
55	Dihydroergotamine			
56	(-)-asperlicin C			Joshi et al. 2020
57	Oriciacridone F			
58	Adlumidine		Inhibition of TMPRSS2	Vivek-Ananth et al.
59	Qingdainone			2020
60	Oxoturkiyenine			
61	3α,17α-cinchophylline			
62	Cathepsin L			
63	Caffeine		Inhibition of ACE2	Mohammadi et al. 2020
64	Nicotine			
65	Pseudojervine			Cheng 2020
66	Protopine		Inhibition of RdRp	Pandeya et al. 2020
67	Allocryptopine			
68	6-acetonyldihydro chelerythrine			
69	3'- methyl isoliquiritigenin	Phenolics and Polyphenols	Inhibition with M^{pro}, PL^{pro}, and RdRp	Khalifa et al. 2023
70	7-hydroxyflavan			
71	7-hydroxyflavanone			
72	7-hydroxyflavan-3-ol			
73	7-methoxy-3,4'-methylenedioxyflavan-3-ol			
74	7-hydroxy-3',4'-methylenedioxyflavan			
75	2',4'-dihydroxy-3'-methyl-3,4-methylenedioxychalcone			
76	Isoliquiritigenin			
77	*p*-nitrophenol		Inhibition of M^{pro}, PL^{pro}, and RdRp	
78	Quercetin		Inhibition of S-Adenosylmethionine, M^{pro}, RdRp, ACE2, and interaction with hACE2-S, Nsp16	Giordano et al. 2023; Omotuyi et al. 2020

(Continued)

TABLE 23.1 *(Continued)*
Plant Secondary Metabolites and Their Target Proteins in SARS-CoV-2

Sl No.	Compound	Class	Activity Against SARS-CoV-2	References
79	**Rutin**		Inhibition of envelope protein E, M^{pro}, PL^{pro}, ACE2	El-Mordy et al. 2020, Giordano et al. 2023
80	**4-(2-hydroxyethyl) phenol**		Inhibition of PL^{pro}	Srinivasan et al. 2022
81	**4-Hydroxybenzaldehyde**			
82	**Methyl 3,4-Dihydroxybenzoate**			
83	**Luteolin**		Interaction with RdRp and inhibition of M^{pro}	Munafo et al. 2022, Rakshit et al. 2021
84	**Isoquercetin**		Inhibition of M^{pro}	El-Hawary et al. 2022
85	**Hesperidin**		Interaction and binding with $3CL^{pro}$, S protein, and RBD-ACE2	Attia et al. 2021, Mohammadi et al. 2022, Utomo et al. 2020
86	**Catechin**		Inhibition of M^{pro} and S protein	Ghosh et al. 2021
87	**Dieckol**		Interaction and inhibition of RBD/ACE2, $3CL^{pro}$	Al-Khafaji et al. 2021, Gentile et al. 2020
88	**8,8-Bieckol**			
89	**6,6-Bieckol**		Interaction and inhibition of S protein/TMPRSS2	Al-Khafaji et al. 2021
90	**Phlorofucofuroeckol B**			
91	**3′-Methoxydaidzin**		Inhibition of $3CL^{pro}$	Harisna et al. 2021
92	**Genistin**			
93	**Neobavaisoflavone**		Inhibition of S2 protein	
94	**Methylophiopogonone A**			
95	**Epigallocatechin gallate**		Interaction with $3CL^{pro}$, S-RBD, PL^{pro}, RdRp, ACE2	Mhatre et al. 2021
96	**Theaflavin digallate**			
97	**Epigallocatechin**		Inhibition with $3CL^{pro}$	Ghosh et al. 2021
98	**Gallocatechin**			
99	**Catechin gallate**			
100	**Hydroxymatairesinol**		Interaction with PL^{pro}	Allam et al. 2021
101	**Sesamin**			
102	**Sesamolin**		Interaction with RdRp	
103	**Ferulic acid**			
104	**Genkwanin-6-C-beta-glucopyranoside**		Interaction with Nsp10	Mohammad et al. 2021
105	**Chrysin**		Interaction with Nsp9	Mohammad et al. 2021
106	**Methyl 3,4,5-trihydroxybenzoate**		Inhibition of protease enzyme	Sherif et al. 2021
107	**Punicalin**		Inhibition of S protein and TMPRSS2	Suručić et al. 2020
108	**Punicalagin**			
109	**Isorhamnetin**		Inhibition of M^{pro} and interaction of ACE2 receptor	Vicidomini et al. 2021, Zhan et al. 2021
110	**Kaempferol**		Inhibition of M^{pro} and RdRp	Shaldam et al. 2021
111	**Myricetin**		Inhibition of M^{pro}	Liu et al. 2021
112	**Hesperetin**		Inhibition of S protein, ACE2, and interaction with TMPRSS2	Cheng et al. 2021, Utomo et al. 2020
113	**Naringenin**		Inhibition of M^{pro}, TPC2, and interaction with S protein	Abdallah et al. 2021, Clementi et al. 2021, Utomo et al. 2020
114	**Chlorogenic acid**		Inhibition of M^{pro}	El-Gizawy et al. 2021

(Continued)

TABLE 23.1   *(Continued)*
Plant Secondary Metabolites and Their Target Proteins in SARS-CoV-2

Sl No.	Compound	Class	Activity Against SARS-CoV-2	References
115	Gallic acid		Inhibition of M^{pro}, Inhibition of Furin	Alrasheid et al. 2021, Suručić et al. 2020
116	p-Coumaric acid		Inhibition of M^{pro}, RdRp	Shaldam et al. 2021
117	Sinapic acid		Inhibition of envelop E protein	Orfali et al. 2021
118	Hydroxytyrosol		Interaction with S protein	Paolacci et al. 2021
119	Apigenin		Interaction with Nsp10, Nsp9, and inhibition of M^{pro}	Mohammad et al. 2021; Omotuyi et al. 2020
120	Pectolinarin		Inhibition of M^{pro} and S protein	Tallei et al. 2020
121	Rhoifolin			
122	Epicatechin			Frengki et al. 2020
123	Herbacetin		Inhibition of M^{pro}	Jo et al. 2020
124	Apigenin-7-O-rutinoside		Interaction with $3CL^{pro}$	Albohy et al. 2020
125	Didymin			
126	Prunin		Interaction of Nsp16/10	
127	Acaciin			
128	Isosakuranetin		Interaction with ACE2-PD	
129	Acacetin			
130	Salvitin			
131	Isosakuranin		Interaction with RBD-S protein	
132	Theaflavin 3-O-gallate		Inhibition with $3CL^{pro}$	Bhatia et al. 2020
133	Punicalagin			
134	Protocatechuic acid 4-O-glucoside			
135	Mearnsitrin III		Inhibition with $3CL^{pro}$	El-Mordy et al. 2020
136	Quercetin 3-O-α-L-rhamnopyranoside			
137	Thymol		Inhibition of S-RBD	Kulkarni et al. 2020
138	Carvacrol			
139	Eucalyptol			
140	Geraniin		Inhibition of S-RBD	Arokiyaraj et al. 2020
141	Kaempferitin			
142	Kaempferol 7-O-rhamnoside			
143	Tectochrysin		Inhibition of RBD/ACE2	Omotuyi et al. 2020
144	Galangin		Interaction of PD of S protein	Utomo et al. 2020
145	Myricitrin		Inhibition of M^{pro}	Cherrak et al. 2020
146	Quercitrin			
147	Theaflavin			Jang et al. 2020
148	Eriocitrin		Inhibition of RdRp	Puttaswamy et al. 2020
149	Curcumin		Interaction with SARS-CoV-2 protease, ACE2	Mohammadi & Shaghaghi 2020, Utomo et al. 2020
150	Theaflavin		Inhibition of RdRp	Lung et al. 2020
151	Dimethoxy-4-hydroxyacetophenone	Acetophenones	Inhibition of M^{pro}, PL^{pro}, and RdRp	Khalifa et al. 2023
152	2,4-dihydroxyacetophenone			
153	2,4-dihydroxy-6-methoxy-3-methylacetophenone			
154	2,4,6-trimethoxyacetophenone			

(Continued)

TABLE 23.1 *(Continued)*
Plant Secondary Metabolites and Their Target Proteins in SARS-CoV-2

Sl No.	Compound	Class	Activity Against SARS-CoV-2	References
155	β-sitosterol 3-O-β-glucopyranoside	Steroid	Inhibition of M^{pro}, PL^{pro}, and RdRp	Khalifa et al. 2023
156	Chebulagic acid	Tannins	Interaction with M^{pro}	Du et al. 2021
157	Ellagic acid		Interaction with M^{pro}, RdRp, Nsp9	Muhammad et al. 2021, Pandeya et al. 2020
158	Friedelin	Terpenoid	Interaction with S protein, and inhibition of $3CL^{pro}$ and PL^{pro}	Tungary et al. 2022, Diniz et al. 2021
159	3-β-friedelanol		Inhibition of $3CL^{pro}$ and PL^{pro}	Diniz et al. 2021
160	3-β-acetoxy friedelane			
161	Epitaraxerol			
162	Saikosaponin A			
163	Saikosaponin B2			
164	Saikosaponin C			
165	Saikosaponin D			
166	Tanshinone IIA			
167	Tanshinone IIB			
168	Methyl tanshinonate			
169	Cryptotanshinone			
170	Tanshinone I			
171	Dihydrotanshinone I			
172	Rosmaraquinone			
173	Celastrol			
174	Pristimerin			
175	Tingenone			
176	Iguesterin			
177	Ferruginol			
178	Dehydroabieta-7-one			
179	Sugiol			
180	Cryptojaponol			
181	8-β-hydroxyabieta-9(11),13-dien-12-one			
182	7-β-hydroxydeoxycryptojaponol			
183	6,7-dehydroroyleanone			
184	3-β,12-diacetoxyabieta-6,8,11,13-tetraene			
185	Pinusolidic acid			
186	Forskolin			
187	Cedrane-3-β,12 diol			
188	α-cadinol			
189	Betulinic acid			
190	Betulonic acid			
191	Glycyrrhizin		Interaction with $3CL^{pro}$ and PL^{pro}	Giofrè et al. 2021
192	Deacetylnomilin			
193	Ichangin			
194	Ichangesin			
195	Nomilin			
196	Limonin			

(Continued)

TABLE 23.1   *(Continued)*
Plant Secondary Metabolites and Their Target Proteins in SARS-CoV-2

Sl No.	Compound	Class	Activity Against SARS-CoV-2	References
197	**Obacunone**			
198	**Oleanolic acid**			
199	**Oleanolic aldehyde**			
200	**Betulinic acid**			
201	**β-amyrin**			
202	**Glycyrrhizin**			
203	**Tirucallol**			
204	**Masticadienoic acid**			
205	**β-sitosterol**			
206	**Glycyrrhizic acid**		Interaction with TMPRSS2	Puttaswamy et al. 2020
207	**Fallacinol**	Anthraquinone	Interaction with S protein	Prabhu et al. 2021
208	**Hypericin**		Interaction with M^{pro}	Puttaswamy et al. 2020
209	**4-O-Demethylbarbatic acid**	Depsidone	Interaction with S protein	Prabhu et al. 2021
210	**Bismahanine**	Carbazoles	Interaction with S protein	Puttaswamy et al. 2020
211	**Resveratrol**	Stilbenoids and Bibenzyls	Inhibition of ACE2 complex	Wahedi et al. 2020
212	**Piceatannol**			
213	**Pterostilbene**			
214	**Pinosylvin**			
215	**9,10-dihydrophenanthrene**	Phenanthrene	Interaction with 3CL^{pro}	Yamari et al. 2023
216	**1,5,7,10-tetraazaphen-9-one**		Inhibition of ACE2	Diningrat et al. 2021

Abbreviation notes: main proteases (M^{pro}), papain-like protease (PL^{pro}), RNA-dependent RNA polymerase (RdRp), 3-chymotrypsin-like cysteine protease (3CL^{pro}), spike glycoprotein (S protein), uridine specific endoribonuclease (Nsp15), angiotensin-converting enzyme 2 (ACE2), Transmembrane protease, serine 2 (TMPRSS2), human angiotensin-converting enzyme 2 (hACE2), envelope protein (E), receptor-binding domain (RBD), Two-pore channels (TPCs)

23.4.1   Phenolics and Polyphenols

Identification of the crystal structure of the main replicase enzyme, namely PLpro, helped unravel its key active residues (Günther et al. 2021). This, in turn, helped researchers identify three major compounds, namely, methyl 3, 4-dihydroxybenzoate (HE9), 4-hydroxybenzaldehyde (HBA), 4-(2-hydroxyethyl)phenol (YRL), which fall under the polyphenolic group (phenol derivatives) with potential inhibitory activity (Srinivasan et al. 2022). Most of the phenolic compounds have been shown to target specific proteins of this virus, which are mainly used in host penetration (Das et al. 2020), like 3-chymotrypsin-like protease (3CL^{pro}), main protease (M^{pro}) (Patel et al. 2021a), helicase, nucleoside-triphosphatase (NTPase) and N7-methyltransferase (N7-MTase) (Piccolella et al. 2020, Russo et al. 2020). Similarly, flavonoids are known for possessing antimicrobial, antioxidant, anti-inflammatory, and immunomodulatory properties (Abotaleb et al. 2018, Kopustinskiene et al. 2020, Liskova et al. 2020). Notably, a few studies demonstrated a strong antiviral activity against SARS-CoV-2 (Kang et al. 2012, Ngwa et al. 2020). Among them, 3′-methyl isoliquiritigenin and p-nitrophenol have shown inhibitory effects on the key viral proteins, including M^{pro}, PL^{pro}, and RdRp, showcasing their potential as a versatile inhibitor targeting multiple stages of viral replication (Khalifa et al. 2023). Additionally, apigenin, luteolin, and isorhamnetin, all belong to the class flavonoids, have demonstrated inhibitory interactions with the crucial SARS-CoV-2 proteins, viz., M^{pro}, RdRp, and the ACE2 receptor (Mohammad et al. 2021, Munafo et al. 2022, Omotuyi et al. 2020, Rakshit et al. 2021, Vicidomini et al. 2021, Zhan et al. 2021). Similarly, pectolinarin has shown promising outcomes in inhibiting both M^{pro} and the S protein (Tallei et al. 2020). Two polyphenols, named theaflavin and quercetin also exhibit inhibitory effects against a number of viral proteins, including the main replicase protein RdRp (Giordano et al. 2023, Jang et al. 2020, Lung et al. 2020). Similarly, a study on chlorogenic acid and gallic acid demonstrated inhibitory effects specifically against M^{pro} (Alrasheid et al. 2021, El-Gizawy et al. 2021). Researchers explored the inhibitory potential of p-Coumaric acid against both M^{pro} and RdRp, suggesting its dual-targeting capabilities (Shaldam et al. 2021). A phenylpropanoid, named sinapic acid, on the other hand, exhibited inhibitory effects on the viral envelope (E) protein (Orfali et al. 2021). Additionally, hydroxytyrosol, showed a significant interaction with the S protein (Paolacci et al. 2021), while, curcumin interacts with the SARS-CoV-2 protease (Mohammadi & Shaghaghi 2020). Researchers have identified these compounds through their interactions with key viral proteins, providing a basis for potential therapeutic interventions. Furthermore, flavonol glycosides displayed strong interactions with the active site of the viral RdRp (Puttaswamy et al. 2020). Similarly, flavone glycosides, phenolics, and polyphenolic compounds characterized by aromatic rings and hydroxyl (−OH) groups have demonstrated noteworthy antiviral activity (Khan et al. 2021). These collective findings contribute valuable insights into the inhibitory mechanisms of phenol derivatives by hindering the major functions associated with the viral proteins, emphasizing their potential as lead for the development of antiviral therapies.

23.4.2 TERPENOIDS

Terpenoids and their derivatives have been attributed to the inhibition of 3CL^{pro} and PL^{pro} in SARS-CoV-2. Seven tanshinones and four quinone-methide-based triterpenes, identified and isolated from various plant species, have demonstrated significant inhibition of the proteins 3CL^{pro} and PL^{pro} (Diniz et al. 2021). A few triterpenoids, in addition, have demonstrated substantial interactions with the receptor binding domain (RBD) of the viral S protein, possibly interfering with S and ACE2 interaction, and thereby viral entry (Tungary et al. 2022). Interestingly, a terpene named friedelin was able to interact with S protein, and able to inhibit 3CL^{pro} and PL^{pro} as well (Diniz et al. 2021, Tungary et al. 2022). Additionally, quite a number of PSMs including tanshinone I, celastrol, epitaraxerol, saikosaponin derivatives, pristimerin, ferruginol, etc., showed inhibitory interaction with both SARS-CoV-2 proteases viz., 3CL^{pro} and PL^{pro} (Diniz et al. 2021). A triterpenoidal saponin, named glycyrrhizinic acid could interact with Transmembrane protease, serine 2 (TMPRSS2), however, the interaction requires further validation (Puttaswamy et al. 2020).

23.4.3 ALKALOIDS

Showing effective activity against potential SARS-CoV-2 proteins like RdRp, 3CL^{pro}, and ACE2, PSMs in the class alkaloids show inhibitory results affecting the overall mechanism of host penetration and injection of the viral material inside the host (Wink 2020) (Maurya et al. 2020). Alkaloids from the plant families viz. *Amaryllidaceae, Apocynaceae, Papaveraceae, Asteraceae*, and *Solanaceae* have been reported to possess biological activities against SARS-CoV-2 (Cordell et al. 2001). In a recent study, researchers have explored the inhibitory effects of lycorine and narciprimine on key enzymes associated with viral replication, namely M^{pro}, PL^{pro}, and RdRp (Alrasheid et al. 2021, Khalifa et al. 2023). The findings suggested lycorine's inhibitory impact on all three proteins, while narciprimine and capsaicin demonstrated inhibitory effects specifically against M^{pro}, showing the potential to restrict viral replication. While, metabolites such as fumigatoside E, norquinadoline A, and thebaine target the ACE2 receptors, and likely inhibit the interaction with the S protein and viral entry (Ismail et al. 2021, Youssef et al. 2021), while piperine specifically targets the S protein (Ruan et al. 2020). However, a few alkaloids were also effective in the inhibition of TMPRSS2 (Vivek-Ananth et al. 2020).

23.4.4 ACETOPHENONES

The inhibitory potential of several acetophenones against key viral enzymes, including PL^{pro}, M^{pro}, and RdRp is well mentioned (Khalifa et al. 2023). Among the compounds studied, dimethoxy-4-hydroxyacetophenone exhibited inhibitory effects on all three proteins, suggesting its potential as a multi-target inhibitor. Similarly, 2,4-dihydroxyacetophenone, 2,4-dihydroxy-6-methoxy-3-methylacetophenone, and 2,4,6-trimethoxyacetophenone also demonstrated inhibitory activity against the same viral proteins (Khalifa et al. 2023).

23.4.5 OTHER PSMs

Steroids, including β-sitosterol 3-O-β-glucopyranoside, exhibited inhibitory effects against key viral enzymes PL^{pro}, M^{pro}, and RdRp, suggesting its potential as a broad-spectrum inhibitor (Khalifa et al. 2023). Furthermore, tannins such as ellagic acid and chebulagic acid were investigated for their interactions with viral proteins M^{pro} and RdRp (Du et al. 2021, Pandeya et al. 2020), while anthraquinones like fallacinol and hypericin exhibited interactions with the S protein and M^{pro}, respectively (Prabhu et al. 2021, Puttaswamy et al. 2020). Moreover, significant interactions of 4-O-demethylbarbatic acid, a depsidone with the viral S protein were observed (Prabhu et al. 2021). Similarly, bismahanine, a carbazole, also exhibited inhibitory interactions with the S protein (Puttaswamy et al. 2020). These findings collectively contribute valuable insights through which these natural compounds may potentially interfere with crucial components of the SARS-CoV-2 virus, offering promising avenues for further exploration and development of plant-based antiviral therapeutics.

23.5 MECHANISM OF ACTION OF PSMs AGAINST SARS-CoV-2 PROTEINS

The versatility of PSMs lies in their ability to target multiple therapeutic aspects simultaneously (Drašar 2022). Their widespread utilization in treating diverse diseases, such as viral infections and associated complications, highlights their effectiveness in addressing various health concerns. Following the onset of the COVID-19 pandemic, there has been an expedited effort to investigate and comprehend the mechanism of action of PSMs against SARS-CoV-2 (Swain et al. 2021). The main mechanism of the PSMs lies in their potential to modulate inflammatory mediators and pathways in immune response (Meng et al. 2011), as well as inhibit the formation of plaque and adverse effects of cytokine storm by either interacting directly with the viral proteins and disrupting its functions or minimizing the effect of interferons (IFN) and interleukins (IL) (Du et al. 2021, Lin et al. 2020). To gain better insight into the molecular mechanism, research has been undertaken focusing on understanding the interactions of PSMs with critical proteins essential for the survival of the virus through *in silico, in vitro,* and *in vivo* experiments (Anand et al. 2021, Bijelić et al. 2022, Kanchibhotla et al. 2022). The main structural proteins viz., nucleocapsid (N) protein and S glycoprotein present on the surface of SARS-CoV-2, and the accessory protein open reading frame 3a (ORF3a) is involved in the inhibitory interaction through hydrogen and hydrophobic interactions with a number of PSMs (Fam et al. 2023, Majnooni et al. 2021, Yang et al. 2010). Inhibition of the S protein leads to the inability of the virus to penetrate the host, while the intervention of N protein can greatly lower the immune responses in the host (Yang & Rao 2021), while inhibition of accessory proteins, such as ORF3a, which is a potassium channel, can affect the process of viral release (Gorkhali et al. 2021). Inhibition of E protein channels by a novel plant-derived chemical revealed *in vitro* protective activity possibly preventing the internalization of the particles (Xia et al. 2021). Interestingly, interactions of certain PSMs and ACE2 protein were reported, additionally illustrating their association in inhibition of major replicative enzyme and cleavage enzyme viz., RdRp and 3CL^{pro}, respectively of SARS-CoV-2 (Maurya et al. 2020, Pandeya et al. 2020, Wink 2020). Inhibition of RdRp potentially resists its association with the other non-structural proteins (Nsp), leading to blockage of the RNA synthesis process (Gorkhali et al. 2021). Inhibition of viral proteases, such as 3CL^{pro} can significantly reduce the replication efficiency and suppression of host immune system responses (Gorkhali et al. 2021). PSMs, specifically flavonoids, have been shown to act through inhibition or reduction in the inflammatory cytokines such as IL-6, IL-1β, TNF-α, IL-8, IFN-γ, IL-33 leading to inhibition in SARS-CoV-2 induced inflammatory storm (Al-Rikabi et al. 2020, Fei et al. 2019, Guo et al. 2019, Jung et al. 2016, Salaverry et al. 2020). In addition, hydrogen-bond interaction between naringenin and active site residues of M^{pro} of SARS-CoV-2 possibly stops the cleavage of polyproteins involved in replication (Khaerunnisa et al. 2020). On the other hand, most of the phenolic compounds including ferulic acid, chlorogenic acid, and caffeic acid target the 3CL^{pro} protease activity of SARS-CoV-2 (Nguyen et al. 2021, Xiong et al. 2021), and alkylphenol such as ginkgolic acids also shows inhibitory effects by interfering with the replication of the enveloped virus (Borenstein et al. 2020). To be noted, a phenolic derivative, namely, 2-methyl-6-(1-phenylethyl) phenol is a potent inhibitor as it formed an inhibitory hydrogen bond and hydrophobic interaction with Nsp13 (helicase) that possibly interfered with the replication of the virus (Zia et al. 2021). Phenolic compounds such as emodin, resveratrol, and piceatannol showed an inhibitory nature toward interaction between viral S protein and hACE2, possibly prohibiting entry (Singh et al. 2021, Zhou et al. 2020). Similarly, alkaloids, have been demonstrated as potent DNA-interacting agents, inhibiting the major replicase proteins such as S, N, and E proteins and the main protease 3CL^{pro} (Fam et al. 2023, Majnooni et al. 2021, Yang et al. 2010). In addition, a few alkaloids, namely taspine and berberine, have the potential to block Nsp15 (endoribonuclease), which is responsible for the cleaving of 3′ uridines and hiding from antivirals (Kumar et al. 2021). Thus, Nsp15-targeted metabolites can be a potential antiviral agent. Molecular docking studies revealed inhibitory binding activity of alkaloids such as anisotine and solanidine against M^{pro} (Ghosh et al. 2021, Ubani et al. 2020), and ergota and dihydroergotamine against RdRp and 3CL^{pro}, thus interfering with the overall processing and replication of the virus (Gül et al. 2020). Tannins are known for their antiviral properties, thus, they are effective against SARS-CoV-2 3CL^{pro}, TMPRSS2, entry, and plaque formation (Chiou et al. 2022, Du et al. 2021, Wang et al. 2020). 3CL^{pro} and PL^{pro} were major targets of tanshinone derivatives, while shikonin targeted the binding pocket residues of 3CL^{pro} through non-covalent interactions, both belonging to quinone type of PSMs (Lim et al. 2021, Ma & Wang 2022, Zhao et al. 2021). Similarly, certain terpenoids including 6-Oxoisoiguesterin, isoiguesterin, and 20-Epi-isoiguesterinol interact through the H-bond with the active residues viz., GLN189, THR292, and THR24, respectively, of SARS-CoV-2 3CL^{pro}, possibly inhibiting its function (Gyebi et al. 2020). A marine terpenoid, named stachyflin was considered the lead compound against RdRp, which acts through interacting with the active site residues, and thus inferring with the viral replication (Sahoo et al. 2021). Interestingly, evaluation of several combinations of cannabidiol and terpenes could lower the SARS-CoV-2 infectivity through immune-system modulation and significantly reduce the effect of cytokine storms (Santos et al. 2022). An overview of the different classes of PSMs and their target proteins of SARS-CoV-2 is presented in Figure 23.1. Although a number of studies have been carried out to unravel the bioactivity of the PSMs as anti-SARS-CoV-2, further investigation to elucidate the molecular mechanism of action to approve it for further validation is required. For instance, viral response to certain PSMs tends to vary across different cell lines, indicating the requirement of multiple testing to approve it for *in vivo* and clinical trials against SARS-CoV-2 and COVID-19 (Khadilkar et al. 2023)

Nevertheless, although there are US Food and Drug Administration-approved drugs for COVID-19, plant-derived natural products or PSMs are still believed to be potential alternatives with lesser side effects. Although a few PSMs are toxic, the correct dose can act as a remedy, and thus, exploration of plant-based products or metabolites is increasing with time (Mahmood et al. 2020). In addition, there has been evidence of PSMs disrupting cell membranes, interfering with nucleotide synthesis and metabolism, and inhibiting cell communication (Anandhi et al. 2014, Chitemerere &

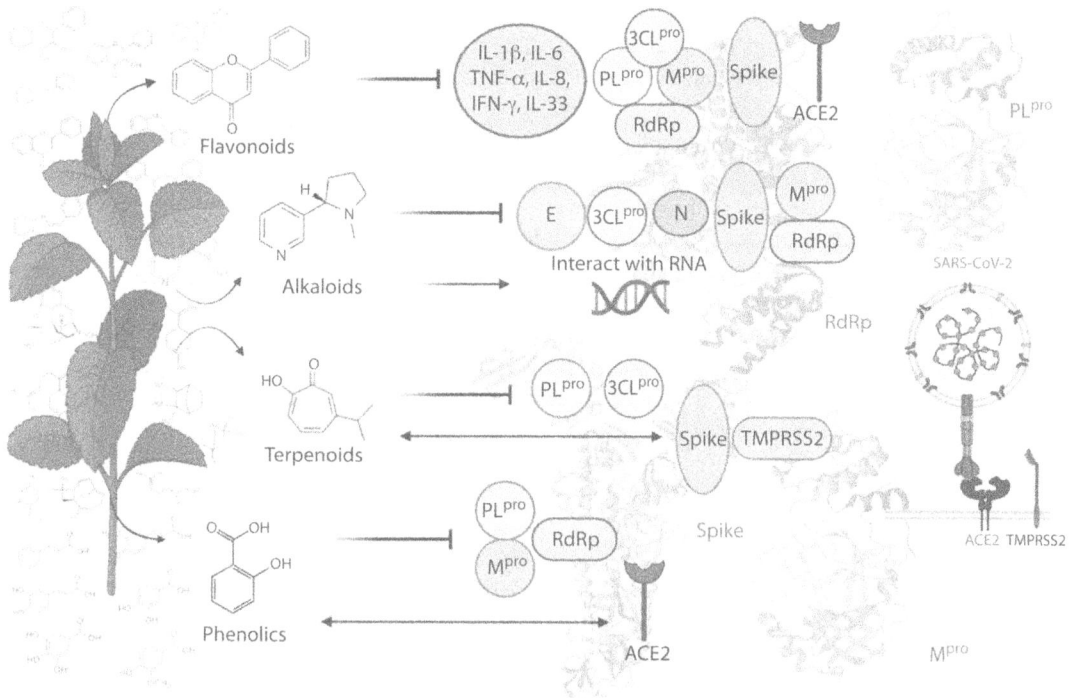

FIGURE 23.1 Schematic representation of different classes of plant secondary metabolites (PSMs) and their target proteins of SARS-CoV-2. The two-sided arrow mark indicates interaction, and the blunt arrow indicates inhibition between PSMs and SARS-CoV-2 proteins.

Mukanganyama 2014). Moreover, close genetic and protein similarities observed among strains of the coronavirus beta family contribute to the rationale for considering PSMs as viable candidates. Furthermore, there have been reported instances of anti-SARS-CoV-2 activity associated with these compounds (Kaur et al. 2021). With the advancement of analytical techniques, isolation and production of lead metabolites from plants have become easier and more convenient (Patel et al. 2021b), and with the evolution of tools such as molecular farming, the production of bioactive PSMs in large-scale through metabolic engineering have overcome the availability limitation (Mohammadinejad et al. 2019). Overall, this can facilitate in deciphering and discovery of novel-plant derived drugs against SARS-CoV-2.

23.6 TECHNOLOGIES AND CHALLENGES

To evaluate the antiviral properties of PSMs, various research groups have employed three approaches, including *in vivo*, *in vitro*, and *in silico*, each contributing valuable insights into the assessment of PSM efficacy against SARS-CoV-2 activity (Ponticelli et al. 2023, Raihan et al. 2021).

23.6.1 *In vivo*

The *in vivo* technique involves a suitable animal model with the desired viral infection, exhibiting clinical symptoms identical to those seen in humans. The prevention and control of SARS-CoV-2 necessitate the search for suitable animal models for viral investigation through preclinical experiments (Yuan et al. 2020). A number of animals currently used as models for COVID-19 research, namely, mice, ferrets, golden Syrian hamsters, cats, dogs, and a few non-human primates, including

African green monkeys, rhesus macaques, and common marmosets for virus characterization and development of antiviral drugs (Rosa et al. 2021). The limitations of the models such as the expression of the ACE2 receptor in the cell surface and other parameters must be examined during experimental design. The most suitable models will vary depending on the specific type of research and should provide authentic and translatable results for humans (Choudhary et al. 2022). There is currently little information available on *in vivo* studies that examine the antiviral activity in animal models infected with SARS-CoV-2 (Jantan et al. 2023). Most of the limitations of this technique are the lack of an animal model capable of reproducing the entire spectrum of the COVID-19 phenotype, and limited lethality and extrapulmonary manifestations. In addition, there are no severe clinical symptoms such as cytokine storm, respiratory failure, or multiple-organ failure, and the inability to replicate comorbidities specific to COVID-19 that already existed, such as diabetes, obesity, hypertension, and chronic obstructive pulmonary disease. Moreover, increased expenses as a result of stricter biosafety level 3 (BSL-3) regulations and variations in the immune systems and anatomy between species are noteworthy (Choudhary et al. 2022).

23.6.2 *IN VITRO*

The *in vitro* assay is an effective tool for assessing the antiviral activity of a prospective repurposed medication against COVID-19 (Nowakowska et al. 2022). An *in vitro* cell line is necessary to comprehend the virus's life cycle, isolate and amplify the virus for additional study, and assess therapeutic compounds (Takayama 2020). Cell culture methods provide a rapid and reliable means of assessing antiviral activity by analyzing the virus's ability to replicate and infect (Panda et al. 2021). Various cell lines are employed for replicating and isolating the SARS-CoV-2 virus, including human airway epithelial cells, Caco-2 cells, Vero E6 cells, Calu-3 cells, HEK293T, and Huh7 (Takayama 2020). Vero E6 cells stand out as the most frequently utilized clone among the mentioned cell lines for the replication and isolation of SARS-CoV-2 (Chen et al. 2023, Ogando et al. 2020). The *in vitro* approach employs three key antiviral tests: plaque-reduction neutralization tests (PRNTs), plaque assays, and real-time polymerase chain reaction (RT-PCR) analysis. However, it is important to note that each of these assays can be labor intensive and time consuming. The initial antiviral assay conducted in biologically relevant cells typically takes 2 to 3 days to complete, while an additional 3 to 4 days are required for plaque counting in Vero cells or for the completion of cell extractions and PCR analysis (Clair et al. 2023). In recent years, plate-based cytometers have shown notable capabilities in high-throughput vaccination screening techniques. These innovative methods offer a valuable platform for assessing potential antiviral medication candidates efficiently (Clair et al. 2023, Lopez-Gomez et al. 2023).

The selection of the *in vitro* method over *in vivo* approaches has several advantages, including cost-effectiveness, the ability to promptly measure product performance, and ethical considerations. This approach eliminates the need for drug testing in animal models or humans, thereby addressing ethical concerns and providing more immediate insights into the measured performance of the product (Polli 2008). However, it has its limitations as well. First, under physiological conditions, the tropism of the cell lines may not precisely mimic the multiplication and impact of SARS-CoV-2 on human organs (Chu et al. 2020). Second, testing was conducted using a homogenous viral isolate, despite reported heterogeneity in SARS-CoV-2 (Chu et al. 2020). Third, ACE2 expression exhibits heterogeneity across the various cell types found in human organs and tissues (Hikmet et al. 2020). Consequently, SARS-CoV-2 infection rates and mechanisms in various cell types may not be fully revealed by employing immortalized or cancer cell lines (Pandamooz et al. 2022). Furthermore, the majority of these human cancer cell lines have mutations linked to tumors, like those in P53, which control viral replication (Ma-Lauer et al. 2016). Concerns are raised about whether the biology of SARS-CoV-2 is recapitulated in a physiological setting by these cancer cell lines. Furthermore, because the majority of these cells come from malignancies, this impacts the transfection efficiency of the virus (Pandamooz et al. 2022).

23.6.3 *In Silico*

In silico studies have emerged as an effective technique in the drug development process for identifying potential PSMs with inhibitory and interaction affinity to SARS-CoV-2 proteins (Mushebenge et al. 2023). In the past century, *in silico* technology has significantly improved the quality of healthcare studies by enabling advanced prediction and analysis capabilities (Moradi et al. 2022). Under current research, techniques such as virtual screening and molecular docking and dynamics, utilizing crystal structures of viral targets from SARS-CoV-2 or related viruses like SARS-CoV, are being explored with PSMs (Agamah et al. 2020). Molecular docking is one of the most popular *in silico* techniques for predicting the relationship between molecules and biological targets. This is achieved through the computation of a ligand's molecular familiarization with a receptor, followed by the utilization of a docking score to determine its correlation (Pinzi & Rastelli 2019). The bioactive PSMs are ranked based on the binding strength/energy and interaction with the target proteins and compared with that of a known inhibitor or approved drugs (Kushari et al. 2023). *In silico* studies provide a solid prediction for drug design, however, it does not guarantee that a molecule would completely block or activate a target protein (Danao et al. 2023). Under the umbrella of *in silico* methods, various tasks are being pursued, including molecular modeling, virtual screening, phylogenetic analysis, machine learning classification of COVID-19 data, development of online tools utilizing COVID-19 acquired data, optimization and antimicrobial peptides construction, design of peptidomimetics, and the development of multi-epitope COVID-19 vaccines (Moradi et al. 2022).

The Nsps of SARS-CoV-2, such as PL^pro, M^pro, and RdRp, serve as primary target molecules in molecular docking and dynamics for the design of antivirals (Hosseini et al. 2021). Additionally, other proteins including S, RdRp, ACE2, RBD, and endonuclease (Nsp15) have been employed as targets in the pursuit of identifying potential repurposed medicines for COVID-19 (Parihar et al. 2022). The outcomes of *in silico* studies reveal robust binding interactions between certain PSMs, such as curcumin, apigenin, chrysophanol, and gingerol, with the S glycoprotein. Additionally, compounds like laurolistine and acetoside exhibit strong binding affinity with the M^pro, suggesting their potential as candidates for anti-SARS-CoV-2 treatments (Kushari et al. 2023). Most recent investigations conducted through an *in silico* approach to delineate the antiviral properties of PSM have been demonstrated. A total of 32 phytochemical constituents (PCCs) were used as ligand molecules against the SARS-CoV-2 M^pro binding site residues (Mukherjee et al. 2022). Among them, epicatechin-gallate, quercetin, and kaempferol showed stable binding affinity to the M^pro active site. Their binding energies were −8.5, −7.6, and −7.3 kcal/mol, respectively, illustrating that epigallocatechin-gallate exhibits the highest affinity when compared to the re-docked inhibitor binding energy. The antioxidants, namely, capsaicin, epicatechin, and ellagic acid significantly bind and interact with M^pro, the most important active site residue, Cys145, and can effectively suppress its activity (Pandey & Verma 2020). In addition, another research group could identify 26 antagonists of SARS-CoV-2 3CL^pro (Samad et al. 2023).

In silico approaches serve as a valuable platform for assessing the activity of potential therapies against molecular targets, enabling the identification of the appropriate candidates for subsequent *in vitro* and *in vivo* studies. By concentrating on selected objectives, this approach effectively reduces the costs associated with laboratory trials, which demand significant financial and human resources (Sadeghi et al. 2021). Moreover, this approach provides valuable information regarding side effects, drug-like properties, and the toxicity of candidate compounds (Agamah et al. 2020). Overall, the physiochemical and pharmacokinetics parameters of the candidate PSMs can be evaluated. However, the results of *in silico* studies required *in vitro* and *in vivo* assessment for validation purposes to use as an alternative for the treatment of COVID-19 (Samad et al. 2023, Danao et al. 2023).

23.7 CONCLUSION

Plants serve as prolific sources of various secondary metabolites with potent antiviral properties. Natural medicines present a safe avenue, considering the extensive ethno-pharmaceutical history of many plant-based remedies with fewer side effects and toxicity, they are regarded as a promising and safe option. The literature summarized in this chapter has demonstrated a number of PSMs that possess bioactivity against the majority of the SARS-CoV-2 proteins, including 3CL^{pro}, M^{pro}, RdRp, TMPRSS2, and Nsp15, as per evidence obtained through *in silico*, *in vitro* and *in vivo* approaches. The PSMs included in the compound classes of phenolics, alkaloids, terpenoids, and a few others have shown possibilities of developing into effective therapeutics agents or drugs against COVID-19. The traditional medicinal use of plant-derived compounds coupled with their observed antiviral properties, provides a strong foundation for further research and development in this zone. The diverse range of PSMs can potentially target emerging variable strains that will enhance novel therapeutic outcomes for COVID-19 treatment and reduce the chances of drug resistance. Designing and developing safe PSM-derived potential antiviral drugs demands rigorous experimentation and exploration as well as an ethical approach to clinical and preclinical trials. There is a need to emphasize the development of validated *in vitro* techniques or strategies that can serve as reliable alternatives to *in vivo* tests for assessing the efficacy of vaccines or drugs. Incorporating both traditional herbal remedies and advanced molecular technologies, this holistic approach holds tremendous scope in the pursuit of effective treatments for SARS-CoV-2, representing a pivotal contribution to the worldwide endeavor for enhanced health and the management of COVID-19. Therefore, full-fledged research on COVID-19 concentrating on the current limitations is required for the development of effective treatment strategies against SARS-CoV-2-related diseases or complications.

REFERENCES

Abdallah, H.M., El-Halawany, A.M., Sirwi, A., et al. (2021). Repurposing of some natural product isolates as SARS-COV-2 main protease inhibitors via *in vitro* cell free and cell-based antiviral assessments and molecular modeling approaches. Pharmaceuticals, 14(3), 1–18. https://doi.org/10.3390/ph14030213

Abotaleb, M., Samuel, S.M., Varghese, E., et al. (2018). Flavonoids in cancer and apoptosis. Cancers, 11(1), 1–13. https://doi.org/10.3390/cancers11010028

Adhikari, B., Marasini, B.P., Rayamajhee, B., et al. (2021). Potential roles of medicinal plants for the treatment of viral diseases focusing on COVID-19: A review. Phytotherapy Research, 35(3), 1298–12131. https://doi.org/10.1002/ptr.6893

Agamah, F.E., Mazandu, G.K., Hassan, R., et al. (2020). Computational *in silico* methods in drug target and lead prediction. Briefings in Bioinformatics, 21(5), 1663–1675. https://doi.org/10.1093/bib/bbz103

Agarwala, P., Bhargava, A., Gahwai, D.K., et al. (2022). Epidemiological characteristics of the COVID-19 pandemic during the first and second waves in Chhattisgarh, Central India: A comparative analysis. Cureus, 14(4), 1–12. https://doi.org/10.7759/cureus.24131

Akram, M., Tahir, I.M., Shah, S.M.A., et al. (2018). Antiviral potential of medicinal plants against HIV, HSV, influenza, hepatitis, and coxsackievirus: A systematic review. Phytotherapy Research, 32(5), 811–822. https://doi.org/10.1002/ptr.6024

Albohy, A., Zahran, E.M., Abdelmohsen, U.R., et al. (2020). Multitarget *in silico* studies of *Ocimum menthifolium*, family *Lamiaceae* against SARS-CoV-2 supported by molecular dynamics simulation. Journal of Biomolecular Structure & Dynamics, 40(9), 4062–4072. https://doi.org/10.1080/07391102.2020.1852964

Aleem, A., Samad, A.B.A., & Vaqar, S. (2023). Emerging variants of SARS-CoV-2 and novel therapeutics against coronavirus (COVID-19). Florida, United States: StatPearls Publishing. http://www.ncbi.nlm.nih.gov/books/nbk570580/

Alfaro, M., Alfaro, I., & Angel, C. (2020). Identification of potential inhibitors of SARS-CoV-2 papain-like protease from tropane alkaloids from *Schizanthus porrigens*: A molecular docking study. Chemical Physics Letters, 761, 1–10. https://doi.org/10.1016/j.cplett.2020.138068

Alipour, Z., Zarezadeh, S., Ghotbi-Ravandi, A.A. (2024). The potential of anti-coronavirus plant secondary metabolites in COVID-19 drug discovery as an alternative to repurposed drugs: A Review. Planta Medica, 90(3), 172–203. https://doi.org/10.1055/a-2209-6357

Al-Khafaji, K., Saygili, E.I., Taskin-Tok, T., et al. (2021). Investigation of promising antiviral candidate molecules based on algal phlorotannin for the prevention of COVID-19 pandemic by *in silico* studies. Biochemistry & Molecular Biology Journal, 7(1), 1–11.

Allam, A.E., Amen, Y., Ashour, A., et al. (2021). *In silico* study of natural compounds from sesame against COVID-19 by targeting M^{pro}, PL^{pro} and RdRp. RSC Advances, 11(36), 22398–22408. https://doi.org/10.1039/d1ra03937g

Alrasheid, A.A., Babiker, M.Y., & Awad, T.A. (2021). Evaluation of certain medicinal plants compounds as new potential inhibitors of novel corona virus (COVID-19) using molecular docking analysis. *In Silico* Pharmacology, 9(1). https://doi.org/10.1007/s40203-020-00073-8

Al-Rikabi, R., Al-Shmgani, H., Dewir, Y. H., et al. (2020). *In vivo* and *in vitro* evaluation of the protective effects of hesperidin in lipopolysaccharide-induced inflammation and cytotoxicity of cell. Molecules, 25(3), 1–11. https://doi.org/10.3390/molecules25030478

Anand, A.V., Balamuralikrishnan, B., Kaviya, M., et al. (2021). Medicinal plants, phytochemicals, and herbs to combat viral pathogens including SARS-CoV-2. Molecules, 26(6), 1–28. https://doi.org/10.3390/molecules26061775

Anandhi, D., Srinivasan, P.T., Kumar, G.P., et al. (2014). DNA fragmentation induced by the glycosides and flavonoids from *C.coriaria*. International Journal of Current Microbiology and Applied Sciences, 3(1), 666–673.

Arokiyaraj, S., Stalin, A., Kannan, B.S., et al. (2020). Geranii herba as a potential inhibitor of SARS-COV-2 main 3CLPro, spike rbd, and regulation of unfolded protein response: An *in silico* approach. Antibiotics, 9(12), 1–16. https://doi.org/10.3390/antibiotics9120863

Assefa, Y., Gilks, C.F., Reid, S. (2022). Analysis of the COVID-19 pandemic: Lessons towards a more effective response to public health emergencies. Globalization and Health, 18(1), 1–13. https://doi.org/10.1186/s12992-022-00805-9

Attia, G.H., Moemen, Y.S., Youns, M., et al. (2021). Antiviral zinc oxide nanoparticles mediated by hesperidin and *in silico* comparison study between antiviral phenolics as anti-SARS-CoV-2. Colloids and Surfaces B: Biointerfaces, 203, 1–7. https://doi.org/10.1016/j.colsurfb.2021.111724

Atzrodt, C.L., Maknojia, I., McCarthy, R.D.P., et al. (2020). A guide to COVID-19: A global pandemic caused by the novel coronavirus SARS-CoV-2. The FEBS Journal, 287(17), 3633–3650. https://doi.org/10.1111/febs.15375

Bhar, A., Jain, A., & Das, S. (2022). Natural therapeutics against SARS CoV2: The potentiality and challenges. Vegetos, 36(2), 322–331. https://doi.org/10.1007/s42535-022-00401-7

Bhatia, S., Giri, S., Lal, F.A., et al. (2020). Identification of potential inhibitors of dietary polyphenols for SARS-CoV-2 M protease: An in silico study. Tropical Public Health, 1(1), 21–29.

Bhuiyan, F.R., Howlader, S., Raihan, T., et al. (2020). Plants metabolites: Possibility of natural therapeutics against the COVID-19 pandemic. Frontiers in Medicine, 7, 1–26. https://doi.org/10.3389/fmed.2020.00444

Bijelić, K., Hitl, M., & Kladar, N. (2022). Phytochemicals in the prevention and treatment of SARS-CoV-2-clinical evidence. Antibiotics, 11(11), 1–16. https://doi.org/10.3390/antibiotics11111614

Bleasel, M.D., & Peterson, G.M. (2020). Emetine, Ipecac, Ipecac alkaloids and analogues as potential antiviral agents for coronaviruses. Pharmaceuticals, 13(3), 1–9. https://doi.org/10.3390/ph13030051

Bok, K., Sitar, S., Graham, B.S., et al. (2021). Accelerated COVID-19 vaccine development: Milestones, lessons, and prospects. Immunity, 54(8), 1636–1651. https://doi.org/10.1016/j.immuni.2021.07.017

Borenstein, R., Hanson, B.A., Markosyan, R.M., et al. (2020). Ginkgolic acid inhibits fusion of enveloped viruses. Scientific Reports, 10(1), 1–12. https://doi.org/10.1038/s41598-020-61700-0

Borquaye, L.S., Gasu, E.N., Ampomah, G.B., et al. (2020). Alkaloids from *Cryptolepis sanguinolenta* as potential inhibitors of SARS-CoV-2 viral proteins: An *in silico* study. BioMed Research International, 2020, 1–14.

Callaway, E. (2020). The race for coronavirus vaccines: A graphical guide. Nature, 580 (7805), 576–577. https://doi.org/10.1038/d41586-020-01221-y

Cascella, M., Rajnik, M., Aleem, A., et al. (2023). Features, evaluation, and treatment of coronavirus (COVID-19). Florida, United States: StatPearls Publishing. https://www.ncbi.nlm.nih.gov/books/NBK554776/

Chavda, V.P., Patel, A.B., Vihol, D., et al. (2022). Herbal remedies, nutraceuticals, and dietary supplements for COVID-19 management: An update. Clinical Complementary Medicine and Pharmacology, 2(1), 1–13. https://doi.org/10.1016/j.ccmp.2022.100021

Chen, D.Y., Turcinovic, J., Feng, S., et al. (2023). Cell culture systems for isolation of SARS-CoV-2 clinical isolates and generation of recombinant virus. iScience, 26(5), 1–22. https://doi.org/10.1016/j.isci.2023.106634

Cheng, J. (2020). Exploring the active compounds of traditional mongolian medicine Agsirga in intervention of novel coronavirus (2019-nCoV) based on HPLC-Q-Exactive-MS/MS and molecular docking method. ChemRxiv. https://doi.org/10.26434/chemrxiv.11955273.v1

Cheng, F.J., Huynh, T.K., Yang, C.S., et al. (2021). Hesperidin is a potential inhibitor against SARS-CoV-2 infection. Nutrients, 13(8), 1–13. https://doi.org/10.3390/nu13082800

Cherrak, S.A., Merzouk, H., & Mokhtari-Soulimane, N. (2020). Potential bioactive glycosylated flavonoids as SARS-CoV-2 main protease inhibitors: A molecular docking and simulation studies. PLoS ONE, 15(10), 1–14. https://doi.org/10.1371/journal.pone.0240653

Chiou, W.C., Chen, J.C., Chen, Y.T., et al. (2022). The inhibitory effects of PGG and EGCG against the SARS-CoV-2 3C-like protease. Biochemical and Biophysical Research Communications, 591, 130–136. https://doi.org/10.1016/j.bbrc.2020.12.106

Chitemerere, T.A., & Mukanganyama, S. (2014). Evaluation of cell membrane integrity as a potential antimicrobial target for plant products. BMC Complementary and Alternative Medicine, 14(1), 1–8. https://doi.org/10.1186/1472-6882-14-278

Choudhary, S., Kanevsky, I., & Tomlinson, L. (2022). Animal models for studying COVID-19, prevention, and therapy: Pathology and disease phenotypes. Veterinary Pathology, 59(4), 516–527. https://doi.org/10.1177/03009858221092015

Chowdhury, P. (2020). In silico investigation of phytoconstituents from Indian medicinal herb 'Tinospora cordifolia (giloy)' against SARS-CoV-2 (COVID-19) by molecular dynamics approach. Journal of Biomolecular Structure & Dynamics, 39(17), 6792–6809. https://doi.org/10.1080/07391102.2020.1803968

Choy, K., Wong, A.Y.L., Kaewpreedee, P., et al. (2020). Remdesivir, lopinavir, emetine, and homoharringtonine inhibit SARS-CoV-2 replication in vitro. Antiviral Research, 178, 1–5. https://doi.org/10.1016/j.antiviral.2020.104786

Chu, H., Chan, J.F., Yuen, T.T., et al. (2020). Comparative tropism, replication kinetics, and cell damage profiling of SARS-CoV-2 and SARS-CoV with implications for clinical manifestations, transmissibility, and laboratory studies of COVID-19: An observational study. The Lancet Microbe, 1(1), e14–e23. https://doi.org/10.1016/S2666-5247(20)30004-5

Clair, L.A., Chan, L.L., Boretsky, A., et al. (2023). High-throughput SARS-CoV-2 antiviral testing method using the celigo image cytometer. Journal of Fluorescence, 2023, 1–10. https://doi.org/10.1007/s10895-023-03289-x

Clementi, N., Scagnolari, C., D'Amore, A., et al. (2021). Naringenin is a powerful inhibitor of SARS-CoV-2 infection in vitro. Pharmacological Research, 163, 1–4. https://doi.org/10.1016/j.phrs.2020.105255

Cordell, G.A., Quinn-Beattie, M.L., & Farnsworth, N.R. (2001). The potential of alkaloids in drug discovery. Phytotherapy Research, 15(3), 183–205. https://doi.org/10.1002/ptr.890

Cucinotta, D., & Vanelli, M. (2020). WHO Declares COVID-19 a pandemic. Acta Biomedica, 91(1), 157–160. https://doi.org/10.23750/abm.v91i1.9397

Danao, K., Nandurkar, D., Rokde, V., et al. (2023). Molecular docking: Metamorphosis in drug discovery. E.S. Istifli, (ed). Molecular docking-recent advances. IntechOpen. https://doi.org/10.5772/intechopen.105972

Das, A., Ahmed, R., Akhtar, S., et al. (2020). An overview of basic molecular biology of SARS-CoV-2 and current COVID-19 prevention strategies. Gene Reports, 23, 1–18. https://doi.org/10.1016/j.genrep.2021.101122

De Sá, É.R.A., Costa, A.N., Costa, R.K.M., et al. (2021). In silico study of the interactions of Pilocarpus microphyllus imidazolic alkaloids with the main protease (M^{pro}) of SARS-CoV-2. Molecular Simulation, 47(1), 74–87. https://doi.org/10.1080/08927022.2021.1873321

De, R., Pandey, N., & Pal, A. (2020). Impact of digital surge during Covid-19 pandemic: A viewpoint on research and practice. International Journal of Information Management, 55, 1–5. https://doi.org/10.1016/j.ijinfomgt.2020.102171

Dhama, K., Sharun, K., & Tiwari, R., et al. (2020). COVID-19, an emerging coronavirus infection: advances and prospects in designing and developing vaccines, immunotherapeutics, and therapeutics. Human Vaccines & Immunotherapeutics, 16(6), 1–7. https://doi.org/10.1080/21645515.2020.1735227

Dhamija, N., & Mangla, A. (2022). Plant secondary metabolites in antiviral applications. A.K. Sharma, A. Sharma (eds). Plant secondary metabolites. Springer. https://doi.org/10.1007/978-981-16-4779-6_15

Dias, D.A., Urban, S., & Roessner, U. (2012). A historical overview of natural products in drug discovery. Metabolites, 2(2), 303–336. https://doi.org/10.3390/metabo2020303

Diningrat, D.S., Sari, A.N., Harahap, N.S., et al. (2021). Potential of Hanjeli (*Coix lacryma-jobi*) essential oil in preventing SARS-CoV-2 infection via blocking the Angiotensin Converting Enzyme 2 (ACE2) receptor. Journal of Plant Biotechnology, 48(4), 289–303. https://doi.org/10.5010/jpb.2021.48.4.289

Diniz, L.R.L., Pérez-Castillo, Y., Elshabrawy, H.A., et al. (2021). Bioactive terpenes and their derivatives as potential SARS-CoV-2 proteases inhibitors from molecular modeling studies. Biomolecules, 11(1), 1–19. https://doi.org/10.3390/biom11010074

Drašar, P.B. (2022). Plant secondary metabolites used for the treatment of diseases and drug development. Biomedicines, 10(3), 1–2. https://doi.org/10.3390/biomedicines10030576

Du, R., Cooper, L., Chen, Z., et al. (2021). Discovery of chebulagic acid and punicalagin as novel allosteric inhibitors of SARS-CoV-2 3CLpro. Antiviral Research, 190, 1–7. https://doi.org/10.1016/j.antiviral.2021.105075

El-Gizawy, H.A., Boshra, S.A., Mostafa, A. (2021). *Pimenta dioica* (L.) Merr. bioactive constituents exert anti-SARS-CoV-2 and anti-inflammatory activities: Molecular docking and dynamics, in vitro, and in vivo studies. Molecules, 26(19), 1–22. https://doi.org/10.3390/molecules26195844

El-Hawary, S.S., Ali, T.F.S., Abo El-Ela, S.O., et al. (2022). Secondary metabolites of *Livistona decipiens* as potential inhibitors of SARS-CoV-2. RSC Advances, 12(30), 19505–19511. https://doi.org/10.1039/d2ra01306a

El-Mordy, F.M.A., El-Hamouly, M., Ibrahim, M.T., et al. (2020). Inhibition of SARS-CoV-2 main protease by phenolic compounds from *Manilkara hexandra* (Roxb.) Dubard assisted by metabolite profiling and *in silico* virtual screening. RSC Advances, 10(53), 32148–32155. https://doi.org/10.1039/d0ra05679k

Elshafie, H.S., Camele, I., & Mohamed, A.A. (2023). A comprehensive review on the biological, agricultural and pharmaceutical properties of secondary metabolites based-plant origin. International Journal of Molecular Sciences, 24(4), 1–20. https://doi.org/10.3390/ijms24043266

Eslami, N., Aghbash, P.S., Shamekh, A., et al. (2022). SARS-CoV-2: Receptor and co-receptor tropism probability. Current Microbiology, 79(5), 1–13. https://doi.org/10.1007/s00284-022-02807-7

Fam, M.S., Sedky, C.A., Turky, N.O., et al. (2023). Channel activity of SARS-CoV-2 viroporin ORF3a inhibited by adamantanes and phenolic plant metabolites. Scientific Reports, 13(1), 1–13. https://doi.org/10.1038/s41598-023-31764-9

Fei, J., Liang, B., Jiang, C., et al. (2019). Luteolin inhibits IL-1β-induced inflammation in rat chondrocytes and attenuates osteoarthritis progression in a rat model. Biomedicine & Pharmacotherapy, 109, 1586–1592. https://doi.org/10.1016/j.biopha.2018.09.161

Frengki, F., Putra, D.P., Wahyuni, F.S., et al. (2020). Potential antiviral of catechins and their derivatives to inhibit SARS-CoV-2 receptors of Mpro protein and spike glycoprotein in COVID-19 through the *in silico* approach. Jurnal Kedokteran Hewan, 14(3), 59–65. https://doi.org/10.21157/j.ked.hewan.v14i3.16652

Fuzimoto, A.D., & Isidoro, C. (2020). The antiviral and coronavirus-host protein pathways inhibiting properties of herbs and natural compounds - additional weapons in the fight against the COVID-19 pandemic? Journal of Traditional and Complementary Medicine, 10(4), 405–419. https://doi.org/10.1016/j.jtcme.2020.05.003

Garg, S., & Roy, A. (2020). *In silico* analysis of selected alkaloids against main protease (Mpro) of SARS-CoV-2. Chemico-Biological Interactions, 332, 1–12. https://doi.org/10.1016/j.cbi.2020.109309

Gasmi, A., Shanaida, M., Oleshchuk, O., et al. (2023). Natural ingredients to improve immunity. Pharmaceuticals, 16(4), 1–28. https://doi.org/10.3390/ph16040528

Gendrot, M., Andréani, J., Boxberger, M., et al. (2020). Antimalarial drugs inhibit the replication of SARS-CoV-2: An *in vitro* evaluation. Travel Medicine and Infectious Disease, 37, 1–6. https://doi.org/10.1016/j.tmaid.2020.101873

Gentile, D., Patamia, V., Scala, A., et al. (2020). Putative inhibitors of SARS-CoV-2 main protease from a library of marine natural products: A virtual screening and molecular modeling study. Marine Drugs, 18(4), 1–19. https://doi.org/10.3390/md18040225

Ghosh, R., Chakraborty, A., Biswas, A., et al. (2021). Identification of alkaloids from *Justicia adhatoda* as potent SARS CoV-2 main protease inhibitors: An *in silico* perspective. Journal of Molecular Structure, 1229, 1–8. https://doi.org/10.1016/j.molstruc.2020.129489

Giofrè, S.V., Napoli, E., Iraci, N., et al. (2021). Interaction of selected terpenoids with two SARS-CoV-2 key therapeutic targets: An *in silico* study through molecular docking and dynamics simulations. Computers in Biology and Medicine, 134, 1–13. https://doi.org/10.1016/j.compbiomed.2021.104538

Giordano, D., Facchiano, A., & Carbone, V. (2023). Food plant secondary metabolites antiviral activity and their possible roles in SARS-COV-2 treatment: An overview. Molecules, 28(6), 1–29. https://doi.org/10.3390/molecules28062470

Gorkhali, R., Koirala, P., Rijal, S., et al. (2021). Structure and function of major SARS-COV-2 and SARS-COV proteins. Bioinformatics and Biology Insights, 15, 1–32. https://doi.org/10.1177/11779322211025876

Govender, K., & Chuturgoon, A. (2022). An overview of repurposed drugs for potential COVID-19 treatment. Antibiotics, 11(12), 1–16. https://doi.org/10.3390/antibiotics11121678

Guerriero, G., Berni, R., Muñoz-Sánchez, J.A., et al. (2018). Production of plant secondary metabolites: Examples, tips and suggestions for biotechnologists. Genes, 9(6), 1–22. https://doi.org/10.3390/genes9060309

Gül, Ş., Ozcan, O., Aşar, S., et al. (2020). *In silico* identification of widely used and well-tolerated drugs as potential SARS-CoV-2 3C-like protease and viral RNA-dependent RNA polymerase inhibitors for direct use in clinical trials. Journal of Biomolecular Structure & Dynamics, 39(17), 6772–6791. https://doi.org/10.1080/07391102.2020.1802346

Günther, S., Reinke, P., Yaiza, F., et al. (2021). X-ray screening identifies active site and allosteric inhibitors of SARS-CoV-2 main protease. Science, 372(6542), 642–646. https://doi.org/10.1126/science.abf7945

Guo, H., Li, M., & Xu, L.J. (2019). Apigetrin treatment attenuates LPS-induced acute otitis media though suppressing inflammation and oxidative stress. Biomedicine & Pharmacotherapy, 109, 1978–1987. https://doi.org/10.1016/j.biopha.2018.07.022

Gupta, S., Singh, V., Varadwaj, P.K., et al. (2020). Secondary metabolites from spice and herbs as potential multitarget inhibitors of SARS-CoV-2 proteins. Journal of Biomolecular Structure and Dynamics, 40(5), 2264–2283. https://doi.org/10.1080/07391102.2020.1837679

Gurung, A.B., Ali, M.A., Lee, D.K., et al. (2020). Unravelling lead antiviral phytochemicals for the inhibition of SARS-CoV-2 Mpro enzyme through *in silico* approach. Life Sciences, 255, 1–13. https://doi.org/10.1016/j.lfs.2020.117831

Gyebi, G.A., Ogunro, O.B., Adegunloye, A.P., et al. (2020). Potential inhibitors of coronavirus 3-chymotrypsin-like protease (3CLpro): An *in silico* screening of alkaloids and terpenoids from African medicinal plants. Journal of Biomolecular Structure & Dynamics, 39(9), 3396–3408. https://doi.org/10.1080/07391102.2020.1764868

Harisna, A.H., Nurdiansyah, R., Syaifie, P.H., et al. (2021). *In silico* investigation of potential inhibitors to main protease and spike protein of SARS-CoV-2 in propolis. Biochemistry and Biophysics Reports, 26, 1–13. https://doi.org/10.1016/j.bbrep.2021.100969

Henss, L., Auste, A., Schürmann, C., et al. (2021). The green tea catechin epigallocatechin gallate inhibits SARS-CoV-2 infection. Journal of General Virology, 102(4), 1–8. https://doi.org/10.1099/jgv.0.001574

Hikmet, F., Mear, L., Edvinsson, A., et al. (2020). The protein expression profile of ACE2 in human tissues. Molecular Systems Biology, 16(7), 1–16. https://doi.org/10.15252/msb.20209610

Hosseini, M., Chen, W., Xiao, D., et al. (2021). Computational molecular docking and virtual screening revealed promising SARS-CoV-2 drugs. Precision Clinical Medicine, 4(1), 1–16. https://doi.org/10.1093/pcmedi/pbab001

Ismail, E., Shantier, S.W., Mohammed, M.S., et al. (2021). Quinoline and quinazoline alkaloids against COVID-19: An *in silico* multitarget approach. Journal of Chemistry, 2021, 1–11. https://doi.org/10.1155/2021/3613268

Izcovich, A., Siemieniuk, R. A., Bartoszko, J.J., et al. (2022). Adverse effects of remdesivir, hydroxychloroquine and lopinavir/ritonavir when used for COVID-19: Systematic review and meta-analysis of randomised trials. BMJ Open, 12(3), 1–12. https://doi.org/10.1136/bmjopen-2020-048502

Jain, C., Khatana, S., & Vijayvergia, R. (2019). Bioactivity of secondary metabolites of various plants: A review. International Journal of Pharmaceutical Sciences and Research, 10(2), 494–504. https://doi.org/10.13040/IJPSR.0975-8232.10(2).494-04

Jamal, Q.M.S. (2022). Antiviral potential of plants against COVID-19 during outbreaks-an update. International Journal of Molecular Sciences, 23(21), 1–25. https://doi.org/10.3390/ijms232113564

Jamshidnia, M., Sewell, R.D.E., & Kopaei, M.R. (2022). An update on promising agents against COVID-19: Secondary metabolites and mechanistic aspects. Current Pharmaceutical Design, 28(29), 2415–2425. https://doi.org/10.2174/1381612828666220722124826

Jang, Y.S., Lee, D.E., Ju, D.U. et al. (2023). antiviral effects of secondary metabolites from *Jatropha podagrica* leaves against the Pseudotyped virus of SARS-CoV-2 omicron. Plants, 12, 1–14. https://doi.org/10.3390/plants12233942

Jang, M., Park, Y.I., Cha, Y.E., et al. (2020). Tea polyphenols EGCG and theaflavin inhibit the activity of SARS-CoV-2 3CL protease *in vitro*. Evidence-Based Complementary and Alternative Medicine, 2020, 1–7. https://doi.org/10.1155/2020/5630838

Jantan, I., Arshad, L., Septama, A.W., et al. (2023). Antiviral effects of phytochemicals against severe acute respiratory syndrome coronavirus 2 and their mechanisms of action: A review. Phytotherapy Research, 37(3), 1036–1056. https://doi.org/10.1002/ptr.7671

Jayaweera, M., Perera, H., Gunawardana, B., & Manatunge, J. (2020). Transmission of COVID-19 virus by droplets and aerosols: A critical review on the unresolved dichotomy. In Environmental Research, 188, 1–18. https://doi.org/10.1016/j.envres.2020.109819

Jo, S., Kim, S., Kim, D.Y., et al. (2020). Flavonoids with inhibitory activity against SARS-CoV-2 3CL^pro. Journal of Enzyme Inhibition and Medicinal Chemistry, 35(1), 1539–1544. https://doi.org/10.1080/147 56366.2020.1801672

Joshi, R.S., Jagdale, S., Bansode, S., et al. (2020). Discovery of potential multi-target-directed ligands by targeting host-specific SARS-CoV-2 structurally conserved main protease. Journal of Biomolecular Structure & Dynamics, 39(9), 3099–3114. https://doi.org/10.1080/07391102.2020. 1760137

Jung, U., Cho, Y.Y., & Choi, M.S. (2016). Apigenin ameliorates dyslipidemia, hepatic steatosis and insulin resistance by modulating metabolic and transcriptional profiles in the liver of high-fat diet-induced obese mice. Nutrients, 8(5), 1–16. https://doi.org/10.3390/nu8050305

Kanchibhotla, D., Subramanian, S., & Ravi Kumar, R.M., (2022). An in-vitro evaluation of a polyherbal formulation, against SARS-Cov-2. Journal of Ayurveda and Integrative Medicine, 13(3), 1–6. https://doi.org/10.1016/j.jaim.2022.100581

Kang, S.Y., Kang, J.Y., & Oh, M.J. (2012). Antiviral activities of flavonoids isolated from the bark of *Rhus verniciflua* stokes against fish pathogenic viruses *in vitro*. The Journal of Microbiology, 50(2), 293–300. https://doi.org/10.1007/s12275-012-2068-7

Kashyap, P., Thakur, M., Singh, N., et al. (2022). *In silico* evaluation of natural flavonoids as a potential inhibitor of coronavirus disease. Molecules, 27(19), 1–23. https://doi.org/10.3390/molecules27 196374

Kaur, N., Singh, R., Dar, Z., et al. (2021). Genetic comparison among various coronavirus strains for the identification of potential vaccine targets of SARS-CoV2. Infection, Genetics and Evolution, 89, 1–15. https://doi.org/10.1016/j.meegid.2020.104490

Khadilkar, A., Bunch, Z.L., Wagoner, J., et al. (2023). Modulation of *in vitro* SARS-CoV-2 infection by *Stephania tetrandra* and its alkaloid constituents. Journal of Natural Products, 86(4), 1061–1073. https://doi.org/10.1021/acs.jnatprod.3c00159

Khaerunnisa, S., Kurniawan, H., Awaluddin, R., et al. (2020). Potential inhibitor of COVID-19 main protease from several medicinal plant compounds by molecular docking study. Preprints. https://doi.org/10.20944/preprints202003.0226.v1

Khairi, L.N.H., Fahrni, M.L., & Lazzarino, A.I. (2022). The race for global equitable access to COVID-19 vaccines. Vaccines, 10(8), 1–14. https://doi.org/10.3390/vaccines10081306

Khalifa, M. F., Fahim, J. R., Allam, A. E., et al. (2023). Studies on the nonalkaloidal secondary metabolites of *Hippeastrum vittatum* (L'Her.) Herb. Bulbs. ACS Omega, 8(30), 26749–26761. https://doi.org/10.1021/acsomega.2c07886

Khan, T., Khan, M.A., Mashwani, Z., et al. (2021). Therapeutic potential of medicinal plants against COVID-19: The role of antiviral medicinal metabolites. Biocatalysis and Agricultural Biotechnology, 31, 1–15. https://doi.org/10.1016/j.bcab.2020.101890

Khanna, K., Kohli, S.K., Kaur, R., et al. (2021). Herbal immune-boosters: Substantial warriors of pandemic Covid-19 battle. Phytomedicine, 85, 1–20. https://doi.org/10.1016/j.phymed.2020.153361

Kopustinskiene, D.M., Jakstas, V., Savickas, A., et al. (2020). Flavonoids as anticancer agents. Nutrients, 12(2), 1–25. https://doi.org/10.3390/nu12020457

Koritala, T., Hussain, A., Pleshkova, Y., et al. (2021). A narrative review of emergency use authorization versus full FDA approval and its effect on COVID-19 vaccination hesitancy. Infezioni in Medicina, 29(3), 339–344. https://doi.org/10.53854/liim-2903-4

Kulkarni, S.A., Nagarajan, S.K., Ramesh, V., et al. (2020). Computational evaluation of major components from plant essential oils as potent inhibitors of SARS-CoV-2 spike protein. Journal of Molecular Structure, 1221, 1–11. https://doi.org/10.1016/j.molstruc.2020.128823

Kumar, S., Kashyap, P., Chowdhury, S., et al. (2021). Identification of phytochemicals as potential therapeutic agents that binds to Nsp15 protein target of coronavirus (SARS-CoV-2) that are capable of inhibiting virus replication. Phytomedicine, 85, 1–10. https://doi.org/10.1016/j.phymed.2020.153317

Kumar, N., Sood, D., Van Der Spek, P.J., et al. (2020). Molecular binding mechanism and pharmacology comparative analysis of noscapine for repurposing against SARS-CoV-2 protease. Journal of Proteome Research, 19(11), 4678–4689. https://doi.org/10.1021/acs.jproteome.0c00367

Kushari, S., Hazarika, I., Laloo, D. et al. (2023), An integrated computational approach towards the screening of active plant metabolites as potential inhibitors of SARS-CoV-2: An overview. Structural Chemistry, 34, 1073–1104. https://doi.org/10.1007/s11224-022-02066-z

Lim, C.T., Tan, K.W., Wu, M., et al. (2021). Identifying SARS-CoV-2 antiviral compounds by screening for small molecule inhibitors of Nsp3 papain-like protease. Biochemical Journal, 478(13), 2517–2531. https://doi.org/10.1042/bcj20210244

Lin, T., Luo, W., Li, Z., et al. (2020). Rhamnocitrin extracted from Nervilia fordii inhibited vascular endothelial activation via miR-185/STIM-1/SOCE/NFATc3. Phytomedicine, 79, 1–9. https://doi.org/10.1016/j.phymed.2020.153350

Liskova, L., Koklesova, M., Samec, K., et al. (2020). Flavonoids in cancer metastasis. Cancers, 12(6), 1–30. https://doi.org/10.3390/cancers12061498

Liu, H., Ye, F., Sun, Q., et al. (2021). *Scutellaria baicalensis* extract and baicalein inhibit replication of SARS-CoV-2 and its 3C-like protease *in vitro*. Journal of Enzyme Inhibition and Medicinal Chemistry, 36(1), 497–503. https://doi.org/10.1080/14756366.2021.1873977

Lopez-Gomez, A., Pelaez-Prestel, H.F., & Juarez, I. (2023). Approaches to evaluate the specific immune responses to SARS-CoV-2. Vaccine, 41(43), 6434–6443.

Lung, J., Lin, Y., Yang, Y., et al. (2020). The potential chemical structure of anti-SARS-CoV-2 RNA-dependent RNA polymerase. Journal of Medical Virology, 92(6), 693–697. https://doi.org/10.1002/jmv.25761

Mahmood, N., Nasir, S.B., & Hefferon, K. (2020). Plant-based drugs and vaccines for COVID-19. Vaccines, 9(1), 1–16. https://doi.org/10.3390/vaccines9010015

Majnooni, M.B., Fakhri, S., Bahrami, G., et al. (2021). Alkaloids as potential phytochemicals against SARS-CoV-2: Approaches to the associated pivotal mechanisms. Evidence-Based Complementary and Alternative Medicine, 2021, 1–21. https://doi.org/10.1155/2021/6632623

Ma-Lauer, Y., Carbajo-Lozoya, J., Hein, M.Y., et al. (2016). p53 down-regulates SARS coronavirus replication and is targeted by the SARS-unique domain and PL[pro] via E3 ubiquitin ligase RCHY1. Proceedings of the National Academy of Sciences of the United States of America, 113(35), E5192–E5201. https://doi.org/10.1073/pnas.1603435113

Malekmohammad, K., & Kopaei, M.R. (2021). Mechanistic aspects of medicinal plants and secondary metabolites against Severe Acute Respiratory Syndrome Coronavirus 2 (SARS-CoV-2). Current Pharmaceutical Design, 27(38), 3996–4007. https://doi.org/10.2174/1381612827666210705160130

Manoharan, Y., Haridas, V., Vasanthakumar, K.C., et al. (2020). Curcumin: A wonder drug as a preventive measure for COVID19 management. Indian Journal of Clinical Biochemistry, 35(3), 373–375. https://doi.org/10.1007/s12291-020-00902-9

Markov, P.V., Ghafari, M., Beer, M., et al. (2023). The evolution of SARS-CoV-2. Nature Reviews Microbiology, 21(6), 361–379. https://doi.org/10.1038/s41579-023-00878-2

Maurya, V.K., Kumar, S., Prasad, A.K., et al. (2020). Structure-based drug designing for potential antiviral activity of selected natural products from Ayurveda against SARS-CoV-2 spike glycoprotein and its cellular receptor. VirusDisease, 31(2), 179–193. https://doi.org/10.1007/s13337-020-00598-8

Ma, C., & Wang, J. (2022). Validation and invalidation of SARS-CoV-2 papain-like protease inhibitors. ACS Pharmacology & Translational Science, 5(2), 102–109. https://doi.org/10.1021/acsptsci.1c00240

Mégarbane, B., & Scherrmann, J.M. (2020). Hydroxychloroquine and azithromycin to treat patients with COVID-19: both friends and foes? Journal of Clinical Pharmacology, 60(7), 808–814. https://doi.org/10.1002/jcph.1646

Meng, X.Y, Zhang, H.X, Mezei, M, et al. (2011). Molecular docking: A powerful approach for structure-based drug discovery. Current Computer-Aided Drug Design, 7(2), 146–157. https://doi.org/10.2174/157340911795677602

Mhatre, S., Naik, S., & Patravale, V. (2021). A molecular docking study of EGCG and theaflavin digallate with the druggable targets of SARS-CoV-2. Computers in Biology and Medicine, 129, 1–9. https://doi.org/10.1016/j.compbiomed.2020.104137

Mohammad, A., Alshawaf, E., Marafie, S.K., et al. (2021). Molecular simulation-based investigation of highly potent natural products to abrogate formation of the nsp10–nsp16 complex of SARS-CoV-2. Biomolecules, 11(4), 1–15. https://doi.org/10.3390/biom11040573

Mohammadi, M., Yahyapour, Y., Nasrollahian, S., et al. (2022). A review on herbal secondary metabolites against COVID-19 focusing on the genetic variants of SARS-CoV-2. Jundishapur Journal of Natural Pharmaceutical Products, 17(4), 1–11. https://doi.org/10.5812/jjnpp-129618

Mohammadi, N., & Shaghaghi, N. (2020). Inhibitory effect of eight secondary metabolites from conventional medicinal plants on COVID_19 virus protease by molecular docking analysis. ChemRxiv, 1, 1–11. https://doi.org/10.26434/chemrxiv.11987475.v1

Mohammadi, S., Heidarizadeh, M., Entesari, M., et al. (2020). *In silico* investigation on the inhibiting role of nicotine/caffeine by blocking the S Protein of SARS-CoV-2 versus ACE2 receptor. Microorganisms, 8(10), 1–14.

Mohammadinejad, R., Shavandi, A., & Raie, D.S., (2019). Plant molecular farming: Production of metallic nanoparticles and therapeutic proteins using green factories. Green Chemistry, 21(8), 1845–1865. https://doi.org/10.1039/c9gc00335e

Moradi, M., Golmohammadi, R., Najafi, A., et al. (2022). A contemporary review on the important role of in silico approaches for managing different aspects of COVID-19 crisis. Informatics in Medicine Unlocked, 28, 1–9. https://doi.org/10.1016/j.imu.2022.100862

Muhammad, S., Hassan, S.H., Al-Sehemi, A.G., et al. (2021). Exploring the new potential antiviral constituents of *Moringa oliefera* for SARS-COV-2 pathogenesis: An *in silico* molecular docking and dynamic studies. Chemical Physics Letters, 767, 1–11. https://doi.org/10.1016/j.cplett.2021.138379

Mukherjee, S., Sharma, D., Sharma, A.K., et al. (2022). Flavan-based phytoconstituents inhibit Mpro, a SARS-COV-2 molecular target, *in silico*. Journal of Biomolecular Structure & Dynamics, 40(22), 11545–11559. https://doi.org/10.1080/07391102.2021.1960196

Mukhtar, M., Arshad, M., Ahmad, M., et al. (2008). Antiviral potentials of medicinal plants. Virus Research, 131(2), 111–120.

Munafo, F., Donati, E., Brindani, N., et al. (2022). Quercetin and luteolin are single-digit micromolar inhibitors of the SARS-CoV-2 RNA-dependent RNA polymerase. Scientific Reports, 12(1), 1–9. https://doi.org/10.1038/s41598-022-14664-2

Mushebenge, A.G.A., Ugbaja, S.C., Mbatha, N.A., et al. (2023). Assessing the potential contribution of *in silico* studies in discovering drug candidates that interact with various SARS-CoV-2 Receptors. International Journal of Molecular Sciences, 24(21), 1–17. https://doi.org/10.3390/ijms242115518

Narkhede, R.R., Pise, A.V., Cheke, R.S., et al. (2020). Recognition of natural products as potential inhibitors of COVID-19 main protease (Mpro): *In-silico* evidences. Natural Products and Bioprospecting, 10(5), 297–306. https://doi.org/10.1007/s13659-020-00253-1

Nasim, N., Sandeep, I.S., & Mohanty, S. (2022). Plant-derived natural products for drug discovery: Current approaches and prospects. The Nucleus, 65(3), 399–411. https://doi.org/10.1007/s13237-022-00405-3

Nguyen, T.T.H., Jung, J.H., Kim, & M.K. (2021). The inhibitory effects of plant derivate polyphenols on the main protease of SARS coronavirus 2 and their structure–activity relationship. Molecules, 26(7), 1–15. https://doi.org/10.3390/molecules26071924

Ngwa, W., Kumar, R., Thompson, D., et al. (2020). Potential of flavonoid-inspired phytomedicines against COVID-19. Molecules, 25(11), 1–10. https://doi.org/10.3390/molecules25112707

Nowakowska, A., Choi, H., Park, K., et al. (2022). *In vitro* antiviral activity of remdesivir against SARS-CoV-2 and its variants. Journal of Bacteriology and Virology, 52(4), 149–159. https://doi.org/10.4167/jbv.2022.52.4.149

Ogando, N.S., Dalebout, T.J., Zevenhoven-Dobbe, J.C., et al. (2020). SARS-coronavirus-2 replication in Vero E6 cells: Replication kinetics, rapid adaptation and cytopathology. The Journal of General Virology, 101(9), 925–940. https://doi.org/10.1099/jgv.0.001453

Omotuyi, I.O., Nash, O., Ajiboye, B.O., et al. (2020). *Aframomum melegueta* secondary metabolites exhibit polypharmacology against SARS-CoV-2 drug targets: *In vitro* validation of furin inhibition. Phytotherapy Research, 35(2), 908–919. https://doi.org/10.1002/ptr.6843

Orfali, R., Rateb, M.E., Hassan, H.M., et al. (2021). Sinapic acid suppresses SARS CoV-2 replication by targeting its envelope protein. Antibiotics, 10(4), 1–13.

Ozyigit, I.I., Dogan, I., Hocaoglu-Ozyigit, A., et al. (2023). Production of secondary metabolites using tissue culture-based biotechnological applications. Frontiers in Plant Science, 14, 1–28. https://doi.org/10.3389/fpls.2023.1132555

Pal, D., & Lal, P. (2023). Plants showing anti-viral activity with emphasis on secondary metabolites and biological screening. D. Pal (ed). Anti-viral metabolites from medicinal plants. reference series in phytochemistry. Springer Cham.

Panda, K., Alagarasu, K., Patil, P., et al. (2021). *In vitro* antiviral activity of α-mangostin against Dengue Virus Serotype-2 (DENV-2). Molecules, 26(10), 1–13. https://doi.org/10.3390/molecules26103016

Pandamooz, S., Jurek, B., Meinung, C.P., et al. (2022). Experimental models of SARS-CoV-2 infection: Possible platforms to study COVID-19 pathogenesis and potential treatments. Annual Review of Pharmacology and Toxicology, 62, 25–53. https://doi.org/10.1146/annurev-pharmtox-121120-012309

Pandeya, K.B., Ganeshpurkar, A., & Mishra, M.K. (2020). Natural RNA dependent RNA polymerase inhibitors: Molecular docking studies of some biologically active alkaloids of *Argemone mexicana*. Medical Hypotheses, 144, 1–6. https://doi.org/10.1016/j.mehy.2020.109905

Pandey, A.K., & Verma, S. (2020). An *in-silico* evaluation of dietary components for structural inhibition of SARS-Cov-2 main protease. Journal of Biomolecular Structure and Dynamics, 40(1), 136–142. https://doi.org/10.1080/07391102.2020.1809522

Panigrahi, G.K., Sahoo, S.K., Sahoo, A., et al. (2023). Bioactive molecules from plants: A prospective approach to combat SARS-CoV-2. Advances in Traditional Medicine (ADTM), 23(4), 617–630. https://doi.org/10.1007/s13596-021-00599-y

Paolacci, S., Kiani, A.K., Shree, P., et al. (2021). Scoping review on the role and interactions of hydroxytyrosol and alpha-cyclodextrin in lipid-raft-mediated endocytosis of SARS-CoV-2 and bioinformatic molecular docking studies. European Review for Medical and Pharmacological Sciences, 25(1 Suppl), 90–100. https://doi.org/10.26355/eurrev_202112_27338

Parihar, A., Zafar, T., Khandia, R., et al. (2022). *In silico* analysis for the repurposing of broad-spectrum antiviral drugs against multiple targets from SARS-CoV-2: A molecular docking and ADMET approach. Research Square. https://doi.org/10.21203/rs.3.rs-1242644/v1

Patel, M., Pandey, S., Kumar, M., et al. (2021b). Plants metabolome study: Emerging tools and techniques. Plants, 10(11), 1–24. https://doi.org/10.3390/plants10112409

Patel, B.M., Sharma, S., Nair, N., et al., (2021a). Therapeutic opportunities of edible antiviral plants for COVID-19. Molecular and Cellular Biochemistry, 476(6), 2345–2364. https://doi.org/10.1007/s11010-021-04084-7

Patgiri, P.R., Rajendran, V., & Ahmed, A.B. (2022). Clinico-epidemiological profiles of COVID-19 elderly patients in Guwahati city, Assam, India: A cross-sectional study. Cureus. https://doi.org/10.7759/cureus.24043

Piccolella, S., Crescente, G., Faramarzi, S., et al. (2020). Polyphenols vs. coronaviruses: How far has research moved forward? Molecules, 25(18), 1–18. https://doi.org/10.3390/molecules25184103

Pinzi, L., & Rastelli, G. (2019). Molecular Docking: Shifting paradigms in drug discovery. International Journal of Molecular Sciences, 20(18), 1–23. https://doi.org/10.3390/ijms20184331

Polli, J.E. (2008). *In vitro* studies are sometimes better than conventional human pharmacokinetic *in vivo* studies in assessing bioequivalence of immediate-release solid oral dosage forms. The AAPS Journal, 10(2), 289–299. https://doi.org/10.1208/s12248-008-9027-6

Ponticelli, M., Bellone, M.L., Parisi, V., et al. (2023). Specialized metabolites from plants as a source of new multi-target antiviral drugs: A systematic review. Photochemistry Reviews, 22, 615–693. https://doi.org/10.1007/s11101-023-09855-2

Prabhu, S.S., Sathishkumar, R., & Kiruthika, B. (2021). Computational screening and molecular docking of lichen secondary metabolites against severe acute respiratory syndrome-CoV-2 main protease and spike protein. Asian Journal of Pharmaceutical and Clinical Research, 14(12), 100–104. https://doi.org/10.22159/ajpcr.2021.v14i12.43227

Puttaswamy, H., Gowtham, H.G., Ojha, M.D., et al. (2020). *In silico* studies evidenced the role of structurally diverse plant secondary metabolites in reducing SARS-CoV-2 pathogenesis. Scientific Reports, 10, 1–24. https://doi.org/10.1038/s41598-020-77602-0

Raihan, T., Rabbee, M.F., Roy, P., et al. (2021). Microbial metabolites: The emerging hotspot of antiviral compounds as potential candidates to avert viral pandemic alike COVID-19. Frontiers in Molecular Biosciences, 8, 1–31. https://doi.org/10.3389/fmolb.2021.732256

Raj, K., Kaur, K., Gupta, G.D., et al. (2021). Current understanding on molecular drug targets and emerging treatment strategy for novel coronavirus-19. Naunyn-Schmiedeberg's Archives of Pharmacology, 394(8), 1383–1402. https://doi.org/10.1007/s00210-021-02091-5

Rakshit, G., Dagur, P., Satpathy, S., et al. (2021). Flavonoids as potential therapeutics against novel coronavirus disease-2019 (nCOVID-19). Journal of Biomolecular Structure and Dynamics, 40(15), 6989–7001. https://doi.org/10.1080/07391102.2021.1892529

Rosa, R.B., Dantas, W.M., do Nascimento, J.C.F., et al. (2021). *In vitro* and *in vivo* models for studying SARS-CoV-2, the etiological agent responsible for COVID-19 pandemic. Viruses, 13(3), 1–29. https://doi.org/10.3390/v13030379

Ruan, Z., Liu, C., Guo, Y., et al. (2020). SARS-CoV-2 and SARS-CoV: Virtual screening of potential inhibitors targeting RNA-dependent RNA polymerase activity (Nsp12). Journal of Medical Virology, 93(1), 389–400. https://doi.org/10.1002/jmv.26222

Russo, M., Moccia, S., Spagnuolo, C., et al. (2020). Roles of flavonoids against coronavirus infection. Chemico-Biological Interactions, 328, 1–13. https://doi.org/10.1016/j.cbi.2020.109211

Sadeghi, M., Miroliaei, M., Fateminasab, F., et al. (2021). Screening cyclooxygenase-2 inhibitors from Allium sativum L. compounds: *In silico* approach. Journal of molecular modeling, 28(1), 24. https://doi.org/10.1007/s00894-021-05016-4

Sahoo, A., Fuloria, S., Swain, S.S., et al. (2021). Potential of marine terpenoids against SARS-CoV-2: An *in silico* drug development approach. Biomedicines, 9(11), 1–22. https://doi.org/10.3390/biomedicines9111505

Sahu, K.K., Mishra, A.K., & Lal, A. (2020). Trajectory of the COVID-19 pandemic: Chasing a moving target. Annals of Translational Medicine, 8(11), 694–694. https://doi.org/10.21037/atm-20-2793

Salaverry, L.S., Parrado, A.C., Mangone, F.M. et al. (2020). *In vitro* anti-inflammatory properties of *Smilax campestris* aqueous extract in human macrophages, and characterization of its flavonoid profile. Journal of Ethnopharmacology, 247, 1–20. https://doi.org/10.1016/j.jep.2019.112282

Samad, A., Ajmal, A., Mahmood, A., et al. (2023). Identification of novel inhibitors for SARS-CoV-2 as therapeutic options using machine learning-based virtual screening, molecular docking and MD simulation. Frontiers in Molecular Biosciences, 10, 1–17. https://doi.org/10.3389/fmolb.2023.1060076

Santos, S., Barata, P., Charmier, A., et al. (2022). Cannabidiol and terpene formulation reducing SARS-CoV-2 infectivity tackling a therapeutic strategy. Frontiers in Immunology, 13, 1–12. https://doi.org/10.3389/fimmu.2022.841459

Sarwar, Z., Ahmad, T., & Kakar, S. (2020). Potential approaches to combat COVID-19: A mini-review. Molecular Biology Reports, 47(12), 9939–9949. https://doi.org/10.1007/s11033-020-05988-1

Shaldam, M.A., Yahya, G., Mohamed, N.H., et al. (2021). *In silico* screening of potent bioactive compounds from honeybee products against COVID-19 target enzymes. Environmental Science and Pollution Research, 28(30), 40507–40514. https://doi.org/10.1007/s11356-021-14195-9

Sherif, Y.E., Gabr, S. A., Hosny, N.M., et al. (2021). Phytochemicals of *Rhus* spp. as potential inhibitors of the SARS-CoV-2 main protease: Molecular docking and drug-likeness study. Evidence-based Complementary and Alternative Medicine, 2021, 1–15. https://doi.org/10.1155/2021/8814890

Singh, N.A., Kumar, P., Jyoti, et al. (2021). Spices and herbs: Potential antiviral preventives and immunity boosters during COVID-19. Phytotherapy Research, 35(5), 2745–2757. https://doi.org/10.1002/ptr.7019

Soy, M., Keser, G., & Atagündüz, P. (2021). Pathogenesis and treatment of cytokine storm in COVID-19. Turkish Journal of Biology, 45(SI-1), 372–389. https://doi.org/10.3906/biy-2105-37

Srinivasan, V., Brognaro, H., Prabhu, P.R., et al. (2022). Antiviral activity of natural phenolic compounds in complex at an allosteric site of SARS-CoV-2 papain-like protease. Communications Biology, 5(1), 1–12. https://doi.org/10.1038/s42003-022-03737-7

Suručić, R., Tubić, B., Stojiljković, M. P., et al. (2020). Computational study of pomegranate peel extract polyphenols as potential inhibitors of SARS-CoV-2 virus internalization. Molecular and Cellular Biochemistry, 476(2), 1179–1193. https://doi.org/10.1007/s11010-020-03981-7

Swain, S.S., Panda, S.K., & Luyten, W. (2021). Phytochemicals against SARS-CoV as potential drug leads. Biomedical Journal, 44(1), 74–85.

Takayama, K. (2020). *In vitro* and animal models for SARS-CoV-2 research. Trends in Pharmacological Sciences, 41(8), 513–517.

Tallei, T.E., Tumilaar, S.G., Niode, N.J., et al. (2020). Potential of plant bioactive compounds as SARS-CoV-2 main protease (Mpro) and spike (S) glycoprotein inhibitors: A molecular docking study. Scientifica, 2020, 1–18. https://doi.org/10.1155/2020/6307457

Teoh, E.S. (2016). Secondary Metabolites of Plants. E.S. Teoh, (ed). Medicinal Orchids of Asia. Springer Charm. https://doi.org/10.1007/978-3-319-24274-3_5

Teoh, S.L., Lim, Y.H., Lai, N.M., et al. (2020). Directly acting antivirals for COVID-19: Where do we stand? Frontiers in Microbiology, 11, 1–18. https://doi.org/10.3389/fmicb.2020.01857

Tomas, M., Capanoglu, E., Bahrami, A., et al. (2022). The direct and indirect effects of bioactive compounds against coronavirus. Food Frontiers, 3(1), 96–123. https://doi.org/10.1002/fft2.119

Tungary, E., Wahjudi, M., & Kok, T. (2022). Secondary metabolites of various Indonesian medicinal plants as SARS-CoV-2 inhibitors: *In silico* study. Media Pharmaceutica Indonesiana, 4(2), 136–146. https://doi.org/10.24123/mpi.v4i2.5255

Twaij, B.M., & Hasan, M.N. (2022). Bioactive secondary metabolites from plant sources: Types, synthesis, and their therapeutic uses. International Journal of Plant Biology, 13(1), 4–14. https://doi.org/10.3390/ijpb13010003

Ubani, A., Agwom, F., RuthMorenikeji, O., et al. (2020). Molecular docking analysis of some phytochemicals on two SARS-CoV-2 targets: Potential lead compounds against two target sites of SARS-CoV-2 obtained from plants. BioRxiv. https://doi.org/10.1101/2020.03.31.017657

Utomo, R.Y., Ikawati, M., & Meiyanto, E. (2020). Revealing the potency of Citrus and Galangal constituents to halt SARS-CoV-2 infection. Preprints. https://doi.org/10.20944/preprints202003.0214.v1

Uvarova, E.A., Belavin, P.A., & Deineko, E.V. (2022). Design and assembly of plant-based COVID-19 candidate vaccines: Recent development and future prospects. Vavilov Journal of Genetics and Breeding, 26(3), 327–335. https://doi.org/10.18699/vjgb-22-39

Vaou, N., Stavropoulou, E., Voidarou, C., et al. (2021). Towards advances in medicinal plant antimicrobial activity: A review study on challenges and future perspectives. Microorganisms, 9(10), 1–28. https://doi.org/10.3390/microorganisms9102041

Vicidomini, C., Roviello, V., & Roviello, G.N. (2021). *In silico* investigation on the interaction of chiral phytochemicals from *Opuntia ficus-indica* with SARS-CoV-2 Mpro. Symmetry, 13(6), 1–15. https://doi.org/10.3390/sym13061041

Vivekanandhan, K., Shanmugam, P., Barabadi, H., et al. (2021). Emerging therapeutic approaches to combat COVID-19: Present status and future perspectives. Frontiers in Molecular Biosciences, 8, 1–10. https://doi.org/10.3389/fmolb.2021.604447

Vivek-Ananth, R., Rana, A., Rajan, N., et al. (2020). *In silico* identification of potential natural product inhibitors of human proteases key to SARS-COV-2 infection. Molecules, 25(17), 1–28. https://doi.org/10.3390/molecules25173822

Wahedi, H.M., Ahmad, S., & Abbasi, S.W. (2020). Stilbene-based natural compounds as promising drug candidates against COVID-19. Journal of Biomolecular Structure & Dynamics, 39(9), 3225–3234. https://doi.org/10.1080/07391102.2020.1762743

Wang, S.C., Chen, Y., Wang Y.C., et al. (2020). Tannic acid suppresses SARS-CoV-2 as a dual inhibitor of the viral main protease and the cellular TMPRSS2 protease. American Journal of Cancer Research, 10(12), 4538–4546.

Wink, M. (2020). Potential of DNA intercalating alkaloids and other plant secondary metabolites against SARS-CoV-2 causing COVID-19. Diversity, 12(5), 1–10. https://doi.org/10.3390/d12050175

World Health Organization, Weekly epidemiological update on COVID-19 - 25 January 2022. (2022). Retrieved from https://www.who.int/publications/m/item/weekly-epidemiological-update-on-covid-19---25-january-2022, Accessed on 27th June, 2024.

World Health Organization (WHO), Tracking SARS-CoV-2 variants. (2023). Retrieved from https://www.who.int/activities/tracking-SARS-CoV-2-variants, Accessed on 17th June, 2024

Xia, B., Shen, X., He, Y., et al. (2021). SARS-CoV-2 envelope protein causes acute respiratory distress syndrome (ARDS)-like pathological damages and constitutes an antiviral target. Cell Research, 31(8), 847–860. https://doi.org/10.1038/s41422-021-00519-4

Xiong, Y., Zhu, G.H., Wang, H.N., et al. (2021). Discovery of naturally occurring inhibitors against SARS-CoV-2 3CL[pro] from *Ginkgo biloba* leaves via large-scale screening. Fitoterapia, 152, 1–9. https://doi.org/10.1016/j.fitote.2021.104909

Xu, Z., Ullah, N., Duan, Y., et al. (2023). Plant secondary metabolites and their effects on environmental adaptation based on functional genomics. Frontiers in Genetics, Section: Genomics of Plants and The Phytoecosystem, 14, 1664–8021. https://doi.org/10.3389/fgene.2023.1211639

Yamari, I., Abchir, O., Mali, S.N., et al. (2023). The anti-SARS-CoV-2 activity of novel 9, 10-dihydrophenanthrene derivatives: An insight into molecular docking, ADMET analysis, and molecular dynamics simulation. Scientific African, 21, 1–11. https://doi.org/10.1016/j.sciaf.2023.e01754

Yang, C.W., Lee, Y.Z., Kang, I.J., et al. (2010). Identification of phenanthroindolizines and phenanthroquinolizidines as novel potent anti-coronaviral agents for porcine enteropathogenic coronavirus transmissible gastroenteritis virus and human severe acute respiratory syndrome coronavirus. Antiviral Research, 88(2), 160–168. https://doi.org/10.1016/j.antiviral.2010.08.009

Yang, H., & Rao, Z. (2021). Structural biology of SARS-CoV-2 and implications for therapeutic development. Nature Reviews Microbiology, 19(11), 685–700. https://doi.org/10.1038/s41579-021-00630-8

Yang, L., Wen, K.S., Ruan, X., et al. (2018). Response of plant secondary metabolites to environmental factors. Molecules, 23(4), 1–26. https://doi.org/10.3390/molecules23040762

Yepes-Pérez, A.F., Herrera-Calderon, O., Sánchez-Aparicio, J., et al. (2020). Investigating potential inhibitory effect of *Uncaria tomentosa* (Cat's Claw) against the main protease 3CLpro of SARS-CoV-2 by molecular modeling. Evidence-based Complementary and Alternative Medicine, 2020, 1–14. https://doi.org/10.1155/2020/4932572

Youssef, F.S., Alshammari, E., & Ashour, M.L. (2021). Bioactive alkaloids from Genus *Aspergillus*: Mechanistic interpretation of their antimicrobial and potential SARS-CoV-2 inhibitory activity using molecular modelling. International Journal of Molecular Sciences, 22(4), 1–23. https://doi.org/10.3390/ijms22041866

Yuan, L., Tang, Q., Cheng, T., et al. (2020). Animal models for emerging coronavirus: Progress and new insights. Emerging Microbes & Infections, 9(1), 949–961. https://doi.org/10.1080/22221751.2020.1764871

Zhan, Y., Ta, W., Tang, W., Hua, R., Wang, J., Wang, C., & Lu, W. (2021). Potential antiviral activity of isorhamnetin against SARS-CoV-2 spike pseudotyped virus in vitro. Drug Development Research, 82(8), 1124–1130. https://doi.org/10.1002/ddr.21815

Zhao, Y., Du, X., Duan, Y., et al. (2021). High-throughput screening identifies established drugs as SARS-CoV-2 PLpro inhibitors. Protein & Cell, 12(11), 877–888. https://doi.org/10.1007/s13238-021-00836-9

Zhou, Y., Hou, Y., Shen, J., et al. (2020). Network-based drug repurposing for novel coronavirus 2019-nCoV/SARS-CoV-2. Cell Discovery, 6(1), 1–18. https://doi.org/10.1038/s41421-020-0153-3

Zhou, Y., Rahman, M.M., & Khanam, R. (2022). The impact of the government response on pandemic control in the long run—A dynamic empirical analysis based on COVID-19. PLoS ONE, 17(5), 1–21. https://doi.org/10.1371/journal.pone.0267232

Zhu, H., Wei, L., & Niu, P. (2020). The novel coronavirus outbreak in Wuhan, China. Global Health Research and Policy, 5(1), 1–3. https://doi.org/10.1186/s41256-020-00135-6

Zia, M., Muhammad, S., Shafiq-urRehman, et al. (2021). Exploring the potential of novel phenolic compounds as potential therapeutic candidates against SARS-CoV-2, using quantum chemistry, molecular docking and dynamic studies. Bioorganic & Medicinal Chemistry Letters, 43, 1–10. https://doi.org/10.1016/j.bmcl.2021.128079

Index

Note: Page numbers in **bold** and *italics* refer to tables and figures, respectively.

For Product Safety Concerns and Information please contact our EU
representative GPSR@taylorandfrancis.com
Taylor & Francis Verlag GmbH, Kaufingerstraße 24, 80331 München, Germany

9 781032 590301